Ocean Space '85

Ocean Space Utilization '85

Proceedings of the International Symposium Nihon University, Tokyo, Japan, June 1985

Volume 2

Editor: W. Kato
Associate Editors: B.C. Gerwick, Jr., M. Homma,
R. Lenschow, O.T. Magoon

With 351 Figures

Springer-Verlag
Tokyo Berlin Heidelberg New York

W. Kato
Department of Architecture
College of Science and Technology
Nihon University
Kanda Surugadai
Chiyoda-ku, Tokyo
101 Japan

ISBN-13: 978-4-431-68286-8 e-ISBN-13: 978-4-431-68284-4
DOI: 10.1007/978-4-431-68284-4

Softcover reprint of the hardcover 1st edition 1985

Preface

Ocean development has conventionally been targeted at the exploitation of natural resources, however this trend is gradually changing: Ocean space has itself come to be regarded as a precious resource. Since problems associated with energy, food supply, and population will become even more crucial over the coming years, ocean space is being reevaluated as a means for providing solutions in many of these areas. The future prosperity of mankind will to a large measure be dependent on the manner in which the resources of the oceans are utilized.

To develop ocean space effectively, international cooperation is essential since a global perspective is paramount. It is on this basis of cooperation and communication that the present symposium was organized, bringing together experts from all over the world to report current findings in ocean development and to consider areas of future research. The material covered here deals with all aspects of these areas which are now of such of vital importance.

M. Takanashi

Symposium Chairman
President of Nihon University

Symposium Committees

Chairman

M. Takanashi (Nihon University, Japan)

Vice-Chairman

M. Homma (The University of Tokyo, Japan)
W. Kato (Nihon University, Japan)
S. Motora (Nagasaki Institute of Applied Science, Japan)
K. Okamura (Japan Marine Science and Technology Center, Japan)
M. Sato (Nihon University, Japan)

Overseas Honorary Members

C.L. Bretschneider (University of Hawaii, USA)
J.P. Craven (University of Hawaii, USA)
B.C. Gerwick, Jr. (University of California, USA)
R. Lacroix (A L'ecole Nationale des Ponts et Chausses, France)
R. Lenschow (University of Trondheim, Norway)
O.T. Magoon (American Shore and Beach Preservation Association, USA)
J.C. McCall (NOAA Data Buoy Center, USA)
C.C. Mei (Massachusetts Institute of Technology, USA)
J. Penzien (University of California, USA)
A.F. Richards (FUGRO B.V., The Netherland)
O.C. Zienkiewicz (University College of Swansea, UK)

Organizing Committee Chairman

K. Shibata (Nihon University, Japan)

Steering Committee Chairman

W. Kato (Nihon University, Japan)

Executive Committee General Chairman

K. Kawanishi (Nihon University, Japan)

Secretary General

J. Motooka (Nihon University, Japan)

International Reviewing and Editorial Board

O. Saijo (Nihon University, Japan)
T. Sakou (Tokai University, Japan)
M. Sakuta (Nihon University, Japan)
R. Tanaka (Japanese Association for Steel Pipe Piles, Japan)
N. Tosaka (Nihon University, Japan)
Y. Yamanouchi (Mitsui Engineering and Shipbuilding Co., Ltd., Japan)
K. Yoshida (The University of Tokyo, Japan)
O.C. Zienkiewicz (University College of Swansea, UK)

Working Group of Nihon University

Chairman:
T. Nishimura

Secretaries:
K. Kamizono, K. Masuda

Members:
K. Hotta, A. Ito, T. Kawanishi, T. Kondo, A. Kuroyanagi, G. Shimizu, S. Togawa, N. Yokouchi

Table of Contents

Special Session

3. Review of Ocean Space Utilization Projects

SAGA I

A Long Range Autonomous Submarine

H.G. Delauze

COMEX S.A., 13275 Marseille Cedex 9, France

I. INTRODUCTION

In 1987 the SAGA I will be the world's largest commercial submersible ever launched.

Basically, it is a mobile underwater habitat situated in the heart of a highly autonomous and long range submersible. 28 meters long, with an external hull displacement of 550 tons, SAGA I is made up of two compartments ; inside one of which a navigation and support crew of six men can live at atmospheric pressure, the other where four to six divers can live in saturation with the possibility of locking out, as from a classical diving bell, to carry out underwater work.

The SAGA I will be the only vessel able to release divers and various tools at a depth of 450 meters, or robots up to 600 meters, and this whatever the surface conditions may be, or even in areas inacessible to traditional surface support vessel such as the fringe regions under ice packs.

The submersible will have sufficiently large energy reserves to undertake a 150 nautical mile submerged trip (the distance between Stavanger and Cormorant oil field), for its divers to work on site night and day for one week and then for it to return to its base without surfacing.

The "surface independent" or "all underwater" solution could be the only answer to the dramatic growth in size and cost of surface vessels able to perform operations in deep, rough seas or even ice covered areas.

II. THE SAGA I PROJECT

For these reasons, COMEX launched the SAGA I project at the end of 1982. The technical and financial responsibilities are now shared with IFREMER (ex CNEXO) the French National Agency for Ocean Development.

The French Ministry of Industry and Research supply a large amount of financial backing through the Hydrocarbon Industry Support Funds.

The SAGA project also has a European dimension in that the E.E.C supports financially some of the technological developments used within the project.

This project will has a value of more than 250 Million French Francs and extends over 4 years distributed as follows :

- Late 1982 : Launching of the program
- 83 - mid 84 : Detailed engineering studies

- mid 84 - late 85 : Fabrication, construction, assembly
- First half 86 : Sub-systems testing and quay testing
- Second half 86 : Sea trials and final testing
- 1987 : Operational demonstration

III. TECHNICAL DEVELOPMENTS

The SAGA I uses part of the submersible "ARGYRONETE", the construction of which was started by Captain COUSTEAU in the late 60's then stopped because this very ambitious project was too far ahead of its time in respect of the technical means available and the offshore industry's requirements.

The all underwater concept can be put at work today thanks to significant industrial and reliable progress in several key technologies, essential for future underwater work :

- Closed-circuit Stirling engine able to produce up to 75 Kw of power for long duration. Two such Stirling plants, developed by United Stirling in cooperation with Kockums (Sweden), will be installed onboard the SAG I thus providing safe, reliable and noiseless operations.
- Cryogenic storage of the combustive, oxygen, allowing energy to be stored in the smallest possible space.
- Extensibe use of composite materials to lighten the vehicle.
- Closed-circuit diver's breathing and heating apparatus.
- Computerized navigation and instrument system.

IV. POSSIBLE APPLICATIONS

The submersible will be able to carry out a variety of subsea tasks in oil and oil-related activities : geophysics, geotechnics, survey, etc.

- Maintenance of wellhead and subsea production systems
- High-resolution seismic operations on possible drill sites
- Unreworked sample-taking by rotary coring
- Static or dynamic penetrometry
- Construction assistance in hyperbaric welding operations
- Pipeline surveys and bathymetric surveys
- Operation of robot vehicles and systems : ROVs, remote telemanipulators, module replacement or repair, etc.

V. THE FUTURE

The major challenges of the SAGA project are not exclusively technical. They are also psychological and "cultural" since this project intends to demonstrate the operational feasibility of a concept, that appears as impossible as deep human diving in the late 60's - now an industrial reality down to 450 meters.

This is why a large portion of the program will be devoted to operational demonstrations, in actual conditions and in varied areas, of the major tasks listed above.

After 12 months of operational demonstrations in 1987, the future of SAGA I is already ensured. In October 1984 COMEX and IFREMER signed a deal with two Canadian companies - ISE (International Submarine Engineering) of Vancouver, BC and ECS (Energy Conversion Systems) of Ottawa, Ontario to form a new company, ISTS (International Submarine Transportation Systems).

ISTS is a Canadian company, owned 50/50 by the French and Canadian partners.

The objective of ISTS is to integrate into the SAGA I, as from 1988, a nuclear power plant, presently being developed in Canada, in order to create a new submersible : the SAGA N.

The SAGA-N submersible will be an experimental base for remote controlled vehicles and a subsea support for activities in extreme conditions, a development that could revolutionize Ocean Science and commercial exploration and exploitation of the sea, that could do for the ocean what the Space Shuttle is doing for space.

SAGA I EXPLODED VIEW

N°	LEGEND
1	ATMOSPHERIC COMPARTMENT
2	MAIN PILOT CABIN
3	LIVING QUARTERS
4	OBSERVATION PILOT POST
5	RESCUE BELL AND R.O.V. PILOT POST
6	STIRLING ENGINES
7	CRYOGENIC OXYGEN STORAGE
8	DIVE CONTROL ROOM
9	DIVERS' COMPARTMENT
10	DIVERS' TRANSFER TRUNK

An Analysis of Offshore Man-Made Islands

T. Matsumoto
Coastal Development Institute of Technology, Hayabusacho, Chiyoda-ku, Tokyo, 102 Japan

SUMMARY

This paper introduces some of our research results concerning man-made offshore islands. We will discuss the various types and characteristics of offshore man-made islands, their development in Japan, and one very interesting application, the sea city.

1. INTRODUCTION

Japan is surrounded by ocean. Much of its 380,000 sq. kilometers is covered by steep mountains, leaving only roughly 25 percent of its territory habitable. Mostly concentrated in coastal regions, this area must support rapidly developing business activity, and cities are becoming overcrowded. Fortunately, however, Japan's 200-nautical-mile zone encompasses nearly 4,510,000 sq. kilometers, more than 10 times its land area.

It is no exaggeration to say that Japan's future depends on how these vast sea territories are used. Offshore man-made islands are one very important aspect of their development.

2. THE DEVELOPMENT OF MAN-MADE OFFSHORE ISLANDS IN JAPAN

The concept of man-made islands in Japan was first mentioned in a report issued in 1977, by the Transportation Technology Council, an advisory organ established by the Minister of Transport. This report was titled "An Inquiry into the Technological Aspects of the Construction of Structures on the Surface of the Ocean, and Policies Necessary for their Construction." Since then, although the phrase "man-made island" has been used in various ways, a definition more precise than "an island artificially created in an offshore ocean area, well separated from land," has yet to be developed.

Man-made offshore islands are not a new phenomenon in Japan. Dejima Island was constructed off Nagasaki, Kyushu, in 1636, and was the base for a Dutch trading firm which flourished as the West's sole window to feudal Japan.

Gunkanjima, or Battleship Island, a small island 1.2 kilometers in circumference and 80,000 sq. meters in area, lies 20 kilometers west of Nagasaki. Named for its resemblance to a battleship, it was created by erecting a 10-meter wall around existing natural islands and outcrops, in order to mine coal from the ocean floor. At its peak, it supported mining and housing facilities, and a population of 5,000 residents. Operations on the island continued from 1891 until its closure in 1974, when it was returned to its natural state.

From 1960-1975, the fast pace of Japan's economic growth was accompanied by rapid expansion of harbor and special ocean industry areas, especially in the Tokyo and Osaka bays, and the Inland Sea. Artificial islands were created in response to the high usage rate of shoreline and surface areas.

Famous among them are Higashi-Ogishima and Daikoku Pier in Tokyo Bay, and Port Island and Rokko Pier in Osaka Bay.

In Omura Bay, Nagasaki Prefecture, the Nagasaki Airport was built, using Mino Island, about one kilometer from shore. Construction of the Kansai International Airport in Osaka Bay is scheduled to begin soon.

Other noteworthy man-made island projects are Kansai Electric Co.'s Gobo thermal power plant in Wakayama Prefecture, and the Kanda Earth and Sand Disposal Facility constructed in Suonada, Fukuoka Prefecture, by the Fourth Harbor Construction Bureau of the Transport Ministry. The former is a 350,000-sq.-meter island created for a coal-fired thermal power plant, and the latter is a 1.2 million-sq.-meter island used for the disposal of earth dredged from shipping lanes.

Presently, much research on the various kinds of man-made offshore islands is being carried out in Japan by government agencies, including the Ministries of Transport and International Trade and Industry, and by private foundations.

From 1980 to 1983 the Ministry of Transport, in conjunction with Keidanren and the Kozai Club, undertook a feasibility study for construction of a man-made island, 2-5 kilometers offshore, housing a coal-fired power plant. In 1984 and 1985, further feasibility studies were launched on various types of islands in each of Japan's six sea areas.

The Research Institute for Ocean Economics has completed studies of plans to build offshore cities. These islands will contain power plants using nuclear energy, coal, or LNG. One idea is to use high-temperature coolant water from nuclear power plants in fish farms for flounder and flatfish.

Research is currently under way on an offshore city with a four-tier deck to be constructed on the open sea, which was proposed by a private research organization in 1983. This ultra-modern city would feature a 24-hour international airport, an advanced INS (information network system), and a population of over one million people.

3. TYPES AND CHARACTERISTICS OF MAN-MADE ISLANDS

The many different functions of man-made islands make it difficult to speak of them generally. However, generalizations can be made concerning the special characteristics of different types according to construction and concept. Man-made offshore islands are classified as follows:

3-1 Undersea mining bases

As was stated before, Battleship Island was constructed for the purpose of mining coal from the sea bed. Today, however, the most numerous and largest in scale of these projects are offshore oil drilling platforms.

In Japan, the only such platforms in actual operation are the Aga Platform in the Japan Sea and the Joban Platform in the Pacific Ocean. But as the Ministry of International Trade and Industry is presently actively pursuing resource development tests, we can expect this number to increase gradually.

3-2 Harbor installations

Two types of man-made islands are used in harbor installations. The first type is extensions made from shore to allow more efficient use of the shoreline or water surface. Not suprisingly, these islands are most often located in large harbors of big cities. Daikoku Pier in Yokohama and Port Island in Kobe are typical examples.

The second type are used for the construction of harbor facilities in deep water. They are usually built as ports to handle very large quantities of one kind of goods; for example, as a loading or distribution base for crude oil, coal, or iron.

There is no such installation in Japan, but Robert Bank Island, located south of Vancouver and designed for shipping coal, is one such example.

3-3 Airport installations

There may be many reasons for locating an airport on a man-made island: to lessen the effect of noise or air pollution, or because there is no appropriate land site. Nagasaki Airport and Kansai International Airport are typical examples, both sited on reclaimed land. Alternatively, many floating-type airports have been proposed. The English seadrome plan, and the American FLAIR plan are well-known examples.

3-4 Recreation and leisure installations

No recreation-oriented island has been built in Japan, but two unique examples are the island in Tampa, Florida, where special habitats have been created for fowl and other wildlife, and the Hart-Miller Island in Baltimore, which features forests, natural beaches, fishing piers, and swamps.

3-5 Industrial installations

A cooperative study carried out by the Ministry of Transport, Keidanren, and the Kozai Club in response to the development of man-made islands describes a three-stage process in the use of ocean space: 1) employing reclaimed land for industrial purposes; 2) constructing man-made islands near shore with a multiple-purpose, i.e., for both industry and cities; 3) creating functional areas on and around man-made islands built offshore.

Man-made islands of the second stage are found in Tokyo Bay. Ukishimacho, Chidoricho, Ogishima, and Higashi-Ogishima are all good examples. Third-level islands are still in the research stages. Examples of projects in other countries are those planned for the North Sea off Holland, and the NFS/RANN plan in America.

3-6 Electric power plant sites

There are many reasons for situating power plants offshore. Land is often difficult to acquire; nuclear power plants require a buffer zone due to the danger of radiation leaks; coal fired plants produce a great many air pollutants and coal transports require a deep berth.

One example is the Gobo thermal power plant located off the Pacific ocean coast of Wakayama Prefecture. More widely known is the planned nuclear power plant off the New Jersey coast of America. Many such construction projects are currently being studied in Japan.

3-7 Island cities

Man-made islands used as cities can be divided into two groups.

The first type are those in which the direction of the existing city's development was toward the sea. In this case, a man-made island was built on the sea, and development of the city continued onto the island. Port Island in Kobe is a typical example. Other such islands can be found in Tokyo Bay.

The second type is an island built as an independent, functional, fully integrated urban unit. No islands of this type have actually been constructed, but many are being studied. Among them are the Pilkington Sea city in England, the Hawaii and Triton Sea City in the U.S., and Kikutake's plan for Sagami Bay. The Marine Information City scheme being studied now by a certain private research organization also belongs in this category.

4. AN APPROACH TO SEA CITIES

As was shown in 3) above, each type of offshore man-made island is a complicated undertaking. Using the results of studies carried out by our organization, we would like to investigate further some aspects of sea cities.

4-1 The beginning of a sea city

Japan has a very complex coastline, with numerous coves, inlets, bays and inland seas. According to ancient records, land reclamation projects were under way in both the 7th and 9th centuries.

According to estimations by the Ministry of Transport's Bureau of Ports and Harbors, over the past 100 years, about 200 sq. kilometers have been reclaimed from Tokyo Bay, 50 sq. kilometers from the Osaka Bay, and an overall 900 sq. kilometers for the entire nation. Whether or not they are aware of it, many Japanese now live on land reclaimed from the sea. New land created in this manner is in response to the needs of both the cities and the nation. Similarly, future projects are expected to pay due regard to the sea's eco-system.

Recently, a new type of man-made island is appearing in Japan. Port Island, built from 1966-1981 in Kobe harbor, is a unique man-made island. This 4.36 million-sq.-meter modern island terminal boasts 12 container ship berths, and 16 liner berths. In the 1.2 million-sq.-meter center of the island, a modern city has been built, capable of supporting 20,000 people. It has high-rise housing units, schools, a hotel, an international auditorium, and several parks.

Specialists are probably expecting a new frontier in the 21st century, a sea city far outstripping Port Island located much farther out to sea. A futuristic sea city such as this may use the sea's untapped potential to its fullest. The prospect also excites the imagination of engineers, but obviously its realization would require a massive pooling of human talent and intelligence.

4-2 Growth of sea cities

There are two distinct approaches to building sea cities on man-made islands. The first is the Port Island type, where the land-based city was developing toward the sea. In order to continue this development without interruption, an island was constructed, and the city spread to the new area. This type may be called a coastal sea city. Such cities necessarily supplement the functions of the mother city, and hence are built close.

The second type, a highly independent integrated sea city, may be called the offshore type. As has already been proposed in several excellent published studies, these islands may be located anywhere from several kilometers to hundreds of kilometers offshore.

Several examples of the coastal variety can be seen in the Tokyo and Osaka Bays, and it is likely that more will be built. Offshore islands, proposed in various excellent studies, are still a dream. In real terms, there is a large gap between the two types.

In the development of space, goals, such as sending a man to the moon, were attained step by step, with hard work and money. Likewise in the field of sea development, rather than striving for the ultimate dream of an offshore city, it seems more likely that emphasis will be placed on the more attainable goal of additional coastal island cities. As additional islands are constructed, they can be further integrated, and their usefulness and design can be refined. An example of this growth process can be seen in Diagram I.

Diagram I: The Future Evolution of Man-Made Offshore Islands

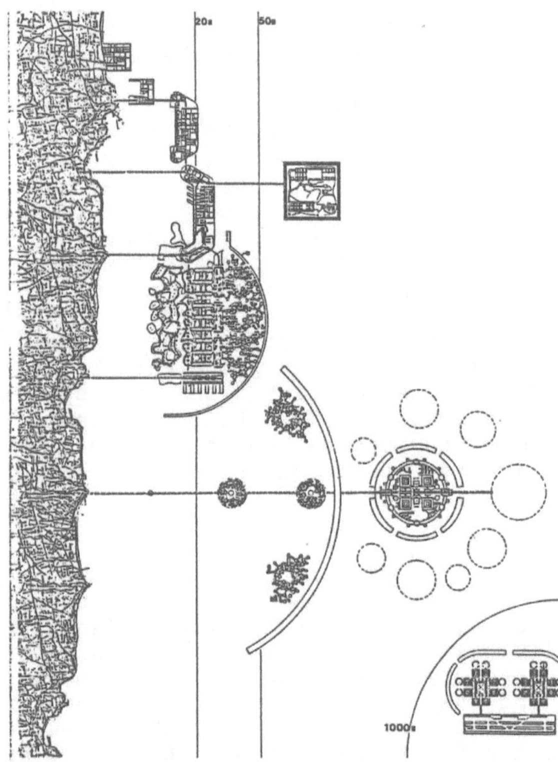

Man-made islands are created for many different reasons: to complement existing facilities, expand the shoreline, or to stand as a symbol.

Planning and construction methods will continue to grow more polished and flexible. At some point in the future, an island which looks totally natural, thick with trees, may in reality be a buffer protecting housing units, hotels, businesses, and a yacht harbor in the interior.

Eventually, coral reefs, atolls, or island chains may appear. Experimental cities self-sufficient in water, food, and energy may some day appear in international waters.

Further research will be extremely important to map out clearly the growth process of sea cities on this type of offshore island.

4-3 A new perspective

Why build a sea city? Can you get people to live on a sea city? Why can't you use a natural island instead of building an artificial one? Can we afford it? These questions are likely to confront new proposals to build man-made offshore islands.

Supporters could cite the frontier spirit and romance involved in building a new city environment on the sea or the opportunity for creating a totally new social system unfettered by existing systems. These arguments, however, do not seem sufficiently persuasive in light of the above objections.

It is certain that the offshore sea city should not be built as a solution to society's short-term needs. In considering the new sea cities, a radically new value system for time, things, and space appears necessary. This perspective should include at least the following main points.

1) Projects should be undertaken as a means of offering new opportunities rather than coping with present social problems.

2) Worthwhile objectives aimed at opening new frontiers should be determined.

3) The sea's resources should be used to the fullest. To this end, rather than considering how they can best supplement land-based needs, they must be viewed in the context of the sea. Optimum integration of the needs of both the land and the sea is an ultimate aim.

4) An international and global awareness is favored over nationalistic self-interest.

5) The sea must not be treated as something to be plundered but as one of society's most important resources.

In order to bring about the realization of this type of offshore man-made island, it is important to study thoroughly the best way to attain a national consensus. To this end, national policy and education, as well as a capable and strong leader, are necessary.

5. ACKNOWLEDGEMENTS

The portion on sea cities covered in the latter half of this paper is based upon the results of research carried out using a grant from the Japan Shipbuilding Industry Foundation.

This research was performed by a committee headed by Japan's preeminent scholar in this field, Dr. Okamura. The authors would like to express their sincere gratitude for the committee's cooperation.

Conception of Offshore Fishery Base

T. Soh

College of Industrial Technology, Nihon University, Izumicho, Narashino, Chiba, 275 Japan

1. INTORDUCTION

Recently, the fields to be studied for the marine have been spreaded, in the various phases, from the resource development to the space utilization. However, on the other hand, the importance for ocean, which is provided by the marine, is recognized again form the problem caused by the population pressure and the lack of resources, and nationalism is risen in terms of maintaining the territories and resources. As a result, many restrictions appear on the fisheries. Therefore, considering that the fisheries in our country are positioned in these problem, the fishery operation within 200 miles EEZ in foreign countries is severely controlled and the fishery itself, specially, is required to be improved in its organization.

However, it is very hard to find its resolution because the problems about the fishery in our country are significantly affected by the historical tradition and the local uniqueness.

Therefore, for the countermeasure, the arrangement of the coastal fishery and the increase and cultivation of the marine resource have been considered as the urgent subject, and then those have been positioned as the actual request which is associated with the policy of the area development and arrangement which intends to reactivation of the coastal area. From recognition of such situation, "Marine-police Concept" and etc., have been produced. This concept is established based on the development of the industry by means of the resource utilization at the coastal area and of the arrangement of the local main cities and their outskirts in their function. Also, the "Marine Launching Concept" and the coastal consolidated industry, which enhances the fishery industry at the specific coastal area, by the Fishery Agency, are about to be carried out.

These marine arrangement and fishery control are intended to control the resources in the entire marine. In addition, it is considered that the conversion from the "Opened Sea" to the "Controlled Sea" is implied. Thus, it has been studied that the new fishery system should be developed through the "Marin Fishery Base", floating on the ocean, based on the full recognition of such current requests and the national intents.

2. NEEDS OF FISHERY BASE

The traditional utilization (collecting/catching fishery, shipping business, and etc.,) and the modern utilization (cultivating fishery, resources at the sea bottom, and etc.,) at the coastal area have been currently complicated. Thus, the fishery base creates the order for the newly consolidated utilization of the coastal area.

Through the modernization of the marine product industry, the activation can be attempted based on the reorganization of the coastal fishery, off-shore fishery and in-shore fishery, and the marine product industry at the coastal fishing village.

The creation of the fishery system in considered to improve the productivity, convenience and effectiveness for the fishery, and to divide the production process functionally.

The adjusting function of the catch of fish and the shipping amount is to be maintained and the market is to be stabilized at the production field or the end area so that the marine products may be smoothly distributed and the distribution loss be made to the minimum.

The effect results to the associated industries, which accompany to the construction of the offshore fishery base, are to be enhanced, and the coastal area is to be developed.

Therefore, it is necessary not only to consider the utilization of the coastal area as the conventional extension of the inland, as well as considering as one impact to resolve various problems contained in the marine product industry, but also to approach from the sea side based on the sea.

It is supposed that the dilemma can be relaxed by effectively extracting the possibility, which the ocean contains, instead of attributing the contradiction at the land side to the sea.

3. CONCEPT IN PLANNING

3-1. Feasibility for Location and Region

The base must be the consolidated location such as the strongpoint. The characteristic of the fishery production activity at the fishery community in the coastal fishery area which is base and its region must be familialized to consider the adjustable plan. Establishment of the location should be considered in association with the region under the optimized condition. However, various actualized factors are included. Therefore, the right location should be selected with some extent of allowance.

3-2 Control for Standard and Function

The standard should be studied based on the fishing boats, fishing type, the fishing method, and etc., which are operated at the region to be studied, and it should be considered as the problem involving the existing fishing action.

In addition, in the base, some more extra spare space or clearance than that of the usual time is required in order that the catch of fish and the replenishment are safely performed and the safety is maintained even at the stormy weather.

Also, because of complication of various functions, the consolidation and multiplication of the functions should be highly attempted.

The spare space is required as the method to change the congested functional device space to the safety and comfortable environment format.

3-3 Relation with Conbination

The offshore fishery base is the floating structure stabilized on the ocean as well as the combined facilities accumulating various functions. It provides the equipment to maintain the interface with the transporting equipment such as the fishing boats and helicopters and in the fishery industrial complex.

Some of those functions provide the high combination and proximity while the others provide the property which is in opposition or to be separated. Therefore, those equipment should be totally assembled by consolidating and arranging the space under the sea surface, the space with high proximity to the land at the sea level, the space above the sea level, and the relationship between those spaces.

Depending on the above three conceptions, the right location is selected through the "Feasibility for Location and Region", the base scale is estimated from the "Control for Standard and Function", and the planning is proceeded considering the arrangement of facilities from the "Relation with Multiplication.

4. SELECTION OF RIGHT LOCATION

The realization of the fishery is the minimum requirement as the condition of location for the offshore fishery base.

The conditions of the fishery to be realized are existence of:

1 Fishes to be caught,

2 Fishery workers, and

3 Fishing methods.

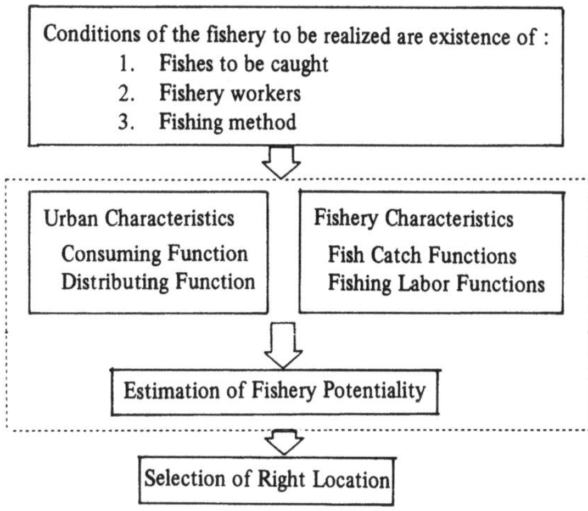

Assuming that these three conditions are arranged, the zone to be selected as the right location of the base is selected using the index called as the fishery potentiality. The fishery potentiality discussed in this report consists of two characteristics: urban characteristics and fishery characteristics.

The urban characteristics indicate mainly the consuming and distributing function. The stronger the urban characteristics in the area are, the more the amount to be consumed and distributed physically becomes. This means that the activities in the cities are active.

The fishery characteristics indicate mainly the fish catch and fishing labor function. The stronger the fishery characteristics in the area are, the more the terristorial landing amount, fishery population, and operated fishing boats become. This means that the fishery harbor is utilized actively.

Therefore, the good fishing area must exist for the fishery potentiality. Not only the fish catch and fishing labor functions but also the consuming and distributing functions at the background must be provided completely. The higher the completeness is, the stronger the fishery potentiality is.

In the specific selection, the following three indexes are established for the evaluaion method.

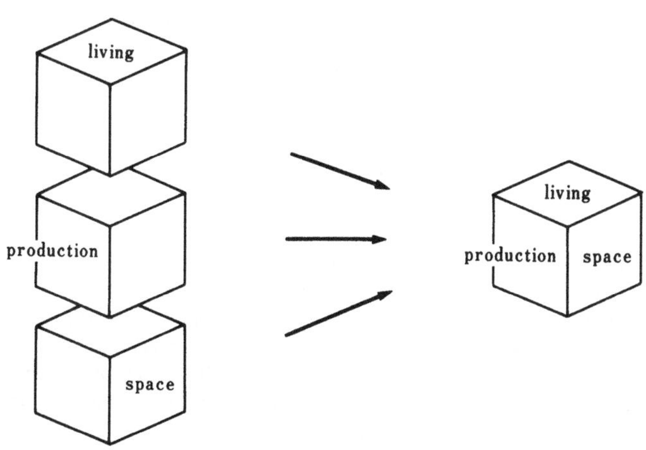

1 "Living System" consisting of residence, society, and economic phases.

2 "Production System" consisting of fish catch, fishing labor and distribution phases.

3 "Space System" consisting of point, line, and surface phases. Thus, the selecting study is performed by mean of the evaluation systems consisting of these three indexes.

First, the items to be evaluated should be established in order to select the coastal area to be the optimum location. Therefore, the urban characteristics item to be evaluated from the view evaluation axis to find the background with the urban structure which can receive the offshore fishery base and the fishery characteristics item to be evaluated from the view evaluation axis to find the fishing area with the fishing potentiality are extracted, respectively.

Upon extracting the items to be evaluated, twenty-three items have been selected as the items to be evaluated from the above-mentioned living, production, and the factor structure between items has been obtained. In addition, as the representative items to be evaluated, four for the urban characteristics and six items for the fishery characteristics are selected, respectively. The results is shown in Table 1. The right location has been selected from the total evaluation by means of the "Zone Classification Method" and the "Deviation Method" using those representative items to be evaluated.

Table 1 Representative evaluation item

Urban characteristics representative evaluation item	Fishery characteristic evaluation item
Total population Number of families DID population Handling amount at fish market per year (in value)	Population fishery harbor area Number of cooperatives Number of fishery organizations Number of fishery harber Total number of utilized fishing boats territorial landing amount

Table 2 Coastal rigions to be optimum location at offshore fishery base

cace	Large coastal region	Prefecturers	Names of coastal region
1	Coastal region of Hokkaido	Hokkaido	Nemuro Kushiro
2	Northan region of Pacific Ocean	Aomori	Northern coast section
3	Northern region of Pacific Ocean	Miyagi	North section Center section South section
4	Northan region of Pacific Ocean	Ibaragi	Northern section of Kashimanada Southern section of Kashimanada
	Middle region of Pacific Ocean	Chiba	Choshi Kujikuri

When the selected for these four cases shown in Table 2 are considered, it is found that either small coastal regions are located at Choshi and northward area which faces to the Pacific Ocean and that the sea area with high potentiality of the fishery values is located at the circumferential sea area of the selected area. From the point of view of the current coastal fishery in our country, we can, now, recognize that results of this study proves that the coastal area selected by this study coincides with the sea area from Choshi offering to Sanriku offering where currently most excellent fishery ground is expanded.

5. SCALE OF FACILITIES

For the procedures of the facility scale establishment on the offshore fishery base, the facility scale characteristic are extracted and the model formula is obtained to establish the facility scale by studying the adaptability to the plan using the scale of the current fishery harbor as the index.

The current fishery harbor facility site shows the significant difference depending on the uniqueness of every fishery harbors. That is, because of the presence of the speciality accumulated with various factors by the topographical conditions and the geographical details, it is difficult to define the causal relation between all the facility scale and the territorial landing amount in certain sample, and the low reliability is provided. The sample, which is provided with more than certain reliability through the recurrence analysis for each type of facility in the fishery harbor included in the region to be analyzed, is selected and, thus, the result is used as the estimating model.

Actually, the facility scale is classified into as follows:

1 Site area for the functional facilities

2 Site area for the water area faciliteis

For the first area 1 , the estimation is made from the relation between the current site area and the territorial landing amount, and for the second area 2 , the total tonnage of the fishing boats which utilize most the anchorage is obtained from the relation between the total number of fishing boats utilizing most the anchorage per day, the total tonnage and the territorial landing amount on the anchorage area, and then the model fishing boat utilizing the anchorage is established to estimate the anchorage area. The results are shown in Table 3.

Table 3

Main functional facilities	Estimated area (m^2)
Cleaning/Reparing facilities	34,602
Fishing tools/Maintenance facilities	121,302
Good disposal place	21,400
Marine products process/preservation facilitiecs	158,788
Open space facility	26,584
Supply facilities	16,760
Living environment facilities	72,226
ANchorage facilities	14,300
Total area	465,962
Anchorage area	200,000

*Note: The offshore fishery base introducing facility contain facilities other than above-mentioned ones.

6. ARRANGEMENT OF FACILITIES

The functional facilities can fully perform their functions only if their arrangement is defined by the facility site, attribute, and etc., and the arrangement is properly made. The facility attribute, which defines the arrangement, can be classified by the type of objects (such as persons, boats, marine products), the relation with other facilities, and etc. When considering the arrangement planning, the mutual relation between the flow of the objects and the facilities is the main estimating standard and then the proximity and the isolation for every facilities can be established.

When the arrangement planning is performed using this method, the relation between the single facilities is emphasized, and a contradiction occurs in the total relation of positions. Thus, in this study, in order to perform the arrangement planning balanced as a whole, the proximity between facilities has been analyzed from the condition of utilization in the current fishery harbor. For the arrangement of functional facilities, it is found that the facilities with the strong proximity have the characteristics to collect together from the analysis of the mutual relation by function on the degree of the mutual proximity on thirty-seven functional faiclities for three hundred and fifty-four fishery harbors within nation-wide fishery harbors. From that result, the model is established for the arrangement planning of the facility function. This is shown in Figure 3.

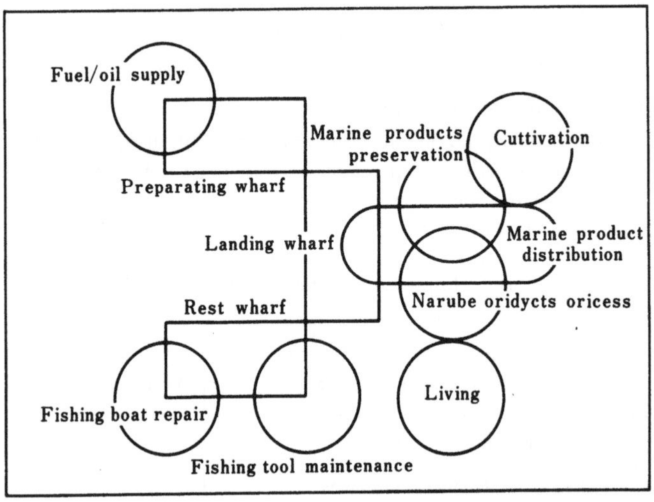

Proximity/spatial characteristics concept diagram
by function of fishing harbor functional facilities

In addition, when the concentrative nature of the facility in the offshore base in taken into account, it is considered that the rationalized arrangement of the anchorage area would enhance the concentrative nature of the offshore fishery base.

7. MODEL OF OFFSHORE FISHERY BASE

The facility required for the offshore fishery base is different a little from that for the fishery harbor located at the coastal area in its content. However, most of functional facilities can be regarded as the same as the facility in the fishery harbor. When considered that the facility of the fishery harbor itself on the fishery production is common to that of the harbor, it is regarded that the type of the fishery functional facility contains the most of theme. The difference between the offshore fishery base and the fishery harbor is as follows : the fishery harbor is established in the close relation to the background area, while the offshore fishery base must be independent of others, constructed itself, and functioned. This difference can be cleared from the fact that the offshore fishery base must be provided with the facility other than that for the fishery harbor. That is, those facilities are the mobile floating structure, the power station in the base, the equipment converting the sea water to fresh water, and etc. In addition, here, these unique facilities of the offshore fishery base are established as the spare space in the facility planning.

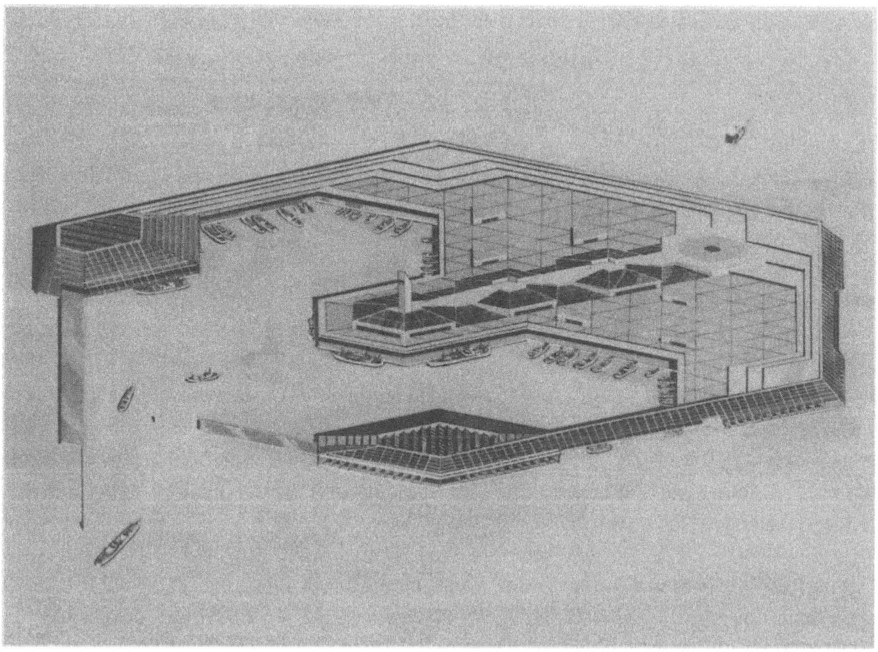

The proposed models are established as follows:
1 Goods distributing center 2 Fishing material maintenancefacility 3 Preservative processing facility for the marine products 4 Site for the multi-purpose 5 Supplying facility 6 Welfare facility for fishery workers 7 Mooring facility 8 Water area facility. The facility associated with the fishery is not always important even if its size is larger than others. It is necessary that the facility should be considered based on the above-mentioned various characteristics, the extent facing to the coast, the convenience, the intensive property, and etc.

Therefore, the arrangement planning for the offshore fishery base is established depending on the space construction in which the distributed and flatted function are enhanced and accumulate. The base consists of three decks such as the upper, middle, and lower layers. The middle deck is the same level with the sea level, and arranged with the landing wharf, the wharf for whipping the marin products, the facility for supplying oil/water and for reparing fishing boats, the marine product process facility, and the functions associated with distribution and supply. The upper deck consists of the elements with high fixing nature such as the welfare facility for the boat crews, the lodging center, and the research center. The lower deck in the sea consists of the elements with high remaining nature such as the same facilities as the middle deck, for example; the marine product process facility, the facility for preserving the marine products, the supplying facility, and the power plant facility.

8. SUMMARY

The image for the offshore fishery base can be summarized as follows:
1 The base is located at the sea area within 20 through 50 sea miles from land, communicates with the facility at land through the satellite, collects the observed data of the weather/ sea weather, and provides the information service to the fishing boats to attempt to establish the safety and the high effectiveness for the operated fishing boats.

2 The fishing boats to be studied will be not only the small fishing boats of 20 through 50 tons but also the dedicated fishing boats with the high mobility.

3 The base performs the supplying service including supplying oil, water, and ice to the fishing boats, and is the supply center which developes the saving of energy, the labor saving, and the labor condition as the relay base which allows the fishing boat crews to take a rest and shift smoothly.

4 In order to maintain the safe sailing of the ships, the base provides the monitor/rescue center for the emergency operation at the sea area patrol/emergency and the floating dock to become the maintenance center with the repair/maintenance functions.

5 The base is the multi-purpose complex with the fishery harbor function, the processing function of the marine products, and the preservative function to become the harbor center.

6 The base maintains the price formation and the distribution system with the consuming market directly and is the delivery center to function as the distribution base.

7 The base becomes the process center with the factory which processes and merchandizes the marine products.

8 The base becomes the cultivating center which provides the function responsible for the role of the seedling/seed production, the stocking and controling fry, the studying of mode of life for the marine living things, and the development study to the cultivating facility for the ocean form, the fish preserves, and etc., that the offshore cultivation in enlarged.

9 The base becomes the study center which provides the place for the technique and the ocean engineering study which are generated when the base is constructed.

The above-mentioned items are the image of the offshore fishery base. These eight function of the centers are combined and the consolidated fishery base is formed.

9. CONCLUSION

This study mainly introduces the approach from the planning studylike side as the feasibility study to construct the offshore fishery base. By proceeding this study, we can find the sea area with the potential fishery value from specially current urban and fishery characteristics. This method allows us to catch the condition of the city formation based on the fishery by associating the city with the fishery village which are previously considered separately for the study objects, and to clarify part of the current fishery construction. Based on this result, we found the possibility to utilize the unused fishing area after arranging the fishery harbor and to activate the adjacent fishing area and harbor. In addition, as the issue of the fishery harbor arrangement, we can catch the spatial/potential characteristics to clarify the principle. It is considered that the above-mentioned matters can be evaluated as the basic data when the offshore fishery base is planned in the future.

Technical Session:
Coastal Zone Utilization
and Management

1. Coastal Zone Management

Urban Waterfront Revitalization

T.R. Hudspeth
Environmental Program, University of Vermont, Burlington, VT 05401, USA

ABSTRACT

This paper provides an historical overview of the rise and fall...and rise again... of waterfronts in American cities. After examining the role of waterfronts in early American history, it traces factors contributing to the decline of urban waterfronts beginning in the late nineteenth century and factors contributing to the rediscovery of waterfronts since the Second World War. Finally, it highlights the importance of citizen participation in planning for revitalization of urban waterfronts.

ROLE OF WATERFRONTS IN EARLY AMERICAN HISTORY

When Europeans first landed on the edges of North America, they settled at river mouths or where harbors protected them from stormy seas and tides. In time, these settlements became active seaports through which came materials, goods, and people from around the world. The country's commercial well-being was tied to the fisheries, shipbuilding yards, and brisk coastal shipping trade of its Atlantic seaports until the Industrial Age. Systems of canals built to form linkages between larger navigable waterways and to avoid rapids, falls, and erratic river courses helped in America's growth and prosperity by opening up the back country in New England, in the Middle Atlantic states, and at the prairie's edge. During the latter half of the nineteenth century, the development of urban America and the pre-eminence of certain cities over others depended largely on their combination of transportation systems; those places which combined railroads and port or river access became dominant.

The waterfronts were the center of activity of these urban areas. They had similar characteristics in most of the cities that developed along the water's edge. There were large wooden docks, behind which were warehouses, shops, offices, bars, restaurants, and hotels. Small industrial plants turning out finished goods from the raw materials unloaded at the adjacent docks were usually found nearby. Businesses catered to townspeople, traders, and seamen from far-off lands. Newspaper and telegraph offices were normally close by. Churches and bawdy houses frequently existed side by side. In colonial America, a great percentage of trade was in illegal goods that avoided various taxes and tariffs; as a result, smuggled and pirated merchandise was also to be found. In short, the waterfronts were exciting, bustling places, full of life, turmoil, and legal and illegal activities. They were the focal points of cities, around which communities grew up.

DECLINE OF WATERFRONTS

Beginning with the Industrial Revolution and continuing right up until very recently, however, waterfronts have declined. Economic and technological forces have contributed to the decline. Changes in the use of urban waterfronts have resulted primarily from four factors: (1) changes in land transportation modes, (2) changing economic role of the central city, (3) technological changes in marine transportation and cargo handling, and (4) steadily decreasing waterborne passenger travel.

Changes in Land Transportation Mode

Since the late nineteenth century, coastal shipping of goods has yielded considerably to rail, and later, truck transport. With the increasing dependence on rail and truck commerce, many industrial concerns shifted their operations away from the waterfront to rail spur lines and major highway interchanges to move the goods more efficiently. The city began to expand further from the water's edge, and the waterfront was no longer the hub of activity.

Between the Civil War and World War I, railroads became the prime mover of goods and people. Vast railyards exempted all other activities at seaport waterfronts and at ports along the Mississippi, Missouri, and Ohio Rivers. The decline of our nation's railroads, in turn, and the rise of the interstate highway system has contributed to the gradual emergence of trucking as the major means of transportation of goods. More and more, freeway corridors have been located across derelict waterfronts, severely limiting access to the waterfront by pedestrians.

Changing Economic Role of the Central City

Many local businesses traditionally located on urban waterfronts have succumbed to centralized mass production. Changing space requirements in manufacturing and mass production technologies have led to the shift of this activity to the suburbs. In many places the economic role of the central city has shifted from manufacturing to highly specialized services, thus eroding the traditional link between urban industries and shipping activity. Consequently, the importance of waterfronts in contemporary industrial operations has declined.

A large number of port areas located next to the central business districts of large cities no longer meet the requirements of modern shipping and transportation technology. Cargo and passenger shipping firms, railroads, warehouses, port-related industries, and other traditional users of urban port areas have changed their activities with urban areas to meet changing requirements. As a result, they no longer require the piers, terminals, and land areas of traditional urban ports situated next to central business districts. As these firms departed or reduced the extent of their activities in port areas, the urban waterfronts were abandoned or became occupied by marginal economic enterprises. Waterfronts were used for less desirable functions including many which were not associated with port functions at all -- highways, airports, water supply facilities, waste water treatment plants, energy facilities, landfills and incinerators, and low-rent warehouses. In a vicious cycle, neglect and underutilization of the area further contributed to undesirable social activities and led to its being perceived "as a place with a bad smell and a bad rumor and good to stay away from" (Balchen and Linville, 1971).

Technological Changes in Marine Transportation and Cargo Handling

The phenomenal growth in size of bulk petroleum carriers has created a need for new and specialized terminals which cannot and need not be located at traditional urban waterfront sites. The largest crude oil carriers are so large that they cannot enter any U.S. ports. Increasingly, offshore terminals are being constructed which connect with onshore refineries and tank farms via sub-marine pipelines. The petrochemical complexes and oil tank farms are located several miles from waterfront areas. The obsolescence of many old fuel terminals, siltation of navigation channels, and fears of oil spills have contributed to this trend.

The movement, storage, and processing of hazardous bulk liquids such as liquified energy gases and toxic chemicals is also being displaced to concentrated locations beyond urban waterfronts. The public has become much more aware of the effects on human health of accidental releases of combustible, radioactive, toxic, carcinogenic, mutagenic, and teratogenic agents, and has questioned the compatibility of such activity in heavily concentrated areas.

The most significant change in cargo handling and shipping has been the shift from break-bulk operations to containerization, which allows goods to move more efficiently. With containerization, goods may be shipped directly from a manufacturing plant to a port and then to a point of distribution without breaking bulk, or packing and unpacking the cargo.

The container revolution has created major land use changes in waterfront areas (Marcus et al, 1980). Containership terminals require fewer berths and less frontage on shipping channels than the old ones, but considerably greater acreage for back-up facilities. They are highly land-intensive operations. While a break-bulk cargo vessel needs only a few acres of waterfront property--basically a pier and warehouse--a single containership requires from 12 to 35 acres of land for sorting and stacking purposes. Containerships tend to be much larger than break-bulk cargo vessels, and to carry much greater cargoes. They make fewer port calls per round trip, stopping only at major ports which serve as load centers to funnel cargo between the port and inland hinterland. Consequently, the amount of cargo loaded and unloaded at a single port call is considerably greater with a containership than with a break-bulk cargo vessel. Also, containerships stay docked at the terminal only 18 to 36 hours, whereas break-bulk cargo vessels spend several days at each port. Therefore, the ports for containerships must have enough room to handle all the cargo that will be loaded and unloaded for each arrival.

Few of the older port cities, especially along the East Coast, are able to provide the acreage needed for container terminals. Consequently, newer container facilities are generally developed in areas remote from traditional urban waterfronts where requirements for back-up space, less expensive land, deeper and wider shipping channels, and access to good highway and rail transportation can be met. So traditional finger piers and much of the break-bulk cargo vessel system are being rendered obsolete as a smaller number of specialized container ports are receiving most of the cargo-shipping business.

Steadily Decreasing Waterborne Passenger Travel

A fourth factor leading to the decline of urban waterfronts is related to the decline in the movement of people by waterborne transportation. People who formerly traveled by boat shifted to planes, and ferries--whose terminals had formerly brought life, excitement, and business to many urban waterfronts--were replaced with bridges.

> The tourists who once flocked to ocean liners to see their loved ones off now drive to airports, and tourism by boat has become a luxurious anachronism best suited to warm climes. Even the charm of a ferry ride has given way to the expedience of a quick trip across a bridge (Harney, 1979).

Decreasing waterborne passenger travel, together with changes in land transportation modes, the changing economic role of the central city, and technological changes in marine transportation and cargo handling, have rendered the traditional uses of urban waterfronts functionally or economically obsolete, and have resulted in a decline of urban waterfronts and in the loss of their positions as the economic, commercial, and industrial centers of cities. They have forced cities to focus their attention inland and turn their backs on the areas where they were born, their waterfronts:

> The common urban waterfront is hardly approachable, much less swimmable, encrusted with wharves, switching yards, sewer outfalls, and other industrial barnacles. It is the true civic outcast, the ghetto of ghettos, familiar only to longshoremen, sanitary engineers, and carp (Cowey, 1979b).

REDISCOVERY OF WATERFRONTS

However, in the 1960's, 1970's, and 1980's, tremendous interest has been shown in transforming our nation's waterfronts from desolate, non-productive areas into the active, vital portions of the cities that they were when first established. Such terms as "Cinderella Syndrome" (Cowey, 1979a), "Ugly Duckling Tale" (Weil et al, 1980), and "transforming a sow's ear into a silk purse with some money in it" have been used to describe this renaissance.

Over thirty percent of the 520 cities in the United States with populations between 25,000 and 50,000 and seventy percent of the 415 cities with populations over 50,000 are located on the edge of a river, lake, bay, or ocean (U.S. Department of the Interior, Heritage Conservation and Recreation Service, 1979). As many of these cities have sought to improve their images as good places to live, work, play, and visit, they have re-examined their waterfronts, have recognized that they are special and important places, and have undertaken revitalization efforts:

> Cities across the nation are beginning to realize that their waterfronts, whether on the ocean, a lake, or a river, are an asset. They're being eyed as potential sources of new income, as means of bringing life back into ailing downtowns, as recreation or residential areas, all depending on the needs, on the nature of the waterfronts, or on the citizens themselves (U.S. Department of the Interior, Heritage Conservation and Recreation Service, 1979).

Presently more than fifty cities have completed waterfront revitalization projects, and another fifty have projects planned or under construction (Lehmann, 1966; U.S. Department of the Interior, Heritage Conservation and Recreation Service, 1979). These range from small-scale improvements at the end of the street to linear parks connected with hike-and-bikeways to multipurpose developments combining recreation, housing, and commercial components.

A variety of innovative funding strategies and resourceful and imaginative approaches to urban waterfront revitalization are being tested by many cities. While most of the projects are multipurpose in nature, some are single purpose. Some are initiated by private groups such as downtown business associations, professional organizations, and corporations; others by governmental leaders. They are funded by federal, state, city, and private sources, and usually a combination. Some, such as Boston and San Francisco, involve extensive adaptive reuse of existing structures, while others, such as Baltimore, involve tearing down and clearing away old structures and rebuilding from scratch. Some feature "showcase" redevelopment projects, while others do not.

As a result of these efforts, rotting wharves and piers and pilings, blighted and abandoned buildings, deteriorating warehouses, scrap yards, and vacant lots are being replaced by a mix of structures, uses, and resources:

*Town houses, condominiums, apartments, hotels, motels, and boatels
*Shops, stores, malls, commercial facilities, marketplaces, convention centers, and office buildings
*Fishing piers and wharves, marinas, boat launches, and boat harbors
*Park and recreation facilities, playfields, open green spaces, picnic areas, children's playgrounds, promenades, walkways and trails for strolling, hike-and bikeways, observation towers
*Museums, aquariums, maritime exhibits, historic ships, arts and crafts centers, theaters, outdoor amphitheaters, galleries, sculptures
*Tourist attractions and sites for festivals and performances.

Why are our urban waterfronts being rediscovered now? Increasing numbers of Americans have traveled abroad, where waterfront amenities are more often the rule than the exception. They have seen for themselves what others (Doyle, 1979; Landscape Architecture, 1974; Mann, 1973; Morrell and Singer, 1978; Simpson, 1976; and Wynne, 1981) have described:

> In Europe, waterfront areas were incorporated into the fabric of the city life prior to the advent of industrialization. As a result, citizen access and use of the water's edge as an amenity were woven closely into the European style of urban life (Morrell and Singer, 1978).

The treatment of urban waterfronts in many cities in Western Europe has provided us with models for improving our own. Creative approaches by our neighbors to the north in Toronto, Quebec City, Montreal, Vancouver, and Victoria have also made us more aware of possibilities for our waterfronts. The viewing of the Tall Ships under Operation Sail as part of our Bicentennial celebration in 1976 and again in 1980 by millions of Americans also made us more aware of our urban waterfronts. Designation of 1980 as "Year of the Coast" focused attention on coastal concerns, including urban waterfronts.

Increased environmental awareness and concern for quality of life over the past two decades account for much of the current interest in urban waterfront revitalization. Several interrelated factors have contributed to these concerns: (1) urban revival, (2) clean-up of waters, (3) urban recreation, (4) coastal zone management, (5) historic preservation, (6) energy concerns, (7) economic development, and (8) conservation of natural resources.

Urban Revival

Having experienced a letdown in expectations after moving to outlying areas, many suburbanites are returning to urban settings closer to employment centers and cultural activities. This movement back to center city areas--especially by middle-class professionals, retired people, and young-- offers viable alternatives to the high costs of energy and transportation, the lack of social interaction, and the increase of social problems in suburban communities. Also, developers are expressing great interest in making adaptive use of urban commercial and residential structures of considerable age and character as the costs of securing suburban building sites increase. Redevelopment of urban areas makes especially good sense because many public services are already in place. High inflation rates, slow growth and even no-growth policies and practices, improvement of safety in downtown areas, improvement of mass transportation systems, and increased costs coupled with shortages of gasoline, have all played a role in revitalizing core cities. So has development of programs which show greater sensitivity to human needs than the wholesale clearance, demolition, and razing characteristic of urban renewal programs in the 1950's and early 1960's. As a result, development trends have gone full circle and attention is again being focused on the key portions of our city centers which have fallen into decay.

Clean-up of Waters

The campaign to clean up our nation's waters has gradually resulted in improved water quality in and near many of our coastal and riverfront cities. Controlling discharge into waterways and constructing waste water facilities has helped us to meet the clean water goal for 1983 set in the Federal Water Pollution Control Act of 1976 for most of the waters in our country to be "fishable and swimmable."

Urban Recreation

Demand for recreational opportunities has increased rapidly in response to increases in population, disposable income, leisure time, and mobility. This demand is especially acute in urban areas where population concentrations are increasing at the same time that the availability of suitable land and water areas to accommodate that demand is decreasing. The Urban Parks and Recreation Recovery Program, passed in 1978, provided funding for urban recreation areas, including decaying urban waterfronts, open space along rivers, and canals fallen into disuse for commerical transportation conduits. Worries about price and availability of gasoline have resulted in even greater use of close-to-home recreation in urban areas.

Coastal Zone Management

In the Coastal Zone Management Act of 1972, Congress established a policy to preserve and enhance the nation's coasts for future generations by guiding protection and development along the nation's 95,000 miles of shorelines. The Act declared a national policy

> to encourage and assist states to exercise their responsibilities in the coastal zone through the development and implementation of management programs to achieve wise use of the land and water resources of the coastal zone, giving full consideration to ecological, cultural, historic, and aesthetic values as well as the needs for economic development (Davenport, 1978).

That legislation has provided 35 coastal states and territories with matching grants to plan and implement their own coastal management programs, which must include guidelines on priority of uses within particular areas of the coastal zone. The funds that have been available through the Act to date have enabled states to address: flooding and erosion problems; fisheries management; energy facility siting; commercial and industrial development on the coasts; public access, private ownership, and recreational development; pollution problems in coastal waters; and waterfront development. Attention has been focused mostly on maintenance of coastal areas in their natural state and preservation of wetlands.

Amendments to the Act in 1976 required state management programs to develop

> "a planning process for the protection of an access to public beaches and other public coastal areas of environmental, recreational, historical, aesthetic, ecological, or cultural value." Although the language does not specifically refer to urban waterfronts, these are the very characteristics found in most in-town waterfront areas. The regulations go on to say that "special attention should be given to recreational needs of urban residents for increased shorefront access" (Harney, 1979).

Historic Preservation

City waterfront areas are old sections, often full of history. In renewing them, cities are finding precious relics of their beginnings and earliest days. In many cases, preservation and restoration of such historic riches are part of waterfront renewal programs. (Whitman et al, 1964).

Restoring historic homes, adapting older buildings to new economically viable uses, and recreating traditional cultural events in these areas of picturesque natural settings, distinctive architecture, colorful heritage, and rich history add to the amenity value of urban waterfronts for residents and to their attraction value for tourists and other visitors. The high costs of new construction make preservation--once considered merely quaint and "in"-- economically practical. Tax breaks available to owners of properties listed in the National Register of Historic Places contribute to interest in recycling old buildings.

Adaptive uses of old structures related to historic seaport and riverport uses--such as lighthouses, canneries, ropewalks, warehouses, chip chanleries, and custom houses--to new economically-viable uses--such as restaurants, shipping pavilions, guest houses, craft shops, museums holding old objects and documents--which do not alter significantly the external appearances creates an excellent setting for re-enactment of historic events; music, theater, and fine arts performances; and fairs and festivals, such as sailboat races and blessing of fishing fleets.

Energy Concerns

The movement back to the city and reuse of old buildings instead of building new ones are both practices which relate to concern for energy. Improving access to waterfronts close to urban centers reduces traveling, thereby conserving energy.

Economic Development

As previously mentioned, there has been a gradual shift of commercial activity away from urban waterfronts. Rail and truck transport have eliminated much of the small scale waterborne traffic that long dominated port cities. And much of the shipping that survives, such as hauling fuel and containerized cargo, requires such vast facilities that few old harbors can handle it. Consequently, miles of shoreline are ripe for development. Revitalization creates jobs in the construction and service sectors and expands the property tax base. When such revitalization includes historic preservation strategies, there are direct economic benefits and spinoffs; employment gains attributed to rehabilitation's labor-intensive nature; savings in resources and in building and demolition costs; protection of investments in basic infrastructures such as streets, sewers, lighting, and public facilities; shorter completion time; fewer delays caused by community opposition; strengthened tax base; and improved business climate.

Conservation of Natural Resources

Revitalizing urban areas instead of constructing new developments helps to conserve the natural environment by preventing urban sprawl and the filling of valuable yet fragile wetland ecosystems, especially tidal marsh estuarine ecosystems. Some of the recreation and open space uses incorporated into urban waterfront revitalization projects help to reduce or mitigate the impact of natural hazards and offer non-structural flood control and shoreline protection.

CITIZEN PARTICIPATION IN URBAN WATERFRONT REVITALIZATION

Participation by citizens in the planning process is essential for any wise environmental planning, and urban waterfront revitalization projects are no exception. Citizens for whom waterfront projects are planned must play an active role if the projects are to enjoy long-term success:

> Underlying successful...programs is a community's shared vision of a waterfront resource tied into the economic and cultural life of the [city]. Achieving community consensus on this goal requires dialogue among many separate groups and interests, which must work together to inspire broad community support (Clark et al, 1979).

Many projects which have failed or been delayed have not involved citizens in the planning process. Public reaction to waterfront projects in which they have been excluded from the planning process has been characterized by heated debate, political controversy, and even litigation.

Citizens have played crucial roles in stopping highways from cutting off waterfront areas from the water in Boston, New York, Baltimore, Chicago, New Orleans, and San Francisco; and in converting a road along the Willamette River in Portland to esplanade. They have opposed energy sitings in highly urbanized waterfront areas in northern New Jersey (Manogue, 1980; Morell and Singer, 1979), and pushed for beautification projects in San Antonio, Brooklyn, and many other places. They have also helped to provide significant involvement of neighboring residents in many cases. Urban waterfront revitalization often results in the creation of new neighborhoods; yet redevelopment plans are usually drawn up with the exclusive participation of business interests unless citizens push for involvement of neighboring residents as well.

Citizens seek to be a part of the decision-making process in a wide range of issues dealing with urban waterfront revitalization:

> The remarkable appearance within the past decade of active local citizen groups intervening in the public decision-making process concerning the siting of large facilities, changes in land use, and other developments of substantial impact on neighborhoods has also been evident on the waterfront. Because access to the waterfront is perceived to be an important goal to more people than the residents of the waterfront community only, waterfront citizen groups can expect to receive support from a wide base throughout an urban area. Proposals to site oil or liquefied gas marine terminals, build tunnels or bridges, erect high-rise condominiums, build or rebuild

waterfront highways, expand airports, site water treatment facilities, or, in fact, make virtually any conceivable change in waterfront land use evoke opposition from voluntary citizen groups. Such groups, although they do not have authority to force concessions from the responsible public agencies controlling permits or funding like port authorities or urban development authorities, can nonetheless greatly influence the development or renewal of waterfront land through political suasion or, occasionally, legal action. It is not uncommon for citizen groups to serve as permanent watchdog committees, moving on to new issues as old ones are resolved and accumulating considerable expertise on waterfront matters among their members (U.S., Committee on Urban Waterfront Lands, 1980).

Urban waterfronts constitute a limited resource. There is fierce competition among industrial, commercial, residential, recreational, defense, transportation, and other land uses. Some are incompatible and result in conflicts, such as heavy industry, pertochemical and energy storage facilities, or marine transportation terminals in the same neighborhood as residences. Others are overlapping and can coexist in mixed use arrangement. It is an arduous task to get all participants involved to agree on what the waterfront should look like. Because water is common property, the disposition of shorelands is often considered to be a public interest issue. But the public interest often conflicts with private property rights, and the expectation of a developer for an appropriate economic return.

But arduous as the task may be, it is one of the most essential tasks in planning for urban waterfronts. Yet, while literature on urban waterfront revitalization is abundant (Hudspeth, 1980), treatment of citizen participation in planning urban waterfront revitalization projects is scant and almost non-existent. What little has been written about this topic consists primarily of case study accounts of citizen action to oppose proposed projects (Manogue, 1980; Morell and Singer, 1979).

Countless questions are raised when dealing with the issue of citizen participation in waterfront revitalization: Who should involve citizens (public officials, planners, architects, developers)? What is the best way to orchestrate constituency involvement? Should everybody be involved? Who should make decisions regarding the allocation of urban waterfront land? How should the priorities of neighborhoods, cities, metropolitan areas, and regions be traded off? Who should be accepted as representative of the interests of each group? Who should benefit from waterfront development? Should recreational facilities be local or regional? Should housing be planned primarily for present residents of the area or should new groups be provided for? What weight should be given to preserving space for future development?

But underlying all these questions is the fundamental question: How do you find out what people want? And how do you present information in easily-understood forms that enable direct comparison of different alternatives? The issue of selecting a technique for involving citizens that is appropriate for urban waterfront projects is very important.

SUMMARY

In the United States, almost every major city is located on a waterway. Over the past two decades many of these cities have undertaken efforts to revitalize their waterfronts. At one time, waterfronts were the focal points of cities. Total communities with heterogenous ways of life grew up around waterfronts. However, economic and technological forces led to the decline of urban ports as the locus of shipping. Trains and, later, trucks moved goods over lands, and the cities began to expand away from the water's edge. Residential areas grew up away from the waterfront, leaving water-related industries, shipping, and fishing to take over the waterfront. The waterfront became an area frequented by a small segment of the population divorced from the rest of the community--the subject of many novels and a Marlon Brando movie-- and cities turned their backs on them. But in recent decades, "the urban waterfront is being given a second chance. Suddenly, after years of neglect, America's coastal and riverfront towns and cities are rediscovering their waterfront areas" (Harney, 1979). Several factors relating to urban revitalization and conservation of natural resources account for this rediscovery and change in status of waterfronts.

Citizen participation in planning urban waterfront revitalization projects is essential. Yet little attnetion has been devoted to this topic, and a need exists for research on techniques for facilitating citizen involvement.

REFERENCES

1. Balchen, B. and J. Linville (1971) The city waterfront: Ending an era of neglect. Nation's Cities 9(4): 8-19.

2. Marcus, H., N. Taneja, and P. Roberts (1980) The impact of changes in transportation technology on the use of land in harbor areas. Pp. 163-175 in U.S., Committee on Urban Waterfront Lands, Urban Waterfront Lands. Washington, DC.: National Academy of Science.

3. Harney, A.L. (ed.)(1979) Reviving the Urban Waterfront. Washington, D.C.: Partners for Livable Places, National Endowment for the Arts, and Office of Coastal Zone Management.

4. Cowey, A.B. (1979b) Urban waterfronts: The forgotten Z in CZM. NationalOceanic and Atmospheric Administration Magazine, January): 10:13.

5. Cowey, A.B. (1979a) The urban coast from a national perspective. Coastal Zone Management Journal 6(2-3): 135-165.

6. Weil, P., F. Fifis, and K. Goodrich (1980) The waterfront: America's ugly duckling tale. Parks and Recreation 15(12): 20-27, 57.

7. U.S. Department of the Interior, Heritage Conservation and Recreation Service, Water Resources Section (1979) Urban Waterfront Revitalization: The Role of Recreation and Heritage. Volume I: Key factors, needs and goals. Volume II: Eighteen case studies. Washington, DC.: U.S. Government Printing Office.

8. Lehmann, R.A. (1966) The principles of waterfront renewal: A summary of experiences in fifty American cities. Landscape Architecture 56(4): 286-291.

9. Doyle, P.J. (1979) People on the waterfronts of Europe: A preview for Cleveland Lakefront State Park. The Cleveland Foundation Quarterly (Fall): entire issue. Cleveland, OH.

10. Landscape Architecture (1974) U.S. runs last along the waterfronts. Landscape Architecture 64(2): 73-75.

11. Mann, R. (1973) Rivers in the City, New York City, NY.: Praeger.

12. Morell, D.L. And G.L. Singer (1978) The urban coastal zone: Challenge of redevelopment. Pp. 66-80 in Coastal Zone '80: Proceedings of the Second Symposium on Coastal and Ocean Management. New York City, NY.: American Society of Civil Engineers.

13. Simpson, A. (1976) Waterfront precedents. Central Waterfront Planning Committee, City of Toronto. Toronto, ON.: City of Toronto Planning Board.

14. Wynne, G.G. (1981) Europe makes waves on Alexandria's waterfront. Urban Innovation Abroad (April). Washington, DC: Council for International Urban Liaison.

15. Davenport, R. (1981) Waterfront utilization: Putting the pieces together. Pp. 19-29 in K. MacDonald (ed.), Waterfront Development: A Reader. Boston, MA.: Coalition of Northeast Municipalities.

16. Whitman, I.L., R.M. David, and S.E. Goldstone (1964) Renewal of waterfront areas presents special problems, brings special opportunities. Journal of Housing 5 (June): 236-255.

17. Clark, J., C. Wilson, and G.L. Binder (1979) Small Seaports: Revitalization through Conserving Heritage Resources. Washington, DC.: The Conservation Foundation.

18. Manogue, H. (1980) Citizen groups: New and powerful participants in urban waterfront revitalization. Pp. 212-240 in U.S., Committee on Urban Waterfront Lands, Urban Waterfront Lands. Washington, DC.: National Academy of Science.

19. Morell, D.L. and G.L. Singer (1979) Energy facility siting in the urban coastal zone: Compatible or not? Coastal Zone Management Journal 6(2-3): 215-231.

20. U.S., Committee on Urban Waterfront Lands, Environmental Studies Board, Committee on Natural Resources, National Research Council (1980) Urban Waterfront Lands. Washington, DC.: National Academy of Sciences.

21. Hudspeth, T.R. (1980) Urban waterfront revitalization employing visual preference as a strategy for managing visual resources and facilitating public participation. Pp. 33-50 in Coastal Zone '80: Proceedings of the Second Symposium on Coastal and Ocean Management. New York City, NY.: American Society of Civil Engineers.

Ocean Utilization and Economic Conflicts in the Coastal Zone

R.H. Charlier

IFAQ-Vrije Universiteit Brussel, Pleinlaan, B1050 Brussels, Belgium

ABSTRACT

It has been a common occurrence, over the last decades, for conflicts and competition to develop, among existing and potential users of the coastal zone. An agonizing choice has been forced in many instances upon governing bodies, even private enterprises, involving frequently considerable economic, social and ecological impact.

The increasing economic importance of the coastal zone and the growing pressures to which it is subjected create an urgent need, especially for nations in the process of industrialization, to first study, then manage the coastal zone. Management and protection of the coastal zone should be harmoniously interwoven. To this end, the geologist has a primordial role to assume. This paper, prepared in part for a governmental agency, examines the combined scenario of geological and economic activities.

INTRODUCTION. EXTENT AND ECONOMIC SIGNIFICANCE.

Coastal zones have a primordial importance from an economic viewpoint because 80% of the world population lives within 50 km of the sea. A thorough knowledge of the littoral fringe is thus indispensable for any planification or exploitation procedure. This littoral fringe is part of the continental shelf geologically defined as the continent's submerged part extending seawards from the shoreline and gradually plunging toward the 200 m. isobath. We regard the continental shelf in this paper, as a geologic concept corresponding to a physical reality, and exclusive of any legal definition.

The maritime facade is conceived, herein as made up of a terrestrial and of an oceanic area, the on-shore and near-shore sections. It is indeed within this zone that the major share of commercial and industrial activities take place, a situation we may presume will perdure until the end of the current decade. The on-shore section poses little problems for inventories and its constituing elements subject to variations have been determined; it is often the site of traditional activities such as agriculture, animal husbandry, tourism, which locate in a space directly concerned by economic considerations and whose tangible results are immediately perceived at the regional or national level.

The marine sector is more difficult to handle, because

it is subjected to change which can only be truly identified and quantified through a more specialized scientific approach.

Besides the traditional vocations such as fisheries, more recent activities are located in the coastal zone; necessitated by economic pressures, they often demand a more sophisticated technology.

THE CONCEPT OF MANAGEMENT AND USE OF THE LITTORAL.

Management and use of the coastal zone require pluridisciplinary investigations which, though they all belong to the domain of oceanology sensu largo, necessitate however knowledge and actions which exteed considerably the field of the disciplines generally considered as "marine sciences". Beyond the boundaries of physics and fluid dynamics, of water chemistry, of geological phenomena, of climatological characteristics, the manager must include what is frequently labeled economic oceanography and which encompasses human influences, the exploitation of biological, mineral and energy resources, and the regional or even international relations, their corollaries. (1)

Contemporary man looks more frequently towards the sea; this leads to a strong stress on coastal zones and a sharp rivalry between potential users. Scientific research dealing with the coastal zone affects a changing character; it was traditionally directed towards basic research, fundamental problems, such as floculation, adsorption, advection, or stratification and mixing of water masses. Economic, social and political demands justify a greater emphasis of research towards topics of direct application leading to tangible results as well for the land as for the marine sector. Exploitation of biological resources, once an artisanal endeavor, is now industrialized in our regions and will become such in a close future in developing countries. With it come construction of more sophisticated harbor facilities, motor ships and electronic equipment research shops, building of factories and cold storage wharehouses. The exploitation offshore and in the open ocean of mineral resources, launches construction of treatment and transformation factories, refineries, oil and gas pipelines terminals. Additionally, electricity plants, in search of cold water, want to implant themselves in the coastal zone.

If physics, chemistry, biology, geology, engineering contribute to the understanding of the coastal zone, so does to a great extent what we may call the complex of environment sciences. Appointment and management of that zone require as a prelude an investigation of its present state : are we faced with an area wherein only natural pollution has had an impact or is it one already threatened or even endangered ? Both water quality and shoreline stability must be studied. Indeed pollutions are either of natural or of anthropic origin; the first are generally in harmony with nature, such as the Amazone's alluvial discharges and accumulations along mangrove coasts. On the other hand large human concentrations along coasts, frequently worsened by a seasonal or year-round tourism, often contribute to a potentially irreversible deterioration of littoral waters quality. Anthropic pollutions evidently include municipal, industrial and even nuclear discharges.(2)

Required counter-action encompasses construction and enlargment of water processing stations, strict limitation of the quantity of dumped waste, examination for and elimination of nocive substances, study of the rates and mechanisms of sedimentation. Not only does it behoove to take into account the consequences on the very site, but the impact must be examined on a regional scale, which implies a survey of coastal currents and their transport capacity, and the characteristics of the mixing of waters. A mere descriptive report will not do, it must be accompanied by a quantitative study allowing to estimate the consequences on a space and time scale.

North Sea riparan states have paid rather little heed to the management of their coastal zones; however a determined pressure is gaining momentum to protect and contain the future development of tourism, habitat, industry, and to insure coast protection. The causal activities in each country are practically non-coordinated and the European Economic Community does not intervene, leaving to local authorities decisions and measures. (3) In Belgium management rests with the national government although area utilization falls within the municipal jurisdiction. Local administrations control coastal areas in Denmark, with industrial development the more advanced along the Baltic coastline. The Netherlands have a long tradition of coast protection and have perhaps the best management of a well-defined littoral fringe; if responsibilities are a local concern, administrations take largely into account public opinion and insure population consultation before undertaking new projects. Yet, Norway is the only country that has a national law applicable to clearly defined coastal zone. The situation in the Federal Republic of Germany is extremely complex : the Federal, State and Local governments each share in management responsibility. As for Great Britain, it does not have a specialized administration that cares exclusively for the coastal zone : countries plan and districts carry out and administer littoral policies. (4)

In many countries, particularly in the Third World, the demand for water, the pressure of industrial expansion -- no matter how modest --, the search for oil and gas and minerals at short distances off-shore, have already brought about important pollutions detrimental to recreation facilities and living space, a situation worsened where the State has no shoreline eminent domain.

Already the location of several industries, the coastal environment is further stressed by private and public investors anxious to develop large harbor zones; the eventual economic rent tempts Third-World countries to allocate their littoral fringes to a total industrial-and-port occupance. In Africa, for instance, where in the past this fringe was often exclusively used for agriculture, animal husbandry, fisheries and trade, and perhaps a very modest harbor activity, one can already witness the rapid implant of industry, transport, manutention. A warning bell was already tolled, not so long ago, by the Centre for Social and Economic Studies of the Third World (Mexico) : it is high time to pay heed to the hereto unspoiled coastal zones. Such action can only occur through an understanding of the prevailing geologic, geographic, social and economic conditions, followed-up by a healthy and realistic management program.

Undoubtedly an economic conflict exists between those who champion the total utilization of the coast for industrial, mining and commercial expansion, and the partisans of a "balanced" occupance between on the one hand industry, and on the other tourism and habitat. It has already been more than a decade that among others Brahtz brought this matter to the fore. (5)

"In the development of oceanographic science and technology, close cooperation should be assured between a wide variety of fields including the development of the technology required to safeguard the environment". (6)

The United States'Coastal Zone Management Act (CZMA) mirrors a national concern to harmonize the demands of urbanization, recreation, industry and energy development in the littoral fringe, and recognizes the recurrent incompatibility of these various uses. The Act underscores the nefarious role of anarchic development generated by repeated pushes. More than 56 million hectares are affected by this legislation; it provides a financial inducement to regional jurisdictions to set up rules aiming at solving opposing requests involving the coastal zone, to establish a priority order for its use, and to determine by whom and how final decisions will be made.

ACTIVITIES, KNOWLEDGE AND MANAGEMENT.

Among the major activities currently carried out and established in the zone are fisheries and aquaculture, harbors and littoral industries, power plants, retrieval of marine aggregates and tourism. Obviously as coastal zone management requires in-depth understanding of this zone, the field of investigation is particularly large. This results not only because of diversity and complexity of the phenomena typical of the interface zone, but also because it constitutes, at the present, a preferential attraction terrain of man towards the sea.

Research is frequently oriented in function of practical objectives of immediate significance : adaptation, exploitation, protection; it behooves to re-establish a just equilibrium between knowledge and use. To attain this aim it is necessary to call upon several scientific disciplines part of oceanology, for instance fluid physics and dynamics; tides, waves and currents; meteorology; water masses, geochemistry and marine chemistry; marine biology; qualitative and quantitative water analysis; inter-action phenomena; and sedimentology. The last becomes a primary topic where sandy or muddy coasts are involved as there is a need to determine transport of material and sedimentary processes, and to estimate the rate of coast aspect modification.

Contemporary evolution is but a moment of a modification whose characterization most often proves to be very delicate and cannot always be extrapolated in time. In the adaptation perspective of an evoluating zone, an examination of the immediate past will provide clearer information on the future situation than the mere observation of present characteristics.

ECONOMIC DEVELOPMENT.

The attempt at economic development that is inherent to adaptation and preservation of the littoral can thus effectively be carried out only if a policy of fundamental and pluri-disciplinary research is concurrently pursued. Such a global approach will allow to understand the especially complex and varied processes which unfold in the coastal zone and will provide the basic information needed for a scientific and technical foundation for a coherent management of the littoral.

As any other, the ocean resource is not unlimited; it is the more important because it represents a major food source for man who will more often have to call on it for the raw materials which are all but exhausted on the land. Pollution and contamination of marine waters is not man-caused, it is a natural phenomenon, but it has been intensified by the more extensive activities of man, the exponential growth of population, the introduction of non-degradable chemical compounds and, especially for coastal regions, the warming up of waters and discharges of municipal garbage and waste.

These changes of the "social" milieu require that a realistic policy be proposed that will favor a balanced use of the different oceanic zones, if it is to be avoided that an ocean that has become polluted becomes a menace to man, where the ocean has always been beneficial to humanity.

BASIC PRINCIPLES OF A STUDY.

Every coastal oceanological study, no matter which major discipline is being used nor what the final aim is, can be efficiently carried out only if researchers and technicians keep in mind two essential basic principles : it is necessary to quantify phenomena and to grasp the meaning of phenomena on a space and a time scale. The evolution of beaches for instance is on a time scale. (7)

A first step is the formulation of a "health report" of the coast; three types of zones can be defined :

(1) zones that are still intact, able to undergo adaptations in a more or less near future;;

(2) endangered zones, because they are particularly fragile, such as estuaries, deltas, lagoons or sandy shores in constant evolution;

(3) already attacked zones, victims of chemical pollution due to industrial or urban discharges, or of inconsiderate adaptations (sites of camping grounds, construction at sea's edge conducive to erosion, etc...)

Based upon the zone under study and the adaptation or intervention type being considered, two types of action can be undertaken; taking France as a prototype one can envision.

(a) continuous actions

- type "Réseau National d'Observation de la Qualité du Milieu Marin" : survey and control water quality in clearly bounded zones;
- type "Surveillance écologique du Blayais" aim first at establishing an ecological reference state prior to an adaptation, then to keep an eye on the milieu's evolution

after such adaptation (implant of a nuclear powerstation near Blaye)
- case of an unconsolidated littoral, thus in constant evolution, continued surveillance actions can also be envisioned in the form of a sedimentary dynamic study (taking of beach profiles at regular intervals, determination of volumes of eroded or accreted sediments, etc...)

(b) punctual actions

- impact study prior to implementation of any new adaptation (something that, for instance was not done for the construction of the Rance River plant, but has been done with an eye upon building a Severn River tidal power plant)
- impact of various discharges upon the quality of water which could lead to ensuing ecosystems modifications, even when keeping in mind that the marine environment can "absorb" far better than its land counterpart, all "aggressions" by diffusing and dispersing them;
- impact of new construction near the edge of the sea on the shoreline's evolution;
- intervention in case of milieu degradation :
  - fight against pollution : limitation of discharges, purification stations, et alia.
  - counteraction against coastal erosion : construction of defense structures.

Finally the development of protected areas is necessary. One must define "control zones", watch over their integrity, in order to safeguard a part of the natural patrimony, so that it can be transmitted to future generations, and also to preserve the equilibria that are useful, for instance fish fray zones.

Which factors carry an influence upon the coastal zone ? They are varied and numerous : dredging, commercial and navigation activities, discharges, military activities, tapping of sea energy, minerals extraction, some others.

CONCLUSIONS. INTERDISCIPLINARY UNDERTAKING. CONFLICTS.

Any study dealing with the coastal fringe and the inner continental shelf will be interdisciplinary, but its base will be geological, geographical and oceanological. It will include a substantial "economic" contribution. Once completed it may provide an important priming for the launching of a region in a developing country.

Because of the complexity of problems encountered, the coastal environment constitutes an outstanding milieu wherein the different scientific disciplines able to help in its knowledge and development interpenetrate and cannot be dissociated. It is indispensable that any problem pertaining to a coastal zone be studied through the joint efforts of teams made up of physicians, biologists, geologists, chemists, marine science specialists in a true pluridisciplinary endeavor.

Beyond matters physical and geological, exploitation of a coastal zone triggers economic and social consequences. Whether of offshore or open ocean origin, transformation, transportation, marketing of an ocean product will have an outfall upon life conditions and social order. Economic conflicts between various types of users of the coastal zone are often unavoidably inevitable and deep. Wither industrialization or tourism-recreation is repeatedly a coastal zone management's dilemna.

Upon these considerations places the international impact : beyond the primary pollutions, changes in the social fiber of a State influence to-day life and work conditions in an adjoining State; economic territorial conflicts flare up; changes in a commercial vocation may plunge a neighbouring region into an economic depression. Cooperation in counteracting beach erosion and coastline retreat must be on a regional or national level. A one sided effort is doomed to limited effectiveness. (8)

It is thus necessary to view any action of maintenance, adaptation, conversation, remedy, counteraction in the framework of international cooperation. This implies for industrialized nations with technological proficiency to transfer to less developed countries know-how through technical aid and personnel.

In short, the pluridisciplinary enquiry is required to understand the economic and social impact and to assess the international outfall of any action taken in a coastal area. Some problems overtake the size of a country : sea management is among them. A decision structure at the regional, perhaps national level must be envisioned. Beyond the immediate interests of each country, we must learn to manage the earth and its resources.

References

(1) Cariani, O. : L'Europe et les ressources de la mer. Paris, Ed. Georgi 1973.
Osgood, E. et al. : Towards a national oceanpolicy : 1976 and beyond. Washington, D.C., Johns Hopkins University Press 1976
California Advisory Commission on Marine and Coastal Resources. The review of the California comprehensive ocean area plan : Sacramento, CA. 1972
California Coastal Zone Conservation Commission : Geology; findings adopted on September 17, 1974 and November 19, 1974. San Francisco, The Commission 1972.
California Coastal Zone Conservation Commission : The marine environment; findings adopted July 23, 1974 : San Francisco, CA, The Commission 1974.
Senten, J.R. and Charlier, R.H. : Composition of waters and sediments in the Antwerp Harbor. Envir. Geology 5,3 (1984) 159-165

(2) Haulot, A. : Les écosystèmes saisonniers. Bordeaux, Doctoral thesis, University of Bordeaux I 1973
Charlier, R.H. : Give the earth a chance : Int. J. Environm. Studies IV, (1970) 77-89.
Charlier, R.H. and Vignaux, M. : Towards a rational use of the ocean. Proc. U.S. Naval Inst. 100, 4/854 (1974) 26-41.

(3) Brown, E.D. : Sea use planning in the North Sea : the legal framework. Proc. Ann. Conf. The Law of the Sea Inst. (The Haghe) (1978) 436-480

(4) Cuyvers, L. : Het beheer van onze Zeeën : Antwerpen-Amsterdam, De Nederlandse Boekhandel 1981.
Charlier, R.H. : Frange Littorale : appréhension géologique et

stratégie d'aménagement. Bull. Inst. Géol. Bass. Aquit. 14 (1976) 181-192

(5) Aubert, M. : Pollution marine et aménagement des rivages. Nice, CERBOM. 1973
Beeson, M. : A California town debates tourism. Place 12 (1983) 1-4
Brahtz, J.F.P. (editor) : Coastal zone management : multiple use with conservation. New York, John Wiley 1972
Charlier, R.H. : An assessment of the oceanic, geologic and economic significance of the littoral fringe. Int. Ocean Develop. Conf. (Tokyo, Aug. 5-8, 1975) III, 5, E-3007 (1975) 277-284
Fischer, J.C. : L'intégration sito-écologique des aménagements littoraux. Aménagement et Nature 39 (1975) 23-24
Ketchum, B.H. (editor) : The water's edge : Critical problems of the coastal zone. Cambridge, MA, Massachusetts Institute of Technology Press 1972

(6) Addres by Yoshinosi Ihara at the Symposium on the Human Environment and Coastal Saving, Tokyo, August 5-8, 1975

(7) Dempsey, K. : Impact of Dana Point Harbor, California on littoral sediment transport. Fullerton, California State University (Thesis) 1982

(8) Charlier, R.H. : Shiftings sands : Long Island Business III, 12 (1956) 8-16
Kaufman, W. and Pilkey, O.H., Jr. : The beaches are moving : the drowning of America's shoreline : with a new epilogue. (Living with the Shore) : Durham, N.C., Duke university Press 1983.

A Direction and Policy for Coastal Zone Management in Japan

S. Shikada, M. Mino, K. Haji, T. Moriuchi, A. Arakawa, H. Nishijima and E. Nagasaki

National Agency of Land Development and Planning, Kasumigaseki, Chiyoda-ku, Tokyo, 100 Japan

Abstruct

In Japan, people have constructed many local communities and cities on the plains near the ocean, and have possessed a deep relationship, either directly or indirectly, with the sea.

Presently, a variety of human activities such as those related to the fishery industry, maritime transportation, the manufacturing industry, recreation and leisure, the construction of cites and so on are diffused throughout the coastal zones of Japan.

Because much effort is being made involving surveys and research on problems of science and technology related to marine development, utilization, and conservation, various activities in the country's coastal zones will become more active in the future. Therefore, the promotion of comprehensive use will become necessary.

As a matter of fact, designs for the utilization of various abundant resources (including area space) in coastal zones as the nucleus of promoting regional development are being formulated within local public bodies. (refer to next table)

However, on the other hand, there exist situations where the environment of parts of the coastal area has deteriorated or competition between existing utilization fields and new utilization fields have arisen.

Considering these facts, the formulation of a coastal zone utilization plan can be considered as effective means to promote the comprehensive use of coastal zones. Therefore, in our efforts derected at determining an appropriate plan, we have identified the concepts of a standard guidance principle, which will be used as a reference when actually formulating the plan. Especially, a broad range of data regarding coastal zone utilization have been collected and organized, and ideas concerning the functions, basic policies, and items to be considered when preparing the plan regarding the notion of coastal zone utilization and the coastal zone utilization plan have been arranged.

However, it must be mentioned that the items stated here are not the official views of the National Land Agency.

Meaning of the Coastal Zone Utilization Plan

Basic Concept

It is believed that the following item should be dealt with as the basic concept when utilizing coastal zones.

Because the coastal zone utilization plan will realize these basic concepts, it is necessary to keep these basic concepts in mind when formulating the plan.

a. Consideration of coastal zones as national land space en bloc with the ocean

Coastal zones are spaces which include both the land and ocean situated on either side of the coastline. Because the development, utilization, and conservation of various resources (space will also be considered as a resource) existing in such areas interact with

Maritime Utilization of Local Public Entities in Japan

Local Public Entities	Plan, Design, Survey and Others	Preparation Date	Note
Hokkaido	"Survey of Plans and Current Maritime Problems of Developmental Projects in Hokkaido"	March 1981	Survey of Problems and Other Aspects of Maritime Developmental Projects
City of Rumoi	"Report on the Rumoi City Marinepolis Plan"	March 1983	City Construction Designs Based on the Most Efficient Use of the Sea
Oma-cho, Aomori Prefecture	"Basic Scheme for the Development of the Shimokita Region"		Establishment of Marine Farms and Development of Marine Energy Based on the Most Efficient Use of the Sea
Kuji City, Iwate Prefecture	"Comprehensive Developmental Scheme for Kuji Bay"		Development of an Industrial Housing Complex in areas Surrounding Kuji Bay, and Improvement of Fishing Grounds by Artificial Fish Reefs
Fukushima Prefecture	"Survey Report on Measures to Realize Construction of the Fukushima Prefecture Marine Recreation Base"	March 1982	Examination of the Contents to Prepare the Marine Recreation Base and Measures to Realize the Base
Toyama Prefecture	Interim Report of the Toyama Prefecture Survey Committee for Maritime Comprehensive Utilization Measures"	January 1982	Survey Report to Promote Comprehensive Measures for Marine Development (Reports of the Basic Survey Committee, Marine Energy Committee, Marine Products and Environment Committee, and Coastal Zone Utilization Committee)
City of Toyohashi	"Bay City Complex 2001"		Aiming at International Economic and Cultural Exchange Primarily Through the International Trade Fair Site
Wakayama Prefecture	"Basic Survey of Maritime Science Technology Research Facilities in Wakayama Prefecture	March 1982	Part of the Considerations of the "Kuroshio Wakayama 21 Plan" Aiming at Promoting Marine Development as well as Research and Development of Marine Science Technology
Hiroshima Prefecture	"Basic Scheme of the Ocean of the People of Hiroshima" "Basic Plan for the Preparation of the Ocean of the People of Hiroshima"	March 1982	Make efforts to Prepare an Ocean for the People of Hiroshima Where They Can Use, Love, and Become Intimate, With the Ocean, and Also Understand the Nature and Culture of the Inland Sea.
Oita Prefecture	"Basic Plan of Marinepolis"	April 1983	With the Promotion of Coastal Fishery as its Core, This Plan Aims at Regional Development by the Most Efficient Use of the Sea
Nagasaki prefecture	"Draft Proposal of the of the Basic Scheme for Marine Development and Utilization in Nagasaki Prefecture" "Note 1-3 On Marine Utilization in Nagasaki Prefecture"	March 1974 1977-1979	Examinations of Utilization Plans Emphasizing the Pursuit of the Direction Forward a Balanced and Rational Use of the Sea
Okinawa Prefecture	"Survey on Marine Utilization and the Possibility of Marine Development	March 1981	Consideration of Marine Utilization to Promote Development in Okinawa Prefecture
City of Shimizu	"Marinepolis Shimizu"	1982	Comprised by the 3 Schemes of the Shimizu Marine Development Scheme, Shimizu Food Belt Scheme, and Shimizu Civil Port Island Scheme, and Also Centered Around the Plan of the Shimizu Harbor LNG Base, This Plan Aims at Wide-ranging Marine Utilization

both land and sea, it is necessary to consider this space as land space en bloc with the ocean.

b. Promote advanced utilization of national land resources while maintaining a balance between development and conservation

In coastal zones, the development, utilization, and conservation of resources greatly affect the utilization of other resoureces, and many of these effects are one-sided. Therefore, it is necessary to comprehensively utilize the abundant resources by restricting the aforementioned effects to a minimum and maintaining a balance with the environment.

c. Consideration of coastal zones as important spaces for present and future Japanese people

Coastal zones must be utilized from the perspective of the Japanese people as a whole. Because there is a great possibility of new coastal zone utilization due to new technological developments, it is necessary to view the plan from a long-term perspective.

Objective of the Coastal Zone Utilization Plan

The coastal zone utilization plan will be prepared in order to deal with various regional problems. The following objectives can be cited as examples:

a. Regional development

Efforts will be made for regional development through development and utilization of untapped development potentials of coastal zones, where a variety of uses are possible such as industry, energy-related facilities, the mining industry, fishery industry, and recreational activities.

b. Utilization of coastal zone resources

Numerous varieties of resources exist in the coastal zone. From the viewpoint of land resource management, efforts will be made for advanced and varsatile use by preventing inidividual and desorderly development, utilization, and conservation of these resources.

c. Environment conservation

In many instances, the development and utilization of coastal zone resouces have significant effects on the environment, and it is very important to keep these effects at a minimum in order to protect the natural scenery and various ecomsystems, to make possible the versatile utilization of coastal zones, and also to secure the new forms of use in the future. Environment conservation is an item of particular importance in the plans for the very highly used coastal zones surrounding inland bays.

Parts of these objectives overlap with one another. Therefore, it is believed that the actual plans will be prepared by combining these objectives.

Functions of the Coastal Zone Utilization Plan

The coastal zone utilization plan has the following functions, and it is necessary to promote the actualization of basic concepts of coastal zone utilization by formulating plans in which these functions can be fully displayed.

a. Coordination of coastal zone utilization

In cases where the individual utilization of coastal zone resources (including space) is being conducted separately as well as inefficiently, the aforementioned plan functions to actualize efficient utilization by the overall integration of various types of use.

b. Regulation or promotion of coastal zone utilization and development

The plan functions to restrain disorderly development and utilization and regulates such undertakings to the required minimum standards. On the other hand, versatile forms of use will be possible, and the plan also functions to promote uses which had been restricted up until now or other new related ones.

c. Various adjustments

It is believed that it will become necessary to make various adjustments between regions or utilization fields. These arrangements may be made during the process of formulating the plan, or are also possible by clarifying the system or procedure of

conducting plan adjustments, adjustment principles, or various standards of judgement. It is then possible to carry out rational and efficient adjustments.
Basic Policies when Formulating the Coastal Zone Utilization Plan

The following items can be considered the basic policies when formulating the Plan:

Advanced Utilization

The following 2 directions can be considered as attempts toward advanced utilization.

a. Advanced utilization making the most efficient use of the characteristic of coastal zones

Types of use appropriate for various coastal zones differ significantly by location, effects of the oceanic and atmospheric weather, and location, form, and condition of the coastline of coastal zones. Therefore, it is necessary to pursue those forms of utilization which involve the advanced use of coastal zone characteristics.

Some forms of such uses are maritime recreation making the most of the natural scenery, promotion of cultivated fisheries making the most of the fishing ground environment, and establishment of industrial production and distribution centers making the most of social assets of the region.

b. Advanced utilization by multiple use

A variety of needs tends to be concentrated in coastal zones. Therefore, it is necessary to respond to these diverse needs appropriately by promoting multiple use. It can be said that the multiple accumulation of various urban functions by man-made islands and redevelopment of urban coastal areas illustrates one future direction.

Balance between Development and Conservation

It is necessary to fully consider the balance between advanced use and that space which should be conserved. Therefore, such a balance needs to be examined from the following viewpoints:

a. Scale and necessity of use of coastal locations

In cases where coastal locations will be altered or exclusively owned due to developmental usage, the scale and necessity of use must be appropriate. Thus, scales and the like will be considered from the following perspectives:

- Scale of use is reasonable in relationship to the capacity of coastal zones in terms of development.
- Satisfactory balance exists between supply and demand.
- No other suitable location which can be subsituted for the aforementioned exists.

b. Environmental conservation

Planned measures regarding the environment and the relevant preventive measures related to the effects of development on the environment are necessary. The following items can be given as examples:

- Some of the tidelands, algae grounds, recreation sites, nature parks, and scenic areas are outstanding forms of the natural environment. Conservation must be given top priority in these places.
- Positive creation of the environment and environmental improvements will be promoted, represented by such examples as man-made tidelands, man-made beaches, and the promotion of seawater exchange.
- Measures for the prevention of usage effects and improvement of use which will become the original source of pollution due to changes in the ocean current which are caused by changes in the ocean, will be implemented.
- Preventive measures will be implemented in cases where organisms are affected.

c. Antidisaster measures

The coastal areas of Japan face various effects of nature from the ocean such as tidal waves, high tides, and coastal erosion; therefore, development, utilization and conservation which take these effects into consideration must be implemented. Anti-earthquake measures must also be kept in mind.

Public Access

Coastlines have been important as space to carry on our daily lives and space for recreation for many people. It is believed that this demand will increase significantly in the future; for example, when considering such developmental use as roads, coastal industry complex projects, coastal residential sites, various types of urban redevelopment, facility arrangements in which unnecessary monopolized use of coastlines exist should be avoided to secure public access. When there are many forms of monopolized use, it is desirable that the residents can secure the use of the coastlines by establishing man-made beaches or coastal parks.

When establishing coastal conservation facilities in places such as coastal parks, it is believed that these facilities will be more effective by constructing structures where people can develop a fondness for water, such as structures where people can easily stand on the water's edge or wade in the water.

Requests to Individual Plans Made by Various Sectors

The major objectives of various developmental usage plans are occasionally industrial development or harbor development, or individual use purposes. Among these objectives and in common with these various plans are guidelines which are effective for improving the plans. The coastal zone utilization plan requests that individual development plans positively incorporate these themes. The utilization plan will also incorporate efforts to make adjustments. The following can be cited as examples of such guidelines:

a. To improve the quality of services for coastal zone residents, employees, and users by positively incorporating parks and green zones into the plans
b. To improve the scenic impact by modifying the coastal zone scenery or urban coastal scenery
c. To improve the water fronts of regions as well as make efforts to revitalize the regions by the redevelopment of harbors, urban redevelopment, and coastal industry development through scrap-and-build methods.
d. Civil engineering technology is occasionally used to promote seawater exchange for the purpose of promoting water quality improvement or increasing the productivity of marine products. Such technology should be positively improved to make it effective.

Utilization Adjustment considerations from a Long-term Perspective

Long-term prespectives are necessary for future coastal zone usage because coastal zone use relies heavily on future technological developments, and also because there are possibilities of new forms of usage resulting from advancements in technological development.

Consolidation of Coastal Zone Data

When formulating the plan, it is desirable to draw up discret usage plans through collecting as much information as possible regarding the concerned coastal zone, systematically analyzing as well as consolidating information at all times, and conducting new surveys. In addition, consideration on the yet unknown possibilities should be made.

Sphere of Areas Considered to Be Coastal Zones

Coastal zones are considered as a comprehensive space including land areas as well as the ocean located on either side of the coastline; however, the sphere of areas, the object of the plan, must be clarified to a certain degree. It must be determined how to consider coastal zones within the aforementioned sphere based on the regional conditions and which cannot be determimed uniformly. Nevertheless, when determining the sphere, considerations from the following viewpoints are necessary:

a. Social and economic radius of the region located around the coastal zones
b. The ocean area of the coastal zone as one of the whole, which is determined by the conditions of the oceanic and atmosheric weather, coastal geography and so on
c. Sphere of the land area affected by various factors of the ocean such as the atmoshere or winds
d. Water quality of the ocean area, and also the sphere of the land area that affect the ocean area by the movement of earth and sand, and the like

Reference

〔1〕National Land Agency; "Survey and Research on the Formulation of a Coastal Zone Utilization Plan", March, 1984.

Citizen Participation in Environmental Policy Making in the Coastal Area

Some Lessons from a Cross-National Study

T. Abe[1], T. Tsurutani[2], N.P. Lovrich[2] and J.C. Pierce[2]

[1] College of Humanities and Sciences, Nihon University, Mishima, Shizuoka, 411 Japan
[2] Department of Political Science, Washington State University, Pullman, WA 99164-4880, USA

ABSTRACT

Two critical gaps are suspected to separate the citizenry from the policy elites in postindustrial society: in policy-relevant knowledge and in values concerning economic growth and ecology. They are said to prevent rational policy making. A cross-national study recently undertaken indicates, however, that the gaps feared are actually of the kind that support the proposition that citizen participation in environmental policy making is imperative.

INTRODUCTION

In postindustrial society the prospect of a widening gap between the few and the many in levels of policy-relevant knowledge is of increasing concern, for, if real, the gap would affect the quality of democracy. And if such knowledge tends to replace wealth as the principal source of power, as some analysts suggest (e.g., Bell and Touraine), we need to ascertain how it is in fact disbributed. This, however, is only part of our concern. Power is ethically neutral, for it is a means to an end. The end, in turn, is a function of values of those who use that power. We should therefore also find out what kinds of values are held by those who have that knowledge.

While some analysts suspect a de-democratizing trend in the distribution of power, as alluded to above, there are others who fear a hyper-democratizing tendency. The concern here is not with the distribution of knowledge but with what is viewed as a critical variance between policy elites and the public in values. In this view, policy elites possess some approximation of appropriate values that points them in the direction of restraint and future-oriented concern in resource utilization and preservation of nature. These elites, however, are critically inhibited from following an appropriate course of action by the dynamics resulting from the expanding popular demands. This popular appetite for self-gratification has been abetted by the acceleration of redistributive politics and rapid economic growth. This view is the basis of the worry about the "governability" of democracy (Crozier et al. and Janowitz). The logical extention of this view is that in the final analysis, democracy is incorrigible and that the remedy needed to save the masses from themselves is to replace democracy by some sort of authoritarian rule capable of enforcing rational preservationist policy.

These twin concerns (the putative knowledge gap and the alleged variance in values between policy elites and the public) prompt us to investigate the extent to which they are indeed justified. This paper explores these concerns on the basis of findings from an empirical cross-national study undertaken in 1982-84 (1). The subject of the paper is of great importance to this distinguished symposium since ocean resource exploration ought to be undertaken with utmost caution and restraint to avoid what one eminent human ecologist called "the tragedy of the commons" (Hardin). Put simply, our query is: are the people in our democracies capable of participating in policy making for proper approaches and uses of ocean-based resources in the future?

CONCEPTUALIZATION OF POLICY-RELEVANT KNOWLEDGE

Ultimate decision making perforce remains in the hands of the few whom we refer to as "policy elites" -- experts, politicians, bureaucrats, and others with preferential access to power. This is so not because these few actors are inherently superior to the citizenry but because they happen to occupy certain strategically compelling locations in the decisional process. In a democracy they are proxies for citizens in those locations, there to do certain kinds of work that citizens do not have time and/or inclination to do. In other political systems the advantage those few enjoy is not limited to the strategic locations but also extends to control over the dissemination of "intelligence" (information and knowledge). A democracy delimits the means by which policy elites can withhold intelligence from the public by such devices as freedom of information, freedom of the press, etc. The citizenry has not only the right but also the responsibility to know as much as its proxies do about public affairs. A society becomes more democratic as the elite-mass gap in knowledge is narrowed. In this paper we present certain data for two major democracies -- Japan and the United States.

The "general public" is represented in our study by the sample of residents surveyed. The "policy elites" consist of elected and high appointive officials and policy-area experts. The "interests/organizations" contain three types of respondents: members of environmental groups taking part in efforts to protect environmental values over the course of the past decade; members of civic groups devoted to "good government" in the broadest sense that became active in attempts to educate the public when issues of environmental protection came to the public agenda; and "economic interests" understood as business and labor organizations with a pecuniary stake in the economic development of local resources.

Policy-relevant knowledge in our study involved two separate measures, one constituting a basic ecology science quiz designed to assess the respondent's command of basic scientific knowledge in ecology, and the other entailing an assessment of his familiarity with technical terms employed in the discussion of environmental problems. Table 1 deals with basic scientific knowledge.

Table 1 shows a considerable variation in levels of knowledge between the public and the policy elites in both nations. But is this variation significant? We think not. The general public in the table is quite undifferentiated. It consists of rank-and-file citizens of whom relatively few are active and many engage only in formal political acts (e.g., voting). Some are even indifferent. The type of "citizen" input that produce impact on both the manner and the content of policies is made primarily by those "participatory" citizens organized into politically relevant groups. It is they that count disproportionately in democracy when it comes to concrete government policies. They are, in our survey, represented by environmental and civic groups. Though "economic interests" are included under the rubric of "interest/organizations," they are not of the same popular and voluntary cross-stratal "citizen" nature as the environmental and civic groups. Thus there is reasonable justification for identifying environmental and civic groups as constituting the key "participatory public" specified above.

Table 1. Distribution of Knowledge of Ecology Science (2)

Level	General Public		Interests/Organizations: Environmental		Civic		Economics		Policy Elites: Officials		Experts	
	JA	US	JA	US	JA	US	JA	US	JA	US	JA	US
Low 0	12%	14%	1%	8%	2%	13%	1%	14%	8%	12%	4%	0
1	21	23	8	18	6	12	4	17	8	8	9	4
2	32	27	26	24	27	18	13	30	23	21	16	5
3	22	26	42	30	44	44	38	30	46	49	24	64
High 4	12	9	24	20	22	13	45	8	15	10	48	26
Total	99	99	101	100	101	100	101	99	100	100	101	99
N	649	522	238	66	158	77	192	92	48	67	388	72
Mean	2.0	1.9	2.8	2.4	2.8	2.3	3.2	2.0	2.5	2.4	3.0	3.1

This participatory public, by virtue of its experienced involvement and activist orientation, maintains a closer strategic proximity to the channels of policy making. To the extent that there ought to be some correlation between participation and knowledge, these groups should be expected to demonstrate higher levels of knowledge than the undifferentiated public. Table 1 indicates they in fact do so. In Japan, the participatory public is even superior in this sense to "officials" and not far behind "experts." A similar situation obtains in the United States.

Knowledge of technical terms used in policy debate, too, pertains to the question of whether and to what extent the public should participate in policy making. These terms are important in that they constitute the vocabulary of valuation, deliberation, and decision making. We now take a look at this dimension.

Table 2. Distribution of Technical Term Familiarity (3)

Level	General Public		Interests/Organizations: Environmental		Civic		Economic		Policy Elites: Officials		Experts	
	JA	US	JA	US	JA	US	JA	US	JA	US	JA	US
High (9-12)	46%	29%	71%	48%	66%	51%	66%	36%	92%	71%	89%	93
Medium (13-16)	42	38	25	37	28	45	28	45	6	22	10	5
Low (17-27)	12	33	4	15	6	4	6	19	2	7	1	2
Total	100	100	100	100	100	100	100	100	100	100	100	100
N	673	464	223	54	151	73	151	87	47	55	386	68
Mean	13.2	14.9	11.5	12.9	11.7	12.4	11.7	13.5	10.0	11.8	10.4	9.8

The policy elites in both nations are clearly more knowledgeable of technical terms than the general public, but the participatory public in each is clearly closer to the policy elites than to the general public. Again, this suggests a correlation between participation and knowledge.

KNOWLEDGE APPLIED TO WHAT VALUED ENDS?

Even if the public is knowledgeable, substantive popular participation would, some analysts argue, still inhibit effective and rational policy making because the public's differential situational location confines it to the kind of perspective and values that are not congruent with imperatives of societal necessities and collective interests. Here, the main concern is not with the public's "ignorance" but rather with its allegedly narrow, short-range self-interests regarded as inimical to the larger interests of the collectivity. The fundamental difference between the few and the many in this view lies not in wealth, nor in knowledge but critically in values or "dispositions." But, is this concern empirically justifiable? We shall now consider this query by examining certain issue-specific values. The first of such values is a "New Environmental Paradigm" (NEP). This construct incorporates "such concepts as limits to growth, steady-state economy, a society in balance with nature, and the rejection of anthropocentric exploitation of the natural world" (Watts and Wandesforde-Smith, p. 349-50). It is indeed the kind of value orientation which those observers worried about the "governability" of democracy impute to the policy elites and which they imply are wanting among the masses. Table 3 shows the distribution of this value among those different groups in our survey.

Table 3. Distribution of "NEP" Attitude (4)

Level	General Public		Interests/Organizations: Environmental		Civic		Economic		Policy Elites: Officials		Experts	
	JA	US	JA	US	JA	US	JA	US	JA	US	JA	US
Low (6-21)	30%	35%	32%	7%	28%	12%	26%	46%	13%	38%	19%	18
Medium (22-24)	28	33	46	19	39	31	37	35	45	24	43	29
High (25-30)	42	32	22	74	33	57	37	19	42	38	38	53
Total	100	100	100	100	100	100	100	100	100	100	100	100
N	631	479	220	58	149	68	188	83	45	55	376	66
Mean	23.1	22.7	22.5	26.3	23.1	24.9	23.4	21.6	24.1	22.9	23.8	24.6

Table 3 indicates some discernible variations in the distributions of support for the NEP. Overall, in mean scores, the two nations are quite alike. However, when the degrees of concentration of NEP attitude are investigated, an interesting variation can be noted. In Japan, the general public and officials contain the highest concentration of such orientation, while in the United States the similar phenomenon is most pronounced among the participatory public and experts. The participatory public in Japan is strangely muted in this regard, showing the lowest concentration of NEP commitment. In mean scores, which perhaps are the most useful indicators of aggregate group orientation, Japan's participatory public is fundamentally similar to the general public and to the policy elites as well. In contrast the American participatory public stands out as somewhat more favorable to NEP than either the general public or the officials. NEP, however, is but part of the ongoing transformation of values in postindustrial society. Another important aspect of this transformation relates to environmental pollution, the most widespread and visible of the negative impacts of economic growth. Table 4 indicates the distribution of pollution sensitivities among the groups in our survey.

Table 4. Distribution of Index of Number of Conditions Defined as Pollution (5)

Index	General Public		Interests/Organizations: Environmental		Civic		Economic		Policy Elites: Officials		Experts	
	JA	US	JA	US	JA	US	JA	US	JA	US	JA	US
Low (≤29)	17%	30%	12%	26%	17%	17%	19%	41%	11%	28%	19%	13%
Medium (30-34)	34	34	17	25	30	29	47	29	35	35	35	46
High (≥35)	49	36	61	49	53	54	34	30	54	37	46	41
Total	100	100	100	100	100	100	100	100	100	100	100	100
N	663	458	230	55	156	65	191	76	44	51	383	61
Mean	33.9	32.2	35.5	33.8	34.4	35.1	32.9	30.9	34.9	32.9	33.7	33.7

This table shows that in both nations the general public is nearly as pollution-sensitive as the policy elites and that the participatory public is more sensitive than the policy elites. Both the general and participatory publics display a greater sensitivity to pollution danger than economic interests do. If such sensitivity is a vital element of relevant policy making, then it would seem obvious that popular participation becomes important in that the public, general and participatory, must act as a restraining force upon the policy elites and economic interests. Negative externalities perpetrated in postindustrial society are in large measure the function of insensitivity toward pollution dangers, insensitivity compounded by the greed for self-gratification and profits.

Corollary to pollution sensitivity is the concern with ecological preservation. Table 5 shows the distribution of preservationist orientation among our groups.

Table 5. Distribution of Preservationist Orientation (6)

Level	General Public		Interests/Organizations: Environmental		Civic		Economic		Policy Elites: Officials		Experts	
	JA	US	JA	US	JA	US	JA	US	JA	US	JA	US
Low 1	1%	1%	0%	2%	1%	4%	0%	1%	0%	0%	0%	0
2	6	10	7	0	4	0	6	19	7	12	3	15
3	57	61	54	24	60	6	78	67	62	63	68	58
4	34	26	37	67	34	0	16	13	31	24	28	28
High 5	2	2	3	7	2	91	0	0	0	2	1	0
Total	100	100	101	100	101	101	100	100	100	101	100	101
N	642	483	229	66	152	54	190	88	42	59	383	69
Mean	3.3	3.2	3.4	3.8	3.3	4.7	3.1	2.9	3.2	3.2	3.3	3.1

In both the United States and Japan, the public, general as well as participatory, manifests a slightly higher mean tendency toward preservation than do the policy elites and economic interests. When we consider the magnitude of concentration of strong preservationist orientation (combination of 4 and 5 levels) the distinctiveness of the public (again both general and participatory) from the policy elites and especially the economic interests becomes much sharper. Both the general and participatory publics show significantly higher degrees of preservationist concentration than the other groups and in the United States the gap between the participatory public on the one hand and the policy elites and economic interests on the other is impressive.

CONCLUSION

That the policy elites have higher levels of knowledge than the general or participatory public is not surprising. After all it is their business to be more knowledgeable than those whom they represent either directly or indirectly. There are strong reasons for supposing, however, that the knowledge gap between the elites and the masses can be significantly reduced. Democratization of education and advances in information and communication technologies have clearly contributed to the narrowing of the traditional gap between the elites and the public. Impetus for knowledge, however, is not limited to education or mass media. One analysis of data collected through our research indicates a consistent correlation between seriousness of concern and knowledge. It finds that the degree of concern with relevant policy issues is "the best predictor" of the level of knowledge when compared to such conventional variables as education, occupation, and income (Pierce et al., p. 12). This finding reflects the critical potency of motivational or situation-specific causes of knowledge, the very same causes that underlie the policy elites' levels of policy-relevant knowledge. Policy elites who are interested in promoting environmental preservation should attempt to raise the popular awareness of the seriousness of a given policy issue "through appropriately designed strategies of information dissemination." (ibid.)

On the important question of the values giving direction to the use of power (knowledge), we find that, in mean scores, the difference between the public and the policy elites is rather small. It is interesting to note that the difference falls in favor of the general and/or participatory publics than the policy elites in two of the three measures. In the one simple measure which perhaps represents the quintessential value to be promoted in the postindustrial age (Table 5), both the general and participatory publics clearly surpass the policy elite (let alone the economic interests) in the magnitude of concentration of high commitment (levels 4 and 5 combined).

Our findings concerning valuational variance between the policy elites and the citizenry suggest the necessity -- both intrinsic and instrumental -- of substantive popular participation in environmental policy making. This conclusion applies to the exploration and utilization of ocean-space resources, in two senses. One is that the policy elites should be assisted, at times even checked, by the public in the making of relevant policies. This is the case not so much in determining specific technologies to be used as in delimiting the substantive parameters of decisional goals. The other is that conflict between economic interests and policy elites may be intervened in by the public, especially the participatory public. To the extent that postindustrial society is permeated and substantially sustained by these economic interests, these interests remain potent politically. The policy elites, to the extent that they endeavor to resist the power of these groups, would then be well advised to seek the support of the public in general and participatory public in particular in their attempts at rational policy making consistent with the public interest. It must be remembered that it was the desire for growth and profit that led to an exponentially rapid exploitation of nature, in the process inflicting irreparable damage to ecology and that it was only when voices in the public were raised with vigor and skill that policy makers began to address the issue against the resistance from those economic interests. The public, then, should be viewed as an indispensable party to a coalition that alone can withstand the temptation of growth and profit and to obviate the risk of incalculable harm to our new frontier of resource exploration, the ocean space. The public is a vital force in our attempt to avoid the "tragedy of the commons."

Notes

1. Research supported by the Japan-U.S. Friendship Commission (through the Institute of International Education), Washington State University, Nihon University, the Ohta Foundation, and the Japanese Ministry of Education. The survey was conducted in the City and County of Spokane, Washington and the contiguous cities of Mishima, Numazu, and Fuji in Shizuoka Prefecture. For details of survey methods and logistics involved in our study, see Lovrich et al. in References below.

2. The ecology science quiz used for this table was taken from Maloney et al. in References below.

3. The technical term familiarity test is explained at length in Lovrich et al.

4. The New Environmental Paradigm measure was developed by Dunlap and Van Liere in References below.

5. For details of this multi-item additive scale, see Lovrich et al.

6. Ibid.

References

1. Bell, D. The coming of post-industrial society. New York: Basic Books 1974.

2. Dunlap, R. and Van Liere, K. The "new environmental paradigm": a proposed measuring instrument and preliminary reports. Journal of Environmental Education 9 (1978) 10-19.

3. Hardin, G. The tragedy of the commons. Science 162 (1968) 1243-8.

4. Crozier, M., Huntington, S., and Watanuki, J. The crisis of democracy. New York: New York University Press 1975.

5. Janowitz, M. Social control of the welfare state. New York: Elsevier 1976.

6. Lovrich, N. P., Pierce, J. C., Tsurutani, T., and Abe, T. Knowledge and action in environmental politics. A paper delivered at the 1984 annual meeting of the Western Political Science Association, Sacramento, April 12-14, 1984.

7. Maloney, M. et al. A revised scale for the measurement of ecological attitudes and knowledge. American Psychologist 30 (1975) 787-92.

8. Pierce, J. C., Lovrich, N. P., Tsurutani, T., and Abe, T. Policy relevant information and public attitudes. Forthcoming in Journal of Soil and Water Conservation.

9. Touraine, A. The post-industrial society. New York: Vintage Books 1971.

10. Watts, N. and Wandesforde-Smith, G. Postmaterial values and environmental policy change. Policy Studies Journal 9 (1980) 346-58.

The Coastal Zone Atlas as a Potential Tool for Citizen Participation in Japanese Coastal Area Planning

H.A. Shapiro

Department of Environmental Planning, Osaka Geijutsu University, Minami-Kawachi, Osaka, 585 Japan

ABSTRACT

In this paper, the bio-physical as well as some of the socio-cultural and political conditions related to Japan's coastal area, its use and management etc. are introduced. Then the nature of coastal information in that nation is briefly discussed. Much of that information is either inaccessible or incomprehensible to environmentally-concerned coastal citizens wishing to actively participate in coastal area decision-making. The concept of the coastal atlas is then introduced with a description of some examples. The coastal atlas is suggested as one possibly effective tool for helping the people in their effort to actively and constructively participate in coastal zone management, planning and protection. Finally, work in progress on one such atlas for the Osaka Bay area is briefly introduced, and some of the difficulties foreseen facing such work are outlined.

INTRODUCTION

Japan is both a coastal and maritime island nation on the western edge of the Pacific Basin. It is able to boast one km. length of coastline for every $10km^2$ of land area.[1] Furthermore, it contains over $10km^2$ of sea surface area (in its 200-mile zone) for every 1 km^2 of land area. Yet within its confines of 377,765km^2 of land area, 66.9% of which are steeply sloping, forested and virtually uninhabitable mountainland, live some 120-million people. Over 40% of this number live in the nation's coastal cities and towns where over half of Japan's industrial productive capacity is likewise located.[2] Nearly 10,000km^2 of Japan are flood-prone coastal plains on which 17.3% of the nation's people live.[3] Under these circumstances, the desire for more "habitable",i.e. flat, land is understandably strong. The mountains offer one possible source of such land assuming they can be flattened, but this is expensive, often locationally inconvenient, wrought with hazards and quite foreign to the flatland-loving Japanese people. The other source is located at the edge of the sea which is readily accessible, offers little resistance and is generally convenient to existing settlements. It is also more suitable to the people's cultural instincts. However, the nature of such coastal places is environmentally poorly understood. In this paper, an attempt will be made to show how a coastal zone atlas could be used to better understand the opportunities and limitations of the coastal environment, serve as a basis for coastal zone management and environmental planning, and aid in increasing the opportunity for constructive citizen participation in the coastal area management and planning process under extremely difficult and complex political and cultural conditions.

THE COASTAL CONDITIONS OF JAPAN

The great geological and climatic variability of Japan have combined with natural erosion and sedimentation processes to form Japan's 33,000km long coastline. As of 1976,[4] some 59.6% of the coast was classified as natural (neither the beachline nor water area adjacent to it have been altered by man), 19.2% was judged to be semi-natural (the water area off the beachline remains natural,but the beachline has been altered by structures) and 21.2% was classified as artificial (both the beachline and water area off it have been altered by man-made structures). Land uses have been likewise classified into three types.[4] As of the same year,54.7% were natural uses (orchards,woods,beaches or cliffs,etc.), 21.2% were farmland and rest,24.1% were industrial, urban and related uses. In both coastal conditions and uses, with the spread of coastal landfill projects, the figures for both natural coastal length and natural uses have been in steady decline, while those for artificial coast and industrial-urban related uses have been rising. In the 33 years between the end of World War II and 1978, a total of 113,350 ha of coastal land had been created by landfill nationally, over three times that created before 1945. Of this postwar fill area, some 52,000 ha has been for agricultural purposes, over 35,000 ha has been for industry and the rest,nearly 26,000 ha, was for other uses.[3]

The beauty of the Japanese coast is attested to by the fact that no less than 50 marine parks may be found there.[5] Its biological productivity can be seen in the fact that marine fisheries provide almost all of the nation's domestic fish needs with coastal and nearshore areas providing over 3/4 of the total.[6] The importance of the coast for the nation's economy is evident from the fact that over half of all of Japan's economic activity is located there.[7] Also, over 14.5 million sports fishermen,[8] some 150 million beach users annually[7], 1,087 commercial ports,[9] and 2,846 fishing ports,[10] along with 12 of the nation's 15 cities of a half-million or more people[11] are all located along Japan's heavily used coastline.

Administratively, the ultimate authority of Japan's coastal area management is the central government with much of its administrative power delegated to the prefectural governments and a handful of major municipal governments. All administrative bodies manage the coast in a vertical sectionalized manner according to numerous,equally sectionalized national level laws.[12] Japan's coastline is jurisdictionally divided between the Ministry of Transportation(M.O.T.), the Ministry of Construction (M.O.C.) and the Ministry of Agriculture,Forestry and Fisheries (M.A.F.F.), 51.2%, 23.1% and 25.7%,respectively.[13] Each ministry tends to approach the solution of problems within its jurisdictional area sectorally and without collaboration with other ministries or agencies, unable to depart from the mission authorized to it by law even though the need to do so may be recognized. This sectional centralism[14] has been considered to be a unique feature as well as a major limitation of Japan's approach to coastal area management.[7]

COASTAL INFORMATION SYSTEMS

It goes without saying that wise and effective coastal zone management and planning must be based on reliable scientific information about the elements,processes and their interactions in coastal and ocean ecosystems. Such information can be obtained by a large variety of means, for instance by remote sensing from satellites and aircraft platforms etc. for large areas,[15] or by detailed field studies of smaller areas or of areas requiring highly specific information. Remote sensing is particularly useful in providing data on the landward portion of the coastal zone, especially land use and land cover information. This kind of information is essential in determining things like coastal watershed sediment yields. It is also useful in the analysis of non-point air pollution sources, among others. It is also applicable in gather information on the characteristics of bays and estuaries. For instance, it can be used to monitor water temperature, salinity, turbidity, chlorophyll content and total sediment loads. Its use for detecting and/or monitoring pollutants as well as current tidal dynamics has also been suggested.[15] It has been used in the landward part of California's coastal strip to observe interrelated physical, biological and cultural phenomena such as land use changes,traffic trends, and recreation and other impacts on ecosystems.[15]

Perhaps the most commonly used method is aerial photographic mapping and monitoring. Extensive formal training is not necessary to do air photo interpretation, it provides much needed high spatial resolution imagery, and it is readily available. These advantages can be expected to make this particular data collection method the one most used in the foreseeable future.

Non-photographic sensors on aircraft platforms exist but are not nearly as available as is aerial photography. Also satellite-based multi-spectral scanners produce image data, but such data is limited to large area studies,such as offshore waters,coastal plains,watersheds or estuaries and bays.[15] Landsat and Seasat are typical examples. All of these methods are part of the data collection element in any coastal management system. Once collected, this data has to be stored so that it can be readily retrieved in forms that are useful to planners, decision-makers, scholars and others. Such forms tend to be maps, graphs, charts and/or tables to mention a few. Computerized information management is being increasingly called on to fulfill these tasks.[16,17,18] But whatever the form of the information, it has to be understandable to its users. Thus, technical terms will be necessary if the user is an expert, but if the user is the general public, technical terms will need to be explained or translated into common terms; Quantitative data will need to be converted into commonly used units. Furthermore, the recipient needs to be informed of the reliability of the data as well as its limitations. By documenting all sources, validations may be done as needed. This helps to protect objectivity in information systems. Whether the information system is manual or computerized, it has to be flexible to continue to be effective.[19]

COASTAL INFORMATION IN JAPAN

A wide variety of information about the nation's coast exists in Japan. Statistics on coastal contidions,projects and costs,plans nationwide and prefectural can be found in the Japan Coastal Society's Manual.[20] Then, each national government agency has information about the coast relative to its own jurisdictional limitations. The largest collection comes from the Ministry of Construction, specifically the National Geographic Society which publishes all government maps,including information on coastal topography, geology, land use and public works as well as past hazards and even land ownership, though not every place is necessarily covered nor in every category. Next is the Marine Safety Agency which also publishes coastal condition maps, sea bottom condition and topographical maps as well as information of navigation channels and tides. The Fisheries Agency puts out information about fishing grounds and fishing ports. The Ministry of Transportation has information on coastal erosion, recreational facilities,especially marinas, and designated port and harbor areas. The National Meteorological Agency handles information on wind speed and direction, temperature, sunlight and the like. Finally, the Environment Agency has information on national park areas and bird protection areas. The Rivers Bureau of the M.O.C. has its own set of coastal statistics as well.[13] In addition, each agency has a quasi-academic society which publishes regular journals for its members. For instance, the Ports and Harbors Coastal Society (M.O.T.) publishes a journal called Harbor Hazards, while the Japan Ports and Harbors Society (M.O.T.) publishes a magazine entitled Ports and Harbors. The Japan Coastal Society (M.O.C.) puts out a journal called Coast, and so on. There are research institutes which have been doing long-term research on the nation's coast. For instance, there is the M.O.T.'s Economic Research Center and the private Japan Research Institute, both in Tokyo.

To this author's knowledge, the only effort to date to systematically compile a large amount of this widely scattered information was M.O.C.'s National Coastal Area Conditions Survey of 1974.[21] In that survey, seven government agencies centered on the M.O.C. cooperated to compile 24 items of information on the coast within an area 2km wide,1 km landward and 1 km seaward of the mean high tide line. This survey was expanded in 1977 to include more detailed information on the coastal water area out to 50m in depth.[22] This data was recorded as coded mesh data. Unfortunately, most of the above mentioned information is not readily available to the general public except for certain maps,such as the Maritime Navigation Maps and selected Coastal Condition Maps which may be purchased through selected distributors. Without influential connections and/or special permission, little of the rest is easily accessible, especially in areas of potential controversy,such as large landfill projects etc. Government societies such as those mentioned above are in principle closed to all but government officials and members of affiliated organizations.

There is as yet no equivalent of the U.S. Freedom of Information Act in Japan,so great pains are necessary for the ordinary person or even scholar to obtain most kinds of coastal information. Efforts are under way in some areas to establish open information programs,but it is a slow process.[23]

In addition, the information in the previously mentioned M.O.C. Coastal Area Conditions Survey is in a form such that even if it were accessible, it would be nearly useless to anyone but trained experts. Such conditions make it nearly impossible for coastal citizens to actively and constructively participate in the nation's coastal zone management and planning process.

RISING COASTAL CITIZEN AWARENESS AND MOVEMENTS

Japan had had a long history as a centralized authoritarian state with enforced conformity.[24] Authority was a given , taken for granted as an unavoidable part of the natural order,[25] and obedience to that authority had to be unconditional. Such a traditional political environment did not allow democracy to develop in Japan. Not even did the concept of "rights" exist in that nation until a word for it was invented after its Dutch equivalent during the Bakumatsu Era (around 1860).[26] It was not until the post-World War II Japanese Constitution of 1947 developed by General MacArthur's staff and calling for the guarantee of fundamental human rights and the development of democratic institutions that democracy became a firm part of the political culture of Japan. Nevertheless, having been imposed through military occupation and not from the poeple's demand for a role in decision-making, the extent to which it has taken root remains a point of controversy.

However, with the postwar spread of urban blight, nuclear weapons and environmental pollution etc., the people began to organize as never quite before into what are called shimin undo (citizens' movements). Despite the diversity of their demands, they all are very unhappy about the government's industry-first policy of high economic growth and are working to get citizen-first policies put into effect. They question the government's decision-making process and hope to eradicate the elitist bureaucratic mentality of the nation and replace it with direct,continuous,citizen participation in policy making.[25] They are not nationally unified nor do they necessarily oppose the entire established system. Rather, they have been said to represent a new political trend seeking genuine popular sovereignty and democracy with full citizen participation in government and access to information,both still lacking in modern Japan.[25]

In Japan, policy decisions are made at the national level and flow down through the prefectural and local governments to the people. Citizen movements seek to change this flow so that decisions originate with the people and are reflected in national policies and actions. Because they go against the grain, so to speak, of the traditional role of conservative Japanese and threaten to usurp the power of the up-to-now all powerful bureaucrats, they are,needless to say, both scorned and feared.

True, there are precedents for these citizens' movements especially since Japan came out of its self-imposed isolation nearly 120 years ago (in 1868 to be exact). It took the postwar constitution and the negative impacts of rapid economic growth to bring them to their present stage of development, the environmental issue being a major concern of almost every citizens' movement today.[27] For the purpose of this paper, a few representative coastal citizens' movements will be mentioned here.

At the height of Japan's high economic growth period in the late 1960's, landfill for industrial uses was rampant especially along the Pacific coast. In the city of Wakayama at the eastern end of the Seto Inland Sea, one of the first "direct action" types of movements developed. It culminated in a 33-day hunger strike by its leader,Mr. Kazuya Ujita, in the spring of 1969 as a last resort to stop landfill of the City's last natural beach by the neighboring giant Sumitomo Iron and Steel Corp. The success of Mr. Ujita's action and his movement stand even today as a unique example of the citizens' resolve and determination to protect the local natural coastal environment of Japan.[28,29]

At the national level, there is the Irihama-ken undo (the Coastal Access Rights Movement) which began in the city of Takasago on the northeastern coast of the Seto Inland Sea in 1973. The coast of that city, once famous for its fabled beauty, had in the course of rapid industrialization been almost completely lost to landfill and was subsequently made entirely inaccessible to the city's people. In the fall of 1973, a number of concerned citizens led by Mr. Hiroshi Takasaki, gathered and began what became Japan's first national coastal citizens' movement. It is aimed at the legal recognition of the people's coastal access rights, the protection of the coastal environment from pollution and environmental destruction, the restoration of lost coastlines, the recovery of lost or denied public coastal access, and repeal of the Landfill Law,among others.[30] Through its members' tireless efforts, this movement has achieved the construction of a small artificial beach and fishing pier along the Takasago City coast, and has stimulated such projects elsewhere by its example. However, it has yet to achieve its primary goal, the legal recognition of coastal access rights.

The last example to be presented here is that of a regional coastal citizens' movement, the Peoples' Alliance to Protect the Seto Inland Sea Environment. It began in 1972 and consisted of representatives from 23 coastal movements (with some 80,000 members) from throughout the Seto Inland Sea region. Over the last decade or so, this Alliance in conjunction with the Japan Scientists' Association has conducted surveys,published documents of a wide variety, and placed pressure on the government to establish and strengthen the Seto Inland Sea Environmental Protection Act.[31,32] At present, the Alliance consists of 43 movements (over 200,000 members) from throughout the region. Its leaders have used elections, the most popular method of influence among activists,[27] to try to elect representatives who reflect their views on environmental issues to political office.[33] Its activities in the most polluted and most extensively disturbed part of the region, the Osaka Bay estuary[34] at its eastern-most end, have been conspicuous and important. It has conducted three major surveys of the Bay since 1981 aimed at both collecting and organizing information on the Bay as well as increasing coastal awareness and concern among the region's people. It is this work which is particularly in terms of the theme of this paper. See Fig.1 for the locations mentioned here.

THE COASTAL ATLAS

Protecting, restoring, maintaining and/or properly developing coastal resources requires scientifically reliable and easily understandable information as has been mentioned. Much of this information often exists already, but its whereabouts may be unknown resulting in wasteful and often expensive duplication. And even if its whereabouts is known, its contents may not be understand nor its significance recognized making it of little use in coastal zone management and planning. One effective way to help eliminate these problems is to develop an inventory of the conditions of the coastal environment in question in the form of a coastal zone atlas.

Many examples, some dating from before the U.S. Coastal Zone Management Act of 1972, can be found in the U.S.. For instance, as far back as 1969, the State of Texas began compiling an Environmental Geologic Atlas of its coastal zone. Each volume consists of an explanatory text and nine folded maps, such as environmental geology, man-made features and water systems, environmental and biological assemblages, and current land uses to mention some. All maps are in color and include a coastal area extending from the inner continental shelf to some 40 mi. inland.[35] Because of their small scale (1:125,000 and 1:250,000), small scale problems cannot be easily nor sufficiently studied.

In 1972, Florida published a Coastal Zone Management Atlas that includes four maps of each coastal county at 1:125,000 scale. These maps depict preservation, conservation and development zoning areas and include coastal features considered important for that category of zoning.[36] Here again, their small scale limits them to regional scale evaluations.

Hawaii began to publish its marine atlas entitled "Bays and Harbors" in 1974. The atlas shows areas proposed for harbor construction and emphasizes the marine environment and pilotage. However, physical data on most of the area is very limited.[37]

In 1977, Massachusetts published an Atlas of Resources as part of its Coastal Zone Management Program. The text includes citizens' desires for future use of the coastal zone and outlines coastal area problems. The Atlas maps are at 1:40,000

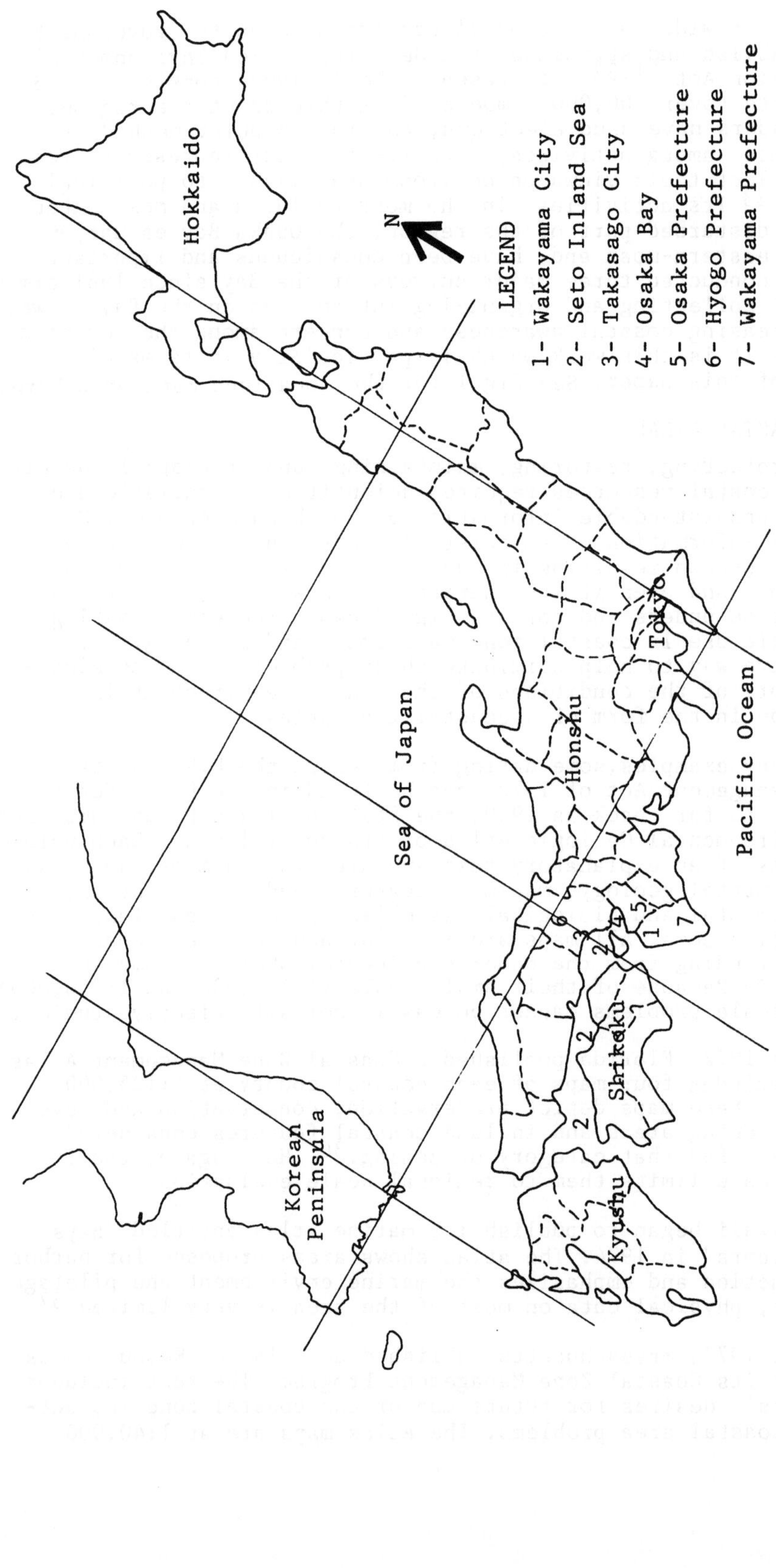

Fig.1. The Location of Places in Japan in the Order Mentioned in the Paper

scale(reduced from 1:24,000). Colored and symbolic overlays show significant resource areas, such as dunes,contaminated and clean shellfish, the 100-year floodplain, and areas of critical erosion or accretion. The data included came mainly from government sources and from citizen participation.[38]

Finally, in 1976 Washington State began producing a Coastal Atlas to improve its coastal zone related decision-making process. Washington evaluated the needs of potential users of the Atlas and various ways to fulfill those needs,and then it developed its Atlas maps at 1:24,000 scale and all in color. It used computer techniques to generate four-color separation plates. This automated data base[39] can be updated and completerevised rapidly and inexpensively in the future. The Atlas consists of information on coastal drift, coastal slope stability, coastal geology, sand and gravel resources, coastal flooding, land use and land cover, and critical faunal and floral areas.[40] It is the only state of those mentioned here to provide for future revisions of its atlas. This can be thought to reflect the high cost of revision. Atlases need to provide for updating due to the inclusion of previously unknown information and due to changes in old information. The scientific studies necessary to provide new information could come from various sources,including developers or independent consultants, academic institutions, government agencies and even citizen groups. The Washington Atlas is an example of one that can be used and understood by the public,planners, decision-makers, developers and conservationists.

Whatever the form, an atlas is a static document that depends on continuous monitoring and surveillance of environmental change, both natural and man-caused, to keep it up to date and make it a useful tool for coastal zone management, coastal planning and environmental assessment, and for coastal citizen participation etc.

TOWARD A COASTAL ATLAS FOR THE OSAKA BAY AREA

The Osaka Bay area is the center of Osaka Basin and is located at the extreme eastern end of the Seto Inland Sea (see Fig. 1). The Basin contains all of Osaka Prefecture,southeastern Hyogo Prefecture,a bit of northern Wakayama Prefecture and of course the Bay (see Fig. 2). The Bay is bordered by 20 cities and towns. The population of the Basin is something over 12,000,000, about 10% of the entire nation, of which nearly 2/3 live in the coastal cities and towns. The area of the Bay itself is approximately 1,700 km^2 and covers over 1/3 of the area of the Basin.[42] This combined with the fact that so many people live, work and recreate etc. near the Bay necessarily implies an intimate and immediate interrelationship between man and the Bay in terms of their influence upon each other.

The main force influencing Osaka Bay today is landfill. It is being done on a continuous basis and has been so since Japan's postwar recovery was completed in 1955. The present rate of fill is something over 200 ha annually on the average.[34] Fig.3 shows the distribution of landfill in the Osaka Bay area

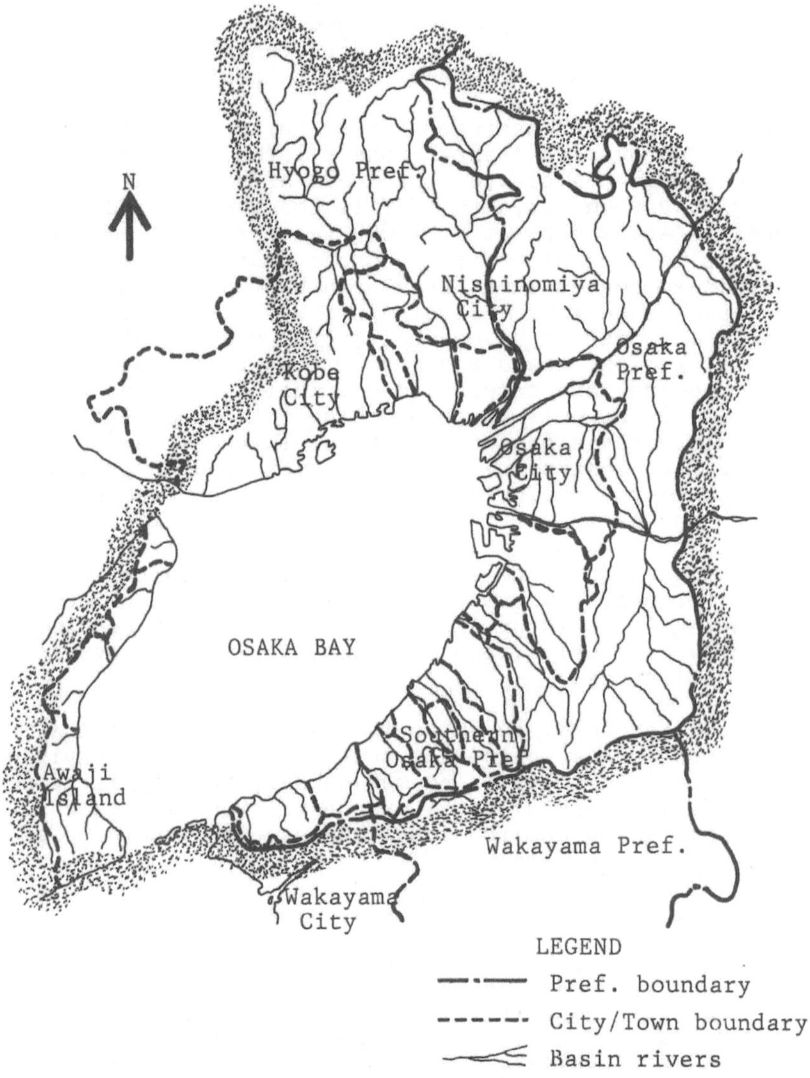

Fig.2. Osaka Basin and Its Coastal Cities and Towns

Fig.3. Landfills Completed, In Progress or Planned in Osaka Bay
(Units; Hectares)

Location / Year	Amagasaki, Nishinomiya and Ashiya	Kobe City	Osaka	Awaji Isl.	Total
1955-1965	-	429	1,708	-	2,137
1966-1975	315	646	3,119	-	4,080
1976-1980	43	144	1,170	63	1,420
*until 1985	467	534	1,250	178	2,429
**1985-1995	910	1,300	5,720	-	7,930
***Up to mid 21st Cent.	....	......	....	......	. 13,000
Total	1,735	3,053	12,967	241	30,996

Notes: 1) 1 hectare = 10,000m^2 = 2.47 acres

2) * fills in progress (see Ref.46)

3) ** Ref.: Osakawan Kowan Keikaku no Kihon Koso no Gaiyo(Outline of the Basic Concept for Osaka Bay Port Planning), 3rd MOT Bureau, Kobe: (Sept. 1984)

4)*** See Ref.48.

from 1955-1980 by coastal area. It is clear from this figure that more than half of the area of fill was done during the high economic growth period (1965-1975).

Before filling began on a large scale, the Bay's waters were clean,and there were 36 sandy beaches[43] backed by blue-green pines (the famous hakusa seisho scenery revered by Japanese since ancient times). The Bay was a source of abundant, high quality fish and marine living resources. After filling,mainly for industry, almost all of the natural beaches were obliterated, the air and water were polluted, the quality of fish changed for the worse,[44] and a great deal of coastal access was prohibited by the port, industrial and other fill land users who became the private owners of the newly filled land,[45,46] to mention but a few. In recent years, there has been a very slight increase in swimmable coastline length due to the construction of some artificial beaches. However, these have been or are being built at great public expense as well as at the expense of other natural environments,such as river beds, mountains and sea bottoms etc. from which sand is taken to build these new beaches by landfill.[47] No let up is in sight in the inertia to fill the Bay area as attested to,for instance,by a very recent proposal by Bay area economic leaders to continue large scale filling operations for the next 100 years[48] (see Fig.4). At this rate, the future of the Bay as a viable ecosystem,as well as the health and welfare of the Bay area's citizens face an imminent threat as can be imagined from the series of maps shown in Fig.5, the probable shape of the Bay assuming the continuation of large scale filling operations and plans.[49,50]

Unlike the late '60's and early '70's when pollution activities were strongly opposed by citizens across the nation, with the national and local economies in recession, many people's concern for pollution and the environment has tended to wane with commodity prices and personal savings etc. being given priority.[51] However, as the impacts of environmental destruction and pollution again become more directly experienced, citizen motivation to participate in actions related to improving environmental quality has started increasing again in some places, the Osaka Bay area being one of them.

In recent years, some of the more environmentally-sensitive citizens and scholars of the Bay area have been working to try to increase coastal awareness and concern among the population. For instance, in February of 1981, they held a major symposium in Osaka concerned with the problems of the Seto Inland Sea in general and those of Osaka Bay in particular.[52] That same year on May 31st, they jointly undertook the first of three surveys of Osaka Bay's coastal area. Some 450 citizens and scientists participated in a combined land and sea survey of the southern Osaka Bay area expected to be directly influenced by the proposed New Kansai International Airport (K.I.A.P.). Among the sea area events were a water quality survey, a fish variability survey, a study of on-sea wind speed and direction, while on shore surveys of coastal organisms, cultural relics and the changing awareness of the sea by coastal citizens etc. were

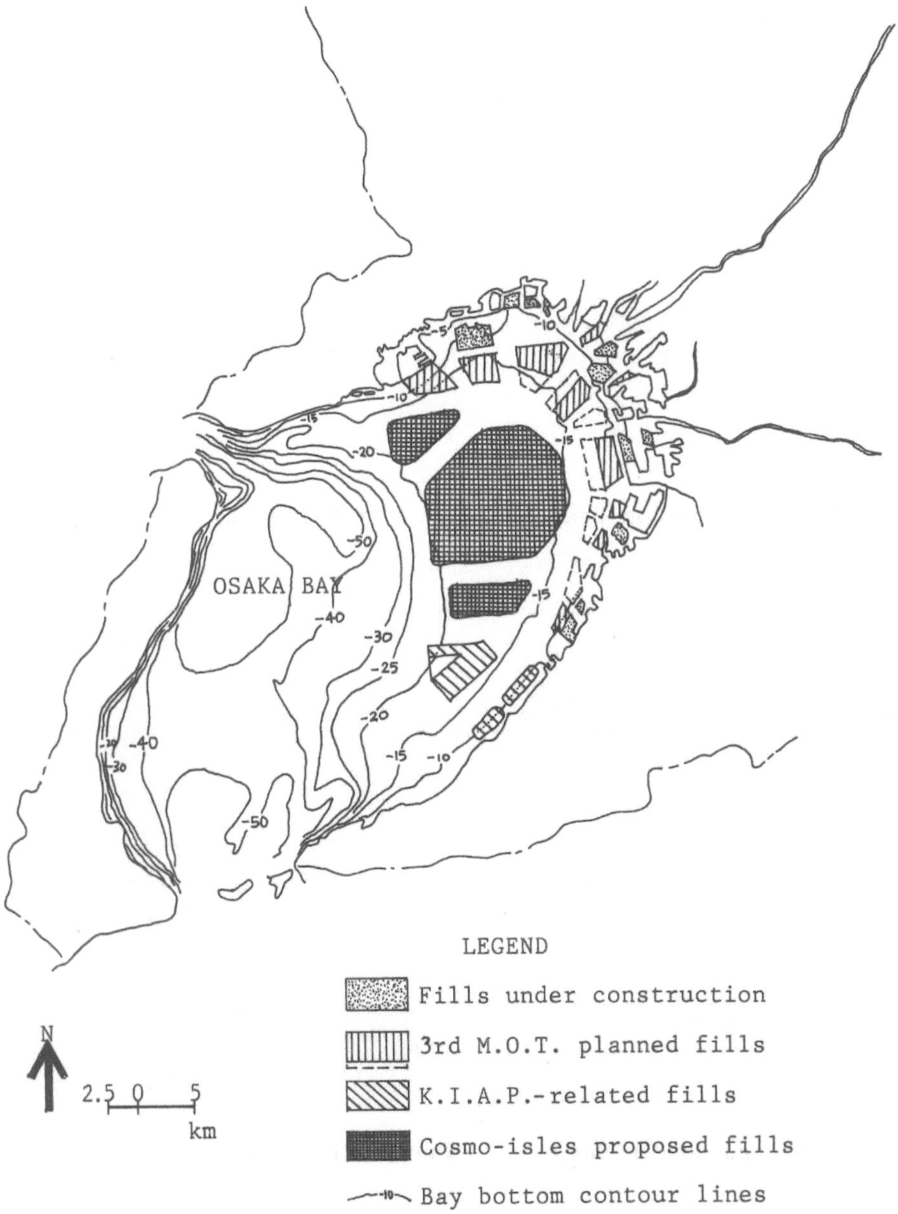

Note: see Refs. 48, and Outline of the Basic Concept for Osaka Bay Port Planning. 3rd MOT Bureau, Kobe: Sept. 1984.

Fig.4. Proposed Large Scale Landfills for the Osaka Bay

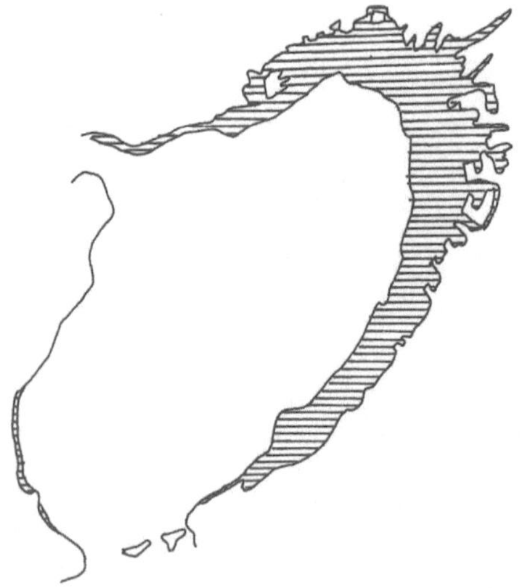

(1) If Filled to 15m Deep

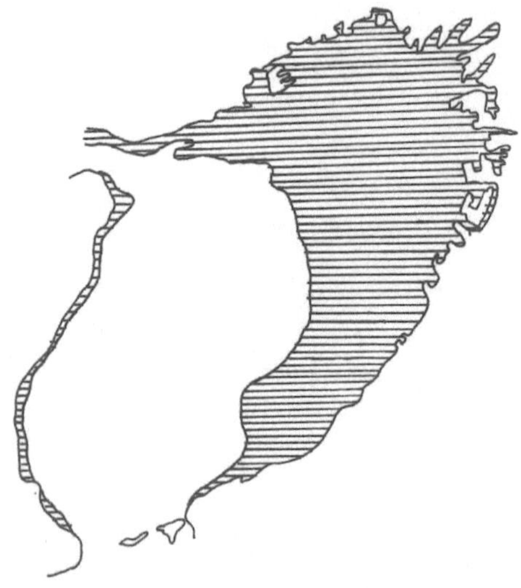

(2) If Filled to 20m Deep

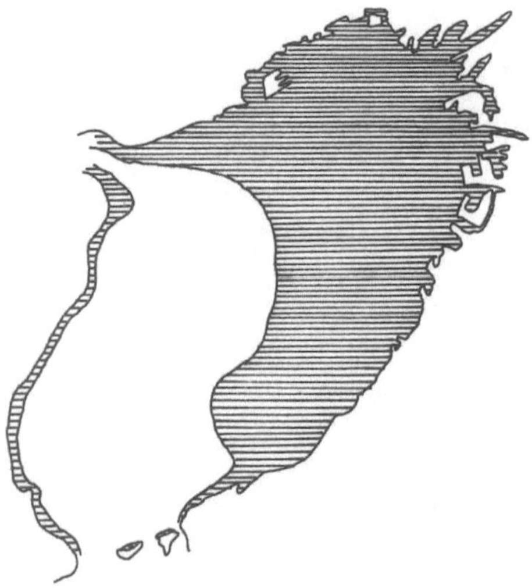

(3) If Filled to 25m Deep

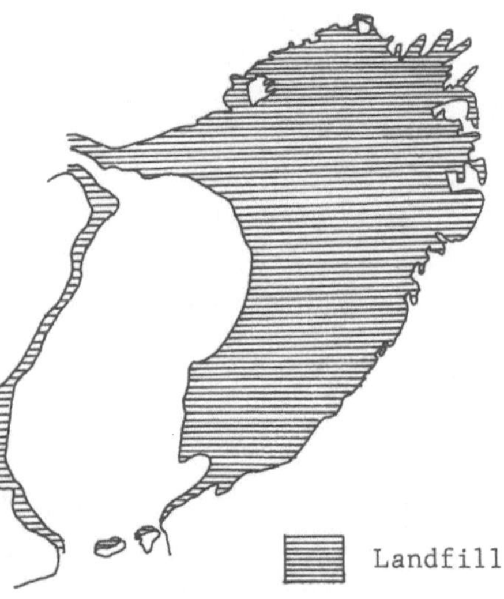

(4) If Filled to 30m Deep

Fig.5. Changes in Osaka Bay Shape Assuming Fills 15-30m Deep done. Government police boats harassed the surveying citizens whose boats had neared the exact site of the new airport 5 km offshore.[53,54]

The 1982 survey took place on June 26th and 27th and consisted of a continuation of the studies done the previous year but expanded to cover not only the southern Osaka Bay area but also the northern coast off of Kobe City where numerous giant fills have been done, are in progress or are planned. Also, the northeastern coast near Osaka City with similar conditions was viewed. Several hundred people from all over the Bay area participated.[55]

The most recent and by far most extensive survey so far was conducted during the summer of 1983 by nearly 1,000 people from all three Bay area prefectures. In addition to a continuation of the work started previously, a detailed survey of coastal conditions and of coastal access of the entire Osaka Bay area coastline followed by mapping of the results at 1:25,000 scale were done. The coast was classified as natural, semi-natural, artificial but accessible, and artificial but with access impossible or prohibited. The results revealed that 58.2% of the Osaka coast, 34.9% of the surveyed Wakayama City coast and 32.2% of the Hyogo coast were of the artificial-inaccessible type, almost all of it in the most heavily populated areas.[45]

There is quite a bit of information about Osaka Bay and its coastal area, but it is widely scattered, alot of it is unknown but to a few, and much of it is inaccessible to the average citizen especially in areas of planned fill and/or pollution, not unlike conditions elsewhere in Japan. The information collected and/or developed by the coastal citizens themselves, though compiled into reports in a simplified easy-to-read and understand form, is not yet fully coordinated nor complete. Lacking access to desirable government information, if citizens' pleas are going to be paid attention, more than a few reports on existing conditions here and there, anti-pollution demonstrations and/or symposiums etc. will be needed to facilitate and develop educated opposition and constructive alternative plans and proposals. It is for this reason that this author, a participant himself in these citizens' surveys, has recommended the systematic development of a comprehensive ecological inventory in the form of a coastal zone atlas for the entire Bay area.[56] The project would involve first the identification of the coastal zone of the Bay area using the method shown in Fig.6. In the first stage, the primary coastal zone would be the object of study. Such a coastal zone identification study has already been done as a student project under the author's guidance,[57] and as such it is in need of review and improvement,but it could serve as a guideline for early atlas work. It is recommended that the primary coastal zone be divided into 12 blocks each including one or two entire cities (see Fig. 7). Such a breakdown should enable coastal citizens and scientists to work within a common framework and develop and/or collect needed information about their respective cities. This information might best be fed to a common center, perhaps at the author's university research institute, for final coord-

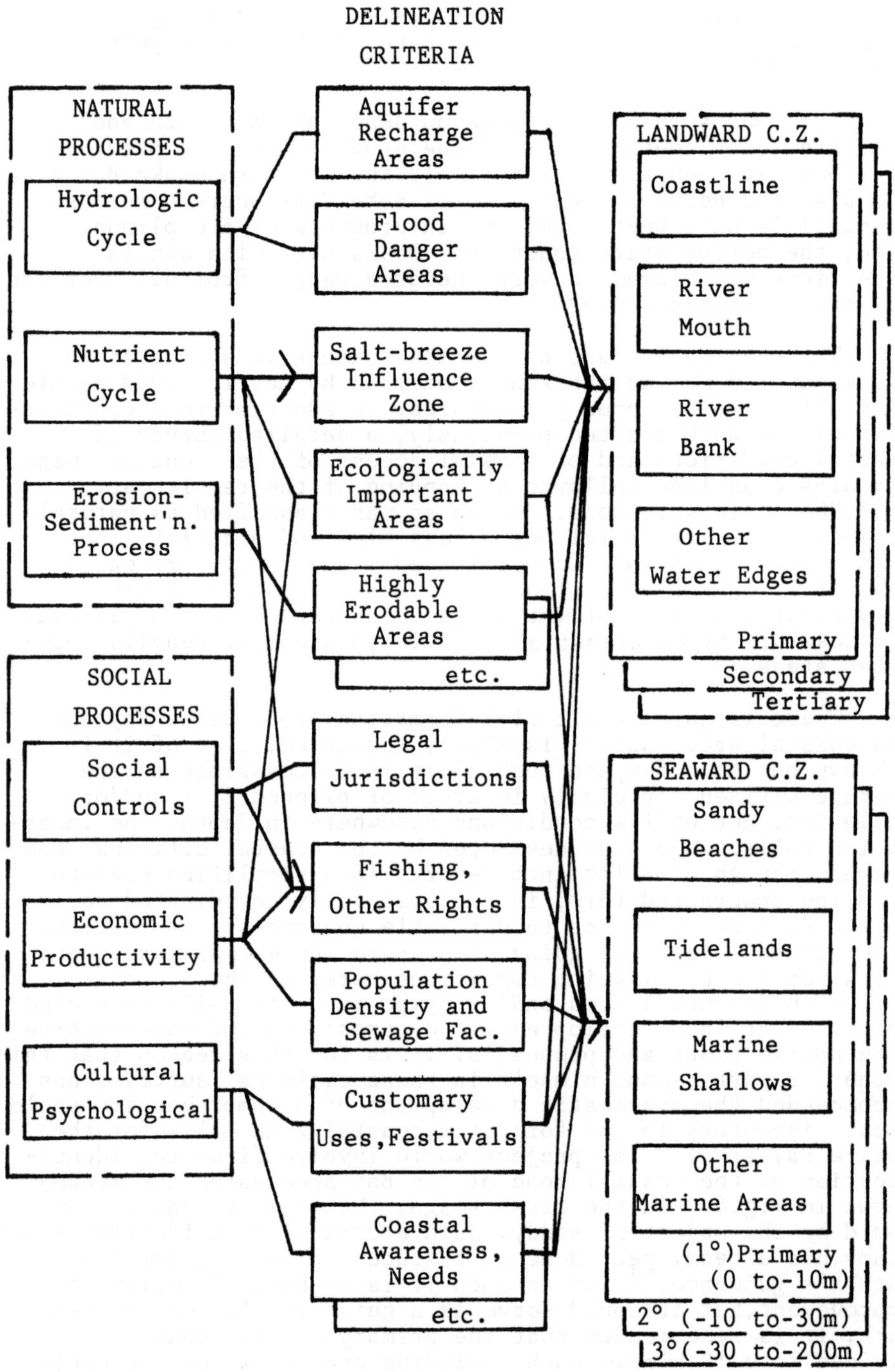

Fig.6. A Coastal Zone Delineation Process For Osaka Bay

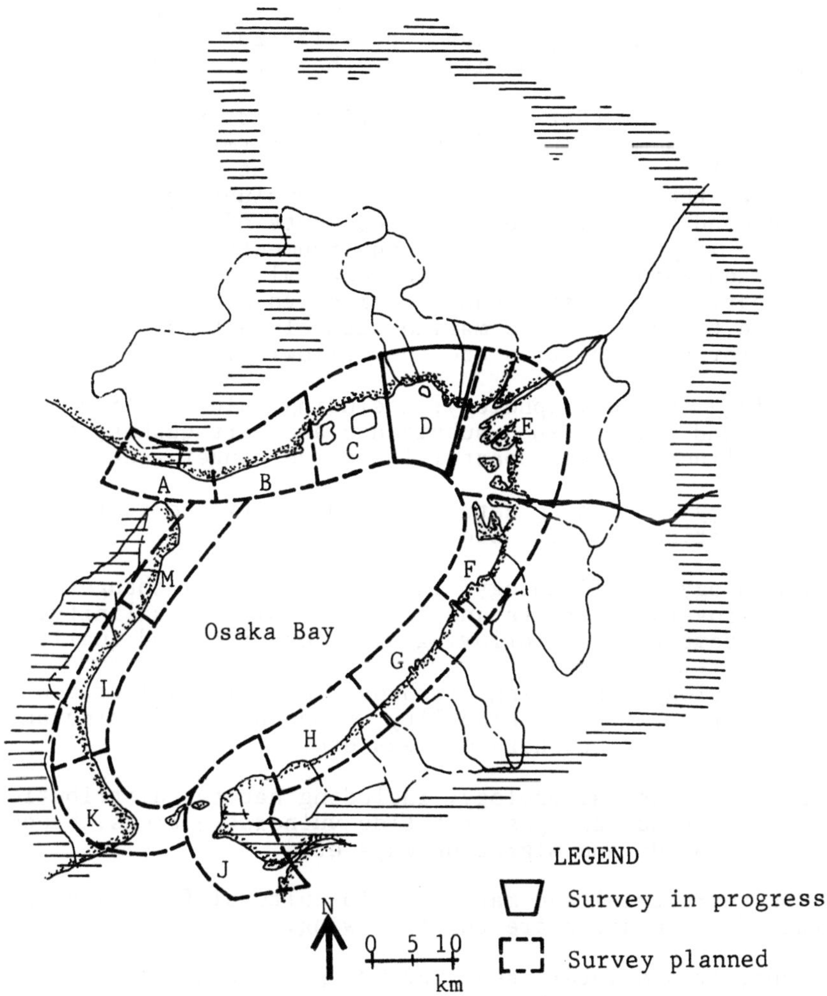

Fig.7. Suggested Survey Blocks for Coastal Zone Atlas

ination, compilation and publication in the form of a Coastal Zone Atlas for Osaka Bay series, one atlas for each block or one atlas for each surveyed item for the entire Bay area coast.

It is expected that the people will need an accessible atlas to serve a multitude of purposes. A formal survey of citizens on this item remains to be done, but informal inquiries suggest that the atlas would be used to evaluate government plans, to conduct environmental assessments of both government and citizen proposals, and for the environmental planning of land, air and water uses within the coastal zone seen as an integral part of the Osaka Basin region, to mention but a few. Some of the terrestrial and marine natural factors considered necessary for such an atlas are the following:

1. Geology and geomorphology: including vulnerability to and damage from subsidence and earthquakes, coastline changes, coastal landslide danger areas etc.

2. Pedology: including natural productivity and self-purification capacity etc.

3. Hydrology and limnology: including seasonal changes, coastal currents and drift, turbidity, salinity, water temperature, inflowing water quantity and quality etc.

4. Meteorology: including airsheds, inversion vulnerable areas, fog vulnerability, typhoon hazard vulnerable areas, flood hazard areas etc.

5. Plant and animal ecology: including habitat distribution, seasonal changes, red tide vulnerable areas, breeding grounds and migration ways etc.

Among the terrestrial and marine socio-cultural factors considered necessary for the atlas are the following:

1. Land, air and water uses: including changes past to present and planned changes etc.

2. Distribution of pollution sources: including both aerial and point sources etc.

3. Distribution of rights: including land ownership and fishing rights etc.

4. Coastal awareness: including changes from past to present awareness basinwide etc.

5. Land price: including changes over time etc.

6. Population distribution and density: including changes over time and place etc.

7. Distribution of cultural assets, festivals and customary uses etc.

8. Distribution of accessible coasts and access routes etc.

9. Distribution of administrative jurisdictional areas

10. Distribution of citizen needs and desires, satisfactions and dissatisfactions with respect to the Bay and its coastal uses etc.

It is expected that as specific problems arise in the coastal zone, other factors may need to be included. It is recommended that all information be prepared on maps of an equal scale, such as 1:25,000, and in color where that makes understanding easier, and that all maps be provided with explanations in a form understandable to all prospective users. Bi-lingual (English and Japanese) explanations could encourage international understanding and exchange as well as advice. Of course, all information would have to be carefully documented for easy verification. The overall format of the atlas might well resemble the Washington Atlas mentioned earlier,and the Osaka Bay Atlas should be likewise automated so that it could be kept easily and inexpensively up-to-date. Information from government sources could be used, but experience has shown that it needs careful verification especially when involving areas outside normal jurisdictions. However, it is expected that academic and citizen sources will likely comprise the bulk of atlas data, but these too will need verification to maintain objectivity.

This author has begun some preliminary work in the development of a model coastal zone atlas for the Osaka Bay area, in particular for Block D which includes the city of Nishinomiya. Mapping work at 1:25,000 scale has been started on terrestrial and marine geomorphology, recorded flood damage areas, past and present land and water use, and coastal conditions and accessability. Mapping by hand and in color is being done, and bi-lingual titles and legends are being prepared. Government information was used for the geomorphology and floods of record with the marine area of the former extensively revised due to gross inaccuracies discovered upon verification. Published information on past land use was used while an air photograph analysis is being done to develop the present land use information. Field confirmation remains to be done however. The coastal conditions and access maps developed by the citizens in the 1983 survey referred to before[45] are being used directly. It is thought to be one good way to show the people how their efforts will fit into the atlas-making work and help encourage their further cooperation. We hope to be able to publish the work we are doing in final atlas form and continue our work on other natural and socio-cultural factors thereafter. Effort is now being made to obtain some domestic funding for this work, and every opportunity is being taken to solicit citizen and scientific cooperation in this work. It can only be hoped that this model atlas project will serve as an example of what can and should be done if the citizens are going to be better able to fulfill their desire to actively participate in and influence the coastal zone decision-making process in an educated way, by constructive planning and continuous environmental monitoring and surveillance.

SOME FORESEEN DIFFICULTIES

This effort at atlas-making is not expected to be without difficulties though. For instance, the author's call for any and all kinds of coastal-related information at several citizens meetings has so far produced little. Spreading the word sufficiently and persuading others in the region to come forward with coastal information will be an important task of the leaders of related citizens' movements as well as of cooperating scientists. We hope that our preliminary efforts in final atlas form will provide an example of sufficient interest to encourage the yet to come cooperation and support.

Another difficulty is expected to involve funding for this atlas-making effort. Without sufficient funding, it will be difficult if not impossible to continue atlas-making. The initial cost of such work, especially if it is automated, could prove to be very expensive. In principle, an effort of this sort with its many public benefits should be the responsibility of government or at least funded by it with adequate technical support. In a sense, without greater government cooperation in citizen efforts like this, the present rift between the people and those who are supposed to be representing them will very likely continue to widen. Government support,understanding and cooperation would greatly aid in helping to make realistic and effective use of the information and plans based on it that the citizens develop. Unfortunately, no such government support or cooperation can be expected any time soon.

The coastal zone information center suggested before could become a depository for not only citizen and scientists' coastal information but for government information from all branches of at present sectionally divided bureaucracy too. Work through that information center could provide a common data bank for all concerned with coastal zone management, planning and protection etc.,something that doesn't exist yet.Here too,unfortunately, it would be unrealistic to expect government cooperation in such an effort due mainly to the basic distrust or fear that development-oriented government officials have toward citizens who just might disrupt government plans.

Another difficulty could be that of maintaining interest in and support of coastal zone atlas-making over the considerable amount of time expected to be necessary to complete the work for the entire Bay area coastal zone. In this age of speed, slowness, for whatever the reason, is often seen as akin to non-necessity. The necessity will surely increase over time, but unless the atlas-making effort can be speeded up, this may not be recognized. Calls for the full range of coastal information to assess and/or develop alternatives to specific coastal area plans could on the one hand stimulate atlas-making efforts, but on the other hand, if atlas-making has not yet reached completion in specific problem areas, unless information is available even if not yet in atlas form, plans might well go ahead without the needed objective data base as they have before. This could encourage pessimism about the work and drive citizens back to their former opposition-without-alternatives methods.

Unlike many places in the U.S., there are practically no citizens or scientists in Japan who can sustain themselves doing only citizen movement work,i.e. there are almost no full-time citizen movement activists in Japan today. This means that most efforts involving monitoring, data collection and surveys etc. not to mention the atlas-making work itself might well have to be done almost entirely after hours. This could slow the whole process even more. These have been but a few of the difficulties seen facing coastal zone atlas-making in the Osaka Bay area, things likely to happen almost anywhere in Japan where such efforts are attempted by coastal citizens and sympathetic scientists alone.

CONCLUSION

Despite these and other difficulties, the reality of democracy in postwar Japan with its demand for a more active role of the people in decision-making cannot be ignored. Environmentally concerned citizens will need easily understandable yet reliable scientific information on the coastal zone and elsewhere if they are going to be able to participate effectively and constructively in coastal matters. In this paper, the coastal atlas form of ecological inventory was suggested as one means of fulfilling their desire and helping them in their effort to protect, restore and properly use the coastal environment of Japan now and in the future. Judging from what has been happening to the Japanese coast over the last two decades or so, it seems clear that coastal zone policy-making and planning can no longer be left to the "elite" of the central and prefectural governments nor to the business and finance communities. Not only is the Japanese economy involved, but also so is its cultural and natural environments. The people's voice at all levels in regard to coastal matters must be heard, and the coastal atlas,compiled and used properly,should aid in making this happen by creating a systematic, objective and verifiable basis for constructive citizen participation in coastal affairs. It can serve as a common base for conflicting interests to rationally discuss their differences in terms of the intrinsic opportunities and constraints of the object of their demands, the coastal ecosystem,and aid all parties in more environmentally sensitive adaptations to that both valuable and extremely vulnerable environment.

References

1. Aramaki,M.: Ikiteru Nagisa (The Living Coast). Tokyo: Sanshodo 1978.

2. Nagao,Y. (ed.): Engan-iki Keikaku Shiko Nyumon (Introductory Thoughts on Coastal Zone Planning). Tokyo: Japan Ports and Harbors Association 1982.

3. Nagao,Y.: Coastal Zone Planning and Management in Japan. Dept. of Civil Engineering,Kyoto University 1984 (mineograph).

4. Midori no Kokusei Chosa (Nature Conservation Survey). Tokyo: Environment Agency 1976.

5. Marine Parks in Japan. Tokyo: Marine Park Center of Japan 1975.

6. 200-kairi jidai no Taio-suru Kaiyo Kaihatsu Riyo ni-kansuru Kenkyu (Basic Research on Orderly Ocean Development in the 200-mile Era). Tokyo: Japan Research Institute 1978.

7. Inoue,S.: Toward a Framework for Coastal Zone Management. University of Washington M.S. Thesis 1977 (unpublished).

8. Japan Sports Fishing Survey. Tokyo: Japan Sports Fishing Association 1973.

9. Tanigawa,T; Iijima,S.; Nagasaki,S. (eds.): Atarashi Kai-iki Riyo Gijutsu Benran (Manual of New Coastal Use Technology). Tokyo: Fuji Technosystem 1978.

10. Fishing Ports in Japan. Tokyo: Fishery Agency 1978.

11. Census of Japan. Tokyo: Bureau of the Census,Prime Minister's Office 1970.

12. Shapiro,H.A.: Coastal Area Management in Japan: An Overview. Coastal Zone Management Journal 12(1) (1984) 19-56.

13. River Bureau (ed.): Kaigan Tokei (Coastal Statistics). Tokyo: Ministry of Construction 1978 (annual).

14. Samuals,R.J.: The Politics of Regional Policy in Japan. Princeton,N.J.: Princeton University Press 1983.

15. Stow,D. and Estes,J.: Status of Remote Sensing for Coastal Zone Management. Coastal Zone '83. Vol.I (1983) 565-574.

16. Bailey,A. and Colwell,R.: The Chesapeake Bay Information System (BIF) for Environmental Storage and Access. Proc. CODATA Conf., Warsaw, Poland (oct. 1982).

17. Dolan R.; Hayden,B.; Rea,C.; May,S.: Design of a Coastal Information System. Report of the Dept. of Environmental Science (1980).

18. Domanski,P.; BIOS a Data Base for Marine Biological Data.

J. Plankton Research, 3(3) (1981) 475-491.

19.Weyl,P.K.: Simple Information Systems for Coastal Zone Management. Coastal Zone Management Journal 9(2)(1982) 155-182.

20.Kaigan Benran (Coastal Manual). Tokyo: Japan Coastal Society, annual.

21.Zenkoku Kaigan-iki Genkyo Chosa (National Coastal Area Conditions Survey). Tokyo: Ministry of Construction 1974.

22.Kaigan-iki Genkyo Chosa Hokokusho (Coastal Area Present Condition Survey Report). Tokyo: Ministry of Construction River Bureau and National Geographic Institute 1977.

23.Open Information. Mainichi Daily News (Nov.29,1982).

24.Richardson,M.: The Political Culture of Japan. Berkeley: University of California Press 1975.

25.Koschmann,V.J. (ed.): Authority and the Individual in Japan: Citizen Protest in Historical Perspective. Tokyo: University of Tokyo Press 1978.

26.Henderson,D.F.: Law and Political Modernization in Modern Japan. Princeton,N.J.: Princeton University Press (1978) 387-456.

27.McKean,M.: Environmental Protest and Citizen Politics in Japan. Berkeley:University of California Press 1981.

28.Matsumoto,H.: Umi ga Achira-e Shinde Iku, Hansto 33-nichikan no Kiroku (Our Sea is Going to Die: Record of a 33 Day Hunger Strike). Osaka: 1970 (unpublished).

29.Shapiro,H. and Zanic,T.: Coastal Citizens Movements in Japan and California. Coastal Zone '83. Vol.II (1983) 1803-1822.

30.Shapiro,H.A.: The Coastal Access Rights Movement in Japan. Coastal Zone Management Journal 8(1) (1980) 1-43.

31.Akaha,T.: Conservation of the Seto Inland Sea. Coastal Zone Management Journal 12(1) (1984) 83-136.

32.Nishimura,T.: Umetate to Setouchi-no Hakai (Landfill and the Destruction of the Seto Inland Sea). Seto Naikai Journal 2 (Nov.1976) 5-20.

33.Hyogo Governor Wins 4th Term. Mainichi Daily News (Nov.2, 1982).

34.Japan Scientists Association Seto Inland Sea Committee (ed.): Umetate Jigoku-no Setouchi Engan (The Landfill Hell that is the Seto Inland Sea Coast). Kyoto: Hotsubunsha 1985.

35.University of Texas at Austin: Environmental Geological Atlas of the Texas Coastal Zone. Texas: Bureau of Economic Geology (7-volumes) 1972-1978.

36. Florida Coastal Zone Management Atlas:A Preliminary Survey Analysis. Florida: Florida Coastal Coordinating Council 1972.

37. Marine Atlas of Hawaii- Bays and Harbors: Seagrant Miscellaneous Report UNHI-SEAGRANT-MR-74-01. Hawaii: University of Hawaii 1974.

38. Massachusetts Coastal Regions and an Atlas of Resources. Massachusetts: Massachusetts Coastal Zone Management Program 1977.

39. Coastal Zone Atlas of Washington: Automated Data Base, An Introduction. State of Washington: Dept. of Ecology 1980.

40. The Washington Coastal Atlas-An Overview. State of Washington: Dept of Ecology 1976.

41. Schoen,R.: The Role of Earth Science in Coastal Zone Management. Coastal Zone '78. Vol.III (1978) 1589-1601.

42. Shapiro,H.A.: Comprehensive Environmental Management and Planning for the Osaka Bay Coastal Zone and Bayshore Cities (I). Numata,M.(ed.): Integrated Ecological Studies in Bay-Coast Cities I. Chiba: Shobunsha (1979) 78-91.

43. Hayashi,K.: Kaisui-yokujo to Setouchi Kaihatsu (Swimming Beaches and Development in the Seto Inland Sea). 10th Setouchi Symposium. Osaka: (Feb. 14-15,1981) 109-115.

44. Umekawa,T.: Osakawangan-no Gyogyo Genkyo (Present Conditions of Commercial Fishing in the Osaka Bay Coastal Area). Seto Naikai Journal 11 (Aug.1981) 31-33.

45. Koyama,E. and Ito,D. (eds.): Ima Osakawan-wa(What's the Condition of Osaka Bay Now?). Osaka and Kobe: Citizens' Alliance Considering Osaka Bay Problems (1983).

46. Nishikawa,E.: Osaka-wan Umetate Kaihatsu to Shin-kuko Kei-Keikaku (Osaka Bay Landfill Development and the New Kansai International Airport). Airport Problem Joint Research Council (ed.): Genso-no Kyodai Kaihatsu (The Dream of Gigantic Development). Tokyo: Shiobunsha (1981) 58-71.

47. Konishi,K.: Jinko nagisa no Fukeizaigaku (The Diseconomy of Artificial Beaches in Japan). 5th National Coastal Access Rights Symposium Proc. (Part 3) (Aug. 1983).

48. Kansai Economic Support Council: 21-seiki no Chiiki Kaihatsu Koso (Regional Development for the 21st, A Concept). Osaka: Nov. 1984.

49. Shapiro,H.A. and Isobe,Y.: Ecological Planning, Its Method and Application,Part II. Kenchiku Bunka Journal 32(367) (May 1977) 23-152.

50. Shapiro,H.A.: Geographical and Related Aspects of Landfill for Transportation and Industry in the Japanese Coastal

Environment: Especially in the Seto Inland Sea Area. 5th Pacific Science Association Inter-Congress Proceedings. Manila: (Feb. 1985) (in press).

51. Yoshii,H.: Environmental Perceptions and Beliefs of Chiba City Residents. NUmata,M.(ed.) Chiba Bay-coast Cities Project,IV. Chiba: Shobunsha (1982) 11-39.

52. Japan Scientists Association Setouchu Council (ed.): 10th Setouchi Symposium. Osaka: (Feb. 1981).

53. Koyama,E. and Ito,D. (eds.): Ikiteru ka? Mizu wa...Sakana wa... (Is Osaka Bay Still Alive? What About Its Water and Fish?: Record of the May 31,1981 "Let's Take a Good Look at Osaka Bay-Campaign). Osaka and Kobe: (Aug. 1981).

54. Kono Me De Miyo, Osakawan 5-31, Osakawan Chosa ga Watashitachi ni Shimeshita Mono (Let's Take a Good Look at Osaka Bay: What the May 31,1981 Survey Showed Us). Seto Naikai Journal 11 (Aug. 1981) 6-59.

55. Kono Me De Miyo, Osakawan Dai-2 Engan Jumin Tentai Shukai (Let's Take a Good Look at Osaka Bay; The 2nd Coastal Citizens Survey). Seto Naikai Journal 12 (Aug,1982) 66-74.

56. Shapiro,H.A.: Shigen Mokuroku (An Ecological Inventory for the Osaka Bay Coastal Zone). Koyama,E. and Ito,D.(eds.): Ima Osakawan wa (What's the Condition of Osaka Bay Now?). Osaka and Kobe: Citizens Alliance Considering Osaka Bay Problems (1983).

57. Osaka Bonchi Seitai-gakuteki Chiiki Keikaku Hokokusho (Ecological Planning for the Osaka Basin). Osaka: Osaka Geijutsu University,Ecological Planning Research Institute (1979-1981) 1-45 (unpublished).

Technical Session:
Coastal Zone Utilization
and Management

2. Coastal Zone Planning

An Activities Allocation Model Considering Interactive Effects for Coastal Zone Planning

Y. Nagao and T. Morikawa

Department of Transportation Engineering, Kyoto University, Yoshida Honmachi, Sakyo-ku, Kyoto, 606 Japan

ABSTRACT

In order to plan the efficient utilization of coastal zones considering preservation of the environment, an activities allocation model in consideration of their interactive effects is proposed in this paper. Here, agglomeration economies are considered as a positive interactive effect, and external diseconomies are treated as a negative effect. This model is formulated by the n-person non-cooperative games and solved by computer simulation. The allocation plans made by this model are evaluated from the view points of efficiency and equity. As a case study, this model is applied to actual coastal waters and its usefulness is examined.

INTRODUCTION

Since Japan has little flatland suitable for residential and productive activities, coastal zones have been utilized for those activities. Recently, the demand for coastal waters as well as lands has rapidly increased. However, since there is no rational methodology of the coastal zone use planning, an appropriate planning of those zones has not been developed. As a result, random development of coastal zones yields to an environmental disruption and sprawling of the zones. Under these circumstances, a methodology for rational coastal zone use planning is strongly required. For the rational planning of coastal zone use, the effective utilization of limited space and the decrease of the negative interactions among activities should be considered. In order to establish this methodology, it is necessary to develop an activities allocation model which considers the interactive effects explicitly.

This research aims to build an activities allocation model which considers the interactive effects among activities explicitly, and to propose a method of coastal zone planning. This model is a simulation model. Here, the interactive effects mean the following two concepts. First, the positive interactive effect is agglomeration economies which occurs when similar activities gather. Next, the negative one is external diseconomies mainly caused by environmental pollution. It occurs when incompatible activities gather.

BASIC CONCEPTION AND PREMISES FOR MODELLING

(1) Basic Conception for Modelling

The activities allocation model proposed in this paper is a simulation model formulated by the n-person non-cooperative games. This model is characterized by treating the positive and negative interactive effects among activities explicitly. The reason for treating them is that coastal waters are susceptible to environmental change on account of their geographical and ecological distinction. This model is formulated on the basis of the following recognition of the use of coastal waters.

Each of the activities demanding space in coastal waters wants to locate at an advantageous point to itself, namely, the point with high locational potential value of the activity. However, since coastal waters are of limited space, there should be competition among activities in locating. And if there are interactive effects among activities, the locational potential of a point will be changed according to the location of the adjacent points. Thus, the condition that each activity competes with each other for maximizing its locational potential value can be recognized as the state of the n-person non-cooperative games. That is to say, if the locational point is regarded as a strategy, the potential value of that point can be regarded as the payoff of the strategy. Further, the condition that a payoff is changed by the interaction of other activity's location can be considered the essence of the games. Namely, in the games a payoff of a strategy varies according to other players' strategies.

On the other hand, the planner will aim to utilize the coastal waters efficiently and impartially among activities. The efficiency and the impartiality are expressed in this model as follows. First, for efficiency each activity predicts other activities' strategies stochastically and chooses the strategy which brings the maximum expected payoff. Next, for impartiality this game has a rule that when two or more activities want to locate at the same point, the activity which relatively has fewer points with high potential values can locate there.

(2) Premises and Assumption for Modelling

1) The concerned area is coastal waters where no plan of utilization is fixed, and is divided into squared unit grids where natural and social conditions can be regarded as uniform.
2) Mixed use of a unit grid is prohibited, that is, a grid is used for only one activity.
3) The area demanded by each activity is apriori given by an appropriate forecasting.
4) The potential values of each grid for concerned activities are scored by professional evaluations.
5) When an activity is allocated at a certain grid, its interactive effects influence the surrounding eight grids.
6) The interactive effect value is defined as the change of potential value. The value activity l at grid j affecting on activity k at grid i is given by Eq.(1). This equation is based on the gravity model, and that value is assumed to be inversely proportional to the square of the distance

between two grids. The numerator of Eq.(1), f^{kl}, denotes the coefficient of the interactive effect, which represents the effect value at a unit distance. The denominator r(i,j) represents the distance between grid i and grid j.

$$a^{kl}(i,j) = f^{kl}/\{r(i,j)\}^2 \qquad (1)$$

THE FORMULATION OF THE ALLOCATION MODEL

(1) The Outline of the Model

As mentioned before, this activities allocation model is formulated as the n-person non-cooperative games where each activity plays the part of a player. Hereafter, in order to correspond to the terms used in the theory of the games, we will refer to an activity as a player and refer to a potential value as a payoff, respectively.

Each player obtains one grid at each play. The process of obtaining a grid is as follows: each player decides his strategy of the play, every player presents his strategy simultaneously, conflicts of strategies are managed if there are any, and the grid given to each player is determined. After one play finishes and every player obtains his grid, the potential values of the grids adjoining those grids change, owing to their interactive effects. The above-mentioned play shall be done over again until the demand of all activities is satisfied, and eventually an allocation plan will be made. Naturally the demand varies from activity to activity, so the activities which satisfy their demand drop out of the game. Fig.1 shows the above-mentioned flow of allocation planning.

(2) The Procedure of the Simulation

The procedure of the simulation of the activities allocation will be explained in accordance with the flowchart shown in Fig.2.

1) Initial potential values

The potential values which are obtained by means of the potential analysis are used for the first play.

2) Calculation of weight

Here, weights mean relative potential values which sum up to a unit. Namely, the weight of activity k and grid i at h-th play is given by

$$w_h^k(i) = p_h^k(i)/\sum_i p_h^k(i) \qquad (p_h^k(i) \text{ is a potential}) \qquad (2)$$

3) Calculation of expected payoffs

As stated before, in choosing his strategy, each player predicts other players' strategies stochastically, calculates the expected payoff for each strategy, and chooses one of them which brings the greatest expected payoff. That is to say, each player calculates expected payoffs on the assumption that other players would take mixed strategies and the above-mentioned weights are regarded as the probability distribution of the mixed strategies. As will be mentioned in detail at 7), if there is a conflict of strategies, the player giving that grid the greatest weight among all players can receive it. Therefore, the expected payoff of player k and grid i at h-th play can be calculated as follows:

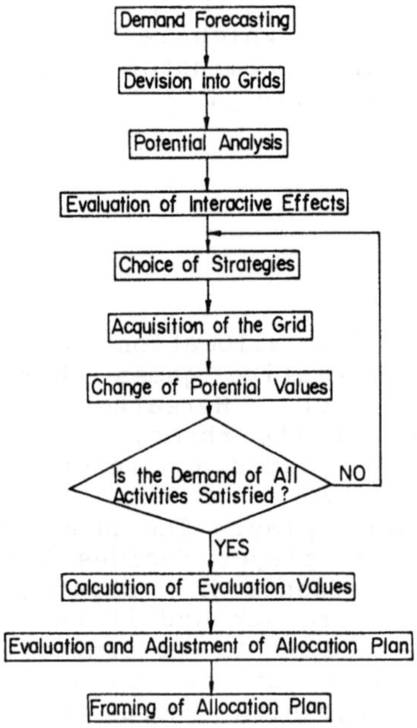

Fig.1 Flowchart of Framing an Allocation Plan

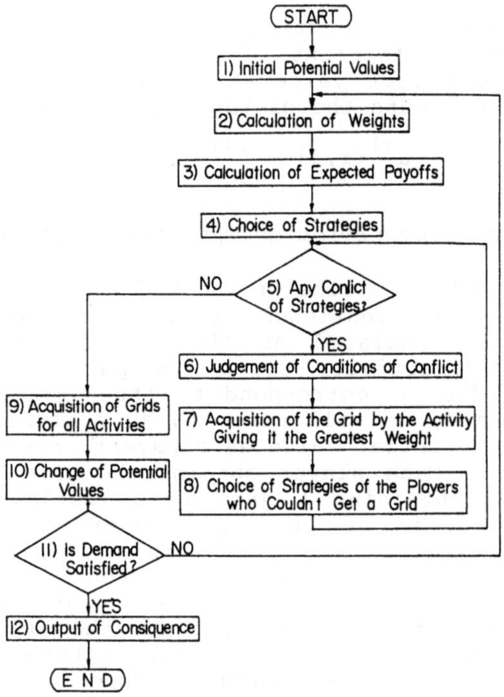

Fig.2 Procedure of Simulation

$$e^k_h(i) = p^k_h(i) + \sum_{l} \sum_{j} w^l_h(j)\, a^{kl}(i,j) - \sum_{l \in L} w^l_h(i)\, p^k_h(i) \qquad (3)$$
$$\scriptstyle l \neq k \quad j \neq i$$

where

$$L = \{l \mid w^k_h(i) < w^l_h(i)\} \qquad (4)$$

4) Choice of strategy

The strategy of player k at h-th play is that which brings the greatest $e^k_h(i)$ in Eq.(3).

5) Judgement of whether there is any conflict of strategies

A conflict of strategies means that two or more players take the same strategies. If there is a conflict, go to 6), and if not, go to 9).

6) Judgement of a condition of a conflict

When there is a conflict, how many activities participate in it at how many grids should be searched.

7) Acquisition of the grid by the activity giving it the greatest weight

If two or more players conflict for a grid, the player who gives it the greatest weight among them can receive it. The reason is that the player who gives it a smaller weight has relatively more grids with high potential values.

8) Choice of strategy of the player who couldn't get a grid

The players who couldn't get grids at 7) have to change their strategies. They calculate the expected payoffs again considering the interactive effects of the grids allocated at 7) and choose their strategies, then go back to 5).

9) Acquisition of grids for all activities

When there is no more conflict, all grids to which all

activities are allocated at that play are determined.

10) Change of potential values

When all allocated grids at that play are determined, all potential values around them change owing to their interactive effects. These changed values shall be used at the next play. Therefore, the potential value of activity k and grid i at h+1-th play is,

$$p^k_{h+1}(i) = p^k_h(i) + \sum_l \sum_j a^{kl}(i,j)\ \delta^l_k(j) \tag{5}$$

where

$$\delta^l_h(j) = \begin{cases} 1 : \text{when activity l was allocated to grid j at h-th play} \\ 0 : \text{otherwise} \end{cases}$$

EVALUATION OF ALLOCATION PLAN

In this paper the allocation plan made by the above-mentioned model is evaluated from two points of view; namely, efficiency and equity. As a measure of efficiency the sum total of potential values obtained by all activities is used, and as that of equity a degree of satisfaction is proposed for each activity. A degree of satisfaction of activity k is given in Eq.(6).

$$S(k) = \{PSUM(k) - PMIN(k)\}/\{PMAX(k) - PMIN(k)\} \tag{6}$$

where PSUM(k) is the sum total of potential values activity k received after the allocation finishes, PMAX(k) and PMIN(k) are those values in relation to an advantageous allocation and a disadvantageous one for activity k, respectively. An advantageous allocation for activity k is the allocation that activity k takes his favorable grids in advance and then the remaining activities are allocated by the proposed model. A disadvantageous allocation is to the contrary. It can be considered that the allocation with equalized S(k) in regard to every activity is equitable.

A CASE STUDY

As a case study, the model proposed in this paper is applied to the coastal waters in the Osaka Bay area. The concerned area is the coastal waters at a depth of less than about 20m, and divided into small grids (about 1 km^2). The concerned activities are the following four activities: recreational, industrial, fishery and residential type activity. Demanded space of each activity is calculated as 24, 40, 32 and 30 grids for recreational, industrial, fishery and residential type activity, respectively. The potential analysis has been conducted for each activity[1)]. For example, Fig.3 shows the consequence of recreational type activity. The coefficient of interactive effects, f^{kl}, namely the change of potential values that activity l affects activity k at a unit distance is defined as shown in Table 1.

On the basis of the above premises, activities were allocated by the proposed model. The map of consequence is shown in Fig.4 and two measures, namely, the sum total of potential values, PSUM, and a degree of satisfaction, S(k),

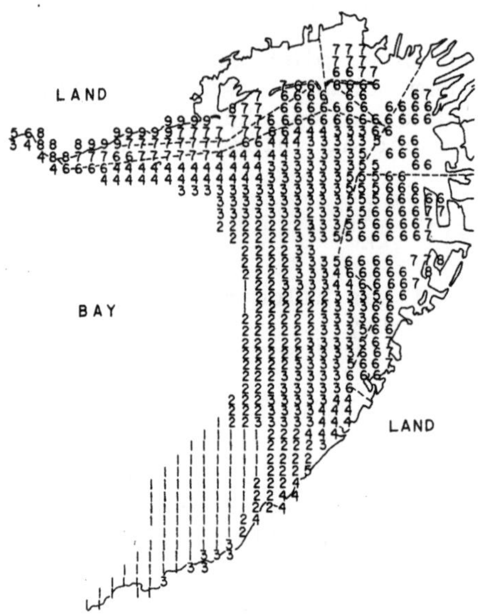

Fig.3 Potential Map of Recreational Type Activity

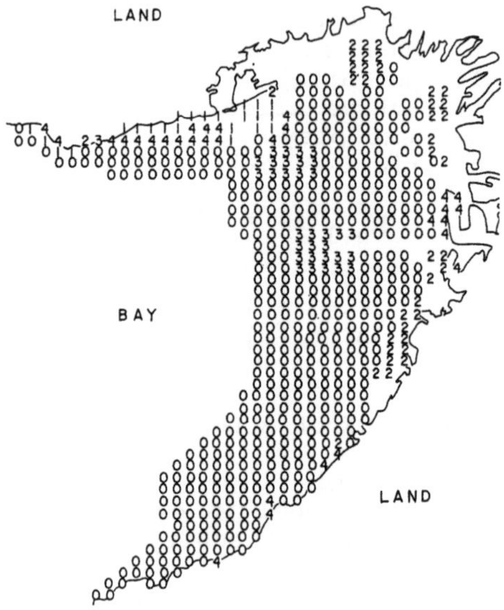

Fig.4 An Allocation Plan

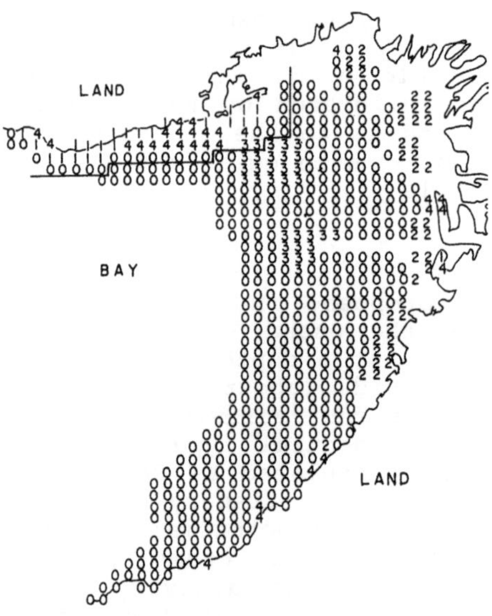

Fig.5 An Allocation Plan under a Use District

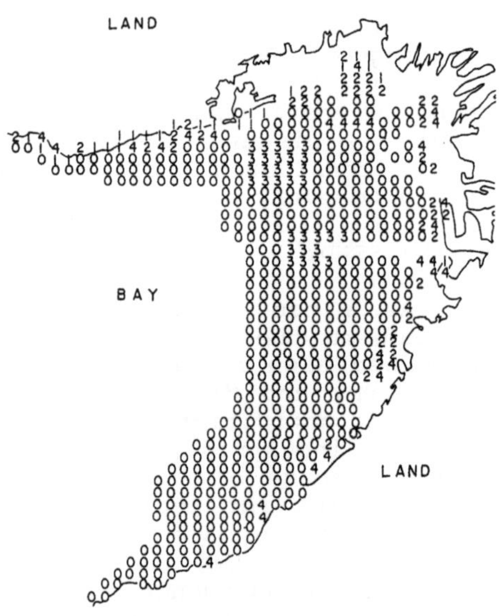

Fig.6 An Allocation Plan without Consideration of Interactive Effects

Table 1 Coefficients of Interactive Effects

k \ l	Recre.	Ind.	Fish.	Resid.
Recre.	0.3	-0.4	0.0	0.0
Ind.	-0.2	0.4	-0.2	-0.2
Fish.	0.1	-0.4	0.2	-0.2
Resid.	0.2	-0.4	0.0	0.2

Table 2 Evaluation Values of the Allocation Plan

k	Recre.	Ind.	Fish.	Resid.
PSUM(k)	206.5	358.5	312.3	251.0
PMAX(k)	218.4	388.4	314.4	281.8
PMIN(k)	183.6	322.0	314.4	197.6
S(k)	0.66	0.55		0.63
PSUM	1128.3			

Table 3 Evaluation Values under the Use District

k	Recre.	Ind.	Fish.	Resid.
PSUM(k)	209.5	356.5	313.7	250.6
S(k)	0.74	0.52		0.63
PSUM	1130.3			

Table 4 Evaluation Values without Consideration of Interactive Effects

k	Recre.	Ind.	Fish.	Resid.
PSUM(k)	191.0	318.0	288.0	238.0
PSUM	1035.0			

are shown in Table 2. In Fig.4, the grids on which 1,2,3 and 4 are printed express those to which recreational, industrial, fishery and residential type activity were allocated respectively. The grids on which 0 is printed express those to which no activity was allocated. In Table 2, the degree of satisfaction of fishery activity is blank because on account of no conflict at the grids to which fishery activity was allocated the value of PMAX shows the same value of PMIN in Eq.(6) and the degree of satisfaction cannot be calculated.

Next, as an alternative which accepts the intention of the supply side, the activities were allocated on the condition that a use district is assigned. Fig.5 shows the map of the consequence. In this figure the top-left area from the line is regarded as the environmental preservation area, so industrial type activity cannot be allocated there. The evaluation values are shown in Table 3.

It can be seen from Table 2 that the degrees of satisfaction are fairly equalized. Probably the reason is that this model originally aims to equalize all activities to some extent. That is to say, when there is a conflict at a grid, the rule that the activity giving the greatest weight can obtain it works as the activity whose remaining girids have relatively small potential values has priority.

If the evaluation values in Table 2 are compared with those in Table 3, the sum total of the potential values in Table 3, namely on the condition that a use district is assigned, is greater than that in Table 2, namely without a use district. However, the degree of satisfaction without a use district is more equalized, so it is considered that industrial type activity would not be satisfied with this assignment of the use district.

Finally as the sensitivity analysis of the interactive effects, activities were allocated without consideration of the effects, and the consequence is shown in Fig.6 and Table 4. As shown in this figure, every activity was allocated so scatteringly that there would be problems of efficiency and environment. Further, the sum total of the potential values is fairly small, so we can justify the significance of considering the interactive effects in allocating activities.

CONCLUDING REMARKS

In this paper, an activities allocation model in coastal waters is proposed considering the interactive effects among activities. The results we obtained through this research may be summarized as follows:

1) By means of the activities allocation model considering both the positive and negative interactive effects, which have hardly been considered so far, efficient allocation plans can be made. Since this model is aided with a computer, it can be fit for an enlargement of the concerned area and an increase of the activities.

2) In most of former planning, equality of interests groups concerning the plan has not been thought much of, but efficiency has. By dint of equalizing degrees of satisfaction proposed in this paper, the plan which also aims at equality of interests groups may be able to be made.

Finally, the following points are left as future problems:

1) It occurs in this model that the potential value of the grid an activity had obtained is reduced on account of other activities adjacently allocated. Therefore, the allocation plan by this model is not always the most efficient one.

2) In using this model, it is necessary to evaluate the interactive effects precisely. However, the study of this field has hardly been done so far, so further study is necessary.

Reference

1. The 3rd District Port Construction Bureau of Ministry of Transport: A report of an investigation of proper utilization planning in waters of Osaka Bay, 1979.

Marine Recreation

An Advancing Technology for Future Ocean Space Technology

T. Morash

Department of Recreation and Leisure Studies, California State University, Northridge, CA 91330, USA

ABSTRACT

Major trends and new programs in Marine Recreation are examined, including floating cities, underwater parks and hotels, and aquaculture, as well as effects of stress on humans living and working in the underwater environment.

MARINE RECREATION DEVELOPMENT

The demand has never been greater for the development and management of marine recreational activities and facilities, as increasingly larger numbers of participants seek out water-oriented activities. In turn, the growth increases the already substantial socio-economic-cultural factors that enhance the recreational and economic values of the beaches, shorelines, harbors and marinas, underwater parks and preserves and the adjacent waters. Our coastal areas and shorelines are attracting various competing users, thus, emphasizing the need for careful management of existing facilities, and to an even greater degree, of planning for new or expanded activities to balance the potentials of overcrowding against the quality of the recreation experience, while maintaining a positive cost-benefit ratio and an increased quality of life. Millions of people owe their livelihoods to the sea and many more are enjoying the recreational aspects of the sea.

It is essential, for common understandings, that a review of the current status of Marine Recreation be undertaken, including recent areas of expansion, such as, recreational use of purified wastewater ponds, expanding underwater park systems, fish farming for recreational purposes, recreational programs for off-shore "floatels," and boating and SCUBA programs for the handicapped. Utilization of artificial reefs for off-shore aquatic centers to provide camping, picnicking, recreational fishing, and underwater photography and exloration all hold great potential.

Growth of user-conflicts, as the number of users increases, is to be expected. We are an oceanic society and must wisely plan for the use of the oceans without destroying them. There are changing trends to be studied; inventories of the shorelines and beaches to be developed; assessment of the full potentials of the oceans for marine recreation programs and increasing tourism participants; changing trends in marina designs to handle larger demands for boats that are trailer-towed and need temporary storage; and marine education programs for the general public as well as our youth are but a reminder of the task that lies ahead.

The sound of the surf and the vastness of the sea make coastal waters very special. One of the main stated reasons people recreate is to relax and find relief from stress and tension; however, when the individual seeking relaxation through the recreation activity finds that recreation spaces overcrowded, expectations are thwarted and the stress/tension levels rise, rather than offering a pleasant, relaxing experience.

When facilities in a given area are not attractive or are heavily loaded, the user will move on to more desirable facilities in other areas because geography is not a barrier and outdoor recreation cannot be viewed as a local matter. High quality facilities are generally the recipient of excessive demands, while lesser facilities are under-utilized. Distance becomes a prime factor in the consumption of marine recreation activities, even when they are "free" because the user must find transportation to the activity and this involves monetary considerations. One must also calculate the time factor involved for time and money are usually not unlimited. The user must then calculate and measure the satisfaction level of the anticipated activity against all other possible users to which time and money might be used.

Wise management of the resources is dependent upon adequate research to determine just what it is the user desires in the way of activities in order to develop a sound management plan that has balanced the potential of overcrowding against a cost-benefit ratio to handle the allocations of space. Such perceptive management is essential to sustaining high quality in marine recreation programs that are serving increasingly larger numbers of participants in a wider variety of activities than ever before -- and all related to significant economic planning, physical site implications and potential multiple-use conflicts. Planning for these marine activities is socially, emotionally, and economically most important in today's complex world of pressures. Coastal shorelines with great recreation potential also are attractive for various competing economic uses. Private residential developments, hotels, townhouses, refineries, pipelines, terminals, and marinas are all drawn to coastal land where land costs are measured by the frontage foot. Conflicts among user groups require perceptive planning based upon sound research, which at this point in time has not been adequately developed.

Unless this process is carefully and precisely carried out, beach recreation could be adversely affected, as Gardner M. Brown states, "by conventional forms of pollution from the shore causing closure of beaches to water contact sports, changing coastal land use and development may exclude beach usage; sea-wall construction or dredging of channels may modify currents and sand deposition and degrade beach resources. It is thus necessary to understand the recreational value of the beach as a recreation site in order to compute the benefits of protecting a beach from these adverse impacts. Beaches provide other services besides recreation, for example, protection from flooding and erosion of land areas behind the beach. These protective functions may be degraded by a variety of activities including over-extensive development and utilization of the beach for recreation, alteration of sand movements, and changes in land use. Such factors must be considered in the determination of the total economic value of beaches.

Needs and desires for different types of recreation services are being expressed openly and includes swimming, sunbathing, snorkeling, underwater photography, fishing, sailboarding, jet skiing, diving and wreck diving, surfing, boating and racing, sailing, water skiing, and wind surfing. Each of these sports requires adequate spacing of areas and freedom from crowding that otherwise could destroy the value of the recreational experience being sought.

Underwater Parks and Reserves. One of the interesting developments in marine recreation has been the growth and development of underwater parks and reserves which were first established in the 1970's on the Monterey Peninsula in California and at the John Pennecamp Coral Reef Park in Florida. The objectives of such programs provide protection of the resources while permitting participation in marine activities and nature interpretation programs. Canada, Japan, Australia, Israel, the Philippines, and Mexico have also established underwater parks which emphasize recreation, education, and tourism.

The State of California Underwater Parks provide us with some data as to the economic impacts from this form of marine recreation. Currently, there are over 550,000 certified divers in California, representing 37% of all certified divers in the nation; and 20,000 new, active divers become users of the oceans resources annually. Standard and Poor's Industry Survey indicated these divers spend over $30 million on related services; i.e., instruction, boat trips, tours, and round-trip transportation to underwater sites. A 1980 poll of leisure activities in the United States indicated that one percent of all Americans participate in SCUBA diving and twice that number (2%) snorkel.

The State of California provides eleven marine park units and one inland underwater unit at the present time, with an additional 23 units under consideration. Facilities range from primitive to those with separate divers' facilities, lockers and dressing areas. At the Point Lobos Underwater Unit, divers may use an underwater guide to animals frequently seen. Printed in full color with concise descriptive data, the flip charts are plastic, laminated for full waterproofing, and are carried on the dive as an excellent educational tool to enhance and enrich the total diving experience.

Historical Reserves. An even more specialized form of underwater park is Truk Lagoon, one of two existing historical reserves protecting shipwrecks. Truk Lagoon Historical Monument is located in the Caroline Islands (a U.S. trust) in the South Pacific. The other historical reserve is Fathom Five Underwater Park in Tobermory, Ontario, Canada, with its boundaries both in Lake Huron and Georgia Bay. This 45 square mile park, established in 1970, contains 18 shipwrecks that are easily viewed through crystal clear water, plus a dive center and interpretive center. It draws over 250,000 non-diving visitors annually, and provides one of the few sources of revenue to the Island.

Role of Marine Recreation. Marine Recreation, a relatively new ocean specialization to many, is a viable and natural partner with the hard sciences due to its common base, the oceans. In addition, there are shared interests in the impacts people make on the marine environment and the need for research and information as to the impacts and relationships between marine resources and recreational/leisure users. The Marine Technology Scoiety has recognized such factors and has established a Marine Recreation Committee within its operational structure in order to provide greater cooperation and sharing of research and common concerns.

Aquaculture. The heated water from power plants and industries warrants further in-depth study of the utilization of that warmed water for increasing the temperature of reservoirs, lakes, and holding ponds. The addition of treated wastes from water treatment plants could provide increased nutritional materials in the form of nitrogen and other needed elements so that fish production could further reduce the plant's operational costs.

It is becoming apparent that the trend is to recognize that sludge can be a resource of economic value rather than a costly and useless waste product. It becomes important to ascertain the differences between use and disposal and waste and salable product. There is a growing market for the conversion of pollutants/wastes into highly salable products, and further, that comprehensive waste management can be a most desirable environmental, economical asset. As cheap energy disappears, natural biological systems will become increasingly valuable and this quality must be recognized in order to protect and enhance those natural systems.

Additional research is needed to determine the possibilities of combining such processes as described with the development of future park sites planned around wastewater lagoon systems to provide economic benefits to wastewater treatment operators and recreation departments now hard pressed to find available land. Numerous recreational activities including swimming and fishing could be offered, along with play areas, picnic areas, and green lawns watered and fed by treated water through reclamation processes and wise management.

It is not the intent of this author to focus on Aquaculture; however, such a venture should be encouraged in our estuaries, bays and shores. The Japanese have already proven the economic efficiencies of shifting from the harvesting of the ocean's natural production of foods to the actual "farming" of seas in a controlled environment.

Fish farming is also tied into marine recreation programs in several Pacific region countries and states such as Taiwan, where Tilapia and Pacu from the Amazon are transferred to recreational fishing areas. Some power plants have already created warm water lagoons, lakes, and fish ponds to provide increased revenue from fish production and recreational fishing access to the general public.

Wesley Marx, in "The Oceans," states that "marine development could serve to enhance, restore, and enrich the environment. Guided by this vision, oil platforms might undergo metamorphosis into high-rise marine suburbs. Below deck, fish farms might flourish, nourished by ocean sewage outfalls whose output would be as carefully blended as a puree from a food processor."

<u>Floating Hotels.</u> Another user of marine recreation services is the petroleum and natural gas corporations involved in the off-shore drilling processes. Semi-submersible "Floatels" of 600-person capacity have been built for Chevron U.K. at a cost of approximately $54.4 million. The self-propelled, twin hulled units were built in Sweden; and contain a medical center, theater, and recreation facilities. Europe's newest $120 million, seven-story "Floatel" has been opened at the Norwegian Ekofish Oil Field, 180 miles out in the cold North Sea for the Phillips Petroleum Company. The 160 ton structure has 106 rooms, cinema, chapel, dining room, coffee shop, hospital, and recreation facilities for their 1,000 workers.

<u>Training Programs.</u> As the off-shore rigs have continued to grow in size, with larger numbers of workers to operate the equipment on a 24-hour basis, so has the need for recreation services. Working in relatively confined areas and extreme weather conditions tends to contribute to the human factors of stress, tension, and boredom. It has become apparent that the need is great for "on-board" recreation programs provided on the larger rigs by trained recreation personnel. The demand for specially trained personnel to handle programs on such units will reqire training not previously provided by the colleges and universities.

Marine Recreation courses have been developed to provide understanding in coastal zoning and access laws; use conflicts; modes of behavior involved with isolation, confinement, boredom and anxiety; the sociology of the small group syndrome; the effects of stress, work and play; specialized programs; laws and behavior in an enclosed community, and human alterations of the environment. The Department of Recreation and Leisure Studies at California State University, Northridge offers a course in Marine Recreation Management which provides professional training, laboratory work, and practical experiences.

Mankind's expansion of scientific exploration of the frontiers of space and the oceans for both utilization and potential colonization produces growing concerns with the impacts on those environments and on those who will be and are "working" in those habitats. Tasks may not always be accomplished as easily in ocean environments as on land. This obstacle in itself raises concerns about the demand or stress that workers must confront in such situations. The increasingly larger number of oceanic workers creates increased concerns with the amount of stress confronted and the impact that stress imposes on workers and their performance. Much has been learned about stress, its causes, what it is, how it affects task performance, and how to control and reduce the effects it has on task performance. Recreation activities can be an effective method of stress management and easily implemented in any work environment. These activities can reduce stress levels in highly stressed individuals, enhancing the quality of the work environment while inreasing task performance.

Indeed, each time we send people out into these new environments to complete predesigned tasks, we are confronted with new problems not always encountered on dry land. We find that the task cannot always be accomplished as it is done on-shore. Furthermore, we have found that these new environments, as well as the conducting of the tasks in the new situation, create new demands upon those persons working in these settings.

Based on the work of Selye, Stennet, Glass and Singer, we can identify stressors which are generally found in oceanic work situations:

(1) Hours of work--is it an 8-5 job, five days a week, or are the hours consistently varying or fluctuating? For some workers a consistent work schedule can be very distressful, while for others, it is very enjoyable (eustressful).

(2) Shift--people work best at varying times of the day. A person who works best during the day, but who has been assigned to a night shift, undoubtedly will have to contend with some distress.

(3) Separation from family and friends for extended periods of time can be potentially distressful. The extent to which separation will be distressful depends on the degree to which the worker relies on family and friends to enhance his life.

(4) Work conditions--how much noise is present, space availability, degree of dimness or light, air pollutants, color modification due to depth of the working situation, and vibration and motion.
Vibration is usually experienced as a consequence of operating a tool such as pneumatic drill or driving a vehicle which transmits engine vibration to the driver's seat. Of course, in some oceanic situations, vibration is more pronounced (e.g. drilling rig pipe changes, construction vibration, engine vibrations). Not only are there vibrations in oceanic work situations, but also in equipment used in conjunction with oceanic work. Studies by Jacobsen and Klauser (1983) have found the vibrations experienced from rotary winged aircraft by pilots and crew are significant. The results from Jacobsen and Klauser's study suggests that these workers may be submitted to vibration three times the amplitude of a helicopter.

In addition to vibration, motion is a stressor associated with oceanic work environments. The to-and-fro motion of a ship or shore boat, or the roll of an aircraft are among the most common oceanic motion stressors. Physiologically, motion stress is precipitated by rotary acceleration of the head which is registered in the semi-circular canals of the inner ear.

Vibration, however, is a greater potential stressor than motion. The amount of vibration an individual can experince without eliciting a stress response depends upon the vibration's size of acceleration and its amplitude, as well as personal characterisrtics of the individual. Most of the consequences are physical in nature: blurred vision, headaches, shaking (particularly in the hands), and muscle tension (although very low frequencies, such as those found in massage units, may have a relaxing effect on the muscles). Long-term effects may include Reynaud's disease, where hands become swollen and painful.

Physiologically, the results of these stressors are more likely to be reactions to the physiological outcomes than to the stressors themselves. It is the blurred vision, the tensed muscles, the shakey hands, or the appearance of a long-term effect such as Reynaud's disease which serve as precipitators of psychological sress reactions.

(5) A fifth potential stressor is found with those you work with--is there compatibility with fellow workers? Are all workers part of the "group"? Do they need friendship or is it forced upon them? Not being part of a group when such membership is desirable can be distressful, as can undesired friendship.

(6) Responsibilities -- too much work (work overload) or too little (work underload) can either be distressful or eustressful. Not matching the worker's degree of need for responsibilities with that assigned them produces distress.

(7) Traveling on the job -- do the workers enjoy it or would they rather be with their family and friends?

(8) Autonomy or self-control over one's work task and responsibilities -- the loss of autonomy when autonomy is desired is very distressful.

All of these stressors or demands have been found to directly affect stress levels in workers, thereby affecting their task performance. In addition to affecting task performance, these stressors listed have physiological, as well as psychological outcomes ranging from: 1) depression, 2) frustration, 3) anger, 4) impulsive behavior, 5) inability to concentrate, 6) dizziness, 7) emotional alertness (keyed-up), 8) insomnia, 9) hypermotility, 10) pain in the neck or lower back, 11) loss of appetite, 12) increased smoking, use of drugs and alcohol, 13) nightmares, 14) accident proneness, 15) high blood pressure, 16) liver ailments, 17) arthritis, as well as other physiological and psychological disfunctions.

Research results have provided us with mechanisms to manage these stressors. They include matching work shifts with employees' personalities; redesigning work instruments and tools; providing workers with high levls of autonomy or internal control, among others.

<u>Recreation Programs: An Alternative Approach</u>. Even in the best situations, application for such stress management mechanisms has not always been practical, possible, or convenient. Interspersing recreation activities between stressful work tasks has been found to be successful in reducing stress levels and the post-stress decrement (decreased task performance). Recreation activities have the ability to increase task performance while affecting stress levels, and appears to be linked to the notion of optimum level of stress.

Recreation activities were found to stimulate under-stressed workers in a study conducted by Finney (1979). Workers who engaged in recreation activities between unstimulating (repetitive/boring) work tasks reported to experience less anxiety and less post-stress work performance decrement than their counterparts who did not engage in recreation activities. Furthermore, the worker's ability to perform tasks at a consistent level of output was significantly greater than those workers who did not engage in recreation activities.

A second study by Finney (1981) found that workers who were highly stressed and then allowed to engage in recreation activities significantly out-performed their counterparts who did not participate in the activities. Workers who were subjected to a work task providing low levels of internal locus of control, experienced high stress levels subsequent to the completion of the work task. Some workers then engaged in a recreation activity allowing them to exercise a high level of internal control over their participation. Workers who engaged in the recreation activity persisted three times longer on a second frustrating stressful task than those workers who did not engage in the recreation activity.

Additionally, those workers engaging in the recreation activity reported less anxiety while working on the second stressful task than their counterparts. These laboratory results are consistent with research results presented earlier in this paper. As stress levels become too low or too high, there is a direct impact on subsequent work task (lower performance levels).

However, if workers who are experiencing low or high stress levels are allowed to participate in recreation activities, they do not experience the post-stress performance decrement effect. Interestingly, these results are consistent with those reported by corporations throughout the United States, Japan, Germany, and U.S.S.R., among others who have instituted Employee Recreation and Fitness Programs as a means to lower stress levels while increasing work task performance.

Many large and small corporations (Xerox, Chevron, Lockheed, Kodak, Carte Blanche, Los Angeles Times) have reported significant changes in employees' work behavior among those who regularly engage in their Employee Recreation Programs. In addition to increased work performance, these companies have reported decreases in industrial accidents, horse play, absenteeism, turnover, work task errors, and reported anxiety and stress levels.

How do recreation activities assist the human body in maintaining its optimum level of (stress) arousal, thereby providing optimum task performance? On one hand, recreation activities have been found to assist in stimulating a worker that is under-stressed, which is the main condition associated with boredom. This revitalizing function is implied by the generic term "recreation".

On the other hand, recreation activities have also been found to assist the worker in reducing his stress level as a result of experiencing low levels of perceived internal control. Activities such as jogging, progressive relaxation, meditation, and others, which provide the "player" with high levels of perceived internal locus of control are effective mechanisms in reducing stress levels (along with the many characteristics associated with high stress levels discussed in this paper) while increasing task performance.

Furthermore, recreation activities are easily implemented in any work situation or environment. From the simplest form of recreation (day dreaming) to the most complex (those needing facilities), recreation activities are an important component in any work environment. However, the confines of underwater habitats, floating cities and work platforms will require less space-consuming activities.

Regardless of the space constraints, recreation activities play an important role in maintaining high task performance levels while reducing workers' stress levels. This is particularly true of oceanic work environments related to man's continuing search into the world's oceans and underwater habitats.

Colonization. The question arises as to the potential need for colonizing some portions of the world's oceans to help answer the need for new sources of energy, food, and minerals. Colonization of both space and the oceans may be required for other reasons too, such as sociological, psychological and recreational needs, which sets forth interesting challenges at this point in time.

One of the most exciting additions to Disney World in Florida has been the opening of its Experimental Planned City of Tomorrow (EPCOT Center). One of the pavilions is called "Horizons" which shows possible habitats of the future, one of which is an Underwater Habitat displaying buildings and possible modes of underwater transportation.

The Japanese architect, Kyonori Kikutake, working with John Harmon at the University of Hawaii, has developed comprehensive plans for an entire "Floating City" that includes a land-sea use concept using land areas for housing, parks, transportation, and a floating city comprised of hotels, offices, condominiums where people get around easily because moving walkways replace vehicles. Each structural unit of the city will be a half-mile in diameter and are similar to structures now used on offshore rigs in the North Sea. A model has been constructed and is afloat on a bay in Hawaii.

And on the local scene, in 1984, a Southern California entrepreneur has recently announced the opening of the World's First Undersea Resort, located off the coast of Georgetown, Grand Cayman Islands in the Caribbean. Guests will reach the unit, called JULES' HABITAT, by mini-submarine. This first hotel will accommodate four people at one time. SCUBA gear allows guests to explore the colorful coral reefs where they may use the underwater videotape camera to record the adventure. A "wet room" between living areas opens to the sea and allows guests to change into or out of SCUBA gear or use the mini-submarine. Included in the services are personal computers, video games, digital symphonic sound system, and instantaneous satellite communications around the world. Several larger Undersea Hotels are planned for other sites around the world. Costs are targeted at $230 a day including meals and a "merperson" to bring the meals.

When cities become more crowded, especially those along the shorelines, it is anticipated that man will turn to the sea for floating cities, providing residence space with shops and stores, as is currently being designed with models already being tested at Nihon University in Japan and also in other Pacific Basin countries. The future for Marine Recreation appears to be one of steady growth in the next few decades due to the increasing interest shown by the larger number of users of the oceans and shorelines.

As mankind moves into deeper water, drilling constructions of man-made habitats on the ice, and searches the bottom of the seas for minerals, energy and foods to meet the needs of a growing population, midst the threat of declining food production world-wide, the question appropriately arises as to when colonization of the ocean will begin. History tells us that there are moments when a variety of elements come together at a time never before noted and there is a major step forward or it may be lost forever. We may be looking at that opportunity now or in the very near future. Frontiers with physical isolation have existed before and the early pioneers tested themselves and their knowledge against the wilderness. Now we may once again start out anew but with greater knowledge at hand.

A Study on the Effect of the Location of the Sea City as a Solution to Urban Problems

N. Yokouchi[1], S. Sakurai[1] and H. Satoh[2]

[1] Department of Ocean Architecture and Engineering, College of Science and Technology, Nihon University, Narashinodai, Funabashi, Chiba, 274 Japan
[2] Institute of Anthropospace, Ltd., Yoyogi, Shibuya-ku, Tokyo, 151 Japan

ABSTRACT

The significance of sea city construction is not only the creation of a new space, but also lies in solving urban problems with its greatly affecting cities on the continent. We can expect sea city development to be effective especially for the cities whose centers neighbor the sea. This reserch was designed to propose construction of the sea city and then to predict its effect on the cities of the continent through simulation analysis.

INTRODUCTION

Today it is obvious that we have various urban problems in most of the big cities of our country which have vast coastal areas. Especially, with the advance of the spreading phenomenon, those who should be city dwellers are forced to move from the city center area. Urban communities are therefore on the point of collapse. The cause of the present situation originated in an imbalance of city functions, seen as follows. Big cities have developed with an economy-centerd policy, placing little emphasis on the residential environment. From the aspect of urban structure, they are constructed as the one point concentration type which has a core in the city center - that is to say a "monocore city". Imbalances in the city functions of business and residence also demonstrate this trend, as observed in the difference between the populations of daytime and nighttime. Now urban problems are difficult to solve by mere temporary means, such as spreading the urban area into the suburbs or increasing the residential capacity of the city center. To solve urban problems, it is necessary to correct the imbalance of city functions and reorganize the urban structure itself. In order to do so,we suppose it is important for this impact to bring about transformation of urban structure in the area of city centers which are considered to be the nuclei of big cities. As a practical method, we would like to propose the construction of the sea city and define the effectiveness of its location in this study.

PURPOSE OF THIS STUDY

The sea city, construction of which is proposed in this study, is a newly created city space on the coast. It is expected to function with the city center to correct the imbalance of city functions. The sea city not only has many merits - easy selection of location and function, cheap land prices, simple legal rights, etc. - but also can exert a great influence on the city center, which is the root of urban problems, without directly affecting the existing city area. Most importantly the location will not only bring new city space to the coast near the city center, but will function cooperatively with the city center to reorganize the imbalanced urban structure. Nevertheless, if the influence by the sea city is too strong, the location will generate new urban problems. And the location with ease, will tend to develop recklessly, as if it were an urban spread moving toward the sea. Therefore, for the sea city to function effectively, it is important to determine its influence on the city structure by its location. It is also important to determine the effect of changing the current trend of an imbalance of city functions. Moreover, it is important to understand how adoption of various programs will affect the result. The purpose of this study is to determine the effect of the location on urban problems by making prediction models, as well as examining various programs and sizes for the sea city, etc.

OUTLINE OF THIS STUDY

The excessive concentration of businesses in the city center caused by the advantage of centralization has produced various disadvantages for the city center. It has deteriorated the residential environment, bringing about a decline in residential population and thereby lowering the activity of urban communities (cf. Fig.1). This trend is clearly demonstrated by the increase in the difference between the populations of daytime and nighttime. To correct the imbalance of city functions,we think it valuable to increase the residential potential of the city center and form communities there. In other words, the addition of the residential function to the city center will correct the imbalance of city functions, and it is expected that revitalization of communities will retard deterioration of the residential environment and, moreover, improve it. As we mentioned, the sea city with a primarily residential function is expected to form the core of a community. As initiator of the revitalization of the residential function, it corrects the imbalance of city functions. We should ensure that the sea city dose not become involved in the mechanism (cf. Fig.1) formed in the city center. However, if it does become involved, the construction of the sea city will only bring about new urban problems and contribute to the spread of them.

To demonstrate the abovementioned hypotheses, we built prediction models on the effect of location and analyzed them with a simulation in this study. We used the econometric model which is suitable for quantitative and positive prediction. To raise the accuracy of prediction and further define the meaning of these models we made two types of models.

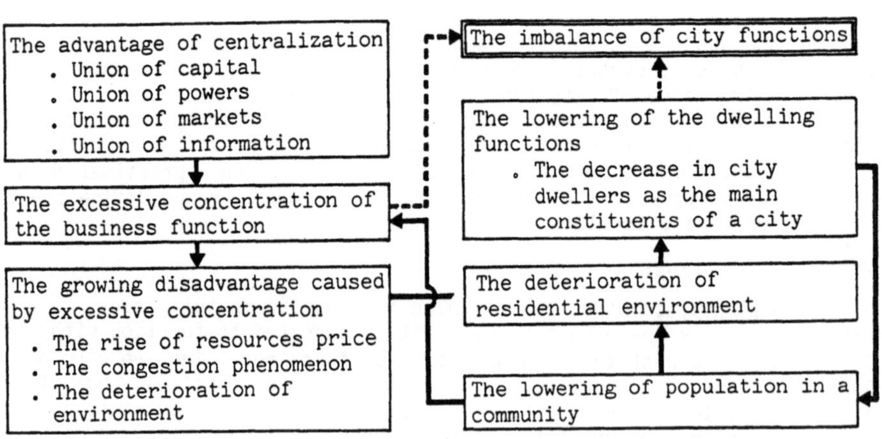

Fig.1. The structure of the exclusion of the dwellers in the city center

One is the prediction model for future trends of the city center on the mechanism producing the imbalance of city functions (Model I). The other is the prediction model for the spreading phenomenon and its relationship to the sea city (Model II). Using each model, we projected the populations of daytime and nighttime, in the cases of having and lacking the sea city. In comparison with each other, we analyzed the feasibility of the sea city as a solution to urban problems. We chose a sample area in Chiyoda ward, Tokyo, where the imbalance of city functions is remarkable. The contents of this study are provided in the flow chart, Fig.2.

THE STRUCTURE OF PREDICTION MODELS ON THE EFFECT OF THE LOCATION OF SEA CITY

In this study, we chose "land use" forming the foundation of urban structure as an exogenous variable, in common with model I and model II. We also chose "nighttime population" and "daytime population" measuring the degree of the imbalance of city function, as endogenous variables. Moreover, in model I, choosing "social stock" as an exogenous variable, we set up the structure shown in Fig.3 on the concept of Fig.1. In model II, we set up the structure shown in Fig.4, choosing "population of sea city" and "land prices" as exogenous variables. On the structure of each model, we set up the items shown in Table 1 and organized the structure relationship as the order of cause and effect (cf. Fig.5, 6). The essential difference between both models is that "land prices" and "floor space" are endogenous variables in model I, but they are exogenous

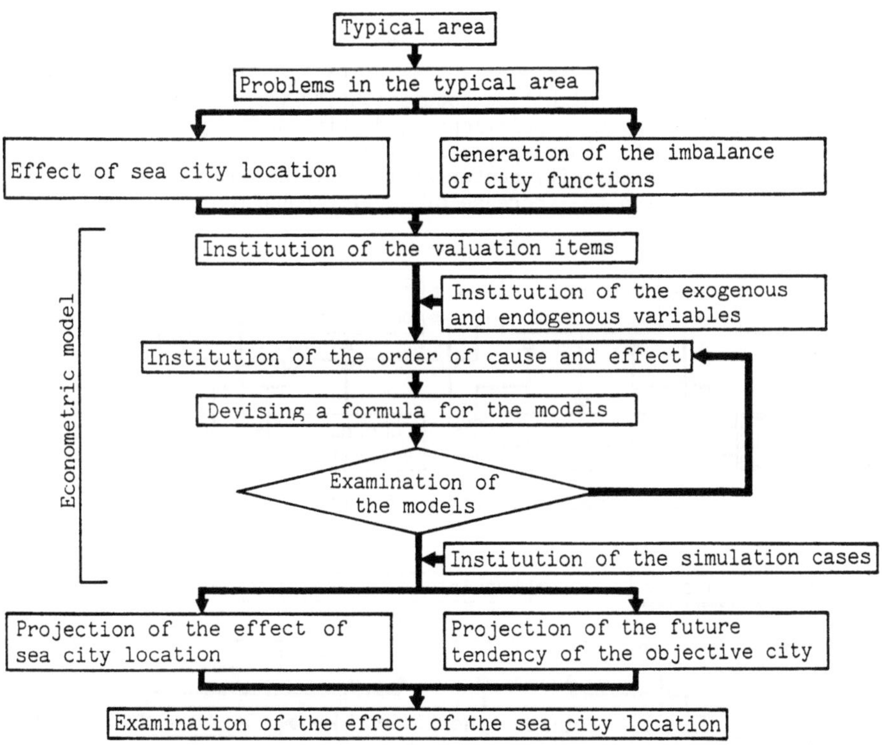

Fig.2. Flow chart of the study

variables in model II. Therefore, it seems that the city activities created by the location of the sea city can retard excessive concentration, the rise in land prices that accompanies it in model II. Estimating the cause and effect order for fourteen years, from 1968 to 1981, simultaneously, we devised a formula. On the results, we should like to show the structural equations and statistical values of each model in Table 2,3. According to them, there are fourteen equations showing the coefficient of determination over 0.9 of sixteen equations in model I and eleven of fourteen in model II. Solving the structural equations as simultaneous equations and inducing the reduced-form equations, we put the whole test and the final test to them. The results of these tests were valuable, so we think both models are persuasive.

SET UP OF THE SIMULATION CASES

The purpose of a simulation analysis is to forecast the effect of constructing the sea city as a solution to urban problems. In this simulation, we would like to analyze the following four cases and forecast the effects for five years, from 1981 to 1986.

- Case 1: the case of allowing the city to develop at the present rate without constructing the sea city
- Case 2: the case of only spreading the city center area and increasing the population by twenty thousand without constructing the sea city
- Case 3: the case of constructing the sea city with a population of twenty thousand
- Case 4: the case of constructing the sea city with a population of thirty thousand

THE RESULTS OF A SIMULATION ANALYSIS AND THEIR IMPLICATIONS

We analyzed case 1, 2 with model I and case 3, 4 with model II and divided the results into "nighttime population","daytime population" and "the difference between the populations of daytime and nighttime" (Fig.7, 8, 9). We would like to consider the results.
In the case of leaving the city center as it is (case 1), the decrease of nighttime population and the increase of daytime population progress

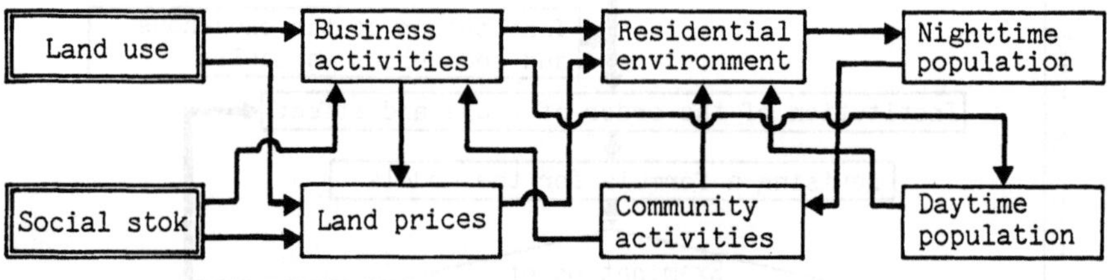

Fig.3. Structure of model I

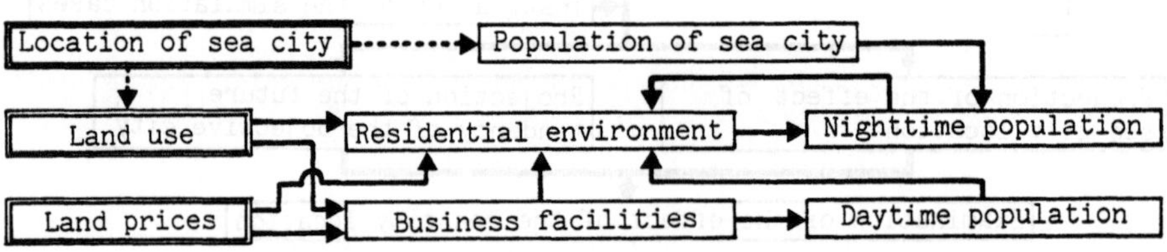

Fig.4. Structure of model II

gradually and the imbalance of city functions is projected increase (The result of this prediction shows a similar tendency, compared with the transition of the surveyed values from 1982 to 1983.). In the case of spreading the city center and increasing its residential function (case 2), the difference between the daytime and nighttime populations is projected to decrease. However, the difference expands little by little until it is thought to be in the same degree as case 1. This, as we mentioned, shows the concentration of business functions and the rise in land prices caused by the spread of the city center which decreased the nighttime population. It also demonstrates the difficulty in solving urban problems only by the spread of city center, even if the residential function is added to it.

In the case of constructing the sea city (case 3, 4), the nighttime population without the population of the sea city is projected to increase by nearly thirty thousand in case 3 and nearly fifty thousand in case 4, compared with case 1. Also the decrease of the daytime population and the shrinking gap between the daytime and nighttime populations are clearly seen in case 3 and case 4. This shows that the increase in nighttime population, caused by the construction of the sea city, stimulated the improvement of the city center area and increased the incoming population. These characteristics should be evaluated as the effects of constructing

Table 1. The table of the valuation items

	Symbol	Valuation items	Selected variable	
			Model I	Model II
1	AFSR	Average floor space of residence	○	
2	BIT	Burdens of inhabitant tax	○	
3	DP	Daytime population	○	○
4	EM	Employees		○
5	ES	Establishments		○
6	FAB	Floor area of business	○	●
7	FAR	Floor area of residence	○	●
8	FARS	Floor area of residence of sea city	●	●
9	HO	Housings		○
10	LAB	Land use area of business	●	●
11	LACP	Land use area of city park		●
12	LAR	Land use area of residence	●	●
13	LARD	Land use area of road	●	●
14	LARS	Land use area of residence of sea city	●	●
15	LPNP	Land use area of city park/Nighttime population		○
16	LPPI	Land prices/Personal income		○
17	LR	Land prices	○	●
18	MI	In migration	○	○
19	MO	Out migration	○	○
20	NP	Nighttime population	○	○
21	OP	Old-age constituency per population	○	○
22	PCE	Personal consumption expended	○	
23	PCPI	Personal consumption expended/Personal income		○
24	PI	Personal income	○	
25	PN	Public nuisances	○	
26	PNNP	Public nuisances/Nighttime population		○
27	PP	Productive age population	○	○
28	PRB	Product of business	○	
29	PS	Population of sea city	●	●
30	TA	Traffic accidents	○	
31	TANP	Traffic accidents/Nighttime population		○

●: Exogenous variable ○: Endogenous variable

the sea city, and thereby the correction of the imbalance of city functions. Compared with case 3, the greater increase in nighttime population is seen in case 4. However, a sudden decrease is seen in the latter half of the projected years, showing an excessive increase in nighttime population causing the business function to concentrate in the city center still more and further downgrade the residential environment. This development suggests that the construction of a big sea city is not always effective. To function effectively, we must know if the sea city has an upper limit in scale and in the case of exceeding it, if the sea city will exert a negative influence on us.

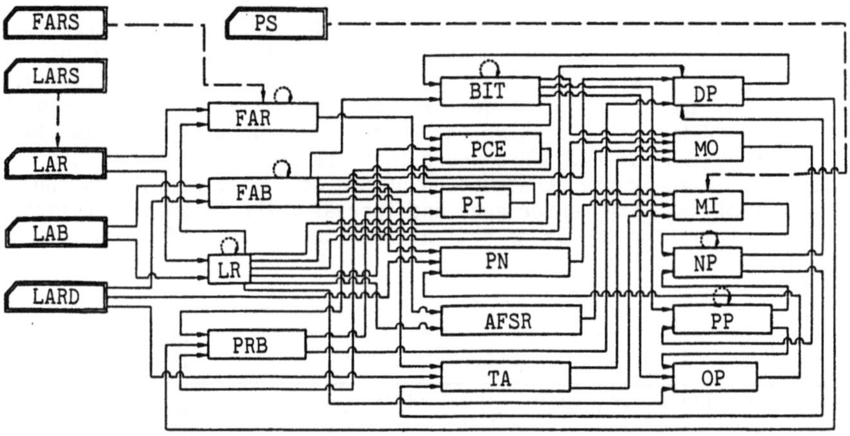

: Exogenous variable
: Endogenous variable
: Relationship of synchronism
: Definition formula
: Relation with a delay

Fig.5. The chart of the order of cause and effect (Model I)

Table 2. The results of projection of model I

Structural equation	Coefficient of determination		
	Partial test	Whole test	Final test
FAR =-27.53+(0.634D-1)LAR +(1.184)FARL-(0.793D-2)LR	0.9798	0.9878	0.9759
FAB =-826.8+(2.142)LAB +(0.564)FABL+(1.640D-1)LARD	0.9327	0.9496	0.9319
LR =-7648 +(23.78)LAR +(21.47)LAB +(0.609)LRL	0.8717	0.9038	0.9475
PRB =-767.9-(3.106)DP +(1.036)FAB +(20.07)PCE	0.9847	0.9875	0.9759
BIT =-1.889+(0.315D-2)DP +(1.027)BITL+(0.132D-3)FAB	0.9555	0.9683	0.9674
PCE = 665.9+(20.23)PI +(12.31)BIT +(0.588D-2)LR	0.9859	0.9826	0.9760
PI = 1.195+(0.113D-2)PRB -(0.180D-2)FAB	0.9757	0.9653	0.9892
PN = 1.292+(4.190)OP -(0.446D-2)LARD-(0.571D-3)FAB	0.5256	0.5781	0.7549
AFSR= 11.83-(0.271D-3)LR -(0.140D-1)FAR	0.9567	0.9780	0.9536
TA =-1.763-(0.378D-2)LARD+(0.854D-3)FAB +(0.333D-1)PP	0.9620	0.9698	0.9294
DP = 886.8+(0.392D-1)FAB -(2.341)PP -(0.547D-2)PRB -(0.297D-1)LR	0.9259	0.9286	0.8922
MO = 5.868-(0.749D-1)AFSR+(2.956)TA +(0.169D-2)LR -(1.189)BIT	0.9752	0.9851	0.9159
MI =-2.253+(6.019)PN +(5.811)TA +(0.927D-3)LR	0.9263	0.9739	0.9150
NP =-17.09+(1.527)PPL -(4.121)MI -(0.180)PP	0.9475	0.9715	0.9223
PP = 49.66-(1.320)BIT +(3.044)MO +(0.796D-1)PPL	0.9711	0.9798	0.9762
OP = 0.196-(0.152D-2)PP +(0.223D-2)BIT +(0.151D-4)LR	0.9879	0.9789	0.9679

As we mentioned, the location of the sea city seems effective for correcting the imbalanced of city functions. Thus we believe the effect of the sea city's location is understood by defining the difference by the effects in a scale.

CONCLUSION

It has been a long time since the construction of a sea city was first proposed. Once the sea city was treated an ideal city,but its construction

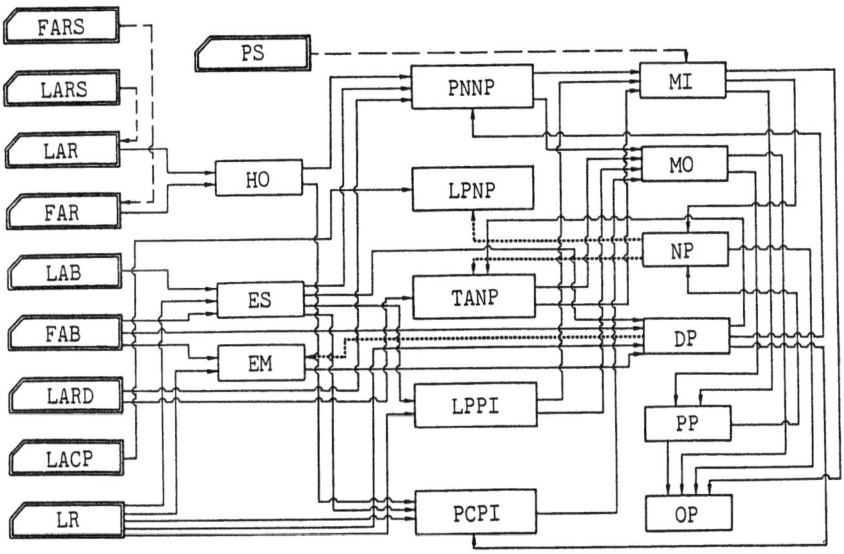

Fig.6. The chart of the order of cause and effect (Model II)

Table 3. The results of projection of model II

Structural equation	Coefficient of determination		
	Partial test	Whole test	Final test
MI =-3.475-(230.0)PNNP+(700.0)TANP+(0.323D-2)LPPI	0.9305	0.9565	0.7432
MO =-0.815-(51.39)TANP+(749.1)PNNP+(0.139D-1)LPPI +(0.135D-2)PCPI	0.9615	0.9708	0.8829
NP = 46.19+(0.283)PP +(4.661)PI	0.9184	0.9706	0.8202
DP = 251.1-(0.268D-1)LR +(0.892D-1)FAB +(0.607)EM +(1.197)ES	0.9081	0.9702	0.7973
PP = 42.71+(2.345)PI +(2.349)MO	0.9804	0.9735	0.9930
OP = 0.338+(0.903D-2)MI -(0.434D-2)MO -(0.206D-2)PP -(0.107D-2)NP	0.9858	0.9932	0.9802
PNNP= 0.007-(0.565D-4)LARD-(0.847D-4)HO +(0.229D-3)ES +(0.549D-5)DP	0.4534	0.5669	0.4692
LPNP= 0.776+(0.181D-2)LACP-(0.431D-2)NPL	0.9824	0.9853	0.9545
TANP=-0.038+(0.123D-4)LARD+(0.353D-4)DP +(0.232D-3)NPL	0.9239	0.9468	0.6156
LPPI= 1337 +(0.322)LR -(0.307)ES	0.4387	0.2733	0.7188
PCPI= 3141 -(0.153)LR +(3.062)HO -(12.50)ES -(3.298)DP	0.9832	0.9860	0.9622
HO =-14.46+(0.183D-1)LAR +(0.183)FAR	0.9950	0.9958	0.9912
ES = 70.43+(0.218)LAB +(0.513D-1)FAB -(0.199D-2)LR	0.8562	0.8921	0.5728
EM =-210.8+(0.356D-1)LR +(0.246)FAB +(0.607)ES +(1.560)DPL	0.8465	0.9055	0.0890

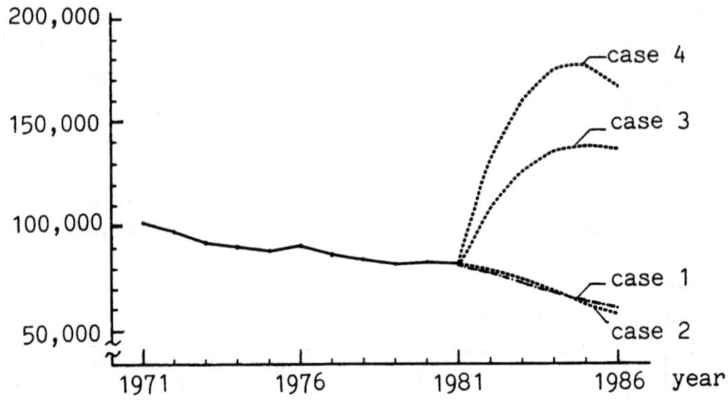

Fig.7. The result of projection of nighttime population

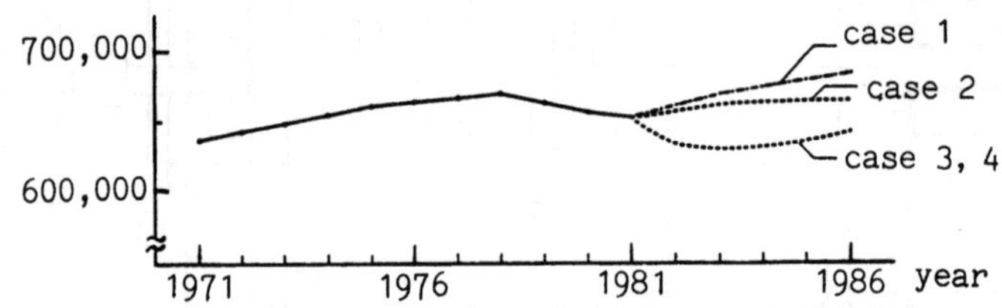

Fig.8. The result of projection of daytime population

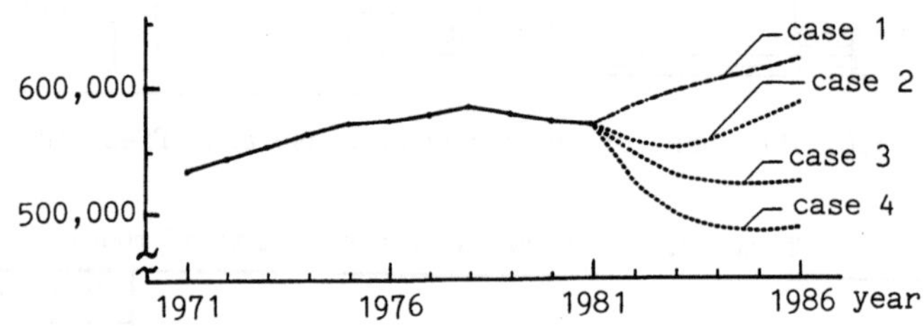

Fig.9. The result of projection of the difference between daytime and nighttime populations

has recently become technologically possible. Consequently, actual examples such as "Kobe port island" and others have been seen and new projects are being proposed. Yet there are some projects whose purpose is only to create new space. To be sure, it is fascinating to note that we can see the growth of a vast, new area near the overpopulated city and solve part of the urban problems by transferring some of the city's activities from the city center to the sea city. Nevertheless, the easily begun construction involves a danger of further generating urban problems by the spreading phenomenon. To effectively construct sea cities, it is imperative to define the social meaning of a sea city and carefully develop the concrete theory of the planning method. In this study, by projecting the effect of the sea city's location, we showed the implications of it, as a solution to the urban problems of the present age. Moreover, we grasped the difference in effect by programs in this study. We believe that the results of this study will play an important role in sea city plans in the future.

A Theoretical Study of the Ocean Space Utilization Planning for the Urban Coastal Water

H. Miyazaki[1], T. Kondo[2], A. Kuroyanagi[2] and S. Araya[2]

[1] Maebashi City College of Technology, Kamisatori, Maebashi, Gunma, 371 Japan
[2] College of Science and Technology, Nihon University, Kanda Surugadai, Chiyoda-ku, Tokyo, 101 Japan
[3] Seibu Marketing Information Service Co., Ltd., Ikebukuro, Toshima-ku, Tokyo, 170 Japan

Abstract

This paper has the purpose of establishing a new technique useful in connection with the process of determining the right functional disposition in coastal zone. The Technique proposed herein regards the land area and the sea area, that hold the coastline between then, as an unfined coastal zone,[1)] systematizes the multi-attribute structure of values referring to the functional disposition of the coastal zone as an integrated social system, and establishes a method to devise solutions to the trade-off problems that occur in the case of multi- dimensional utilization of the coastal zone, as well as trade-off problems between development and conservation. Furthermore, a space value evaluation diagram to draw up the master plan of the water area utilization in the Tokyo Bay is proposed here, by using the technique in question.

Introduction

Cities located in coastal zones of Japan have grown to megalopolis proportions, and they constitute the economic information, cultural and political centers of the regions that they belong to. It must be borne in mind however, that the phenomenon of explosive demographic concentration lies in the background of the said urban growth. As a consequence of the said phenomenon, the Japanese coastal cities are nowadays plagued with such urban problems[2)] as land shortage, energy shortage, environmental pollution,etc., because the degree of concentration of urban component elements of every kind surpass by far the permissible proportions. The said distortion brings about the expansion of the through the enlargement of the sphere of action by urbanization of the inland areas. In reality however, the space available for urbanization is limited by the geographical conditions peculiar to Japan, and furthermore problems of various kinds, represented by the sprawl and doughnut phenomena, are coming to the fore as a consequence of the chaotic enlargement

Spaces facing the sea have very high economic pervasiveness in view of their geographical characteristics, and also from the technical standpoint they can be regarded as easy-utilization spaces. On the other hand, it must be borne in mind that they are places encompassing an abundant and complicated ecological system, and are furthemore places blessed with excellent natural environment. As can be seen, the spaces of the land area and sea area, divided by the shoreline, are very important from every standpoint. uch being the case, various kinds of approaches can be considered in connection with the invigoration of the utilization of coastal zones. In connection with the problem of redevelopment of the land area side however, it is obvious that the status quo of the land utilization in the land area is at a deadlock, and the authors think that the invigoration of the coastal-zone-utilization through the effective use of the sea area facing the megalopolises is an available means to find a solution to the impasse. Under the circumstances, this paper proposes the scheme which consists of invigoration the coastal zone through the establishment of this technique for effective utilization of the sea area.

Objective of the study

This study is aimed at establishing a new technique useful in connection with the process of determining the right functional disposition in coastal zones. This technique regards the land area and the sea area, that hold the coastline between them, as an unified coastal zone, and handles both conservation and development in an integrated way, by making the most of the peculiarities of this space available for multi-dimensional and multipurpose utilization, and by complying with the requirements related to the natural characteristics as well as the local and ecological characteristics of the coastal zone in question. The existence of such facts as multi-dimensional antagonistic goals, uncertainties without reappearability, effects on areas, individuals and groups of various kinds, and long-term effects on the superstructure of the society as a whole, corresponding to a given functional disposition, must be borne in mind when cosidering the utilization of coastal zone. Under the circumstances, in this study the coastal area is regarded as a social system, the multi-attribute structure of values referring to the functional disposition of the coastal zone is systematized, and then the techniques aimed at devising the right solution to the trade-off problems between the multi-dimensional forms of space utilization as well as the trade-off problems between conservation and development are duly examinded and established. Furthermore, this study is also aimed at proposing the correct functional disposition and the space value evaluation diagram to draw up the master plan for utilization of the water area.

Method of study

The Tokyo Harbour (Figure 1), comprising its northern sea area located to the north of the line interconnecting Cape Kannonzaki and Cape Futtsu and the land area consisting of 453 cities, towns and villages within a sphere of 100Km from the center of the Tokyo Metoropolis, and having a high demand of multipurpose sea area utilization, is selected as object of study. The 2KmX2Km quadrilaterals in accordance with the national standerd mesh are regarded as dimensional units for water area utilization.

the functions taken into consideration in connection with the water area utilization are the residential area, industrial area, commercial area, urban facilities area, park, green and open area, fishing ground area, port and harber area, and sea route area. The evaluation analysis is carried out by drawing up, on the basis of the utility theory[3] the evaluation tree diagram referring to the seven functions shown in (Figure 2). When a given Function is arranged, the utilities the land area gives the sea area in connection with each function, and the utilities the sea area gives the land area must be maximized for the land cover function in question to realize and to integrate in the area in question. Such being the case, the composite utility function corresponding to each one of the land cover function in question are established by introducing the concepts of utility independence and preferentiol independence in the utility theory. These functions have the following forms.

$$U_{i:1\sim7}=\sum_{z=1}^{344}\sum_{j=1}^{453}\left(\frac{1}{T}\right)^{2}\frac{\left(\prod_{v=1\sim3}\left(1+KK_{j}^{i}F_{j(x)}^{i}\right)-1\right)}{K}$$

U_i : Utility function by land cover function.

F_x : Dimentional utility function.

i : Mesh number.

T_j : Distance between mesh and land area.

K : Scale composition coefficient.

j : 453 cities, towns and villages.

The composite utility function values of 344 water area meshes are calculated for each land cover function, by means of the aforestated functions. The obtained results are shown in (Figure 3). The utility values by function, calculated from the space utilization that interacts with both land area and water area, take into due consideration both the utilization function and the characteristics of each space. In connection with the water space however, its supply is fixed, and as a matter of fact there is conflict between the various functions as a result of the overlay of the utility contour maps of the 7 functions, as can be seen from the utility conter map (Figure 4). Such being the case, the technique to solve the trade-off problem between the various functions is examined in the following.

(1) If the mutual space utilization degree of each block of the water area mesh is regarded as 1, the space utilization degree calculated for the sea area in question becomes 344. This value can be interpreted as water area utilization supply degree of the Tokyo Bay.

(2) The utility value by function of each mesh mentioned above is interpreted as water area utilization demand from the standpoint of mutual influence between sea and land.

(3) The water area utilization demand by function is calculated by the following expression, by using the aforestated results.

Sea area utilization demand by function=(Utility value of each land cover function)X(Number of water area mesh blocks)/(Utility values by function)

(4) The location demand of each land cover function, calculated with the aforestated function, is 38 mesh blocks for residential area, 39 mesh blocks for industrial areas, 39 mesh blocks for urban facilities area, 39 mesh blocks for parks, green and open spaces, 35 mesh blocks for fishing ground and 39 mesh blocks for harbour sea routes.

(5) The utility values by mesh and by function calculated above are introduced in each mesh in decreasing order, and the process is repeated until fulfilling the value of demand by utilization function.

(6) The demand value by function is supplied to the 344 mesh blocks of the Tokyo Bay by means of the aforementioned method. There are two distinct alternatives regarding the disposition of function in the 344 mesh blocks, the disposition of a single function and the disposition of composite functions. The standpoint adopted in this study is that the higher the value of the mutual utility between land and sea of the utilization function in the coastal zone, the higher shoud be the priority regarding the introduction of the function. Such being the case, the introduction scale in the mesh becomes 1.0 in the case of single function. Furthermore, when a composite function utilization is picked up, the function disposition within the mesh and its scale are determined by assuming a introduction function of 1.0 in the mesh, and by carrying out the proportional distribution of each utility of the picked up functions. The Tokyo Harbour Water Area Utilization Basic Map shown in (Figure 5) is drawn up by repeating this method for each function.

Results

As can be seen from (Figure 5), the areas off Shinagawa Yokohama and Kawasaki, as well as the center of the bay, are suited as residential area, urban facilities area and commercial area. Therefore, the location of single functions is impossible in these mesh blocks, but the mixed location of residential area, urban facilities area and commercial area is possible. The mesh blocks of Funabashi City, Futtsu City and from the offings of Yokosuka to the center of the bay are suited as industrial areas. As for the sea routes and harbour areas, the mesh blocks from Chiba City to the offings of Kisarazu City are available for their location as single functions The mesh blocks from Kisarazu City to the offings off Futtsu City are suited for location of fishing grounds as a single function. The park, green and open space areas are assumed to accompany residential areas, and it became obvious that they should be located at the offings of Futtsu City, in the form of single functions. Generally speaking, most of the coastal background cities at the west side of the Tokyo Bay are specialized in terms of economic environment factors and living environment factors, and furthermore they outnumber by far the coastal zone of the east side in demographic terms. On the other hand, most of such coastal cities as Chiba, Ichihara, etc., of the east side of the Tokyo Bay, are specialized in terms of production environment. Furthermore, the central part of the sea area is the zone with the highest development capacity from the standpoint of natural environment. This is the summary of the results of this study.

Discussion

The technique to draw up the scheme of utilization of water areas is examined in this study, by using the utility theory. When preparing the model, the coastal zone is interpreted as an integrated social system. and the mutual effect the land area and water area exert on each other, as well as the nature of the sea area, are taken into due consideration. The result obtained by this model,which is expressed in terms of directivity of the water area utilization, can be regarded as a valid results from the standpoint of utility of water area utilization in coastal zone. It must be borne in mind however, that these results reflect the status quo of the coastal zone, but it does not identify its future image.

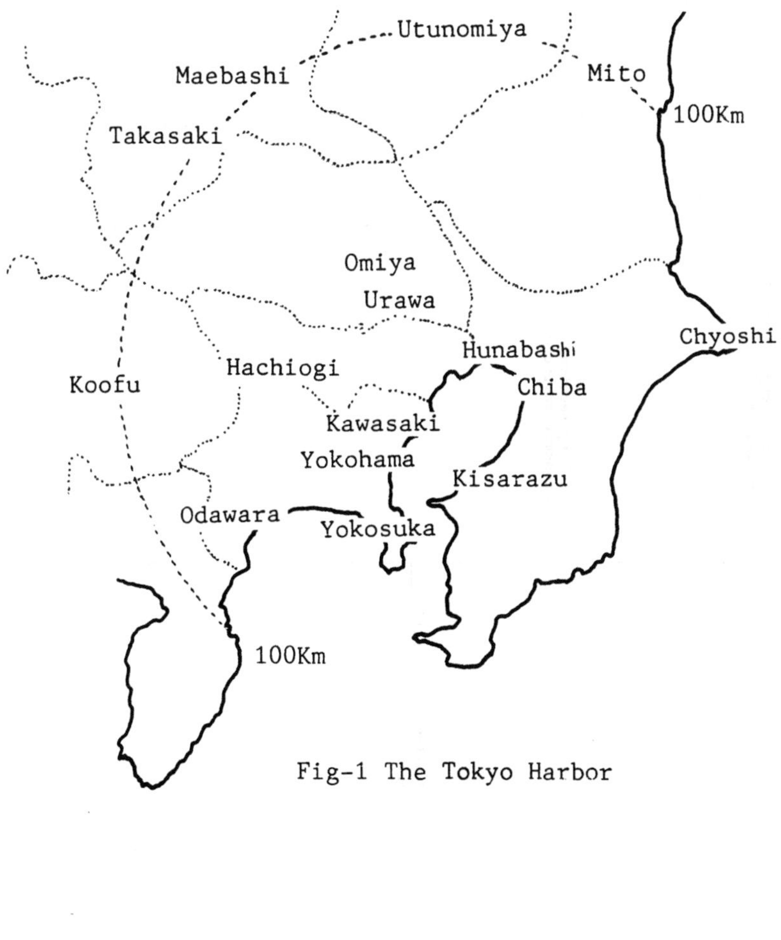

Fig-1 The Tokyo Harbor

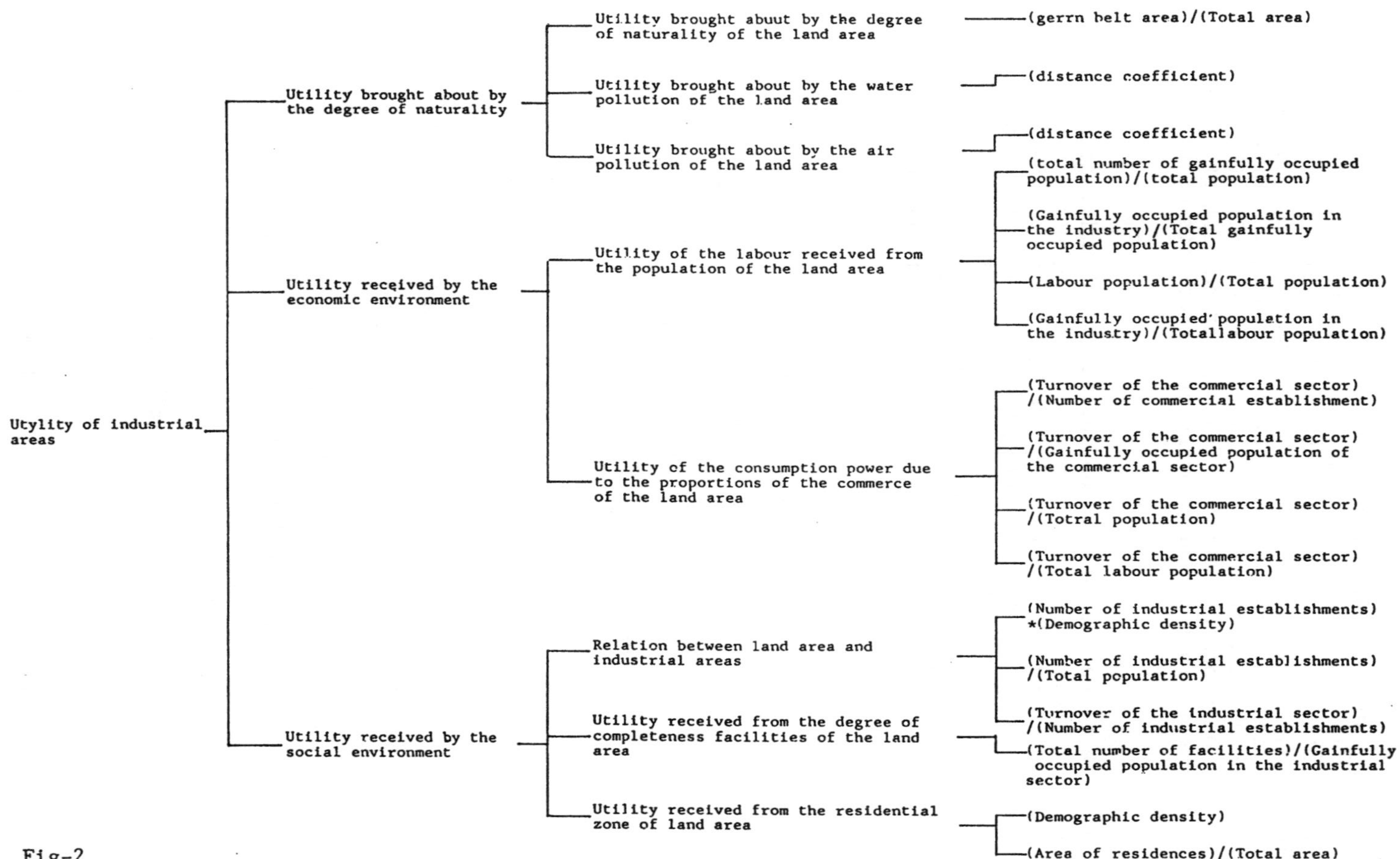

Fig-2
The evaluation tree diagram (Utility of industrial areas)

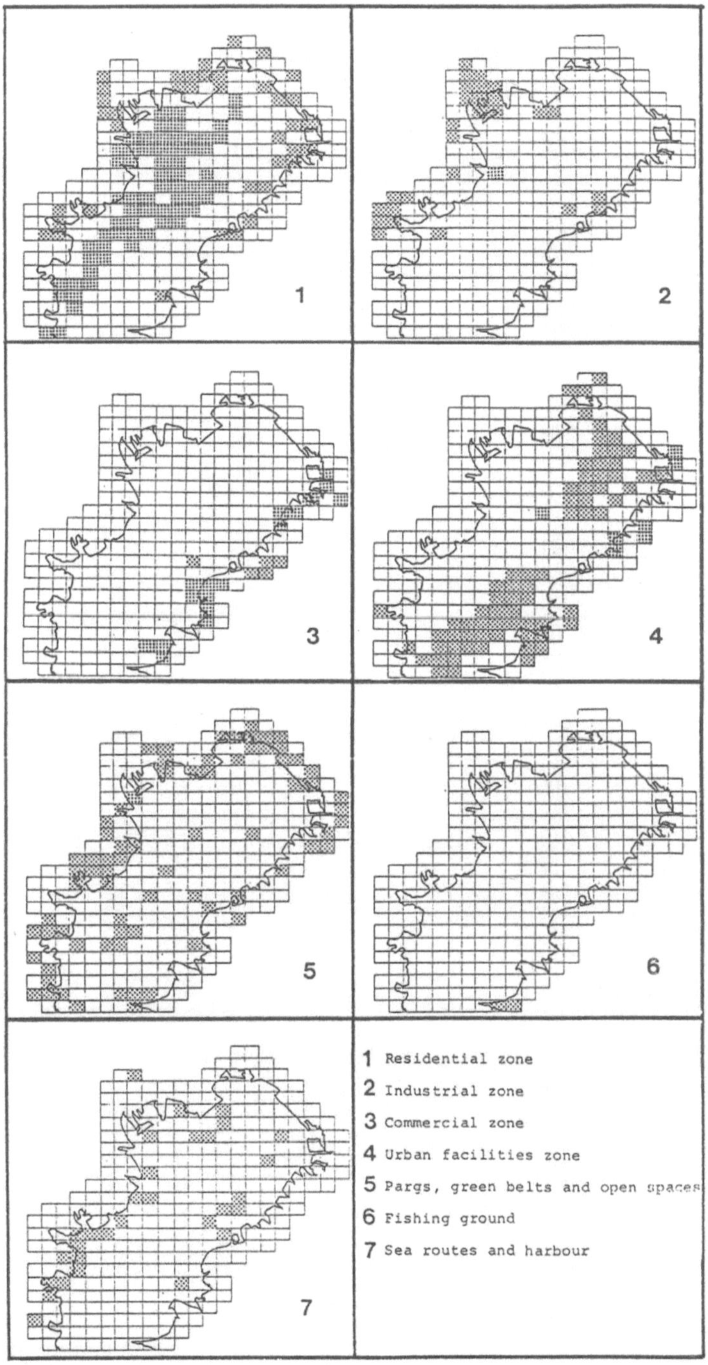

Fig-3
The composite utility function values of 344 water area meshes are calculated for each land cover function

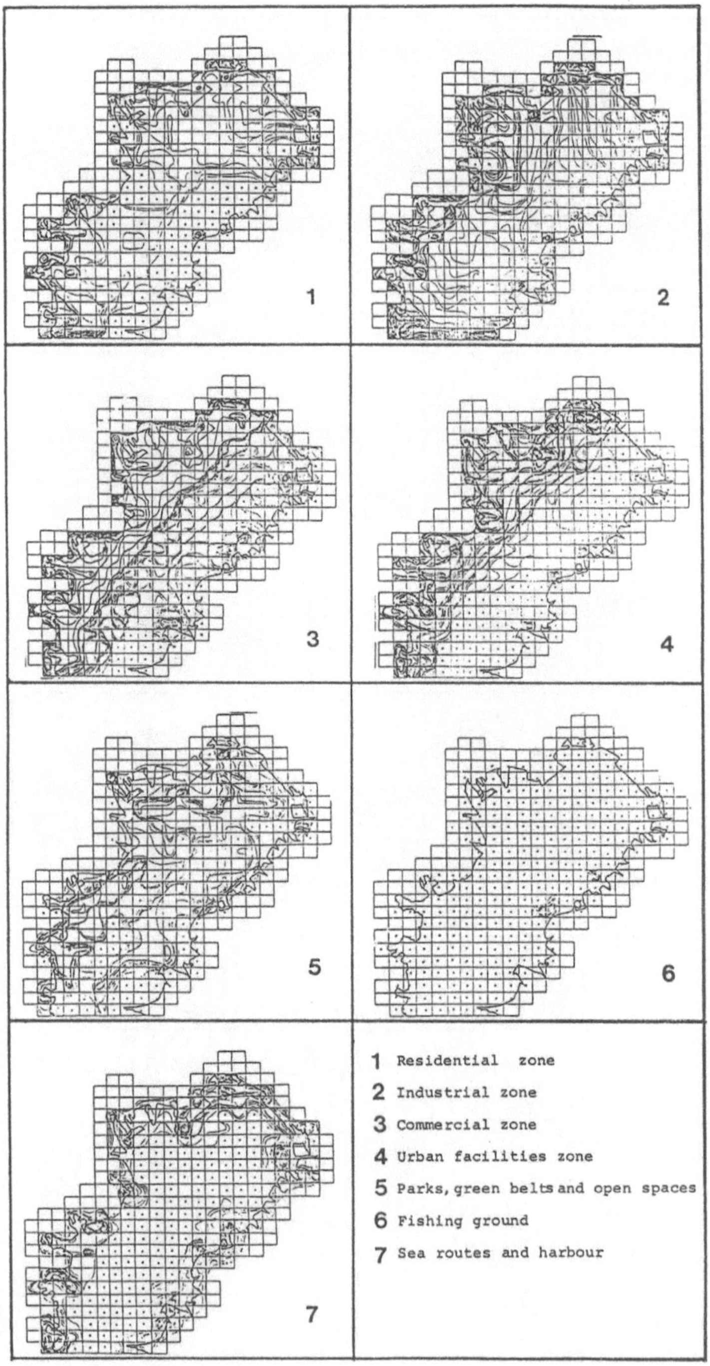

Fig-4
The utility conter maps of the 7 land cover function

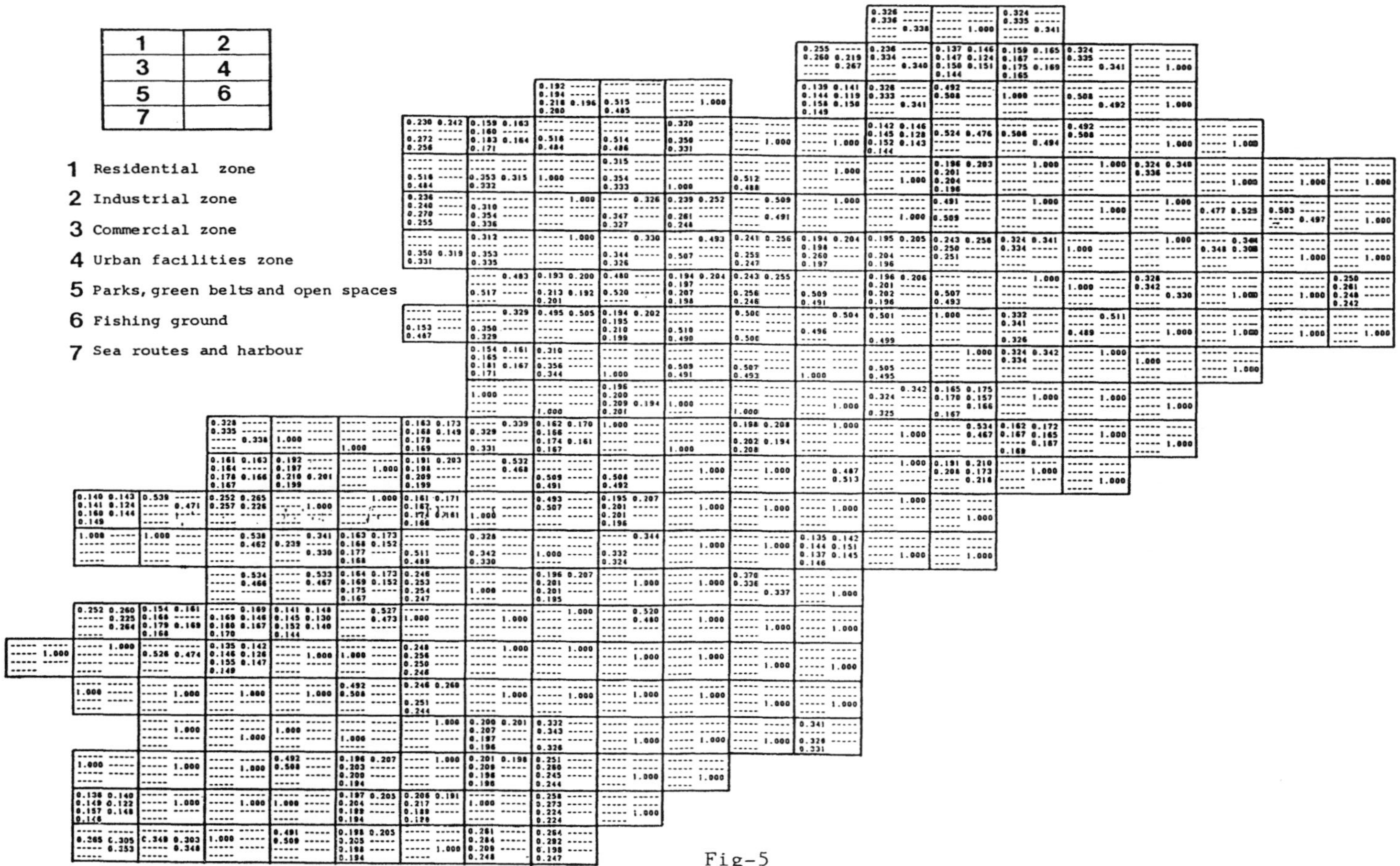

Fig-5
The Tokyo Harbour Water Area Utilization Basic Map

REFERENCES

1) US. The Coastal Zone Manegement Act of 1978 sec 303

2) National Land Agency: "The 3rd Report of National Land Planning & Development in Japan", Association of Land Planning, Tokyo Japan,(1978)

3) Decisions With Multiple Objective Preferences and Value Tradoff By Ralph L. Keeney & Howard Raiffa 1976

Waterfront Environment Planning for Urban Coastal Area

Y. Arai

Construction Division, Ports and Harbours Bureau, Ministry of Transport, Kasumigaseki, Chiyoda-ku, Tokyo, 100 Japan

[Abstract]

Several waterfront green zones in urban ports and coastal areas have been selected as the subject of case study in this paper.

The background of the study is the problem that artificial structures have replaced rich and pleasant natural coastline of Japan as a result of ports and harbours construction, expansion of reclaimed land and building of facilities for the prevention of disasters.

Recently, however, the citizens' demand for the restoration of natural environment and the creation of pleasant and enjoyable space on the waterfront of urban communities has become stronger.

Thus, various plans have been implemented to create such waterfront environment in many urban ports.

In this paper, the conditions of desirable waterfront environment and the direction of the compilation of the manual for the creation of waterfront pakrs are examined through case studies.

It is hoped that this paper will be of some help to create rich and pleasant waterfront environment.

1. UTILIZATION OF COASTAL ZONE AND GREENERY ON THE WATERFRONT

Japan, consisting of four major islands, has the land area of 370,000 km^2 and the total coastline extension of 34,000 km. Topographically, mountains and hills occupy larger part of its land: inhabitable land is only one-fourth of the total area, most of which is distributed along the coast. Many of the cities are concentrated in this zone where active city life takes place. The cities and local municipalities with coastline amount to 1017, which is 31 % of all the cities and local municipalities in Japan. The number of population in this zone is 61 million, accounting 52 % of Japan's total population. This zone yeilds the industrial shipment of 124 trillion yen, which is 58 % of the nation's total industrial shipment.

Majority of the cities on the coast own ports and harbours. Thus, cities with ports and harbours sustain the economic and urban activities of this nation. The number of ports in Japan are : specially designated major ports, 18, major ports, 114 and local ports, 962, respectively.

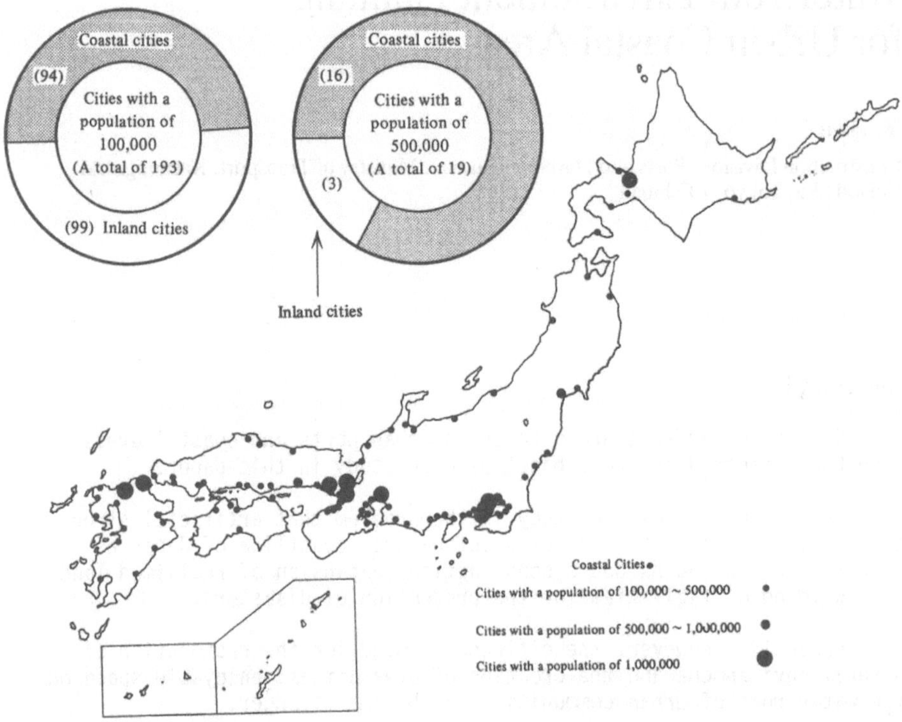

Fig.-1. Distribution of Coastal Cities (1980)

Since ports and harbours, usually located in the front of the cities, occupy expansive space where distribution and production activities are carried out, citizens cannot freely enter their ground. They practically keep citizens away from the sea. In addition, development of ports and/or cities accelerates the reclamation of the sea, and artificial structures are constructed along the coast to protect the national land from natural disasters. This tendency further discourages citizens from familiarizing themselves with the sea.

Recently, however, the desire to enjoy marine activities has become stronger among citizens, and the direction is toward the creation of the coastline and ports and harbours that allow easy access to, and contact with the sea. Hence, more and more projects are adopted to create water front environment that abounds in nature, providing space for seaside enjoyment. The creation of green zones within port ground was legally acknowledged in 1973 : it is relatively recent in the long history of ports and harbour construction. Therefore, as of 1983, only 16 % of all the Japanese ports and harbours own green zones which expand over 10 % of their open ground.

Characteristically, recently-created green zones tend to be quite spacious and serve multiple functions and purposes. The functions that are expected of the green zones are diverse. For example, they may be regarded as a contact point where citizens explore and enjoy the sea and ports, as space for citizens' recreational activities, as an asset to enrich the living environment in urban space, and as the symbol of urban re-development. Characteristics and uniqueness of the seaside and port-side space are effectively utilized in many of the facilities within the green zone. Recent tendency is that facilities are constructed to provide citizens with space to come into contact with the sea, enjoy marine activities, and to preserve and re-create nature on the waterfront. There are also facilities that display unique history and cultural climate of the port. Especially remarkable are the increase of plans of such facilities as fishing piers, revetments and promenades that allow citizens to enjoy the view and surroundings on the waterfront, man-made seashores where citizens can touch the water, re-created natural milieu equipped with features unique to the seashore, open squares where various events may be held, and structures that symbolize the ports.

2. COMPILATION OF THE MANUAL FOR THE IMPROVEMENT OF WATERFRONT GREENERY IN THE COASTAL ZONE

1) Special Features of the Coastal Zone in the Urban Sapce

Special features of the coastal zone in urban ports, in view of the creation and improvement of green zones, are enumerated below:

* It faces the sea. This means that the area can offer seaside enjoyment to citizens for one thing, but that it is also exposed to severe meteorological conditions.
  Also note that nature abounds on the seashore or, if not, it has the potential, as is manifested by the hypothesis that the very first life on earth was born in the ocean.

* Currently, the form of utilization of the port ground is highly specialized.
  As compared to other urban space, the ports and harbours are utilized for extremely specialized purposes of industrial production and distribution. Hence, the unique view is created. On the other hand, however, most of the space within the port is not suitable for citizens' leisurely walk nor for various activities in urban life.

* The port has a long history of development.
  The development of the port city and that of the port are closely related, and the development has taken place over long period of time. This is the reason for the existence of rather mal-functioning port facilities or only historically significant structures in the central part of the city.
  Such space, however, has high potential of playing an important role in rebuilding both the city and the port.

* It is vast and flat space beyond human scale. Compared to residential areas and areas of other human activities, the scale of port ground and reclaimed land is unproportionately huge. Faced with this vastness, citizens often feel rejected. On the other hand, it can be positively regarded as the expanse of land hardly obtainable in urban areas.

2) Expected Functions of the Coastal Green Zone in Urban Space

The coastal green zone in urban space is expected to function in various ways. Roughly divided, there are two categories. The first category of expected roles derives from the functions of, and activities in the coastal zone itself and the second category of demands originates from the city behind it.

According to some questionnaire surveys, the green zone is essentially grasped as open space in the latter category since some examples of the expected roles are sports facility, greenery and multi-purpose open square. On the other hand, the expected roles in the former category are closely related to the natural characteristics of the coast, e.g., sea bathing, viewing of the sea and promenade along the seashore, visiting cultural facilities such as marine museum and fishing.

3) Compilation of Manual for the Creation of the Coastal Green Zone

In order to promote rebuilding and improvement of port cities, taking the characteristics and expected functions of the coastal green zone into consideration, the manual for the creation of the waterfront greenery needs to be compiled.
The contents of the manual are roughly outlined below:

(1) Determination of Planning Principle (Type and purposes)

(2) Arrangement of the Green Zone (Where and how)

(3) Space Composition (Number and type of facilities)

(4) Design and Construction of Individual Facilities
(Technical information regarding bathing beach, fishing pier, historical park, etc.)

(5) Method of Maintenance and Management (Planting, etc.)

For successful compilation of the manual, case study of typical waterfront green zones across the nation should be carried out.

3. STUDY OF TYPICAL GREEN ZONES ON THE WATERFRONT

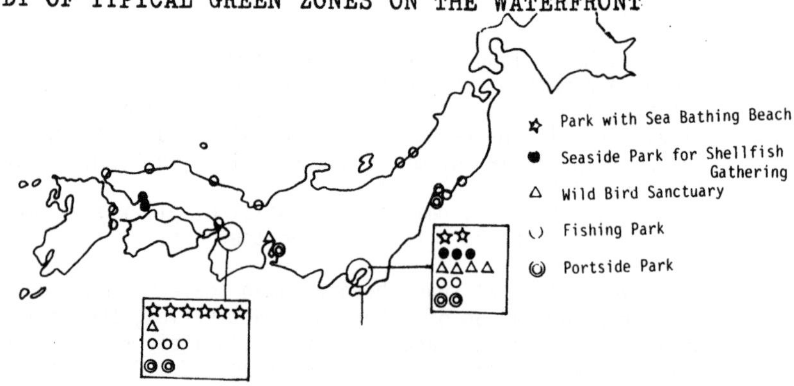

Fig.- 2. Distribution of typical man-made waterfront parks

(1) Park with Sea Bating Beach

The parks in this category have a sea bathing area on the restored beach: some are equipped with not only a bathing beach but also facilities for related recreational activities.

A typical example of this type is Takasago Marine Park in Takasago City, Hyogo Prefecture. In this park, the landscape with white sand and green pine trees has been restored on the man-made beach and on the island. Other examples of large-scale parks are Inage Beach (240,000 m^2) and Makuhari Beach (410,000 m^2) in Chiba Prefecture.

The keypoint in constructing this type of park is to ascertain that the restored beach is suitable for swimming, equipped with necessary facilities, and that it is maintained appropriately. Additionally, large-scale parks in this category should be designed so that they can be visited and enjoyed throughout the year.

A unique marine park with a bathing area has been created within the Port of Osaka. The bathing area is a natural swimming pool, so to speak, realized by enclosing part of the port. Clean water is supplied from the open sea. The purpose of this facility is to make sea bathing possible in the area where that particular enjoyment is otherwise unattainable due to water pollution and other adverse environmental conditions, and to diversify the recreational activities at the sea. Essentially important in maintaining this facility is to keep the water constantly clean by having good system of water supply and drainage. Considering that the number of bathing beaches falls short of the demand around large cities, it is quite feasible that the construction of this type of swimming facility becomes inevitable in some cases.

Fig. 3 Takasago Marine Park (Hyogo Prefecture)

(2) Seaside Park for Shellfish Gathering

Sand beaches and tidal flats are artificially created where shellfish gathering and related seaside recreational activities can be enjoyed. Naturally, shellfish must be able to inhabit and breed on those beaches. To attain this goal, the temperature and quality of the sea water are carefully studied through environmental assessment. Furthermore, the sand quality, the beach profile and the thickness of the sand layer, etc. are determined in reference to the coastal ecosystem. This process differentiates the creation of the seaside park for shellfish gathering from that of the beach for bathing. Marine Park in Yokohama City and Makuhari Beach in Chiba Prefecture are the successful examples, and many more are expected to be made in various locations across the nation.

Shellfish Gathering Profile of Beach

Fig. 4 Man-made Beach (Yokohama City)

(3) Wild Bird Sanctuary

The wild bird sanctuary may be created by protecting the natural habitat of wild birds on the coast and at the river estuary, or by artificially providing tidal flats and other conditions on the restored seashore on the reclaimed land. The wild bird sanctuary can be a place where citizens enjoy not only the observation of wild birds but also the natural environs that are unique to the seaside.

The keypoints in making wild bird sanctuary are, firstly, to secure the environment suitable for the habitation of wild birds, and secondarily to provide the area where citizens can enjoy wild birds without disturbing the ecosystem of the sanctuary. The maintenance and management of the sanctuary in harmony with the nature's order and mechanism are another important keypoint.

In the wild bird sanctuary, the feeding station, the resting place, the refuge spot, and the breeding area, etc. must be properly provided. Each of these facilities must be set up in a place where conditions such as the tidal wind, the wave, the tidal height and bottom sediments, etc. are favorably secured.

Fig. 5 Wild Bird Sanctuary (Osaka City)

Currently, six wild bird sanctuaries in the coastal zone are being either created or planned across the nation. Some are quite spacious as in the case of Kasai Marine Park (442.5 ha) and Osaka Wild Bird Observatory (19.3 ha). The green zone of this type is expected to be more in demand since it is located on the coast which is quite close to the urban center, and since it serves to diversify the types of green zones in the city.

(4) Fishing Park

The park for fishing and related recreational activities may be specially made for that purpose, or part of port and seaside facilities may be utilized. Fishing has always been very popular, and it will continue to be so in the years to come.

In many instances, fishing facilities (mostly piers) are constructed exclusively for that purpose. But there are also cases where facilities with other functions are utilized for fishing as mentioned earlier. The improvement of fishing facilities is expected to rapidly progress both in quality and quantity in the future.

The conditions of the good fishing facility are safety, comfortableness and attractiveness. The most important of all is that the facility is safe and offers good catch.

Fig. 6 Fishing Park (Kobe City)

(5) Port-side Park (Symbol Park, Park with Historical and Cultural Assets and Park for Seaside Events, etc.)

Various types of parks fall under this category. There are parks where citizens and port workers take a rest and learn about and enjoy the port, and parks that symbolize the port and the port city. There are also parks characterized by their historical background and cultural assets.

Representative examples of the port-side park are Yamashita Park in Yokohama Port, and recently constructed North Park of the Port Island in Kobe. Visitors can enjoy the view and atmosphere of the port in the port-side park. Furthermore, the existence of the park contributes to enhance the attractiveness of the port and its vicinity.

The keypoints in planning this type of park are to ascertain that the path from the city to the park is an attractive prelude and that the park commands excellent view of the port and the sea.

The port usually has a long history as it has developed with the city. This is the very reason why old port facilities sometimes remain in the central part of the city. Currently, 803 pieces of historical port facilities are identified in 246 ports throughout the nation. More than half of these pieces are from before the Edo Period. In recent years, these facilities are regarded not only as the core for the rebirth of the port itself but also as the historical and cultural symbol of the port city. They have begun to play an important role in the project to revitalize and re-create unique communities in many local municipalities.

Also note that, taking advantage of the extensive space and unique atmosphere of the port, various events are held. Port festival and fireworks exhibition have long been popular events often held in the port. According to statistics, 270 events were held in 99 ports across the nation in 1983. Ships often take part in these events.

In many of the port cities, these port side events are now acknowledged as an effective means of community's revitalization project. Tendecny is such that the waterfront zone within the port is now regarded as the arena of these events, and its improvement has begun to be examined in many local communities.

Fig. 7 Port-side Park (Yokohama City)

[Conclusion]

The need for the creation of waterfront pakrs in urban ports is storongly felt. Only half of the requested budget for this particular category of public works is actually allotted at present while the amount of budget request is expected to increase in the years to come. The actual investment was approximately 13.8 billion yen in 1984.

This paper deals with some representative waterfront green zones so-far constructed. The study is only half finished: it is hoped to be completed through the analysis of many more cases and the examination of trials and errors in the implementation of plans.

Technical Session:
Coastal Zone Utilization
and Management

3. Development Project and Planning

The Port Island Development in Kobe City as an Urban Complex Artificial Island

S. Waki

Engineering Department, Port and Harbor Bureau, Kobe City Government, Kanocho, Chuo-ku, Kobe, 650 Japan

1. FOREWORD

The Port Island is a marine cultural city complex that has been established in the center of Kobe Harbor and is equipped with modern port facilities and functional urban facilities designed to meet the needs of the 21st Century.

Since its completion, the Port Island's 12 container berths and associated facilities have come to play a pivotal role in one of Japan's largest and most representative ports, while its urban facilities have been specifically designed to create a new city space that will be closely linked to Kobe's future development.

These facilities include housing projects, hospitals, centers for international cultural exchanges and fashion industry events, etc. as well as parks and green areas.

Today, some 20,000 people live on the island, while about 30,000 are engaged in various activities there on a regular basis. A new transit system, the Portliner, connects the new city space with the downtown Kobe area.

The mammoth construction job of building the Port Island extended over 15 years and was completed in 1981. The total cost was ¥530 billion (approx. $2.2 billion). The project presented numerous challenges at all levels, in design and planning, in engineering, in financing, etc. These challenges were met by pioneering a new vision of building a city of the future and adopting innovative ideas that would fulfill this vision.

2. THE BIRTH OF THE PORT ISLAND CONCEPT

Ever since its opening in 1868, efforts have continually been made to improve and expand Kobe Harbor's Port facilities. In spite of these efforts, the rapid increase in freight and ship arrivals that accompanied the exponential growth of the post-war Japanese economy resulted in a chronic shortage of berths.

Moreover, the revolution in marine transport brought about by the development of container ships created an urgent demand for the construction of new large-scale wharves with numerous berths.

On the other hand, the population of Kobe, which stood at 980,000 in 1955, had grown to 1,220,000 by 1965. Ninety percent of these people were living in the area south of the Rokko mountain range, a scant 10 percent of the total city area of 5,410,000 square kilometers.

Natural geographical limitations made it necessary for Kobe to create an urban area equipped with new functional capacities to allow further expansion.

Because the water line of Kobe Harbor was almost completely occupied by existing port facilities, the only viable solution was to reclaim some of the water area in order to create an artificial island where new wharves could be built.

It was these circumstances which led to the birth of the Port Island concept.

The construction site and shape were determined in a port and harbor plan that was established in 1967, priority being given to the planning of the port facilities to be built on the outer perimeter of the proposed man-made island.

3. THE PORT ISLAND PROJECT

3.1 Outline

The construction of the Port Island which comprises a surface area of 436 hectares entailed the moving of 80 million cubic meters of earth. Construction commenced in 1966 and was completed in 1980 at a cost of ¥530 billion.

The purpose of the project may be summarized as follows:

(1) The construction of modern harbor facilities capable of handling the quantative increase and qualitative changes in marine transport brought about by containerization.

(2) The construction of a new city environment to provide Kobe with enhanced urban integrated facilities for its citizens. (The project resulted in the creation of two new urban spaces, the Port Island itself, and the site from which the earth used in reclamation was removed,

which has been named New Town.)

(3) To make intensive public capital investments in areas which will result in both increased employment and economic growth.

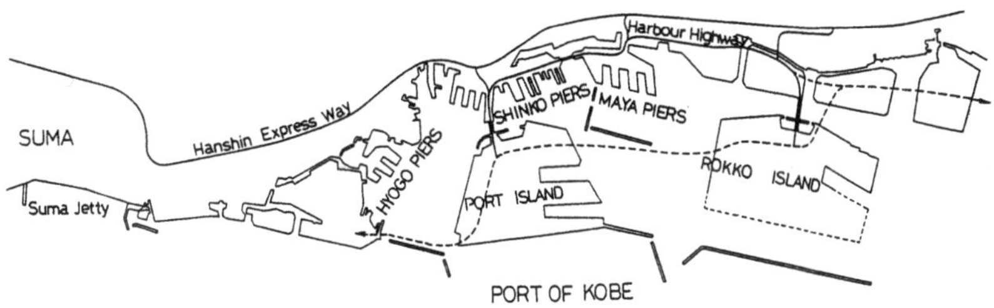

(4) To provide a new base for cultural, economic, and informational exchanges on an international scale.

(5) To play a leading role in the creation of cities of the future by actively adopting innovative concepts and technologies.

Construction responsibilities were assumed as follows:

Ministry of Construction -- Breakwaters and seawall.

Port Development Authority -- Container berths and liner berths.

Kobe Municipality -- Reclamation, roads, bridges, new transit system, parks and green areas, water supply and sewers, hospitals and other public facilities.

Private industrial groups -- The construction of facilities on purchased or leased sites located on the new island.

3.2 Harbor Installations

The already existing piers were built on diluvium which provides relatively good stability. However, it was discovered that alluvial deposits increased in thickness in the southerly and easterly directions to a depth of -12 to -13 meters at the Port Island construction site. The bearing stratum in places was at -30 meters which seriously affected the costs involved in building structures.

- Seawall -

A seawall was built at the southern waterline of the Port Island. This wall serves to protect the Port Island from large waves during typhoons and also greatly facilitated reclamation operations and construction of berthing facilities by shielding the work site from wind waves.

Because construction needed to be done rapidly yet result in a secure foundation, a gravity-type structure using caissons was adopted. Large-scale caisson production was effectuated by employing a method whereby the caissons were built on land, then hoisted by large floating cranes to the work site. To enable rapid and economical foundation improvement, a displacement method was adopted. This dicision resulted in the development of new technology related to the use of large floating cranes and large grab dredgers.

- Container Terminals -

Because container terminals were unknown in Japan at that time, planning for them required far-reaching research in order to overcome the absence of information on dimensions and construction methods, and on the systems used in such structures. After collecting the necessary date and consulting with shipping companies, construction plans for 12 berths were developed on the basis of the following standards:

- Size of ships to be moored : 35,000D.W.T.
- Water depth : -12 meters
- Length of berths : 300 meters
- Berth Area : 300 by 350 meters
- Container cranes : 2 at each berth with a rated output of 30.5 tons. A gravity type structure using caissons, as used in case of seawall construction, was adopted.

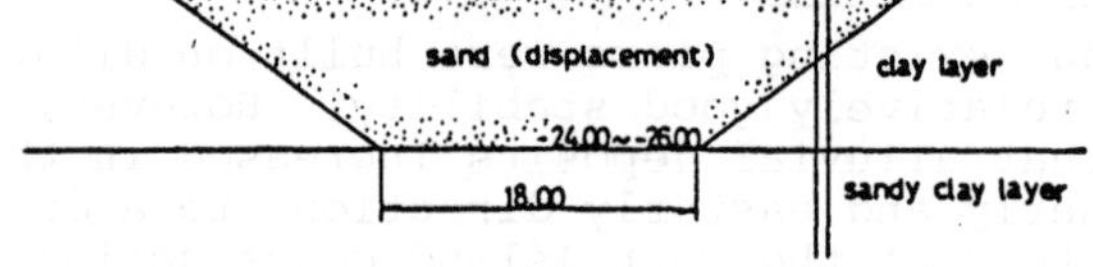

It was realized that unified management of the berthing facilities and adjoining freight handling facilities would be necessary to fully rationalize the transport system based on container shipping. For this reason, a centrally controlled system whereby each berth would be rented by an individual shipping firm for its own use was adopted. Such a system, however, conflicted with the principle of the public use of port facilities as prescribed by the Port and Harbor Law. For this, and other financial reasons, which are dealt with below, a new law was enacted. This statute introduces the concept of enabling port development authorities to operate as an enterprise in order to meet the perceived national demand for a rapid improvements in harbor facilities for overseas trade.

In conformity with this concept, the Kobe Port Development Authority has built 15 liner berths in addition to the container berths. Each of these berths has a length of 200 meters with a water depth of -10 meters. The crown elevation of the surrounding wharves is +4 meters.

- Roads -

It was understood that a bridge connecting the Port Island with the mainland was absolutely essential. Thus the Kobe Ohashi Bridge was built out from Shinko Pier No. 4. This structure was especially designed to link the island to mainland streets and highways without obstructing or hindering the movements of large ships.

It is a three-span, double-deck continuous arch bridge with an overall length of 322 meters and eight traffic lenes.

A new method using floating cranes placed on large concrete blocks was developed for the construction of this bridge. This method was later effectively employed to erect other bridges in the Kobe Harbor area, such as Rokko Ohashi Bridge.

It was also understood that for the port installation to function adequately it was necessary to create an efficient road system to handle the traffic volumes generated by the operation of the new facilities. For this reason, the construction of a four-lane Harbor Highway was undertaken to connect the island not only with the existing wharves but with roads and expressways in the hinterland as sell. A 4.2 kilometer section of the Harbor Highway between the Maya Piers and the Port Island is now open to traffic. When completed, it will reach Rokko Island which is now under construction.

3.3 Functional Urban Facilities

It was perceived that if Kobe City was to move into the 21st Century as an international information-oriented metropolis, modern port facilities must be built, industries which uniquely characterize the area must be promoted, and areas where its citizens can enjoy their leisure time must be established. To achieve this, the Port Island was designed to constitute a qualitatively new form of urban development, which entailed, not only the building of port facilities but also quality-of-life environments to provide satisfactory housing, work and rest facilities for its inhabitants.

(Basic Objectives of the Land Utilization Plan)

(1) To create a new urban focal point that will serve as a unique gateway to the world.

(2) To create a town with a young feeling provided with greenery, quality housing and cultural facilities that will attract active people.

(3) To create a new form of urban management that will allow the integrated operation of the Port Island's various facilities. The Island's functional urban facilities which are surrounded by the port facilities are divided into the following five areas:

- Distribution Business Area -

Provides truck terminals, repair and maintenance facilities for ship board equipment, and shipchandler, shops for stores and supplies.

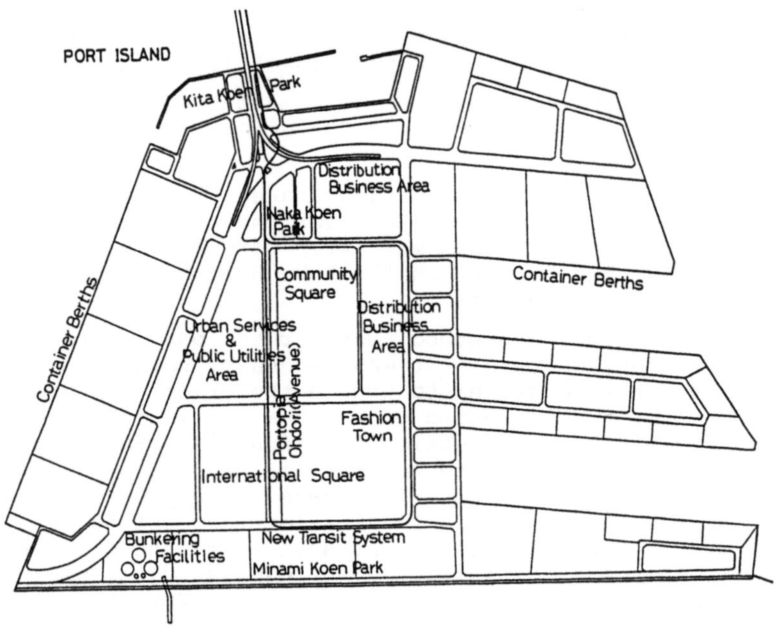

- Urban Services and Public Utilities Area -

Provides various services and utilities serving the Port Island as well as existing urban areas: i.e. hospitals, telephone exchanges, transformer stations, etc.

- Community Square -

In the original conception, housing on the Port Island was to be for the benefit of port workers. However, uniformly constructed housing does not encourage the creation of an active, multi-stratra community. Moreover, housing in existing urban areas had become excessively crowded, while newly developed areas on the fringe of the city, being located at a distance from the work centers, did not allow full utilization of existing urban facilities. For this reason, the plan was developed to build well-equipped housing on the island that would be near to the city center and allow residents to enjoy the sight of the mountains, the sea and surrounding greenery.

The housing units were designed for rational arrangement of living quarters and minimum street congestion. There are about 6,000 units, housing a population of 20,000. Because of the island's close proximity to the city center, a high-density, high-rise housing structures were adopted. The housing was provided both by the Housing and Urban Development Corporation and private housing companies in Kobe, resulting in an abundance of styles and types. In order to preserve the orderly appearance of the streets lining the central green area, subdivisions were sold by competitive bid to private enterprises.

- International Square -

Convention facilities were considered to be a key element for a city to prosper in the 21st Century.

To meet this requirement, the Port Island has at its core:

(1) A base for the creation of a distinctive "Kobe culture" that will highlight the city's international character and provide integrated information services functions that are presently concentrated in Tokyo.

(2) Facilities for the development of informational, cultural and fashion-oriented activities of an international character that will allow a departure from the traditional emphasis on heavy and chemical industries.

(International Trade Center)

To help make the Port Island a base for economic activities of a global nature and to strengthen its role as a gateway to the world, an international conference center, an information center on ports and harbors, trading house offices, exhibition and trading facilities for imported goods and hotels have been provided.

(Fashion Town)

With the aim of establishing Kobe as a city of fashion, a mall area called Fashion Plaza and other streets which have fashion retail outlets from Kobe and other parts of Japan and from abroad at their core, have been created.

(Exotic Town)

In order to further enhance the exotic and international character of Kobe, efforts are being made to provide a site to be used primarily for the sale of imported goods and various foreign foods and beverages.

(Citizens' Plaza)

Located at the center of International Square, this plaza is a broad, open street area leading from Portopia Avenue to Fashion Town and Exotic Town. It is the focal point for both local and international cultural and informational events, i.e. assemblies, festivals, bazaars, concerts and outdoor exhibits, etc.

- Parks and Green Areas -

The Port Island was conceived as a modern urban center which would provide a pleasant living environment where port and urban activities were to be carried out side by side. In order to attain this objective, the Marine Park complex was created comprised of Kita Koen, Naka Koen and Minami Koen parks linked by green connecting pathways for pedestrian traffic. A secondary objective was to provide a quality urban environment overall by upgrading parks and greenery in built-up areas. The idea evolved to provide a large park near the city center because of the difficulty in procuring adequate space for parks and green areas in overcrowded built-up districts.

The connecting pathways have been built 3 meters higher than the roadways to shield the housing units from vehicular traffic, and to provide pleasant places for strolling and relaxation.

The environment and appearance of the Port Island is not maintained by laws or ordinances. Instead, agreements to maintain environmental standards were made between Kobe Municipality and the participating enterprises when the land sales contracts were signed.

This is believed to be an effective means to control the land utilization within the scope of the existing system of private ownership of property.

3.4 <u>The New Transit System</u>

The Port Island line of the New Transit System is a new medium-distance, medium-volume transportation system whose facilities create a middle link between railways and buses. It connects the Port Island with the city center.

This system was adopted following extensive research and studies launched in 1971. It is designed to serve as the main transit artery for the Port Island and is part of long-term urban planning reaching into the next century.

Providing the principal means of transportation for people who live and work on the Port Island, Port Island line trains run on elevated tracks over a distance of 6.4 kilometers. The trains are equipped with virtually noiseless rubber tires, and operations are automatically controlled and monitored by a computer.

The infrastructure was constructed by Kobe Municipality as a public undertaking, while the system itself is operated by a newly-formed third sector organization that provides the vehicles, signalling devices, communication and other facilities.

4. IMPLEMENTATION PROBLEMS

4.1 Technical Problems

- Reclamation Method -

In the past, reclamation work in Kobe harbor was carried out almost entirely by transporting earth in dump trucks and unloading it at the designated area. At one time this was viable method because the distances between the loading sites and the waterline were short and transportation could be done entirely on land.

However, this method created various problems such as noise, vibration, and dust caused by the trucks, and differential settlement of the ground where earth had been removed.

Because the Port Island site was not only situated in an outlying waterarea but required an enormous volume of earth, 80 million cubic meters, to be transported, an overhead belt-conveyor and push-barge system was adopted for the sake of efficiency and economy.

Taking into consideration the subsequent land utilization and the distance to the waterline to be covered by belt conveyor, an earth removal site in the hills of Suma ward in the western part of the city was selected. The earth was transported by overhead conveyor to a jetty on Suma beach where it was loaded on push-barges which carried it the 20 kilometers to the Port Island site.

The outstanding features of this include:

(1) Prevention of traffic congestion problems in built-up area;

(2) Realization of large-scale earth moving for reclamation purposes at minimum time and cost;

(3) Utilization of easily-steerable, safe push barges in heavily trafficked Kobe harbor;

(4) Utilization of hopper barges that did not disturb the clay layer of the sea floor making possible the formation of an even foundation.

Pusher Barge

Suma Jetty

Unloader

Spreader

The water area at the Port Island site had a depth of -12 meters, allowing direct dumping from the barges up to the -2 meters level. At that point, newly-developed bucket-wheel unloaders were employed to empty the barges. Continuous transportation and reclamation operations were made possible by the use of shiftable conveyors, stackers and spreaders.

This reclamation method has proved to be revolutionary in many ways. It is now being applied to Rokko Island project where 170 million cubic meters of earth have been moved to date.

- Others -

Consolidation settlement is a problem that inevitably arises on reclaimed ground. According to a standard model for reclamation in a one year period, the estimated settlement for Port Island should total 3.8 meters, reaching 80 percent degree of consolidation (3 meters) three years after completion of the reclamation work.

The reclamation records of different portions of the Port Island differ because of the large area involved and because reclamation had to proceed from the wharves in towards the interior. That is why special measures to deal with differential settlement were instituted for the elevated roads and tracks. For instance, although the elevated tracks for the new transit system have a pile foundation, single beams were used wherever possible to avoid stress caused by settlement. In the curves portions of the track, continuous beams were used, but measures were taken to ensure that they will withstand a 5 centrimeter differential settlement. Corrective actions using jacks are being considered for other cases where differential settlement may create problems. Pile foundations are used for large structures, but, to counteract the effects of differential settlement, steps to promote settlement have been adopted using either the preloading method, whereby the sand used in reclamation is provisionally diverted, or the sand-drain method.

4.2 Financial Considerations

The challenge of building the marine city, the Port Island, came not only in the area of new technological development, but also in the financial field. It was not a small task to procure the vast funds required for construction.

- The Establishment of the Port Development Authority -

Large investments were required to provide for the timely construction of new wharves capable of meeting new demands brought about by the rapid trend towards containerization. This imposed a heavy burden on the national government as well as the port management body.

The Port Development Authority was established to relieve some of this burden. It was based on the entrepreneurial concept of business profitability whereby funds were borrowed on a long-term basis from private sourves, their redemption being assured by income gained from charges for the use of the facilities to be provided.

The financial structure is as follows: Ten percent of the total investment came from the national government and another 10 percent from the City of Kobe, the port management body. Loans procured from the Treasury's investment and loan program constituted 40 percent, while another 40 percent came from private sources.

The redemption of the borrowed funds will be carried out by the issuance of bonds by the Port Development Authority which will collect usage fees from companies having exclusive use of the wharves.

- Kobe City Funding -

Kobe City's contribution to the Port Island project amounted to ¥86.6 billion, the procurement of which also presented a major problem.

A decision was reached to seek about half of the necessary amount, ¥44.8 billion, issuing bonds, while another ¥37.4 billion was procured from West Germany in the form of mark bonds. The first series of the German bonds, with 100 million marks, was issued in 1968, followed by floatations in similar amounts in 1969, 1971 and 1972. Their redemption will be made through proceeds arising from land sales.

The sale of the Port Island land progressed smoothly as a result of its close proximity to the city center and the provision of urban facilities allowing varied forms of land utilization. Land sales have made it possible to cover fund shortages that arose in spite of the bond issues. Moreover, in order to lessen costs, linkages were established between the development of the earth removal areas and the reclamation area while efforts were made to develop new technologies.

5. POSTSCRIPT

Kobe Harbor, as one of the most representative Japanese foreign trade ports, plays a vital role in Japan's economy. In 1983, it handled the largest amount of containerized freight in the world, some 2,300 tons (1,350,000 TEU).

Moreover, it is a primary base of operations for feeder services to East and Southest Asia and one of the main ports of the Asian-Pacific region.

This can be attributed to the port facilities that the Port Island provides.

In addition, as the pace and volume of activities at its functional urban facilities increases, the Port Island project is coming close to reaching its original objectives. It is producing indirect benefits such as the attention generated by the International Conference Center, hotels, and the International Exhibition Hall, increased human contact at international conferences and hotels, the collection of business information at trade exhibits and the promotion of trading transactions. Also, openings are developing for the entry of information industries into the area precisely because the inland provides excellent access to information.

The sequel to the Port Island is neighboring Rokko Island which will cover an area of 580 hectares. Construction began in 1972.

The knowledge and expertise gained in the construction of the Port Island is this serving to sustain our continuing efforts to build a city that will be ideally suited to meet the needs of the future.

Prerequisites for Ports and Harbors in the 21st Century

Casc Study of Bcppu Port

Y. Yabe[1] and T. Maeda[2]

[1] Planning Division, The Fourth District Port Construction Bureau, Ministry of Transport, Tatezaki-cho, Shimonoseki, Yamaguchi, 750 Japan
[2] Environmental Planning Section, Ecoh Co., Ltd., Higashi-Nakanukicho, Tsuchiura, Ibaraki, 300 Japan

ABSTRACT

Any future port and coastal area plan in Japan must take into account emerging social trends, namely; the internationalization, the aging population and the socio-economic maturation. A development plan for Beppu Port will be discussed here as an example to demonstrate how to incorporate these future requirements.

The basic concept in the plan is to develop a port having a close social relationship with the nearby community offering high quality services to all the users, both foreign and domestic, while giving due consideration to the cultural aspects of the local community.

The proposed port plan is to create a facility which is not only a transport center but also will provide a space for tourism, cultural activities and marine leisure activities. The plan is to include space for leisure activities such as shopping, yachting, swimming, fishing, etc. in the vicinity of the port which would meet the demands expected to arise in the 21st century.

1. Background of Beppu Port

Beppu International Tourist Port is situated in the innermost part of Beppu Bay located in the north eastern part of Kyushu as shown in Fig. 1.1. Since its development in 1871, the port has prospered as a key base for tourist routes connecting the Seto Inland Sea and tourist spots in Kyushu. Behind Beppu Port lies the International Tourist City of Beppu known for its numerous hot springs and sightseeing spots. Visitors to the city number as many as 12 million a year. Beppu Port has been improved to provide an access for ferry traffic besides conventional port cargos as one of the major ports serving the prefecture.

1.1 Population and Industry

The population of Beppu City is about 136,000 as of 1980 and is expected to reach 168,000 by 1995. The major industry is tourism, and the city has grown as a tourist town for many years. In 1980, the industrial composition by sector is; primary industry. 2.4%, secondary industry. 16.1%, and tertiary industry, 81.5%. Tertiary industry which includes the service industries is expected to increase further in the future.

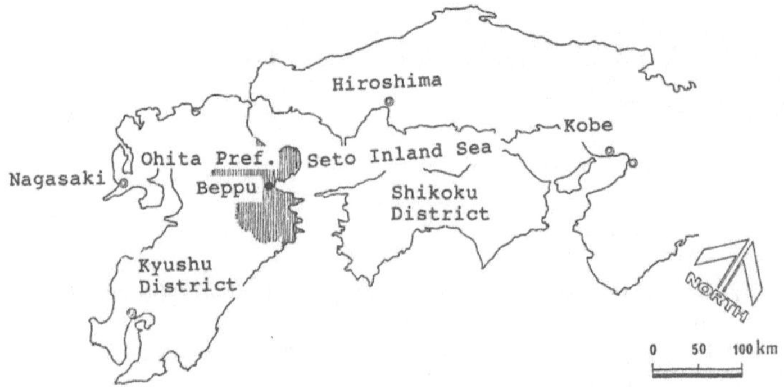

Fig. 1.1 Location map

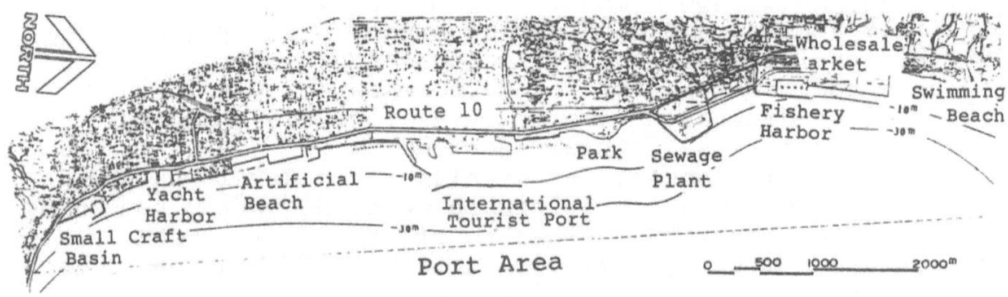

Fig. 1.2 Use of coastal area

1.2 Utilization of the Coastal Area

As shown in Fig. 1.2, the area along the coast of about 10 km within the project area is utilized for various purposes. The area is broadly divided into two sub-areas; a downtown area and a coastal area by the national highway No.10 running from south to north. The facilities in the coastal area are not connected well by road and therefore not efficiently utilized at present. The coastline consists of about 1.4 km beach (0.8 km natural sand beach, 0.4 km artificial beach, 0.2 km rocky beach), about 0.7 km stepped seawall and about 2.8 km port facilities, about 1.0 km fishery port facilities and about 4.0 km armor block seawalls.

1.3 Outline of Beppu Port

Beppu Port which is designated as "A major port" occupies about 10 km of the coastline of Beppu City and has facilities including 990 m of breakwater and, as a cargo/passenger facility, 2 x 4.5 m deep berths, 4 x 5.5 m deep berths, 1 x 7.5 m deep berth, and 1 x 12 m deep berth. Additionally, there are a fishery port, a basin for small craft, a yacht harbor,etc. The facilities for yachts and their leisure boats such as fishing craft need to be expanded to meet the increasing demand. More than 90% of the cargo handled through the port is automobiles carried by ferries amounting to 2,700,000 tons in 1980. The cargo volume is forecast to continue to increase. The number of passengers through the port sharply decreased and is still decreasing after the Sanyo Shin Kan Sen of Japanese National Railway opened its services in 1975.

1.4 Tourist Recreation

1.4.1 Tourist statistics.[1)]

Beppu City has rapidly developed as a tourist city during a period of the high economic growth since 1955, and the number of visitors to the city has also increased sharply as shown in Fig. 1.3. After the Oil Shock in 1973, the number levelled off to the order of 12,000,000. When the changes in the number of tourists are analysed according to stay period, overnight tourists have shown a gradual decrease from a peak of 6,650,000 in 1965. The number of day visitors has been recorded as more than the nightly total since 1972. While the monthly number of tourists exceeds 1,000,000 in six months from spring to autumn, seasonal fluctuation has not been found significant.

1.4.2 Tourist Origins [2)]

In 1982, the origins of tourists were classified as Ohita Prefecture (4,720,000), other areas in Kyushu (6,100,000), and Shikoku (930,000). All three have show an increase, while the visitors from other areas farther away has remarkably decreased from 4,400,000 in 1977 to 670,000 in 1982, indicating a clear trend. Foreign tourists doubled from about 18,000 in 1977 to about 36,000 in 1982 in five years. 65% of foreign visitors were from Asia, mainly Taiwan, Hong Kong and Korea.

1.4.3 Tourist Facilities - Oceanic Recreation

Beppu City provides various facilities for tourists which include 670 inns and hotels with hot spas (total accommodation capacity; 45,000 persons). The city is surrounded by tourist spots such as mountains about 1000 m high, plateaus, valleys and lakes. In the area, observation facilities equipped with a cable car and ropeway, golf courses, a horse back riding club, a wild life zoo, a botanical garden, an art gallery, a museum, leisure parks and sports facilities are available. As for oceanic recreation, the city welcomes a total of about 70,000 bathers a year and about 100 leisure craft are available for fishing and cruising. Recently wind surfing has become increasingly popular however a suitable beach is lacking. The facilities in the project area for marine leisure activities have not been fully developed to allow active participation by interested parties.

2. Social Trends toward the 21st Century

Japan is entering a period of rapid social change as it approaches the 21st century. The growth brought by development of modern industry is on the verge of ending, and a society matured on the heritage of a past economic prosperity is emerging. The Economic Planning Agency in Japan cites three principal social trends of "aging population", "internationalization" and "maturation" which Japan should pay special attention to in the 21st century. In improving a port in Japan, it is essential to accurately assess future trends on formulating a development plan. We discuss the three social trends in the following.

2.1 Aging Population

The population of Japan is forecast to reach a peak of 130 million in the beginning of the 21st century and then begin to taper off. The age structure then will become that of an elderly society in the year 2000 where 15.6% of the total population will be 65 years old or over (about 20 million) a similar state in present Europe as shown in Fig. 2.1.
Japan's figure is expected to exceed that of Europe in or about 2015 when it will be 20% or more.[3)]

2.2 Internationalization

Stagnation of industrialized countries and steady growth of developing countries is expected in a future international economic climate. Japan is expected to maintain her relatively high economic growth rate, with its GNP increasing against the world total GNP rate; 6% in 1970, 10% in 1982, and 12% in 2000 (20% for US and 12% for USSR)[4)]. Interdependence among countries will expand through a promotion of economic cooperation, thus accelerating the trend for further internationalization. This trend is to develop not merely with respect to economy but in cultural exchanges as transportation and communication means advance. These international exchanges will make remarkable progress in the near future and to keep pace with it, even a rural area would be required to cope with the forecast situation of close connection with the world.

2.3 Maturation

Having caught up with advanced countries in the race for industrialization, the economy of Japan is now reaching maturity in various aspects. A socio-economic structure is expected to shift from a material and industry oriented one to an information and knowledge oriented one. This will head toward a society relying fully on densely organized information services. Instead of seeking materialistic affluence, society will find greater values in mental and spiritual comfort as the sense of value continues to diversify. In such an enriched society, roles where cultural and leisure activities play a part will assume an increasing importance, demanding policies which attach more emphasis on social activities rather than economical activities.

3. Coping with the Mature Society

Beppu City terms itself an international tourist city or international tourist port, and has resources such as hot springs and seashores. We shall focus our discussion on the coastal area and ways of coping with future society in particular how to effectively utilize the space for people and goods and to utilize for oceanic recreation in the project area.

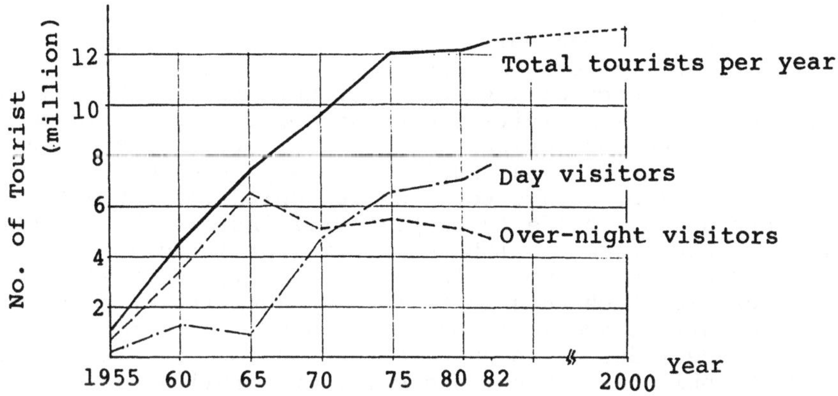

Fig. 1.3 Number of tourist

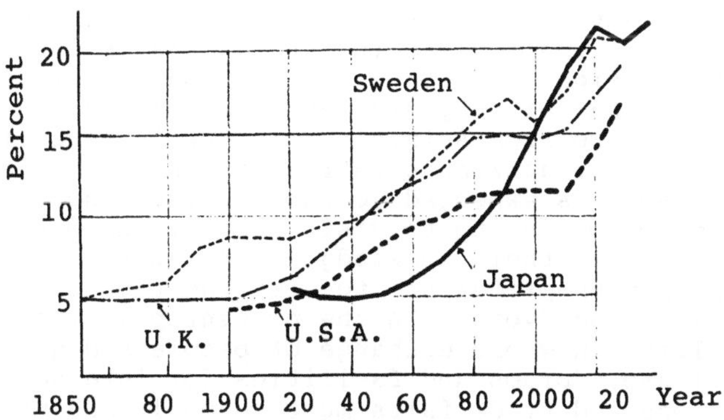

Fig. 2.1 Aging population in the world (population over 65 years of age)

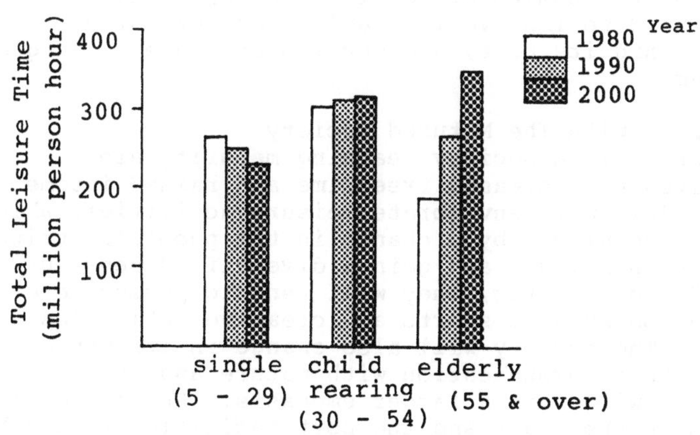

Fig. 3.1 Demand for leisure by life stages

3.1 Coping with the Aging Society

Two major human obligations of labor and child rearing are controlling factors on leisure acitivities in human lives. The presence or absence of either affects the mode and content of leisure activities considerably. Fig 3.1 shows the estimated demand for various leisure activities broken down by life stage.[5] In the year 2000, "single people" will decrease to 90% of todays total while "child rearer" will remain unchanged. "Elderly people" will increase by 1.8 times of that of the present, with their time for leisure exceeding those of the other groups. Thus the "elderly society" will bring significant changes to recreation activities, thereby requiring various new ways and measures suited to their demands. The modes of oceanic recreation conceivable for the "elderly" could be fishing, sketching or strolling on a sunny beach.

3.2 Coping with Internationalization

A future international exchange will become qualitative and diversified by development of information network and airline services. In the tourism industry, this is reflected in the number of foreign tourists visiting Japan; it increaced about 2.3 times to that of 10 years ago (about 1,000,000 in 1982), 43% of whom were from Asia.[6] With the increase in income of newly industrializing countries in Asia, the number of visitiors to Japan is expected to further increase. The tourists are now beginning to visit rural towns off the beaten track, deepening their knowledge of culture , and enjoy meeting other people in preference to visiting famous tourist spots such as Nikko and Kyoto. In the present polarized era of internationalization where exchange of people and goods is expected to increase, forums and facilities for international events, conventions and trade fairs must be built and the know-how to manage these facilities must be acquired. Highly sophisticated management techniques for the above facilities meeting an international standard must be developed. Even a rural town must be equipped with the competitive capacity to live through the coming era of internationalization. Beppu City will need to improve its facilities for marine recreations and its port, and develop software to adequately manage them.

3.3 Coping with The Matured Society

Changes in a society reaching maturity are characterized by increased free time and raised income standard. This will envigorate leisure activities, which will further be encouraged by a change in the people's attitude to respect the individual and gain a diversified sense of values. To be specific, they will tend to pursue more hobbies and recreation such as sports and creative activities spending more time. The society will also change in its mental structure where communication with nature and human relationship will gain a larger emphasis. Recreational facilities on the coast and the port facilities should be planned to provide sufficient space with amenities but giving thorough consideration to the surroundings with respect to, appearance and high quality service.

Table 4.1 Problems in Coastal Area and Their Solutions

	Problems	Counter-measures	Main Facilities	B/F*	A/P	I	M
COASTAL AREA	Inadequate connection, among facil-ities	Securing the route for connecting facilities, improving facilities	Promenade, bicycle paths	○	○		
	Division of the area by Highway No.10 to the coast and the downtown	Road Connec-tion between the coastal area and the downtown	Pedestrian decks, underpaths	○			
	Lack of facilities for oceanic recreation	Securing the land, expand-ing and constructing the facili-ties	Artificial beach, fishing facilities, observation decks	○	○		○
PORT AREA	Increasing cargo volume in a tourist port	Separation for people and commodi-ties, securing the land	Commodity distribution port	○			
	Improving the interna-tional port as a gate from the sea	Creating space with amenities.	Facilities for special events, shopping mall, oceanic museum	○		○	○
	Lack of leisure port for yacht and fishing	Securing the land, expanding and constructing facilities	Marina				○
TOURIST REC.	Stagnation in the increase in the number of tourist	Promotion of international events	Facilities for international events and convention hall	○		○	
	Decreased attractive-ness in conventional tourism resources and facili-ties	Developing diversified facilities and events to meet the future needs	All-season hot spring swimming pools, artificial surfing facilities				○

*: B/F- Basic Facility, A/P- Aging Population
I- Internationalization, M- Maturation

4. Problems in the Coastal Area and Countermeasures

We have discussed the problems and their possible solutions in the Beppu coastal area. Table 4.1 shows the major facilities proposed for the project in relation to the aforementioned major issues of againg population, internationalization, and maturation. Coastal facilites where people can enjoy pleasant oceanic recreation, a promenade, an artificial beach, a fishing facility, an observation deck, etc. are planned as shown in the table. In the port area, there are such important facilities as a space for various events, a shopping mall, an aquarium, etc.

5. Conclusion

1) An improvement plan for a coastal area involving tourism and industry must take into account the future social trends such as aging population, internationalization, and maturation of the society.

2) A coastal area favored with a mild climate is an ideal place for the elderly to live, and is also suitable as an area to offer various types of recreation and cultural activity. The rapid increase in leisure time of the elderly must be taken into account when improving coastal area.

3) Internationalization is expected to develop, and even a rural city in Japan will be required to be internationally competitive in the near future. Beppu, which is an international port, must offer satisfactory facilities and services to fully develop its potential.

4) As this nation approaches its maturity, people's sense of values undergo various changes and achieve sophistication. Development of a coastal area should be planned to meet the diversified and changing demands and provide safe and pleasant facilities.

Reference:

1),2) Tourist Statistics, Beppu City, 1982

3),4) Japan in the year 2000 - to be ready for an internationalized, mature and aging society; Economic Planning Agency, Japan 1983

5) Report of Survey on Promotion of Local Industries by Developing Oceanic Reactivations; The 5th District Port Construction Bureau, Ministry of Transport, Japan 1984

6) Statistics of Foreign Vistiors to Japan; Department of Research & Data Processing, Secretariat to the Minister, Ministry of Transport, Japan 1982

Research on the Development of Deep Water Port of Taiwan District

H.S. Hou* and C.F. Wang**
Institute of Harbor and Marine Technology, Taichung District, Taiwan 435, Republic of China

ABSTRACT

Taiwan is now served by five international ports, two of which were newly opened in the 1970's. They have been under continuous expansion in recent years to meet the needs of the island's burgeoning foreign trade. Due to recent trend of vessel developed to huge load-capacity, the existing port of Taiwan district is inconvenient to the ULCC or VLCC. For decreasing the shipping coast and accomodating the cargo handing of over 100,000 DWT, the deep water port is urgent to be developed in order to promote the competent capability of port cargo handling.

This research is divided into two parts, at first, the prediction of cargo handling and the cargo sources which are adapted to the huge vessel shipping need to be studied for the development of deep water port. The next, then the oceanographical data for the design and the planning of deep water wharf and breadwater need to be surveyed around this Island for possible site of deep water port.

This study deals with the related problems of deep water breakwater construction and discusses how to apply the the present design and construction method to the deep water wharf and breakwater. Further analyses are conducted for this research.

---

*Deputy Director and Principal Senior Researcher, Institute of Harbor and Marine Technology
Professor, Institute of Naval Architecture, National Taiwan University, Taipei, R.O.C.
Project Adviser, Ministry of Communication, Taipei, R.O.C.

**Associate Researcher, Institute of Harbor & Marine Technology

1. Oceanographic data of the proposed deep water ports

Taichung, Kaohsiung, Keelung and Tansui totally 4 ports are tentatively new port will be possibly pinpointed as the deep water port.

The collection of data on site conditions is continued during this phase of this study.

With respect to geotechnical survey activities borings and sub-bottom seismic investigations are started to obtain a more detailed insight into construction possibilities at the various sites. The degree of detail will, however, be adapted to the requirements at this stage of the project; it should be recognized that probably a detailed geotechnical survey for specific areas needs to be carried out once the deep water port master plan has been finalized, and the engineering studies required for the detailed design and implementation have to be undertaken. This final phase is however beyond the scope of the project described in this proposal.

During this phase the topographical and hydrographical surveys will be completed; construction materials, methods and costs will be investigated. The geotechnical surveys will be completed for all four sites as is required to enable a proper conceptual design.

Data collection on site conditions will not be restricted to site surveys. In order to obtain an insight into the spatial variations of the conditions on each of the sites, computations will be made with the mathematical models available to DHL. The items to be studied in this way may vary from site to site. It is foreseen that for each of the sites computations on the following subjects need to be made:

Keelung	:	refraction of waves through currents current conditions.
Tan-Sui	:	current conditions wave refraction and diffraction.

Taichung : current conditions.
Kaohsiung : current conditions.

2. Layout and structural planning of the proposed deep water port.

2.1 Introduction

The selection of the feasible deep water port site will be made by comparing conceptual master plans for the four port site options. Drawing-up a master plan is a multidisciplinary process with many interrelated activities; its main stages are:

1. Determination of technical port requirement based on cargo forecast figures and selected ship sizes.
2. Collection and elaboration of environmental data.
3. Layout and structural planning.
4. Determination of total construction costs.

The methodology to be followed for the layout and structural planning is described in the next two paragraphs.

2.2 Layout planning.

Layout planning is basically the determination of the required dry and wet areas including their shape and location within the port area.

The main boundary conditions which have an influence on the planning of the optimal layout of a port are:

a. Required technical facilities such as:

- water areas for manoeuvring and required depths,
- number of berths with berth lengths and water depths,
- storage sheds and open storage areas,
- industrial areas,
- administration and service buildings,

- roads and other transport access,
- number and types of cargo handling equipment,
- floating equipment and service craft,
- port services such as fuel, water, electricity and sewerage,
- auxiliary port services.

b. Environmental conditions such as :

- wave climate and current velocities,
- water level variations,
- wind climate,
- morphological processes (sedimentation/erosion),
- sub-soil conditions and availability of fill and other construction materials.

All relevant required technical facilities and environmental conditions will be taken into account in order to determine an optimal layout for each considered deep water port location.

It should be pointed out that the optimal layout has to compromise between the various requirements; however, the basis for the selected layout will be save ship manoeuvring and cargo handling at the lowest possible price.

The environmental conditions are the governing boundary conditions for the layout of the entrance of the port. The impact of a particular environmental condition on the technical, and consequently on the economic feasibility of a layout depends completely on the considered location.

E.g. for the Taichung location there will be a strong relationship between the layout of the port entrance and the morphological system balance.

Consequently for this location the authors' attention will be focused on the expected sedimentation in the harbour and access channel as well as the instability of the adjacent coast lines.

For the Tan-Sui and Kaohsiung location also some influence of the morphological system balance on the layout may be expected.

A flexible approach will be followed and the authors will concentrate their sutdies on those aspects which are the most important for a given location.

The possible constraints created by wave penetration in the port and increased current velocities due to the harbour entrance layout, will be taken into account.

The quantification of most of the environmental influences will be determined by means of readily available mathemathical models which allow for the evaluation of various alternatives within a reasonable time and money budget.

The phenomena to be studied, in this phase, by means of mathematical models are :

- current patterns and current velocities,
- wave refraction and diffraction,
- sedimentation and erosion,
- ship manoeuvring.

Current patterns.and current velocities have to be known because they are input in the other three mentioned mathematical models. By means of wave refraction and diffraction the near shore wave climate is calculated from the deep water wave conditions. The near short wave climate has to be known for:

- computation of the sedimentation and erosion,
- computation of the wave attack on the breakwaters and wave penetration in the harbour,
- ship manoeuvring evaluation and downtime computation.

Sedimentation of materials in access channels and ports can lead to considerable costs of maintenance dredging. In many cases maintenance dredging cannot be avoided. However, in the process of determination the layout of the port entrance, the expected costs of maintenance dredging should also be taken into account.

Ship manoeuvring simulation models are a valuable tool for investigating the optimal approach direction, the required width of the access channel and the required stopping length which often has a major influence on the breakwater length.

2.3 Structural planning.

Structural planning will be undertaken for all the envisaged port constructions. The designs will be made in such detail as required for reliable cost estimates.

Again, a flexible approach will be followed by paying extra attention to the most critical construction items of each deep water port location.

E.g. for the Keelung site a breakwater will be a major component of the total port construction costs. Consequently, also the availability of building materials (quarry inspections) and the construction method will have to be considered in some detail in this phase of the study in order to provide reliable cost estimates. For the Kaohsiung location extensive reclamation works may be required. So much attention will have to be paid to this aspect of the port construction.

2.4 Conceptual master plans for the four port locations.

For the four considered deep water port locations conceptual master plans will be developed.

These master plans will be based on information and insight obtained during previous activities, being:

- cargo forecasts,
- ship sizes and expected developments,

- rail and road systems and planned improvements and new constructions,
- layout of the port,
- structural design of the port components.

Much attention will be given to a flexible approach in the master plans which allows adaption to unforeseen developments.

3. Evaluation of alternative port sites

At the end of the second phase the following data will be available for each of the five alternative deep water port solutions:

- conceptual deep water port master plan i.e. layout and structural planning;
- port investment, operational and maintenance costs;
- additional inland transportation infrastructure investment and maintenance costs;
- total transport costs in market prices over a ten year period for incoming and outgoing cargo flows.

One last conversion is required at this stage, namely the conversion of the total transport costs from market prices into shadow prices, which reflect the true costs of transport to the country.

Furthermore, care should be taken to avoid double counting. It may be necessary to look into the economic costs for some investment projects in inland infrastructure, if special conditions would require an additional investigation.

Thereafter it is possible to compare the alternative solutions by way of calculating the net present value of the total investment outlays, maintenance costs and annual transport costs over a minimum period of ten years, against the economic rate of discount.

Preference should go to the port alternative with the lowest total discounted costs, but it may happen that the differences between the alternatives are so small that they cannot be considered significant for projects of the financial magnitude of ports.

Therefore the evaluation of the deep water port alternatives shall be completed with an extensive sensitivity analysis, which will be aimed at all variables introduced in the preceding stages of the study, most notably the transportation simulation model but also the poet investment and maintenance costs estimates, which can be reasonaly expected to have a major impact on the outcome of the study; the analysis will result in the identification of the variables which exert a major influence on the study's conclusions.

Before the formulation of the study's recommendations, the authors will be able to discuss with the Ministry of Communication about the socio-economic consequences of the investigated deep water port solutions in order to evaluate the impace that the inclusion of these consequences would have on each alternative's attractiveness.

The study's recommendations at the end of the second phase shall fully take into account the findings of the sensitivity analysis with the Ministry of Communication regarding socio-economic factors.

4. Detailed planning phase of the proposed deep water ports

4.1 General

Once a site for the deep water port has been selected the conceptual master plan, drawn-up for that particular site in the second phase, will be optimized and further developed into a detailed master plan for the entire deep water port.

In order to be able to draw-up such a detailed master plan the design for the maritime works should be worked out more in detail and the feeder facilities of the international ports should be evaluated.

4.2 Preliminary engineering designs of maritime works

Preliminary engineering designs will be prepared for the civil engineering components of the port. Outline specifications, indicating the required main types and capacities of equipment will be given as well as a precise definition of the works to be carried out, including an estimate of the capital cost and future maintenance cost of constructions, including capital and maintenance dredging.

The collection of additional environmental data may be required for the detailing of some of the constructions. However, the costs for the collection of such data is not incorporated in our offer.

The entrance channel and breakwater are such important components that an improper design may jeopardise the feasibility of the entire port. Therefore in this phase of the study it is envisaged to carry out an additional nautical study in which a number of runs with a ship manoeuvring simulator on real time base is incorporated.

By statistical processing of the executed manoeuvres the required minimum dimensions of the protected area in the harbour and the optimum layout of the harbour entrance can be determined.
Also a relation will be defined between the harbour layout and the anticipated downtime.
The hydraulic stability of the design breakwater will be investigated by means of a physical model.

Especially in the case of extreme design conditions, as may be expected at some locations, a physical model study and the envisaged evaluation of the structural integrity of the armour units is believed to be necessary for the determination of a reliable design.

4.3 Results of Detailed Engineering Planning

The results of the detailed engineering will be the draw-up of a total outline of the future deep water port. A detailed layout will be available showing all constructions and required facilities.

For all major constructions, preliminary designs will be made which will be shown on plan views and on cross sections. Possible construction methods will be described together with realistic implementation schedules and required construction equipment.

The detailed master plan will indicate how the development of the port is foreseen and will also describe the advised sequence of activities with respect to the improvement of the existing port and the new to be built constructions.

5. Evaluation of cargo transportation

5.1 Commodity/cargo projections.

Commodity projections will be undertaken for a number of import and export products, based upon the statistics available from the harbours or the BOFT. Projections will be specified at a level lower than the national level, most probably at the prefecture level or a combination of prefectures depending upon the availability of data and the existing transport pattern.

Projections will thus take into account the regional differences in economic development. As far as Taiwan itself is concerned, projections will cover the the commodities from their origin to the port, or from the port to their final destination.

The projections, especially after 1995, will also take into account the prospects for transshipment and free trade zones as well as Hong Kong's political position.

Special consideration will be given to commodities like coal, which have proved in the "Port Development Study" to be of utmost importance.

Commodity flows will be converted into cargo flows at each point in the transport chain where that would be required. It is evident that containerization will play a large role in this respect and it will receive attention accordingly.

5.2 Transport costs.

Transport costs will be determined for each chain in the transportation link, starting with the shipping freight rates which for at least some commodities or cargoes will be based on a study of trends in shipping. Costs will be specified, if relevant, per commodity/ cargo, ship size, route and port. Likewise, the costs of transport through the port will be determined on a commodity/cargo basis, also distinguishing if relevant, between ports and ship sizes.

Finally, the simulation will require a separate calculation, at least for a number of commodities, of the transport costs by road, rail and sometimes coastal feeder system, in as far as not yet available with the Transportation Planning Board.

5.3 Rail and road systems.

It should be noted, that it requires a major effort to isolate the effects on the road and rail system as caused by port development from those originating in the normal development of domestic transportation needs. In order to prevent the study from becoming a general transport study, it is imperative that the Ministry of Communication actively participates at this point in the study by providing projections of road and rail system developments as well as projections of the domestic transportation needs for both the first ten years and second ten years stage of the study.

On the basis of this information, their estimate of the road and rail system's capacity and their projections of commodity/cargo flows to and from the ports, the authors will be able to determine which bottlenecks will arise in the domestic transport system.

The Ministry of Communication will then be involved for a second time in the study to advise on the possibility of removing the apparent bottlenecks and to indicate which additional investment would be required to that purpose.

Road and rail transport can be expected to be strongly competing. Special attention shall be paid to the modal split and to the factors which influence this. In this respect the tariff policy of the TRA, and specially its long term development, shall have to be determined as reliably as possible.

5.4 Coastal shipping.

The Ministry of Communication has shown the planning in the potential role of coastal shipping for coal and containers in Taiwan.

The authors will certainly take the possibilities of coastal shipping into account, but only if relevant for this study and only as far as the five existing international ports, the existing terminals of power stations, and the new deep water port is concerned. More especially, the authors consider the potential contribution of a coastal feeder system, either by barges or by self-loaders, to the distribution of large quantities of coal most promising and deserving of further investigation.

In addition, the present coastal shipping movements will be fully considered in the projections to be undertaken as described earlier in this methodology.

5.5 Data collection transportation model.

The collection of data for the transportation simulation model is a major task, which for a complex subject like the planning of a deep water port should go hand in hand with the formulation, testing and adjustment of the model. The availability of data on one hand influences the structure of the model, while on the other hand the experience and insight gained while elaborating its concept will have a large impact on the data requirements.

Although two experienced economists will be fully occupied during the second phase of the study with the data collection and subsequent financial and economic analysis - supported by the efforts of other term members in the field of port investment costs estimates - the full cooperation of the Ministry of Communication will be required in order to finalize the study within the proposed time limits.

6. Detailed planning of transportation.

6.1 General.

In this phase of the study the efforts of the authors can be concentrated on the selected location for the deep water port allowing for much more detailing of the various subjects than in the previous phase.

The main activities to be carried out are:

- To work out preliminary engineering designs for all maritime works.
- To investigate the feeder facilities of the international ports including the draw-up of terms of reference for relevant additional studies on this subject.
- To draw-up a detailed master plan for the entire port including cost estimates.

For all the maritime works construction methods and implementation schedules will be further developed.

6.2 Preliminary engineering designs.

All structural components of the port will be designed more detailed than in the second phase.

If required by extreme design conditions the following activities will be carried out with respect to breakwater design:

- Engineering design of the breakwater, taking also in account:
    - availability of construction materials such as sand gravel and rock;
    - construction method and implementation schedule;
    - required floating and land based equipment.
    - stability during construction.

- Optimization of the length of the breakwater by nautical and morphological studies.
- Investigation of the hydraulic stability of the armour layer by a two or three dimensional physical model (as far as not carried out in the second phase of the study).
- Investigation of the geotechnical stability of the breakwater by menas of mathematical models.
- Investigation of the structural integrity of the selected armour units.

The layout of the port entrance will be optimized by means of the earlier mentioned nautical and morphological studies. The nautical studies will encompass also a study on a ship manoeuvring simulator being a mathematical model of the ship's manoeuvring behaviour on real time basis and steered by a pilot and an helmsman.

The ramifications of the computed sedimentation and erosion will be technically and economically assessed.

Reclamation works will be optimized by means of detailed soil balances. The possible use of dredged material, excavated from the entrance channel and port basins, as fill material will be investigated.

Required equipment will be established and implementation schedules will be determined.

Quay construction and shore protection works will be worked out in detail, including construction methods; the stability of slopes will be investigated.

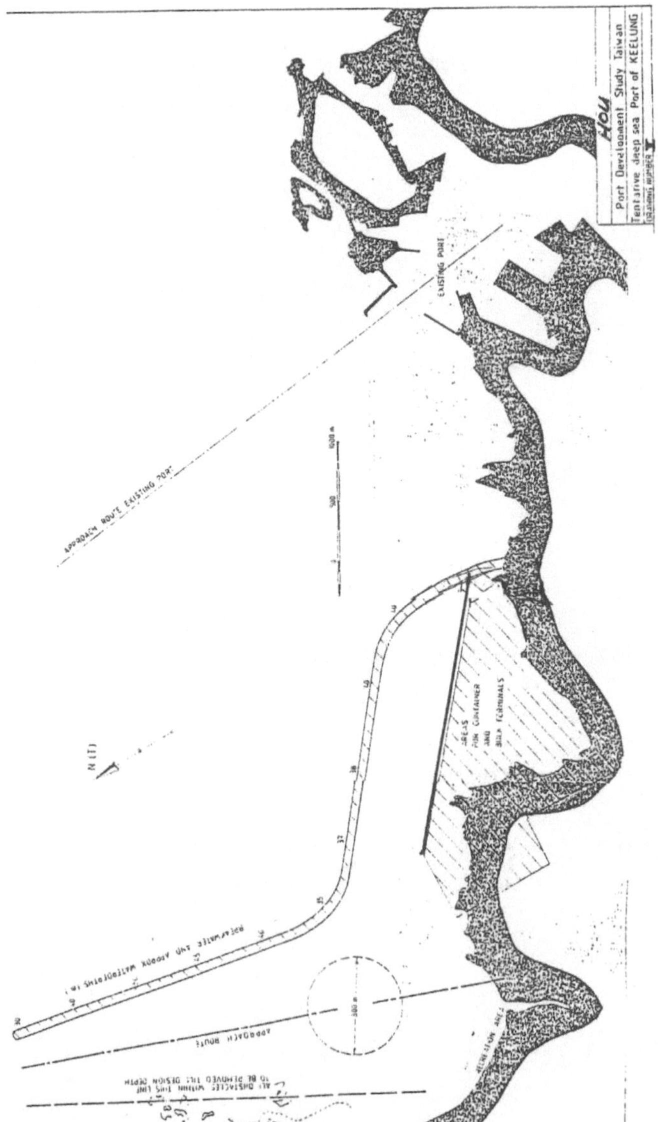
Port Development Study Taiwan
Tentative deep sea Port of KEELUNG
EXISTING PORT
APPROACH ROUTE EXISTING PORT
AREAS FOR CONTAINER AND BULK TERMINALS
APPROACH ROUTE

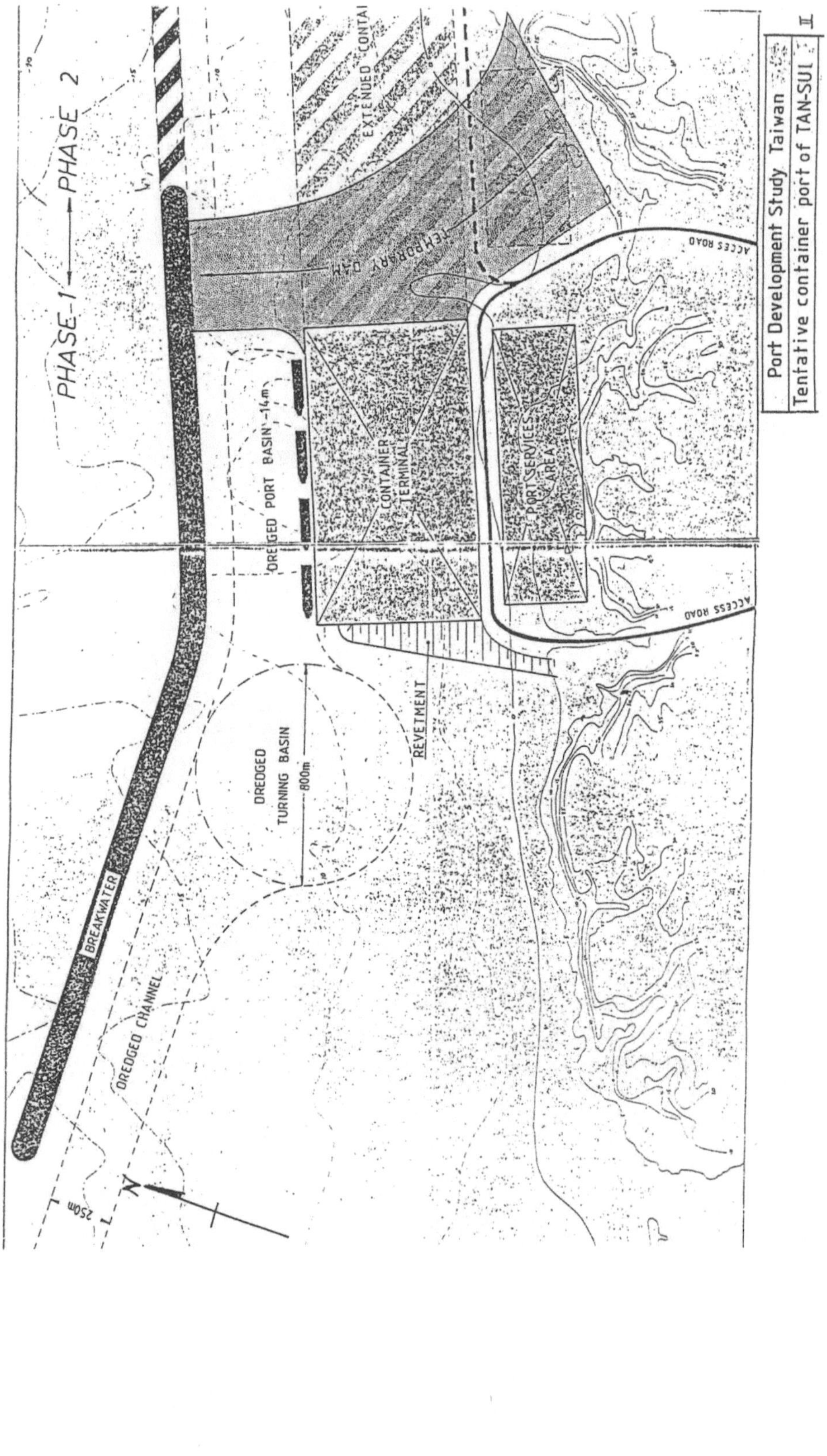
PHASE-1 ——— PHASE 2
EXTENDED CONTAI
TEMPORARY DAM
ACCESS ROAD
ACCESS ROAD
DREDGED PORT BASIN -14m
CONTAINER TERMINAL
PORT SERVICES AREA
REVETMENT
DREDGED TURNING BASIN
800m
BREAKWATER
DREDGED CHANNEL
250m
N
Port Development Study Taiwan
Tentative container port of TAN-SUI
II

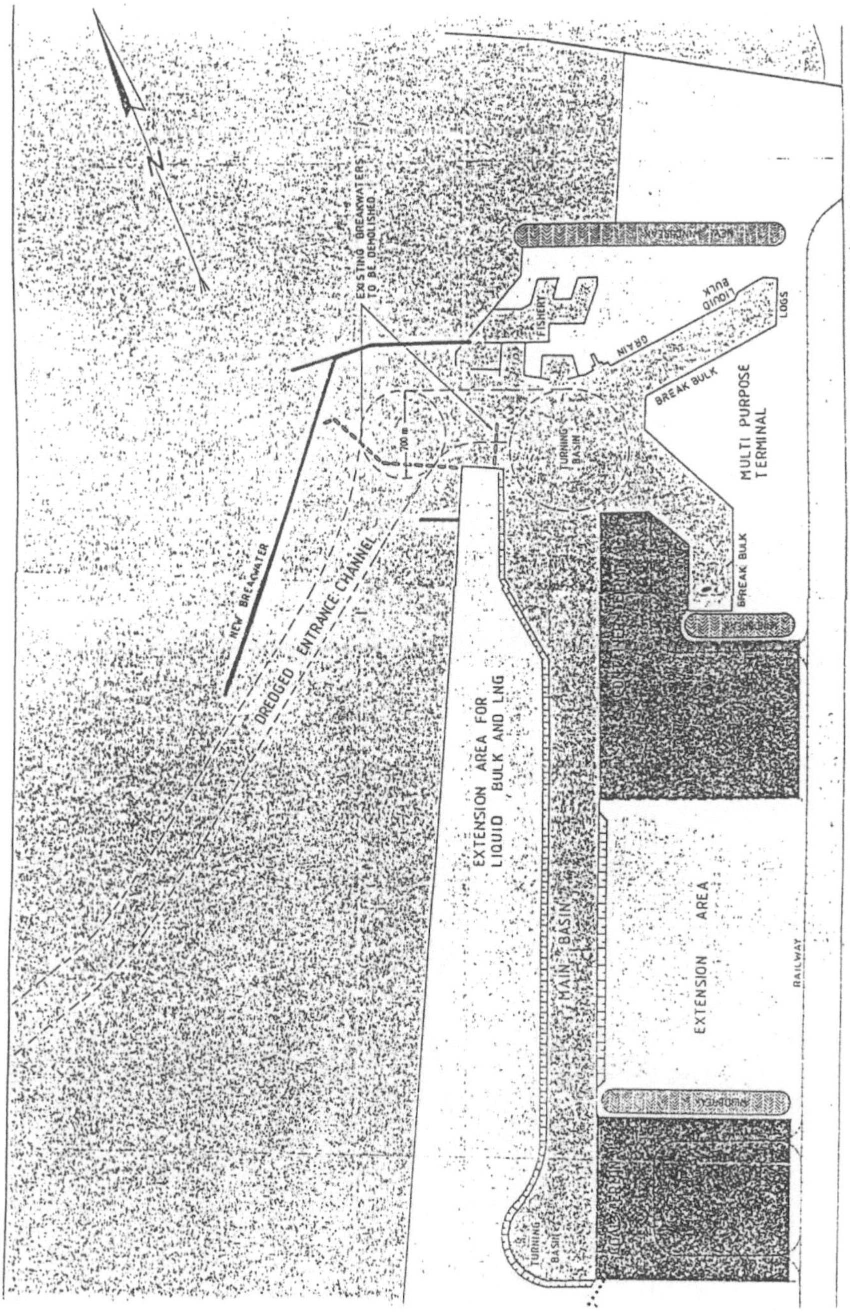

Port Development Study Taiwan
Modification port development plan Port of Taichung III

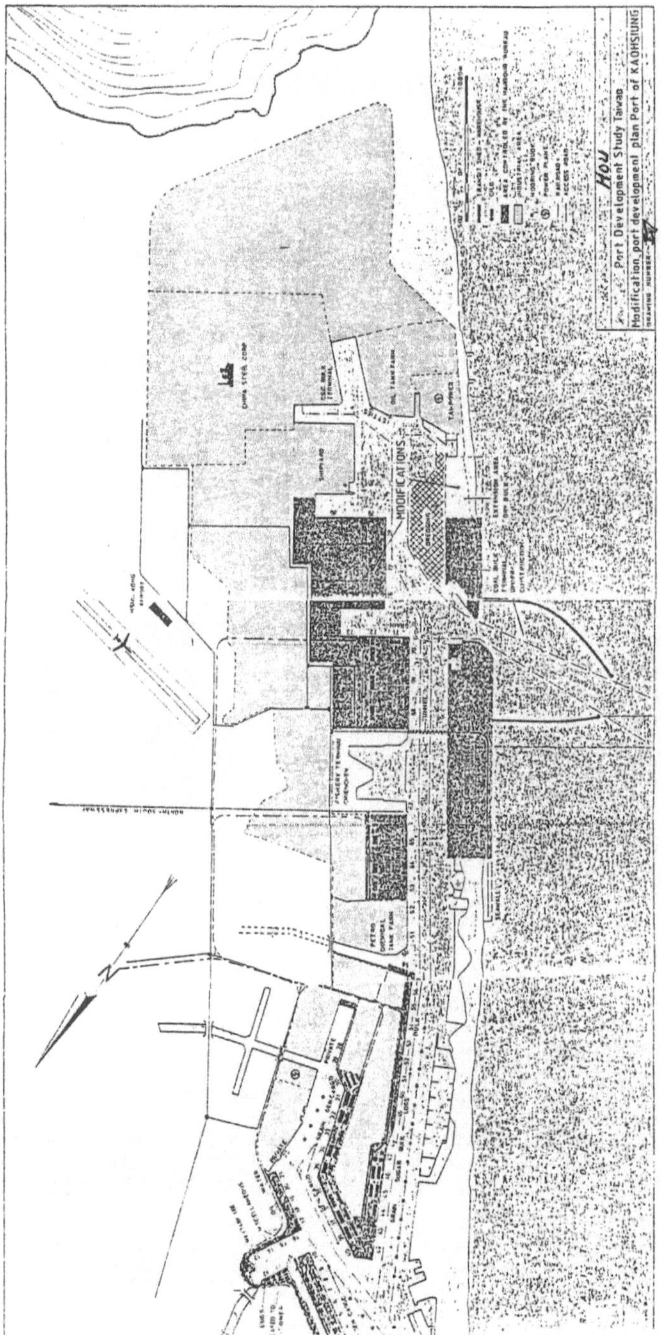

The 2nd Entrance

New Concept of Offshore Fishery Base for Effective Utilization of 200 Miles EEZ

A Method of Select the Position of Japanese Ocean Space

A. Kashimura, A. Kuroyanagi, T. Miyazaki, T. Kondo, T. Soh and W. Kato
Department of Oceanic Architecture and Engineering, College of Science and Technology, Nihon University, Narashinodai, Funabashi, Chiba, 274 Japan

ABSTRACT

The fisheries in Japan are approaching the important turning point and entering the phase at which the fisheries production system matching with the marine environment should be created by changing the point of view for the location of fishery harbors from land to sea. Therefore, it is proposed that the method for selecting appropriate locations to create the offshore fishery bases as the new fishery system beyond the coastal region by installing the offshore fishery bases, avoiding the various social implication with the coastal fishery village areas without being affected from land.

INTRODUCTION

The importance of the ocean is recognized again from the problems caused by the extreme population pressure and the lack of resources, on the other hand, nationalism is risen in terms of maintaining the territories and resources, and the age of spliting ocean is being reached. As a result, many restrictions appear on the fisheries. Therefore, considering that the fisheries in our country are positioned in these problems, the fishery operation within 200 miles EEZ in foreign countries is severely controlled, and the fishery itself, specially, is required to be improved in its organization (1).

This study, therefore, predicts the possibilities of new fishery systems (2) by which new offshore fishery bases (3) are installed avoiding the troubles with the coastal fisheries without being affected from land to develope the activation with priority given to the coastal fishery. And it is intended to develope the methodology on the following two items:

1) To propose the method to select the appropriate location through the relative evaluation from the point of view of the city and fishery characteristics.
2) To select the coastal region at which the marine fishery bases are located.

SELECTION OF THE COASTAL REGION TO BE TARGETED

The selection of the coastal region at which the marine fishery bases are located should be determined by performing the relative evaluation for the coast line and the coastal area according to some standards. Therefore, the conception of the coastal are in a block which is called as the "coastal region (4)", classified large and small coastal region defined by the

Fishery Agency, the Ministry of Agriculture, Forestry and Fishery and the "small coastal region" (111 regions) are employed as the area to be evaluated.

SELECTION OF THE ITEMS TO BE EVALUATED

The representative items to be evaluated are selected to list the optimum coastal region at which marine fishery bases are to be located, after considering the significance of data of the "City characteristics item to be evaluated" extracted from the point of view of finding the background area with the urban structure to be acceptable, and of the "Fishery characteristics item to be evaluated" extracted from the point of finding the fishing ground with the potentiality of fishery. The consolidated evaluation is made using these representative evaluation items through the "Zone classification method" and the "Deviation method".

Upon extracting the items to be evaluated, three indexes are established: Living function (residence, society, and economy), Production function (fisheries, labor, marketing), and Space function (point, line, and area). The city characteristics and the fishery characteristics are assumed to be expressed using the items regarding "Living function and production function" and "Living function and Space function", thus 32 items are selected, and the representative evaluation items are selected.

First, the analysis and significance are examined (t examination) using the coefficient of correlation: r, for the following two conditions; 1) the coefficient of correlation: $|r| \geq 0.3$ and 2) the significance level is allowed at 0.1%. And then, the item relation which satisfies these two conditions is extracted. As a result, 23 items with specially high degree of association between items are extracted, and the factor structure for each "characteristics" is obtained. After classification of items, three items for the city characteristics and six items for the fishery characteristics can be extracted. In addition, it is found that two items, such as "area composition ratio (others)" and "population composition ratio by industries (primary industries)" cannot be positioned as the city characteristics. (see Tables 1 and 2)

ZONE CLASSIFICATION METHOD

In "some areas" located along the shore line, the range area of the characteristics maintained by those areas is not known, but it should be clearly known. Therefore, in the continuously adjacent small coastal region along the shore line, individual fishery potentiality is calculated, and new one territory is established by combining those small coastal region. This zone classification method is to calculate and evaluate the fishery potentiality of such established combined-small coastal region.

Upon calculation of the fishery potentiality index, the total distance of the shore line is not constant. Therefore, the retention amount and their standard amount, which are held by every items composing the characteristics, should be compared to obtain the condition whether the retention amount is proper quantities against the standard amount or not. For that reason, the indexing (5) should be performed to calculate the fishery potentiality index (see calculating equation).

Table 1 City characteristics representative evaluation items

City characteristics items to be evaluated

Items	Category	Representative
City scale (1)	Total population Area composition ratio (Housing site) Number of families DID area	* *
City scale (2)	DID population Population density	*
Distributing function	Handling amount at fish market per year (in value) Handling amount at fish market per year (in quantity)	*
Items to be deleted	Area composition ratio (others) Population composition ratio by industry (Primary industry)	

Table 2 Fishery characteristics representative evaluation items

Fisheries characteristics items to be evaluated

Items	Category	Representative
Fishery harbor adjacent area	Population of fishery harbor area Number of fish market	*
Fishery worker	Number of cooperatives Population engaged in fishery	*
Fishery village	Number of fishery organizations Number of families	*
Fishing method	Total number of utilized fishing boats Number of registered fishing boats Total tonnage of utilized fishing boats	*
Fishery harbor scale	Territorial landing amount Total tonnage of registered fishing boats	*
Point, Line	Number of fishery harbor Total extended distance of shore line	*

Calculating Equation for Fishery Potentiality Index

$$S_{ij} = K_i \times L_j + \alpha \qquad I_{ij} = \frac{F_{ij} - S_{ij}}{S_{ij}} + 1.0$$

$$T_j = \frac{\sum_{i=1}^{4} a_i \times I_{ij}}{\sum_{i=1}^{4} a_i} \qquad G_j = \frac{\sum_{i=5}^{10} a_i \times I_{ij}}{\sum_{i=5}^{10} a_i} \qquad TL_j = \frac{\sum_{i=1}^{10} a_i \times I_{ij}}{\sum_{i=1}^{10} a_i}$$

$$TT_{jk} = \frac{\sum_{k=1}^{n} \left(T_{j+k-1} / L_{j+k-1} \right)}{k} \qquad GT_{jk} = \frac{\sum_{k=1}^{n} \left(G_{j+k-1} / L_{j+k-1} \right)}{k}$$

$$TTL_{jk} = \frac{\sum_{k=1}^{n} \left(TL_{j+k-1} / L_{j+k-1} \right)}{k}$$

- Sij : Standard amount of i item against coastal line total distance Lj of the small coastal region
- Ki : Regression coefficient in i item
- Lj : Coastal line total distance in j small coastal region
- α : Constant
- Iij : Index value against standard amount of i item in j small coastal region
- Fij : Retention amount of i item in j small coastal region
- Tj : City characteristics total index in j small coastal region
- Gj : Fishery characteristics total index in j small coastal region
- TLj : Fishery potentiality index in j small coastal region
- ai : Weight coefficient in i item (see Table 3)
- TTjk : City characteristics total index in k combined small coastal regions with starting at j small coastal region
- GTjk : Fishery characteristics total index in k combined small coastal regions with starting at j small coastal region
- TTLjk : Fishery potentiality index in k combined small coastal regions with starting at j small coastal region

Table 3 Weight coefficient

Items to be evaluated	Weight coefficient
1. Handling amount at fish market per year (in value)	0.6454
2. Total population	0.7593
3. Number of families	0.7575
4. DID population	0.6044
5. Population of fishery harbor area	0.6475
6. Number of cooperative members	0.5582
7. Number of fishery organizations	0.6775
8. Number of fishery harbor	0.4858
9. Total number of utilized fishing boats	0.5524
10. Territorial landing amount	0.5662

Every small coastal regions are evaluated by means of "Zone classification method" according to the following selection standards:

Standard 1 - Extract the combined small coastal regions with the fishery characteristics total index $\geqq 0.9$, for the fishery potentiality provided that the small coastal regions adjacent to the good fishing area should be found.

Standard 2 - Extract the combined small coastal regions with the city characteristics total index $\geqq 0.9$.

Standard 3 - Extract the combined small coastal regions with the fishery potentiality index $\geqq 0.9$.

Standard 4 - The small coastal regions with extremely high fishery potentiality index due to high affection by the city characteristics total index (that is, the small coastal regions have low dependency on the fishery because the large city is located at background area) are rejected because, it is considered that the fishery potentiality is essentially low.

Standard 5 - This study is intended for the coastal fishery, therefore the small coastal regions, which depend on the deep sea fishery, are rejected.

The combined small coastal regions to satisfy the above conditions could not be found. As a result, seven small single coastal regions are selected as shown in Table 4.

Table 4 Results selected by the zone classification method

Code	Large coastal regions	Prefectures	Small coastal regions	Fishery potentiality index
014	Northern region of Pacific Ocean	Aomori	Northern east section	0.94319
022		Fukushima	North section	0.93095
025		Ibaragi	North section of Kashimanada	1.42432
027	Center region of Pacific Ocean	Chiba	Choshi Kujukuri	1.28202
061	Setonaikai coastal region	Ohita	Buzenkai	0.98985
062	East China coastal region	Fukuoka, Saga, Nagasaki	Ariakekai	1.03857
105	Northern region of Japan Sea	Toyama	Toyama	1.43376

DEVIATION METHOD

The deviation method is used to obtain the active condition of the small coastal regions by showing the specialized condition for the small coastal region in percentage using the climograph because it is required to obtain the deviation when the range area of characteristics for "some area" located at the shore line is limited. The small coastal regions to be object are selected by complementing the results selected with the zone classification method each other. The small coastal regions have been able to be classified into four groups using the climograph as shown in Figure 1.

Group (1) - "City characteristics/Fishery characteristics small type" means the type that most of the representative evaluation items is less than 1.0%, the city scale at background is very small, and the small business condition is provided in the fishery. (50 small coastal regions)

Group (2) - "City characteristics specialized type" means the type that the city characteristics are highly specialized comparing with the fishery characteristics. This is found at the small coastal regions with large city of its background, and at the small coastal regions adjacent to the large city. Their dependence on the fishery is very low, that is the fishery potentiality is seemed to be very low. (14 small coastal regions)

Group (3) - "Fishery characteristics specialized type" means the type that is small background city scale and the collected small coastal regions of the local fishery villages. However, such group types of small coastal regions have the high possibility of existence of the small coastal regions with the high dependence on the fishery and the high fishery potentiality. (31 small coastal regions)

Group (4) - "City characteristics/Fishery characteristics filling type" means the type that provides the city scale, to some extent, at the background, the city function, and the high fishery potentiality. Therefore, the possibility of location is very high. (16 small coastal regions)

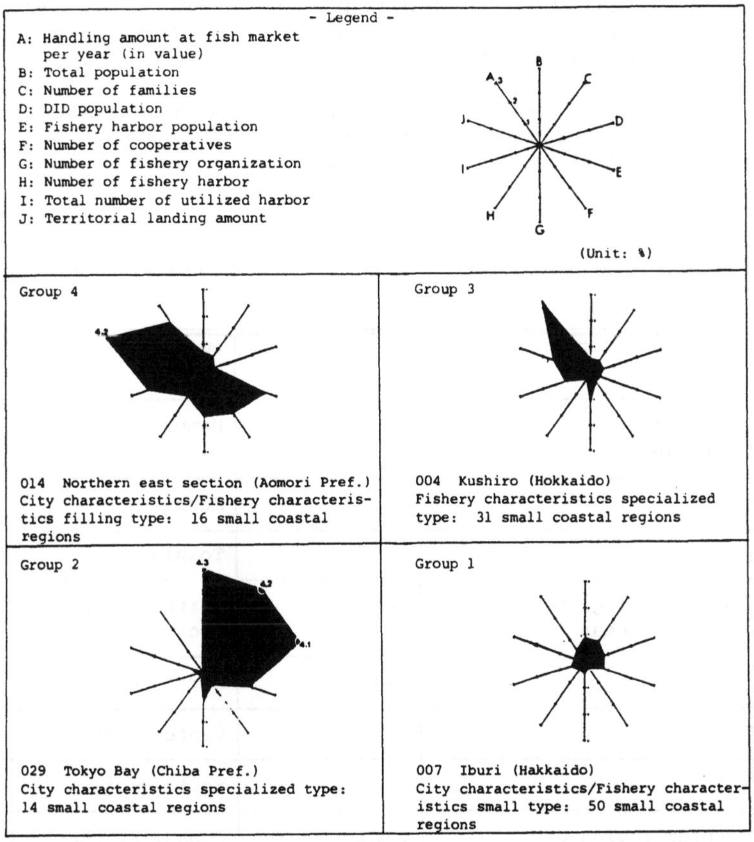

Figure 1 Group results by the deviation method

The small coastal regions are selected from the climograph by group, which, as a rule, belong to the group type of the "City characteristics/Fishery characteristics filling type".

Also, the standard 5 which has been established by the zone classification method should be applied to the deviation method to select the optimum small coastal regions to be located. Assuming that the small coastal regions with the high fishery potentiality exists even in the small coastal regions belonging to the "Fishery characteristics specialized type" as well, 12 small coastal regions have been selected, specially, for the "Territorial landing amount" and the "Handling amount at fish market per year (in value)" as shown in Table 5.

GENERAL EVALUATION

The selected results of the optimum coastal region to be located are shown in Table 6.

The case 1 is the combined small coastal regions of Nemuro and Kushiro (Hokkaido). Both of them belong to the "Fishery characteristics specialized type" with the deviation method. However, they show the extremely specialized condition in their "Territorial landing amount" and "Handling amount at fish market per year (in value)". In addition, the territorial landing amount at Kushiro harbor shows the 1st rank for every year. This is the season why they are selected as the combined small coastal regions.

The case 2 is for the northern east section (Aomori Pref.), which is the small coastal region selected by means of the above-mentioned two methods due to the circumferential sea area being considered as good fishing ground.

Table 5 Selected results with deviation method

Code	Large coastal regions	Prefectures	Small coastal regions	Growth type
003	Coastal region of Hokkaido	Hokkaido	Nemuro	2
004			Kushiro	2
008			Toshima	1
014	Northern region of Pacific Ocean	Aomori	Northern east section	1
019		Miyagi	North section	2
020			Center section	2
021			South section	1
027	Center region of Pacific Ocean	Chiba	Choshi Kujukuri	2
048	South region of Pacific Ocean	Ehime	Uwa	2
058	Setonaikai coastal region	Hiroshima	Hiroshima	1
082	East China coastal region	Fukuoka, Saga Nagasaki	Ariakekai	1
092	West region of Japan sea	Yamaguchi	Yamaguchi	1

Table 6 Selected results of optimum coastal regions to be located

Case	Large coastal regions	Prefectures	Combined small coastal region to be object
1	Coastal region of Hokkaido	Hokkaido	Nemuro Kushiro
2	North region of Pacific Ocean	Aomori	Northern east section
3	North region of Pacific Ocean	Miyagi	North section Center section South section
4	North region of Pacific Ocean	Ibaragi	North section of Kashimanada South section of Kashimanada
	Center region of Pacific Ocean	Chiba	Choshi Kujukuri

The case 3 is for the north, center, and south sections in Miyagi Prefecture. For the north section, its fishery characteristics total index = 0.99. Both north and center sections belong to the "Fishery characteristics specialized type" and their "Territorial landing amounts" are specialized. In addition, the south section belongs to the "City characteristics/ Fishery characteristics filling type". These three small coastal regions have been selected as the combined coastal regions considering that Sendai City located behind those sections is highly evaluated in its marketing and distributing condition.

The case 4 is for the north and south sections of Kashimanada in Ibaragi Prefecture and for Choshi Kujukuri in Chiba Prefecture. Among them, the territorial landing amount for the north section of Kashimanada is about 220,000 tons per year. This is the 111th throughout the country and the 8th in the small coastal regions, and means about 2% of total catch of fish in our country provided that its total amount is about 11,000,000 tons. As a result, the north section of Kashimanada has been selected. and Choshi Kujukuri has been selected because it is the small coastal region selected by both previously-mentioned two methods. In addition, the south section of Kashimanada located between those two small coastal regions has been selected considering the geographical condition, and then these three small coastal regions are selected as the combined small coastal regions.

Some small coastal regions, which were selected by the "Zone classification method" and the "Deviation method" but not selected as the optimum small coastal region to be located, are rejected because the main fishery type of their small coastal region is the cultivating fishery which is not an object for this study and it is found by the deviation method that the small coastal region selected by the zone classification method belongs to the "City characteristics/Fishery characteristics small size type" group.

When the selected results for these four cases are considered, it is found that either small coastal regions are located at Choshi and northward area which faces to the Pacific Ocean and that the sea area with high potentiality of the fishery values is located at the circumferential sea area of the selected area. From the point of view of the current coastal fisheries in our country, we can, now, recognize that the results of this study proves that the sea area selected by this study coincides with the sea area from Choshi offing to Sanriku offing where currently most excellent fishery ground is expanded.

CONCLUSION

From the point of view of the problem of "200 Miles EEZ", this study analyzes the coastal area through two characteristics of "City" and "Fishery" in order to obtain the optimum coastal region to be located, assuming the "Offshore Fishery Base" which contributes to the effective utilization of the coastal area in our country. In addition, this study provides two methods, such as "Zone classification method" and "Deviation method", and then selects four cases of coastal regions to be object. Depending on the validity of this result, the validity for the point of view such as the "City characteristics" and the "Fishery characteristics", and the validity for the methodology such as the "Zone classification method" and the "Deviation method" can be also recognized as the results of this study.

References

(1) Takemasa Miyazaki; "Positive Studies on Forming the Environment of the Coastal Fishery Village Area", Thesis for a Degree for its Application, March 1983.

(2) Takeo Kondo, Wataru Kato, etal.; "Multipurpose Floating Platform for Ocean Resources Development", 10th Meeting UJNR MFT, October 1980.

(3) Junjiro Motooka, Wataru Kato, etal.; "General Studies on the 200 Miles EEZ in Our Country", Research Special Committee specified by President of Nihon University, March 1983.

(4) Statistics and Information Dept. Economic Affairs Bureau, Ministry of Agriculture, Forestry and Fishery, "The 6th Fishery Census", December 1982.

(5) Tadakuni Taniguchi; "On the Evaluation of Nine Types of Cities about Medical Facilities and Manpower", Transactions of the Architectural Institute of Japan No.309, November 1981.

A Case Study on a Project by a Fisheries Cooperative

I. Sakai[1] and K. Mitsuhashi[2]

[1] All Japan Fishing Port Association, Kasumigaseki, Chiyoda-ku, Tokyo, 100 Japan
[2] Disaster Prevention and Coastal Protection Division, Fisheries Agency, Ministry of Agriculture, Forestry and Fisheries, Kasumigaseki, Chiyoda-ku, Tokyo, 100 Japan

ABSTRACT

Fishery is not so big an industry among today's modern industries. But in many small communities, like local towns and villages in Japan, it is still a key industry for developing those areas.

This report introduces a typical project which has brought a certain development to a community by efforts of a fisheries cooperative, based on fishery, through integrated measures combining fishery and resort. The project suggests many things about the way of development in local areas, especially in coastal places where fishery is a major industry.

In the near future of Japan, fisheries cooperatves should play more important roles for developing local areas because coordinate activity connecting fishery with other industries will become very common in local areas.

INTRODUCTION

In the field of "Ocean Development", an effective use of fishery resources is an important one for people living at the seaside.

In Japan, about 400 thousand fishermen are engaged in coastal fishery and over 1.3 million people of fishing families live along the coast of 34 thousand kilometers. The number of fishing hamlets which form towns and villages reaches seven thousands and most of them are scattered along the coast of intricated configuration.

In such a local community, fishery is a key industry to support their daily lives, and development of fishery at the site is directly connected to the development of the place. But, in today's situation in Japan, free fishing operation is unable to be carried out due to fisheries regulations which aim to keep reproducing potentiality of marine products and also to prevent conflicts among fishermen or fishermen with other people.

At present, all the coast of Japan is covered with designated fishing zone based on the regulations, especially in near shore area, most part of it is shared with each cooperative for predominant use of each coast. In this situation, the way of development of a fishery oriented community should satisfy the following.

(1) To raise fisheries potentiality in a limited area covered with fishing rights which belong to a fisheries cooperative in the community.

(2) To shift their fishing operation from low price fish to higher price ones.

(3) To develop new industries based on fishery.

In this report, a project which satisfy the above three items is introduced as a model case of local development based on fishery.

A CASE STUDY ON A PROJECT IN A LOCAL TOWN OF JAPAN

Waita is a hamlet, of 1,200 people mainly engaged in fishery and agriculture, which belongs to Toyoura town, west end of Honshu(main island) of Japan. Many people has come to the hamlet through the year round for visiting "MARINPIA KUROI", a site studied in this report.

MARINPIA KUROI is a seaside resort of 84,000m ,providing rest and recreational facilities scattered in woods and on shore, targeting to families, young groups and aged people. In 1983, the resort, including a fishing pond for leisure fishing, accepted 200 thousand people and had sales of 1,000 million yen(4 million dollars), which shows that it has become a leading industry in a hamlet the same as culture fishery of 880 million yen(3.5 million dollers) sales done on the water in front of the hamlet.

These industries have been planned, raised funds, and managed by Kuroi Fisheries Cooperative under the basic concept of "to create a harmonized environment and to establish local people's industries through combination of people and the sea". Actually, they are an exellent result of fishermen's effort in a hamlet and also shows how a small community is developed through the efficient use of natural resources inherited from ancestors.

Fig.1. MARINPIA KUROI, main building(left) and tennis court(right)

Back Ground Required by the Project

About 20 years ago, in Waita hamlet, fishery was a leading industry cooperatively done by the people living there, but, during the following decade, the hamlet's fishery nearly died out caused by shortage of fishermen to operate fishery and decrease of fish coming from offshore.

The rapid decrease of fishermen depended on rapid development of Japanese economy which had yielded a great difference of income between urban people and local ones and attracted the lattter especially the young to big cities. Decrease of fish had depended on fishing operation vigorously done on offshore waters, because the operation would have intercepted fish schools heading to Waita shore.

As a resuscitative measure of fishery and of people's recalling from urban areas, Kuroi fisheries cooperative decided, in 1966, to start a fish culture which had not been accepted in any other fisheries cooperatives because of its uncertainty on commercial base. Of cource, members of the cooperative were anxious about many things, paticularly the following three items.

(1) The proposed water area for fish culture had not covered enough against rough waves.
(2) Increasing water pollution would bring heavy loss of fish.
(3) How to go about finding good customers who could buy such a large amount of high priced fish constantly.

In spite of the above mentioned anxieties, the cooperative started because it would be the only way to restore fishery of Waita.

The cooperative experienced long hard days to have damages by destructive waves, sea pollutions and stagnant sales of fish. After half a decade of effort, the fish culture has become a key industry to yield big profit.

Fig.2. Fish culture field(left) and MARINPIA's beach(right)

An idea to construct MARINPIA KUROI originally occurred from two phases, one was to create an organization accepting large amounts of cultured fish constantly, and the other was to find an effective use of lands belonging to the cooperative.

When the fish culture had been started, the cooperative had tried to sell the fish to consumers directly adding ordinary way to sell them at fish market, but it had been hard to sell large amounts of fish constantly. To improve such a difficult situation, an admission fee charging fishing pond was opened on the corner of fish culture field. The result was a big success with large numbers of fishing lovers coming and the balance of fish culture rapidly improving. In addition to it, a seaside restaurant was constructed to satisfy the people's demand of instantly enjoying the taste of fish which they themselves hooked up just a moment before. Of course, the restaurant, by the request of guests visiting the site, could also serve various kinds of dishes of fish caught from the sea or cultured in the field. This idea has also contributed a certain improvement of balance to the cooperative.

From the successful result of fishing pond above, MARINPIA KUROI was expected to be a big consumer of fresh fish and a selling place of processed fish products supplied by the cooperative. From the other side, MARINPIA was thought out for utilizing the lands which was bought back from real estate agents in order to protect the water front of Waita from pollution and to keep the environment inherited from ancestors.

As a combination of these two demands, MARINPIA KUROI was planned and developed. The transition of circumstances and countermeasures adopted for improving the situations are shown in Fig. 3.

Present Situation

Outline of MARINPIA's facilities and management are shown in Fig. 4 and Table 1 to 3, which explains the project including fish culture and fishing pond contributing largely to the town's development.

Key to Success

The success of the project depends mostly on its originality and continuous efforts done by the fisheries cooperative.

The key to success are listed as follows.

(1) Clear idea in order to make a basic conception of the project, which are, a) presentation of beautiful scenery on and around the site for visitors particularly for persons enjoying sports, b) adjusted use of water front of the site for marine sports, and c) expansion of stable consuming route for cultured fish, and processed fish products.

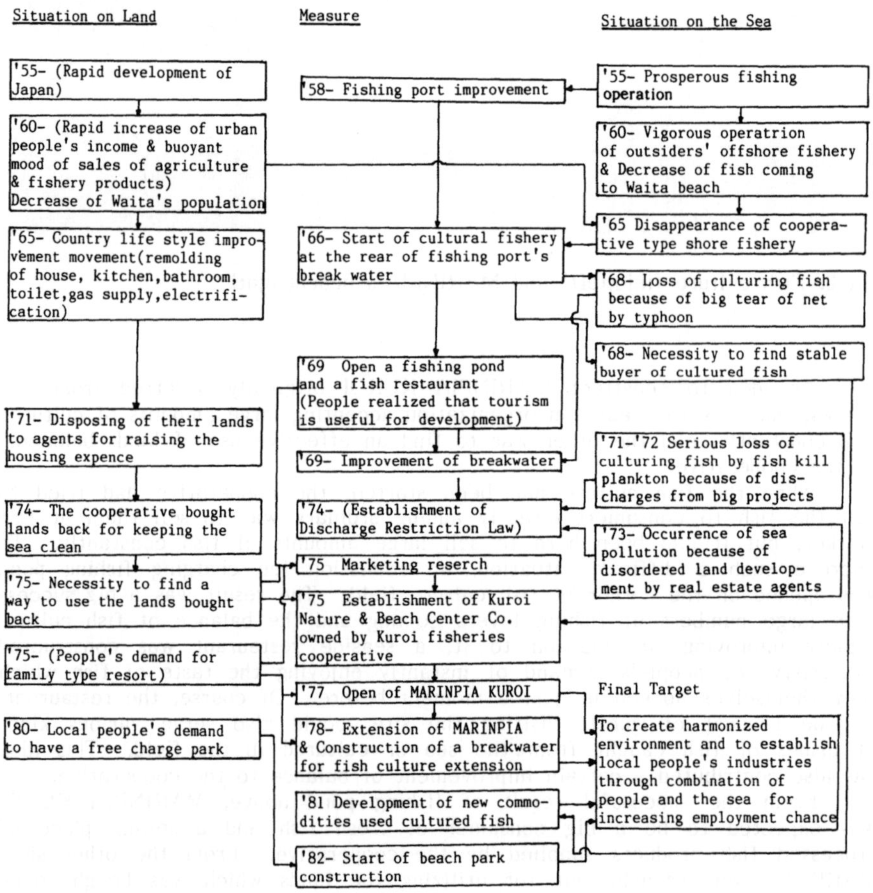

Fig.3. Flow chart of Waita hamlet's development

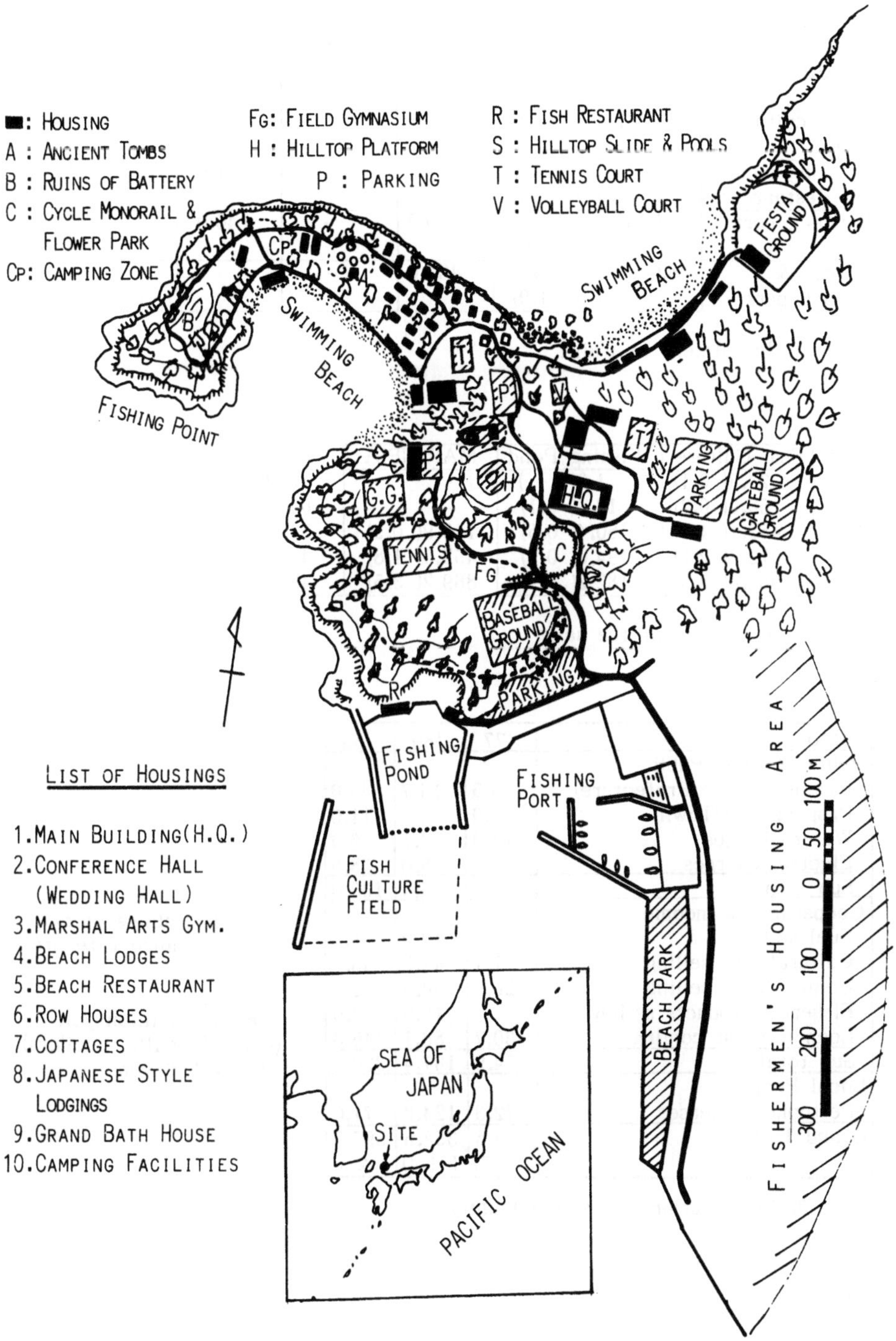

Fig.4. Layout of facilities in Waita

Facilities	Year	Cost	Subsidy	Semi-gov. finance	Coop.	Others
MARINPIA	'75-	2,845.0 (11.4)	-	938.0	1.000.0	907.0
Fish culture field	'65-'82	175.0 (0.7)	110.5	9.9	55.0	-
Fishing pond & Restaurant	'69	9.6 (0.4)	-	-	9.6	-
Fishing port	'58-'82	220.0 (0.9)	180.0	-	40.0	-
Coast & Beach park	'63-	480.0 (1.9)	480.0	-	-	-

Table 1. Investment in Waita

year	Fish Culture	Fishing Pond	MARINPIA	Fishery	Total
'67	24.6(0.1)	1.3(-)	-	72.0(0.3)	96.9(0.4)
'72	60.9(0.2)	13.1(0.1)	-	64.0(0.3)	138.0(0.6)
'77	455.1(1.8)	160.4(0.6)	113.9(0.5)	53.0(0.2)	782.4(3.1)
'80	672.3(2.7)	194.0(0.8)	455.5(1.8)	45.0(0.2)	1,366.8(5.5)
'83	880.6(3.5)	182.2(0.7)	869.2(3.5)	43.0(0.2)	1,975.0(7.9)

Table 2. Transition of the sales

Sector	'77	'80	'83
Public & Semi-Public			
Restaurant tax(Prefecture)	0.5	16.7	45.0
Property tax(Town)	2.9	6.7	11.0
Water charges	13.0	4.1	6.0
Electric charges		18.0	38.0
Sub total	16.4	45.4	100.0
Cooperative's Income			
Fuel selling	-	12.0	8.0
Cultural fish selling	17.7	42.7	46.0
Commodities selling	26.5	34.6	112.0
Fisheries products selling	-	10.0	14.0
Loan interest coming	30.7	88.1	146.0
Sub total	74.9	187.4	326.0
Local People			
Personnel expenses	72.3	124.8	171.0
Grand Total	163.5 (0.7)	357.6 (1.4)	597.0 (2.4)

Table 3. Contribution to the community

*All the table's units and exchange rate are as follows.

unit : 1 million yen
(unit) :(1 million dollers)
1 $ = 250 yen

(2) Careful feasibility study for planning appropriate facilities based on geographical condition of one hour cruise by car from Kita-Kyushu shi, a big city with a population of one million.
(3) Paying full attention for keeping good circumstances and protecting the sea from pollution.
(4) Small but intricated configuration serving various types of beaches consisting of sand, rocks, and their mixture.
(5) Adoption of membership system which has contributed to keeping visitors constantly to decrease financial burden required for construction and management.
(6) Full utilization of public and semi-public sector's subsidizing and lending system for developing local industries.
(7) Active advertisement and fresh planning like a jazz festival in open air which was the first festival done around the area.
(8) Step by step development of facilities adjusting to financial condition.

DIRECTION OF LOCAL FISHERY DEVELOPMENT IN JAPAN

In order to develop fishery and increase fishermen's income, the following themes should be considered.
(1) Extension of fishing ground.
(2) Increase of fish density in unit space.
(3) Saving the cost of fishery.
(4) Raising the price of fish.
(5) Combining fishery with other industries.

The situation of these themes classfied by fishery sectors, according to their fishing operation scale, is shown in Table 4. In this table, size of each circle means the probability to realize the theme in each sector.

Theme / Sector	(1) F. g. extension	(2) Increase f. density	(3) Down the cost	(4) Up the price	(5) Combine with others
World wide	⬤	•	⬤	●	•
Nation wide	•	●	●	●	•
Local scale	•	⬤	●	•	⬤

Table 4. Basic conception of possibility to develop each theme

In local fishery, the circle in column(1) is very small because of severely restricted setting of fishery zone in shore, and the efforts to improve it should be intended for the other four themes. For this reason, following measures may be adopted for local fishery improvement. They are
a) artificial discharge of fish fry, b) creation of artificial fish grounds,
c) promotion of fish culture, d) Improvement of fish distribution system,
e) creation of better environment for fishermen, f) development of new industries combined with other industries, and g) improvement of fishing ports for supporting these measures.

Most of the above mentioned measures will require cooperative works of people in various kinds of fields, and a fishery cooperative should be the key organization to coordinate and to manage these matters. Especially, in the field of combining with other industries, fishery cooperative will be able to have a leadership for developing the new industry through connection with local people and outer industries. In many local places, activities which combined fishery and other industries have been started, paticularly in the field of tourism by fisheries cooperatives. For example, management of fishing jetties and fishing ponds, cruising of glass bottom boat, observation of tamed natural fish, management of under water observation tower, and fish restaurants etc. have been done.

THE ROLE OF FISHERIES COOPERATIVES FOR OCEAN DEVELOPMENT

Many items are considered as a target of ocean development. From the view point of utilization, a) materials composing the sea, b) many forms of life, c) built-in energy, d) wide and deep space may be considered.

Most of the above mentioned items will conflict with the activities of fishery, because fishing rights of fisheries cooperatives in Japan legally authorized based on historical back grounds, and an acceptable proposal to fishermen is a key for smooth starting of a project on the sea.

There are two ways to compensate a certain restriction of fishing activity for an ocean project, one is to pay money to fishermen, the other is to provide funds for promotion plans of the area. In many cases ,the former has brought unhappy lives causing temporary income of big money without purposes, and the latter may be recommendable in Japan.

Generally, local area's fishermen claim to satisfy followings as a compensation. They are a) to live continuously in their home town, b) to work on the sea or seaside with a certain connection to fishery, c) to have a job managed by themselves, d) to make their own town develop by the compensation money.

A case study described in this report will fit the above mentioned requirement, which means that it is very suggestive for smooth starting of an ocean development project near shore.

Fisheries cooperarives, in future, will be expected to play more important role for promoting communities through coordination of fishing activities with other on the sea or on land activities, and for keeping the sea clean with the prospect that the sea is the common property of people.

CONCLUSION

In this report, the authors described a successful local development project and the roles of fisheries cooperatives for developing communities and the sea. Finally, the importance of infrastructures such as fishing ports should be emphasized, because today's prosperity of Waita would not be brought about without the existence of the fishing port, the situation shown in Fig. 3. The success of a project will be brought on the basis of long and steady efforts to develop fundamental facilities done by the people concerned.

The authors wish to express their sincere gratitude to Mr. Ichitaro Masuda, Chairman of Kuroi Fisheries Cooperative, and his staff for the presentation of the data used in this report and detailed explanation on the project.

They also wish to their thanks to Mr. Sadaji Miyoshi and his staff of Fishing Port Division of Yamaguchi Prefecture, and the staff of Toyoura Town for their enthusiastic supporting works.

A Fundamental Consideration for the Fishery Environmental Impact Evaluation Using the Matrix Network Method

T. Otsuki

Engineering Research Institute, TOA Harbor Works Co., Ltd., Anzen-cho, Tsurumi-ku, Yokohama, 230 Japan

ABSTRACT

One example of the fishery environmental impact evaluation using the Matrix Network (M.N.) method is presented and discussed. The M.N. method is thought as method that can quantatively predict and evaluate the impact scales, which various development actions in the sea will have an influence on the fishery environment and marine ecosystem. And, it is said that the characteristics of this method are found in being able to clarify the causal relations of impacts using the matrix at each step, and in the simplicity of calculation that the final prediction values are obtained by calculating their matrices of impact transmissions in order.

INTRODUCTION

Japan, the coastal zone is utilized to a high degree for the coastal fishery, mari- and aqua-culture, etc., therefore, there are very severe criticisms against the environmental disturbances brought about by the large scale development actions in bay or shallow sea. But, it goes without saying that we must promote to utilize effectively the coastal zone in the future. Consequently, prior to the design of the development and utilization plans, it is strongly required to consider the prevention of impacts which are predicted for the surrounding environments, specially, the fishery environment.

However, when we are going to carry out the environmental assessment and prepare its report, necessary informations are not enough and not arranged well under present conditions. Still less, it is said that there are few attempts to evaluate the impacts for the fishery environment.

Then, in this paper, one example of the fishery environmental impact evaluation using the M.N. method is presented in accordance with the following procedure.

PROCEDURE OF THE PREDICTION AND EVALUATION

The environmental changes brought are not always possible to come about simultaneously and unitarily, and it is common that the causal relations, which the other side of environment are changed by one side changing, exist.

Therefore, the M.N. method shows the whole processes as a net-work formed from the matrices between the two factors or phenomena, in order to indicate and predict-evaluate successively what impacts and by what degree the environmental changes, brought by the appearances of the development actions, go to spread over the final objectives. Namely, it is thought that it is desirable that the method proposed in here ① can grasp the whole without omissions, and ② can examine the prediction -evaluation values at every steps of impact transmission processes, to clarify the significance of them obtained finally and enhance the exactitude higher, and also ③ can have the flexible evaluation system which easily introduces the new knowledge.

STEP 1 : RECOGNITION OF ENVIRONMENT

Such actions as developments of harbors have some influences on the surrounding environments, and these impacts are actualized as the changes of environmental conditions. And it is thought that these processes and results of changes are observed and recognized as various environmental phenomena.

Extraction of the Development Actions (D.A.):

Development actions which are predicted to have influence on the environment among many different ones in the plan proposed, we specially regard as the causal elements of impacts and call the Development Actions (D.A.). And, we make it a rule to clarify beforehand the kind of works, scales, terms, etc., of these D.A. extracted.

Extraction of the Environmental Impact Factors (E.I.F.) and Environmental Impact Phenomena (E.I.P.):

The parameters which indicate the environmental changes and phenomena themselves, we call the Environmental Impact factors (E.I.F.). And, the environmental changes and phenomena which are secondarily or further indirectly brought by these E.I.F., we call the Environmental Impact Phenomena (E.I.P.). Namely, it is thought that these E.I.P. are formed from their phenomenal forms (E.I.P. in here) and component elements (E.I.F. in here).

For example, when such E.I.P. as the water pollution is observed, the turbidity or SS which is the parameter of the change of its water quality, is the E.I.F. so-called.

Extraction of the Environmental Component Factors (E.C.F.) and Environmental Component Phenomena (E.C.P.):

Though the new development actions are not brought into existence, actually there is already one environment in its own way in the sea area and many environmental phenomena must be observed. Therefore, such phenomena must be distinguished from the E.I.P. here. Namely, we call such phenomena the Environmental Component Phenomena (E.C.P.) and also call their component elements the Environmental Component Factors (E.C.F.).

Besides, as there are many unknown problems under present conditions, for extracting them it is thought that one method is to carry out the opinionaire and hearing research for the men of knowledge and experience and experts.

Extraction of the Evaluational Objective Phenomena (E.O.P.):

It is thought that there are many cases which the D.A. have direct influence on our lives and health, however, in this paper it is dealt with the cases which the impacts have finally influence on them by way of E.I.P. and E.C.P.. And, such environmental impact phenomena which should be taken up as the objectives to evaluate finally, we call the Evaluational Objective Phenomena (E.O.P.).

Namely, in this paper, the natural environmental aspect in the fishery environment is specially given attention and the fish catch of each fishery kind and fish species, which is thought for a point of contact with its social-economical one, is taken up as the E.O.P..

Extraction of the Impact Limiting Factors (I.L.F.):

It is thought that the particular circumstances and characteristics in the sea area have great influence on the ways and degrees of impacts, in every processes of impacts appearing. Therefore, we take up them as the limiting factors of impacts and call the Impact Limiting Factors (I.L.F.). Namely, the I.L.F. is like the catalyzer in chemical reactions and has the functions to increase or decrease the scales of impacts and the degrees of changes of E.I.F. and E.C.F.. However, it is said that the E.I.F. or E.C.F. itself also has the possibility to be the I.L.F..

STEP 2 : ESTABLISHMENT OF THE EVALUATIONAL OBJECTIVE RANGE

It is the important work to establish the objective range of evaluation in an initial step of the environmental assessment. And, it involves four different aspects as follows.

Regional Range:

It is desirable that the objective sea area is established in a safety view point to be of wider range, in consideration of the fishery actions, meteorological conditions, sea states, and topographical situations, etc..

Generally, when the fishery is in full activity and it is necessary to evaluate in detail, it is desirable to establish the evaluational meshes as small as possible. But, when the fishery activity is not so great, it may be careless to establish the large ones.

Time Range:

It is generally thought that there are three points of time for evaluation as follows.

a Step of Plan: A point of time which is not such a long distant of time, between just before and about three years before the plan is implemented.

a Step of Work: A point of time which the work reaches the zenith.
a Step of Operation: A point of time which the operational conditions come to be stable , between one and three or five years after the operation is started.

Environmental Range:

Owing to the results of previous researches, it is necessary to select objective environmental items carefully. For example, when the impacts for the fishery environment are predicted, it is desirable to select the factors and phenomena connected with the fish catch of each fishery kind and fish species.

Evaluational Range of Impacts:

The evaluational range of impacts must clearly be determined beforehand. Namely, it is the problem that we take up only negative impact or both impacts of negative and positive. But, when the plan is framed, it is generally thought that we must do best to minimize the impacts of negative in the first place.

STEP 3 : MAKING THE P MATRIX

The matrix which shows quantatively the causal relation between D.A. and E.I.F. is called the P matrix (Fig.1). And the prediction for the impact scale of E.I.F. caused by D.A. are done using the vector $\{p\}$ which is obtained by the prediction simulation based on the elements in each column direction in P matrix. Namely, when some D.A. appear simultaneously, it is the vector $\{p\}$ that gathers synthetically the impacts by their appearances.

$$P=\text{A.D.}\overset{\text{E.I.F.}}{\begin{bmatrix} p_{11} & p_{12} & \cdots & p_{1\ell} \\ p_{21} & p_{22} & \cdots & p_{2\ell} \\ \vdots & \vdots & & \vdots \\ p_{k1} & p_{k2} & \cdots & p_{k\ell} \end{bmatrix}}$$

$$\{p_1 \quad p_2 \quad \cdots \quad p_\ell\} \; \{p\}$$

Fig.1. P matrix and vector $\{p\}$

STEP 4 : MAKING THE Q MATRIX

The matrix which shows quantatively the causal relation between E.I.F. and E.C.F. is called the Q matrix. And the impact scales of E.C.F. caused by E.I.F. are indicated using the objective and non-dimentional Environmental Impact Indexes (E.I.I.) which regard the present level as 0 and the maximum impact as 1.

$$Q=\text{E.I.F.}\overset{\text{E.C.F.}}{\begin{bmatrix} q_{11} & q_{12} & \cdots & q_{1m} \\ q_{21} & q_{22} & \cdots & q_{2m} \\ \vdots & \vdots & & \vdots \\ q_{\ell 1} & q_{\ell 2} & \cdots & q_{\ell m} \end{bmatrix}}$$

$$\{q_1 \quad q_2 \quad \cdots \quad q_m\} \; \{q\}$$

Fig.2. Q matrix and vector $\{q\}$

The Q matrix, as shown in Fig.2, is able to be made by substituting the representative prediction vector $\{p\}$ to the function f_{ij} $(1\leqq i\leqq l,\ 1\leqq j\leqq m)$ between E.I.F. and E.C.F.. And its intensive vector $\{q\}$ is, to predict the impacts complicated by plural factors, obtained by multipling each E.C.F. in the column direction in Q matrix using the following formula. $q_m=1-\pi_{i=1}^{l}(1-q_{im})$

And these vectors $\{q\}$ are regarded as the impact scales which the E.I.F. have influence on the E.C.F..

STEP 5 : MEASURING THE INTERACTION BETWEEN THE TWO E.C.F.

It is thought that it is desirable to correct the impact scales obtained before by the sea states, that is to say, the existences and degrees of interactions between the two E.C.F., to evaluate them based on the actual environmental conditions.

Then, in the first place, we make it a rule to clarify the functional relation between each E.C.F., using the results of previous researches and studies. And, in the next place, we clarify quantatively the impact scales between them, using the structural interactions, as follows. Namely, in the present state it is thought to be rather exactitude and practical that these interactions are quantatively clarified by the method of graphic theory, based on the ranking of interactions which is obtained from the results of the opinionaire and hearing researches for the men of knowledge and experience and experts.

When we grasp the impact scale from A to B of E.C.F., we first take up all ways possible to arrive at B from A which are not overlapping with other ways on the way. And they are to be regarded as the effective ways in due order of smaller number of routes. And also, as to all of these effective ways their impact scales are made clear from their functional relations obtained, and the results added of them are regarded as the impact scales of A to B. For example, when there are n ways possible to arrive at B from A and their numbers of routes are r_i $(i=1,2,\cdots,n)$, the impact scale of A to B is obtained by the following formula using the function f between the two E.C.F. and the vector $\{q\}$.

$$\sum_{i=1}^{n}\pi_{j=1}^{r_i} f_{ij,j+1}(q_{i,j}) \qquad (0\leqq f\leqq 1)$$

Besides, the impact scale of A to A is given by the degree of feedback from other factor and of own changes. And, the former is obtained by above-mentioned calculation method, and the latter is obtained by adding 1 to it, since it is thought that it is equivalent to the impact scale in the case which the number of route is 1. That is to say, when there are effective n ways of feedback and the numbers of their routes are r_i $(i=1,2,\cdots,n)$, the impact scale of A itself is given by the following formula.

$$\sum_{i=1}^{n}\pi_{j=1}^{r_i} f_{ij,j+1}(q_{i,j})+f_{11,1}(q_{1,1})$$

In here, $f_{11}=1$

As above-mentioned, on the occasion of clarifing quantatively the impact scales between each E.C.F., we make it a rule to use respectively, ① the number of effective ways for representing the importance of the factor's role in the structure composing of E.C.F., ② the number of routes of effective ways for representing the impact scales and the differences of impact derivery time, and ③ the value which the impact scale 1 of the factor itself is added to it obtained from ① and ② , for representing the impact scale of feedback.

STEP 6 : MAKING THE R MATRIX

The matrix which shows quantatively the interaction betwwen the two E.C.F. , we call the R matrix. And, we make it a rule to obtain the prediction value r_{ij} by the intensive vector $\{q\}$ obtained before, and to make the R′ matrix in the first place. And, in the next place, concerning the E.C.F. which will have large influence on the fish catch of E.O.P., to reach 1 of its maximum prediction value, the R′ matrix is standardized using a coefficient α and the R matrix so-called is made. Namely, the impact of E.I.F. is larger and larger, the fish catch must draw near zero, and at that time the prediction value of E.C.F. must get the maximum value 1.

In R matrix, as shown in Fig.3, the maximum value obtained by scanning toward the row is regarded as the vector $\{r\}$, and the R matrix is predicted and evaluated using this value. And in this paper, we specially take up four factors; the distribution of biological resources, the volume of them, the catching means, the fishery operation scales; which are thought to have direct influence on the final objectives, and we make it a rule to pay attention to the vector $\{r\}$ obtained from these factors.

$$R = \text{E.C.F.}\overset{\text{E.C.F.}}{\begin{pmatrix} r_{11} & r_{12} & \cdots & r_{1m} \\ r_{21} & r_{22} & \cdots & r_{2m} \\ \vdots & \vdots & & \vdots \\ r_{m1} & r_{m2} & \cdots & r_{mm} \end{pmatrix}}$$

$$\{r_1 \quad r_2 \quad \cdots \quad r_m\} \; \{r\}$$

Fig.3. R matrix and vector $\{r\}$

STEP 7 : WEIGHTING OF THE E.O.P.

Fishery kinds and fish species taken up as the final objectives, are weighted beforehand using the coefficient β, γ at every evaluational meshes, based on the fishery statistical materials, etc., in the sea area. For example, when the fish catch of trawlnet fishery forms 16.7 % of the whole fish catches in the objective sea area, we obtain the rates of the trawlnet one at every evaluational meshes in case of setting up the whole one as 100 and the trawlnet one as 16.7. And these rates are used as a coefficient β for weighting the trawlnet fishery. Accordingly, if these coefficients in every meshes are summed up, they come to give 16.7 as a whole in the sea area. Furthermore, in the same way of the case of coefficient β , we make it a rule to obtain the coefficient γ as the weighting value for each fish species.

In this way, at the time of obtaining the weighting values,

in this paper we use the present fish catch in the objective sea area, however, this is nothing but an implication of the set up their fish catches themselves as the criteria for prediction and evaluation, in this M.N. method.

STEP 8 : MAKING THE S MATRIX

The matrix which shows quantatively the causal relation between E.C.F. and E.O.P., we call the S matrix. And, the S matrix is made from the following formula, as shown in Fig.4, using the maximum vector $\{r\}$ and the coefficient β, γ .

$$[S]=\{r\}\cdot\beta\cdot\gamma$$

And, concerning four factors taken specially in here, we make it a rule to obtain the maximum vector $\{s\}$ in their column direction and predict-evaluate the impact scales for the fish catches of each fishery kind and fish species, at every evaluational meshes.

E.O.P.

$$S= \text{E.C.F.}\begin{bmatrix} s_{11} & s_{12} & \cdots & s_{1n} \\ s_{21} & s_{22} & \cdots & s_{2n} \\ \vdots & \vdots & & \vdots \\ s_{m1} & s_{m2} & \cdots & s_{mn} \end{bmatrix}$$

$$\{s_1 \quad s_2 \quad \cdots \quad s_n\} \quad \{s\}$$

Fig.4. S matrix and vector $\{s\}$

STEP 9 : PREDICTION AND EVALUATION

The calculation flow to predict-evaluate using this M.N. method is shown in Fig.5. Such a train of matricies abovementioned is made on every evaluational meshes in the sea area, at each point of times.

Analysis of Sensitivity:

On the occasion of impact evaluation, it is necessary to clarify distinctly in advance what the values obtained finally means. Then, in this paper, they are regarded as the impact moduli which the fishery environments in the present condition - especially, the fish catches - are influenced. And we make analysis of sensitivity about the relation between them, for making clear how the input value of prediction vector $\{p\}$ is equivalent to the output one of impact modulus vector $\{s\}$.

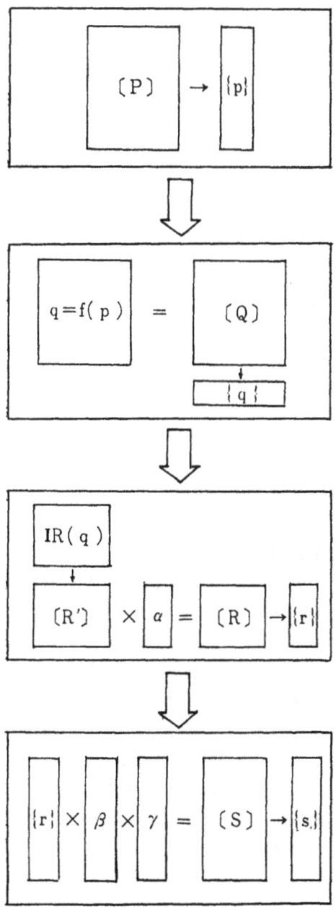

Fig.5. Calculation flow using the M.N. method

Prediction and Evaluation:

In Fig.6, one example of results predicted and evaluated about the impact moduli for the fish catches of trawlnet fishery at the peak time when the dredging and reclamation are carried out in T sea, is shown. Such an impact modulus is obtained at each evaluation time, and is compared with the result predicted when the fishery environment makes progress as it is without the proposal plan, or with the result predicted about its alternatives.

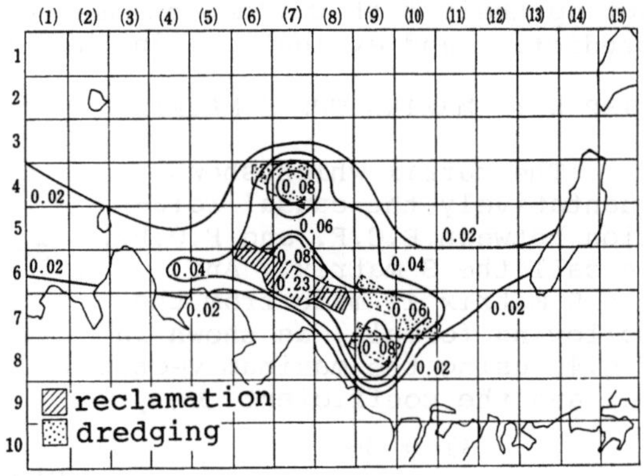

Fig.6. Impact moduli for the fish catches of trawlnet fishery (%, for the whole fish catches)

DISCUSSION

Although the prediction-evaluation values are obtained at a standpoint in safer view point, there may be some problems in using them as the absolute values. However, it is thought that they are fully fit for the relative evaluation of alternatives or each evaluational time and sea area.

Besides, in order to be predicted and evaluated with more scientific exactitudes, it is thought that we must carry out to clarify the causal relations and their environmental structures among these factors and phenomena and to discuss about this M.N. method itself further.

References

1. T.Otsuki : Various environmental impact evaluation methods. Musashino-Shobō, Tokyo 1979.
2. T.Otsuki : The fishery environmental impact evaluation using the M.N. method. Environmental Information Sci., Vol.11, No.2 p.70-78, No.3 p.53-64, 1982.
3. J.C.Sorensen : A framework for identification & Control of resource degradation & Conflict in the multiple use of the coastal zone. College of Environmental Design, University of California, 31 pp. 1971.
4. L.B.Leopold, et al.: A procedure for evaluating environmental impact. Geological Survey Circular, 645, 13 pp. 1971.
5. N.Dee, et al.: An environmental evaluation system for water resource planning. Water Resource Research, Vol.9, No.3, p.523-535, 1973.

Technical Session:
Coastal Zone Utilization
and Management

4. Coastal Process

Changes in Littoral Processes due to Construction of Nearshore Structures

F. Gerritsen and V. Dayananda

Department of Ocean Engineering University of Hawaii at Manoa, Honolulu, HI 96822, USA

ABSTRACT

A method is proposed to verify the change in coastline form due to the construction of nearshore structures. The method involves the calculation of wave diffraction, refraction, leeward of the structure and is based on the condition that the gradient of the driving forces for the littoral currents must be zero.

INTRODUCTION

Construction of nearshore structures in areas with strong littoral drift generally causes changes in the coastal form close to the structure and in most cases it also disturbs the equilibrium of coastal sediment distribution in the region of littoral drift. Prediction of these changes in the coastal configuration is an important aspect in the design of the project.

Along natural shorelines the planform of a beach is often determined by rocky headlands at either one or both extremities of the beach considered (1). The configuration of the planform is governed by the wave climate to which the beach is exposed to. Prediction of changes to be expected due to changes imposed on the nearshore wave climate is a valuable input for the total area planning.

This paper deals with the planform of a stable beach in a man-made or natural situation, where wave diffraction and refraction play an important role in determining or changing the littoral process. Such conditions usually prevail at the downdrift side of coastal structures.

A method, presently available in determining the effect of coastal structures, such as groins and breakwaters on the behavior of the shoreline is based on work of Pelnard-Considère (2), which is based on a solution of the partial differential equation:

$$\frac{\partial x}{\partial t} = \frac{-s}{h}\frac{\partial^2 x}{\partial y^2} \tag{1}$$

in which x and y are the coordinates perpendicular and parallel to the beach (x positive in landward direction). Refinements of this approach has been developed by Bakker (3) and Grÿm (4)

This method generally does not take wave diffraction into consideration and therefore is not applicable to areas downdrift of breakwaters.

This paper deals with wave behavior and littoral drift aspects downdrift of man-made structures or natural headlands. The method is also applicable to coastal lagoons (natural or man-made) behind reefs or offshore breakwaters.

Firstly the methodology will be developed. It is based on the presumption that in a stable shoreline configuration the longshore gradient of the forces that drive the littoral current is zero.

Secondly two examples will be discussed where the proposed method is applicable.

METHODOLOGY

In this analysis the longshore littoral drift is related to the wave induced coastal currents which in turn are related to hydraulic driving forces (and their gradients) acting on the body of water within the surf zone. Effect of winds or tides are not considered. The method applied is valid for a regular wave train approaching the beach from a known direction. It does not take into account wave irregularity or directional spreading.

The principal cause contributing to shoreline changes treated here is the change in the longshore transport. It is realized that in many instances erosion or deposition can result due to changes in onshore/offshore transport caused by changing wave conditions. These changes are considered as dynamic modulations, which can be superimposed on the changes that arise due to changes in the rate of longshore transport.

The calculations are based on stationary conditions. Effects of edgewaves and surfbeat are not included.

It is important to point out that the problem at each particular site is unique because of the specific boundary conditions such as coastal configuration, offshore topography and wave climate.

The wave refraction and diffraction is calculated both outside and inside the surf zone. Snell's law is used for the refraction. In the diffraction calculations a method proposed by Wiegel 1962 (5) is used to obtain wave directions and the Cornu spiral is used for calculating the values of Kd.

Stability conditions for a beach.

A beach will obtain a stable configuration if the gradient of the longshore littoral drift along the beach is zero. This condition can be obtained when the drift equals zero, which is the case when the beach receives no supply from adjacent areas, or when the drift along the shoreline is constant.

It is assumed that the driving force of the littoral drift is the littoral current. Stability will then be obtained when the average longshore current equals zero or has a constant value along the beach. In both cases the condition must be met that the gradient of the driving forces along the beach must be zero or that the forces in this direction must be constant. The constant may be zero when no resultant drift occurs along the beach.

In the following the resultant driving forces on a column of water in the surfzone will be analysed and the criterion of stability defined. The driving forces consist of the various components of the radiation stress, acting on a column of water and the forces induced by a slope (along the beach) of the mean water level.

Driving forces inside the surfzone.

The driving forces inside the surfzone can be categorized as follows:

- components of the radiation stress,
- pressure forces,
- bottom shear stress,
- lateral shear stress.

These forces will be evaluated for a midsurf location, halfway between the breaker point and the shoreline.

Since the problem deals with the surfzone, shallow water approximations may be utilized.

For oblique waves the radiation stress tensor may be expressed as follows:

$$S = \begin{pmatrix} S_{xx} & S_{xy} \\ S_{yx} & S_{yy} \end{pmatrix} = E \begin{pmatrix} \cos^2\alpha + \tfrac{1}{2} & \tfrac{1}{2}\sin 2\alpha \\ \tfrac{1}{2}\sin 2\alpha & \sin^2\alpha + \tfrac{1}{2} \end{pmatrix} \qquad (2)$$

where S_{xx} is the radiation stress component in the x-direction, S_{yy} the component in the y-direction and S_{xy} a shear type component in both x and y directions.

The angle α is the angle between the local depth contour and the wave crest or between the wave ray and the orthogonal on the local depth contour. The x and y directions (respectively perpendicular and parallel to the contour lines) vary along the beach because of the changes in beach direction, but since these changes are usually small no serious errors are introduced in this way.

The forces associated with the above listed phenomena can be evaluated. It is to be emphasized that the component of interest here is the longshore component (in the y-direction). The various forces are:

a. Gradient of the radiation stress component S_{yy} in the y-direction $\left(-\frac{\partial S_{yy}}{\partial y}\right)$

b. Gradient of S_{xy} in the x-direction $\left(-\frac{\partial S_{xy}}{\partial x}\right)$

c. Pressure force from the gradient of the waterlevel associated with the variation of wave set-up in the y-direction.

Neglecting the mean shear stress the wave set-up is obtained from the differential equation:

$$\frac{\partial S_{xx}}{\partial x} + \rho g(h + \bar{\zeta})\frac{\partial \bar{\zeta}}{\partial x} = 0 \qquad (3)$$

where h is the depth with respect to the still water level and $\bar{\zeta}$ the mean wave set-up.

At the midsurf point, the wave set-up, $\bar{\zeta}$ is usually small compared to the depth, h.

Equation (3) may therefore be reduced to

$$\frac{\partial S_{xx}}{\partial x} + \rho g h \frac{\partial \bar{\zeta}}{\partial x} = 0 \qquad (4)$$

The pressure force from the gradient of $\bar{\zeta}$ in the y-direction may be expressed by

$$- \rho g h \frac{\partial \bar{\zeta}}{\partial y} = + \frac{\partial S_{xx}}{\partial y} \tag{5}$$

d. Following a bottom contour in the y-direction the depth remains constant and the resultant pressure force from the bottom is zero.

e. The bottom shear stress in the y-direction is a function of the local littoral current and the orbital bottom velocities induced by waves.

The criterion imposed is that the longshore current in the y-direction is constant (F_f).

f. Evaluating the driving at the mid-surf point, where $\partial V/\partial x = 0$, the lateral friction force may be neglected.

The total driving force F_t can then be expressed by:

$$F_t = - \frac{\partial S_{yy}}{\partial y} - \frac{\partial S_{xy}}{\partial x} + \frac{\partial S_{xx}}{\partial y} + F_f \tag{6}$$

Incorporating the bottom friction force in the constant, the condition of equilibrium becomes

$$- \frac{\partial S_{yy}}{\partial y} - \frac{\partial S_{xy}}{\partial x} + \frac{\partial S_{xx}}{\partial y} = C \tag{7}$$

For a special case (no littoral drift supply to the beach section):

$$- \frac{\partial S_{yy}}{\partial y} - \frac{\partial S_{xy}}{\partial x} + \frac{\partial S_{xx}}{\partial y} = 0 \tag{8}$$

Assuming that at the midbreaker point the wave height is determined by depth only, the first term may be neglected, which gives

$$\frac{\partial S_{xx}}{\partial y} - \frac{\partial S_{xy}}{\partial x} = 0 \tag{9}$$

as a condition of stability for that case.

Using the linear expressions for S_{xx} and S_{xy}:

$$S_{xx} = E\,(\cos^2\alpha + \tfrac{1}{2}) \tag{10^a}$$

$$S_{xy} = E/2 \sin 2\alpha \tag{10^b}$$

the condition of stability can be developed into a relation between the gradient of the wave height in the y-direction ($\partial H/\partial y$), beach slope m and the angle of wave approach α.

For this derivation the following assumptions are introduced:

$h = -mx$, (straight beach slope m)

$H = \gamma h = -m\gamma x$, (constant γ)

$c = \sqrt{g(h + H)} = \sqrt{gh(1 + \gamma)}$

The resultant equation is:

$$\frac{\partial H}{\partial y} = \frac{m\gamma}{2(2\cos + 1)} \left(2 \sin 2\alpha + \frac{\cos 2\alpha}{\cos \alpha} \frac{\sin \alpha_0}{c_0} \sqrt{gh(1+\gamma)} \right) \qquad (11)$$

which is depicted graphically in Figure 1.

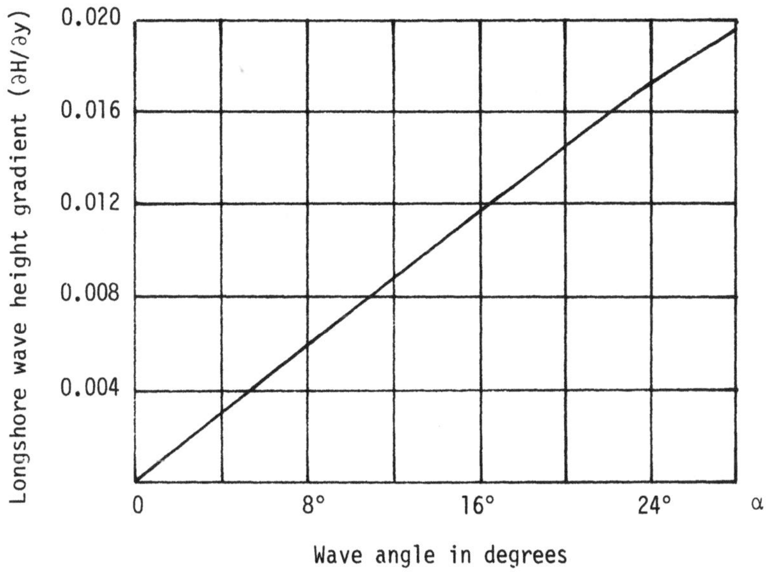

Figure 1. Relation between wave height gradient and angle of wave approach.

The procedure to arrive at a stable beach slope is now an iterative process. Based on an assumed beach configuration equation (1) is verified. Changes are then made into the assumed beach configuration until the conditions of equation (11) are satisfied.

EXAMPLES

Two examples are given to which the above described procedure has been applied.

Beruwela Harbor, Sri Lanka.

Beruwela harbor is a small fishing harbor on the West Coast of Sri Lanka. In order to provide better shelter for the local fishing fleet two breakwaters (A and B, see Figure 2) were built. Due to a change in wave refraction and diffraction leeward of the breakwater rapid shoaling developed immediately leeward of the harbor and erosion of the coast further away from the harbor.

Figure 2 shows the shoreline before and after construction. The change in shoreline configuration could be verified using the method described in this paper. Figure 3 shows the developed shoreline in and near the harbor. Reference is also made to (6).

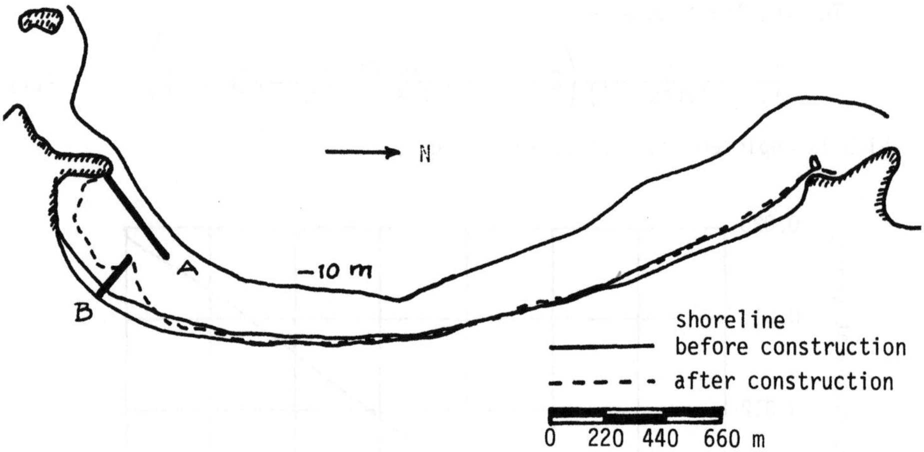

Figure 2: Beruwela Harbor, Sri Lanka.
Shoreline before and after construction.

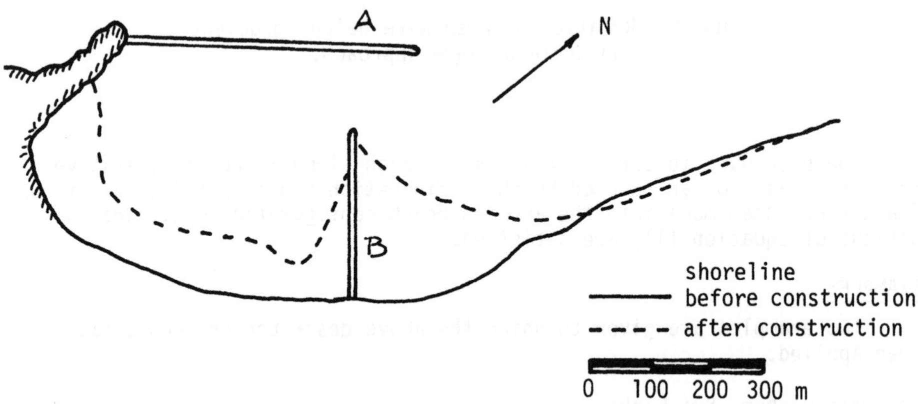

Figure 3: Beruwela Harbor, Sri Lanka.
Sedimentation inside and outside harbor.

Waikiki Beach, Hawaii.

In the middle of Waikiki Beach a section of beach is protected by an offshore breakwater; part of it is relatively high and does not allow wave overtopping under normal conditions, and parts of it have elevations slightly below M.L.L.W. and allow penetration of waves into the lagoon behind the offshore breakwater.

The situation is sketched in Figure 4 and 5. In Figure 5 the 0 and -0.5 m contours are shown, as obtained from a survey and from calculations. In the northern part of the beach the calculated beach form agrees reasonably well with the observed shoreline. At the southern end the agreement is less satisfactory due to the effect of a second gap in the seawall (see Figure 4) which causes the development of a communicative flow condition between the two gaps, giving rise to a different shoreline configuration.

CONCLUSIONS

A method has been developed to predict a stable shoreline configuration downdrift of a natural headland or man-made structures and inside a lagoon.

In applying the method to two specific cases, the harbor of Beruwela in Sri Lanka and man-made lagoon at Waikiki Beach, and in comparing calculated values with observations, useful results were obtained.

Acknowledgements.

The authors wish to thank the students in Ocean Engineering who assisted in field measurements for the example cited at Waikiki and also to the office staff of the department for assistance given in preparing this paper.

References.

1. F. Gerritsen. Beach and Surf Parameters in Hawaii. UNIHI-SEAGRANT-TR-78-02.
2. Pelnard - Consideré, R. Essai de theórie de l' évolution de sable et de galets: IV Journeés de l' Hydraulique, Question III, rapport 1, Juni 1956.
3. W.T. Bakker. The dynamics of a coast with a groin system. Proceedings of the Eleventh Conference on Coastal Engineering, London 1968.
4. W. Grijm. Theoretical forms of shorelines. Proceedings of Ninth Conference in Coastal Engineering, Lisbon 1964.
5. R.L. Wiegel. Diffraction of waves by a semi-finite breakwater. Proceedings of the ASCE, 1962.
6. H.V. Dayananda and F. Gerritsen. Sedimentation studies on a small harbor on a coast of high littoral drift. Proceedings of the 1st International Conference in Coastal and Port Engineering in Developing Countries, March 1983, Colombo, Sri Lanka.

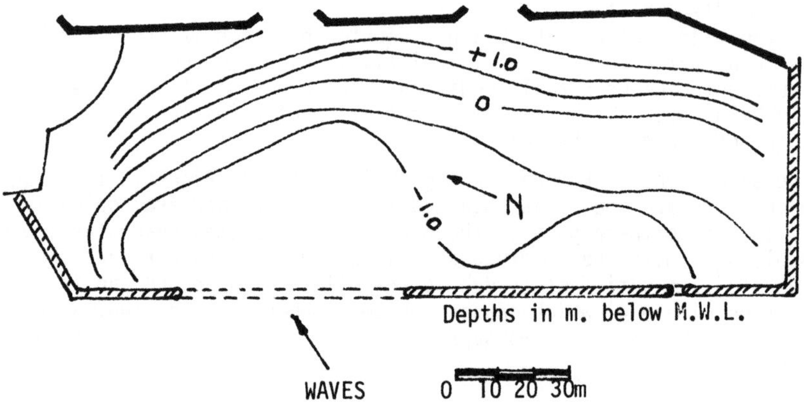

Figure 4: Lagoon behind offshore breakwater
Kuhio Beach, Waikiki, Hawaii

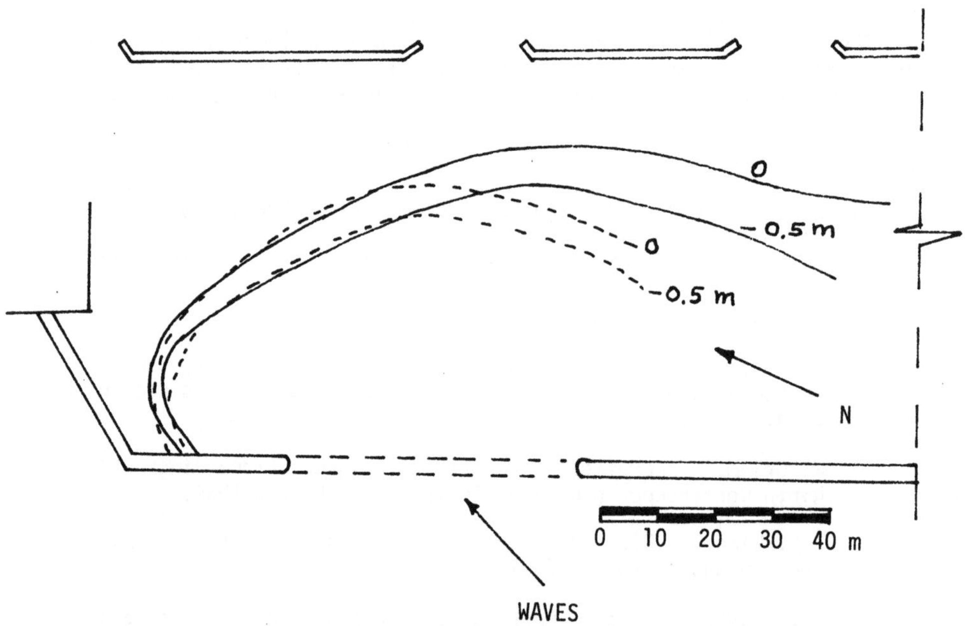

Figure 5: Lagoon behind offshore breakwater
Observed and calculated depth-contours

Long-Term Prediction Model of Phosphorus in Bed Material

T. Horie

Purification Hydraulics Laboratory, Marine Hydrodynamics Division, Port and Harbour Research Institute, Ministry of Transport, Nagase, Yokosuka, 239 Japan

ABSTRACT

A numerical model is developed on the bottom phosphorus behaviors which are closely related with internal production of marine organics. The variation of O-P settling rate and dissolved oxygen are assumed by the record from 1880 to 1983. T-P and I-P in the sediment, PO_4-P in interstitial water, and PO_4-P release rate from sediment surface are simulated to the natural values in a specific point. According to the simulation result, PO_4-P release rate slightly exceeds observed values in summer times, but the seasonal variation and the annual release rate of PO_4-P display agreeable values with the observed. Furthermore, T-P and O-P concentrations of the sediment, and interstitial PO_4-P approach to an asymptotic value to observed values. This model is expected to offer powerful means for the long-term prediction of phosphorus or the estimation of sediment treatment works.

INTRODUCTION

Organic pollution in seawaters is mainly due to the effluent, the internal production, and the release from the sediment. These processes cause great influence to the seawater quality and interact complicatedly with each other. Release flux from the sediment is affected by sediment property, temperature, dissolved oxygen, accumulated substances supplied from seawater, etc. According to the in-situ or laboratory tests, the influence of accumulated substances on release flux is not negligible, and sometimes plays an important role. For good understood of the water quality distribution in the sea, the mechanism of release from the sediment should be analyzed with connection to the influence of accumulated substances. From the above reason, a mathematical model on phosphorus behavior is formulated, based on the data of water quality, sediment, interstitial water and release rate, which are obtained at off-Nishinomia in Osaka Bay.

MATHEMATICAL MODELING

All the essential materials composing organic body, as well as the seawater temperature and illumination, have the possibility to behave as the factors which are influential to the marine eutrophication. Among chemical substances, nitrogen, phosphorus, carbon, iron, manganese, vitamin, etc. are more influential. Of these elements, iron and manganese are sufficiently supplied from the sediment. Therefore, the eutriphication is more sensitive to the content of nitrogen and phosphorus.

With the consideration of water quality distribution in Osaka Bay, the following assumptions are adopted to the model:

i) PO_4-P release (R) is caused by the dispersion in interstitial water,

ii) PO_4-P concentration in interstitial water (C*) varies with bacterial decomposition and dissolution of sedimental organic substances, adsorption to sediment particles, and release into seawater,

iii) PO_4-P is allocated with a certain ratio α into interstitial water and sediment particles,

ix) PO_4-P production in the sediment is caused by the decomposition of newly accumulated substances and existing sediment,

x) O-P supply from seawater to the sediment is carried by settling substance, which is regarded as conservative,

xi) phosphorus moves in the sediment within a certain depth, and

xii) PO_4-P in the overlying seawater mixes with that of the lower seawater in a certain exchange ratio.

The mathematical model formulated by the above assumption is illustrated in Fig.1.

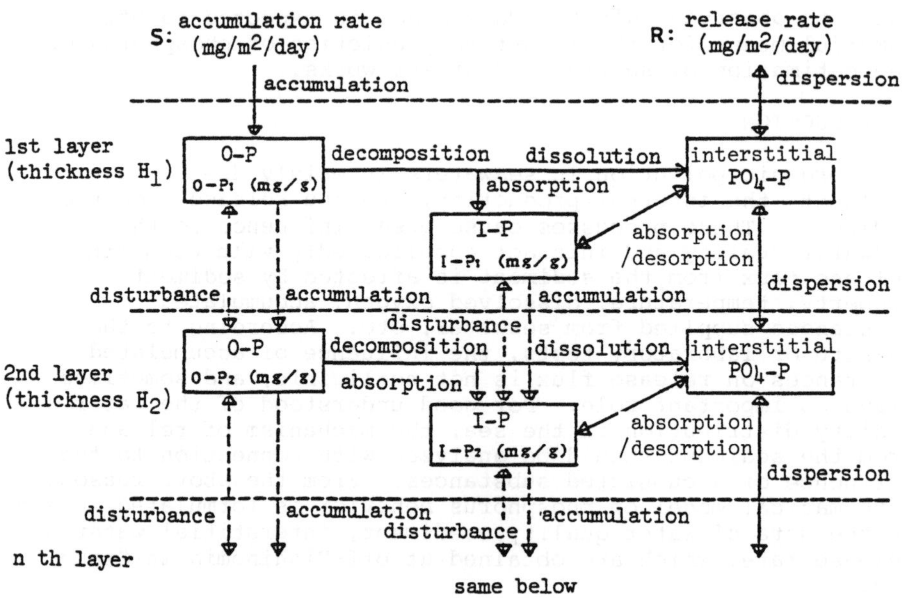

Fig.1 Illustration of the model on phosphorus behavior in sediment

FORMULATION OF EACH PROCESS

Effective thickness of sediment and layer partition

Owing to the local variation of the value of each parameter in the model, the subsequent analysis is made on the sediment samples obtained at a certain number of points in Osaka Bay. According to laboratory tests, the release rate is high in the layers less than 50 cm from the surface, and quite low in deeper layers. Matsumoto et al[1] estimate the accumulation rate of 0.18 to 0.61 cm/year in the inner part of the bay, and annual mean of 0.41 cm by means of ^{210}Pb vertical distribution in the sediment. From these data, the sediment thickness effective to phosphorus release is taken as 60 cm, which corresponds to the accumulation time for about 150 years. Furthermore, this effective thickness of the sediment, 60 cm, is vertically divided into seven sublayers as shown in Fig.2, considering the vertical distribution of T-P, PO_4-P production rate, water content, bioturbation, etc.

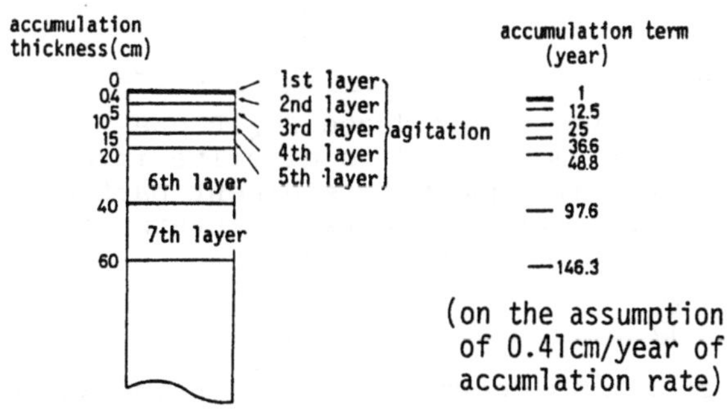

Fig.2 Layer partition of sediment

Settling and accumulation

Accumulation rate is related with the settling rate as

$$T = S \times \frac{1}{\text{sediment ratio}} \times \frac{1}{\text{unit weight}} \times 10^{-4} \times 365 \qquad (1)$$

where, T = accumulation rate (cm/year) and S = settling rate ($SS.g/m^2/day$).

By substituting the sediment ratio = 0.29, the unit weight = 1.12 g/cm^3 and T=0.41 cm into equation (1), S is calculated by 3.9 $SS\text{-}g/m^2/day$. By assuming the settling SS for t days as ΔS_t, the stock in the first layer as $S1_t$, and allowable capacity of SS in the first layer is counted in the following manner:

$$\begin{aligned} & S1_t = S1_{t-1} + \Delta S_t && (\text{ for } S1_t < TS1 \,) \\ & \left.\begin{aligned} & S1_t = TS1 \\ & S2_t = \Delta S_t' \\ & \quad \Delta S_t' = S1_t + \Delta S - TS1 \end{aligned}\right\} && (\text{ for } S1_{t-1} + \Delta S_t \geqq TS1) \end{aligned} \qquad (2)$$

The SS stock in other layer is counted in a similar manner. In the analysis, only SS is related with the accumulation rate, whereas O-P and I-P are transported by SS. O-P and I-P contents in SS are treated as to vary with dissolution and adsorption/desorption.

PO_4-P production

PO_4-P production is considered to be originated from both fresh deposit and old sediment for the first layer,and from only old sediment for the second through fourth layers. PO_4-P production rate for the fresh deposit,r_F, was obtained:

$$r_F = 0.017 \times (\mathrm{OP/SS} - 0.016) \times 1.05^{T-25} \tag{3}$$

where, OP/SS = O-P in 1 g of deposit (mg-P/g-SS) and T = temperature (°C). For old sediment, PO_4-P production rate,r_S, is also obtained as follows:

$$\left.\begin{array}{ll} r_S = 0.0016 \times (\mathrm{OP/SS} - 0.08) \times 1.06^{T-25} & (\text{ for 0 to 5 cm in depth }) \\ r_S = 0.0014 \times (\mathrm{OP/SS} - 0.13) \times 1.06^{T-25} & (\text{ for 5 to 10 cm in depth }) \\ r_S = 0.00035 \times (\mathrm{OP/SS} - 0.14) \times 1.06^{T-25} & (\text{ for more than 10 cm }) \end{array}\right\} \tag{4}$$

Adsorption and dissolution of produced PO_4-P

A certain amount of produced PO_4-P due to the dissolution of organic matters is assumed to dissolve into interstitial water and the rest to adsorp to sediment particles. By defining allocation factor α between interstitial PO_4-P,C_*,and adsorped I-P, IP, α is expressed as:

$$\alpha = \frac{C_*}{\mathrm{IP}} \tag{5}$$

This factor α is dependent on the temperature and the dissolved oxygen. α for 0 to 5 cm in depth is expressed as

$$\alpha = 11.5 \times 0.748^{\mathrm{DO}} \times 1.02^{T-20} \tag{6}$$

Movement of substance in sediment

Regarding the physical property of sediment in the layer less than 15 cm in depth as homogeneous and by assuming that SS contained in one half of the agitation layer moves by 15 cm in distance for four years, the sediment in this layer is completely mixed[2]. Thus agitation coefficient is estimated.

O-P in accumulated SS gradually reduces its amount with the time elapse and determines the vertical distribution of O-P in the sediment. To the undisturbed lower layer, deposit (ΔS_n), O-P in the sediment(P_n), SS(S_n), and PO_4-P production rate(r_n) after n-th day are estimated as

$$\left.\begin{array}{l} P_n = P_{n-1} - r_{n-1} \times S_{n-1} + \mathrm{OP} \times \Delta S_n \\ S_n = S_{n-1} + \Delta S_n \\ r_n = f(P_n/S_n, T_n) \end{array}\right\} \tag{7}$$

where, OP = O-P in deposit and T = temperature.

The dispersion of interstitial PO_4-P between layers is expressed as a general form of the function of vertical concentration gradient. By assuming that temperature compensation factor of dispersion coefficient is equivalent to that of molecular dispersion coefficient of cation or anion[3], that is, 1.03^{T-25}, the dispersion coefficient at arbitrary temperature, D_T, is expressed by

$$D_T = 1.41 \times 10^{-4} \times 1.03^{T-25} \ (m^2/day) \qquad (8)$$

Water ratio and unit weight of sediment

Vertical distribution of water ratio, w, and the unit weight, V_d, of sediment at off-Nishinomia have been obtained by laboratory test as

$$\left.\begin{aligned} w &= 1 - 0.268 \times z^{0.0689} \\ V_d &= 1.03 + 0.168 \times z^{0.0801} \end{aligned}\right\} \qquad (9)$$

where, z = distance below the sediment surface (cm).

CALCULATED RELEASE RATE AND VERTICAL DISTRIBUTION OF PHOSPHORUS IN SEDIMENT

Computational conditions

For the verification of the model, the values of O-P settling rate and DO in the bottom water are assumed for the period of previous one hundred years. The settling rate of phosphorus is assumed to be propotional to the phosphorus in the seawater. The long-term variation of phosphorus has been estimated on the basis of data observed by the Kobe Marine Observatory in 1934-35, 1952-53, and 1974-79, and the data obtained by prefectural governments after 1979. This result is shown in Fig.3.

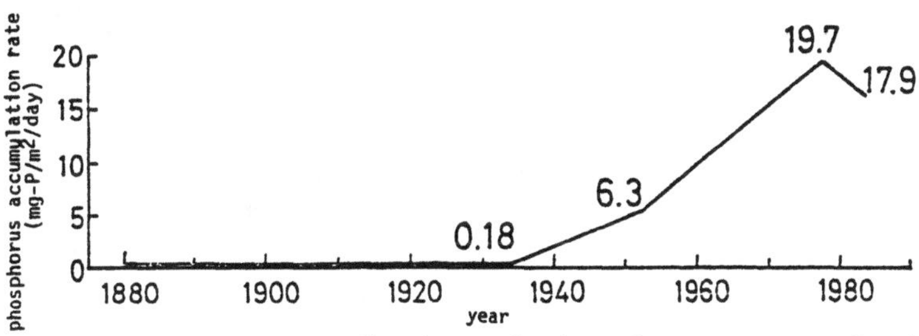

Fig.3 Long-term variation of phosphorus accumulation

Furthermore, the annual variation of O-P in the lower layer has been obtained by the simulation results due to a two level nutrient-cycle model[4]. Also, DO has been obtained for the almost same period with that for O-P. DO in the lower layer is, therefore, given as

$$DO = DO1 - (DO1 - DO2) \times \beta \qquad (10)$$

where, DO1 = DO at the salinity of 31.6 ‰ and a given temperature, DO2 = observed DO (see Fig.4), and β = DO reducing factor to the standard year.

β is decided by the records as shown in Fig.5.

The annual temperature variation is shown with the broken line in Fig.6.

Results

Numerical simulation was executed for the period over one

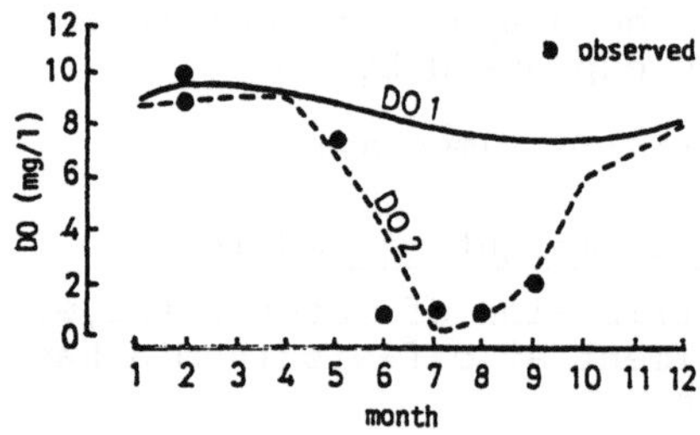

Fig.4 Observed DO and saturated DO

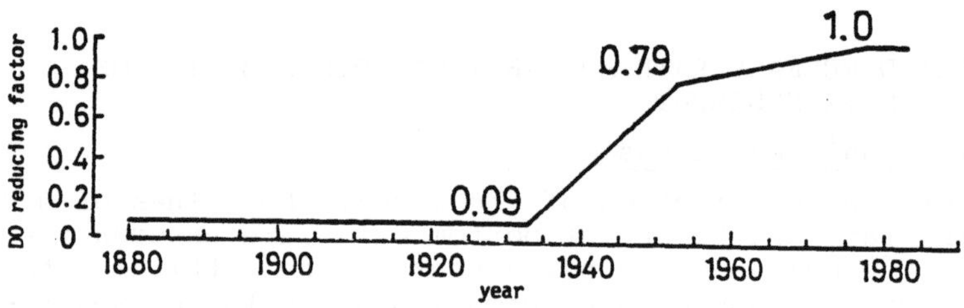

Fig.5 Long-term variation of DO reducing factor

hundred years from 1880 to 1983 by applying the above conditions and time increment $\Delta t = 0.1$ day[5]. The calculated annual variation of PO_4-P release rate is shown in Fig.7 for the most recent year of 1983.

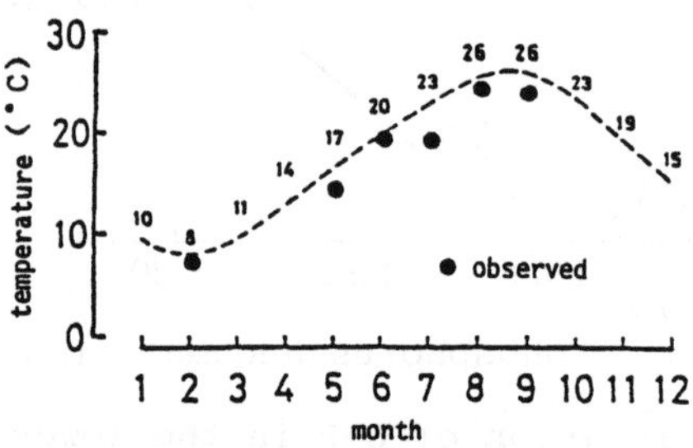

Fig.6 Seasonal variation of temperature

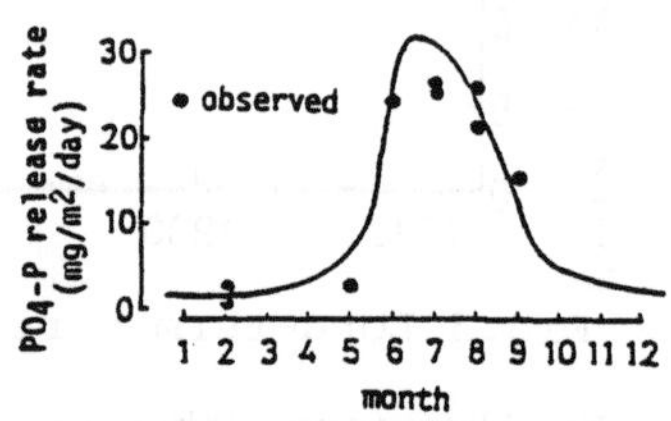

Fig.7 Seasonal variation of release rate

The calculated release rate is slightly greater than the observed value, but the annual variation generally agrees with the observed one. The release rate integrated with the time range of each year is shown in Fig.8. PO_4-P release has a close relation with the settling load of O-P as shown in Fig.3, and rapidly increases after 1950. The observed annual release from 1979 to 1981 is in the range of 3 to 4 g-P/m^2/year, and the calculated value also lies on this range.

The vertical distribution of calculated T-P, O-P, I-P in the sediment and interstitial PO_4-P are shown with the observed values in Fig.9. According to these results, the vertical distribution of calculated phosphorus in the lower layers from the first to the fourth well agrees with the observed values, with the exception of interstitial PO_4-P. The calculated phosphorus lower than the fifth layer is about one half of the observed value; the result of calculation seems to be rather sensitive to initial values, iteration times, and other parameters.

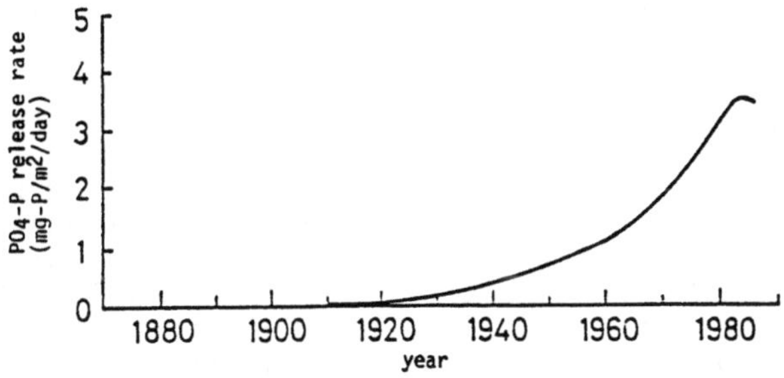

Fig.8 Long-term variation of release rate

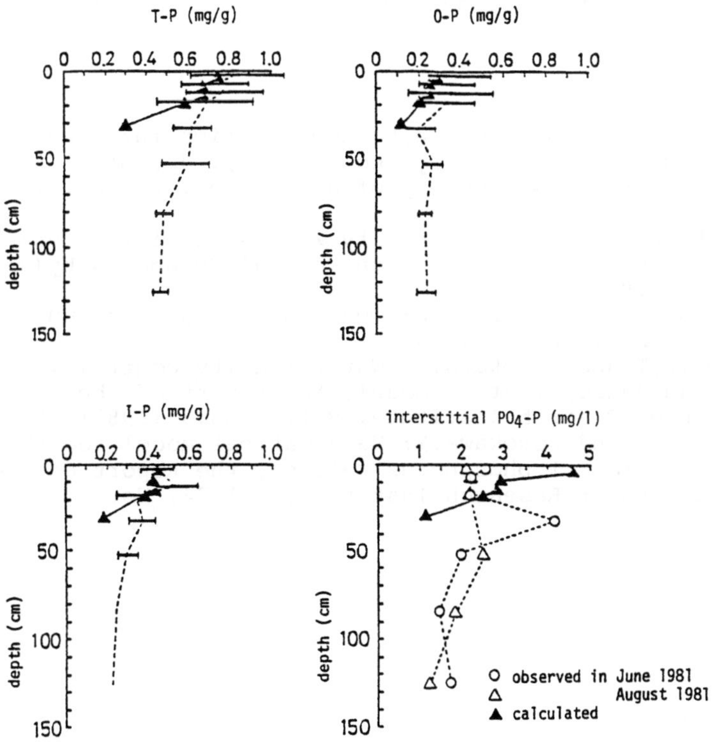

Fig.9 T-P, O-P, I-P in sediment and interstitial PO_4-P

CONCLUSION

i) the annual variation of PO_4-P release rate and the integrated release rate over the year are well simulated to observed values with this model,

ii) the vertical distributions of T-P, O-P, I-P in the sediment are well simulated in the upper layer, but in the lower layer they are underestimated because of the effect of initial condition, iteration times, etc. Interstitial PO_4-P is overestimated in the upper layer,

and

iii) this model can apply to the long-term prediction of phosphorus or to the estimation of sediment treatment works.

ACKOWLEDGMENT

This investigation was financially supported by the Third District Port Construction Bureau and the Environmental Protection Division of the Bureau of Ports and Harbours, Ministry of Transport. Most of field data applied here were obtained by the Third District Port Construction Bureau , Ministry of Transport. I express my great gratitude to Prof.H,Nakanishi and Associate Prof.M,Ukita of Yamaguchi Univ., Prof.T,Horiguchi of Tokyo Metropolitan Univ. and Associate Prof,T,Morioka of Osaka Univ. who gave a great deal of most appropriate instructions on the fulfilment of this investigation.

References

1. Matsumoto,E and Yokota,S: Accumulation rate and heavy metal pollution in Osaka Bay sediments(J), Journal of oceanographical Society of Japan, Vol.18, No.2(1981) 108-115.
2. Kikuchi,T; Benthos activity with special reference to bioturbation, Bulletine on Coastal Oceanography(J), Vol.18, No.2(1981) 66-77.
3. Kadokai,S and Noriki,S; Oceanographical Chemistry(J), Sangyo Tosyo Pub. Ins., (1983).
4. Horie,T and Hosokawa,Y; Water-quality controlling factors in an eutrophication model, the Report of the Port and Harbour Resarch Institute, Vol.22,No.3 (1983) 159-205.
5. Horie,T and Hosokawa,Y; Mathematical model for the prediction of phosphorus release, the Report of the Port and Harbour Research Institute, Vol.24, No.1 (1985).

Environmental Assessment of Waste Water Disposal from Sewage Treatment Plant

A Case Study

A. Murota and K. Nakatsuji
Department of Civil Engineering, Osaka University, Yamadaoka, Suita, Osaka, 565 Japan

ABSTRACT

A case study of the environmental assessment related to the construction of sewage treatment plant in shallow coastal waters is described. On the basis of results of field surveys and hydraulic model tests for examining the actual flow structure in the coastal sea region, a numerical simulation of the flow and diffusion process has been performed to make an assessment of developmental effects on the coastal environment.

INTRODUCTION

Recently, Japan has launched many extensive reclamation programs for nuclear plants, treatment of waste and airport in the coastal sea region. Such developments, including discharges of waste water would have such major impacts on the coastal environment that the assessment of their effects must be made in advance. A significant problem to be resolved is the flow and diffusion process of sea water in the proper coastal sea region. Since it is so complicatedly affected by many elementary processes with different time and space scales (for example, irregular wave, tidal current, wind, density flow and constant current component), it is very important to find out the elementary physical process that dominates the actual fluctuations in coastal environment by field surveys and fundamental experiments, and to develop the adequate modeling for the assessment.

In this paper, the writers selected the reclamation project for New-Tarumi sewage treatment plant in Kobe City, Japan, as 'a case study' in order to consider the overall study of hydrospheric environmental assessment and to discuss how the assessment connected with its development were carried forward.

PROJECT DESCRIPTION

The old town of Kobe City lies from east to west located between the Rokko Mountains and Osaka Bay in the Seto Inland Sea. With increasing development of large-scale new towns in the inland area, the construction of a new tertiary treatment plant, which would be capable of 245,000 m^3/day for 400,000 peoples, was necessary in the old town area because of the above-mentioned topographical condition. In 1974, the Kobe City Authorities planned the reclamation of 360,000 m^2 land required (250m offshore and 1,400m shoreline) in the coastal sea region which is shallower than 10m in depth.

The concerned coastal sea region is located near the Akashi Straites in the northern part of Osaka Bay as shown in Fig. 1. It is characterized by the strong reciprocating tidal current through the deep narrow straites and the periodically appearing circulation, which is caused by interactions of the tidal current with topographycal conditions. As a consequence, an unique tidal pattern is always observed, in which the tidal current goes westwards along the coastline for 8 hours during one tidal period. Such a current pattern is estimated to constitute the predominant factor for the mixing and diffusion process in the coastal sea region.

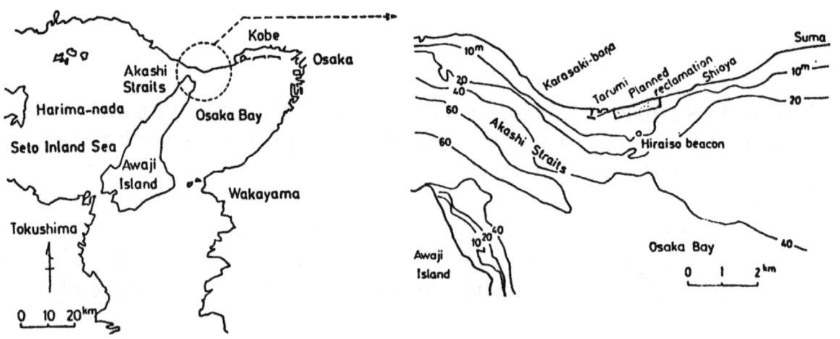

Fig.1. Site for New-Tarumi plant at Osaka Bay

PROCEDURE FOR ENVIRONMENTAL ASSESSMENT

When making an assessment of the developmental effects on the coastal environment, it is necessary to examine the flow and diffusion characteristics of the concerned sea region by means of field surveys before anything else. In particular, it is very important to evaluate what kind of physical process is fundamental in making a prediction test or a simulation model to describe the actual phenomena. For this purpose, field surveys are indispensable, as they enable the determination of the time and space scale of the fundamental phenomenon.

A flow chart, describing the making of an assessment of the flow and diffusion process in the coastal sea region is shown in Fig. 2. The assessment described herein was performed according to this flow chart.

FIELD SURVEYS

The City Authorities carried out field surveys in the concerned coastal sea region twice in 1973 and 1975 in addition to the regular four measurements per year. The results are summarized as follows.

Tidal current

Figure 3(a) shows the surface layer flow pattern of the constant component of tidal flow. It is characterized by the existence of a circulation at Shioya-Tarumi coast offshore of the reclamation point. It is strongly reflected by the coun-

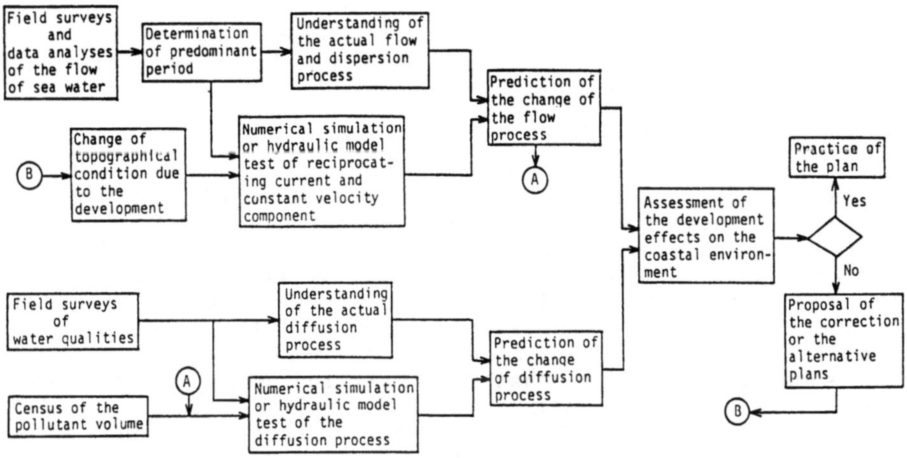

Fig.2. Procedure for prediction and assessment in coastal environment

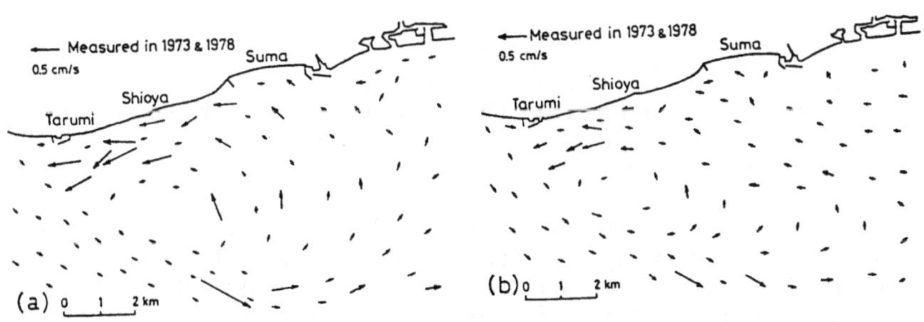

Fig.3. (a) Constant velocity component of tidal flow, and (b) tidal current at edd tide at Tarumi coast

terclockwise rotating flow distinctly observed in the tidal current at the ebb tide shown in Fig. 3(b). Since old times, fishermen said that there is no ebb tide in Shioya-Tarumi coast. That is, the sea water is believed to flow westward during one tidal period regardless of the reciprocating action of tidal currents at the Akashi Straits. It is made clear from the measured records of the flow of sea water that it flows westward for about 8 hours while it flows weakly eastward for 4 hours. This unique flow pattern is estimated to originate in the formation of the wake region which is due to the peaked topography of Karasakibana headland on the west of the reclamation point. As a consequence, as the sea water of clear water quality in Harima-nada Bay, which flows toward the east through the Akashi Straits, goes around and purifies the sea water at Suma coast adjacent to the concerned coastal sea region, its area is the only recreational beach for sea bathing in Osaka Bay.

On the other hand, there exists a clockwise circulation peculiar to Osaka Bay at the southern-side of the concerned sea region. Its flow is caused by interactions of the large scale tidal currents with the sea bottom topography in Osaka Bay. Both circulations have important effects on the diffusion process of water qualities as well as the flow pattern of constant current component.

Diffusion coefficients

In order to evaluate the diffusion phenomena in the concerned sea region, an airplane observation of many drifting plates on the water surface was performed on Oct. 9, 1975. Lagrangian analyses of relative displacements between drifting plates made it clear that the convective contributions due to the tidal current and the counterclockwise rotating flow are predominant in the dispersion process and that the diffusion coefficients are of the order of 10^4 cm^2/sec at the most. It was sometimes observed that the drifting plates flowed in a single line, like a walk of ducks. In such a case, the diffusion coefficient evaluated becomes a negative value, which is physically meaningless. It indicates that upwelling and downward currents exist and that the coastal area has a three-dimensional complexity and active variations, affected by the geometrical irregularity of the sea bottom and coastal line.

Water quality

Every season, measurements of plane distributions of water qualities have taken frequently in addition to point-measurements for evaluating their variations. With respect to the plane distribution of the COD-concentration, it is 1.4 - 1.6 ppm at the eastern area while it is less than 0.8 ppm at the western one. This survey makes clear that the water quality distribution at the concerned sea region is determined by the intense mixing of the clear sea water through the Akashi Straits and the dirty water of Osaka Bay. Near the reclamation point, it is of 1.0 - 1.2 ppm, which is lower than the standard value of 2.0 ppm at Osaka Bay. As a whole, the concerned coastal sea region is considered to have much dispersive activities.

MODELING OF FLOW AND DIFFUSION FIELD

According to the results of field surveys, the following items must be taken into consideration at least in modeling the flow and diffusion process :

1) to grasp what degree the counterclockwise circulation would be influenced by the geometrical change due to the reclamation project.
2) to treat tidal currents as unsteady phenomena having a fluctuating cycle of less than one day.
3) to clearly estimate an ever-changing pattern of water qualities.

In order to understand the complicated flow characteristics, a numerical simulation is not enough to depict the local change of currents in the immediate vicinity of the reclamation point, so a hydraulic experiment must be used supplementarily to give some additional explanations. Therefore, hydraulic model tests of the steady flow were used to evaluate the geometrical effects of the reclamation on the tidal currents, while numerical simulations were used to predict the flow and diffusion process over a wide sea region.

Hydraulic model test

In order to correctly describe how the geometrical shape changes due to the reclamation, affects the formation of a wake region, hydaulic model tests of a steady flow were performed in a 27m x 4m hydraulic basin for some tidal ranges. Its scale is 1/1000 horizontally and 1/160 vertically. An example is shown in Fig. 4. It is a case of the ebb tide in which the effects of reclamation are most apparently observed. The following results were obtained by the hydraulic model experiments : (1) In the ebb tide, the tidal current trends to flow off the shore to divide the existing circulation into halves due to the reclamation and the Karasakibana headland. (2) With regard to the whole tidal range, the geometrical shape change due to the reclamation has little influence on the tidal pattern in the concerned coastal sea region.

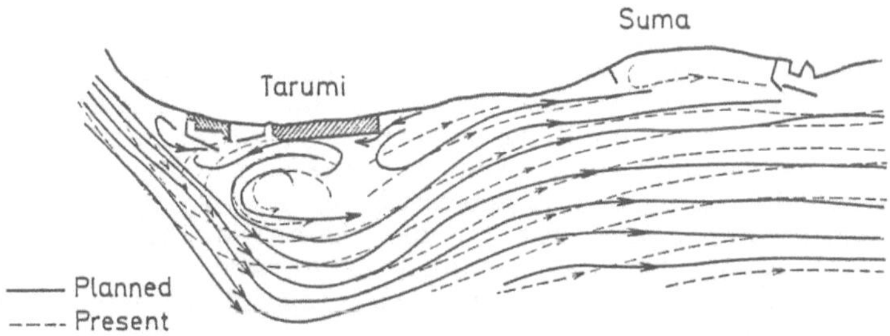

Fig.4. Changes of tidal currents during ebb tide

Numerical simulation

Since the tidal flow at the concerned sea region is active and complicated and the time scale of the simulation is less than a month, it is not necessary to consider the three dimensionality of the stratification. The equations governing such flow and diffusion field will then be the shallow water equations : these are well known and need not be restated in this paper.

A range of pollutant dispersion discharged from the sewage treatment plant can be estimated within a radius of 3.5 km by means of Nitta's formula, which enables the aim for the model test domain to be fixed. The model schematization is given in Fig. 5, in which its boundaries encounter the open sea in three directions due to the convenience of boundary conditions. With regard to the nature of the tidal flow equation solved, two boundary conditions appear to be necessary. When the large scale clockwise circulation at Osaka Bay is simulated, the two flow components of the tidal current are specified at the offshore boundary, Bs, while, when the reciprocating tidal current through the Akashi Straits is simulated, the water surface level is specified at the eastern and western boundaries, Be and Bw. Consequently, as the numerical model became more unstable and divergent, an alternative direction implicit (ADI) method was used as a numerical scheme in order to avoid the computational instability. The basic equations were discretized to conventional rectangular grids of 200 m and the time increment of 30 seconds by means of the finite difference methods.

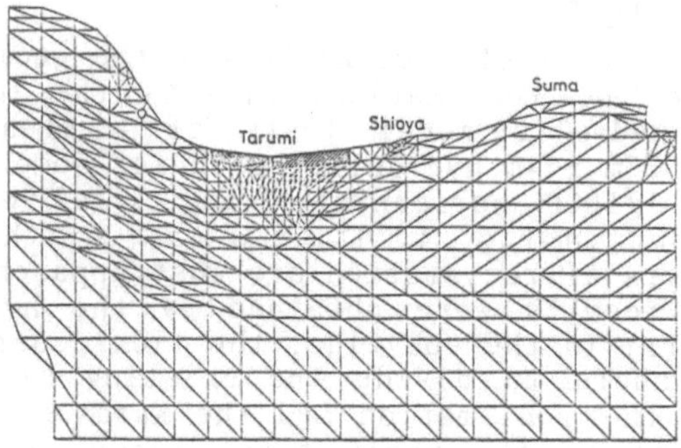

Fig.5. Computational area schematization : finite elements

In the diffusion simulation, the finite element method was applied to evaluate the details of local diffusion process in the vicinity of the outfall, in which arbitrary shaped finite elements were used to describe complex and localized phenomena. Its network was made up of 805 triangular elements of 75m to 600m, which is schematized at Fig. 5. It is, of course, well arranged after due consideration of tidal currents. Besides, the pollutant discharged from the sewage treatment plant and rivers was distributed uniformly within the proper element.

<u>Examples of computation</u>

Figure 6 shows the tidal current and the distribution of COD-concentration at 2 hours after the ebb tide. The discharged outfall is located at the distance of 100m offshore from the reclamation point and the discharged waste water is of 3.26 ppm. The counterclockwise rotating flow, which can be correctly simulated, transports the discharged water in the offshore direction in a tongue-shaped way. This is probably due to the clockwise circulation given at the offshore boundary, Bs, which was expressed in terms of a superposition of tidal harmonic constants based on field surveys.

When examining the tidal currents in the whole tidal range, the geometrical change due to the reclamation, has little effect on the fluctuations of tidal patterns as compared with the case of the present geometrical condition, with the exception of the ebb tide. One of the distinct features in the reclamation case is that the counterclockwise rotating flow moves about 400m in the west-south-west direction at the ebb tide, which is in good agreement with the results of hydraulic model tests. Its scale and its dominant region are, however, not changed.

An ever-changing pattern of the COD-concentration varies with the movements of tidal current. A substantial problem remains in determining what time scale is effective for evaluating the effects of waste water disposal on the coastal environment. After the repeated excursion of more than 60 tidal ranges in such a limited domain, the COD-concentration pattern is determined by the given boundary values. Since this simulation was intended to evaluate the extent dependent of waste water disposal, the computation was stopped when the

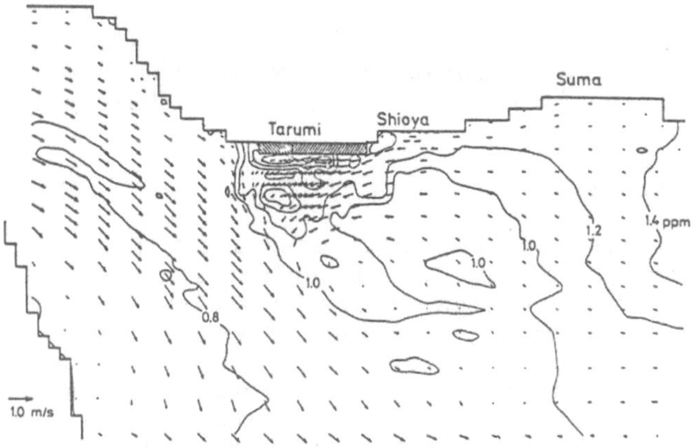

Fig.6. Tidal current and distribution of COD-concentration simulated at 2 hours after ebb tide

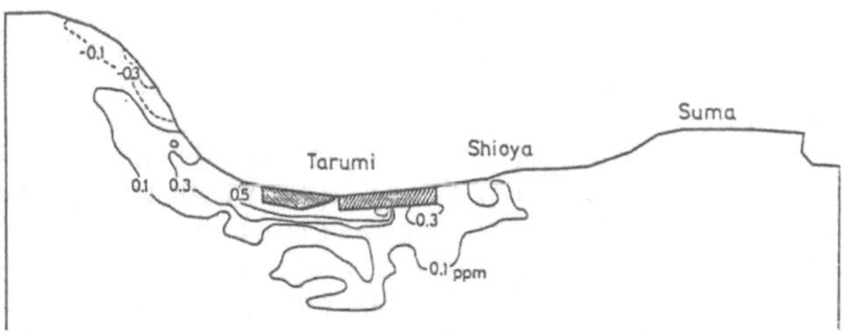

Fig.7. Evaluated extent dependent of waste water disposal

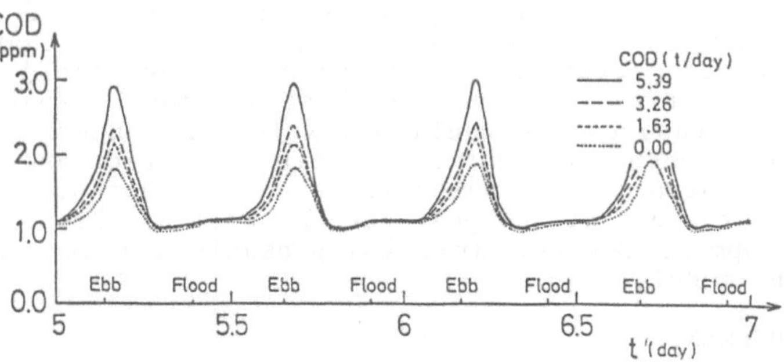

Fig.8. Variation of COD-concentration at Hiraiso beacon

time-averaged COD-concentration pattern of 12.5 hours converges to a degree of less than 5 percent. The comparison of time-averaged pattern between the present situation and the reclamation case is shown in Fig. 7. It becomes clear that the extent of concentration increasing is limited in the immediate vicinity of the outfall.

Figure 8 shows the variation of COD-concentration of Hiraiso beacon, in which the solid line is the present situation and the symbols indicate simulated results for several discharging conditions. Apparently the variation is also true to be coincident with the tidal periodicity for all cases. A peaked value is observed to take place at 1 hour after the ebb tide. It is well reflected in the phenomena that the discharged waste water is stagnant temporarily in the wake region and it suddenly flows offshore accompanied with the growth of the counterclockwise rotating flow in the ebb tide, which results in an periodical variation of the COD-concentration at Hiraiso beacon.

CONCLUDING REMARKS

The detailed environmental assessment made proposals for some improvements of the lay of reclamation land, the location of outfall and the design of discharged system. The Kobe City Authorities, therefore, examined these proposals and made fundamental alternations of the design and layout of the sewage treatment plant for the environmental conservation. In particular, a submerged multiport diffuser was examined by means of hydraulic experiments. After that, the reclamation and the construction of plant were put into practice during 1979 - 1983 along the line of assessment results, and operation was begun in 1984.

The field measurements on Oct. 9 1983 showed that the floating water discharged from the diffuser had salinity of 30.8 millesimal at the water surface in no tidal current advection. It corresponds to a dilution of 1 to 22. These measurements verified that the initial dilution of 1 to 7 demanded for the environmental conservation was obtained over all tidal range and that the construction of plant had no direct impacts on the coastal environment.

In this paper, the writers do not go into the details of the assessment techniques, but rather emphasize the concepts and the methodology applied. Although obviously some uncertainties still remain in regard to the modeling of physical coastal process, it clears throughout this study that mutual repetition of feedback is required among field survey, hydraulic experiment and numerical simulation. In particular, only field measurements would allow for the understanding of the physical phenomena in the coastal sea region and also for model developments. For this purpose, in the authors' opinion, field measurements at the poststage as well as the prestage of the development project are indispensable for improving the assessment model.

ACKNOWLEDGEMENT

This is one of the environment studies entrusted by the Kobe City Authorities. The hydraulic model experiments were mainly carried out by Dr. T. Hirata. The writers wish to express their thanks to Dr. T. Itoh and Dr. A. Wada of Central Research Institute of Electric Power Industry for their kind discussions in preparing a flow and diffusion model system.

Enhancement of Oceanic Productivity through Application of Municipal Sewage

D.A. Segar, E. Stamman and P.G. Davis
SEAMOcean, Inc., Wheaton, MD 20902, USA

ABSTRACT

While it is widely recognized that the application of sewage, treated sewage, or sewage sludge on agricultural land can be beneficial (primarily as a fertilizer), the disposal of these same materials in the ocean is typically viewed in a negative light. Yet, when sewage sludges and effluents are introduced in a properly controlled manner, they can be of benefit to large areas of the ocean. For example, benthic biomass and growth rates of commercially-valuable species have been shown to be stimulated by the input of sewage sludge and effluents into certain ocean areas. Increases in fish productivity have also been associated with these inputs. The application of sewage sludge may be particularly beneficial in those areas which are naturally nutrient-deficient. Further, the potential problems of effluent or sewage sludge application in the ocean are similar to, although probably less severe than, those encountered with land disposal (i.e., toxics, pathogens, and amount applied to a given area).

INTRODUCTION

For many generations the nutrient (nitrogen and phosphorus) value of treated and untreated municipal sewage and sewage sludge (collectively refered to as sewage) has been used to enhance the growth of plants in agricultural production, home gardens, and forests. Sewage has also been used as a soil conditioner and ground cover to accelerate reclamation of strip-mined and other land. While the public in many nations has serious concerns about the use of sewage for some of these applications, the concept that sewage can beneficially enhance the productivity of land plants is generally accepted. In contrast, the concept that the nutrient and nutritive value of sewage could be beneficially utilized in the ocean under appropriate conditions is equally valid and has been largely ignored. Since the scientific community has ignored this potential beneficial use of sewage in the ocean, there is an almost universal perception that sewage which is applied to the land is beneficially recycled, while sewage applied to the ocean is dangerous and damaging. The inconsistency in these two views is obvious.

Although both land and ocean disposal of sewage have been intensively researched, the attitude difference toward these two management alternatives has resulted in different research hypotheses and approaches being applied to the two alternatives. Research into the land application of sewage, particularly composted sludges, has been aimed at developing the knowledge necessary to optimize the practice for maximum benefit with minimum environmental damage and human health risk. Potential problems, when found, have led to recommendations for lower application rates, changes in composting methods, or other such improvements in management practice. In contrast, research into the ocean application of sewage has been dominated by studies of traditional management practices in a search for adverse environmental or human health effects, with the philosophy that any adverse effect found, no matter how small, would justify a higher level of treatment or even an end to the practice.

It has become clear that there are environmental effects and human health risks inherent in ocean application of sewage. However, these are no more serious than the adverse effects of improper land application, and it is apparent that these adverse effects could be reduced or eliminated by optimizing the methods of ocean application and controlling the quality of sewage, particularly the concentrations of toxic metals and organic compounds entering the sewers.

The principal reason for the treatment of sewage is to remove BOD and suspended solids. The oxidizable particulate organic material (POM) in sewage is the principal source of benefit to the marine food chain when sewage is introduced to the ocean in appropriate amounts and locations. Therefore, if the beneficial use of sewage in the ocean can be properly controlled and managed, the need for expensive high levels of treatment may be eliminated or reduced, and simpler treatment methods adopted to specifically condition the sewage for ocean application.

CONCEPTUAL APPROACH TO BENEFICIAL OCEAN APPLICATION

Municipal sewage essentially consists of a low concentration of natural organic particles (POM) suspended in water. POM contains trace metals and trace concentrations of synthetic organics that are present in all foods, the human body, and human wastes. Most municipal sewage also contains small additional quantities of trace metals and synthetic organics that are derived from other sources, including domestic cleaning products, street runoff, and industrial wastes. Sewage also contains micronutrients, such as nitrogen and phosphorous compounds.

Sewage can be beneficial when applied either to the land or to the ocean, if the micronutrient elements are utilized to provide the growth requirements of plants in nutrient-limited environments. In the ocean, an additional benefit can be realized from sewage since the POM can provide food for appropriate consumer organisms. In obtaining these benefits either on land or in the ocean, it is important to minimize any offsetting adverse effects caused by excessive concentrations of toxic trace metals and synthetic organics, or by microbial contaminants.

SUPPORTING EVIDENCE

There is considerable evidence that the application of sewage to the ocean can lead to substantial increases in phytoplankton and zooplankton productivity within the water column (1,2). This increased productivity, in turn, provides food for an enhanced fish population. For example, Sindermann (3) concluded that increases in marine productivity and abundance of fish stocks in coastal areas, including the Southern North Sea, may well be related to the fertilizing effects of domestic sewage. Sindermann noted that high nitrates and phosphates outside the Thames estuary were derived from the sewers of London, and that the catch of fish per unit area had been found to be double the corresponding catch for the rest of the North Sea, the English Channel, and the Kattegat-Skagerak region. The higher average catch was attributed to nutrients from London sewage.

The extreme temporal and spatial variability of most fish populations and the dominance of other effects, such as overfishing, make studies of the effects of sewage on fish populations difficult. However, it is commonly-known fishermen's folklore that fish are more abundant around sewage outfalls. An intensive field study of fish populations at outfalls in Southern California confirms that substantially increased fish populations do occur in the vicinity of some outfalls (4). The same study showed that, at some of the largest outfalls, fish populations were reduced in a small area immediately adjacent to the outfall, but were increased in a larger zone which surrounded this inner area. These data suggest that the application rate of sewage to the area immediately adjacent to the largest outfalls is too high.

Observations of the benthic fauna in this same area (discussed below) support this conclusion. Since sewage discharges at these California outfalls are confined to depths below the upper layer of intense primary production, it would appear that the observed enhancement of fish populations is due to utilization of the sewage POM in the marine food chain, rather than to enhanced primary production stimulated by sewage nutrients.

Unfortunately, relatively little is documented concerning the availability and nutritive value of sewage POM to pelagic animals. Nevertheless, it is clearly possible for properly-applied, treated sewage effluent and municipal sludge to increase fisheries production in the ocean, either by enhancing primary production through nutrient fertilization and/or by enhancing secondary production through incorporation of the POM into the food chain.

In most areas where sewage is introduced into the ocean, the water column is sufficiently deep that primary productivity is insignificant or does not occur at the bottom of the water column. Therefore, the benthos in these areas relies upon the transport of fixed organic carbon from the euphotic surface layers to the bottom. However, the POM contained in sewage can provide a food source for the benthos, which utilizes it in the same way as the naturally-produced, fixed organic carbon transported to the benthos from the surface layers. Depending upon the characteristics and concentration of the sewage and the location of application, the POM in sewage can be utilized by both benthic filter-feeders and deposit-feeders.

Many municipal wastewater outfalls throughout the world are located on the continental shelf, and the effects of these discharges on the benthos within the dispersal area surrounding these pipelines have been intensively studied. The effects of municipal sludge POM which is transported to the benthos have also been intensively studied in areas where municipal sludge is dumped or discharged. These studies of the benthic effects caused by municipal sludge and treated effluent discharges have been more intensive than studies of pelagic effects, since the benthos are less variable in their distribution and are sessile.

The addition of POM to sediments through ocean outfalls is of particular interest, since it is possible to observe the changes that take place with distance from the fixed discharge point. Application rates of the suspended solids thereby form a gradient, with the highest application rates seen at the discharge point and locations extremely close to it and very low application rates found at locations remote from the discharge point.

Studies of the impact of treated sewage effluent, municipal sludge, and other organically-rich materials on the sedimentary infauna have been carried out in many areas (see 5,6). Observed distributions of the benthos in all of these areas are essentially similar. In sediments with the highest organic carbon content, the diversity of benthic infaunal populations is reduced, although the biomass is usually increased. At certain outfalls where organic loading rates are extremely high, the zone at the point of discharge exhibits both reduced species diversity and reduced biomass. In these organically-rich sediments, species that dominate are usually deposit-feeders, particularly polychaetes and bivalve molluscs. In an area which surrounds this central zone (i.e., the zone characterized by the highest organic carbon content in the sediments), there is always a region of lower organic carbon enrichment where diversity of species is normal, but the biomass is higher than areas farther removed from the highest organic sediments. In these transition zones, suspension-feeding species and their predators dominate although deposit-feeders may be more abundant than in unaffected areas.

These studies support a conclusion that the increased sedimentation rate and the greater availability of organic detritus in areas receiving sewage discharges lead to competitive advantages for organisms with different feeding strategies than those which dominate in slower-accumulating, lower organic sediments and that this effect is responsible for the changes observed in the benthos (e.g., 7). In most of these studies, higher concentrations of trace metals and synthetic organic compounds have been found in the organically-rich sediments, compared to the coarser-grained sediments found farther from the discharge. The available evidence suggests that toxic contaminant levels in these sediments are not sufficiently high to have caused the observed changes in benthic ecology. However, in some areas studied where toxic contaminant concentrations are extremely high, there is evidence that these toxic contaminants may contribute to minor ecological changes in addition to those caused by the sewage POM.

It would appear that the changes that take place in benthic fauna in an area to which municipal sludge POM is applied are systematically related to the application rate. As application rate increases, the first effect is a beneficial one: increased biomass without substantial change in the species composition (species diversity). At higher loading rates, the biomass continues to increase, but shifts in species composition may occur toward deposit-feeders which are, in some cases, thought to be less readily included in beneficial food chains (8). At even higher application rates, both species diversity and biomass may be substantially reduced. These changes are entirely analogous to the effects that would be observed by increasing application rates of municipal sludge or composted municipal sludge to a terrestrial ecosystem. At low rates of application, beneficial enhancement of productivity is observed; some plant species would be inhibited at higher rates; and only resistant plant species would survive at extremely high, continuous application rates.

Segar et al. (5) have discussed observations of enhancement of benthic populations at several sewage sludge dumpsites in the United States and Europe. The observations reviewed by these authors (5) demonstrate that enhancement of growth rates and numbers of commercially and ecologically valuable benthic species can be achieved through application of sewage sludge in appropriate continental shelf locations, if the application rate of sewage sludge is controlled. However, current sewage sludge application practices do not permit such control and excessive application rates are achieved at many sites which result in degraded benthos at the application sites surrounded by areas of enhancement.

The effects of application rate on the marine benthos are perhaps best illustrated by considering the information obtained from the intensive studies that have been performed on the benthos of the Southern California Bight where a series of outfalls discharge treated sewage and some sewage sludge from the Los Angeles - San Diego metropolitan areas. Spaced out along a stretch of ca. 300 km of the Southern California coastline, there are four large and several smaller outfalls which discharge treated sewage effluents and one outfall which discharges sewage sludge into the ocean. These outfalls generally discharge at a depth of about 60 km. The natural benthos surrounding these outfalls is, therefore, generally similar for each outfall. Although there are some differences in the characteristics of the effluents or sewage sludge discharged at different outfalls, the POM discharged from different outfalls does not differ significantly in physical or chemical composition, and trace toxic compounds are found at generally similar concentrations with respect to POM. Therefore, the benthos immediately adjacent to each of these outfalls is subjected to different application rates of essentially the same sewage POM. The differences between the benthos surrounding these outfalls and the changes in benthos with distance from each outfall provide an opportunity to study the effects of application rate in detail.

The benthos has been studied intensively at and around these outfalls (9, 10, 11, 12). Word (12) developed an infaunal trophic index to describe the changes in benthos composition induced by POM loading at these outfalls. The index enabled the benthic populations to be classified into three groups (i.e., normal, changed, and degraded), which represent the increasing influence of POM. Benthos within the normal group are dominated by suspension- and surface detritus-feeders and are found in sediments where BOD and total volatile solids are low. The changed benthos are found in sediments with higher BOD and total volatile solids and have a greater percentage of surface deposit- and subsurface detritus-feeders. However, both biomass and the total abundance are substantially increased compared to benthos in normal or background sediments. Within areas with changed benthos, diversity, as measured by several different indices, lies within the range observed in normal sediments and, therefore, these areas of changed benthos are areas of enhancement.

Benthos within the degraded group are dominated by surface deposit- or subsurface deposit-feeders, with diversity and the total abundance below ranges encountered in normal sediments and biomass above normal ranges. Although some sediments characterized by a degraded benthos have extraordinarily high biomass, these are not the most enriched (i.e., highest volatile solids) sediments. The sediments with the highest volatile solids have very low benthic diversity, a predominance of subsurface deposit-feeders, and a lower biomass than found in less-enriched sediments, although biomass is still high compared to background. Therefore, the degraded benthic group ranges from areas which are enhanced, with only slight indications of sediment overenrichment and ecological instability, to areas which are clearly overenriched and have stressed communities generally accepted as undesirable.

Table 1 summarizes pertinent aspects of benthic population studies in the areas adjacent to the one sludge and five effluent outfalls. It can be seen that the degree of benthic alteration and the area over which altered benthos are found are directly related to the rate of introduction of dissolved solids at each outfall. This rate of introduction provides an approximate measure of the maximum application rate of POM at each outfall since the physical and chemical characteristics of the introduced effluents and the physical oceanography at each outfall are generally similar. The data indicate that, at very low application rates (e.g., Oxnard), measurable enhancement of the benthos does not occur, while at higher application rates (e.g., Point Loma, Orange County, Hyperion 5-Mile), enhancement is observable and the area of enhancement increases with increasing application rate. Since these are point discharges, the actual effect of increasing the solids discharge rate is a) to increase the area receiving a POM application rate above the threshold value needed to cause observable enhancement, and b) to increase the maximum application rate of POM which occurs at the point of discharge. As the total solids discharge rate increases even further, the area of observable enhancement increases, while the maximum application rate at and near the outfall also continues to rise until overapplication causes degraded sediment to appear (e.g., Hyperion 7-Mile). As application rates continue even higher, the degraded area also increases in size (e.g., JWPCP).

Table 1. Benthos of Outfall Region in Southern California

Outfall	Flow $l.d^{-1}x10^6$	Total Suspended Solids $mg.l^{-1}$	Total Solids Input Rate $kg.d^{-1}$	Area of Benthos km^2		Approximate Maximum Sediment Concentrations	
				Degraded	Changed	BOD $mg.kg^{-1}$	Total Volatile Solids $mg.kg^{-1}$
Oxnard	42	98	4,100	0	0	–	–
Point Loma	440	130	56,000	0	4	2.3	6
Orange County	680	180	90,000	0	10	3.1	8
Hyperion 5-Mile	1210	62	75,000	0**	>48***	2	4
Hyperion 7-Mile	17	8100*	138,000	3		>20	33
Joint Water Pollution Control Plant (JWPCP)	1270	220	280,000	9	85	21	16

* Total dissolved plus suspended solids
** None within immediate area. Nearby degraded area is associated with 7-Mile discharge.
*** 7-Mile and 5-Mile areas overlap and changed area extends seaward of study area.

Sources: Refs. 10, 13, 14. Data for 1977 at which time Oxnard and Point Loma discharges were primary treated effluent; Orange County was ca. 90% primary,10% secondary; Hyperion 5-Mile ca. 70% primary,30% secondary; JWPCP was primary plus ca 1% digested sludge.

Managed beneficial use of sewage in the ocean could be achieved by maximizing the area of benthic enhancement while avoiding overapplication. It seems obvious from the data in Table 1 that this can be done by limiting the introduction rate of total solids through each outfall. A reasonably conservative approach might be to limit the discharge rate of total suspended solids at each outfall to about 100,000 kg/day, if the outfalls are not located too close together. Alternatively, to allow for possible differences in hydrographic regimes, input management could be achieved through the use of limits on the maximum BOD or total volatile solids observed near an outfall. Although better data and more elegant application rate management schemes using several parameters are possible, it should be remembered that decisions regarding land application of fertilizers, including sewage, are often made with no more information than is available from Southern California marine outfalls, and these practices are successful.

Several interesting conclusions that can be reached from the data in Table 1 and from observations of fish populations around these same outfalls (see above) are as follows:

1) There is no need for secondary treatment to remove additional POM from the discharges at Point Loma, Oxnard, Orange County, or Hyperion 5-Mile. In fact, such additional treatment would REDUCE benthic biomass and the current beneficial enhancement of fisheries would also be partially lost.

2) The adverse effects of the sludge discharge through the Hyperion 7-Mile outfall could be substantially reduced and probably eliminated by discharging the sludge through two or more outfalls separated by approximately 10 km, or more, along the coast or located at different depths.

3) Additional secondary treatment, which would reduce POM input rates at JWPCP, would reduce, or eliminate, the area of degraded benthos. However, this same result could be achieved, perhaps at much lower cost, by utilizing three or more outfalls separated by appropriate distances or located at different depths. Multiple outfalls would also substantially increase the total area of marine enhancement and prevent the generation of more, difficult-to-manage sludge.

4) In planning sewage management systems in heavily populated coastal areas, the optimum strategy is probably to build several long outfalls which minimize the need for high levels of treatment, rather than to build large centralized treatment plants which utilize only one outfall and, therefore, impose a requirement for high levels of treatment to avoid degraded zones around the outfall.

If ocean application of municipal sludge is to be used for beneficial enhancement of the benthos, the application rate must be optimized to maximize enhancement of biomass, while maintaining a diversity and species composition which ensure that the additional biomass is incorporated in useful food chains. It should be emphasized that this optimization process may not involve choosing an application rate which causes no changes in benthic species composition, since certain compositional shifts may be beneficial (i.e., larger total populations of commercially valuable species may be obtained even though the populations of certain commercial species are somewhat reduced).

A STRATEGY FOR BENEFICIAL USE OF SEWAGE IN THE OCEAN

At present, two methods of disposal of sewage in the ocean are utilized: discharge through outfalls, and dumping of sewage sludge from barges. Discharges through current outfall designs are inflexible and difficult to manage, since the discharge location is fixed and the initial dispersion is generally low (approx. 100 times). However, the discharge depth may be selected to some degree, and the application rate can be controlled by limiting the rate of solids introduction through each outfall, establishing minimum spacings between adjacent outfalls, and, less effectively, by incorporating diffusers at the outfall. Ocean dumping from barges is considerably more flexible since the discharge site can be easily moved and the application rate can be controlled by rotating disposal among locations on appropriate time scales. Ocean dumping provides a high potential for dispersion with initial dilutions of approximately 10,000 times. However, ocean dumping has several limitations, including its higher cost compared to outfalls (which limits its practical application to sludges) and the technological difficulty of discharging the sewage into other than the upper euphotic zone of the ocean where the maximum benefits of the POM may not be realized.

Any strategy for ocean productivity enhancement must take into account the following principles:

1) Municipal sewage POM must be dispersed effectively in order to minimize potential impacts from excessive application rates.

2) The application rate must be controlled to maximize the productivity of valuable species.

3) As with all other management options, the concentrations of toxic trace metals and synthetic organics in the municipal sewage must be minimized through pre-treatment and other toxic control strategies.

4) Areas where beneficial application of sewage can be practiced must be carefully selected. The existing information suggests that at least two types of application areas may be appropriate: mid-shelf areas characterized by sediments with low organic content and a relatively impoverished benthic fauna; and shallow areas with fine-grained muddy sediments, such as those found in the German Bight. Application areas must be located such that bathing beach water quality is not adversely affected by the microbial content of the sewage.

5) Shellfish harvested from application areas should be depurated or cooked to eliminate any potential microbiological hazard to human health.

COMPARATIVE BENEFITS OF SEWAGE IN THE OCEAN AND ON LAND

Application of sewage to the land has several disadvantages, including the potential for: groundwater contamination; contamination of crops with trace metals and toxic organics; build-up of contaminant concentrations in the soil (particularly cadmium and copper) and, consequently, limited application rates and lifetime loadings; high cost; limited reliability of pathogen and odor control afforded by the existing technologies; and adverse public reaction. In contrast, application of sewage to the land has the advantage that the nutrient value is maximized by its use in a single-crop, high-yield system, and it is possible, although probably unrealistic, to envisage land application as a containment approach where the toxic contaminants are contained in a given location for future management.

Ocean productivity enhancement by sewage has several disadvantages, including the inability to contain toxic contaminants for future management, and the presently poor control of target organisms for enhancement which leads to a relatively low yield of harvestable food compared to land application. The first disadvantage is of no consequence if the concentrations of the toxic materials in the sewage are reduced to the lowest practicable levels, and if the material is effectively dispersed and assimilated in the ocean. Productivity enhancement through sewage disposal in the ocean has several major advantages, including: the availability of large areas for application of the increasing quantities of sewage that are being generated by major coastal cities; the lack of long-term contaminant build-up in the sediments; the lack of biomagnification of metals through the marine food chain, as compared to land food chains; the reduced need for expensive high levels of treatment to remove POM and BOD; the reduced need for pathogen control since most pathogens are inactivated in sea water; the probable reduction in human health risk (seafood contamination vs. groundwater and crop contamination vs. air pollution); and the low cost of the management technique.

The use of sewage for ocean productivity enhancement is not without risk and potential disadvantages. However, the same statement is also true for the use of sewage for land productivity enhancement and all other practicable sewage management options. Perhaps the most compelling argument in favor of further research on the potential beneficial uses of sewage in the ocean is the public perception regarding the disposal of sewage, particularly in the developed nations. Although, at present, the public is not in favor of ocean disposal since it considers the practice to be damaging to the ocean, they are much more intensely opposed to any waste disposal on the land or into the atmosphere, both of which are closer to their living environment. Therefore, the beneficial use of sewage in the ocean, properly researched and managed, may be a more acceptable practice for long-term municipal sewage management than land application.

In summary, properly managed ocean disposal of sewage would have several benefits, including the conservation of scarce land resources, the protection of groundwater, the minimization of atmospheric pollution, the minimization of human health risks, reduced treatment costs, and finally ocean productivity enhancement.

References

1. Pike, E.B.; Gameson, A.L.H.: Effects of marine sewage disposal. Water Pollut. Control 69 (1970) 355.
2. Eppley, R.W.; Carlucci, A.F.; Holm-Hansen, O.; Giefer, D.; McCarthy, J.J.; Williams, P.M.: Evidence for eutrophication in the sea near southern California coastal sewage outfalls. Coli. Co-op. Oceanic Fish. Invest. Rep. 16 (1972) 74.
3. Sindermann, C.J.: Effects of coastal pollution on fish and fisheries - with particular reference to the Middle Atlantic Bight. Amer. Soc. Limnol. Oceanogr. Spec. Symp. 2 (1976) 281-301.
4. Segar, D.A.; Davis, P.G.: Contamination of populated estuaries and adjacent coastal ocean —— A global review. U.S. National Oceanic and Atmospheric Administration Technical Report, Rockville, Maryland, In press.
5. Segar, D.A.; Stamman, E.; Davis, P.G.: Beneficial use of municipal sludge in the ocean. Mar. Poll. Bull., In press.
7. Caspers, H.: Ecological effects of sewage sludge on benthic fauna off the German North Sea coast. Prog. Water Technol., 9 (1976) 951-956.
8. Boesch, D.F.: Ecosystem consequences of alterations of benthic community structure and function in the New York Bight region. In Ecological Stress and the New York Bight: Science and Management (G.F. Mayer, ed.), pp.543-568. Estuarine Research Federation, Columbia, South Carolina 1982.
9. Word, J.Q.; Mearns, A.J.: 60-meter control survey off Southern California. Southern California Coastal Water Research Project, El Segundo, California, TM229 (1979) 58p.
10. Bascom, W.: Life in the bottom. Southern California Coastal Water Research Project, El Segundo, California, Annual Report 1978 (1979) 57-80.
11. Word, J.Q.; Mearns, A.J.: The 60-meter control survey. Southern California Coastal Water Research Project, El Segundo, California, Annual Report 1978 (1979) 41-56.
12. Word, J.Q.: The infaunal trophic index. Southern California Coastal Water Research Project, El Segundo, California, Annual Report 1978 (1979) 19-40.
13. Bascom, W.; Mearns, A.J.; Word, J.Q.: Establishing boundaries between normal, changed, and degraded areas. Southern California Coastal Water Research Project, El Segundo, California, Annual Report 1978 (1979) 81-94.
14. Schafer, H.A.: Characteristics of municipal wastewater discharges, 1977. Southern California Coastal Water Research Project, El Segundo, California, Annual Report 1978 (1979) 97-102.

A New Model for a Prediction of Beach Deformation around a River Mouth

T. Sawaragi, J. Lee and I. Deguchi

Department of Civil Engineering, Osaka University, Yamadaoka, Suita, Osaka, 565 Japan

ABSTRACT

In order to predict a beach deformation in a region where a sediment discharged from a river mouth or an on-offshore sediment movement plays an important role, it is generally inadequate to apply a one(single)-line theory.

In this paper, a new three dimensional beach deformation model is proposed. In the model, the change of water depth is expressed by the spatial gradient of the bed load flux and the net vertical sediment flux, therefore it becomes very important to accurately evaluate external forces which cause sediment movements and the limit of suspension.

The applicability of this model was tested by three dimensional experiments with a movable bed, then the authors applied this model to predict a beach deformation which will take place in the field after construction of the large reclamation near a river mouth.

INTRODUCTION

Flow fields around an estuary are complicated because of mixed effects of waves, river discharge, tides and density currents. Therefore, sediment movement near an estuary also becomes very intricate.

The sediment discharged from a river mouth diffuses, falls and resupends by the external force. The movements of such sediment could affect the concentration of the suspended load and the depth change, and finally causes the delta formation and the beach deformation.

Some numerical models to predict the beach deformation around a river mouth have been proposed based on the diffusion model.[1)2)3)] However in most of the model, they only took into account the behavior of the discharged sediment from the river into the coastal area without waves and wave-induced currents.

It is needless to say that waves and wave-induced current in the surf zone are significently responsible for the movement of sediments and the beach deformation.

In the first half of this paper, the authors propose a numerical model which enables the simulation of the beach deformation around a river mouth, caused by the discharged flow from the river, incident waves and wave-induced currents.

The model consists of three steps, the first one is the part in which wave deformations and wave-induced currents around the river mouth are calculated. Then the concentration of suspended sediments are estimated. And finally, the beach deformation is simulated based on the continuity equation for the sediment flux.

In the last half of this paper, the authors applied this model to simulate the beach deformation which will take place by the construction of the large reclamation near a river mouth.

BEACH DEFORMATION MODEL

In this model, the water depth h is defined as a distance between the top of the bed load layer and the still water level as shown in Fig.1.

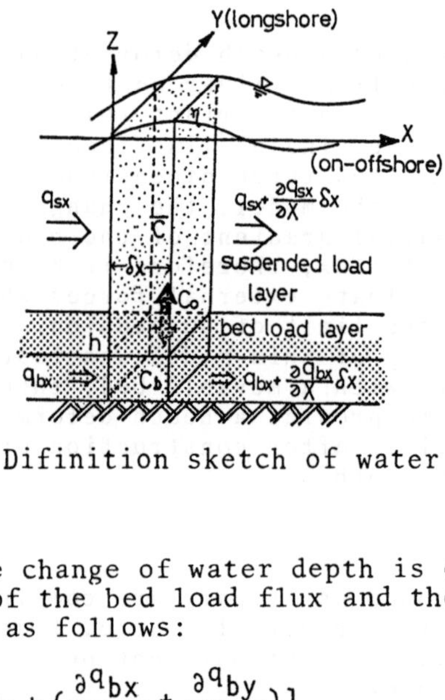

Fig.1. Difinition sketch of water depth

Therefore the change of water depth is expressed by the spatial gradient of the bed load flux and the net vertical flux of sediments as follows:

$$\frac{\partial h}{\partial t}=\frac{1}{1-\lambda}\left[Qs + \left(\frac{\partial q_{bx}}{\partial x}+\frac{\partial q_{by}}{\partial y}\right)\right] \tag{1}$$

where x and y denote the horizontal coordinate in the on-offshore and longshore direction, z is taken upward from the still water level, Q_s is the net vertical flux of sediments (settling flux is assigned a positive), q_{bx} and q_{by} are bed load fluxes in the direction of x and y axis and λ is a porosity.

In order to calculate the change of water depth from Eq. (1), it is required to know the values of Qs, q_{bx} and q_{by}. These values are determined in the following section.

Estimation of Qs

The net vertical sediment flux is expressed as Eq. (2) by using the diffusion coefficient Kz and the settling velocity of sediment w_f.

$$Qs = \left(k_z \frac{\partial c}{\partial z} - w_f c \right)_{z=-h} \tag{2}$$

where c is the time averaged concentration of suspended sediments.

It is generally said that the sediment is brought into suspension in the case where the bottom frictional velocity U_* is larger than the settling velocity w_f[4]. Based on this fact, Eq. (2) can be rewritten as follows:

$$Q_s = \left[-(1-r)c_o w_f \left(1-\frac{u_*}{w_f}\right) - \bar{c} w_f \right] \tag{3}$$

where Co is a concentration at the reference level which is defined as the top of the bed load layer in Fig.1, $\bar{c}$ is the depth and time averaged concentration of suspended sediment, r is the parameter to describe whether the suspension takes place or not(r is zero in case of $U_* > w_f$ and 1 in case of $U_* < w_f$).

As for the reference concentration Co, the following expression obtained from 2-dimensional movable bed experiment is used[5]:

$$c_o = 0.347\, N_c^{1.77} \tag{4}$$

$$N_c = 0.688\, \hat{u}_w^2 / [\, 1.13(\rho_s/\rho - 1) g w_f T]$$

where g is a gravitational acceleration, ρ_s is the density of sediment, ρ is the density of water, T is the wave period and $\hat{u}_w$ is a maximum wave orbital velocity at the bottom. U_* in Eq. (3) is defined by $\sqrt{\tau/\rho}$, where τ is the bottom shear stress. To determine U_*, the following expression for τ is used:

$$\tau = \rho f_w F_b^2 / 2 \tag{5}$$

$$F_b^2 = \frac{1}{2}\hat{u}_w^2 + \frac{2}{\pi}\hat{u}_w (\, U\cos\theta + V\sin\theta\,) + \frac{(U^2+V^2)}{4}$$

where θ is the angle of wave incedence, U and V are the velocity components of depth and time averaged flow in x and y direction and $\hat{u}_w$ is calculated from the linear long wave theory.

$\bar{c}$ is estimated by solving the following time and depth averaged two dimensional diffusion equation of the suspended sediment concentration.

$$(h+\eta)\frac{\partial \bar{c}}{\partial t} + U(h+\eta)\frac{\partial \bar{c}}{\partial x} + V(h+\eta)\frac{\partial \bar{c}}{\partial y} = Q_s + Ts \tag{6}$$

$$Ts = \int_{-h}^{\eta} \frac{\partial}{\partial x}\left(k_x \frac{\partial c}{\partial x}\right) dz + \int_{-h}^{\eta} \frac{\partial}{\partial y}\left(k_y \frac{\partial c}{\partial y}\right) dz \tag{7}$$

where Kx and Ky are coefficients of turbulent diffusion in the x and y direction. In Eq. (6) the shear effect has been neglected under the assumption of the nearly uniform concentration and velocity profile in the vertical direction.

Kx and Ky in Eq. (7) are generally expressed as the product of the representative velocity and length scale. For example,

$$Kx, Ky = a\sqrt{U^2+V^2}\ h \qquad (8)$$

where a is a constant and in the following calculation, we use the emphirical relation a=0.15 after Murray[6].

Estimation of q_{bx} and q_{by}

To evaluate q_{bx} and q_{by}, eq. (9) proposed by Madsen and Grant[7] is used. The applicability of eq. (9) with respect to the rate of time averaged on-offshore sediment transport on a horizontal bottom has already been verified by many investigators.

$$q_{bx}=40\ w_f\ d_{50}\ \tau_x{}^3$$
$$q_{by}=40\ w_f\ d_{50}\ \tau_y{}^3 \qquad (9)$$

where τ_x and τ_y are components of τ given by eq. (5) in x and y direction.

Calculation Of Wave Height,Wave Direction And Wave-Induced Current

In order to determine Qs, q_{bx} and q_{by} mentioned above, it is necessary to know the wave charateristic and wave-induced flow pattern.

In the calculation of wave heights and directions, a wave refraction model is used. Within the wave breaking point(which is determined by the breaker index proposed by Goda)[8], the energy dissipation due to wave breaking[9] is considered. Behind the coastal structure such as jetties and detached break waters, a modified wave refraction model is used to calculate the wave height and direction.

The time and depth averaged flow field due to waves and the river discharge is calculated by usual equations of mass and momentum conservation including the radiation stress in shallow water regions[10].

APPLICATION TO THE FIELD

An area of coastline some 3.6 Km long shown in Fig.2, has been studied. The data of beach topography are interpolated onto the regular grid point(Δs=20m).

Large reclamations are planned to be constructed in the area about 2Km northward from the river mouth. The effect of reclamations on the current and beach deformation are studied under the following storm condition.

a) direction and heights of the incident storm wave

at the offshore boundary: WSW, 3.05m
b) wave period: 5.61 sec
c) discharge rate of the river: 65 m^3/sec
d) duration of wave incidence: 1 hour

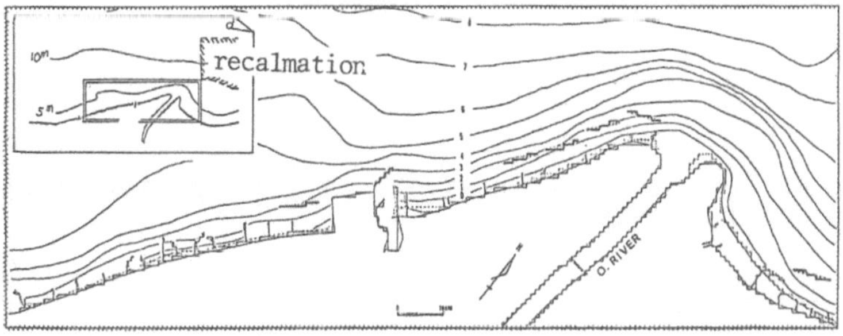

Fig.2. A topography of study area

RESULTS AND DISCUSSION

Computed Results Of The Littoral Current

Fig.3 shows the calculated velocity vector of the time and depth averaged flow before the construction of reclamations. In Fig.3 calculated breaker line is shown by the dash-ed-and-dotted line. From this figure, it is found that strong currents are generated mainly in the shallow water region near the breaker line. At the river mouth, discharged water from the river abruptly changes its flow direction due to the existence of strong longshore current

Fig.4 shows the difference between the velocities of time and depth averaged flow Δv with and without reclamations v and Vo, i.e. Δv=vo-V. For example, in the region where Δv<0,the velocity increases after the construction of reclamations.

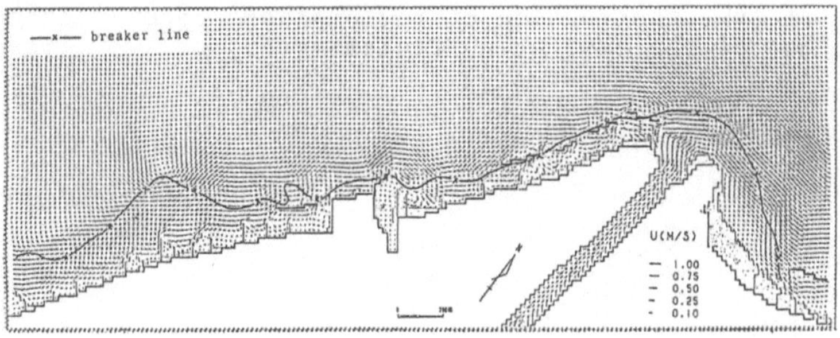

Fig.3. Calculated vector diagram of depth and time averaged flow before construction of reclamations

When there are reclamations, the velocity of the flow decrease in almost all region. However, in the southern coast of the river mouth, there is no significant effect of the

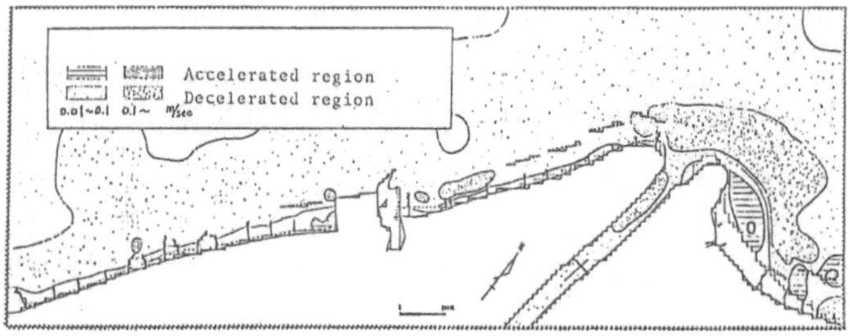

Fig.4. Effects of reclamations on the velocity of time and depth averaged flow

reclamations and Δv is small. While, in the northern coast of the river mouth, the velocity near the shore line increases, and in the region apart from the shore line, the velocity decreases significantly over the wide range. This can be interpreted by the fact that there exists considerable reflected wave from the reclamations as shown in Fig.7. The maximum velocity in front of the river mouth is avout 1.2m/sec when there is no reclamation. However, it becomes about 1.0m/sec after the construction of reclamations. The mean water level at the boundary of the river also rises about 10cm after the construction of reclamations.

Computed Results Of The Depth Change

Fig.5 and 6 denote the distribution of depth change with and without reclamations respectively. In these figures, the eroded and deposited region indicate the area where the water depth changed more than ±1.0 cm/hour from the initial water depth. Whether there exist reclamations or not, the pattern of depth change is almost the same.

Fig.7 denotes the calculated depth change Δh, velocity of the longshore current V, wave height H and the initial beach profile at the representative section A-A shown in the figure. From this figure, it is found that the magnitude of Δh after the construction of reclamations becomes larger than that of the case of no reclamation.

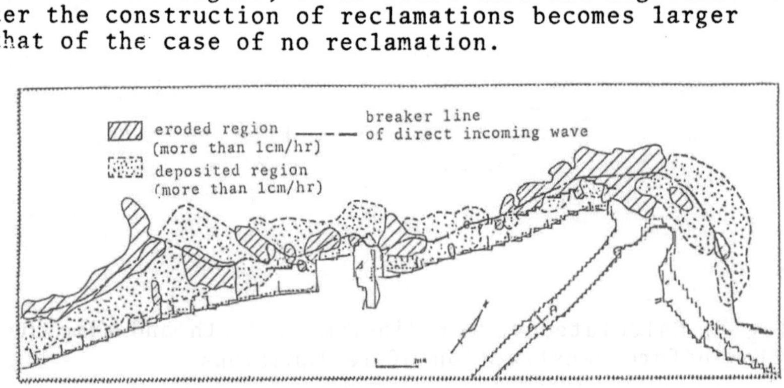

Fig.5. Calculated change of water depth after construction of reclamations

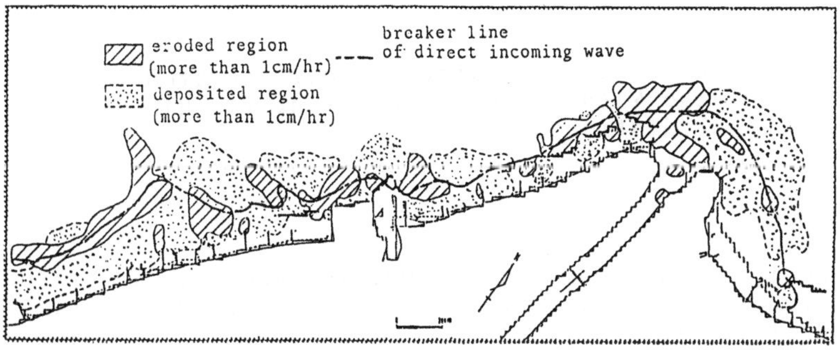

Fig.6. Calculated change of water depth before construction of reclamations

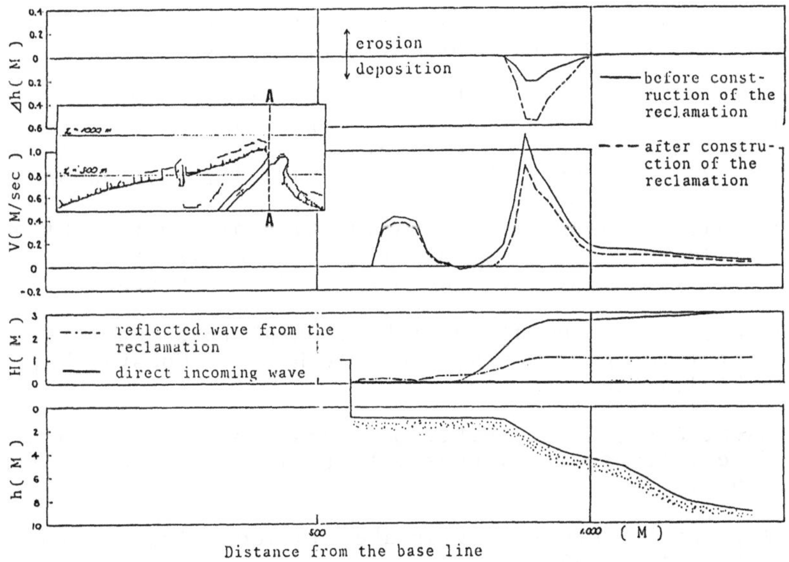

Fig.7. Comparison of depth change Δh, velocity of flow in longshore direction V, wave height H, before and after construction of reclamations along the section A-A at the river mouth

CONCLUSION

The authors proposed a new three dimensional beach deformation model and applied to predict the topographic change which will take place after construction of reclamations in front of the river mouth.

From the calculated results the effect of reclamations on the time and depth averaged flow and the beach deformation are pointed out.

However there still remain uncertainties about the limit of sediment suspension and the rate of bed load transport due to waves and currents.

ACKNOWLEDGEMENTS

The authors wish to express their gratitude to Mr. Hirohata of the Mitsui Kyodo Consultant for his help in numerical calculations.

REFERENCES

1. Nihoul, J.C.J.;Adam, Y.: Dispersion and settling around a waste disposal point in a shallow sea. J. Hyd. Res. 13 (1975) 171-186.

2. Farmer, R.C.; Waldrop, W.R.: A model for sediment transport and delta formation. A.S.C.E. Coastal Sediment'77(1977) 107-115.

3. Özsoy, E.: Suspended sediment transport near tidal inlets. A.S.C.E. Coastal Sediment'77(1977) 914-926.

4. Ashida, K.; Okabe, T.; Fujita, M.: Threshold condition of particle suspension and the concentration of suspended sediments at a reference level. Dis. Pre. Res. Inst. Annual No. 25. B-2. Kyoto Univ. (1982) 401-416. (in Japanese)

5. Deguchi, I.;Sawaragi, T.: Calculation of the rate of net on-offshore sediment transport on the basis of flux concept . Proc. 17th Conf. Coastal Eng. (1984) in press.

6. Murray, S.P.: Simulation of horizontal turbulent diffusion of particle under waves. Proc. 10th Conf. Coastal Eng. (1968) 446-466.

7. Madsen, o.s.; Grant, M.D.: Quantitative discription of sediment transport by waves. Proc. 15th Conf. Coastal Eng. (1976)

8. Goda, Y.: A Synthesis of breaker indices. Proc. J.S.C.E. No. 18 (1970) 339-349.

9. Iwagaki, Y.; Mase, H.; Tanaka, G.: Wave height change of random waves in shallow water. Dis. Pre. Res. Inst. Annual No. 25. B-2. Kyoto Univ. (1981) 509-523. (in Japanese)

10. Sawaragi, T.; Jongsup, L.; Deguchi, I.: A study about near shore currents and three dimensional beach deformation model near a river mouth. Proc. 31th Conf. Coastal Eng. (1984) 411-415. (in Japanese)

Optimization of a Shore Protection Scheme for the West Coast of India

J.R. Weggel and N. Rajendran

Department of Civil Engineering, Drexel University, Philadelphia, PA 19104, USA

ABSTRACT

More than 300 km of the 2000 km-long west coast of India are undergoing severe erosion. The economic importance of India's coastal zone in the States of Kerala, Karnataka, Maharastra and part of Tamilnadu requires that various shore protection measures be investigated to halt the erosion which in some locations exceeds 5 meters per year. Shore protection measures considered were beach nourishment, groin construction and the construction of a rubble-mound seawall. Economic and practical considerations led to selecting the seawall alternative since a source of suitable beach sand was not readily available and the cost of the heavy construction equipment necessary to build groins would have been prohibitive. The shortage of heavy construction equipment, an abundant source of labor and the availability of suitable rock from nearby quarries influence the economically optimum seawall configuration. Hydraulic model tests of the proposed rubble seawall were undertaken to establish the seawall cross-section and to determine the amount of damage caused by various levels of wave attack. The results obtained by an interactive computer program to establish the optimum level of design for the seawall and to determine the most economical cross-section geometry are discussed. The program considers factors such as the size distribution of rock produced by blasting at the quarry, the quarrying costs of producing armor stone of the required size, the cost of transporting and placing the armor stone, and the cost of repairing the seawall in the event design conditions are exceeded.

INTRODUCTION

The heavily populated, 2000 km-long west coast of India is dotted with a number of major and minor ports as well as with other industrial development. In addition, fertile agricultural land lies adjacent to the shoreline where erosion rates of 2 to 5.5 meters/year have been documented. These conditions, caused by waves and tidal flooding, persist along one-sixth of the shoreline in the

States of Kerala, Karnataka, Maharastra and part of Tamilnadu. See Figure 1. The high, steep waves associated with the southwest monsoons occur between mid-May and mid-September and cause severe erosion. Coupled with relatively low elevations along the backshore, monsoon winds and incident waves also cause coastal flooding. Monsoon floods discharging from rivers along this reach of coast interupt the normal longshore sand transport and carry sand offshore aggravating the erosion problem along downdrift beaches. Vast offshore mud flats also exist along part of the west coast. They act to dampen incoming waves, shelter the coast, cause sediment in transport along the coast to be deposited, and cause erosion along downdrift beaches. Existing coastal structures such as groins and breakwaters also contribute to the erosion problem along downdrift beaches. Sediment carried to the coast by rivers during monsoon floods is in the silt-size range and does not contribute much sand to the beaches.

Several alternative shore protection schemes were considered to alleviate erosion and flooding problems along the west coast. Alternatives investigated were beach nourishment with protective dunes; groins, and a rubble-mound seawall or revetment. The preferred solution would have been beach nourishment; however, no suitable source of beach sand was readily available. Groin construction would have required heavy equipment and, in the absence of a wide beach and dunes, would not have provided the desired level of protection. The presence of nearby quarries that could provide suitable, relatively inexpensive rock for seawall construction and the availability of a large labor force led to selection of the seawall alternative.

Hydraulic model tests of various seawall cross-sections were conducted by the Coastal Engineering Section, Karnataka Regional Engineering College (Rajendran, 1979). Seawalls with slopes of 1:5 and 1:6 were investigated both with and without rubble toe-aprons. Toe-aprons 2 m, 3 m, and 4 m wide were studied to determine if they would reduce scour at the seawall toe. The cross-sections studied are shown in Figure 2. The model tests established the armor stone size necessary for stability under attack by the design wave and also determined the level of damage caused by waves exceeding the design wave height.

DESIGN PHILOSOPHY - ECONOMICS

Coastal structures are often designed to withstand wave and water level conditions that are more or less arbitrarily selected. Structures are designed to withstand the 50-year wave or the 100-year wave without consideration for the risk or economic implications such arbitrary design condition selection involves. If too large a design wave is selected, a structure with a high first cost will result; however, it will not require much in the way of expenditures for maintenance and repair because it will rarely be subjected to conditions exceeding its design conditions. See Figure 3. On the other hand, if too small a design wave is selected, the structure will have a low first cost but will need more maintenance and more frequent repairs because its design conditions will be exceeded more often. Somewhere between these extremes, optimum design conditions will exist that result in a minimum annual cost by balancing the structure's annualized first cost against its annual maintenance and repair costs.

This paper describes an interactive computer program written in Fortran 77 to compute the quantities of materials for various seawall geometries designed for various incident wave heights. The program also calculates the annualized first cost of the seawall using cost models for the various components of the seawall and the annual cost of repairing damage expected from waves exceeding the design wave. Specifically, the curves in Figure 3 are quantified with the first cost and annual repair cost given as a function of design wave height.

WAVE ENVIRONMENT - WEST COAST OF INDIA

Data on the return period of waves of a given height are needed to determine the level of damage that can be expected on a seawall when waves exceed the design wave height. Generally, large waves that will cause significant damage to a structure occur infrequently; however, if a structure is designed for a relatively small wave, wave condtions that can cause damage will occur more frequently. The long-term distribution of significant wave heights occuring at the site is needed to determine the probability of damage.

The wave environment along the west coast of India has been described by Dattari (1973) based on 12 months of wave data recorded over an 18 month period in 1968-1969 with a pressure-type wave sensor located in 10 meters of water 1500 meters west of Mangalore Harbor. During the one year of record, the largest wave recorded was 5.4 m high with a zero-crosssing period of 8.9 sec. The maximum significant wave height was 3.44 m with a period of 10.4 sec. The maximum root-mean-squared wave height was 2.50 m. Unfortunately, the short record makes it impossible to determine the return period of large waves without making some assumptions regarding their distribution. Larras (1970) suggested that the maximum annual wave could be related to the wave with a 10 year return period and the wave with a 100 year return period by assuming a logarithmic probability function. Under this assumption, the 10 year wave is 1.39 times the annual wave and the 100 year wave is 1.78 times the annual wave. The resulting distribution of significant wave heights for the west coast of India is given in Figure 4. Recognizing the limitations imposed by the assumptions made in obtaining the long-term wave height distribution, it is fortunate that the economic analysis of the present study is not very sensitive to the distribution. This may not generally be the case however.

COMPUTER PROGRAM DESCRIPTION

Structure Geometry and Stability

Generally, the program uses the equations for rubble structure design given in the U.S. Army Coastal Engineering

Research Center's "Shore Protection Manual" (U.S. Army, 1984). Hudson's equation, given by,

$$W = \frac{wH^3}{K_D(S-1)^3 \cot \theta} \qquad (1)$$

where W = the armor unit weight required for stability,
w = the unit weight of the water,
H = the design wave height,
S = the ratio of the unit weight of the stone to the unit weight of the water,
θ = the angle the structure slope makes with the horizontal, and,
K_D = the stability coefficient,

is used to calculate the armor unit size required for stability. The stability coefficient is entered into the program from a terminal so that the effects of using armor with various stability coefficients can be investigated. For the rubble seawall analysis presented here a stability coefficient of 3.5 (rough quarrystone on a structure trunk) is used for all structure slopes. Slopes varying from 1:3 to 1:6 were investigated. Once the design wave height and resulting armor unit size are specified, the underlayer and core layer sizes are also specified. The thickness of the various layers (structure geometry) and the materials quantities can then be calculated.

Cost Models for Elements of the Seawall

The in-place cost of the armor is assumed to be comprised of three components: a quarrying cost, a transportation cost, and a placement cost. Each of these costs is a function of the armor unit weight and thus of the design wave height.

The quarrying cost is subdivided into a dynamiting cost and a cost associated with selecting and sorting the stones resulting from a blast. This latter cost is assumed to be inversely proportional to the amount of stone produced by a blast that can be used in the size distributions making up the armor layer and the first underlayer. Thus, if only 30% of the stones resulting from a blast can be used as armor or underlayer stone without further blasting, the basic dynamiting cost is divided by .30 to obtain the quarrying cost per tonne. Baird and Hall (1984) discuss the quarrying problem for breakwaters constructed using the Corps of Engineers' recommended armor size gradations and Hudson's equation. They advocate modifying the breakwater cross-section to make maximum use of a given quarry's production. This generally reduces quarry waste and can reduce placement costs but may increase the amount of rock that must be transported to the construction site and thereby increase the costs associated with transportation. In the present analysis the rocks produced in a quarry by a single blast are assumed to have a log-normal size distribution (See Baird and Hall, 1984) and the median size of the distribution can be selected to some degree by modifying the blasting hole patterns. Thus, the median stone size can be selected; however, the blasting cost/tonne of rock produced is assumed to depend on the median size produced as shown in Figure 5. For median weights less than 1 tonne, the blasting cost is taken to be 7 rupees/tonne; for median weights greater than 8 tonnes, the blasting cost is 5 rupees/tonne; between these median weights, the blasting cost decreases linearly from 7 rupees/tonne to 5 rupees/tonne reflecting the lower unit blasting cost to produce larger stones. For the log-normal size distribution assumed, the percent of the size distribution which can be used as armor is shown in Figure 6 as a function of the ratio of armor unit weight to median stone weight. A similar figure can be produced for the underlayer gradation. A maximum of 37% of the distribution (solid line in Figure 6) can be used for armor if the median stone weight produced by the quarry equals the armor unit weight. This assumes that the allowable size gradation of armor varies from 0.75 W to 1.25 W where W is the armor unit weight computed from Hudson's equation. If the allowable gradation is expanded to allow armor weight to vary from 0.7 W to 1.3 W the maximum amount of quarry production useable in the structure armor increases to 46% as shown by the dashed line in Figure 6.

Transportation costs for the armor are based on a fleet of trucks having a capacity of 8 tonnes. Assuming the trucks haul only armor stone gives the variation of the load carried per truck with armor unit weight shown in Figure 7. For armor slightly over 4 tonnes, the truck can carry only a single armor unit and will travel only half loaded. Prudent operations and planning, and the allowable size gradation for armor will in reality smooth out the discontinuity in Figure 7. The cost of operating a truck per kilometer as a function of armor unit weight is shown in Figure 8. The discontinuity at an armor unit weight of 4 tonnes reflects fuel savings realized by operating a half loaded 8 tonne capacity truck between the quarry and construction site. Figure 9 is a synthesis of the information on Figures 7 and 8 giving the cost/km/tonne of transporting armor as a function of armor unit weight.

The placement cost for individual armor units is also a function of armor unit weight; however, total placement cost is a function of the number of armor units rather than the total weight of armor. Placement costs reflect the method of placement: manual labor can be used for relatively small units while increasingly heavier equipment is required as the armor unit weight increases. Figure 10 shows armor unit placement cost as a function of armor unit weight. Armor stone weighing less than 40 kg can be placed by one man and the cost of placement is taken to be 0.5 rupees/unit. Two men are needed to place armor units weighing between 40 and 100 kg, while light, manually operated equipment can be used to place armor weighing up to 2 tonnes and medium, power-driven equipment is needed to place armor weighing over 2 tonnes.

Seawall Damage as a Function of Wave Action

Hydraulic model tests of the revetment were conducted to determine armor layer stability and to determine the damage that would result if waves greater than the design wave height occured. Damage was quantified as the percentage of armor stones moved by the waves and related to the ratio of the wave height causing the damage to the design or zero-damage wave height (expressed as a stability coefficient divided by a zero-damage stability coefficient). The resulting data are shown in Figure 11. Damage tests were conducted for several different wave periods and for structures having different toe-apron widths. This range of testing conditions explains some of the scatter in Figure 11. Generally, damage levels were quite low (less than 5%) for the conditions tested, even for waves much larger than the design wave. This is because the tests were conducted with only a single water level at the seawall toe elevation in the model. For this water depth the surf zone filters the incoming waves so that large waves break seaward of the structure and only smaller broken waves reach the structure. The offshore wave height thus has little influence on the wave height acting on the seawall; rather, it is the water depth at the seawall that determines the wave height. Greater water depths allow larger waves to reach the seawall resulting in greater damage. Additional work is needed to define the return period of various water levels along India's west coast and to define levels of damage in laboratory tests with greater water depths. For the initial analysis presented here, however, the damage function defined in Figure 11 is used.

The annual cost of repairing damage caused by waves exceeding the design wave height is the probability that a given damage-producing range of wave heights will occur times the amount of damage that range of wave heights will cause. Summing the incremental damages caused by each damage-producing range of wave heights results in the percentage damage occuring to the structure on average each year. The cost of repairing the damage is then the percentage damage times a factor proportional to the first cost of the structure. The present analysis resulted in unrealistically low annual repair costs because of the relatively low damage levels predicted by the model tests. Higher, more realistic damage levels resulting from breaking waves supported on higher water levels at the structure will result in higher annual repair costs.

ANALYSIS RESULTS FOR RUBBLE SEAWALL

Some interim results of the economic analysis are shown in Figures 12 and 13. The effect of the distance between the quarry and costruction site on the optimum level of design are illustrated in Figure 12. Distances between quarry and cosntruction site of 20 km, 30 km, 40 km and 50 km were investigated. As expected, increasing the quarry distance increases the per kilometer seawall cost; however, the level of design giving a minimum annual cost per kilometer of seawall does not change. In each case the curves reach their minimum at a design significant wave height of 3.0 m. The estimated return period for a 3.0 m high significant wave is about 1.17 years. The low damage levels for large waves obviously shifts the analysis to recommend lower levels of design; however, a more realistic relationship between damage and waves exceeding the design wave height might shift the minimum toward higher design waves. The armor unit weight required for stability is slightly less than 1 tonne.

Changing the structure slope has a marked effect on which design wave height will give the minimum cost since the required armor size changes significantly with slope. Figure 13 shows that a 1:6 slope gives the minimum cost at a design wave height of 3.0 m. The 1:3, 1:4 and 1:5 slopes give minimum costs at wave heights of 3.0, 4.0 and 3.5 m respectively. The discontinuities at the right side of Figure 13 for the 1:3, 1:4 and 1:5 slopes are caused by the discontinutity at 4 tonnes in the cost curves. For the 1:3 slope the jump to a 4 tonne armor unit occurs between wave heights of 3.75 and 4.0 m. The jump occurs between 4.25 and 4.5 m high waves for the 1:4 slope and between 4.5 and 4.75 m high waves for the 1:5 slope.

CONCLUSIONS

The present analysis suggests that a rubble seawall built on a 1:6 slope to withstand a 3.0 m high significant wave is an economically optimum solution to the erosion problem along the west coast of India. From the model studies this structure with a 2 m toe apron was found to be optimum. Care should be exercised in interpreting the results of the present analysis, however. There are several shortcomings in the present economic model that need to be rectified. The most serious is the lack of hydraulic model test data for water depths above the structure's toe elevation. The damage resulting from larger waves breaking on the seawall will significantly modify the results of the damage model and predict greater annual repair costs. This will shift the recommeded level of design toward higher design wave heights. In addition, other truck-haul transportation cost models should be investigated, specifically models that involve other truck fleets and truck loading plans. Lastly, detailed information on the quarry production is needed to better define the armor quarrying cost model.

REFERENCES

Baird, W.F. & K.R. Hall (1984), "The Design of Breakwaters Using Quarried Stones," Proceedings of the 19th International Conference on Coastal Engineering, Houston, Texas, September 1984.

Dattari, J. (1973), "Waves Off Mangalore Harbor - West Coast of India," Journal of the Waterways, Harbors and Coastal Engineering Division, ASCE, WW1, #9532, February 1973.

Rajendran, N.,(1979), "Design and Testing the Stability of a Seawall for the West Coast," M.Tech. thesis submitted to the University of Mysore, Coastal Engineering Section, Karnataka Regional Engineering College, India, August 1979.

U.S. Army Corps of Engineers, (1984), "Shore Protection Manual," Coastal Engineering Research Center, WES, Vicksburg, Mississippi, 1984.

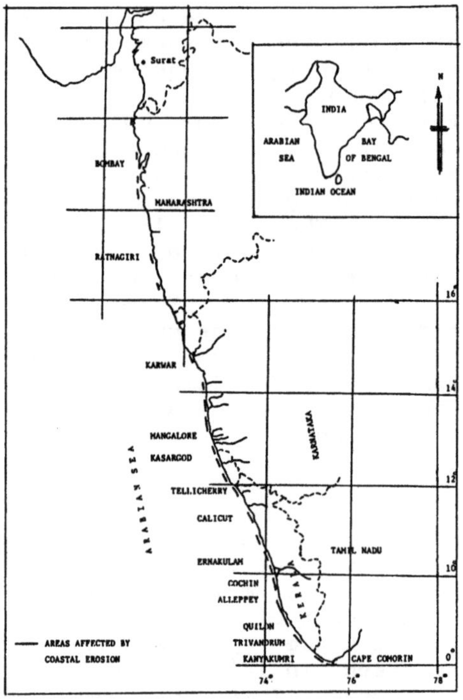

Figure 1 Location Map - West Coast of India

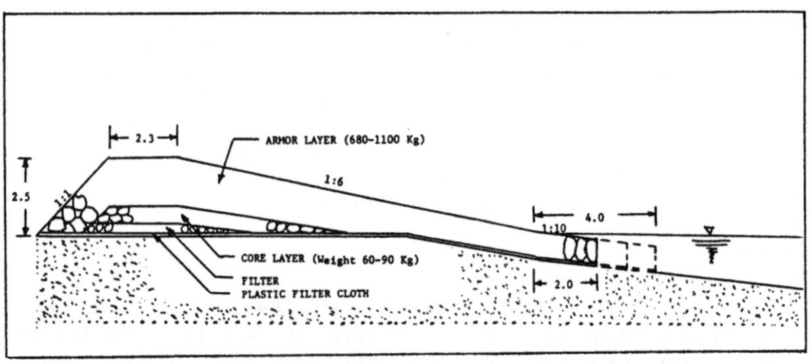

Figure 2 Rubble-mound Seawall Cross-Section

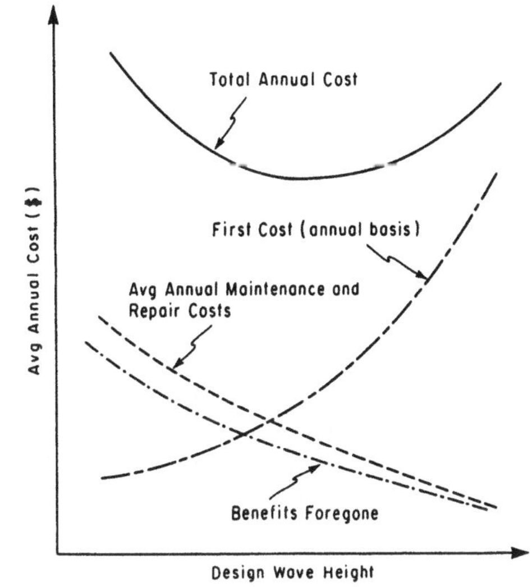

Figure 3 Typical Cost Curves for Seawall - Annualized First Cost and Annual Repair Cost vs. Design Wave Height

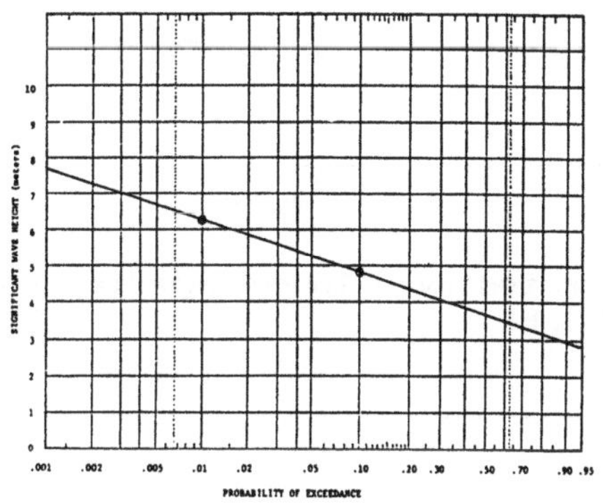

Figure 4 Estimated Return Period of Significant Wave Heights - West Coast of India

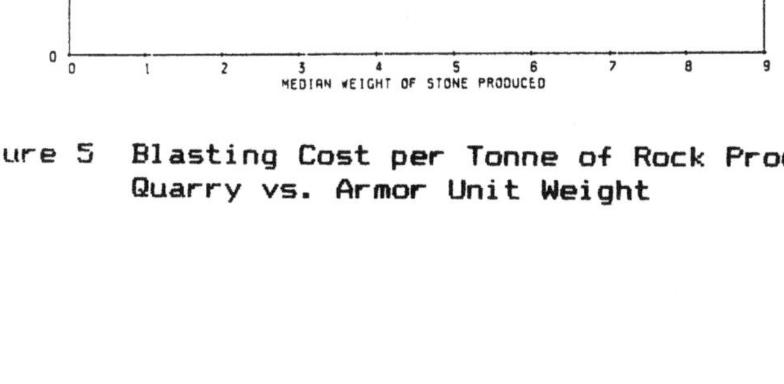

Figure 5 Blasting Cost per Tonne of Rock Produced at Quarry vs. Armor Unit Weight

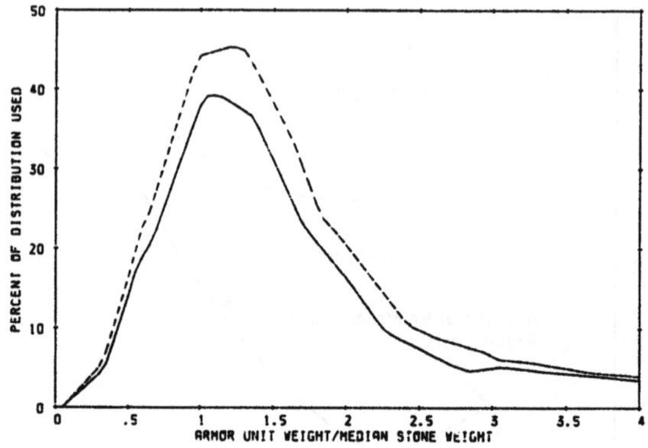

Figure 6 Percentage of Rock Produced by Quarry that can be Used as Armor vs. Ratio of Armor Unit Weight to Median Weight of Stone Produced by Quarry

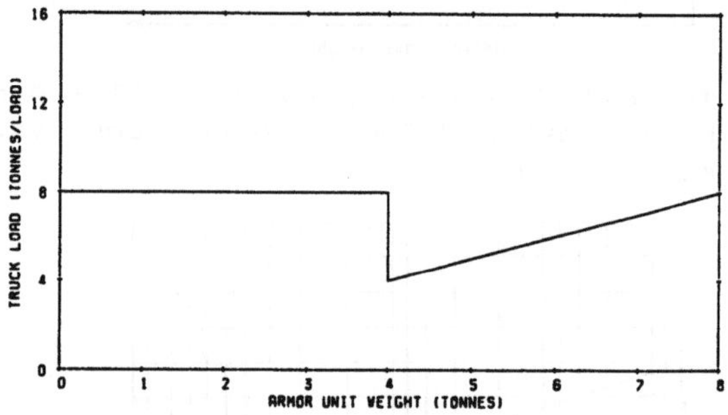

Figure 7 Load Carried by 8 Tonne Capacity Truck vs. Armor Unit Weight

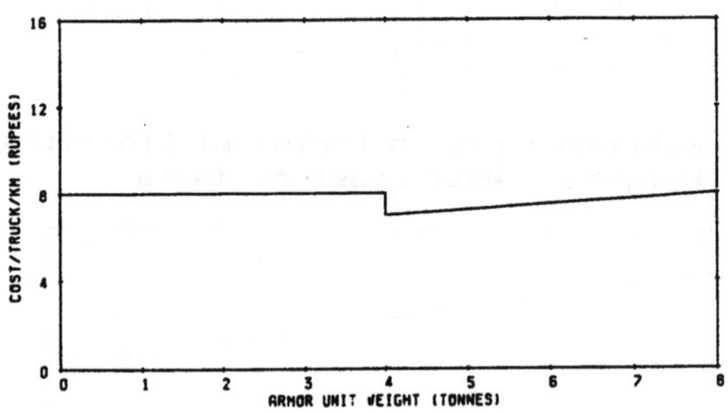

Figure 8 Operating Cost for 8 Tonne Truck per Kilometer vs. Armor Unit Weight

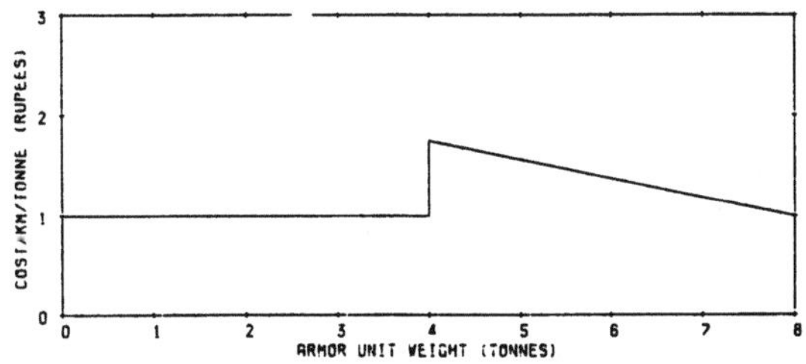

Figure 9 Transportation Cost per Kilometer per Tonne of Armor Stone vs. Armor Unit Weight

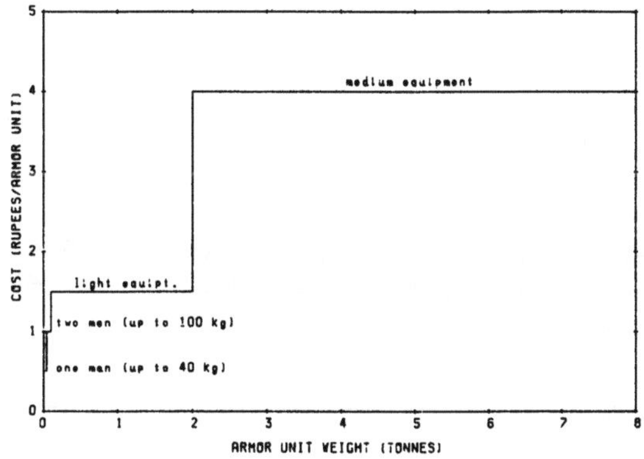

Figure 10 Armor Unit Placement Costs vs. Armor Unit Weight

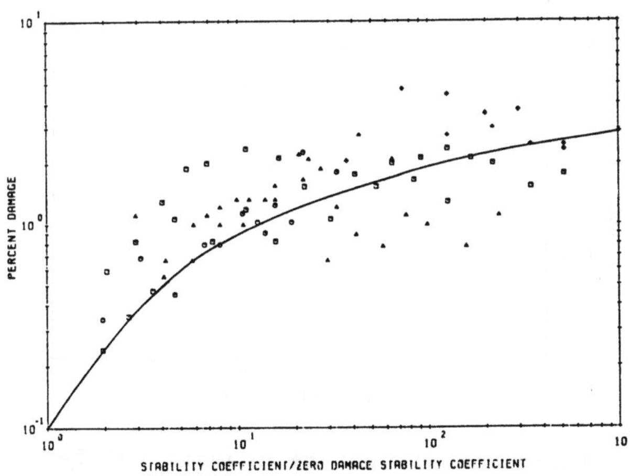

Figure 11 Percent Damage to Armor Layer vs. Ratio of Stability Coefficient to Zero-Damage Stability Coefficient

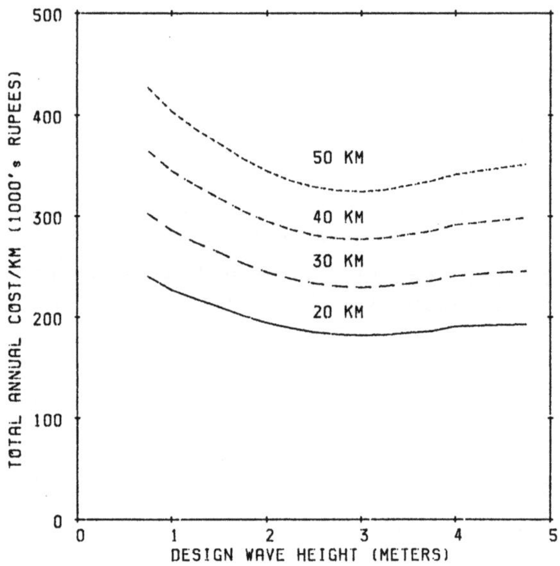

Figure 12 Total Annual Cost of Seawall vs. Design Wave Height for Quarries 20 km, 30 km, 40 km, and 50 km Distant

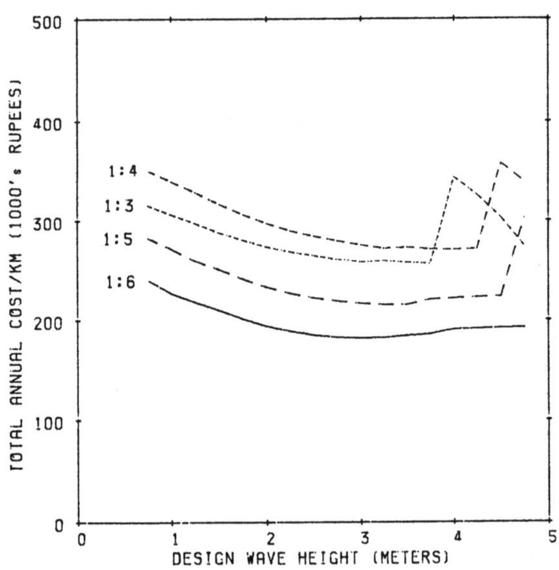

Figure 13 Total Annual Cost of Seawall vs. Design Wave Height for Seawall Slopes of 1:3, 1:4, 1:5 and 1:6

Technical Session:
Coastal Zone Utilization
and Management

5. Environmental and Safety Management

A Global Comparison of Contamination in Populated Estuaries and Coastal Waters

D.A. Segar, P.G. Davis and E. Stamman
SEAMOcean, Inc., Wheaton, MD 20902, USA

ABSTRACT

Many areas of the world ocean which are subject to contaminant inputs from man's waste discharges have been the subject of intensive pollution studies during the last decade. In the United States, the New York Bight Project of the National Oceanic and Atmospheric Administration (NOAA) has been the largest of such marine ecosystem studies. This paper is a brief summary of a study which compared the available infomation concerning contamination and its effects on the New York Bight and on other coastal areas throughout the world (Segar and Davis, in press). These comparisons reveal that problems in many regions of the coastal ocean are quite similar, that nutrient-induced eutrophication and organic loading of sediments in areas with insufficient dispersal of inputs are the most widespread and serious problems, and that inputs of toxic trace metals and synthetic organics to the ocean have had no significant impact on human health or the marine environment, except in a few extreme cases in very small areas of the coastal ocean environment.

INTRODUCTION.

Many of the coastal regions of the world ocean have suffered environmental problems as the quantity of wastes discharged into the ocean from the numerous urban/industrial complexes located at the ocean's edge has approached or exceeded the ability of certain coastal ocean areas to assimilate these wastes. These problem areas often develop because the urban centers are located along the coast where there is safe harbor and protected embayments, but where the dispersive properties of the coastal ocean are consequently restricted. This paper summarizes the results of a larger study on contamination in marine coastal areas throughout the world (Segar and Davis, in press). This larger study cites about 400 references which cannot be listed in the present paper due to space limitations. However, copies of the completely referenced manuscript are available upon request from the authors.

POLLUTION SUSCEPTIBILITY.

Four descriptive parameters were used to categorize the regions of the world ocean that are significantly impacted: i) the population of the watershed region; ii) the degree of industrialization; iii) the water volume of the area receiving waste inputs; and iv) the flushing and mixing characteristics of the affected segment of ocean. The first two factors are easily understood: the sizes of the population and the industrialized base which discharges wastes and storm drain runoff into

Table 1. Pollution susceptibility of selected ocean regions.

Area	Pop.	Industry	Discharge Region Volume	Dispersion and Flushing
New York Bight	3	3	1	2
Lake Erie	3	3	1	2
Seto Inland Sea (Japan)	3	3	1	2
Baltic Sea	2	3	1	3
Mediterranean Sea:	2	2	1	2
Adriatic	3	3	1	3
Saronikos Gulf	3	3	2	3
North Sea	3	3	1	1
Irish Sea	3	3	1	2
S. Calif. Coastal Area	3	2	1	1
Puget Sound	2	2	2	2
Chesapeake Bay	2	2	2	2
Guanabara Bay, Brazil	3	3	3	2
Hudson-Raritan, NY	3	3	3	2
Firth of Clyde, UK	3	3	2	2
Singapore: Harbor	3	3	3	3
Straits of Mallaca	3	2	2	1
Hong Kong: Harbor	3	3	3	2
Adjacent S. China Sea	3	3	1	1
Red Sea	1	1	1	3
Persian Gulf	1	2	1	3
Black Sea	2	2	1	3

Higher numbers indicate a higher susceptibility. Densely populated or highly industrialized areas receive a "3", a high susceptibility. Areas with limited discharge volumes and restricted flushing also rate a "3" for their high susceptibility.

the ocean, either directly or through rivers, strongly influence the potential for the build-up of contaminants to levels at which adverse impacts can occur. The other two factors which determine the extent of anthropogenic impact on the ocean environment are hydrographic factors which affect the degree of dilution of the contaminant inputs. The concentration of contaminants found in a specific region of coastal ocean is determined by 1) the amount of each contaminant introduced into the system, 2) the rate of contaminant dispersal in the coastal water, and 3) the rate at which that segment of coastal ocean is mixed and flushed into the larger volume of the deep, open ocean. Contaminants, including floating rubbish and microbes, only pose a problem to man or to the marine ecosystem if their concentration in the environment reaches a certain critical level. The critical level varies with contaminant, geographical location, and even season. This variation occurs because ecological sensitivity differs between biological communities and biotic life stages, and human uses of the ocean vary regionally and seasonally.

Table 1 lists some representative coastal regions and indicates their potential for developing pollution problems based on the four parameters. Although arguments could be made to change the ranking of an area's factor(s) as assigned in Table 1, the general conclusion is obvious. The regions of greatest concern are enclosed bays, harbors, estuaries, and gulfs sustaining large populations and industries. Next in susceptibility are the large enclosed seas, such as the Baltic and Seto Inland Sea, followed by heavily populated and industrialized areas located on open coastlines with poorly-flushed bottom waters, such as New York Bight, and by dynamically-flushed, semi-enclosed seas, such as the Irish Sea and North Sea. Finally, the areas of least concern include the semi-enclosed seas with small populations and industry, such as the Red Sea and Persian Gulf, and open coastal ocean areas with low populations. However, it is important to recognize that extreme events can lead to severe pollution problems even in regions with low susceptibility. For example, oil spills within the Persian (Arabian) Gulf, uncontrolled because of the Iran/Iraq war, have reportedly led to substantive environmental damage.

Our analysis indicates that the New York Bight is not among the most pollution-susceptible regions of the world ocean. Rather, it is comparable to the large, but poorly flushed, semi-enclosed seas (e.g.,

Table 2. Estimated Contaminant and Natural Inputs to Selected Regions.

		BOD ($\times 10^4$)	N ($\times 10^4$)	P ($\times 10^3$)	Hg	Pb	Cr	Zn	Cd	Total PCB
New York Bight	A	55	14	19	53	2,100	1,600	6,800	110	10
	B	3.2	0.84	1.1	3.1	220	96	400	6.2	0.59
Southern California	A	25	NA	>10	2.8	150	370	840	42	2300
	B	3.6	NA	1.4	0.4	21	53	120	6	330
Mediterranean	A	330	100	360	130	4,800	2,800	25,000	NA	6,000[1]
	B	3.3	1.0	3.6	1.3	48	28	250	NA	60
Seto Inland Sea	A	91	16	17	NA	NA	NA	NA	NA	NA
	B	4.6	0.8	0.9	NA	NA	NA	NA	NA	NA
Baltic Sea	A	41	59	17	29	6,200	900	19,000	250	NA
	B	2.3	3.3	0.9	1.9	>55	NA	NA	NA	NA
Irish Sea	A	37	8.0	22	35	2,400	1,100	10,000	110	NA
	B	4.6	1.0	2.8	4.3	300	140	1,300	14	NA
North Sea	A	110	92	110	47	4,900	4,000	24,000	330	NA
	B	5.0	4.2	5.0	2.1	220	180	1,100	15	NA

A metric tons/yr
B metric tons/ 10^6 persons/yr

[1]Total organochlorines
NA Not available

the Baltic, Mediterranean, Irish and North Seas, and the Seto Inland Sea). Much more susceptible to pollution and its impacts are the smaller embayments with limited dispersion and flushing and with large populations [e.g., the Hudson-Raritan Estuary, Guanabara Bay (Brazil), the Firth of Clyde (Great Britain), Singapore and Hong Kong Harbors, the Saronikos Gulf (Greece), and, if populations continue to grow, Puget Sound and Chesapeake Bay, (U.S.)].

CONTAMINANT LOADINGS.

Comparison of the data indicates that the total annual input of contaminants to the New York Bight is within the range of annual mass loadings to other regions which are similar in terms of pollution susceptibility (Table 2). However, differences in hydrography and geography among the various regions strongly influence the distribution of pollutants in the oceanic environment. The Mediterranean, Baltic, and North Seas are considerably larger than the New York Bight in terms of both area and volume and, therefore, are not analogous to the Bight. The Seto Inland Sea and the Irish Sea are in the same size range as the New York Bight and probably provide more closely matched models for comparison with it.

SEDIMENT CONCENTRATIONS.

Despite considerable differences in input rates and hydrographic, geographic, and geological features among the areas studied, the ranges of contaminant concentrations reported in sediments from the different areas (Table 3) do not show obvious major differences. Those areas where the sediments are naturally fine-grained are found to have the highest metal concentrations, with the lowest concentrations found in coarse sand regions. Areas where metal concentrations are as high as, or higher than, those found in the New York Bight include the Saronikos Gulf, the Solent (Great Britain), the Severn (Great Britain), the Firth of Clyde, Tokyo Bay, the southwest Spanish and Portuguese coasts, and the area close to the 7-mile Hyperion sewage sludge outfall in Southern California. Each of these areas receives high contaminant inputs which

Table 3. Heavy Metal annd PCB concentrations in Selected Marine Sediments (ug/g dry wgt).

	Zn	Cd	Pb	Cu	Cr	Hg	PCB
United States							
New York Bight	4.6-541	<0.47-9.6	5-270	1-320	2-370	0.12-4.9	0.0005-2.2
S. California	44-2900	0.22-140	7-600	9.0-1000	25-1500	0.04-6.1	0.03-12.8
Rhode Island	53-170	0.06-2.5	17-81	26-98	13-81	NA	NA
Mediterranean	2.1-3000	0.02-7.0	2.8-200	3.8-130	1.3-1200	0.01-57	NA
Japan	110-240	0.14-3.5	10-110	23-54	5-180	<0.1-713	0.01-390
Baltic Sea	110-270	NA	25-84	33-78	90	NA	NA
Irish Sea	14-800	NA	10-460	5-200	15-360	0.07-3.3	NA
North Sea	<20->200	0.1-0.8	<10-240	2-50	4.0-41	0.05-0.9	NA
Norwegian Sea	24-2000	0.3-1.5	5.0-11,000	2-2400	NA	90-350	NA
Australia	3-1200	0.15-9.9	4.6-180	2.2-120	1.3-42	NA	NA
Spain & Portugal, Atlantic Coast	6-3100	0.9-4.1	6-1600	2-1400	NA	NA	NA
Bermuda	16-65	<0.25-0.99	6.4-230	7-130	15-484	NA	NA

include various forms of sewage, industrial wastes, and, often, dredged material.

The effects of flushing efficiency on the metal accumulation in sediment can be seen by contrasting the metal concentrations in the New York Bight sediments with those in the Saronikos Gulf. The highest metal concentrations in the Saronikos Gulf sediments are as high as, or higher than, those observed in the most contaminated areas of the New York Bight, despite 1) an Athens-area population about one-tenth that of the New York City region, and 2) a correspondingly smaller amount of industrial effluent which is also discharged from somewhat more dispersed locations when compared to the concentrated discharge into the New York region from Raritan Bay. However, the spatial extent of the contaminated sediments in the Saronikos Gulf is only about one-third to one-half as large as that in the New York Bight. The critical differences between the two regions are the Saronikos Gulf's poorer circulation and dramatically smaller, tidally-induced dispersion which ensure that a greater fraction of the discharged metals remain in the sediments near the disposal site.

Chlorinated pesticide and PCB concentrations vary widely in sediments depending upon the location and the sediment grain size. Both contaminants are found in relatively high concentrations in fine-grain sediments in the New York Bight apex and near British sewage sludge dumpsites. High concentrations are also found near municipal wastewater outfall pipes, such as in the Saronikos Gulf and off Southern California. Fortunately, decreased production and use of DDTs in the United States yave resulted in a decrease in DDT concentrations in wastewater discharges and a resultant decrease in sediment concentrations. Comprehensive data are lacking for the PCB content of sewage sludges dumped in the New York Bight, but the results from different researchers suggest that PCB concentrations have decreased 50% in a period of about 5 years. Sewage sludge disposal is the possible source of elevated DDT and PCB concentrations that have been observed in the Baltic. The highest reported concentrations of PCBs in marine sediments are in New Bedford Harbor, Massachusetts (190,000 ug/g). Extremely high concentrations (up to 390 mg/Kg) have also been reported in water and sediments from various regions of Japan.

WATER CONCENTRATIONS.

The information reviewed for various nearshore and estuarine areas reveals that, while considerable variability exists, the different regions of the coastal oceans of the world do not show obvious differences with respect to trace metal concentrations in the water

Table 4. Range of Heavy Metal and PCB Concentrations in Selected Sea Water Samples [All conc. in ug/l except Hg and PCBs in ng/l]

	Zn	Cd	Pb	Cu	Hg	PCB
United States						
New York Bight	0.5-190	0.0006-46	0.01-8.0	0.024-22	10-240	0.08-80
S. California	1.1-14	0.15	0.6-5.2	1.6-3.0	NA	2.3-36
Mediterranean	0.1-400	0.01-6.9	0.02-14	<0.03-130	9-230	0.2-38
Japan	<0.1-67	NA	NA	0.4-7.7	56-290	600-900
Baltic Sea	1.5-3.5	0.05-6.0	0.05-15	0.5-23	<2-23	0.3-140
Irish Sea	3.8-49	0.03-1.4	0.6-2.9	0.9-17.3	NA	NA
North Sea	0.8-160	0.05-6.2	1.2-10	0.3-21	3.4-600	NA
Norwegian Sea	14-3600	0.01-85	0.9-27	2.2-58	NA	NA
Spain/Portugal Atlantic Coast	17-530	0.08-6.0	2.3-38	1.9-110	NA	NA

column (Table 4). However, in all areas, there is a tendency for the highest metal concentrations to be found in and near the estuarine discharges, particularly in those estuaries that are subject to industrial discharges.

The concentration of PCBs and chlorinated insecticides in the ocean varies greatly between regions. Unfortunately, due to the small concentrations of these contaminants present in seawater and the lack of intercalibration in analytical techniques, most current data are somewhat unreliable. However, in general, reports suggest that these compounds are concentrated in the surface microlayer of the sea (surface slick) and not in the underlying water column.

The available data on concentrations of heavy metals and organic contaminants in marine waters show that concentrations are universally below lethal toxicity levels, except within extremely localized areas which receive large inputs of industrial and sewage discharges.

BIOTIC CONCENTRATIONS.

The many variables which control the concentration of metals in an organism can pose serious problems when comparing contamination among different areas, as illustrated by the experiences of the Mussel Watch program. Metal concentrations in mussels from areas known to be contaminated have been found generally to be higher than those from non-contaminated areas, but other studies have shown that the variability due to factors other than the environmental contamination level have often obscured the regional differences.

In addition, most metals are present in marine waters and sediments in a variety of physical and chemical forms, only some of which are biologically available. Grimanis et al. (1978) have reported elevated concentrations of arsenic (As), but not other metals, in one of two species of fish from the Saronikos Gulf near the Athens sewer outfall, and Halcrow et al. (1973) have reported that the whelk, *Buccinum undatum*, from the sewage sludge dumping area off the Clyde estuary (U.K.) had higher Zn, Ni, and Pb than in a coastal area. However, Grieg et al. (1978) have reported that metal concentrations found in the various species studied in the New York Bight are comparable to those found in similar species in other regions of the ocean.

HUMAN HEALTH CONCERNS.

Apart from the Minamata incident which fueled a world-wide concern for the possible health effects of toxic contaminants released to the marine environment and ingested in seafood, there apparently has been no well documented case of any clinical or subclinical disease in the global human population that can be attributed to an anthropogenic toxic chemical in seafood taken from non-estuarine marine waters. While clinical problems have been caused by many toxic chemicals released into the environment (e.g., Itai-Itai disease possibly caused by ingestion of Cd-contaminated rice, Pb poisoning in chemical industry workers), all instances of such disease (except Minimata) have apparently been caused by ingestion of food grown in contaminated soil or direct human exposure in the work or living environment.

A recent study which examined human mercury levels resulting from the consumption of seafood from Liverpool Bay and the Thames Estuary, both of which receive substantial mercury inputs, concluded that the people were unlikely to be adversely affected by the presence of methyl mercury in the fish they consumed. Even at the highest levels of fish consumption and, therefore, mercury intake, human blood levels of mercury never exceeded the World Health Organization estimated effect level of 200 mg Hg/liter and rarely exceeded one-tenth of this value. While toxic metals other than mercury have even less likelihood of constituting a human health hazard via the marine environment, safe intake limits of cadmium may be exceeded in Hong Kong, where both metal concentrations in seafood and seafood consumption may be relatively high.

ECOSYSTEM HEALTH CONCERNS.

Ecological changes which have been observed include population reductions of some commercially important finfish and shellfish, altered benthic fauna, and fin rot and other fish diseases. These changes have been observed in the New York Bight and in Southern California, the Mediterranean, the Baltic, and, particularly, the Seto Inland Sea. However, changes have not always been detrimental. For example, there is strong evidence that the productivity of the fisheries outside the Thames estuary was enhanced by the nutrients derived from the London sewers.

Laboratory toxicity studies used to infer that ecological damage may take place in certain contaminated ocean ecosystems are questionable, since there has been virtually no unequivocal demonstration that the ecological changes that have been observed in the marine environment were caused by toxic chemicals. The only exceptions to this rule are where discharges of extraordinarily large quantities of a particular toxic chemical contaminant occurred, either through accidental discharges or ocean dumping.

Indeed, most changes in fish and shellfish populations have probably been caused by natural "climatic" fluctuations or by overfishing of certain populations, rather than by toxic chemicals. The same considerations hold true for other observed ecological changes, including fish diseases.

Studies performed of the impact on benthic fauna by sewage, sewage sludge, and other organically-rich materials have revealed a number of similarities. In the sediments with the highest content of organic carbon, the diversity of the benthic infaunal population is reduced, but the biomass is usually increased. The species that dominate in such an environment are typically deposit-feeders, particularly polychaetes and bivalve molluscs. In the area fringing the central portion of the highly organic sediments, there is a region where diversity of species is normal, but the biomass is higher than in areas farther removed from the high organic sediments. In these transition zones, filter-feeding species and their predators dominate.

The "damage" done to the ecology of sedimentary environments by discharges of organic-rich wastes is apparently not permanent. For example, within four years after removal of a municipal sewage discharge from Cannes, the bottom sediments of the Gulf of Napoule that had been "severely altered" by the discharge had recovered such that the infaunal assemblages were entirely normal. Similarly, benthic recoveries were reported in a Swedish estuary impacted by organic waste from a sulfite pulp mill and in the Firth of Forth (U.K.) after a reduction in suspended solids from sewage effluents.

Bottom water anoxic events that have been observed in the New York Bight and in many other locations throughout the world are generally natural phenomena which, in some areas, may be significantly exacerbated by nutrient-induced eutrophication. The large quantities of nutrients, particularly nitrogen compounds, discharged to the ocean primarily from treated or untreated sewage effluents and agricultural runoff, contribute to higher productivity in many coastal regions. In those areas of the coastal ocean where seasonal or permanently low dissolved oxygen conditions occur naturally, the increased oxygen demand caused by transfer of the excess production vertically into these bottom waters has aggravated the low oxygen condition. Consequently, in many areas throughout the world, the frequency and/or severity of anoxic events and associated fish and shellfish kills has increased.

SUMMARY

Although ecological changes have been observed in the vicinity of many marine outfalls and ocean dumpsites, and concentrations of many toxic chemicals have risen in the marine environment over pre-industrial conditions, it has not been established that these elevated concentrations have exerted any significant effects on the marine ecosystems into which they are discharged, except in a small number of restricted areas where discharge rates are extremely high and where flushing and dispersion are exceptionally poor. Furthermore, marine systems have demonstrated a high degree of recuperative ability when major sources of contaminants to a degraded area are reduced. The major impacts of man on the ocean environment are, apparently, those changes in benthic assemblages caused by excessive organic carbon loadings in areas where sewage, treated or otherwise, and other organic-rich discharges are insufficiently dispersed at the point of discharge. In addition, there is strong evidence that eutrophication, induced by nitrogen and other nutrients in treated sewage effluents and agricultural runoff, may be a serious and growing problem in many coastal areas, particularly those that have seasonal or permanent, naturally-low oxygen concentrations in bottom waters close to discharge areas.

For complete literature citations see:

Segar, D.A. and P.G. Davis. (In Press): Contamination of populated estuaries and adjacent coastal ocean- A global review. Special Report Series "Perspectives on the New York Bight", National Oceanic and Atmospheric Administration, NOAA NOS (OMS).

Prevention of Escaping Oil from Mechanical Barriers in the Presence of Water Currents

K. Ueda and Y. Ueta

Marine Engine Department, Ship Research Institute, Ministry of Transport, Shinkawa, Mitaka, Tokyo, 181 Japan

1. ABSTRACT

In the case of accidents of oil tankers and seaside oil refineries, oil control barriers are usually set on the sea surface to prevent spilled oil from spreading, many types of these barriers have been developed so far. However, the conventional oil control devices are not always effective for containing oil slicks on the actual sea surface. Hence, the improvement in the performance of barriers has been required especially for the high values of current velocity, wave height and wind velocity.

This paper presents the fundamental performance of the oil control barriers by the experiment using large-scale water tanks with the oil control systems.

Furthermore, the effect of equipping the conventional barriers with nets is also contained to calm the upper stream and bottom side of oil slicks.

2. INTRODUCTION

At first, The observation is made on the states of oil layers contained by barriers and the streams of water beneath oil layer. Then the limit of oil controlling efficiencies, and the influences of both wave and wind are expermented. Finally, the effect of the method of the prevention of escaping oil from mechanical barriers with the net forward and under the oil layer is examined.

3. FACILITY AND INSTRUMENTATION

To carry out the above experiments, recirculating channels are used. The channel as shown in Fig.1 is mainly used. To measure wind velocity, a Pitot-tube and transducer are used and the measuring point is 20cm above water level. Water current speed is measured by the vane-type speed meter, and the measuring point is 20cm

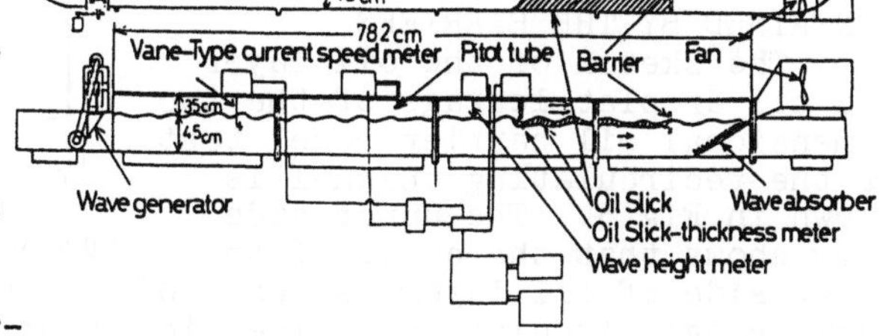

Fig.1. Experimental test tank

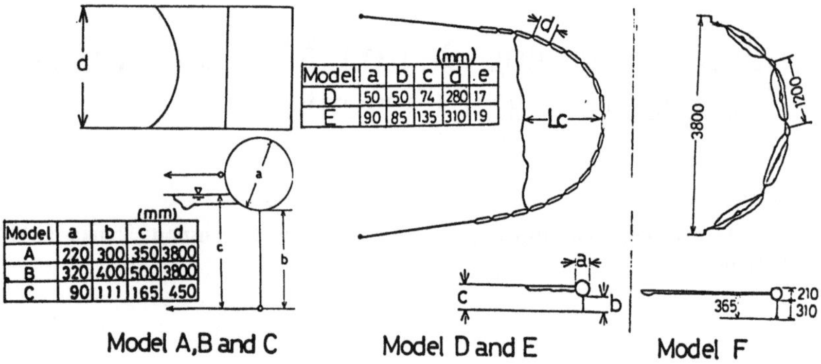

Fig.2 MODEL TEST OIL BARRIERS

(below) water level. The wave height meter is a servo-type and the needle of the wave height meter follows the water surface level or the level of the interface between the water and oil. (Using) this meter, the frequency and the wave height of the interface between the water and oil are measured. To carry out the large-scale model test, the re-circulating channel of the Institute of Ocean Environmental Technology is used. At the test section, this allows area 3.8m wide and 4.1m deep and 60m long to be used. The oil barrier models which are used in the experiment are shown in Fig.2. Flat plates are also used.

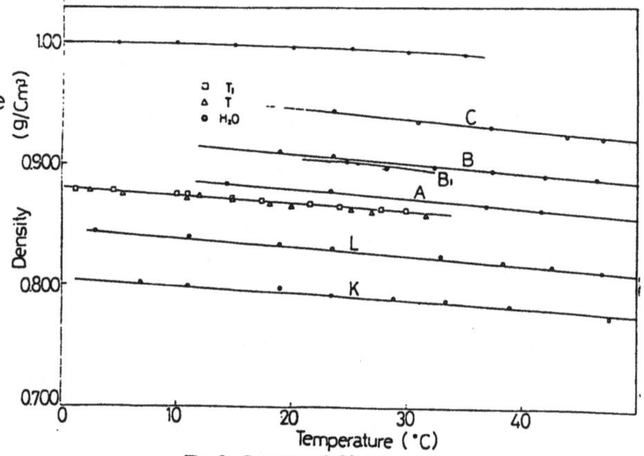

Fig.3 Density of Oils

The oils which are used for this test are shown in Fig.3 and Fig.4. Fig.3 shows the density of the oils and Fig.4 shows the viscosity of the oils.

4. THE BEHAVIOR OF THE OIL LAYER CONTAINED BY THE BARRIERS

The sketch of the oil layer and oil droplet leakage of the two dimensional oil barrier model test at the recirculating channel is shown in Fig.5. The upper side view shows that the curve of the front side of oil layer is shape of "V" and there are two symmetrical circulations. The side view shows that there is

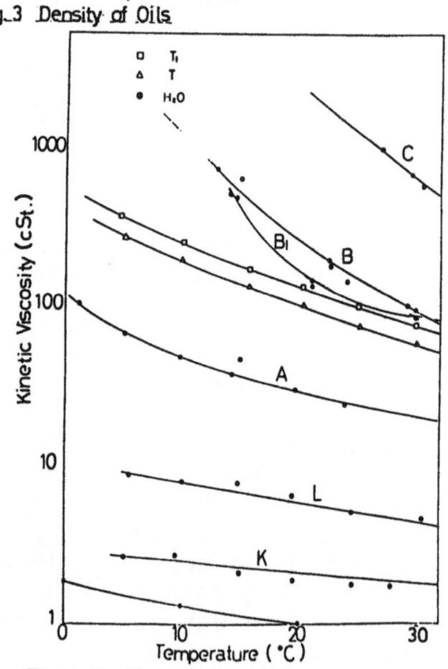

Fig.4 Kinetic Viscosity of Oils

a head wave in front side of the layer. When the current speed becomes fast, oil droplets are torn off the lee side of the head wave and the vortex like B in Fig.5 and the flow like C are observed. Then oil droplets are entrained into the flowing water and escape under the barrier.

The length Lc, which is the oil layer length of the center line, is used as the representative oil layer length because the head waves appear near the center of the oil layer and both walls have little influence on the center line of the oil layer.

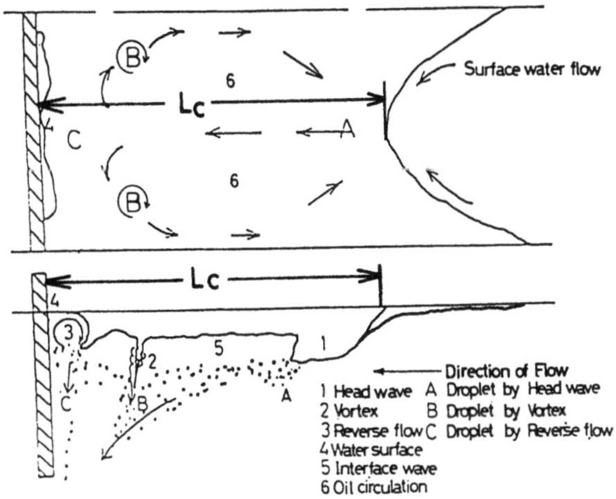

Fig. 5 The behavior of the oil layer contained by barrier

5. THE RESULT OF THE EXPERIMENT

5.1. The Oil Layer by Current

The oil layer on water flow was analyzed by S. Tamiya et al. Concerning these experimental results, they are expressed with the oil layer Froude number "F" and non-dimensional oil layer length "$Lc/\sqrt{S}$" according to Tamiya's analysis.

$$F=\frac{V}{\sqrt{g(1-\sigma)\sqrt{S}}}$$

where V is current velocity (cm/s), g is gravitational acceleration (cm/s^2), σ is ρ_1/ρ_2, ρ_1 is density of oil, ρ_2 is density of water, Lc is the length of oil layer (cm) and S is oil quantity per unit width (cm^3/cm). By the obserbation, the oil layer Froude number, when oil leakage begins to appear is expressed (by) the critical oil layer Froude number "Fc".

Fig.6 shows the relation of F and $Lc/\sqrt{S}$ of oil barrier models concerning different sizes.

No.	Sym	Model	Tank W×D cm	cm	Oil	l	Wave m	Fc
1	▫	P	30×45	5	A	1~8	—	0.77
2	▲	P	30×45	5	B	9	—	0.78
3	●	P	30×45	5	L	5	—	0.84
4	○	P	30×45	5	K	5	—	0.70
5	☆	A	380×430	35	T	500	—	0.75
6	★	A	380×430	35	T	500	3×0.15	0.76
7	○	B	380×430	50	T	500	—	0.79
8	●	B	380×430	50	T	500	3×0.15	0.79
9	■	P	45×45	10	A	12	—	0.75
10	▲	P	45×45	0	T	12	—	0.81
11	◉	P	45×45	10	L	12	—	0.79
12	◎	P	45×45	10	K	12	—	0.89

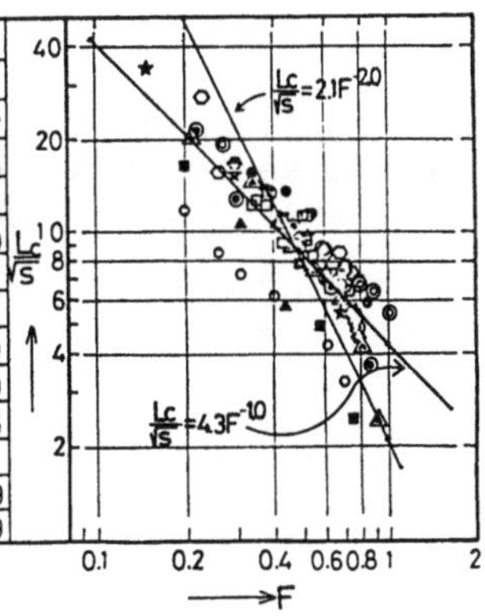

Fig. 6 Relation of F and $Lc/\sqrt{S}$ Concerning Different Model Size

According to Tamiya $Lc/\sqrt{S}$ is expressed with $2.1 \times F^{-2}$, but according to these experiment results, the best estimated line is expressed as follows.

$$\frac{Lc}{\sqrt{S}}=4.3F^{-1.0} \qquad \cdots\cdots(1)$$

In the figure,"P" of the model describes the flat plate. Regardless of the size of oil barrier models, eq. (1) expresses well the state of the oil layers. Some scattering points of Fig.6 seem to be caused by the very thin oil film or very thin dirty film on the water surface. The average of Fc is 0.79. In this case, the influence of the side walls is neglected.

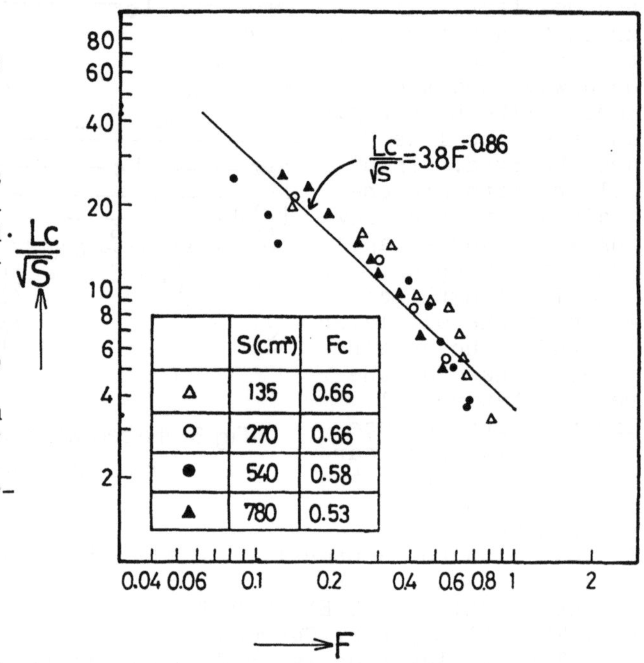

Fig. 7 Relation of F and $Lc/\sqrt{S}$ Concerning Different Oil Quantity (A oil)

5.2. The effect of oil volume.

Fig.7 shows the experiment results concerning the oil volume. The best estimated line of the relation between $Lc/\sqrt{S}$ and F is as follows.

$$\frac{Lc}{\sqrt{S}}=3.8F^{-0.86}$$

Fc is inclined to decrease as the oil volume increses.

5.3. The effect of the oil viscosity.

Fig.8 shows the experiment results concerning the oil viscosity. In the case of low viscosity oil, the oil layer is inclined to become longer. Fc is inclined to decrese

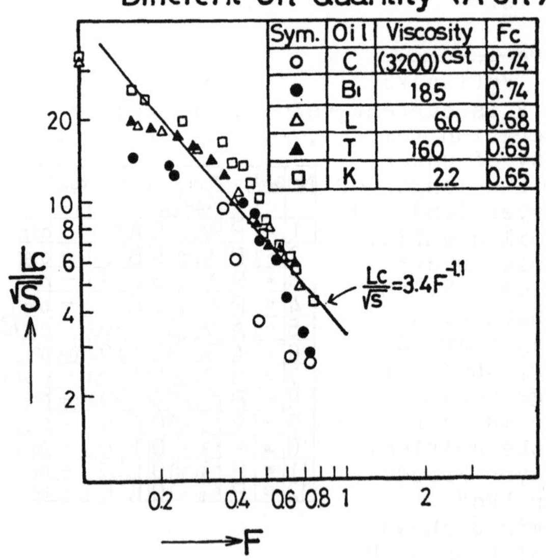

Fig. 8 Relation of F and $Lc/\sqrt{S}$ concerning different viscosity (S=270 cm^2)

as the oil viscosity becomes small.

5.4. The wave height of the oil-water interface near the head wave.

The wave height and the frequency of oil-water interface waves are measured at the point of about 10cm behind the front of oil layer by using the servo-type wave height meter. Fig.9 shows the average of the wave height. From the point where F is about 0.7, the wave height of the oil-water interface near the head wave increases suddenly. This seemes to cause the oil droplet leakage.

5.5. Frequency of the head wave

Fig.10 shows the frequency of the head wave. The frequency of the high viscosity oils, like C-heavy oil increases up to 0.2Hz and the one of low viscosity oils, like kerosene increases up to 6Hz. After that, the oil leakage begins to appear. The high viscosity oils, like C-heavy oil leake as rather large droplets and the oils with low viscosity, like kerosene leake as rather small droplets.

5.6. The deployed oil barrier.

Fig.11 shows the experiment results of the small size of the deployed oil barrier model E. The depth of the model D is too short to contain the oil layer up to Fc. In this case, the oil volume per unit width "s" is expressed with Q/w. Where "Q" is the total volume of the oil and w is the width of the entrance of the

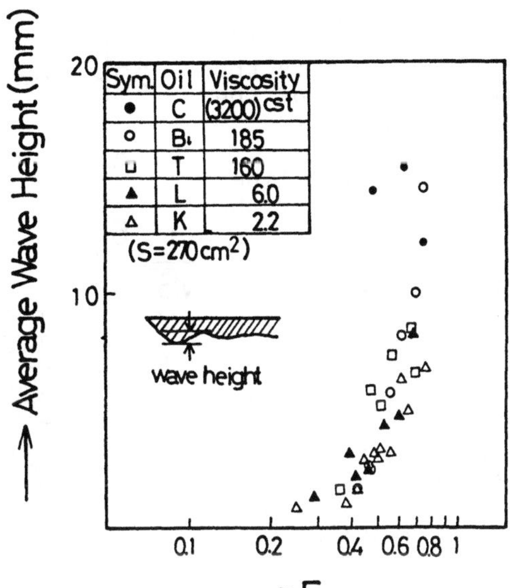

Fig. 9 Average wave height of oil-water interface near Head wave

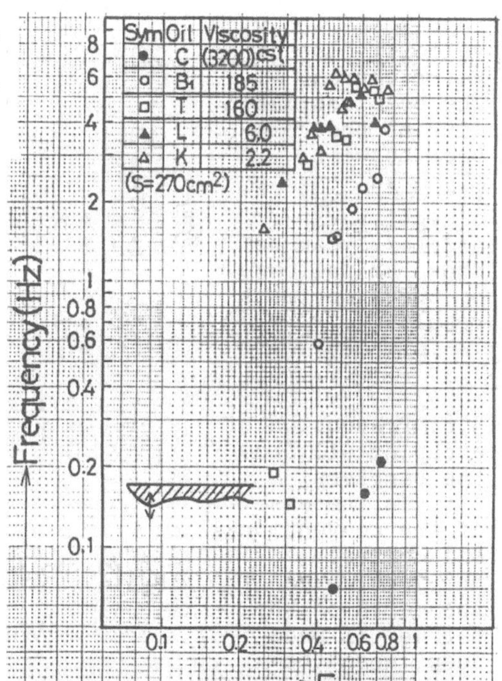

Fig. 10 Frequency of oil-water interface wave near Head wave

deployed oil barrier model. "m" is w/l. "l" is the whole length of the deployed oil barrier model.

When the line of the front of the oil layer is straight, the oil layer is stable. But if "m" is small the line of the front oil layer is inclined to become the shape of V and the oil droplets are inclined to appear. Fc is inclined to decrease as "m" becomes small.

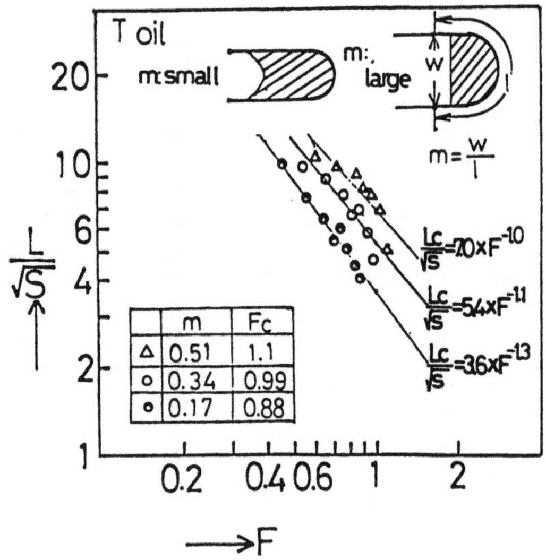

Fig. 11 Relation of F and $L/\sqrt{S}$ concerning the deployed oil barrier (Model E)

5.7. The influence of wind.

The oil layer by wind is shown in Fig.12. "Vw" represents the wind velocity instead of water current V. The best estimated line of the relation between $F(=\frac{Vw}{\sqrt{g(1-\sigma)\sqrt{S}}})$ and $Lc/\sqrt{S}$ can be shown as follows by the experiment result.

$$Lc=135F^{-1.1} \quad \cdots\cdots(2)$$

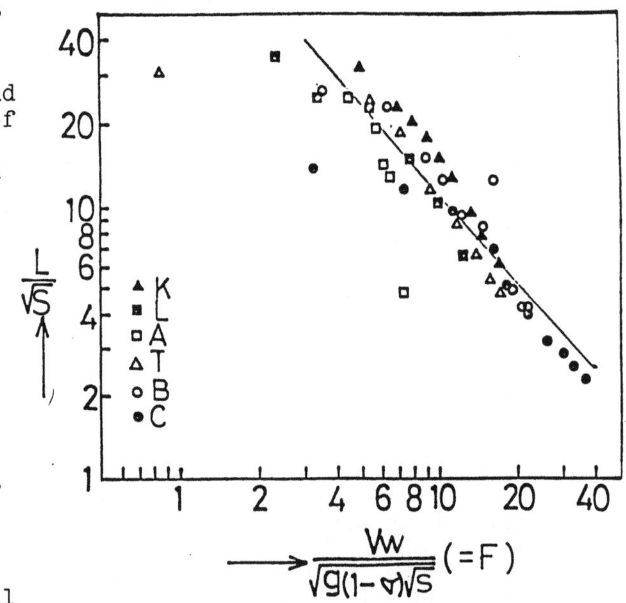

Fig. 12 Relation of F and $L/\sqrt{S}$ concerning wind ($S=270\ cm^2$)

5.8. Prevention of escaping oil from the barriers.

In the case of the conventional oil control barriers, even if the depth of the barrier is made deep, when the current velocity becomes about 50 cm/s, oil droplets appear near the head wave and escape by the vortex under the barrier. So the conventional oil barriers are not always effective for containing oil layers on the actual sea surface. It is difficult to prevent oil from escaping by deploying two or three rows of the oil barrier because of the vortex generation. To prevent oil leakage from the oil barrier, it is necessary to control the vortex generation and the headwave of the front part of oil layer. Concerning this, a method of preventing oil from escaping under the barrier as show in Fig.13 is devised. The effect test result of the double nylon net of 15mm square is shown in Fig.14. This is a comparison of the conventional oil barrier (model F) and the barrier model A with the net. Concerning the model F, when the current velocity is 52.5cm/s and Fc is 0.77, the oil leakage is observed. But concerning the model A with the net, the oil leakage begins to appear when the current velocity is 75cm/s and Fc is 1.1. Using the double brass wire net of 5.5mm square, the effect of each net of this method is examined. Fig.15 shows the result. The net of the front side is very effective.

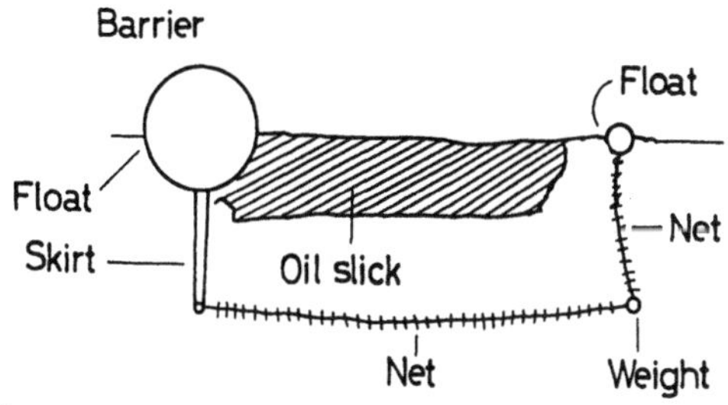

Fig. 13 Prevention of escaping oil from barrier

Sym	Model	Tank	Dep.	Quan	Wave	Fc
		cm	cm	l	m	
○	C	45 x 45	16.5	12	–	0.88
●	C+net	45 x 45	13.5	12	–	
△	C	45 x 45	16.5	12	1.4 x 0.08	
▲	A+net	380 x 430	35	500	–	1.1
◇	F	380 x 430	36.5	500	–	0.77
◆	C+net	45 x 45	15.5	12	1.4 x 0.08	
Oil :T						

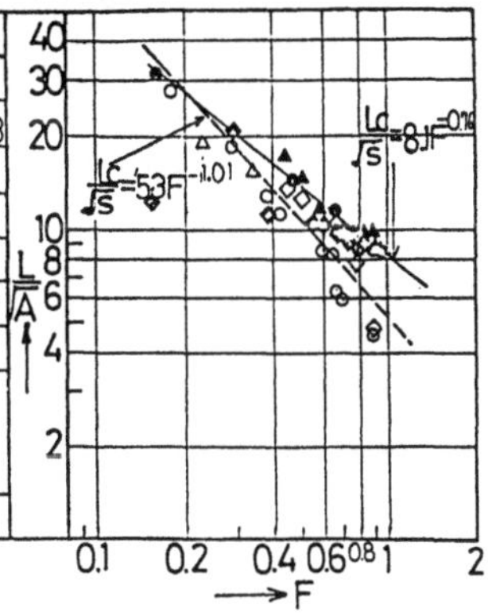

Fig. 14 Effect of preventing method of escaping oil from barriers

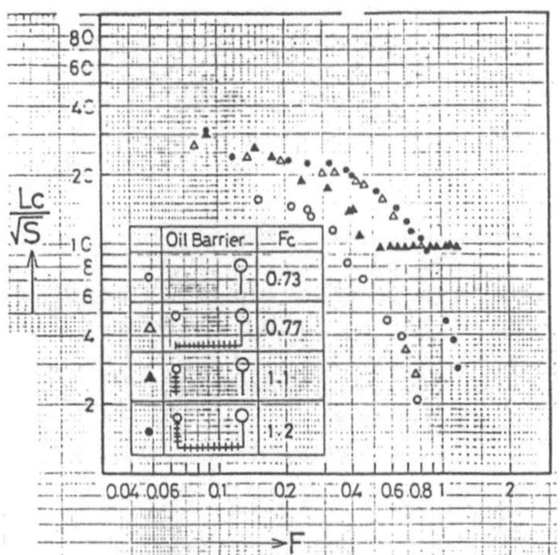

Fig. 15 Effect of Each Net ($S=270cm^2$, Boil)

6. DISCUSSION OF THE RESULTS

The state of the oil layer contained by the barrier can be well expressed with Froude number of oil layer and non-dimensional oil length $Lc/\sqrt{S}$. The relation of F and $lc/\sqrt{S}$ is assumed to be expressed by $Lc/\sqrt{S}=aF^b$, "a" is from 3.4 to 4.3 and "b" is from -1.1 to -0.86 in the case of two dimmensional test. "a" is from 3.6 to 7.0 and "b" is from -1.3 to -1.0 in the case of the deployed barrier test. Concerning Fc which is the value of F when oil leakage appears, Fc is from 0.53 to 0.89 in the case of two dimensional test. Fc is inclined to become small in the case of low viscosity and a large amount of oil. Fc of the deployed barrier test is from 0.88 to 1.1. In the case that m(=w/l) is large, Fc is large too. The wave height of the oil-water interface increases suddenly when F comes near Fc. When the oil leakage is observed, the frequency of the low viscosity oil is from 5Hz to 6Hz and the frequency of the high viscosity oil is about 0.2Hz.

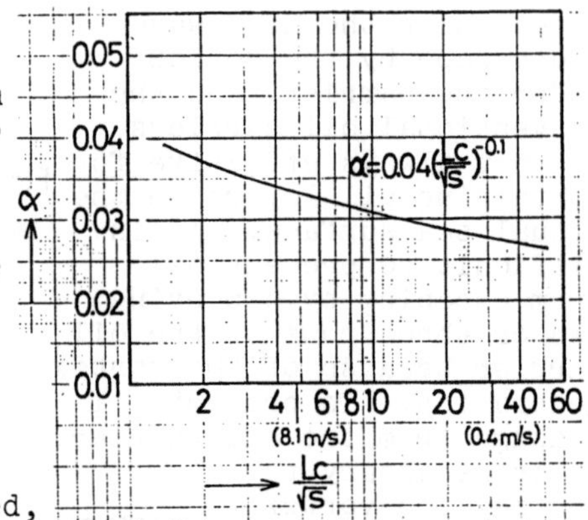

Fig.16 Comparison between wind and water current

Concerning the wave influence on the oil layer, in the case that the oil barrier follows the wave, the influence of the wave is not observed.

Concerning the wind, the influence of water current on the oil layer and the one of the wind velocity are compared. The comparison of eq.(1) and eq.(2) gives

$$\alpha=0.04(Lc/\sqrt{S})^{-0.1}$$

where $\alpha=V/Vw$

Fig.16 shows the relation between α and $Lc/\sqrt{S}$. This means that the influence of the wind velocity on the oil layer is equivalent to the influence of the water current velocity of 2.7%-3.7% of the wind velocity. The velocity shown in the parentheses is the wind velocity in the case of the 12 liter T-oil test.

Concerning the method of the prevention of escaping oil from the barrier, the model A with the net is compared with the model A only. Fc of the model A is 0.75 and Fc of the model with the net is 1.1. The performance is improved by 47%. Fc of the model C with the net is 1.2. By using small mech, the performance is improved by 9%.

7. CONCLUDING COMMENTS

The oil layer contained by mechanical barriers is well expressed with the oil layer Froude number and non-dimensional oil layer length. The oil leakage is observed when the oil layer Froude number is from 0.53 to 1.1. The wind has the same influence on the oil layer as the water current with the velocity which is 2.7%-3.7% of the wind velocity. The method of the prevention of oil escaping from the oil barrier by using net is very effective and the performance is improved by from 47% to 64%.

REFERENCE

1. S. Tamiya, H. Katō, Y. Watanabe and T. Komura : "A treatment of the Equilibrium of oil Layer on Water Flow," Journal of the society of naval Architects of Japan, Vol.135, 1974, pp71-80.

Interaction between Fresh Water and Sediment in Hachinohe Harbor

M. Natori and K. Nikaido

Japan Sediments Management Association, Tsukiji, Chuo-ku, Tokyo, 104 Japan

ABSTRACT

The effects of water quality improvement by removal of sediment are discussed. It has been known that the seawater pollution in Hachinohe Industrial Harbor No. 1 has continued to progress remarkably due to a release of nutrient from pollution sources and accumulated bottom sediment, and the sediment removal was performed as one of the countermeasures. The analysis of the surveyed results of before and after dredging was conducted comparatively in order to know the effects of water quality improvement. With dredging of sediment, the water quality was considerably improved by upgrading bottom sediment that influenced the degree of water quality, and, as a result, the benthos were remarkably rehabilitated.

OUTLINES OF DREDGING WORK

The dredging work was implemented for about 400 days for the periods of November, 1980 and January, 1982. The dredging of a major portion of inner part of the harbor was completed in September, 1980. The dredging materials amounted to about 240,000 m^3 and the mean thickness of dredging materials was approximately 0.7 m. Either grab or pump dredging method was employed depending on water depth, shallow (less than 1 m) or deep. A dredger used for the purpose was a special vessel equipped with apparatus that controls turbidity produced by dredging, and having pump dredging capacity of 500 m^3/h. The driving power was 3,360 horsepower (DE). Two kinds of grab dredger, 2 and 6 m^3, were employed.

ITEMS SURVEYED

The survey in various aspects was conducted on the assumption that dredging could bring about the effect of water quality improvement. The stations for the survey were selected as shown in Figure 1. The items for survey included (1) water quality, (2) sediment, (3) biota, etc.

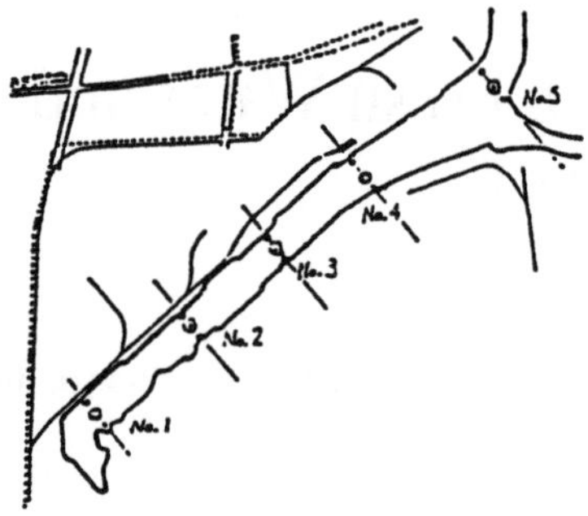

Figure 1. Sampling Stations

Fundamental survey prior to the dredging work and the post-survey to know the dredging effects were conducted in 1979 and in October, 1980, respectively.

EFFECTS OF DREDGING

Change in Sediment

The results of survey for depth before and after dredging are shown in Figure 2. The thickness of sediment ranged between 0.6 and 1.0 m, and the upper layer was almost removed.

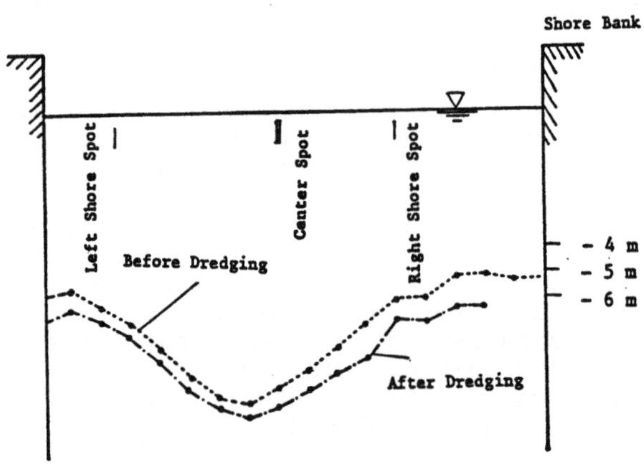

Figure 2. Measurement Results of Depth before and after Dredging (No.4 Station)

The variations of contents of pollutants dredged were studied. With regard to the contents of representative pollutants, such as COD, T-N, T-P and sulphide, the plane distribution before and after dredging is shown in Figure 3(1) ∿ (4).

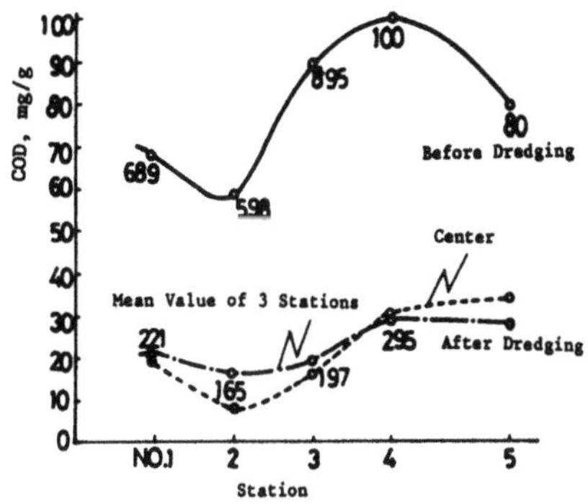

Figure 3(1) Variations in COD at each Station before and after Dredging

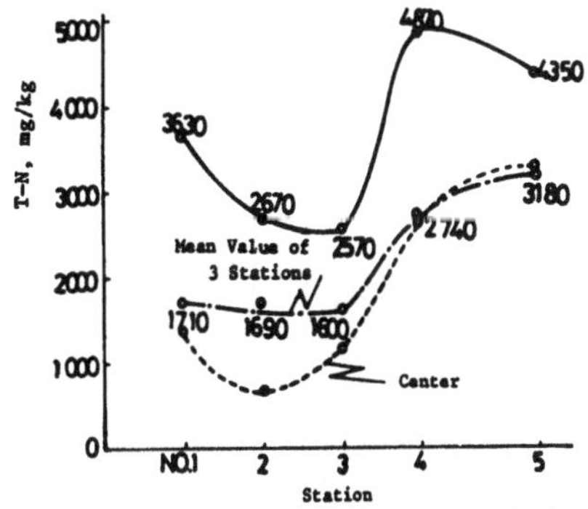

Figure 3(2) Variations in T-N at each Station before and after Dredging

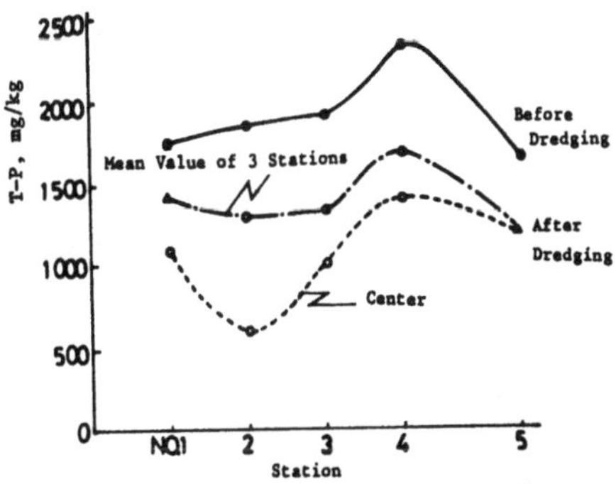

Figure 3(3) Variations in T-P at each Station before and after Dredging

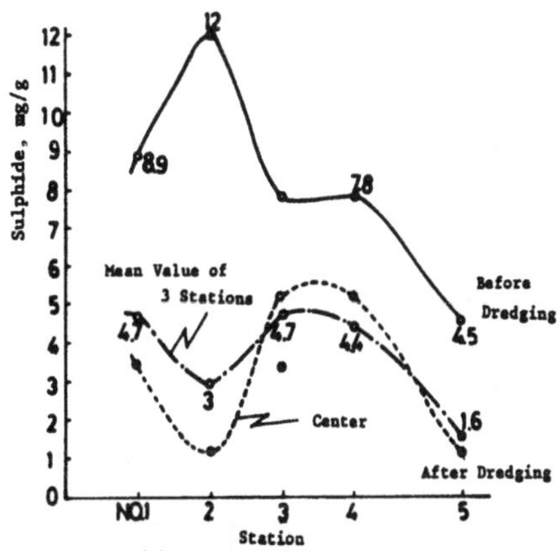

Figure 3(4) Variations in Sulphide at each Station before and after Dredging

Table 1 shows the results of calculation for removal of pollutants before and after dredging. The mean values of removal ratio for all items before and after dredging are 3.113/7 = 0.445. In other words, approximately 55.5% are lowered against quantities of pollutant contents in the upper layer, and it can be said the removal reaches the portion of highly concentrated.

Table 1. Comparative Results of each Station before and after Dredging

		No. 1	No. 2	No. 3	No. 4	No. 5	Remarks	
Water content	before	238.9	124.2	176.2	157.7	155.8	170.56	0.413
	after	69.2	59.4	57.5	72.8	93.4	70.46	
Ignition loss	before	15.1	11.5	16.0	16.4	16.1	15.02	0.57
	after	7.07	6.16	7.30	9.28	12.7	8.50	
COD_{Mn}	before	68.5	59.8	89.5	100	80	79.56	0.29
	after	22.1	16.5	19.7	27.5	28.3	22.82	
Total sulphides	before	8.9	12.0	3.3	7.8	2.0	6.8	0.54
	after	4.7	3.0	4.7	4.4	1.6	2.68	
T-N	before	3.63	2.67	2.57	4.87	4.35	3.62	0.61
	after	1.72	1.70	1.60	2.74	3.18	2.19	
T-P	before	1.75	1.85	1.91	2.32	1.62	1.89	0.71
	after	1.4	1.3	1.4	1.4	1.2	1.34	
n-Hexane extracts	before	23.3	20.3	18.5	21.3	6.08	17.9	0.35
	after	9.2	4.5	6.3	8.2	2.8	6.2	
Mean value of before and after dredging							0.445	

The removal quantities of T-N and T-P are calculated on the assumption that the pollutant containing amo-nts are in proportion to the depth of sediment. 412 ton of T-N and 158 ton of T-P were removed.

Change in Water Quality

The changes in water quality before and after dredging are summarized in Table 2.

Table 2. Comparison of Water Quality before and after Dredging

Item		Before Dredging (in June, Sept.)	After Dredging (in Oct.)
Water temperature		22 ∿ 23.0 (in Sept.)	15 ∿ 160
DO	Surface layer	1∿4 mg/ℓ (in Sept.)	About 6 mg/ℓ
	Bottom layer	5∿8 mg/ℓ	6 ∿ 7 mg/ℓ
SS		Average: 7.52 mg/ℓ (in June)	3.44 mg/ℓ
COD_{Mn}	Surface layer	6.7 mg/ℓ (in Sept.)	4.9 mg/ℓ
	Bottom layer	3.02 mg/ℓ (in Sept.)	3.52 mg/ℓ
T-P	Surface layer	9.48 μg-at/ℓ	5.59 μg-at/ℓ
	Bottom layer	2.62 μg-at/ℓ	1.81 μg-at/ℓ
T-N	Surface layer	2642 μg-at/ℓ	211 μg-at/ℓ
	Bottom layer	416 μg-at/ℓ	91.4 μg-at/ℓ

The vertical distribution before and after dredging is summarized below. Generally, concentration differences of nutrient in the upper and lower layer of the inner harbor, where a degree of pollutant is low, are narrow. The vertical distribution of T-N before and after dredging at No. 4 station is shown in Figure 4, the water quality concentration at upper and lower portion is not seen as that of after dredging. The tendency of plane distribution of water quality is summarized in Table 3.

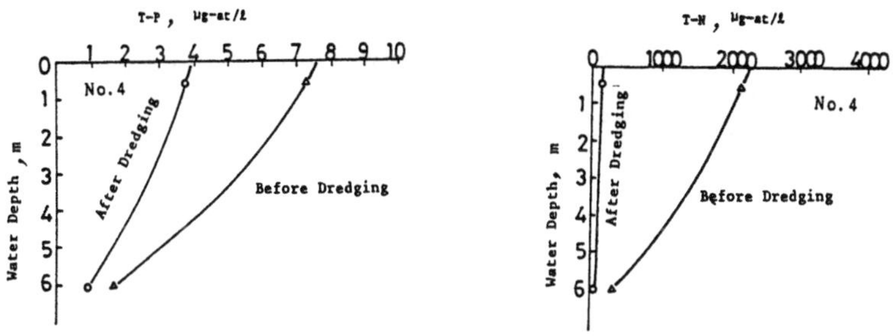

Figure 4. Vertical Distribution before and after Dredging

Table 3. Plane Distribution Tendency of Water Quality before and after Dredging

Item		Fig. No.	Summary
COD	Surface layer	Fig. 5(3)	Varied widely before dredging, but stabilized in the vicinity of 5 mg/ℓ.
	Bottom layer		Tends to vary before dredging, and tends to increase by 25∿4 mg/ℓ towards harbor mouth after dredging.
T-P	Surface layer	Fig. 5(2)	Greatly improved from average 9.5 to 5.5 mg/ℓ.
	Bottom layer		A big difference was not noticed.
T-N	Surface layer	Fig. 5(1)	Reduced greatly from average 2.5 before dredging to nearly 0.02 mg/ℓ after dredging.
	Bottom layer		Reduced greatly from 0.4 before dredging to 0.1 mg/ℓ after dredging.

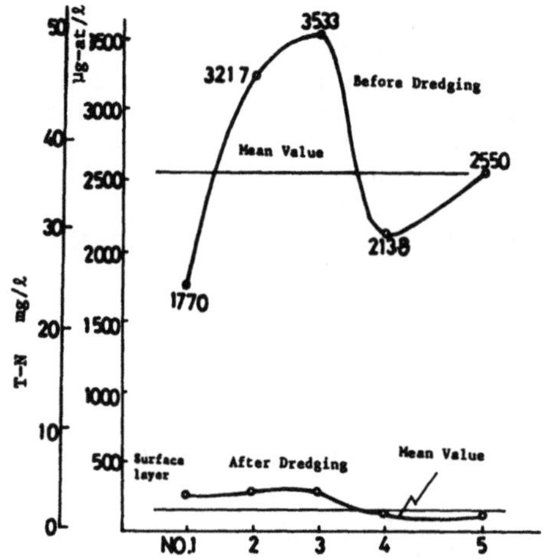

Figure 5(1) Station before and after Dredging, and Total Nitrogen

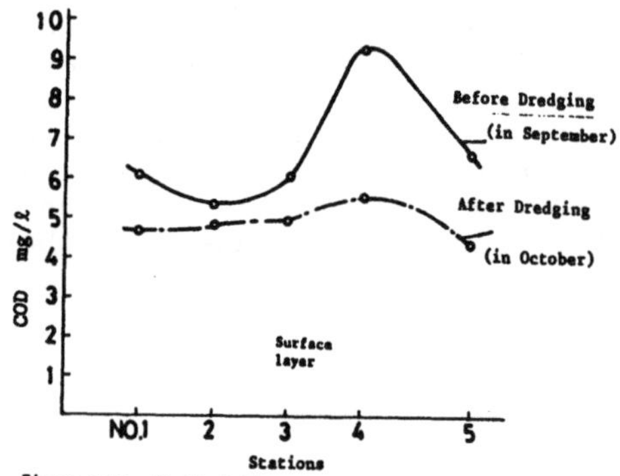

Figure 5(3) Vertical Distribution of COD at each Station

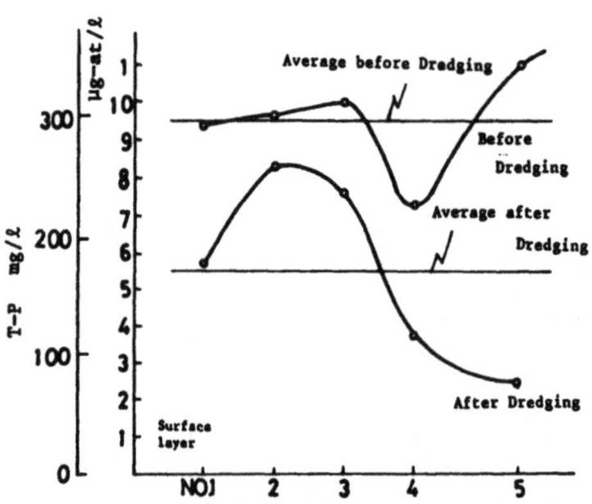

Figure 5(2) Station and Total Phosphorus

Mean values, measured at 5 stations, for T-P and T-N before and after dredging are shown in Table 4. The volume of suspended matters removed was 163.9 kg for T-P and 43.4 t for T-N.

Table 4. Mean Concentration of Nutrient

(μg-at/ℓ)

	Item	Mean Value of 5 Measuring Stations		Mean Value
		Surface Layer	Bottom Layer	
Before dredging	T-P	9.48	2.62	6.05
	T-N	2,642	416	1,529
After dredging	T-P	5.59	1.81	3.70
	T-N	2.11	91.4	151.2

(COD is excluded since value is larger before dredging.)

Change in Biota

Samples of biota were collected from 5 measuring stations each from the left shore, the center and the right shore (total of 15 stations). Kinds of species and population in the center are shown in Table 5 and Table 6, respectively.

From the results of the survey, a study to make a criterion of living organisms and dredging effect was made as follows:

Table 5. Kinds of Species of Macrobenthos

(Species/m²)

Surveyed \ Station		No. 1	No. 2	No. 3	No. 4	No. 5
A (in June) before dredging - 1		2	3	3	5	3
B (in Sept.) before dredging - 2		2	2	2	2	10
C (in Oct.) after dredging	Right shore	6	6	4	4	8
	Central	7	4	8	3	8
	Left shore	8	9	8	7	6
	Total	21	19	20	14	22

Table 6. Total Number (Density) of Macrobenthos

(Significant figure: two figures, population/m²)

Surveyed \ Station		No. 1	No. 2	No. 3	No. 4	No. 5
A (in June) before dredging - 1		1,403	3,910	1.115	189	1,801
B (in Sept.) before dredging - 2		85	607	2,126	208	2,996
C (in Oct.) after dredging	Right shore	8,900	4,800	6,500	1,700	6,600
	Central	14,000	7,200	8,900	1,800	2,800
	Left shore	7,500	12,000	3,500	4,000	3,100

(1) Macrobenthos

a) Number of population is larger after dredging, when compared with that before dredging;

b) kinds of species were changed from 8 species to 20 species after dredging, that is to say, colony is highly diversified; and

c) diversifying index shows more increasing tendency at the end of bay than the mouth of the bay (except for No.4).

Summing up the above, it is considered macrobenthos colonies are greatly influenced by variations of sediment since they inhibit in the sediment. Accordingly, it can be considered that a degree of contamination is low, judging from the high diversifying index which is higher than before dredging.

(2) Meiobenthos

It is obscure since macro- and microbenthos are present in mixture.

(3) Microbenthos

Almost all are dead body of diatomaceae and can inhibit only on surface layer, thus this can not be used for judging dredging effect. What is known from this survey is that, when the microbenthos inhibit in large number, it can be useful to judge a period of living organisms environment. When the benthos is used for index of dredging effect, it is better to utilize macrobenthos.

SUMMARY

As harbor pollution control measures, the dredging work for removing sludge was performed for purification of contaminated seawater at Hachinohe Industrial Harbor No.1 in Aomori Prefecture. In this work, about 240,000 m^3 of sludge were removed. In order to grasp the effect of sediment removal, preliminary and subsequent investigation with regard to water quality, sediment, biota, etc. were carried out.
As a result of water quality survey, suspended solids (SS) was remarkably reduced from 13∿5 mg/ℓ to 7∿2.5 mg/ℓ and from 11∿7.5 mg/ℓ to 8∿2.5 mg/ℓ for total phosphrus (T-P). With respect to sediment, COD was reduced from about 80 mg/g to 23 mg/g, from 15% to 8.5% for ignition loss, and from 8 mg/g to 3.7 mg/g for sulphide.
Meanwhile, kinds of species of benthos was eight, which were increased to 20 species. N-mber of population was 1,400 which increased noticeably to 6,800.
The removal of sediment attributed to the improvement of water quality and living environment of sediment and other organisms, thus resulting in rehabilitation of benthos to a great extent.

REFERENCES

1. Transport Ministry; The Effects of Accumulated Sediment Removal, 1981

Sensitivity of the Kuwait Coastal Environment to Potential Oil Spill Impact

D. Al-Bakri, W. Shublaq, Z. Al-Sheikh and W. Kittaneh
Kuwait Institute for Scientific Research, P.O. Box 24885, Safat, Kuwait

ABSTRACT

Planning for oil spills is becoming a necessity for most of the coastal areas of Kuwait. The important part of planning is the protection of those coastal environments most likely to be seriously damaged by oil contamination in the event of a spill. Based on a reconnaissance field study conducted during May, 1983, the coastal area of Kuwait was subdivided into 10 geomorphic classes each with characteristic morphological and sedimentological features. The relative sensitivity index (RSI), was employed to rate the various geomorphic classes in terms of their sensitivity to potential oil spill damage. A sensitivity map was then developed on which the various coasts of Kuwait were ranked according to their relative sensitivity. Areas with RSI values between 16 and 21 are most likely to be adversely affected by oil spills and should receive priority protection and clean-up countermeasures while areas with RSI values from 4 to 5 are likely to be least affected by oil spills.

INTRODUCTION

Kuwait is situated at the north western corner of the Arabian Gulf, lying between longitudes 46° 30' and 48° 30' east and latitudes 28° 30' and 30° 08' north (Fig. 1). Since the Arabian Gulf is a relatively shallow and semi-closed sea, winds and currents have a pronounced influence on its circulation. The winds affecting the Gulf blow from two main directions, namely the northwesterly wind and, to a lesser extent, the southeasterly. The tides off Kuwait are between mesotidal and macrotidal. Coastal tidal currents vary in direction and velocity along the nearshore area of Kuwait (0.1 to 1.0 m/sec.) The shallow nature of the Kuwait offshore area prohibits the generation of large waves along the shoreline, and waves higher than 50 cm are very rare in the northern area of Kuwait.(1)

The coastal area is considered of special importance to Kuwait since its sea water represents the primary source of freshwater and electricity in the country and due to the fact that most of the urban, industrial, commercial and recreational activities are concentrated along this zone. In addition to that, it is a unique ecosystem and significant nursing ground for many important marine species such as fish and shrimps (the main local food resources).

Since about one quarter of the world's oil supply is produced and shipped through the Arabian Gulf, spills and leaks cause serious environmental problems in the region. In Kuwait, the coastal area is continously subjected to oil pollution resulting from illegal discharge of ballast water, discharge of refineries and petrochemical industries, accidental leaks during loading operations, and other types of spill. Between 1965 and 1976, Kuwait recorded more than 100 spills.[2] Recently, a major spill in the Arabian Gulf began on 2 March, 1983 as a result of serious damage to the wells of the Nowruz oil field, which lies 142 km to the east of Kuwait's mainland shoreline. It was estimated that about 2000 barrels a day were released into the Gulf until the damaged wells were capped almost 12 months later.

In Kuwait, and most of the Arabian Gulf, oil spills occuring in/or near the coastal waters present special problems. In addition to the usual environmental and economic damage, spills may lead to the closure of desalination and power plants thus cut off the country's prime sources of freshwater and electricity.

This study was carried out as an integral part of the effort by the Kuwait Institute for Scientific Research (KISR) to help in the preservation and protection of the coastal environment. The main objective of this work was to develop a sensitivity map on which the shoreline is classified and ranked in terms of its sensitivity to potential oil spill damage, and so permit the concerned agencies to assign priorities to shoreline protection and clean-up countermeasures.

METHODS OF STUDY

The assessment of the senstivity of Kuwaiti coastal environment to oil spill impact began in March, 1983 with a review of the relevant literature and aerial reconnaissance of the coastal zone using a low-flying aircraft. For five days, a ground study was carried out, during which sediment and fauna samples were collected from selected locations to help in the assessment of the physical and biological setting of the shoreline. On the basis of data generated through this reconnaissance survey and information gathered from maps, aerial photographs and published reports[3,4] a geomorphic map of the coastal area was constructed. In this map, the coastline was classified into 10 classes, each with characteristic geomorphological and sedimentological features. The map served as a base for developing the environmental sensitivity map of Kuwait's shoreline.

The principles of the oil spill vulnerability index (VI), developed in the seventies [5,6] were employed to rank the various shoreline types in terms of their vulnerability to potential oil spill impact. The VI is built to reflect the various possibilities of interaction of the shoreline with oil spillage as regards persistance of stranded oil, ease of clean-up, and the extent of the damage to the biological communities. The VI is best used as a guide for comparing the sensitivities of shoreline types not characterized by unique or sensitive features. As shown earlier, Kuwait's coastline is marked by many socio-economic activities that are related to human welfare and safety, and contains unique and sensitive biological features. These factors will strongly influence the ultimate selection of protection and clean-up priorities. Therefore, the VI will not accurately reflect the sensitivity of the shoreline because it ignores the impact of oil spill on the above mentioned factors. To overcome this shortcoming, special features and/or uses for each geomorphic class were defined and then considered, in conjunction with VI, in the assessment of the shoreline sensitivity. In other words, man and all of his activities were considered as important factors in the assessment of oil spill damage on the coastal environment. This approach, considering man as an integral part of the environment, has recently been adopted in environmental shoreline sensitivity studies.[7,8] The special features and/or uses along the coastline of Kuwait are grouped into 3 main categories: sensitive industrial and commercial features, sensitive recreational features; and unique biological features. The environmental sensitivity of the Kuwaiti shoreline was determined by computing the relative sensitivity index (RSI) for each geomorphic class. The RSI was determined by adding the rating values for both the VI and the special features. Then the RSI was employed to construct a sensitivity map to rank the various shoreline types in terms of their vulnerability to potential oil spill damage. Areas with high RSI values are most likely be adversely affected in the event of oil spill and should receive priority protection and clean-up countermeasures.

COASTAL GEOMORPHOLOGY

It has been proven that coastline pollution is greatly a function of coastline geomorphology, exposure to winds and currents, and the type of sediments affected(9). Therefore, it was neccessary to study the coastal morphology and nature of the deposits in order to evaluate the sensitivity of the Kuwaiti coastline to oil spill damage. 500 km of coastlines, including those of the islands, were grouped into 10 classes. The classification proposed by Gundlach and Hayes (5) was used as a guide for studying the coastal geomorphology of Kuwait. The distribution of the various geomorphic classes is shown in Fig. 1 and a brief description of their salient features is given below. Some notes on the general characterisitcs of the biological communities were also included.

Sheltered Muddy Intertidal Flat

These shorelines are located in the most sheltered parts of the Kuwaiti marine environment where the energy level is very low. They are characterized by extensive and gently inclined tidal flats with widths reaching up to 5 km in some areas. The supratidal and upper intertidal areas are covered with muddy sand and sandy deposits. The middle intertidal flat is covered with a mixture of mud and fine to very fine sand while the lower intertidal area is completely covered with mud deposits (silt + clay). The supratidal area is bounded landward by coastal sabkha flats partially covered by vegetation. Biologically, productivity is high in the middle and lower intertidal areas, consisting of large populations of mud skipper fish, crabs, and gastropods.

Mixed Sand and Gravel Beach Fronted by Mud Flat

This shoreline is a relatively sheltered area, protected from waves and strong currents by Bubiyan island. Due to the effect of the tidal current, the energy conditions are relatively higher than those of the sheltered mud flat. It consists of moderately steep (4°-5°) and narrow (5-7m) beach face fronted by a relatively narrow mud flat (200 - 500 m) and bounded landward by coastal sand drifts. The beach is composed largely of a mixture of sand and gravel (0.3 - 4.0 mm grain size). The upper intertidal zone is occupied by oyster banks embedded in a mixture of sand and silt. The middle and lower intertidal areas are covered with mud sediments and moderately rich in molluscs and mudskippers. The biological community of the beach face is limited.

Shelly Sand Beaches Fronted by Mud Flats

This class is characterized by extensive muddy tidal flats fronting beaches and some sand spits or bars, and extensive sabkha flats occupying the backshore. The beach sand is composed primarily of shells and shell fragments. Although this shoreline is developed in a semi-sheltered area, it is subjected to more wave and currents actions than any other part of the northern offshore area. The extensive presence of shell hash along the beach indicates relatively high energy conditions. Biological activities are fairly extensive, consisting mostly of faunal organisms (mainly bivalves of the *Solen sp.*)

Sheletered Rocky Tidal Flat

This area is characterisied by a narrow beach (5-10 m wide) fronted by a wide rocky tidal flat (upto 900 m wide). The beach is composed of medium to coarse sand and bounded landward by coastal ridges and sabkha flats. The rocky area is covered partially by oyster mounds and a thin layer of sand (upto 5 cm thick). This shoreline is found in a semi-closed environment, where moderate waves and currents, caused by the presence of a nearby tidal channel, affect the area. The resident biological community is extensive and diverse, consisting primarily of gastropods (*Cerithidea cingulata)*, oysters, algae and crabs

Artificial Beaches

Narrow (up to 15 m wide) and moderately steep (3-5°) beaches are developed all along the Kuwait City waterfront . The beaches are bounded landward by a heterogeneous fill material used to reclaim large tracts of the tidal area. Consequently, the intertidal zone along this shoreline is either absent or drastically reduced in width (200-500 m when present). The area is characterized by a relatively high energy beach process which is generated by the action of strong currents and limited waves. Beach processes remove sand material from the fill edge to form small beaches at the base of the fill or thin sand patches covering the rocky tidal flat. A limited biological community was observed inhabiting this zone: gastropods, bivalves, tunicates and green algae are dominant.

Exposed, Compacted Tidal Flat

This area is characterized by a straight and open coastline, where the intertidal flat is narrow in most of the area and moderately wide in few locations (reaching up to 400 m), and is largely composed of a hard rocky surface covered in most parts by a compacted layer of fine to medium sand (up to 1 m thick). The upper intertidal zone is covered partially by gravels and boulders embedded in the sand. The beach face is rather steep with slope angle of about 8°, bounded landward by berms and coastal drifts. This area is characterized by relatively high energy conditions as the shoreline is exposed to winds, waves and currents. The biological community is moderately developed, with molluscs being the dominant type.

Eroding Wave-cut Platforms

This shoreline consists of eroding wave benches cut directly into oolitic limestone cliff, 2-8 high. The cliff or rock platform is fronted by relatively narrow and steep beaches (slope angle of 7° - 9°) which are in turn fronted by a wide rock tidal flat (up to 700 m wide) partially covered with oolitic sand. The area is exposed to strong waves and currents, hence it is characterized by high energy conditions. Oncoming waves forcefully reflect back off the rock cliff leaving little or no sediment accumulation at the cliff base and thus allowing little or no sediment accumulation at the cliff base and thus allowing erosional processes to act directly on the cliff. The biological community is diverse but limited in number.

Pocket Beaches and Headlands

This is a cove like coast, consisting of a pocket beach bounded by the two headlands. The beach face is narrow, relatively steep (slope angle of 6-8°) and composed of oolitic material of fine to medium sand sizes (0.06-0.35 mm grain size range). A relatively narrow sandy tidal flat (100 m wide) is fronting the beach face. The intertidal areas off the headland are wide and covered with multiple recurved sand spits and sand bars. The backshore is occupied by sand berms and sand drifts along the northern and middle sectors and by a small oolitic rock cliff (2 m high) along the southern sector. Most parts of this shoreline and particularly the headlands are exposed to high wave energy and a relatively strong longshore current. The biological community is poorly developed.

Exposed Sandy Tidal Flat

This type is marked by a beach sloping at 7°-9°, fronted by a very extensive intertidal flat (reaching up to 2000 m in width). The beach is bounded landward by sand berms in the northern half and wave-cut benches and a small cliff composed of oolite limestone in the southern half. The intertidal area consists of a rocky bottom covered in most places by thick sand deposits and incised by a set of creeks and channels despite the generally flat surface. This shoreline is located in a zone of maximum fetch and subjected to the largest waves in the study area. Megaripples and sandbars developed on the sandy surface are indicative of high wave energy. A strong southward longshore current is responsible for accumulating large masses of sand in this area. The biological communities are moderatley developed, with gastropods and polychaete worms being the dominant types.

Exposed Reefal Flat

Shorelines of the offshore islands of the southern half of Kuwait's territorial water belong to this type. These shorelines are characterised by narrow to extensive oyster or coral reefs developed in the intertidal and subtidal areas of the islands. Some parts of these areas are covered with locally derived fine to medium sand. They are found in an environment of moderate to high energy conditions.

SPECIAL FEATURES

As has been discussed earlier, the coastline of Kuwait is considered of high socio-economic value and contains many unique or sensitive biological features. Such special features will strongly influence the ultimate protection and clean-up priorites because of the public pressure and economic loss associated with the temporary shutdown of services and businesses. Special features or uses for each geomorphic class are defined and briefly described in Table I.

Table I. Special Features of Kuwait's Shoreline

Shoreline Type	Special Features
Sheltered Muddy Intertidal Flat	- Unique biological features: ecologically productive area, containing rare types of organisms such as mudskipper fish and mud crabs, considered representative of the mudflat ecosystem of the Arabian Gulf, some parts allocated for a proposed Nature Reserve, the main nesting, feeding, and breeding grounds for resident and migrating birds, and important nursery ground for fish and shrimps.
Mixed Sand and Gravel Beach Fronted by Mud Flat	- Unique biological features: ecologically productive area, contain rare fauna such as oyster (*Saccostrea cucullata*), and nesting grounds for birds
Shelly Sand Beaches Fronted by Mud Flat	- Unique biological features: ecologically productive area, nursery ground for shrimps, and feeding and nesting ground for migrating birds.
Sheltered Rocky Tidal Flat	- Unique biological features: ecologically productive area and important nursery ground for fish and shrimps. - Sensitive industrial and commercial features: water intakes for two main desalination and power generation plants are located along this coast. - Sensitive recreational features: a marina and boat harbour are present along this coast.

Artificial Beaches	- Unique biological features: nursery rounds for shrimps, and site of an aquaculture research station. - Sensitive industrial and commercial features: the main commercial port of Kuwait, and the water intake for a small desalination and power plant are located on the western end of the coast. - Sensitive recreational features: many marinas, boat harbours, swimming and sailing beaches adjacent to roads and residential areas are present along this coast.
Exposed Compacted Tidal Flat	- Sensitive industrial and commercial features: water intakes for two desalination and power generation plants, cooling and processing water intakes, oil loading terminals and a major importing and exporting harbour are located along the southern sector of the coast. - Sensitive recreational features: marinas, and swimming and sailing beaches adjacent to residential areas are present along the northern sector of the coast.
Eroding Wave-cut Platforms	- Sensitive recreational features: swimming and sailing beaches fronting chalets.
Pocket Beaches	- Sensitive recreational features: beaches adjacent to residential areas in the southern sectors and chalets in the northern parts.
Exposed Sandy Tidal Flat	- Unique biological features: nursery ground for shrimps along some parts of the coast, and an aquaculture research site is located along a tidal inlet. - Sensitive industrial and commercial features: large desalination and power plants, and an oil loading terminal are being constructed along this coast. - Sensitive recreational features: wide swimming and sailing beaches adjacent to chalets.
Exposed Reefal Flat	- Unique biological features: ecologically productive areas, and locally recognized as an important ecosystem largely because of the presence of coral and oyster reef. The islands are important nesting grounds for many types of birds. - Sensitive recreational features: they offer alternatives to the recreational facilities in Kuwait's urban areas.

RELATIVE SENSITIVITY INDEX

The evaluation of the Kuwaiti coastlines sensitivity to potential oil spill impact was carried out by defining the relative sensitivity index (RSI) for each geomorphic class. This index was developed by adding the rating value of VI to those of the special features (Table 2). A rating scale of 1 to 10 was used for VI while a scale of 1 to 20 was employed to rate the special features. In turn the special features rating scale was subdivided as follows: 1-5 for unique biological features (BF), 1-5 for sensitive recreational features (RF), and 1-10 for sensitive industrial and commercial features (ICF). The rating scales was designed to ensure that the special features, in general, and the industrial/commercial features, in particular, will dominate the evaluation processes of determining shoreline sensitivity.

Local experts and scientists were consulted in defining the special features and their relative importance along the coastline of Kuwait. Team consensus was secured for rating the VI and special feature indices. A sensitivity map showing the distribution and ranking of Kuwaiti shorelines in terms of their relative sensitivity to potential oil damage was developed (Fig. 2).

Table 2. Relative Sensitivity Index (RSI) of Kuwait Shoreline

Shoreline Type	% of total shoreline	Special Features Index				VI (1-10)	RSI
		B.F (1-5)	R.F (1-5)	I.C.F (1-10)	Total (1-20)		
Sheltered Muddy Intertidal Flat	50	5	-	-	5	10	15
Mixed Sand and Gravel Beach Fronted by Mud Flat	5	5	-	-	4	9	13
Shelly Sand Beaches Fronted by Mud Flat	6	3	-	-	3	8	11
Sheltered Rocky Tidal Flat	3	4	1	9	14	7	21
Artifical Beaches	5	1	5	8	14	3	17
Exposed, Compacted Tidal Flat	8	-	4	10	14	4	18
Eroding Wave-cut Platform	3	-	3	-	3	1	4
Pocket Beaches and Headlands	6	-	3	-	3	2	5
Exposed Sandy Tidal Flat	5	2	2	7	11	5	16
Exposed Reefal	9	4	2	-	6	6	12

CONCLUSIONS

Applying the RSI, which is the combination of VI and special features indices, in assessing the sensitivity of Kuwaiti coastlines to potential oil spill damage, proved to be a more useful tool than applying the VI alone for such type of assessment. In this approach, the oil spill damage to the natural environment and its impact on man and his sensitive activities were considered in the evaluation process.

Areas classified as being the most sensitive to oil spill damage (RSI=16-21) account for 21% of the total shoreline and include the sheltered rocky tidal flat; exposed, compacted tidal flats, artificial beaches; and the exposed sandy tidal flat. Wave-cut platform, pocket beaches and headlands, comprising of 9% of Kuwait's shoreline, are the least likely to be damaged (RSI=4-5). Within the intermediate category (RSI=11-15) are the sheltered mudflats in the north and the reefal areas around the islands in the south.

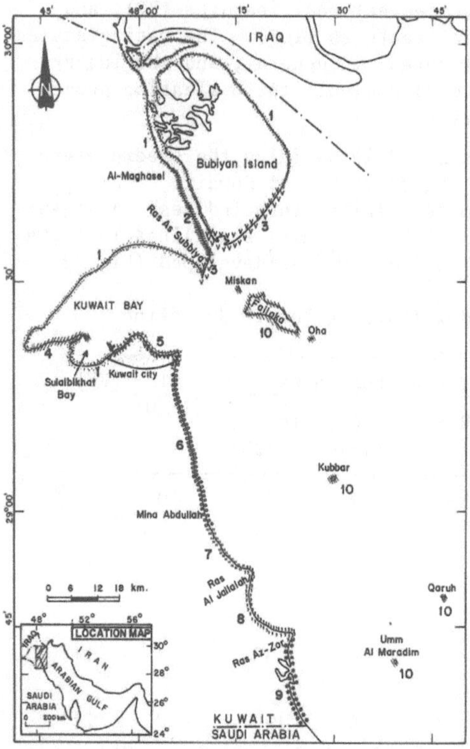

Fig. 1. Location and coastal geomorphology map.

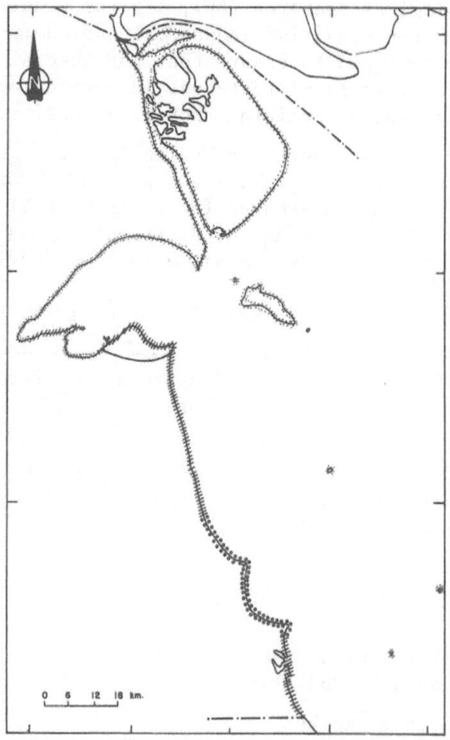

Fig. 2. Sensitivity map of kuwaiti coastline

Fig. 1. Location and coastal geomorphology map.

Fig. 2. Sensitivity map of kuwaiti coastline

1. Sheltered muddy intertidal flat
2. Mixed sand and gravel beach fronted by mudflat
3. Shelly sand beaches fronted by mudflat
4. Sheltered rocky tidal flat
5. Artificial beaches
6. Exposed compacted tidal flat
7. Eroding wave cut platforms
8. Pocket beaches and headlands
9. Exposed sandy tidal flat
10. Exposed reefal falt

Most Sensitive Shoreline

Moderately Sensitive Shoreline

Least Sensitive Shoreline

References

1. Hayes, M.O.: Preliminary investigations of coastal processes and geology of Kuwait city waterfront. Unpublished technical report, Kuwait Municipality (1977)

2. Pickett, R.L.; Partridge, R.M.; Arnone, R.A.; Galt, J.A. The Persian Gulf, Oil natural circulation. Sea Technology (1984) 23-25.

3. Khalaf, F.I.: Intertidal flat sediments of Kuwait, preliminary sedimentological, textural and mineralogical investigation. A technical report No. KISR/PPI 039/EES-RR-R 7804. Kuwait Institute for Scientific Research, Kuwait, 1978.

4. Al-Abdul Razzaq, S.; Khalaf, F.; Al-Bakri, D.H.; Shublaq, W.; Al-Sheikh, Z.; Kittaneh, W.; Al-Ghadban, A.; and Al-Saleh, S.: Marine sedimentology and benthic ecology of Kuwait marine environment. A technical report, Kuwait Institute for Scientific Research, Kuwait, KISR No. 694. 1981.

5. Gundlach, E.R.; Hayes, M.O.: Vulnerability of coastal environments to oil spill impacts. Marine Technology Society Journal. Vol. 12, No. 4 (1978) 18-27

6. Michel, J.; Hayes, M.O.; Brown P.J. Application of an oil spill vulnerability index to the shoreline of Lower Cook Inlet, Alaska. Environmental Geology. Vol. 2, No. 2 (1978) 107-117

7. Lindgren, K.; Norrby, S. Oil and chemical pollution from ships. Min. of Argric., Stockholm, (1980) 201.

8. Owens, E.H.; Ribilliard, G.A. Shoreline sensitivity and oil spills. A reevaluation for the 1980's. Marine Pollution Bulletin, Vol. 12, No. 3 (1981) 75-78.

9. Berne, S.; Marchand, M. amd D'ozouville, L. Pollution of seawater and marine sediments ion coastal area. AMBIO, Vol. 9, No. 6 (1980) 287-293

Acknowledgement

The authors would like to express their appreciation to Dr. F. Khalaf and Dr. H. El-Baroudi for their continous support during the various stages of the work. Thanks are also due to Ms. F. Al-Suderawi for her help and advice, Mr. Paolo Schiannini for editing the manuscript, and Ms. Leena Lobo for providing necessary typing.

Some Problems of Environmental Investigation and Assessment

M. Shinozaki and Y. Shinozaki

Takenaka Technical Research Laboratory, Takenaka Komuten Co., Ltd., Minamisuna, Koto-ku, Tokyo, 136 Japan

1. Introduction

The construction of airports, power plants, waste disposal plants and large-scale industrial bases has often caused noise, vibration, air pollution and water pollution on land, and so these facilities at sea have recently been considered and some of them have already been constructed and been in operation there. However, they are of large-scale or super large-scale and environmental impacts, such as water pollution, deposits of drift sand or coastal erosion, a shift of wind brought by the occupation of an enormous space and changes of scenery, result from these construction works. Also, as a large quantity of waste water is drained with the operation of these facilities, contaminants in traces are eventually accumulated in the sediment and living things, and various problems arise along with it. Since these construction works have much influence on the neighborhood living environment, it is the trend of public opinion, in planning and executing works, to assess the environmental influence in advance (hereinafter referred to as "the environmental assessment"), to estimate the items for influence and their quantity (hereinafter referred to as "impacts"), and that unless they are above the standards for the preservation of environment, the works can not be done. In the environmental assessment, the author is now picking up several points, which are as follows:

(1) To grasp the mechanism of the generation and spread of impacts.

(2) To grasp "the existing condition" and to assume its model.

(3) To make the assessment items and to establish the standards for the assessment.

In order to estimate the impacts of natural environment, it is most important to grasp the mechanism of the existing condition quantitatively, to set up the assessment system incorporated with it, and to select indicators suitable for combining the impact system with the assessment.

This report deals with the necessity of examining the characteristics of the sea area in order to grasp the existing condition, the process of grasping the many-faceted features, water and sediment as a ecosystem and planktons which feed on them, the recognition of the existing condition of phenomena with compound characteristics, and the introduction of study cases.

2. Outlines of Environmental Assessment

Reclaiming land from the sea for constructing an industrial complex is given as an example for outling the environmental assessment.

The construction works are done in order of reclaimation, dredging, embankment, creation and foundation work. And then, port facilities and complex are constructed and start operating.

After a lapse of several years, its effect can be seen there. The items influencing environment, covering from the construction to the operation, are referred to as "the influencing element on environment".[1)] With the progress of the work, the influencing elements have an impact on every moment upon the environmental factors, such as water quality, muddiness, the shift of current and water temperature. Also, the combined factors influence the present phenomena, such as coast configuration, sediment, ecosystem and environment. The impacts on natural environment in human society are the results of compound phenomena and are judged from the standards for the preservation of envirionment. Fig. 1 shows the above mentioned process. By the way, the measures for preserving environment are given by "The Assessment for the Environmental Influence" (The Central Council for Anti-Pollution Measures). The idea is shown in Fig. 2.

3. Mechanism of the Generation and Spread of Impacts

Reclaiming work is given as an example for discussing one of the problems, the generation and spread of impacts. Part of the reclaiming work is modeld in Fig. 3. Fig. 4-1 shows an interaction between environmental impacts and "excavating", a process of the ground improvement, which is part of what is described in Chapter 2. Also, Fig. 4-2 gives a link between digging and dredging in excavating process, and their effects. Impacts are generated by digging and dredging the seabed, a mere micro part in excavating which is a process of the ground improvement necessary for reclaiming work in a macro construction work. Since the impacts are connected with various matters before they are accumulated and revealed as influencing phenomena, it is suggested in Fig. 4 what indicators should be selected to estimate the effects in advance. The flow shows the relaction between water and living things, between sediment and living things and between sediment and marine phenomena (waves, tides and coastal current), and the characteristics of living things there. The flow is the very ecosystem of the coastal zone with compound phenomena, and it is very important but very difficult to grasp the ecosystem. Therefore, it is most effective to classify the ecosystem extensively. A study of various phenomena, which aims at "the classification of sea areas" will be made later.

4. Environmental Indicators and Their Characteristics

To grasp the existing condition and to estimate impacts, a quantitative measurement by means of environmental indicators is necessary. At present, concerning the indicators aiming at the preservation of coastal environment, DO (dissolved oxygen), COD (chemical oxygen demand), pH and SS (suspended solids) are defined as the indicators for water quality and their quantitative standards are established. Table 1 shows the Environment Code. However, as described in Chapter 3, marine phenomena are so complicated that extensive and careful consideration should be given to the indicators for assessment. In examining the indicators, the following three points will especially be important:

(1) Type of indicators: differential or integral.

(2) Information obtained: Local or extensive.

(3) Phenomena grasped: Steady or unsteady.

In (1), water quality, a typical indicator of coastal environment, is of a local and differential type changing momentarily, where as, in the formation process of seacoast, the aspect of the coast in summer and winter reflects the weather and marine phenomena then, and the sediment at the seaside and in the surf zone of beach is classified uniquely according to its grain size resulting grom the accumulation of impacts ranging at least from several weeks to several months, which is considered to be information of a integral type.

In (2), for instance, the expansion of indicators can be seem by following suspended solids peculiar to the river flowing from its mouth into the sea, and the movement of the ocean current and the position of a cold water mass can be obtained by measuring the expansion of temperature.

Also, in (3), phenomena which are supposed to be caused by some familiar impacts like an abnormal development of some kinds of planktons, and the fact that the catches of migrating fishes at the time of periodical migration have relevance to the increase and decrease of sunspots, are examples of unsteady or steady phenomena.

An example of the relation between the impacts of reclaiming work and the indicators for assessment is given in Table 2. Further, the indicators in Table 2 are examined and their characteristics are given in Table 3, taking the abovementioned problems, (1) to (3), into consideration. In Table 3, the view point to assess influencing phenomena is divided into the following two points:

Physical phenomena covering environmental factors and habitats surrounding the physical phenomena.

In addition, as substitute characters for changes of environment, time, space and phenomena are selected, which have two categories, respectively and show the characteristics of the indicators.

5. Characteristics of Sea Areas

In relation to influencing phenomena as an ecosystem, factors on the seabed, such as water, living things and soil of seabed are to be examined.

5.1 Grasp of Sea Areas characterized by Water Quality and the Distribution 300 logical Planktons

Studies of the mutual relationship between the distribution of 300 locgical planktons and exchange of their dominant species and seawater, and marine phenomena like water quality have long attracted the attention of many researchers. In order to survey the rapid pollution in the Inland Sea of Japan, Hirota[2)] et al. have recently clarified specific dominant species and the corresponding marine phenomena as mentioned hereunder.

An example of copepods is shown. The order of their appearance is given ranging from the inner part of the bay, where nutritive salts due to land water contaminated, are highly concentrated, but the salinity is lowered, the waters are shallow and muddy, and the water temperature is greatly changed, to the mouth of the bay, where the concentration of nutritive salts is lowered, but the salinity is highly concentrated, the waters are deep and clear, and the water temperature is slightly changed. The results of the survey on the population by dominant species at each measured point in the S Sea Area are given in Fig. 5 with the axes of species. To analyze this data statistically and to grasp the characteristics of each point, a practical application of the Hayashi's Third Statistical Theory is made. As a result, the classification of sea areas is obtained as shown in Fig. 6.

According to Fig.6, St. 10 and St. 26 show the properties of a bay-type, and St. 9 in the river-mouth is considered to be a stagnation point due to a countercurrent. The ocean-type I is reasonable enough but the ocean-types II and III are contrary to our expectations. This area seems to be a large stagnation with the whirpool, considering the existence of a clockwise flow around the ocean-type III due to the ocean current.

5.2 Characteristics of Sea Areas from the Viewpoint of Mineral Composition of Sediment[3)]

In surveying the coarse mineral composition of sediment in the S Sea Area, sampling stations and the results of the analyses are shown in Fig. 7 and in Table 4, respectively. Based on the patterns of X-ray diffraction analyses, the following sorts of sea areas can be obtained:

(1) St. 2, St. 3 and St. 30 (in the mouth of the H River) have higher relative intensity of albite than any other station.

(2) St. 9 and St. 15 (in the vicinity of the port) have no peak intensity of plagiodase.

(3) St. 103, St. 109, St. 30, St. 26 and St. 24 (in the range of the mouth of the H River and offshore) are closely akin in X-ray diffraction pattern.

(4) St. 13 is similar to the stations in both (1) and (3), and seems to form the border between the sea areas in (1) and (3).

The following inference can be drawn from the above.

The area of influence of the basin of the A River is limited to part of the mouth of the A River and its vicinities. The fact that a different distribution can be seen in St. 9 and St. 15, is due to the effect of dredged earth which accompanied the port construction of several years ago. The reason why the distribution of the sediment with similar geological features to those in the basin of the H River can widely be seen offshore, in St. 103, St. 109, St. 30, St. 26 and St. 24, is that, after the H River drained into the sea, that distribution came up north with the coastal current coming up north. Therefore, it is necessary to give full consideration of these conditions to the area of influence of the H River.

Next, fine-grained minerals, especially clay minerals, are considered. An analysis of clay minerals is shown in Table 5. To sun up, the distribution is characterized by a large relative quantity of chlorite in the coastal zone and much montmorillonite offshore.

In analyzing items on sediment pollution, a greater response of TP (total phosphorus), T-N (total nitrogen), sulphide, COD (chemical oxygen demand) and ignition loss is shown in both St. 26 and St. 24 with a large quantity of montmorillonite. The distribution of heavy metals can scarcely be detected, but a response of T-Hg (total mercury), Cd (cadmium) and Pb (lead) is shown in St. 26 and St. 24.

Another example of the relation between the distribution of clay minerals of sediment and items on sediment pollution is given in the case of Tokyo Bay. Fig. 8, Fig. 9 and Fig. 10 show the distribution of clay minerals, COD and heavy metals, respectively. As shown in these Figures, except the river-mouth, the areas where the montmorillonite content is high, are heavily polluted. However, it was found that the above-mentioned relation does not apply to the sea area having been contaminated excessively, like Tokyo Bay, especially on organic pollution in summer. Judging from the fact that the values of other ignition losses increase in summer, a large quantity of remains of planktons are considered to be heaped up and decomposed violently.

Similar results are obtained concerning the relation between the distribution of clay minerals and environment items on sediment pollution in the case of Osaka Bay and the Inland Sea, too. From the above, it can be considered that the distribution of montmorillonite with high surface activities fulfils the function of the indicators for sediment pollution.

5.3 Grasp of Sea Areas characterized by Sediment
—— Characteristics of sea areas observed from the grain size distribution of sediment ——

Sediment moves due to disturbance and it is called the drift sand phenomenon in coastal zones. In order to clarify the drift sand phenomenon, it is necessary to examine the composition of grain sizes of sediment. Sieve analyses are often done to obtain the following values:[4),5)]

(1) Grain size of central value $Md\phi = \phi_{50}$

(2) Grain size of mean value $M\phi = 1/2(\phi_{84} - \phi_{16})$

(3) Standard deviation $\sigma\phi = 1/2(\phi_{84} - \phi_{16})$

(4) Skewness $\alpha\phi = \dfrac{M\phi - Md\phi}{\gamma\phi}$

where ϕ_{50}, ϕ_{84} and ϕ_{16} are grain sizes corresponding to 50, 84 and 16 percent, respectively, in the cumulative percentage P in the ogive for grain sizes expressed in the ϕ unit.

The results classified on the coast show a curve resembling the normal curve or the error curve in the grain-size distribution curve or the ogive for grain sizes. Though the theoretical endorsement from the hydrodynamic point of view may be an untouched subject, plotting the relation between each of (1) central values, (2) standard deviation or selection coefficient and (4) skewness, and the depth of water, or the combination of parameters of central values, and selection coefficient or skewness, is practical and useful for estimating phenomena.

As an example, the sediment in the northern part of Sagami Bay is given. The relation between grain-size composition in that sea area and the depth of water is shown in Fig. 11 (a). Concerning the relation between the depth of water and the selection coefficient, a boundary line is drawn at a depth of 12 meters in sea water, medium or fine sand is found in the shallower part and soil can be seen in the deeper one, and the selection coefficient is slight in the part shallower than 12 meters as shown in the Figure. In this sea area, because of high selection, waves have much influence on the part deeper than 12 meters, but have little on the depths. From the relation between $Md\phi$ the central value of grain sizes, and $\sigma\phi$, the selection coefficient, it is seen that grains of sand with $Md\phi$ are approximately equal to 2, medium and fine sand, are highly selected. In contrast with this, Fig. 12 shows that, in the S Sea Area, the selection coefficient concentrates on a depth of about 8 and 14 meters and on central values $Md\phi$ equal to 2.5 to 3.0, but is slight in the part shallower than 5 meters. Since the slope of the seabed in Sagami Bay is steeper than that in the S Sea Area and the former has the boundary line at a depth of 12 meters, where as the latter at a depth of only 5 meters, the depth of about 5 meters in the S Sea Area is considered to be the surf zone of beach with a gentle slope and horizontal displacement occurs at this water depth. Further, as shown in Fig. 11 (b), the boundary line of the delta of the Obitsu River in Tokyo Bay is shallower than that of the S Sea Area, and there seems to be a slow current due to the waves.

6. Summary

This report has dealt with the grasp of the existing condition which is most important as a common problem in surveying the environmental assessment and the indicators for it, especially "sea area".

"Outlines of environmental assessment" has mentioned the process intended as a measure at present and the existence of difficult problems in carrying into execution.

"Mechanism of the generation and spread of impacts" has clarified the relation of cause and effect of impacts as a clue to the solution of the problems mentioned above, has arranged it in time series, and has discussed it as a proposal for finding estimated items and the indicators for it.

"Environmental indicators and their characteristics" has taken up indicators and has stated the necessity of studying what the chosen indicators mean to estimate phenomena. Especially, items and categories are given in the indicators in the sea area and the characteristics of each indicator are tabulated.

"Characteristics of sea areas" has shown an example of the S Sea Area in line with a ecosystem view.

For the characteristics of sea areas, "water quality and the distribution of animal planktons" and "the distribution of mineral composition of sediment" have been taken up. And the latter has dealt with the relation between the distribution of clay minerals, and the pollution and contamination of sea areas, and has proposed that the distribution of clay minerals should be an indicator following up a change with passage of time in contrast to water quality, an instantaneous indicator. It has also related with examples that the distribution of coarse minerals is an indicator corresponding to marine phenomena, physical forces, such as currents and waves, where clay minerals are an indicator making use of chemical properties of the surface of the sea.

7. Conclusions

This report is mainly composed of a few examples which the authors have experienced, and careful consideration should be given to things which have not fully been discussed. However, we cannot wait for a perfect solution for the preservation of natural environment. Therefore, it is very urgent to make good use of already-known information and knowledge which are considered to be accumulated with time and to be improved to a higher level. To attain this, it is necessary for us to select information concentrating on the points that are practicable, to construct a theory for solving the problems on the basis of the information, to clarify precise and imprecise points by superposing the solutions of some tentative plans, and further, to concentrate our efforts to study the imprecise points, though it may be commonplace. This report has made no mention of the following; however it is also necessary to aim at the assessment grounded on the value system from the point of estimators, namely, the measure of value taken by both beneficiaries and sufferers. The authors have proposed that the idea that the estimate of the effects of development works is finally related to human life, should be the starting point, that the iving space, a fundamental element of human life, should be grasped as property in "economics", that various kinds of property, such as material, immaterial, social, political and environmental property should be regarded as viewpoints for assessment, and that assessment items should be selected, based on a genera idea of property in line with social indicators, but I have not touched on these matters in this report. However, let me add that the indicators having been examined in this report are selected from among them.

References

1) Hajime Inamura, "Environmental Assessment Method in Port Planning", Technical Notes of the Port and Harbor Research Institute of the Ministry of Transport, No. 235, 1976.
2) Reiichiro Hirota, "Oceanic Life in the Inland Sea of Japan", Vol. 14, pp. 245 - 251.
3) Yukuo Shinozaki et al., "Grasp of Sea Area Environment from the Viewpoint of Mineral Composition of Sediment", Proceedings of the Annual Conference of the Japan Society of Civil Engineers (Résúme), October, 1979.
4) Noriyuki Nasu, Sea Bottom, "Composition of the Earth", pp. 135 - 150.
5) Goda et al., Seashore, Port and Harbor, pp. 301 - 357.
6) Yuji Hirayama, "How To Estimate Abundance", Nikkei-shinsho.

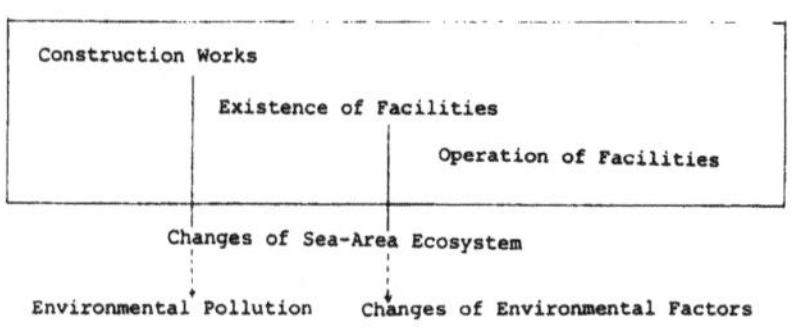

Fig.1 Environmental Assessment System

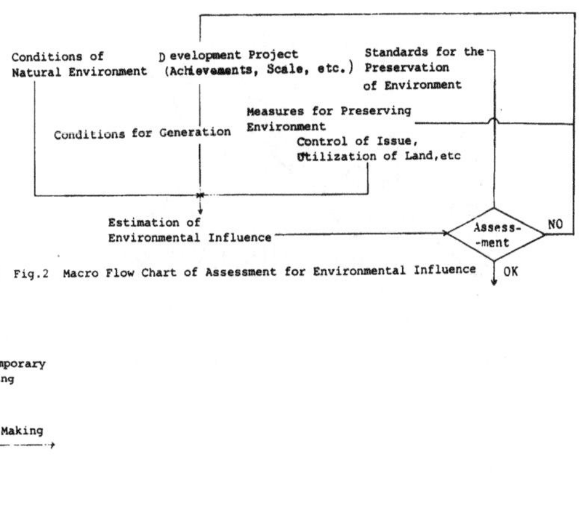

Fig.2 Macro Flow Chart of Assessment for Environmental Influence

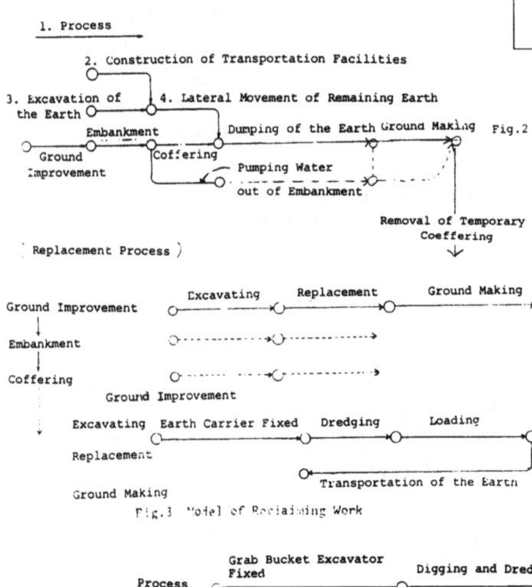

Fig.3 Model of Reclaiming Work

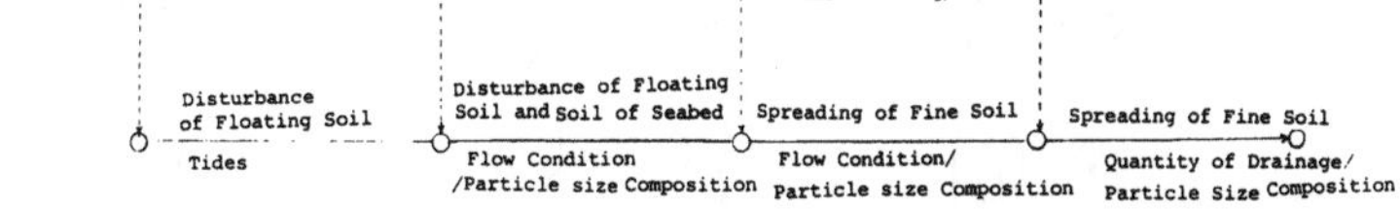

Fig.4-1 Interaction between Excavating Process and Impacts

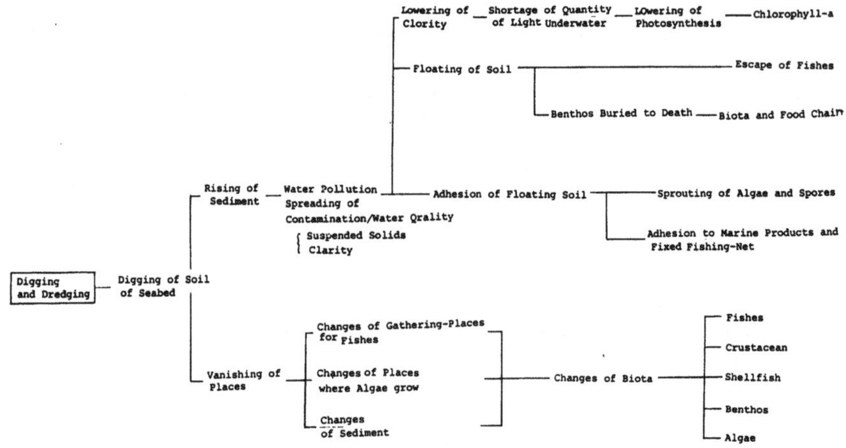

Fig.4-2 Link between Digging and Dredging, and Their Effects

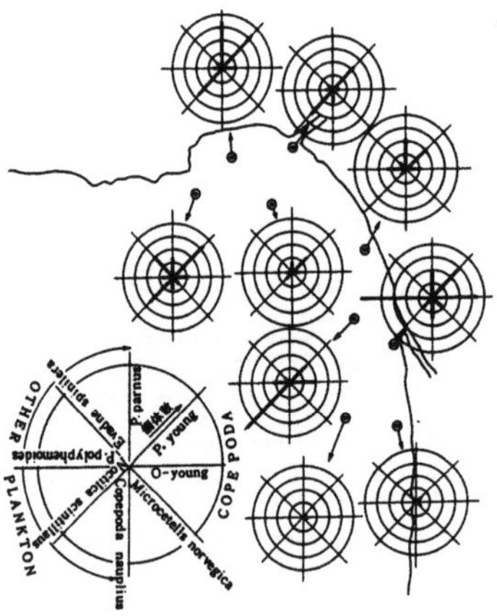

Fig.5 Distribution of Animal Planktons in the S Sea Area

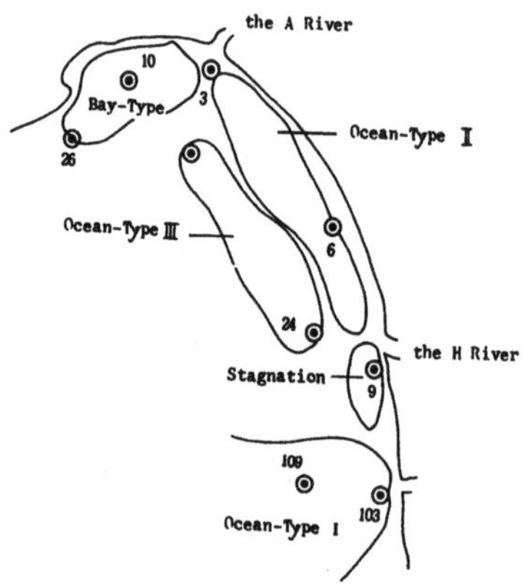

Fig. 6

Classification of Sea Areas for the Distribution of Animal Planktons by the Hayashi's Third Statistical Theory

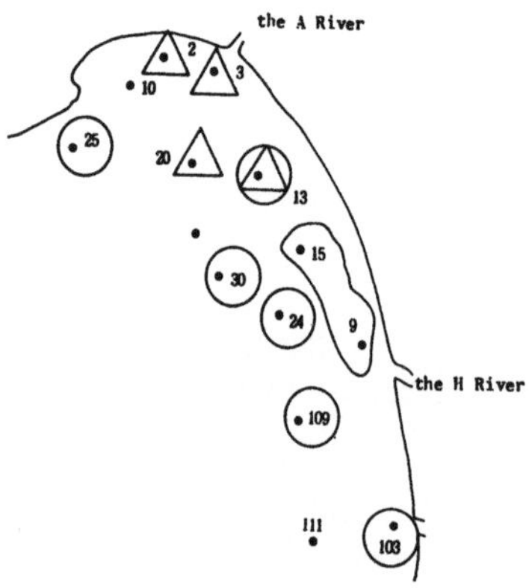

Fig.7 Gathering Stations for Sediment

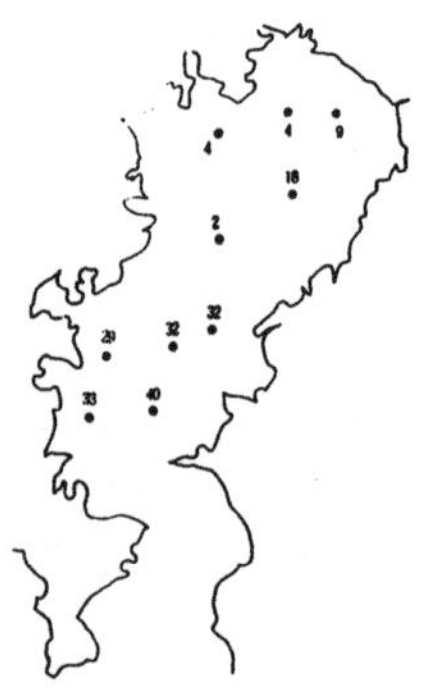

Fig. 8 Distribution of Montmorillonite in Tokyo Bay

Fig.9 Distribution of COD in Tokyo Bay

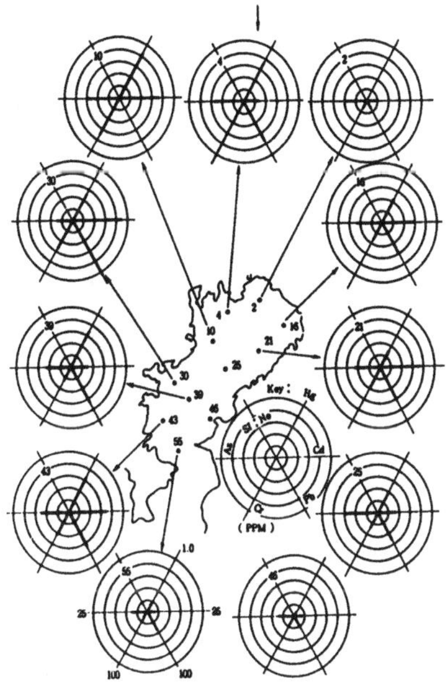

Fig. 10 Distribution of Heavy Metals in Tokyo Bay

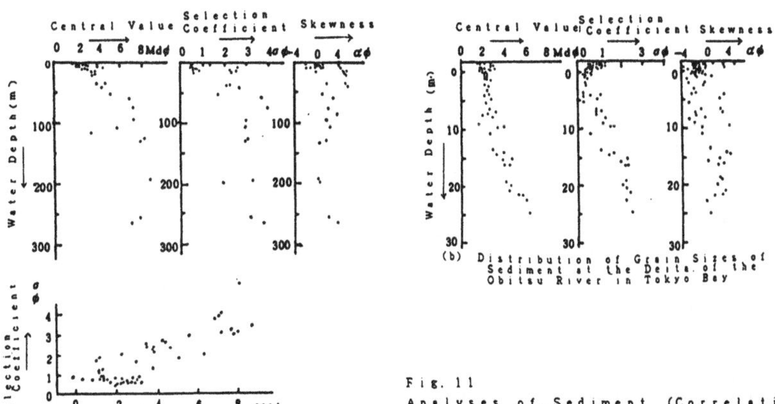

(a) Distribution of Grain Sizes of Sediment in the Northern Part of Sagami Bay

(b) Distribution of Grain Sizes of Sediment at the Delta of the Obitsu River in Tokyo Bay

Fig. 11
Analyses of Sediment (Correlation between Central Value, Selection Coefficient and Skewness, and Water Depth) by Nasu in 1956, and by Nasu and Sato in 1957

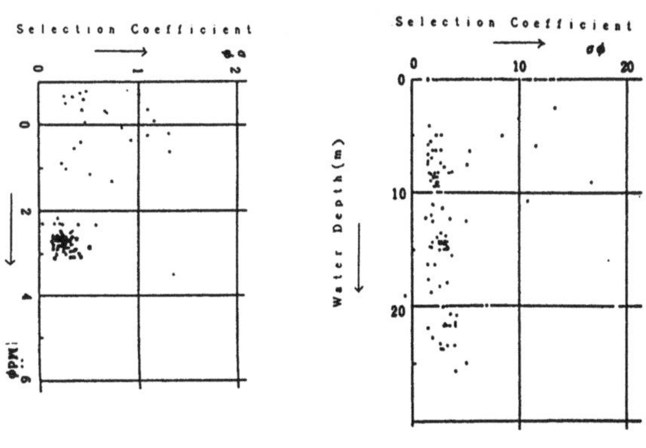

Fig. 12 Analyses of Sediment in the S Sea Area

Table 1 Environment Code "Environment Standards on Water Pollution"
(Decided by Cabinet Meeting, Apr. 1970)

Item / Type	Suitability for Application	Standard Values			Number of Colibacilli in Groups	Extracts with n-Hexane, (Oil)
		pH	COD	DO		
A	First Degree of Environment for Fishery, Bathing and Those shown below D	between 7.8 and 8.3	under 2 PPM	above 7.5 PPM	under 1.000 MPN/100ml	Not detected
B	Second Degree of Environment for and Those shown below C	between 7.8 and 8.3	under 3PPM	above 5 PPM	———	Not detected
C	Preservation of Environment	between 7.0 and 9.3	under 8 PPM	above 2 PPM	———	———

Table 2 Impacts and Indicators for Assessment

Impacts				Indicators for Assessment	
	Influencing Phenomena	Factor Phenomena	Factor Phenomena	Indicators	Indicators
1	Disturbance of Soil and Sand	Muddiness Floating Soil	Water Pollution Suspinded Solids	Standards for Water Quality	Distribution of Floating Soil
2	Deformation of Seabed	Flow Conditions Benthos			Distribution of Living Things
3	Extermination of Fixative Organisms	Attached Organisms to Seabed	Biotas		
4	Construction of Islands	Currents and Flow Conditions	Marine Conditions		
5	Draining Waste Water from Factories	COD SS Nutritive Salts	Water Pollution Water Pollution Water Pollution	Standards for Water Quality	
4	Draining Warm Water from power Plants	Temperature Salinity			
4'	Changes of Marine Conditions	Coastal Sand Hill Coastal Erosion			
2'	Benthos, Nekton Migrating Fishes	Left-Eyed Right-Eyed Flatfish	Food Habit		Food Chain
5'	Increase of Nutritive Salts	Plants and Planktons	Biotas		
1'	Generation of Floating Soil	Benthos		Standards for Water Quality	
4'	Changes of Currents	Stationary Waters			

Table 3 Characteristics of Indicators

Indicators			Dimension & phenomena	Time: Differential Type	Time: Integral Type	Space: Small Area	Space: Large Area	Phenomena: Steady	Phenomena: Unsteady
Investigation	Geospher	Water	COD	O		O			O
			SS	O		O			O
			Water Temp	O		O			O
			Chlorophyll	O		O			O
		Topography	Topography		O	O		O	
			Geology		O		O	O	
		Sediment	Ig Loss		O	O		O	
			Mineral composition		O		O	O	
	Biosphere	Nekton	Fishes, Shellfishes, etc		O		O	O	
			Plankton	O		O			O
		Benthos	Microorganism	O		O			O
			Macro Benthos		O	O		O	
			Fishes		O	O			

Table 4 X-Ray Diffraction Analyses of Mineral Composition

No.	Mineral composition	Marine sediment												Land	
		St-2	St-3	St-26	St-20	St-13	St-15	St-30	St-24	St-9	St-109	St-103	St-111	St-1	St-2
1	Quartz	⧻	⧻	⧻	⧻	⧻	⧻	⧻	⧻	⧻	⧻	⧻	⧻	⧻	⧻
2	Albite	#	#	+*	#	+	+	+	+*	+	+	+	+	+	+
3	Plagioclase	+*	+	+	+*	+*	−	+*	+	−	+*	+*	+*	+	+*
4	Chlorite	+*	+*	+*	+*	+*	+*	+*	+*	+*	+	+*	+*	+*	+*
5	Chloritoid	−	+	+*	+*	−	+*	+*	+*	+*	+*	+*	+*	+*	+*
6	Tremolite	−	+*	−	−	−	−	−	−	−	−	−	−	+*	−
7	Montmorillonite	−	−	−	−	−	−	−	−	−	−	−	−	−	−

Note; Peak intensity of X-ray analysis
卌 very strong ⧻ strong # medium
\+ weak +* very weak − none

Table 5 Analyses of Clay Minerals
(Composition of Fine-grained Minerals)

Data No.	Grain Size	粘 Clay Mineral 物					Non-Clay Mineral			
		K	S	C	M	Mx	Q	F	O	Gy
103	1.18	+?	+	++	+		+++++	++		
S − 2	1.17	+	+	++			+++++	+++++		
1 − 1			+?	+	++		+++	++	+	+?
S − 26	20.52			+	+++		+++	++		
OM		+	+	++	+	?	+++++	++		
S − 13	1.26	+?	+	+	++		+++++	+++++		
S − 9	7.88		+	++	+	?	+++++	+++		
S − 24	8.86	+?	+	++	+++		+++	++	+	+
S − 20	1.60		+	+	+		+++++	++	+	+

Note;
K; kaolinite, Halloysite; + (S − 2), ? (OM)
C; chlorite, Q; quartz,
Mx; mixed layer mineral, O; α-cristobalite,

The Utilization of Ocean Space for Waste Disposal

Changing Perspectives

M.A. Champ and S. Panem

Office of Policy, Planning and Evaluation, U.S. Environmental Protection Agency, Washington, DC 20006, USA

ABSTRACT

With seventy-one percent of the world covered with seawater, should the land carry the burden totally for waste disposal? Why does the ocean have a "protected" or "preferred" status when it comes to the disposal of wastes? What is the concept of assimilative capacity? What are the ocean waste disposal issues? What is comprehensive-integrated waste management? Why does the risk assessment - risk management approach to environmental management coordinate and integrate waste management actions? What issues in the risk assessment process need public debate? What critical oceanographic information is needed to apply the assimilative capacity concept in the decision making process for the utilization of ocean space for waste disposal?

INTRODUCTION

Thoroughout history, wastes have been cast onto the land, into rivers and into the sea. Wastes either floated away, dissolved, or sank, and thereby were out of sight or out of mind. Coastal civilizations have always used the sea for waste disposal. Ships have always thrown their wastes overboard. Fish processing wastes, seafood wastes, fish parts, culls, or inedible or unmarketable species have been discarded into the sea or at sea since the time that man began to fish. Maritime nations have found it essential to maintain shipping lanes and harbors, by dredging the sediments deposited by riverine, estuarine, or coastal processes, with the subsequent discharge or dumping of these materials onshore or in shallow waters to create new lands or directly into coastal waters adjacent to shipping lanes.

The ocean proved the "NIMBY" (Not In My Back Yard) cheap outlet for waste disposal. As nations became aware of the consequences of direct municipal and industrial discharges to waterways, waste treatment procedures and facilities were required. However, this cleaning up of the discharges by per-

---

* The views expressed in this article are those of the authors alone and do not necessarily reflect the views of the U. S. Environmental Protection Agency.

mit systems to decrease the biochemical oxygen demand (BOD) levels, suspended solids, nutrients, created a new problem (sludges). They certainly couldn't permit the dumping or discharge of these sludges back into the waters they were trying to clean up. The unregulated discharge or disposal of contaminants in the marine environment had affected water quality, fishery resources, and marine ecosystems. The problem persists today. The purpose of this paper is to identify the waste disposal issues that should be addressed in developing rational policies and is to discuss the concepts of risk assessment/risk management, and assimilative capacity as they apply to comprehensive waste management and to identity critical oceanographic information that is needed.

THE "PREFERRED" OR "PROTECTED" STATUS OF THE OCEAN FOR WASTE DISPOSAL

In the United States, the factor that has most recently impeded the use of the ocean for waste disposal has been the development of the perspective that the ocean has a "preferred" or "protected" status for waste disposal. This policy developed during the 1970s when national and international attention was focused on the onslaught of increasing waste load discharges and illegal industrial dumping activities on the high seas and major environmental crises were identified in coastal waters. This "pollution" [1] was seen as the harbinger of environmental crises in the U.S. and immediate measures were requested of the government to protect the oceans [2]. At that time there was not a consensus of opinion on how to identify ocean pollution problems from a "source/function perspective.

Therefore, in the minds of many environmentalists, scientists, engineers, members of the public, elected officials, and decision makers, there was a need for the ocean to have a "preferred" or "protected" status policy until more information was available, and that the ocean should only be used for waste disposal as an alternative of last resort. This desire led to the implementation of U.S. ocean dumping regulations [3] and a global regulatory mechanism [4].

From the current perspective of having the benefit of the last fifteen years experience, the "preferred" or "protected" status policy was not inappropriate. It provided time for extensive research on the fate and effects of contaminants disposed of in the ocean [5]. Also, the regulatory process aided the information collection process, because the discharger/dumper industry was required to monitor chemical and biological effects in the field and laboratory and report them to the regulatory agency in a specified time period as part of the permit granting process.

WASTE DISPOSAL ISSUES

A spectrum of perspectives on the use of the ocean for waste disposal in the U.S. can be identified and are summarized here. In the U.S. the current policy reflects the perspective of a "preferred" status and permits are granted on a case-by-case basis. The debunking of such "preferred" status of the ocean has been voiced by Osterberg [6,7,8,9,10]. He believes that "Planet Water" is a better name than "Planet Earth" because 71% of its surface is covered with seawater. Over 99% of all liquid water on earth is saltwater with less than 1% being fresh or drinkable. This small amount amount of freshwater and land support some 4.5 billion people This position holds that the laws that protect the oceans are forcing the most toxic wastes to go on land, placing man in jeopardy by contaminating the limited volume of freshwater. He notes that when considering sources of food, the oceans only provide about 1% [11]. The key question is "what is more critical to mankind -- the 99.4% of the water polluted with salt which provides so few of our needs, or the tiny 0.6% of fresh liquid water that contributes nearly all of our food, fiber and shelter?"[10, 12].

Osterberg calls the ocean natures' trash basket, and uses the case of radioactivity to support his point [7]. There are nearly 500 billion curies of natural radionuclides in seawater today, and the data suggest that marine organisms may receive higher exposures, largely from Polonium -210 than do land animals. Preston and his colleagues [13] have suggested that for every atom of radioactivity that man has added (nuclear bomb testing or nuclear industrial discharge) that nature has added 1000 more. Eisenbud [14] has reported that even though 42% of the fallout from nuclear testing fell into the Pacific Ocean, people living in San Francisco only received 0.2% of the Strontium -90 in their diet from seafood far less than found in milk in Wisconsin. The explanation for this is that the pathways of transporting contaminants from the ocean back to man are diffuse.

Other perspectives have been presented by Goldberg and Kamlet. Goldberg [15] makes an argument that the ocean can be used for waste disposal based on the assimilative capacity of a body of seawater. This capacity is defined "as the amount of a given material that can be contained within it without producing an unacceptable impact on living organisms or nonliving resources." He would determine this amount by the use of titration endpoints, in which "the polluting substances in the discharged material with the water body becomes evident at an endpoint." His examples of these endpoints were the U.S. Food and Drug Administration's (FDA) safe-maximum levels of contaminants in marine resources consumed as foods by man. This model is flawed if the endpoint protected only human health and not the health of the organisms in the ecosystem. It would be similarly flawed, if its use protected only organisms consumed by humans, particularly if the toxic environmental effect was expressed below the established endpoints (for example, if fish containing mercury below the 0.5 ppm standard set by FDA for human consumption exhibited reproductive failures and yet were approved for human consumption). Kamlet's [16] rebuttal to Goldberg's position was an argument for keeping the "preferred" status policy of the ocean and presented four critical perspectives for the disposal of any waste on land, in the air, or at sea:

> The ocean, as a commonly owned (or unowned) resource, is not protected by marketplace and political forces; consequently, ocean disposal should not be permitted for persistent, toxic materials unless disposal in other media has at least marginally greater environmental impacts.
>
> The ocean, as the prototypical dispersal medium, is an inappropriate place to dispose of persistent, toxic materials: the land, which if properly managed is the exemplary containment medium, is, in general, a sounder choice for the management of such wastes.
>
> No waste management strategy can be totally free of risk to health or the environment; management decisions should be based on multimedium comparisons and risk minimization.
>
> In view of the rudimentary ability of marine science to detect, much less correct, problems associated with waste disposal, we cannot prudently rely on a permissive approach based on crude assimilative capacity models in the hope that after-the-fact monitoring and a decades-long response time will ensure that health and the environment are protected.

He fears for the use of the "ocean as a fall-back waste heap for any material deemed too toxic or too controversial to dump on the land." Kamlet argues that "the ocean, unlike the land, is not protected by the market place and political forces that protect private property"...and that the ocean "is nobody's backyard and fish do not vote." He believes that if ocean disposal decisions are made on the basis of the socio-political acceptability of land-based alternatives, the ocean will always be the disposal medium of choice. He calls it the "tragedy of the commons" where the ocean is the "global commons" described by Hardin [17]. Kamlet questions a basic premise: "is it better to put a persistent toxic material in a dispersal medium or in a containment medium?"[16]. For persistent synthetic chemicals, such as PCBs, Kepone, DDT, and the like, he believes that they should be isolated and contained to the fullest possible extent if they cannot be destroyed with high temperature incineration.

Land containment options were thought to have three important characteristics: (1) monitoring can be on a small scale with emphasis on release from the containing system, (2) the persistence of long-term toxicity can be followed by time interval bioassays, and (3) if the containment system is leaking, the waste can be moved and repackaged, providing better "fail safe" mechanisms and permanent isolation from the environment.

The issues identified during the 1970s reflected both scientific and policy concerns: (1) there were great scientific [laboratory and/or field] uncertainties of the short- and long-term effects of ocean dumped wastes, or the potential impact on human health, (2) any large addition of chemicals to the ocean (such as from ocean dumping) was predicted to disrupt the stable chemical and osmotic balance that exists between marine organisms and the seawater, (3) due to the complexity and interdependence in marine food webs, they were considered very delicate and easy to disrupt, (4) as the oceans were a free commodity (no charge for their use) they could be subject to abuse, (5) there was limited ability to take remedial action after the wastes were dumped in the ocean, whereas contamination on land was thought to remain confined and available to future disposal options, and (6) the costs of using the ocean was considered inversely related to incentives for the waste disposer to develop pretreatment, recycling, or land based alternatives.

THE ASSIMILATIVE CAPACITY CONCEPT

Dilution is a central concept in managing waste discharges, but dilution is not a self-purification mechanism since it does not lead to reduction of the constitutent within the environment. However, the importance of dilution mechanisms comes from the reduction of toxicity due to a decrease in waste toxicant concentration to biota during the initial dilution and exposure period [18].

The concept of assimilative capacity probably originated from engineering waste load allociation studies. Webster [19] defines assimilation with a biological emphasis as: "the incorporation or conversion of nutrients into protoplasm that in animals follows digestion and absorption and in higher plants involves both photosynthesis and root absorption." The critical assumption on which the concept of assimilative capacity of the ocean is based is that the aquatic environment can tolerate some wastes, because various physical, chemical, biological, and biochemical processes will degrade these wastes [20, 21]. Assimilative capacity is therefore an ecosystem concept which describes the load of wastes (materials) which can be accommodated without changing the ecological balance of the system. To measure natural assimilative capacity, one must determine the rate of biodegradation of a given waste material in a body of water without altering the distribution or abundance of organisms [22].

The general advantages of using ocean space for waste disposal can be summarized: (1) the costs for land-based waste disposal alternatives are increasing, (2) municipalities are having difficulty in managing the disposal of large volumes of sewage sludges using land-based alternatives, due to public dislike for locating incinerators, composting facilities, and land fills in urban areas, and the saturation of local sludge markets, and (3) there is growing public attention to the environmental and public health hazards from land-based waste management alternatives.

Curiously, policies which support a "preferred" status force the development of environmentally undesirable disposal options. Also it is quite possible that the costs to develop and operate a land-based alternative, may not balance the environmental and public health risks.

COMPREHENSIVE WASTE MANAGEMENT

The key to comprehensive waste management over the long-term is to emphasize control at the source and recycling [23, 24], thereby minimizing the need for disposal, and to assess waste management options from a risk management process [25]. In 1981, the U.S. National Advisory Committee on Oceans and Atmosphere issued a special report on the role of the ocean in waste management strategies and recommended that "...waste should be disposed of in the manner and medium that minimizes the risk to human health and the environment, and at a price that this nation [U.S.A.] is prepared to pay" [26]. The report further identified a major policy problem in that the U.S. environmental legislation of the 1970s resulted in a disjointed process of waste management by medium. For example, disposal from ships and barges at sea is regulated under the Ocean Dumping Act, while pipe discharges (outfalls) into the nation's waterways and oceans fall under the Clean Water Act. The incineration of wastes is regulated under the Clean Air Act when on land, and under the Ocean Dumping Act when at sea. The NACOA report also concluded that waste disposal regulated by this medium-by-medium approach (land, air, or water) would be responsible for shifting the risk posed by individual classes of wastes to the medium of least regulation rather than to the medium of least risk [26]. The regulatory process needs to manage waste disposal from a multi-media perspective, allowing for coherent integration of approaches and alternatives[23,25,26, 27,28, 29]. What is needed are approaches that consider and balance environmental, economic, public health, and social considerations. An approach which could meet all of the above requirements is the application of risk assessment and risk management techniques to the decision making process.

RISK ASSESSMENT AND RISK MANAGEMENT

The task of reducing the total risk in the whole environment from exposure to toxic chemicals is perhaps the most complex managment initiative ever undertaken. The variety and degrees of scientific uncertainty of the available information, and the need to protect a variety of environmental values, as well as human health, all work to complicate the decision making process. Many times regulatory actions unwittingly transfer contaminants from one environmental medium to another via pollution control technology; or uncoordinated regulation of the same industry or the same substance by different regulatory offices or agencies [30].

For example, since 1945, over 65,000 different chemicals have been listed as being in commercial production in the U.S. alone. Approximately 1,200 new chemicals are added to this list every year. Many of these chemicals have not been characterized as to their toxicological potential, a few thousand have demonstrated some toxic effects and are used or released to the environment in such a magnitude that there is concern [30]. Management of toxic chemicals is a critical aspect of waste management. The contamination of clean fine sediments in a harbor with a highly toxic chemical (such as PCBs in New York Harbor) can immediately create an inverse ratio between the available options for disposal and the costs of disposal [23].

The risk assessment/risk management approach uses risk reduction as a decision making tool. This regulatory approach is quite different from developing discharge guidelines or regulations based on available technology and cost, because by nature, the risk approach is an integrative management concept. Risk assessment is distinguished and separated from risk management [31]. Risk assessment is the scientific study to identify the problem (i.e. how toxic, to whom, what-when-where, and why), whereas risk management is deciding what to do about the problem identified and is a decision making process aimed at control strategies, and cost considerations. In the risk waste management process, the key issues that need public debate are the "values" used by the policy and decision makers, acceptability of costs of control measures or alternatives and environmental effects, the acceptability of environmental and public health risks and benefits, and the measure of confidence - the degree(s) of uncertainity in the information used in the decision process. The key to risk managment in waste disposal is to not only balance risk reduction against effects and control resources, but to balance risk against risk and reduce the total risk in the whole environment. These approaches should make regulatory judgments more consistent and more reflective of the state of scientific understanding.

There is not a standard formula for conducting a risk assessment, it is an analytical tool which must be tailor made to the needs of each use. An ocean disposal risk assessment for an environmental effect has to integrate physical, chemical, and biological processes, and identify target organisms, critical pathways, and food webs (to predict the actual level of exposure to an organism and a population and subsequent effects). Risk assessment focuses on two measurable factors: hazard and exposure. For a chemical to cause a risk, it must be toxic (be a hazard) and be present in the ocean at levels above background, that is if it occurs naturally in the ocean. Four steps are required: (1) hazard identification, (2) dose-response assessments, (3) exposure assessment, and (4) risk characterization (degree of uncertainty and assumptions). The uncertainties stem from the extrapolation of effects from hazard and dose response assessments that were conducted in the laboratory to predicting effects in the field, and from laboratory studies of rats and mice to predictions of effects in man. Several tools are available to predict this uncertainty: benefit/cost analysis, risk benefit analysis, and cost-effectiveness analysis.

Conceptually, the risk balancing approach appears to be a more comprehensive and integrative approach for waste disposal management. However, this approach may have a slow acceptance by policy and decision makers, because there is a limited state-of-art knowledge of risk assessment, and there are major questions which we must answer in institutionalizing risk assessment methods into a regulatory decision making framework. These are: (1) Are our abilities to predict the fate and behavior of wastes disposed of with land-based alternatives equal to ocean based alternatives?, (2) If we could accurately predict the effects of both land-based and ocean based alternatives, can we equally or fairly compare these effects, or are we comparing apples and oranges?, (3) Due to the uncertainty of assessing the risks of either land-based or ocean alternatives, if we find that we have made an error in our assumptions because of new additional information, can we recover the waste and remedy the damage?, (4) If increased recycling or better pretreatment is the preferred waste management alternative in the long-run, (because it prevents the generation or the contamination of waste), why consider the risk approach?

CRITICAL INFORMATION NEEDED TO APPLY THE ASSIMILITIVE CAPACITY CONCEPT TO DECISION MAKING FOR THE UTILIZATION OF OCEAN SPACE FOR WASTE DISPOSAL

To develop a risk assessment protocol for the disposal of wastes at sea, there are several unique informational needs, which are specific to physical, chemical, and biological oceanographic processes that play a role in dispersion, dilution, fate, behavior, of wastes and subsequent environmental effects. These are (1) waste characterization with respect to prediction of its reaction in seawater, and disposal methods (engineering technology), with respect to technology that maximizes dispersion or containment, (2) dumpsite selection criteria [32], because some sites are more dispersive than others which may accumulate wastes [33], and (3) due to the uncertainties in an oceanographic risk assessment, a monitoring program is needed to verify the assessment and management predictions [34].

The physical properties of a waste (specific gravity, volatilization, etc.) and the physical oceanographic processes at a dumpsite are the factors that determine the post dumping waste exposure concentrations and time periods to biota. For example waste discharge from a moving vessel at 6 knots obtains an initial dilution of 1:5000, whereas discharge from an outfall achieves an initial dilution of about 1:100. Appropriate dumpsite selection is the key in the pre-assimilative capacity matrix. Because careful site selection can maximize the spatial boundaries of a dumpsite for waste dispersion,therefore decreasing the waste concentration per unit volume of seawater. In sites where waste materials are rapidly dispersed by water movement, there is minimum accumulation, providing a greater degree of assimilative capacity. At highly dispersive dumpsites, long-term adverse effects have not been detected [33, 35]. A semi-quantitative model [36] has been developed for assessing the dispersion at a site being considered for dumping. The model uses dumpsite physical characteristics such as water depth, tidal currents, wave-induced currents, distance from the coast, degree of open water exposure, and seasonal storms to determine local (dumpsite) and regional dispersiveness for sewage sludges for the 8 global sewage sludge dumpsites.

After the waste has been dumped the following factors/ processes have been determined influential to the fate and behavior of wastes dumped at ocean dumpsites [37]: (1) dilution/dispersion processes [settling rates (flocculation/ particle dynamics) vertical mixing rates (within and below mixed layer) - horizontal dispersion rates - mass fluxes to the seafloor - bioturbation (magnitude, depth, rate)], and (2) bioavailability/bioaccumulation/biomagnification processes [- solid/solution chemistry - uptake processes and pathways - depuration kinetics - detoxification processes - laboratory and in situ toxicity assays - fecundity assays].

For two U.S. deep ocean dumpsites, the short- and long-term environmental consequences from dumping municipal and industrial wastes have been projected. O'Connor et al. [35] have estimated waste distribution and biological responses to the ocean dumping of annual volumes of 7×10^6 m^3 of New York/ New Jersey sewage sludges at the 106-Mile site a as first order exercise. Waste dispersion, settling velocity distributions, biological uptake and depuration rates are calculated. Estimates are made of annual rates of sediment contaamination and subsequent bioturbation with release back into the water column. Biota uptake rates and effects in plankton, worms, and fish are also calculated. The second study is an environmental assessment of the pharmaceutical wastes that have been ocean dumped off Puerto Rico [38]. The Puerto Rico waste is currently not being ocean dumped, but disposed via a nearshore ocean outfall.

Risk assessments are being prepared and reviewed for the New York/New Jersey Sewage Sludges, balancing land-based incineration and ocean dumping as the two waste management alternatives. Composting or other land-based alternatives were rejected because they exceed the economic or environmental impacts that were acceptable. Therefore, in the near future, the risk approach for waste management will have its first application and be available for analysis.

REFERENCES

1. Champ, M.A. Canadian Journal of Fisheries and Aquatic Sciences. Vol. 40 Suppl. 2, (1983):5-8.
2. U.S. Council on Environmental Quality. U.S. Govt. Print. Office. Washington, D.C. (1970). 45p.
3. U.S. Environmental Protection Agency.Federal Register. 42 (7):2485 (1-11-77).Federal Register 49(88):19005(5-4-84).
4. Inter-Governmental Maritime Consultative Organization London. 36p. + 1979 Amend. Pub.76.14E.
5. Champ, M.A.; Park, P.K. IFI/Plenum. New York. 1982. 399p.
6. Osterberg, C. The New York Times. 8-9-81
7. Osterberg, C. In: Waste Management-82. Post (Ed.) Vol. 2 Low Level Wastes. Univ. of Arizona,Tucson. (1982):402-414
8. Osterberg, C. IAEA Bull. Vol. 24.No.2. (1982):3-34.
9. Osterberg, C. The Washington Times. 6-7-83.
10. Osterberg, C. In: Ocean Waste Management: Policy and Strategies. Paper # 13. 9 p. IDOS Spec. Symp. (1983).

11. Roels, O. Sea Technology. Vol. 23. No. 8. (1982):63-67.
12. U.S. National Research Council. National Academy Press. Washington, D.C. (1984). 179p.
13. Preston, A., et al. Poc. Royal Society. Ser. E. (B). 72, 41. London. (1972):411.
14. Eisenbud, M. Health Physics. Vol. 40. (1981):429-437.
15. Goldberg, E. D. Oceanus. Vol. 24. No. 1. (1981):2-9.
16. Kamlet, K. S. Oceanus. Vol. 24, No. 1. (1981):10-17.
17. Hardin, G. Science. Vol. 162 (1968):1243
18. Cullen, P. Water News. (9 March/Apr, 1984):11-12.
19. Webster's New Collegiate Dictionary, G & C Merriam Co. Springfield, Massachusetts. (1979). 1531p.
20. Goldberg, E.(Ed.)Proc.Workshop on Assimilative Capacity of US Coastal Waters for Pol. NOAA. Boulder, CO (1979).
21. Champ,M.A.; Hebard,J.F. In: The Manag. of Oceanic Resources-The Away Ahead. Pub. by the Univ. of Southern California, Los Angeles (1981):389-409.
22. Champ,M.A. In: The Law of Sea and Ocean Ind: New Oppor. & Restraints. Johnston & Letalik (Eds.). Law of the Sea Ins. Univ. Hawaii. (1983): 282-295
23. Mayer, G. F. (Ed.). Ecological Stress and NY Bight: Sci. & Mang. ERF, Columbia, South Carolina. (1982). 715p.
24. Farrington, et al. Oceanus. Vol. 25. No. 4. (Winter 1982/83): 39-50.
25. Champ, M. A.; Park, P. K. (Eds). Waste Management in the Ocean. Krieger Pub. Co. Malabar, Florida. (In Press).
26. National Advisory Committee on Oceans and Atmosphere. Gov. Print. Office. Washington, D.C. 103p. 1981.
27. Boxer, B. Marine Pollution Bulletin.15(9)(1984):320-325.
28. U.S. National Research Council. National Academy Press Washington, D. C. (1984). 210p.
29. U.S. National Research Council. National Academy Press. Vol. IX. Washington, D. C. (1978). 187p.
30. U.S. Environmental Protection Agency.Risk Assess.&Management: Manag. the Process. Washington, D.C. (1984). 35p.
31. U.S. National Research Council. National Academy Press. Washington, D.C. 1983. 191p.
32. GESAMP. Reports & Studies No. 16. Inter-Governmental Maritime Consultative Organization. London. (1982). 60p.
33. Norton, M.G.; Champ, M. A.: In: Wastes in the Oceans Series, Duedall et al. 1985 John Wiley & Sons, NY
34 International Atomic Energy Agency. Tech. Doc. 296. Vienna. (1983). 55p.
35. O'Connor, et al. Canadian Journal of Fisheries and Aquatic Sciences. Vol. 40. Suppl. 2. (1983):228-241.
36. Champ, et al. Paper #59. Special Symp. on Contam. Fluxes Through the Coastal Zone. ICES, Copenhagen (1984):27p.
37. Champ, et al: In: Proc. Workshop on Assimilative Capacity of the Oceans for Man's Wastes. Su & Hung (Eds.). SCOPE/ICSU, Taipei, Republic of China. (1983). p362.
38. O'Connor, T. P. NOAA TM NOS 28. Rockville, Maryland. (1983). 119p.

Marine Biotechnology and Pollution Control

S.M. Lyons and S. Panem

Office of Policy, Planning, and Evaluation*, U.S. Environmental Protection Agency, Washington, DC 20006, USA

ABSTRACT

The marine sciences are on the verge of commercial exploitation by the developing biotechnology industry, using techniques of genetic engineering. Application of these new technologies represents an important opportunity for addressing a ubiquitous and persistent problem of industrialization -- toxic waste disposal. As marine biotechnology advances, its application to pollution control in the oceans should be encouraged.

INTRODUCTION

Biotechnology has been defined as the application of biological systems and organisms to technical and industrial processes (1). Included in this definition are such traditional industrial applications as the use of microorganisms in effluent treatment and in the fermentation industry, as well as industrial applications of recombinant DNA and hybridoma technologies. Commercial development of these latter techniques, often described as the "new biotechnology", has occurred in a wide variety of industrial sectors (2). Application of these techniques may have a large economic impact on such areas as the production of new drugs, food, and chemicals, the improvement of agricultural products, and degradation of toxic wastes.

Marine biotechnology, which is the application of this technology to the marine sciences, has begun to attract increasing attention from commercial enterprises (3-11). Recognition of the enormous resources of the oceans has led to the formation of a number of start-up biotechnology companies (12) hoping to exploit these resources, and has also stimulated interest from established companies (13). Most of this attention is directed toward the discovery of new pharmacologically active compounds, the development of specialty chemicals, and aquaculture. Mineral leaching, metal concentration, enhanced oil recovery, and biomass conversion as an alternative source of fuel have also been mentioned as

---

*The views expressed in this article are those of the authors alone and do not necessarily reflect the views of the U. S. Environmental Protection Agency.

potential targets for commercialization. Relatively little attention, however, has been directed toward potential applications of biotechnology to control contaminants in the marine environment. Now that the commercialization of marine biotechnology appears to be imminent, it is important to emphasize the potential of this technology for the detection, monitoring, and control of pollution in the marine environment.

OPPORTUNITIES IN MARINE POLLUTION CONTROL

There are many problems of toxic waste release in the marine environment that are amenable to biotechnological solutions. The direct application of engineered microorganisms at toxic waste sites in the marine environment (and on land), and the use of cellular products of engineered microbes are both likely to be useful for pollution control problems. The use of genetically engineered microorganisms to digest particularly hazardous and recalcitrant toxic chemicals offers great promise for solution of some of our most difficult toxic waste disposal problems. Immobilized enzyme systems have been suggested as a technique to improve waste water treatment and may have other pollution control applications (14). Enzymes can perhaps be identified in bacterial and fungal species that degrade specific toxic compounds, and the genes coding the enzymes cloned into the appropriate host for mass production.

Microorganisms that can tolerate harsh environmental conditions have been found in several extreme aquatic systems. Submarine hydrothermal vents, hypersaline lakes, and hot springs could yield microbial species useful as hosts for specialized waste treatment applications, or as a source of degradative enzymes resistant to extremes of temperature, salinity, or pressure (9, 15).

Biotechnology will also have an impact on pollution control in less direct ways. The development of alternative, nonpolluting ways to manufacture chemicals, fertilizers, and pharmaceuticals is conceivable. Biological pesticides may also substitute for a variety of highly toxic chemical pesticides that are released into the environment in large quantities. The generation of alternative fuels such as methane and ethanol may be enhanced by biotechnological applications to biomass conversion, reducing air and water pollution from fossil fuels. Chemosynthetic bacteria have recently been discovered in submarine hydrothermal vents that use hydrogen sulfide as a source of energy. These may be useful for removing sulfer compounds from coal and oil, which could lessen the impact of acid rain (10). Research on nitrogen fixation and plant hormones may lead to the genetic engineering of agriculturally important plants and/or their microbial symbionts to allow reduction in the use of high levels of fertilizers.

Improvements in detection and monitoring of toxic chemicals may also be achieved through biotechnology. Biotechnology may advance the development of new biosensors, for example a transistor coated with enzymes that undergo a conformational change and send an electrical signal when a particular toxic chemical binds (1). This could vastly improve the sensitivity of detection methods. Metal accumulating bacteria or other indicator bacteria in particular proportions may indicate the presence of specific toxic chemicals. For example, oil-degrading microbial species are enriched in both numbers and proportionately in areas of marine oil pollution (16). These microbes may be exploited by the biotechnology industry to improve the quality of detection and monitoring techniques for toxic substances.

OBSTACLES TO DEVELOPMENT

To date, there has been little popular discussion of biotechnology for control of pollution in the marine environment. This is not unique to marine biotechnology, and may be due, in part, to a common view of the oceans as the "ultimate sink" -- an erroneous public perception that they have a limitless capacity to absorb pollutants. Except for the highly visible catastrophe, such as a major oil spill, there is little public awareness of marine environmental problems.

In the recent past, attention has focused on cleaning up the rivers and lakes of the United States. Now pollution of other non-marine aquatic systems, such as groundwater and acid rain pollution are prominent in public discussion. Since much of the pollution of non-marine aquatic environments will ultimately find its way to the oceans, a secondary effect of reducing toxic waste discharge into non-marine systems is a lessening of marine pollution levels. This may be an especially important route for reducing the levels of some toxic substances in the world's oceans. Exclusive attention to non-marine water pollution, however, neglects the additional benefits that can be derived from targeting primary sources of marine pollution.

One reason for the comparative lack of activity in marine biotechnology for pollution control is most probably the interdisciplinary nature of marine environmental problems. Communication and collaboration between molecular biologists, marine biologists, and pollution control specialists is needed for the development of this application of marine biotechnology. Further, although commercial interest in marine biotechnology is increasing, the term marine biotechnology may not connote pollution control applications to the molecular biologists primarily responsible for the research directions of the biotechnology industry. Technical obstacles may also exist to the development of pollution control biotechnology. The molecular biology and genetics of those microorganisms most likely to be useful for marine pollution control have yet to be adequately explored. For example, there are dozens of genera of marine bacteria and fungi known to have some capacity for degrading petroleum. The bacterial genera most commonly cited that may be useful for controlling petroleum pollution include *Pseudomonas*, *Achromobacter*, *Flavobacterium*, *Acinetobacter*, *Vibrio*, *Bacillus*, *Arthrobacter*, *Nocardia*, *Corynebacterium*, and *Micrococcus* (16).

Yet, the biotechnology industry that has thus far emerged, emphasizes the commercial production of drugs and specialty chemicals, depending on well understood laboratory species such as yeast and E. coli. For most marine pollution control applications, these microbial species are not appropriate because they are not adapted to environments where toxic waste contamination usually occurs. They are intolerant of shifting environmental conditions, and are likely to be susceptible to competition and predation. Therefore, a great deal of basic research needs to be performed to allow the sophisticated manipulations of genetic engineering to be dependably accomplished in other potentially useful microbial species. Consequently, the lead time necessary for commercial exploitation is increased, and the accompanying elevated financial risk may have discouraged companies from pursuing pollution control applications of biotechnology.

Other factors contribute to the perception in some segments of the biotechnology community that the commercial potential of biotechnology for pollution control is not favorable. For many toxic substances there is a natural, albeit slow, process of biodegradation. This process may be augmented by deliberately seeding the toxic waste site with microbes that degrade particular toxic chemicals. Conventional genetic methods are commonly employed to improve the capacity of microorganisms to degrade toxic chemicals, and if these methods are deemed adequate for the job there is little incentive to apply recombinant techniques. A number of firms currently offer microorganisms, some genetically altered and others naturally-occurring, as commercial products for the degradation of toxic substances (17). None of these products, however, is genetically engineered by recombinant DNA techniques. The question is not whether new biotechnological methods can improve currently used microbial products for pollution control, but whether the amount of improvement is worth the investment. Many believe that there is plenty of room for improvement, but are also skeptical about the cost/benefit relationship.

The small start-up companies of the biotechnology industry have concentrated on pharmaceutical, chemical, and agricultural products, which have a large and proven market. If biotechnology firms can successfully develop such products and manufacture them cheaply and reliably, substantial profits are assured. Pollution control applications, however, are not presently perceived as a substantial market for the biotechnology industry, so investment in this area may be too speculative, even for organizations accustomed to high risk. Investment in long-term R & D for pollution control biotechnology poses a level of risk that may be beyond the capacity of small companies with limited resources, which are often under tremendous pressure to bring products to market quickly.

THE PETROLEUM CASE

The control of petroleum pollution has long been used as a popular example of a potential application of biotechnology. In the United States, the first patent for a genetically engineered microbe was awarded to the General Electric Company for a strain of Pseudomonas putida, developed by Dr. Ananda Chakrabarty (18). The microorganism was engineered to rapidly digest many of the components of crude oil. Curiously, this strain has never been developed as a commercial product, which suggests that technical and commercial barriers need to be overcome before genetic engineering will be effectively used for pollution control.

Petroleum in the marine environment derives from a variety of point and non-point sources, including natural seeps, accidental spills, ballast tank washings from large tankers, industrial and municipal wastewaters, urban and river runoff, and precipitation from the atmosphere (16). Large, single event spills from tanker accidents or oil-well blowouts attract a great deal of public attention, but on average contribute only a small fraction of the total marine petroleum burden (16). The impact of genetic engineering on the control of such spills may be limited because rough weather conditions in the open ocean and the need for speed in confining oil spills impose severe constraints on microbial digestion as a practical way of dealing with oil spills. Accidents occur unpredictably and the damage to marine plant and animal species may be immediate if a spill occurs in a coastal area. Generally, any containment effort must occur within at most a few days if it is to be effective. This would necessitate keeping huge quantities of a microbial agent on hand in anticipation of an emergency. Additionally, no single microbe is ever likely to be generated that is capable of rapidly digesting all of the components of the extraordinarily complex mixture that constitutes crude petroleum.

There may eventually be biotechnological applications for cleaning up such spills, but some of the other sources of petroleum in the marine environment are perhaps better prospects for practical solution using biotechnology. Chronic petroleum discharges from specific, predictable sites are better candidates for biotechnological solution. Such systems as oil refinery waste treatment, municipal wastewater treatment, ballast tanks, and drilling well production waters offer the advantage of a contained or semi-contained environment. This allows capture and retention of the toxic components of petroleum. Growth parameters such as temperature, substrate concentration, and pH can be more easily controlled, and the engineered microbes can be protected from competitors and predators. Finally, engineered microorganisms can presumably be more easily prevented from release into the environment in such systems, which may be an issue of concern in some cases.

ENVIRONMENTAL CONSEQUENCES

The deliberate release of genetically engineered microorganisms into the environment is a controversial issue that is currently under intense debate in the scientific and regulatory communities. In the United States, a rationale for federal regulation of the deliberate release of genetically altered organisms was recently proposed (1), and the Organization for Economic Cooperation and Development is also studying the formulation of regulatory guidelines on deliberate release. The potential hazards of releasing genetically altered microorganisms are difficult to assess because of the uncertain effects of new genetic combinations and the exceedingly complex nature of ecological interactions.

Microorganisms are commonly used for biodegradation of toxic chemicals at spill sites and in industrial and municipal wastewater treatment. These include many altered by conventional genetic techniques such as mutagenesis and selection, and constitute an existing data base that may be helpful in the process of risk assessment for genetically engineered microorganisms.

The concern over deliberate release is especially pertinent for pollution control, since using engineered microbes directly on sites of toxic waste contamination would necessarily involve release into the environment and often would require that the microorganism persist in the environment for some period of time. Although some have argued that overly restrictive regulation of direct release of engineered microbes could deter companies from attempting to apply biotechnology to pollution control, it is also possible that well-designed regulations will encourage the safe development of the technology. Determinination of the environmental impact of genetic engineering for marine pollution control involves a balancing of the risks of deliberate release with the benefits for treatment of hazardous wastes.

CONCLUSION

The proprietary nature of most industrial research makes it difficult to identify the level of effort applied to pollution control biotechnology. There is probably more effort than is apparent from published reports in the scientific literature. In fact, the recent Biotech '84 USA conference had several speakers from industry who described their companies' efforts along these lines (19-21). It would seem appropriate as marine biotechnology develops to encourage the use of this technology to address problems of marine pollution.

The marine sciences are on the verge of commercial exploitation by the biotechnology industry. It is important that biotechnological applications to control pollution in the marine environment are not overlooked. It is ironical that one of the first microbes engineered by recombinant DNA techniques for industrial purposes, Chakrabarty's oil digesting Pseudomas strain, was designed as a tool for the control of oil pollution, and yet has never been commercialized. At the time, it was hailed as a breakthrough for pollution control, and since, has often been cited as an example of a socially useful product from recombinant DNA research. Biotechnology represents an important potential opportunity for addressing a ubiquitous, persistent problem of industrialization -- toxic waste disposal.

References and Notes

1. Office of Science and Technology Policy. Proposal for a Coordinated Framework for Regulation of Biotechnology; Notice. Fed. Regis. 49: 50856-50907.

2. Office of Technology Assessment. 1984. "Commercial Biotechnology: An International Analysis", U. S. Congress, Washington, DC.

3. Colwell, R. R. 1983. Biotechnology in the marine sciences. Science 222:19-24.

4. Colwell, R. R. 1984. The industrial potential of marine biotechnology. Oceanus 27:(1)3-12.

5. Colwell, R. R., Pariser, E. R., and Sinskey, A. J., eds. 1984. "Biotechnology in the Marine Sciences", John Wiley & Sons, New York.

6. Curtin, M. E. 1985. Chemicals from the sea. Biotechnology 3:(1)34-38.

7. Goldoftas, B. 1982. High tech for the high seas. Technology Review, July, pp. 74-75.

8. Klausner, A. 1985. Food from the sea. Biotechnology 3:(1)27-32.

9. Regan, D. L. 1980. Marine biotechnology & the use of arid zones. Search 11:377-381.

10. Tucker, J. B. 1985. Biotechnology goes to sea. High Technology, February, pp. 34-44.

11. Weiner, J. 1985. Marine biotech in the Negev desert. Biotechnology 3(1):41.

12. For example, SeaPharm (Princeton, NJ), Ocean Genetics, Inc. (Walnut Creek, CA), Cyanotech Corp. (Woodlinville, WA), and Bio-Polymers, Inc. (Farmington, CT).

13. For example, Phillips Petroleum (Bartlesville, OK), Weyerhaueser (Oregon Aquafoods, subs., Springfield, OR) and Standard Oil of Indiana (Amoco Research Center, Naperville, IL), and in Japan -- Suntory and Mitsubishi Kasei.

14. Owings, D. 1984. Panel discussion: emerging industrial applications. In "Genetic Control of Environmental Pollutants", G. S. Omenn and A. Hollaender, eds., Plenum Press, New York, p. 345.

15. Kobayashi, H. A. 1984. Application of genetic engineering to industrial waste/wastewater treatment. In "Genetic Control of Environmental Pollutants", G. S. Omenn and A. Hollaender, eds., Plenum Press, New York, pp. 195-214.

16. Floodgate, G. D. 1984. The fate of petroleum in marine ecosystems. In "Petroleum Microbiology", R. M. Atlas, ed., Macmillan, New York, pp. 355-397.

17. For example, Petroferm (Amelia Island, FL), Flow Laboratories (Englewood, NJ), Polybac Corp. (Allentown, PA), Sybron Biochemical (Salem, VA), and Occidental Chemical Corp. (Grand Island, NY).

18. Chakrabarty, A. M. 1973. Genetic fusion of incompatible plasmids in Pseudomonas. Proc. Nat. Acad. Sci., USA. 70: 1641-1644.

19. Lundgren, D. G. 1984. Mineral leaching for recovery and waste treatment - an overview. In "World Biotech Report, Vol. 2: USA", Online Publications, Pinner Middlesex, UK.

20. Munnecke, D. M. 1984. Biotechnological approaches to hazardous waste treatment. In "The World Biotech Report, Vol. 2: USA", Online Publications, Pinner Middlesex, UK.

21. Sojka, S. A. 1984. Treatment of hazardous wastewaters by specialized bacteria. In "The World Biotech Report 1984, Vol. 2: USA", Online Publications, Pinner Middlesex, UK.

Additional Readings

1. Adler, R. G. 1982. Biotechnology and genetic engineering - the promise for pollution control. Environmental Forum. September, pp. 6-8.

2. Alexander, M. 1981. Biodegradaton of chemicals of environmental concern. Science 211:132-138.

3. Bourquin, A. W., and Pritchard, P. H., eds. 1979. "Proceedings of the Workshop: Microbial Degradation of Pollutants in Marine Environments." Environmental Protection Agency, Washington, DC. EPA-600/979-012.

4. Genex. 1983. Impact of biotechnology on pollution control and toxic waste treatment. Contract report prepared for the Office of Technology Assessment, U. S. Congress, Washington, DC.

5. Johnston, J. B., and Robinson, S. G. 1984. "Genetic Engineering and the Development of New Pollution Control Technologies", Environmental Protection Agency, Washington, DC. EPA-600/2-84-037. NTIS PB84-148972

6. Kobayashi, H., and Rittmann, B. E. 1982. Microbial removal of hazardous organic compounds. Environ. Sci. Technol. 16:170A-183A.

7. O'Brien, T. 1983. NBS and Industrial Biotechnology: Instrumentation and Measurement Needs. Planning Report 14, National Bureau of Standards, Department of Commerce, Washington, DC.

Ocean Dumping of Industrial Wastes

I.W. Duedall[1], P.K. Park[2] and D.R. Kester[3]

[1] Department of Oceanography and Ocean Engineering, Florida Institute of Technology, Melbourne, FL 32901, USA
[2] U.S. National Oceanic and Atmospheric Administration, Rockville, MD, USA
[3] Graduate School of Oceanography, University of Rhode Island, Kingston, RI, USA

ABSTRACT

The oceans receive industrial wastes from outfall pipes, nonpoint sources, atmospheric fallout, and ocean dumping. No one knows the total quantity of these wastes entering the ocean. What is known, however, is that certain regions of the oceanic environment have been adversely affected by the presence of some kinds of waste materials. While some wastes discharged into the sea can be assimilated by the ocean without causing serious harm, others wastes cannot, and this has led to environmental damage. In order to understand the effects of wastes in the ocean, it is necessary to know (1) the types, quantities, and characteristics of the wastes discharged at sea; (2) the oceanographic conditions of the receiving waters; (3) important natural processes occurring in the ocean; (4) the biological effects of the waste; and (5) the extent of biodegradation of the waste substances. Although dumping of wastes at sea will probably occur into the future, every attempt should be made to minimize the wastes through waste-utilization schemes.

TYPES AND QUANTITIES OF INDUSTRIAL WASTES AND LOCATIONS OF DUMPSITES

Industrial wastes discharged at sea can be broadly classified as acid-iron wastes, coal wastes, incineration wastes, miscellaneous chemicals, and military wastes. Acid-iron wastes are usually mixtures of sulfuric or hydrochloric acid containing compounds of iron or titanium oxide. In the United States, for example, a majority of the acid-iron wastes have been dumped at coastal sites near the states of New York, New Jersey, and Delaware. Deepwater dumpsites, also used for disposal of acid-iron wastes and other industrial wastes, are found in the slopewater off New York. The Puerto Rico dumpsite, a deepwater area for disposal, was recently discontinued as a dumpsite for pharmaceutical wastes. The deepwater dumpsite off New York is scheduled in the near future to receive larger quantities of industrial wastes, including coal wastes.

On a global basis, the greatest concentration of dumpsites exists in Europe (Fig. 1), in areas of the North Sea and the English Channel. France, the United Kingdom, the Federal Republic of Germany, The Netherlands, Belgium, and Spain discharge a total of about 8×10^6 metric tons per year of industrial wastes (Duedall et al., 1983a) compares with about 5×10^6 tons per year of industrial wastes

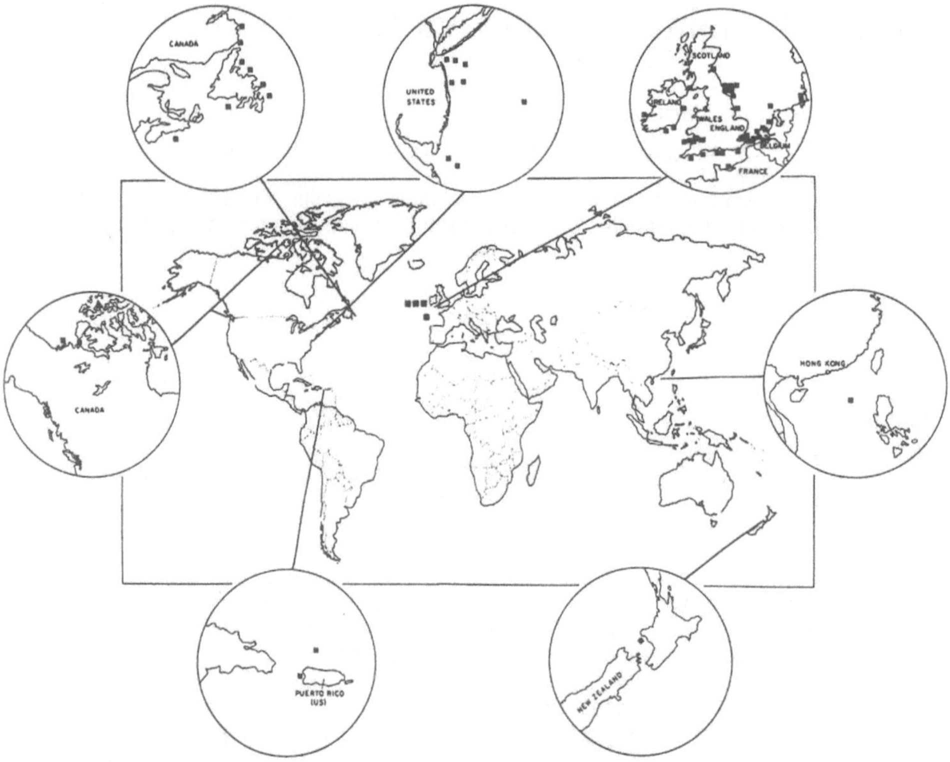

Fig.1. Ocean dumpsites in 1979 used by Contracting Parties to the London Dumping Convention (Duedall et al., 1983a).

discharged by the United States. Dumping-activities information provided in permits issued by member states to the London Dumping Convention (LDC) is available from the International Maritime Organization, a United Nations-based organization. In the United States, the Ocean Dumping Act provides that dumping will only be permitted when the dumping will not unreasonably degrade or endanger human health, welfare or amenities or the marine environment, ecological systems, or economic potentialities. Factors that must be considered when addressing whether or not to dump have been recently discussed by Erdheim (1985):

1. The need for the proposed dumping.

2. The effect of such dumping on human health and welfare, including economic, esthetic, and recreational values.

3. The effect of such dumping on fisheries resources, plankton, fish, shellfish, wildlife, shorelines, and beaches.

4. The effect of such dumping on marine ecosystems, particularly with respect to the transfer, concentration, and dispersion of such material and its by-products through biological, physical, and chemical processes; potential changes in marine ecosystem diversity, productivity, and stability; and species and community population dynamics.

5. The persistence and permanence of the effects of the dumping.

6. The effect of dumping particular volumes and concentrations of such materials.

7. Appropriate locations and methods of disposal or recycling, including land-based alternatives and the probable impact of requiring use of such alternative locations or methods upon considerations affecting the public interest.

8. The effect on alternative uses of oceans, such as scientific study, fishing, and other living-resource exploitation and nonliving-resource exploitation.

9. That in designating recommended sites, the United States Environmental Protection Agency will utilize all feasible locations beyond the edge of the continental shelf.

FATE OF WASTES DUMPED INTO THE OCEAN

Many dynamic processes occurring in the sea either work together or act independently to disperse or degrade wastes that enter the ocean. Broadly speaking, these processes are either physical, chemical, biological, or benthic (Fig. 2). Rapid mixing of wastes with seawater can occur because of advective and dispersive processes; frequently dilutions greater than 10^4 are observed in less than an hour after a dumping event. Chemical processes such as precipitation can lead to the rapid settling of wastes. This is particularly true for acid-iron wastes which, when mixed with seawater, result in the formation of ferric hydroxide particles that can settle quickly from the surface waters where dumping occurs. Biological processes can modify some wastes. For example, organic materials in some wastes may be metabolized by certain kinds of marine organisms.

REUSE OF WASTES

Every attempt should be made to utilize waste rather than dispose of it. Wastes are often concentrated in some metals; such wastes could be recycled to recover the metals. Recycling would have the added benefit of decreasing the rate of depletion of resources. On a global basis, for example, some of our natural resources, are presently becoming substantially depleted. Figure 3 shows the projected hypothetical lifetime of the present-day reserves of some commonly used metals, assuming present usage levels continue; it also shows the significance of recycling to prolong their depletion. Economic factors are obviously quite important. For many resources, it is still cheaper to extract the resource than to reuse it.

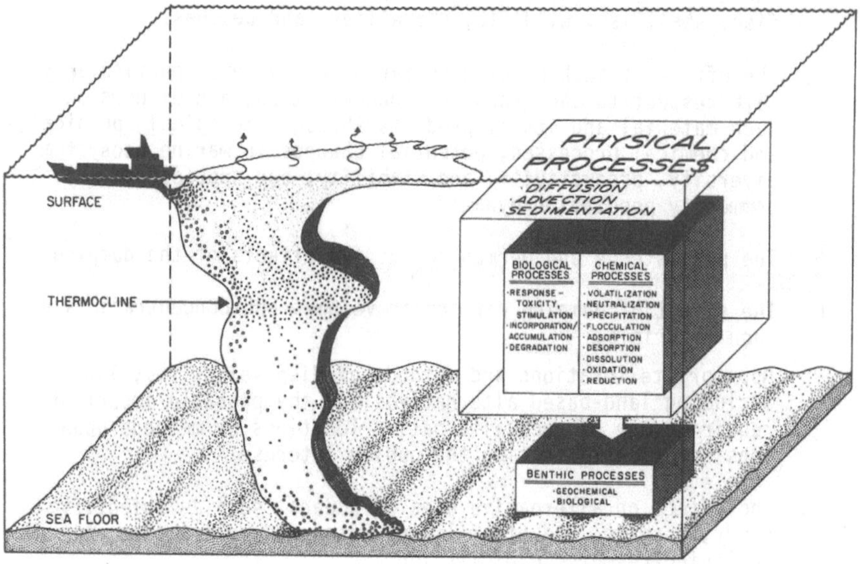

Fig.2. Schematic of the distribution of waste in the water column after a dumping event, and some processes that will affect distribution and fate of the waste (Duedall et al., 1983a).

COAL WASTES: AN INDUSTRIAL WASTE THAT CAN BE USED IN THE OCEAN FOR CONSTRUCTION OF ARTIFICIAL REEFS

Coal wastes (fly ash, scrubber sludge), which is dumped at sea by some countries, may soon emerge as the principal industrial waste destined for ocean disposal. The reason for this is the shift away from oil in favor of coal as the main source for electricity. With coal for electricity comes the production of a huge quantity of waste. For example, a 1000-MW coal-fire power plant equipped with scrubbers will produce about 200 tons per day of calcium salts ($CaSO_4 \cdot 2H_2O$ and $CaSO_3 \cdot \frac{1}{2}H_2O$), 500 tons per day of fly ash, and about 100 tons per day of bottom ash.

What can be done with these wastes? A group of scientists in the United States has been developing a program in which the coal wastes are used for artificial-reef construction. Basically, the method involves combining the ash material with additives and then running the mixture through a concrete block-making machine. The blocks are mass-produced and then transferred to a barge for transport to the sea for disposal as a reef (Fig. 4). On the seabed, the blocks form a random assemblage of crevices suitable for bottom habitat (Fig. 5).

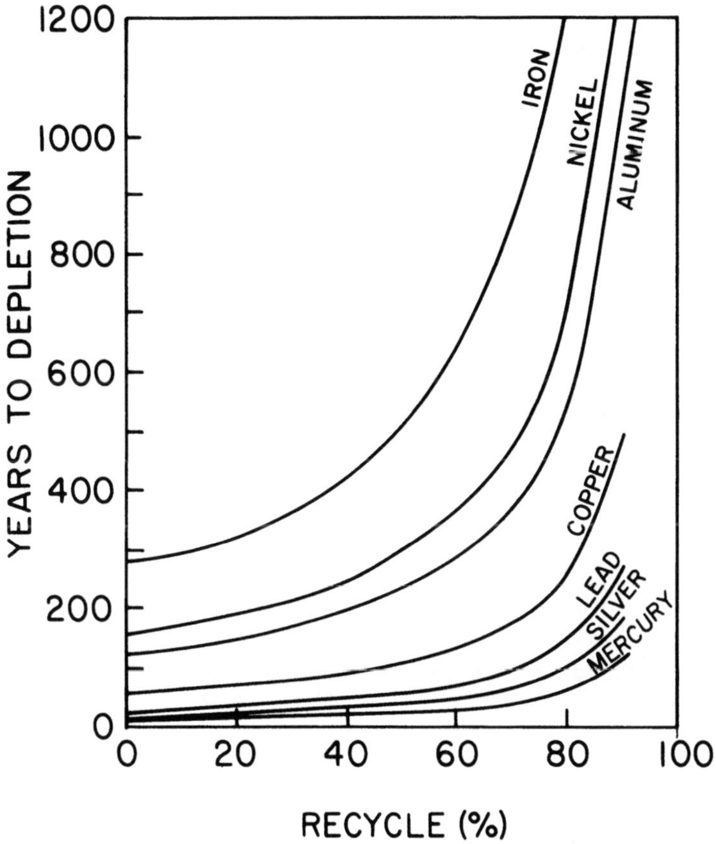

Fig.3. Projected hypothetical lifetimes of the reserve of some common metals and the significance of recyling on prolonging their depletion (Duedall et al., 1983b).

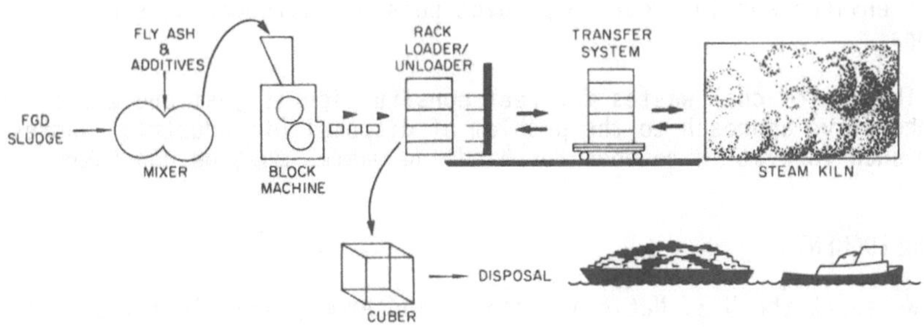

Fig.4. Schematic showing the processing of coal waste for production of blocks for construction of artificial reefs.

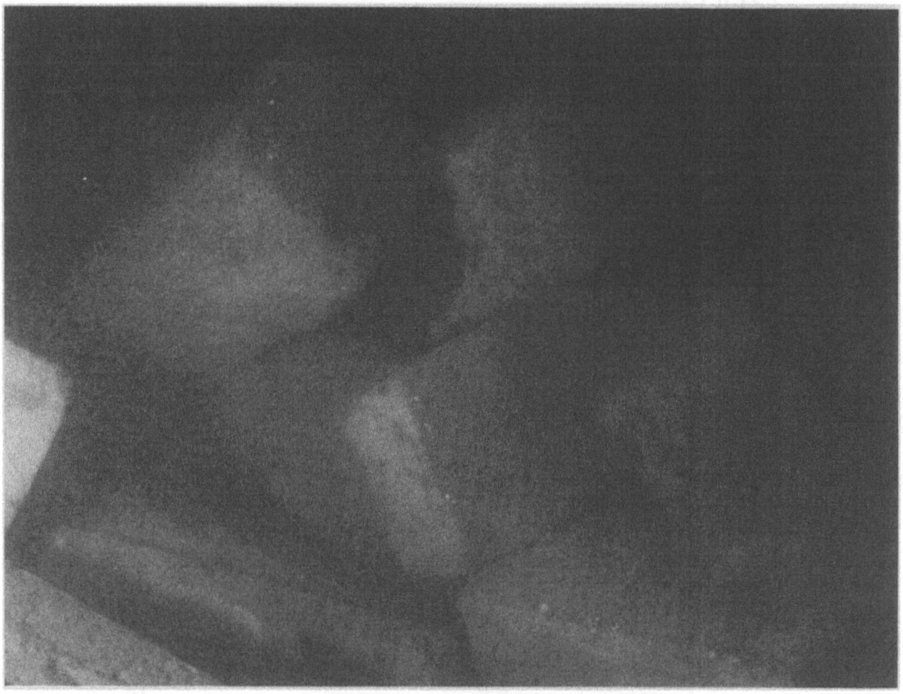

Fig.5. Coal waste blocks on the seabed: an artificial reef.

CONCLUSIONS

The dumping of wastes at sea will continue into the future. Through research we must understand the effects of these wastes on the marine environment in order to predict possible outcomes of such actions.

The use of coal wastes for reef construction is just one example of an innovative approach to the problem of disposal of industrial wastes. Other such innovative methods for handling wastes must be considered.

ACKNOWLEDGMENTS

We thank the U.S. National Oceanic and Atmospheric Administration for financial support.

REFERENCES

1. Duedall, I.W.; Ketchum, B.H.; Park, P.K.; Kester, D.R. Global inputs, characteristics, and fates of ocean-dumped industrial and sewage wastes: an overview. In: Wastes in the Ocean, Vol. 1, Duedall, I.W., Ketchum, B.H.; Park, P.K.; Kester, D.R. (eds.), Wiley-Interscience, New York, 1983a.

2. Duedall, I.W.; Ketchum, B.H., Park, P.K.; Kester, D.R. Scientific strategy for industrial and sewage waste disposal in the ocean. In: Wastes in the Ocean, Vol. 1, Duedall, I. W., Ketchum, B.H.; Park, P.K.; Kester, D.R. (eds.), Wiley-Interscience, New York, 1983b.

3. Erdheim, E., United States marine wastes disposal policy. In: Wastes in the Ocean, Vol. 6, Ketchum, B.H.; Capuzzo, J.M.; Burt, W.V.; Duedall, I.W.; Park, P.K., Kester, D.R., Wiley-Interscience, New York, 1985.

Technical Session:
Coastal Zone Utilization
and Management

6. National and Regional Planning

Planning for the Comprehensive Management of Ocean Space

Y. Ohtahara[1], T. Kondo[2], M. Inoue[3], H. Tsuji[4] and S. Hamada[5]

[1] National Institute of Resources, Science and Technology Agency, Primeminister's Office, Kasumigaseki, Chiyoda-ku, Tokyo, 150 Japan
[2] Department of Oceanic Architecture and Engineering, College of Science and Technology, Nihon University, Hiroo, Shibuya-ku, Tokyo, 150 Japan
[3] Applied Meteorological Engeneering and Consulting Services Co. Ltd., Ueno, Taito-ku, Tokyo, 110 Japan
[4] Technical Research Institute, Ohbayashi Gumi Building Construction Co., Ltd., Shimokiyoto, Kiyose, Tokyo, 204 Japan
[5] Offshore Engineering Department, Kajima Construction Co. Ltd., Motoakasaka, Minato-ku, Tokyo, 107 Japan

1. Introduction

Japan, a nation completed surrounded by the sea, has been able to benefit in many ways from the ocean, including various aspects of material and cultural exchange as well as the exploitation of a variety of food for human consumption from vast ocean resources. Furthermore, the development of Japan's culture has been significantly influenced by the sea. In these aspects, Japan's far-ranging and deep relationship with the sea is intimate and inseparable.

When viewing the many forms of land utilization in Japan, it can be seen that the coastal zones are extensively used as evidenced by the fact that approximately 70 percent of all Japanese cities as well as about half of the highly-populated districts are located in coastal regions. Demand for these coastal land zones can be expected to increase in the future, even leading, in some cases, to requests that zones along the coast be reclaimed.

However, because of the fact that these areas represent a source of vitally important natural resources, coastal zones consisting of the long and narrow tracts of land and sea along the coast of a certain width are extremely precious assets not only for us at the present but also for our children of the future.

Ocean development (utilization) in Japan until the present has had a tendency to lack a basic territorial management principle based on the concept of sea zone utilization. As a result, there is an urgent need to draw up a basic principle in line with the overall status of land in Japan. Furthermore, as seen by the convening of the United Nations' Conferene on the Law of the Sea for signatures of the text held in December 1982, interest in matters related to the ocean have reached the highest level of importance ever.

According to the above then, by treating land space resources here as space also including ocean space, we have attempted to clarify the characteristic features and issues based on the current situation of ocean development (utilization) in Japan, the results of which shall be used to examine ways to utilize ocean space.

2. The Nature of the Ocean and Ocean Space Resources in a More Advanced Manner

Man is attracted to the abundance of nature (in terms of earth science, ecology and sentiment) represented by the ocean, and efforts to harvest this nature shall continue to grow in importance in terms of the future of mankind.

Moreover, being a nation surrounded by the sea on all sides, the 200-mile sea zone of Japan represents more than ten times the national land area. Together with the developmental use of ocean resources, such as marine organisms, efforts to promote the use of ocean space, including the sea surface, undersea and the seabed, by achieving technologies for sea space utilization will become an important part of ocean development in the future.

Table 1 Classification of Ocean Space in Terms of Utilization

- Ocean space resources
  - Socio-economic use (actual resources)
    - Use of land reclaimed from the sea
      - Place for daily living
      - Place for industrial production (including regenerative and dumping processes)
      - Place for storage
      - Place for traffic and transport
      - Place for recreation
    - Use of special ocean characteristics
      - Place for production of organic resources
      - Place for production of ocean energy
      - Place for traffic and communications
      - Place for recreation
      - Place for securing water
  - Ecological use (potential resources)
    - Cultural use (human conditions)
      - Place for marine culture (festivals/rituals/literature/history/the arts)
      - Place for adventure
      - Place for emotional recreation
    - Use of special ocean characteristics (climatical/geographical conditions)
      - Open space (atmospheric circulation)
      - Food chain (place for natural purification)

Table 2 Present State of Technologies Concerning Utilization of Ocean Space

Level of necessity	Necessary measures	Marine organisms resources	Seawater/seabed resources	Ocean energy sources	Ocean space resources
Level I "desire"	Quantity sufficiency/ management	[Cultivation fishery] Cultivation fishery; Plant of seedlings; Acoustic baiting	[Sea water desalination] [Construction materials] Grit, Ballast; [Energy sources] Coal, Oil, Gas		[Reclamation] Airport, Factory, Storing; Electric power station; Marina, Marina, Offshore observation tower
Level II "hope"	Solution for trouble areas; Technological development	[Marine fish cultures] Offshore fishing; Planned production	[Dissolved substance in seawater] [Mineral resources] Manganese, Hydrothermel ore deposit; Uranium	[Clean energy] Sea-thermal power genaration; Wave power generation	[Offshore artificial island] [Marine recreation]
Level III "dream"	Solution for problematic areas; Technological development training	[Marine farm] Marine biofarm		Tidal power generation; Ocean/tidal current power generation; Density differencial power generation	[Underwater base] Marine city; Underwater factory; Underwater city; Undersea park

The term ocean space resources used here shall refer to the atmosphere, hydrosphere and geosphere, the most basic source of resources for human existence and activity. Moreover, since they are three-dimensional in characteristic ---- sea surface, undersea and seabed ----, they are also regarded as a source of resources comprising of sea water and various ecosystems in addition to the ocean space.

3. Characteristic Features and Issues of Sea Utilization in Japan

(1) Traditional use of the sea

Since Japan is comprised of a narrow chain of islands surrounded by the sea, and also because Tokyo Bay, the Inland Sea and many other areas are blessed with calm waters separated from high sea, many cities have created new land by reclaiming the sea areas located to the front. Since many of these cities are all port cities, harbor life has contributed towards their development, and offshore fishery activities have been very actively carried out.

The problems involved in this type of sea utilization may be summarized as follows:

a) Access to the sea has been cut off because a large part of the reclaimed land is for industrial use and thus monopolized by business firms. As a result, the sea has become something remote, not near, for urban dwellers.

b) Plans for new forms of sea use more often than otherwise have caused disputes with those engaged in fishery and other traditional forms of sea use.

c) Due to emphasis on disaster prevention and other needs related to coastal industries, the importance of clean and comfortable water has not been sufficiently taken into account.

d) Because the use of a sea zone has been promoted each time a project has been launched, ideas never took root that would have led to a master plan for the sea calling for regional uniformity, inter-project coordination, and the planned use of all sea zones concerned, as well as their preservation.

(2) Sea utilization based on land theory

Utilization of sea space in Japan has involved the reclamation of shallow sea zones for industial use, urban development, public facilities, and the like. Reclaimed sea zones were those that were suited for coping with functions considered forced out from zones of normal land use.

However, the use of ocean space pursued only on the basis of land theory has resulted in the irreversible problem of increased space consumption, such as the loss of natural seashores and tideland, which are recognized as resources for amenities, and the destruction of ecosystems triggered by the contamination of sea zones.

(3) Single utilization of the sea

Japan's coastal zones, with their high concentration of people and industires, are used extensively. Viewing this in terms of a special unit, we see that the ways in which they are being used are so simple that they can be easily classified into stretches of coastal industries, harbor distribution, marine preservation and marine recreation, each form of utilization ranging from several kilometers to well over ten kilometers. Moreover, looking also at the various types of sea projects noted above, we can see that many are mono-functional-types projects.

4. Necessity of Utilizing Ocean Space

Many areas of the sea remain unknown --- empirically, logically and scientifically --- that it is even unclear whether or not the empirical rules of the present-day social system can be applied. The necessity of ocean space can be divided into three levels.

(1) The level at which the livelihood needs of people have reached the level of "desire" (level I).

The conventional use of the term "necessity" or "need" refers to those that have reached this level, that is, their form and the way needs are satisfied are clear. Thus, the needs themselves are recognized as "objects."

(2) The level at which the needs have reached the level of "expectation" or "hope" (level II).

The needs at this level are not yet the object of any clear desire because the method of satisfying them have not been perfected either technologically or economically. Thus, it is difficult for them to represent an end in themselves.

(3) The level at which the needs are still at the level of "dreams" (level III).

The needs at this level are either unclear or incomplete in terms of both their objects and method of satisfying them. Thus, they are sometimes identified as "illusions."

5. Basic Principle of Ocean Space Utilization in an Advanced Manner

(1) Ocean space as a land resource

If ocean space is treated as land space, for the purpose of orderly development, it would be necessary to have a comprehensive and integrated philosophy of development, as well as a plan for the appropriate use of space. The use of ocean space for complex, multiple purposes will exert a profound effect on the way land is used. Therefore, Japan should make a serious effort to establish and recognize ocean space as part of its land resources.

(2) Complex, multi-purpose utilization

Since the space that can be used effectively in marine development is limited despite the vast ocean area, that Japan possesses, the country must make an effort to use it for complex, multiple purposes. Also, although a serious effort will no doubt be made to exploit the sea zones at greater depths, construction costs for such an undertaking would be enormous. Thus, ocean development would require efforts to promote technological development that take into account both economical and technological efficiency.

(3) Ocean development as a social system

In order to solve such complicated and interrelated problems as those involved in ocean development, it is necessary to adopt a social systems approach that takes into consideration man, space (nature and ecology), and society (multitude of human activities). Without such an approach, among other things, it would be difficult to use the sea zones in an advanced manner, combine utilization of land and that of the sea under one system, and treat ocean space as part of Japan's natural resources. Also, to solve the various problems confronting Japan with a social systems approach, it would be necessary to treat them as a system and develop science and technology designed to resolve their related issues. Since single-purpose technology will not achieve this, complex multi-purpose technology is necessary.

(4) Method of conducting ocean space utilization projects

What is clear about the use of ocean space in the future is that the areas of investigation will increase, every area will be intricately related to all other areas, and the process of carrying out projects will require a longer period. Table 3 shows the areas of investigation in ocean space utilization projects and their basic points of view.

The content of investigation in these areas will differ depending on the respective projects being implemented. However, in whatever form ocean space might be used, a similar, fundamental way of thought should lie at the base.

Futhermore, with regard to the manner by which ocean space utilization projects are being implemented, it is important to present the final actual projection, not as something static but as one that is always related to the production process.

(5) Promotion of a comprehensive policy

It is expected that the need to establish a fundamental project based on a long-term perspective concerning utilization of space in the ocean areas surrounding Japan will increase. If this is the case, a comprehensive policy would be required governing the scientific, technological and institutional measures designed to smoothly and efficiently adjust the scale and characteristic features of Japan's ocean space as part of its land

Table 3 Areas to Be Investigated in the Ocean Space Utilization Plan and Basic Perspective

Objectives	Ocean space is regarded as a source of natural resources. The appropriate utilization of ocean space is essential for establishing a healthy and culturally enriching life environment for the Japanese people. This is important for both Japan and the entire earth.
Resources	Create new possibilities and achieve the goals of the plan by comprehensively utilizing marine resources (geological, ecological, and emotional resources).
Ecology	The ecosystem is a barometer for environmental preservation, and as such will be used to conduct environmental assessments, both attempting to preserve the ecosystem, and, at the same time, effectively use ocean resources by controlling and utilizing the ecosystem (as a precondition, the ecosystem's conditions must be ascertained).
Areas	The ocean space will be used as a three-dimensional space. Furthermore, the coastal and 200-mile offshore zones will be utilized in a way where the ocean and land are integrated.
Time	Much of the utilization of ocean space is carried out over a ten-year or longer period. Along with implementing plans into concrete form in stages, it is also necessary to draw plans flexibly so that, for example, new technology may be continuously introduced. Such a system of process planning will be devised.
Internationalism	A system of international mutual assistance will be established. For example, in order to shift from resource nationalism to resource mutual assistance, free ports, an energy distribution system, and other concepts are being drawn up.
Industry	In addition to the possibility of locating industries ranging from basic industry to the fourth industry (information industry), a new industry will be established involving the planning, construction and management of ocean space.
Technology	Technologies for the utilization of ocean space in an advanced manner (with self-sufficiency as the axis) (1) Technologies for planning, construction and management of ocean space (including deep sea offshore ocean space) (2) Technologies for expanding production (application of such forms of high technology as marine biotechnology) (3) Technologies for the multiple use of ocean energy
Education	To reach a consensus that "the sea grants us rich resources and refreshes our lives," a curriculum designed to promote marine education for people of all ages will be devised. The training of marine education experts will be conducted.
Organization	Establishment of a unified national organization for ocean space management, establishment of a federation of local governing bodies involved in the coastal zones and establishment of a tie and cooperation with the private sector.

resources, restrictions on use, method by which they are used, and the balance between supply and utilization of ocean space available in Japan.

6. Conclusion

Appropriate utilization of ocean space is imperative for establishing a healthy and culturally enriching life environment for the Japanese people. This is important not only for Japan but for the entire earth.

To actively and smoothly promote utilization of ocean space in an advanced manner, the following steps must be taken:

(1) Promotion of related administrative measures

(a) Comprehensive utilization of all land in Japan, including sea zones

(b) Drafting and adopting plans for developing sea zones and examining a management and administrative system for sea zones

(2) Promotion of the related scientific and technological development

(3) Promotion of international cooperation as well as understanding and cooperation by the Japanese people

Reference

〔1〕Science and Technology Agency, Research Institute of Resources, "Research Report of an Advanced Usage for Ocean Space", Report No. 129, January 1985.
—"Kaiyo Kukan no Koudo Riyou ni Kansuru Chyosa"—

〔2〕National Land Agency, "Prospect for the 21st Century in Japan", November, 1984.
—"Nihon 21 Seiki e no Tenbou"—

〔3〕Economic Planning Agency, "Japan in 2000 years" August, 1982
—"2000 Nen no Nihon"—

Coal Transhipment Centers

An Economic Development Potential for the Emerging Micronesian Governments in the Pacific

M.K. Actouka*

Pacific Basin Development Council, University of Hawaii, East West Center, Honolulu, HI 96813, USA

ABSTRACT

The Micronesian governments of the Republic of the Marshalls (RM), the Federated States of Micronesia (FSM, which includes Yap, Truk, Pohnpei and Kosrae), the Commonwealth of the Northern Mariana Islands (CNMI, which includes Saipan, Tinian and Rota), the Republic of Belau (RB) and the Territory of Guam are all aspiring for some degree of economic self-sufficiency. Energy along with capital and raw materials is one of the basic and fundamental resources needed to attain this goal. All of these island governments do not at this point have proven indigeneous sources of oil and coal. However, as early as the 1800's, these Pacific islands were being used by the rim countries as coaling stations. The oil embargo of the 1970's and the unstable cost of fuel oil since have led to various economic and engineering studies aimed at verifying the potential advantages of supertanker ports, transhipment of oil and coal and the utilization of various alternative renewable energy sources. Of these options, this paper will concentrate on Coal Transhipment Center (CTC). The importance of a CTC is based on the concept of security, reliability and accessibility. Strategic and cost effectiveness are also integral to this concept. The CTC can be located on any of the islands. The assumptions are that the CTC through-put of 12 million tons per year and that it will provide at least 100 jobs and a revenue of close to $30 million. The additional benefits of the CTC are the by products such as electricity, fertilizer and construction materials. (1)

INTRODUCTION

The Micronesian islands are under the United Nations Trusteeship Agreement until the Compact of Free Associations agreements are signed. (2) Guam however became an unincorporated territory of the United States in 1950. The Micronesian islands consists of more than 2,000 or more islands and cover an ocean area of about 3,000,000 square miles. The 1980 census showed that the population is young (50% are less than 15 years old) and increasing at a rate of more than 3% a year. (3)

---

*The findings and opinions expressed in this paper are of the author and do not necessarily reflect the position of the Board of Directors. This paper is an expansion of a recently completed study by the author and colleagues.

Table 1. General Characteristics of the U.S. Affiliated Islands

Island	Land Area (sq. mi.)	Population 1980	Fuel Import (Mbbls)	Per Capita Consumption
Guam	212	105,000	9,178	1.47
CNMI	182	16,900	330	0.91
RM	66	31,000	157	0.16
RB	192	12,000	148	0.26
FSM	271	74,000	229	0.09

Source: DOI/DOE Final Report, Territorial Energy Assessment, 1982. (4)

For the last 500 years the islands have been governed by the Spanish, the Germans, the Japanese under the League of Nations Mandate and now the United States of America. This historical heritage has provided at times a common language, government structure and monitary system that is foreign to the Micronesians who have diverse languages, cultures, religions and traditional governments. The people and the land have been scarred by foreign wars, especially World War II, where lives were lost, fortunes taken away and hopes died.

OCEAN REGIME

With the land based natural resources diminishing, more attention and importance have been directed toward the ocean resources and rights. Already exploited are the tuna and other marine resources. Minerals, energy and legal ownership of the resources are now being explored and defined. The United Nations Law of the Sea Conference (UNCLOS) and the subsequent Treaty is a testimony to the importance of the ocean and its resources. The complexity of ownership, management and exploitation and control of this regime have allied some countries, especially coastal ones on one side and the land locked on the other. The rejection of the United Stated of the Treaty and declaration of its own 200-mile Exclusive Economic Zone (EEZ) in 1984 is an outstanding example. (5)

With the UNCLOS and 200-mile EEZ in concert with the current Micronesian ties with the United States, the Northern Central Pacific becomes a vast EEZ. Each of the island groups control the following: Republic of Belau (178,000 sq. mi.), FSM (889,000 sq. mi.), Guam (65,000 sq. mi.), CNMI (237,000 sq. mi.), and Republic of Marshalls (565,000 sq. mi.). This major natural resource has both economic and political implications. Defense interest, however, of the rim nations should not be overlooked. Of immediate benefits are the fishing rights and the potential minerals (manganese crusts), energy (ocean thermal, hydrogen, oil), waste disposal (nuclear waste), taxing sea lanes (oil and coal transhipments, goods and people) and military transports. As Micronesian governments become more economically and politically self-reliant, their sentiments and alliances can shift to other rim countries. With growing concern about nuclear hazards becoming more of a uniting force, the EEZ's of the Pacific Basin countries will virtually block any sea lane between the rim countries (i.e. U.S. and Japan). The Central North Pacific is part of the shipping route between the energy consumers (Japan, Taiwan, Korea and Hong Kong) and the suppliers (Australia, Canada, United States, Mexico and South America). The strategic location of Micronesia in this lane and their political stability could provide a secure and accessible point for the storage and transhipment of oil and coal for the rim as well as some of the basin countries. In the last decade various joint ventures have studied the potential for refineries, oil and coal transhipment and the stockpiling of strategic fuel supplies. (6)

COALING STATIONS

The protected harbors and channels, available land and close proximity of major sea lanes have been the historical basis for the islands becoming coaling stations as early as the 1800's. Reports show that Lelu, Kosrae harbor, was loading coal in 1872 and the Marshalls in 1881. Truk was a Japanese coaling station in 1935. During the Japanese period, coal was being imported to fuel the sugar and phosphate industry on Saipan and Angaur. To support the economic activities of that period, coal imports increased from 5 tons (1921) to 9,000 tons (1935). (7)

Remains of the coal imported during the Japanese Mandate were used by the only blacksmith on Pohnpei as late as 1963. Today, coal is being used or considered for power generation for cement and sugar industries and for power utilities in the Pacific Basin. Discussions, economic and engineering feasibility studies have been going on in the U.S. affiliated islands regarding the possibility of oil-to-coal conversion in the utilities, transhipment and storage facilities. The continued rise and the recent fluctuations of fuel oil has been a major driving factor of this renewed interest in coal. (8)

Currently, all the Micronesian governments are heavily dependent on U.S. federal aid for the operation and maintenance of the power utilities. It is quite common to either reprogram funds or request additional funds to defray additional fuel costs. Fuel costs alone can consume as much as 10% of the total budget of some of the island governments. (9)

Table 2. Electricity Production Revenue Deficits in Micronesia (1980 - 1981)

Government	Cost U.S. $	Micronesian Revenues	Dept. Interior Grants	Deficit
Palau	5,964,151	811,892	3,820,000	(5,152,295)
Yap	2,362,208	285,479	2,083,400	(2,076,729)
Truk	3,737,541	248,645	2,292,000	(3,498,896)
Pohnpei	3,762,532	367,848	2,760,300	(3,394,684)

Source: U.S. Department of Defence, Economic Development in the Marshall Islands, 1984. (10)

The main fuel used in the utilities is diesel and residual fuel and is costing more than the governments' export trades. With coal prices now being competetive, more interest is surfacing in the utilization of coal as an economic development potential.

The power systems in the islands vary from less than a megawatt to 120 megawatt base load on Guam. Though the sizes and capacities are wide spread, power outages, oil supply contract changes and high cost of fuel are common to almost all. The stability and accessibility of coal as a power generation fuel can, in the medium range, provide some solutions to the chronic problems. One solution is by bringing coal into the islands in the establishment of CTC's.

Table 3. Electrical Generation Capacity in the Pacific

Island	Base Load (MW)	Operating Capacity (MW)	Capacity/person (kwh/person)
Guam	120	173	1.64
CNMI	10.5	21.6	1.28
RM	3.1	4.9	0.16
RB	2.2	6.8	0.55
FSM-Truk	1.1	6.0	0.16
-Pohnpei	1.8	2.5	0.11
-Yap	1.0	3.3	0.40
-Kosrae	0.2	0.5	0.09

Source: DOE/DOI Territorial Energy Assessment, 1982. (4)

COAL TRANSHIPMENT CENTERS

The CTC's should be strategically centrally located between the importer and exporters of coal. In addition, some basic infrastructures should already have been in place. They include: telecommunications, air transportation, banking, stable currency, political and social order. These are already part of the reason that over the last decade major multi-national corporations have proposed joint ventures in energy related projects for the islands.

It is evident that the need and interest is there from U.S.A. as well as other rim countries in puting major investment projects in the Micronesian islands. Aside from the fishing industry and the tourist industry, energy as a common demoninator in development is being aggressively expanded. (11)

The basic components of CTC's are: channels and ports, the loading, mixing, unloading and stockpiling equipment. The equipment costs are fixed while channel and port dredging and expansion varies according to the existing conditions and the selected size of CTC. For example, for a site in Tiniam, with

12 million ton through-put, the estimated cost of the CTC is 50 million dollars.

Table 4. Proposed Investment in Energy Projects

Island	Approximate Cost (U.S.$)	Comments
Marshalls	25,000,000	IPSECO, a London based company, has built a 12 MW power plant, and residual fuel storage in Majuro.
Belau	27,500,000	Government of RB and IPECO are negotiation similar contract. In 1974, Japan and Iran studied the feasibility of a Superport.
Rota, CNMI	10,300,000	Government and IPSECO discussed contract for power plant, transmission and fuel storage.
Maug, CNMI	400,000,000	IDC International proposed the construction of a 105,000 storage in 1977.
Tinian, CNMI	50,000,000	Economic and engineering feasibility study for a 10 million through-put CTC. A number of oil companies are also seeking possibilities of oil transhipment facilities.
Guam	300,000,000	Feasibility study of a Superport in 1977.

Source: Government of Guam, Guam Supertanker Port, 1978, U.S. Dept. Defence, Dept. Interior, and Pacific Daily News Articles.

Table 5. Cost Estimate of a CTC on Tinian

Item	Cost in $ million
Site work	3.0
Marine construction	8.0
Foundations	3.0
Three stackers	6.0
Two reclaimers	10.0
One shiploader	10.0
Material handling	10.0

Source: Actouka, et al, Pacific Coal Trade: Economic Opportunities for CNMI, 1983. (1)

The operation of the major equipment at the CTC will require large amounts of electrical power. This can justify the installation of a coal-fired power plant near the CTC to provide economical electric rates and reliable power source. This in conjunction with the dilema that small islands face of a small base load, can be combined to efficiently and technically utilize 25 to 50 Mw steam coal-fired power generators. There are cases, like Tinian-Saipan, Moen-Dublons (in Truk) where power can be transmitted via submarine cables to load centers and preserve the environmental integrity of tourist destinations.

Another advantage of coal-fired power system is that as the island governments look for indigeneous fuel sources (i.e. biomass from energy tree plantations and municipal solid waste (12)) become successful, the conversion from coal fuel fired to biomass is possible technically and economically. Thus the large investment in the coal-fired system will not be lost when biomass conversion does occur.

It is also important to note that the effect of CTC's on the deepening and widening of channel/harbor/turning basin at the existing or new ports will greatly enhance the ability of the containarized large ships to load/unload consumer goods. This type of bulk handling and shipping will reduce the current costs of imported materials and goods. (13)

The CTC size and stockpiling requirements were designed and based on, for illustrative purposes, steam coal import for Japan for 1990. The projection of 63 million ton per year might be too optimistic, but it also gives the extra added advantage that if Japan does not require that much, there is the potential to export to Taiwan, Korea or Hong Kong and still maintain a 90-day emergency supply for one of these countries. The through-put of 12 million tons per day closely approximates 20% of Japan's projected 1990 steam coal import. (14)

The closing of Guam Oil and Refining Company recently left the North Central Pacific with no refining capabilities. The closest refineries are in Singapore and Philippines. This event could be a blessing in disguise for the CTC since Guam is a major defense installation which imports around $300 million worth of oil products per year.

The question of pollution, dust from stockpiling, loading and discharging and burning will be critical as most island environments are delicate. With modern advances in monitoring and control of pollution, CTC's should not pose more environmental risks than oil or other fuel transhipments in the islands. However, in contract negotiations and formulation, there should be clear requirements stated to protect the welfare of the islands.

The cost incurred as a result of the additional handling and stockpiling at the CTC before the consumer ports should not deter the potential developers of CTC. Table 6 shows that there is an actual savings in the transhipment. In addition to the monitary advantages, there are enhanced logistics, space savings at the consumer ports, and the close proximity of strategic stockpiles.

Table 6. Estimation of Transhipment Cost Savings. (U.S.$)

	Direct Shipment	Transhipment		
	NSW-JPN	NSW-CTC		CTC-JPN
Deadweight tons of ships	60,000	100,000	150,000	60,000
Tons of coal shipped	58,200	97,000	145,000	58,200
Days in ports	2.3	2.6	5.1	0.8
Daily cost in ports ($)	18,275	22,582	26,713	18,275
Total cost in port ($)	42,033	58,713	104,181	14,620
Days at sea*	23.7	17.2	17.2	6.5
Daily cost at sea ($)	27,284	33,379	39,173	27,284
Total cost at sea($)	646,631	574,119	673,776	177,346
Total shipping cost ($)	688,664	632,832	777,957	191,966
Average cost per ton($)	11.83	6.52	5.35	3.30
		NSW-CTC-JPN		
Transhipment cost savings		9.82 2.01	8.65 3.18	

*Days at sea are calculated on the round trip basis; Vessel's speed is 15 knots; Distances are as follows:
NSW (Newcastle, NSW.) - JPN (Yokohama, Japan): 4,268 nautical miles;
NSW - CTC (coal transhipment center, Tinian): 3,096 Nautical miles;
CTC - JPN: 1,172 nautical miles.

Source: Actouka, Jenkins and Miklius. (1)

CONCLUSIONS AND RECOMMENDATIONS

The discussion of a CTC in this paper is an expansion of an earlier work that concentrated on one specific site in Micronesia. However, the aim of this effort is to apply the findings of a specific site to other possible sites in the North Central Pacific to improve the chances of implementation and the selection of the best location.

In concluding this brief presentation, a number of points should be highlighted for the coal, utility, transportation industries and governments involved in steam coal business to consider and share potential benefits. First, as oil prices become more unstable and volatile, coal firm prices have contributed to a more sustained growth in the steel and power generation industries in the Pacific rim countries. Second, coal is not a commodity that has uniform qualities that utilities can switch suppliers at short notices. Each power plant needs special standards that ties to a long-term contract with a supplier. These long term contracts (e.g. for 10 years) can be factored into the CTC system and the quality of coal can be attained by blending, cleaning and enhancement processes at the CTC. A CTC located in Micronesia will have three benefactors: The exporter will have an additional stockpiling site and a point of discharge with deep draft and appropriate facilities. The island will gain revenues, jobs, training, reliable energy source and other by-products. The importer will have a secure and close supply of coal for emergencies. These advantages, I believe, are the technical and economic rationale that will make a CTC possible in Micronesia in the near future.

REFERENCES

1. Actouka, M.K.; Jenkins, R.W.; Miklius, W.: Pacific Coal Trade: Economic Opportunities for CNMI, Pacific Basin Development Council Research Institute, Honoulu, 1983.

2. The Office for Micronesian Status Negotiations, Draft Environmental Impact Statement for the Compact of Free Association, 1983; and U.S. General Accounting Office, The Challenge for Enhancing Micronesian Self-Sufficiency, 1983.

3. U.S. Dept. of Commerce and Bureau of Census,: 1980 Population Census for Trust Territory, Guam, American Samoa and Commonwealth of the Northern Mariana Islands, 1983.

4. U.S. Dept. of Interior and Energy, Final Report, Territorial Energy Assessment, 1982.

5. Prodeedings of the 7th International Ocean Symposium, Exclusive Economic Zone 1982, ISS International, Inc. Tokyo, 1983; and, Friedheim, R.L.: Japan and the New Ocean Regime, Boulder and London, Westview Press, 1984.

6. In the last year the local papers have been reporting the take-overs and proposed plans to purchase of the power systems, construction of transhipment of oil fuels. The companies include IPSECO of London, UNELCO of France, Mitsubishi Corp. of Japan, Sea Hunter Corp. of Los Angeles, Shell, Caltex, and others.

7. Kaneda, K.: Political and Economic Characteristics of Japan's Colonies. The Journal of the Kobe University of Commerce, Reprint No. 125, 1939. Hezel, F.X.: Foreign Ships in Micronesia, Trust Territory Historic Preservation and U.S. Heritage Conservation and Recreation Service, Saipan, 1979. Ballendort, D.A.: Secrets Without Substance. The Journal of Pacific History, Vol. 19, No. 1-2, April, 1984.

8. In the State of Hawaii, already all the cement companies have converted to coal. The sugar industry is already starting to convert. The major electric utility is reportedly seriously considering coal fuel use.

9. U.S. Dept. of the Interior, Office of Inspector General, Annual Report, Government of Guam, American Samoa, Marshalls, Federated States of Micronesia, Republic of Palau and the Commonwealth of the Northern Mariana Islands, 1984.

10. U.S.Dept. of Defence, Economic Development in the Marshall Islands, 1984.

11. Various Over-All Economic Development Plans for Guam, CNMI, American Samoa and the Five Development Plans of FSM, Marshalls, and Belau.

12. Actouka, M.K. and Kvandal, S.C.: Municipal Solid Waste Energy Conversion Study for Guam and American Samoa, Proceedings of BioEnergy 84, World Conference and Exhibit, Gothenburg, Sweden, 1984: and Pacific Island Development Program, Energy Mission Report Ponape FSM, East West Center, Honolulu, 1982.

13. U.S. Dept. of Army, Army Corps of Engineers, Draft Interim Survey Report and Environmental Statement, Apra Harbor, Guam, 1982; and Port and Harbor Study for CNMI, 1981. Pike, D.: Coal Transport: An Analysis of Technical and Economic Factors, London, The Financial Times Business Information Ltd. 1982.

14. Borg, I.Y.: Coal, An Option for Power Generation in U.S. Territories of the Pacific, 1981; and Technical Considerations Relating to Use of Coal in American Samoa for Power Generation, University of California, Livermore, California, 1982; and Kimura, T: Demand for Imported Steam Coal in Asian Countries, The Institute of Energy Economics, Tokyo, 1983; and Cwik, L.J.: Austrilian Coal/Environmental Policies: Analysis, Recommendations, Thesis, University of Hawaii, Honolulu, 1983.

Coastal Development and Government's Policies in Singapore

N. Matsumoto

International Division Group, Penta-Ocean Construction Co., Ltd., Koraku, Bunkyo-ku, Tokyo, 112 Japan

ABSTRACT

Due to its limited land resources and rapid economic development, the government of the Republic of Singapore has been conducting extensive reclamation projects in its territorial waters since 1966. By 1982, 3,200 ha of land area has been added by land reclamation, which comprises 5% increase in the total land of the country. In this paper, past and future reclamation projects in the country and the use of reclaimed area have been thoroughly studied. Furthermore, the government's policies against those projects and their background are also discussed.

INTRODUCTION

The Republic of Singapore is composed of one main island and other 50 small islands located off Malay Peninsula. The total area of the country is merely 618 square km (almost equivalent to Awajishima Island of Japan), while the country has a population of 2.4 million which is expected to grow to 2.7 million in 1990. Its economy has been growing strongly with almost 10% of annual growth rate, even after the world-wide depression due to the Oil Embargo.

Under those conditions, the government has given the high priority to land development, and the country has been actively carrying out extensive land reclamation in its territorial waters since 1966.

DRIVING FORCES OF RECLAMATION PROJECTS

One of the important factors that have driven the country to the series of extensive reclamation projects is its limited land resources and the growing population. Although the total land area has increased by 5% from 1967 by means of land reclamation, its population has shown a phenomenal 25% increase since 1967. Accordingly, the gross population density became 4,000 persons per square km in 1982 from 3,400 persons per square km in 1967. The figure in 1982 is more than twice of what was just after World War II. Although the growth rate of population has shown a slight decrease in the recent few years, the trend of population increase is expected to continue in the future. The total population in 1990 is estimated to grow to 2.7 million, and the number is expected to reach 4 million in 2040. The rapid increase of population in the country of limited land has created a keen demand of land.

Another factor of the extensive reclamation is stemmed from the government's policy of industrialization.

During the era of British colony, the country prospered as a freetrade center in Southeast Asia, and at the same time, Singapore played an important role as a British military base in that area. In the 1960's, while the national revenue declined due to the price fall of rubber that was the main trading item for the country at that time and due to the retreat of the British military after Singapore's independence, its industrial sectors showed rapid growth by means of policies of actively inviting foreign investments in manufacturing sectors through its financial strength acquired in the colonial era. In the last decade, the rapid industrialization was shifted to technology-oriented sectors, such as oil refinery, ship-building, electric machinery, etc. Further, in the 1980's, the restructuring of the manufacturing sectors -- phasing out of lower value-added activities and establishing higher value-added and more technology-oriented industries, e.g., computers, periphrals, electronic components, pharmaceuticals, etc. -- is the main trend of the eighties.

Despite the world-wide stagnation due to the Oil Crisis, the economy of the country has grown strongly, and the real GNP growth has been keeping the level 10% in recent years. GNP per capital achieved US$4,340 in 1980 (US$8,920 in Japan), and new investments in manufacturing sectors in terms of gross fixed assets reached S$1.75 billion in 1982, only slightly less than S$1.92 billion achieved in 1981.

Furthermore, thrusted by trade expansions between Western countries and China, Singapore has been increasing its importance as a business center in Southeast Asia for international businesses of Western countries.

The rapid modernization of manufacturing sectors and the growing importance as a business center in Southeast Asia together have created further demand for land in addition to its residential and recreational needs due to the growing population. In that situation, the government of Singapore has been carrying out extensive reclamation projects to meet those ever-growing needs for land.

RECLAMATION PROJECTS IN SINGAPORE

Almost all of reclamation projects in Singapore have been undertaken by the three governmental organizations: the Housing and Development Board (H.D.B.), the Jurong Town Corporation (J.T.C.), and the Port of Singapore Authority (P.S.A.). For the reclamation projects, jurisdiction of the three organizations are divided as follows:

Table-1 Jurisdiction of H.D.B., J.T.C. and P.S.A.

Organization	Purposes	Main Location
H.D.B.	Residential, Commercial Recreational, Educational, etc.	Eastcoast
J.T.C.	Industrial	Southwestern Coast & Offshore Islands
P.S.A.	Transportation (Airport & Harbour)	Changi, Westcoast & Offshore Islands

Up to date, the total reclaimed area completed by the three organizations comes to about 3,200 ha, comprising 5% increase of land in the country, and areas of another 1,000 ha are currently under construction. 87% of the total reclaimed area completed is on the main island, while reclamation of offshore islands accounts for the rest. Table-2 describes the whole reclamation projects both completed and proposed, and their locations are shown in Fig.-1.

From the technical point of view, obtaining good fill materials has been one of the most hard-to-solve problems for those projects. In most cases, low undulatory hills have been excavated to utilize the hill-cut soil for fill materials, such as the cases of Eastcoast Phase I to Phase IV and Phase VI to

Table - 2 Reclamation Project in Singapore

Project	Area (ha)	Employer	Period	Cost (S$ million)	Contractors
Eastcoast Reclamation Phase I	405	HDB	1966-1970	45	Ohbayashi-Gumi Ltd.
Eastocast Reclamation Phase II	53	HDB	1970-1971	10	Ohbayashi-Gumi Ltd.
Eastcoast Reclamation Phase III	67	HDB	1971-1975	23	Ohbayashi-Gumi Ltd.
Eastcoast Reclamation Phase IV	486	HDB	1971-1976	44	China Engineers & Chin Ann (Pte.) Ltd.
Eastcoast Reclamation Phase V	154	HDB	1974-1977	106	Ohbayashi-Penta JV
Eastcoast Reclamation Phase VI & VII	360	HDB	1979-1985	385	Ohbayashi-Rinkai JV
Changi Airport Reclamation	653	PSA	1976-1979	230	Penta-Tao-Rinkai JV
Westcoast Reclamation at Pasir Panjang	-	PSA	1971-1975	12	Several Contractors
Westcoast Reclamation off Westcoast Road	-	HDB	1975-1977	-	Several Contractors
Reclamation of Tuas	400	JTC	1973-1976	12	Tiong Seng Contractors (Pte.) Ltd.
Pulau Seraya off Jurong	175	JTC	1975-1978	35	Loh & Loh-Daloh JV
Pulau Ayer Herbau	120	JTC	1976-1977	21	Kajima Corporation
Pasir Ris Reclamation	44	HDB	1978-1979	5.5	Tiong Seng Contractors (Pte.) Ltd.
Reclamation of Beting Sudong	243	PSA	1976-1978	45	Tao Harbour Works Co., Ltd.
Reclamation of Small Offshore Islands & Reefs	192	PSA	-	54	Several Contractors
Jurong Tuas Reclamation	637	JTC	1984- 1988	602	Penta-Ocean Construction Co., Ltd.
Reclamation of Pulau Sakra & Bakau	155	JTC	1984-1986	44.5	Boka-Ballast JV
Reclamation of Pulau Busing	55	JTC	1982-1984	32.5	Boka-Ballast JV
Reclamation of Punggol	1,152	HDB	1985-1993	874	(Future Plan)
Reclamation off Tekong	400	-	-	-	(Future Plan)
Reclamation at Tekong/Ubin	Unspecified	-	-	-	(Future Plan)

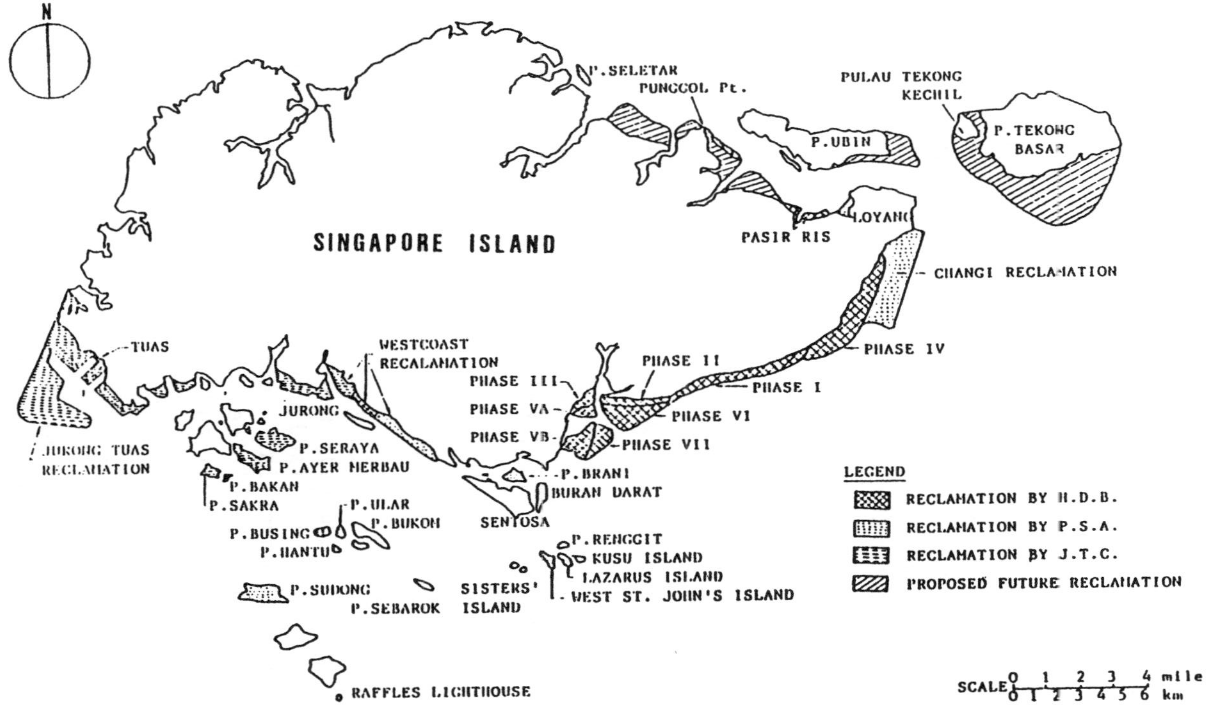

Fig.-1 RECLAMATION PROJECTS IN SINGAPORE

Phase VII, Pasir Ris, Westcoast and Tuas Reclamation. Flattened winning areas of hill-cut soil were, then, developed into residential areas, e.g., Tampines and Bedok Newtown, reservoirs and industrial estate, e.g., Tuas Reclamation. In other cases where hill-cut soil is not available, sea sand has been excavated for fill materials from seaside swamps or from seabeds as the cases of Changi Reclamation and Offshore Islands. Most parts of those hills, however, have been already excavated and earth fill from the land is substantially exhausted, while most of the sand deposits on the seabeds have also been utilized and now are coming to their limits.

LAND UTILIZATION OF RECLAIMED AREAS

As a result of the fast industrialization and the rapid population growth, land utilization of the country has been going through dramatic changes in the recent two decades.

A comprehensive land-use survey conducted in 1982/1983 has shown that the build-up*area of all over the country expanded by 80% from 14,411 ha in 1967 to 25,953 ha in 1982. Table-3 summarizes the results of the survey showing the current land utilizations and their increases over the 14 years.

Reclaimed land is utilized for wide range of purposes according to the previously mentioned three developing authorities. Reclaimed land developed by H.D.B. is used for residential, commercial, recreational and educational purposes, while land reclaimed by J.T.C. is used mainly for industrial development and P.S.A. prepares land for airport and harbour.

(1) Eastcoastal Reclamation

On the eastern end of Eastcoast Reclamation, Changi International Airport is already in operation. (1,131 ha, 68% of the total area of the airport's 1,663 ha, is reclaimed land developed by P.S.A.) Other reclaimed areas developed by H.D.B. are utilized

---

* Build-up area includes residential, commercial, industrial, educational, institutional, warehousing, utilities, telecommunication, transportation and sites under development.

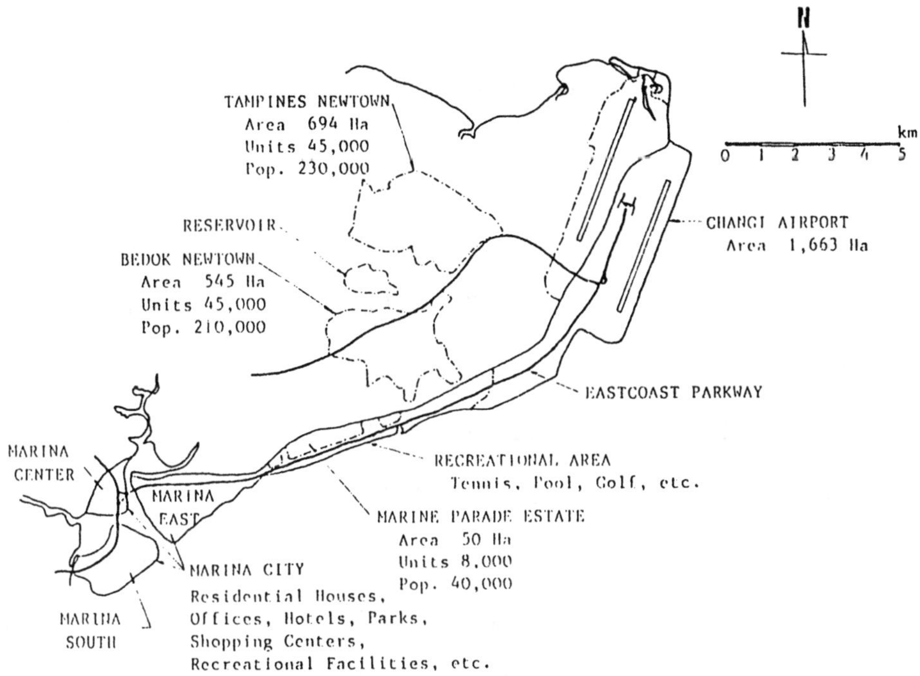

Fig.-2 SINGAPORE EASTCOAST RECLAMATION; DEVELOPMENTS OF RECLAIMED AREAS AND BORROW AREAS

for residential estate, recreational facilities such as tennis courts, swimming pool, golf range, cycling courses, etc, and for highways connecting between the airport and the central business district. After completion of Phase VI and VII, the area is planned for a future new town 'Marina City' featuring residential dwellings, offices, shopping centers, and recreational and educational facilities. Furthermore, new residential estate and reservoirs for water supply are constructed on the flatten hills of winning areas for fill materials as shown in Fig.-2.

(2) Westcoast Reclamation & Reclamation of Offshore Islands by P.S.A.

Reclaimed areas of Westcoast and Offshore Islands developed by P.S.A. are mainly used for warehousing and harbour facilities.

Table-3 Land Use of Singapore

Category	Area (ha) & Percentage Over Total Land	Increase from 1967 to 1982
Residential	8,716 (15.3%)	+ 16%
Transportation	7,457 (13.1%)	+181%
Industrial	3,347 (5.9%)	+359%
Recreational	1,725 (3.0%)	+143%
Institutional	1,364 (2.4%)	+ 60%
Educational	1,142 (2.0%)	+ 85%
Utilities	1,002 (1.8%)	+129%
Commercial	803 (1.4%)	+ 13%
Warehouse	522 (0.9%)	+ 40%
Buildings under Construction	1,602 (2.8%)	+191%
Agricultural	8,101 (14.2%)	- 43%
Water/Wooded Area	5,929 (10.4%)	- 29%
Special Use	4,979 (8.7%)	- 19%
Other*	10,353 (18.0%)	- 1%

* Quarries, Cemeteries and Vacant Land (includes Mrina South)

(3) Jurong, Tuas & Offshore Islands Developed by J.T.C.

Reclaimed land of Jurong, Tuas and Offshore Islands constitutes significant parts of Jurong Industrial Estate which is the largest industrial estate in Singapore, comprising 90% of the country's industrial areas. Industries in the area are classified into four categories : clean, dirty, heavy and light, and factories are allocated in the Estate according to the categories. Dirty industries requiring deep berth, such as petrochemical plants, oil refinaries and oil fired power stations are located on reclaimed Offshore Islands like Pulau Ayer Merbau, Pulau Seraya, Pualu Busing and Pulau Sakra and Bakau. On waterfront sites or reclaimed land, allocated are heavey industries, such as shipbuilding/ship-repairing, oil rig fabrication yards and offshore oil-related equipment industries. In inland areas, J.T.C. has been preparing Standard Factories, ready-built factory complexes, for light and clean industries. Reclaimed land and flattened hills of winning area are utilized for those industries.

JURONG TUAS RECLAMATION PROJECT

At the western extremity of Singapore, a large-scale reclamation project was commenced in March 1984. The project, called Jurong Tuas Reclamation, will yield 637 ha when completed and involves 70 million cubic meter of fill material. The S$602 million reclamation project was awarded to PENTA-OCEAN CONSTRUCTION CO., LTD. of Japan. The 5 year project tendered by J.T.C. is one of the largest civil engineering construction in the world awarded to a single contractor. After its completion, the reclaimed land will be utilized for power generation and industrial purpose.

CONCLUSION

As is discussed above, ever increasing demand for land in Singapore, which is thrusted by its rapidly increasing population and industrialization policies, has been eased by coastal development through extensive reclamation projects. It is no exaggeration to say that the nation is "building up" its own country through the projects.

Land is a strictly limited resource for the country, and reclaimed land has an increasing importance for the government. Authorities, however, are not rushing in the usage of the land. Careful plannings in land use, such as found in combination of high-rise residential buildings, highways and abundant greenery of Eastcoast Reclamation, and in Standard Factories of Jurong Industrial Estate, have created unique scenery that can hardly be found in other countries.

Currently, Jurong Tuas Reclamation and several offshore reclamation projects of Southern Islands are under construction, and Eastcoast Reclamation Phase VI & VII are almost finished. In this October, government approvals were granted to S$874 million reclamation project around Punggol for a public housing estate. We could conclude that the policy of coastal development through land reclamation seeking new land resources will be continued in the future as well in the Republic of Singapore.

REFERENCE

HDB Annual Report 82/83, Housing and Development Board, Republic of Singapore.

PSA Annual report '83, Port of Singapore Authority, Republic of Singapore.

JTC Annual Report 82/83, Jurong Town Corporation, Republic of Singapore.

JTC Annual Report 83/84, Jurong Town Corporation, Republic of Singapore.

1982 Land & Building Use Report of Survey, Ministry of National Development, Republic of Singapore, December 1983.

"The Survey of Fill Materials for Reclamation Projects in Singapore's Territorial Waters," Japan International Corporation Agency, March 1979.

"Jurong Tuas Reclamation Project," Kensetsu-Kogyo Shinbun, p.18, clm.2, September 28, 1984.

"Reclamation Project Approved," The Strait Times (Singapore), p.15, clm.1, October 20, 1984.

Konishi, Teruo, "Jurong Development Projects," Present and Future of Singapore Changi Airport and their surrounding areas, pp.37-41, Osaka Science and Technology Center, March, 1981, Osaka, Japan.

Konishi, Teruo, "Eastcoast Reclamation Projects" Present and Future of Singapore Changi Airport and their surrounding areas, pp.42-48, Osaka Science and Technology Center, March, 1981, Osaka, Japan.

Yokoo, Michio, "Existing Conditions and Major Projects in Singapore," Present and Future of Singapore Changi Airport and their surrounding areas, pp.19-21, Osaka Science and Technolgy Center, March 1981, Osaka, Japan.

"Singapore Reclamation," Construction of Artificial Irlands, pp.515-521, Japan Ocean Development Sonstruction Association Inc., March, 1984, Sankai-do Tokyo, Japan.

Research on the Development of Wai-Shan-Ding Barrier, SW Coast of Taiwan, R.O.C.

H.S. Hou* , C.F. Wang** and T.J. Wang***

Institute of Harbor and Marine Technology, Taichung District, Taiwan 435, Republic of China

ABSTRACT

Due to the abundant supply of sediment from the rivers, the west coast of Taiwan has shown a rapid accretion in history. As a result of this large areas could be reclaimed, such as the delta North of Putai — Wai-Shan-Ding sand barrier. In recent years, however, sediment supply is reduced due to river training and flood control. Consequently, the coastal area is now subject to erosion as the net sediment transport exceeds the supply to the beach. Therefore, shore protection is going to be cultivating the Australian pine tree by the Chia-Yi County Government.

Since the relatively well protected location of the area with regard to both waves and wind, no special navigational problems have to be envisaged. For the point view of whole development, it is easily planned the lee wind side of the delta as a marina, and wharf could be simply constructed as a floating type for sightseeing and cargo handling between this island to offshore island (such as P'eng-Hu or King-Meng, etc.). The main purpose of this marina is used for happy vacation time, such that the wharf is designed for the anchoring of yacht and hydro-foil cruiser, etc., the breakwater is need to constructed convenient to be fishing. Siwwing beach, beach park and some facilities for sightseeing are planned and designed in this offshore area for utility of the coastal management.

INTRODUCTION

In the western coast of Taiwan, Pu-Tai bay has less wind and littoral drift. It is a good base to build a moderate harbor. It has not only to prosper the southwest part of Taiwan, but also to develop a common assistant harbor of Taichung Harbor and Kaohsiung Harbor. The inistry of Communication tries to establish this harbor site as the new port for domestic coastal shipping and even for oversea trade shipping.

For planning the port, the oceanographic condition is very important, especially the wave condition, the wave characteristics of the bay area is quite special, therefore both calculations and field measurements are studies.

Wind waves are computed from SMB method, the wind records of Au-Gu meteorological station are used for forecasting its wave feature. Since the existence of the key of Wai-San-Ding, the wind speed is small (average 5 m/sec), wave height is commonly below 2 m.

* Deputy Director & Principal Senior Researcher, Institute of Harbor & Marine Technology
Profeşsor, Institute of Naval Architecture, National Taiwan University

** Associate Researcher, Institute of Harbor & Marine Technology

***Head, Planning and Construction Division, Chia-Yi County Government

Waves caused by typhoon intrusion are calculated by using Tang's (1970) numerical model. The computed results are used for the design wave of the proposed Pu-Tai New Port.

For studying the wave characteristics in the Pu-Tai Bay, an offshore waverider was installed (at -12 m deep) to measure the sea waves, and the proposed harbor site of KSK cassette type wave meter is set up along the steel of concrete block which was sunk into the sea-bed. The decaying and refracting phenomena are analyzed. For understanding the wave characteristics, the wave spectra analyses are reduced from the measured waves for summer monsoon and winter monsoon respectively.

1. The Wind Condition at Pu-Tai Proposed Harbor Site

There was no field wind observation at Pu-Tai before 1981. To know more about the wind climate at Pu-Tai, the wind records at Ou-Gu which have been observed for more than ten years because Pu-Tai and Ou-Gu are closely located and sheltered both by the Waishanding Bar area are analyzed. While, the wind records of Pu-Tai coastal area observed after 1981, are used to correct those of Ou-Gu. According to the long-term recorded wind data, the winter monsoon lasts from September to March of the next year and the prevailing wind directions are NNE, NE and ENE with average wind speed ranged from 1.6 m/sec to 5.5 m/sec. This is shown in Fig. 1-1. However, as shown in Fig. 1-2 which is the wind rose during 1973 and 1974, the prevailing directions are from NNE, N and NNW, etc. This reveals that winter monsoon has been affected by new constructions located in this area. From the long-term data, also summer monsoon blows during April and September and the prevailing wind directions are NNE and SSW with average wind speed ranged from 1.4 m/sec to 4.8 m/sec. However, the prevailing wind directions are changed and scattered mostly from NNE, NNW, SSW and SSE, etc., during 1973 and 1974, as shown in the figure.

Wind is an important condition for the study of harbor layout and then it is necessary to obtain the wind conditions near a proposed harbor. Therefore, an Anemometer was installed at Pu-Tai coast. The field records of wind condition at Pu-Tai shown, the winter monsoon in December mainly blows the wind from NNE, N and NNW directions, etc., and with the stronger wind speed. In January, the prevailing wind directions are the same as those in December and totally 90% of occured freq. The strongest wind occurs in N direction, and then in NNE direction. These are shown in Fig. 1-5. Also, in February, wind overwhelmingly occurs in NNE, N, NNW directions which are in total 80% of occurred frequency. Wind in N direction occurs most frequent but those in N and NNW directions are strongest. In March, wind comes more often from WNW, NW and NNW directions which totally reach up to 60%. Wind speed is highly reduced such that the occurrence of wind with speed stronger than 5 m/sec is only 17%.

2. Wave Condition in the San Barrier Island

2.1 Monsoon waves at the Pu-Tai harbor

As described in the paragraph of 3.2, the summer monsoon prevails frequently in the directions of NNE, SSW and WSW with the average speed from 4.1 to 4.7 m/sec, and the winter monsoon prevails frequently in the directions of NNE, NE and ENE with the average speed from 5.0 to 5.4 m/sec at the site. Using the statistical results of the wind records in 1973, the significant waves are deduced by SMB method. Wind waves of winter monsoon are prevailing (from October to next March) are always about 1 M high and period is 4-5 sec, and those of SW, WSW summer monsoon waves are prevailing (during April to September) are always below 1 M high. Because the orientation of the Pu-Tai harbor entrance will be WSW, it is not the dominative waves in NNE, NE or ENE direction. Besides, the winds in June and in July occur most frequently in SSW direction, but the induced waves are negligible because of the weak average speed is about 3.6 m/sec.

2.2 Hurricane waves at the Pu-Tai harbor site

Because wave gauges are occasionally stricken to out of order and thus the data about hurricane waves are always lack, the design wave of a proposed harbor is always predicted by using the model typhoons which will most seriously attack the harbor. Referring to the records of the typhoons which hit Taiwan before, the important ones for this area are Typhoon Viola, which hit Taiwan on July 28, 1969 with a maximum wind speed of 15.0 m/sec from SSW direction at Pu-Tai at 5 o'clock on July 27, and Typhoon Elsie, which hit Taiwan on September 27, 1969 with a maximum wind speed of 25.8 m/sec from WSW direction at the site at 8 o'clock on the day. The courses of the two typhoons are shown in Figure 2-13. Here, Typhoon Viola, Typhoon Elsie and Typhoon Nora which occurred in 1973, and Typhoon Elsie which occurred in 1975 are selected as the model typhoons to predict hurricane waves by using the numerical wave-perdiction model for moving fetch typhoon deduced by Dr. Tang (1970). Results show that Typhoon Viola induces a maximum significant wave of 6.6 M in height and 15 sec in period which is shown in Figure 2-3, and Typhoon Elsie (1975) produces a maximum significatn wave of 7.0 M high an 15 sec in period which is shown in Figure 2-2. The hurricane waves are used as the criterion of the harbor planning and design of the proposed harbor.

2.3 Design wave of Pu-Tai harbor site

Pu-Tai harbor site which is located in the southwest coast of Taiwan is sheltered by the Waishanting Barrier Island, therefore, the wind wave height near the site induced by the winter monsoon is always under 1 m and period of 4 to 5 sec and therefore are negligible. Also the wind waves induced by summer monsoon are negligible. The dominative waves of the proposed harbor are induced by typhoon, and therefore, they are uses as the design wave of the harbor which is 7 m of height and 15 sec of period from SSW direction.

3. Coastal Zone Developing Layout

The optimal solution of the Wai-San-Ding Sand Barrier is scheduled as Fig. 3. The plan consists of coastal industrial area, harbor, fishery harbor, power plant, offshore airport terminal, oil-refinery and oil-storage tank swimming beach, yacht port, coastal park and transportation system etc.

The design sections of sea-wall quay and inner breakwater are shown in Fig.4.

Detail planning is still going on and the financial plan is also under way.

4. References

(1) Hou, Ho-Shong, Lee, C.P. Ueng, G.H. & Lin, L.H.

"Research Report of Planning and Design of Pu-Tai New Harbor Master Plan", Graduate Instatute of Harbor and Ocean Engineering, Keelung, Taiwan, R.O.C. Oct. 1980.

(2) Hou, Ho-Shong, Lee, C.P. & Lin, L.H.

"Relationship Between Alongshore Wave Energy and Littoral Drift in the Mid-West Coast at Taiwan", Proceeding of the Seventeenth International Coastal Engineering Conference, March 23 to 28, 1980, Sydney, Australia.

(3) Hou, Ho-Shong

"Researcher of Harbor Site Investigation, Planning and Design of Coastal shipping System in Taiwan, R.O.C.", Special Report No. 7, Institute of Harbor & Marine Technology, Wuchi, Taichung District, Taiwan, R.O.C., November, 1982.

(4) Hou, Ho-Shong, Tseng, Z.M. and Weng, G.H.

"Research on the Wave Decaying and Refracting of the Pu-Tai New Port" J. of CICHE, Nov. 1983.

(5) Tang, L.W. (1970)

"Planning and Design of Coastal Engrg.", Taipei, R.O.C.

(6) Hou, Ho-Shong (1983)

"Planning and Design of Coastal and Harbor Engrg.", Special Report No.9, IHMT, Feb. 1983.

(7) R.G. Dean, E.P. Berek., C.G. Gable and R.J. Seymour (1982)

"Longshore Transport Determined by an Efficient Trap", 18th International Conference on Coastal Engineering, Capetown.

(8) Hou, Ho-Shong, T.J. Liaw and T.J. Wang (1984)

"Littoral Transport Rate along the Wai-San-Ding Sand Barrier." J. of CICHE Nov. 1984.

(9) Hou, Ho-Shong, Z.M. Tseng and G.H. Weng (1984)

"Research on the Way Decaying and Littoral Transport of the Pu-Tai New Port", 19th International Conference on Coastal Engineering, Houston, Texas, U.S.A.

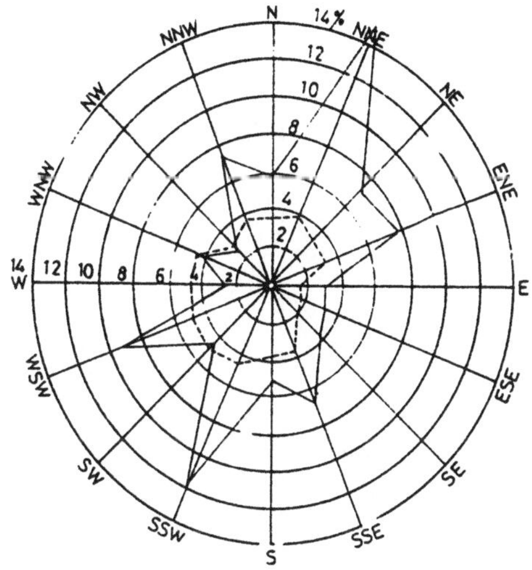

4月～9月 (1963～1965、1968～1972)

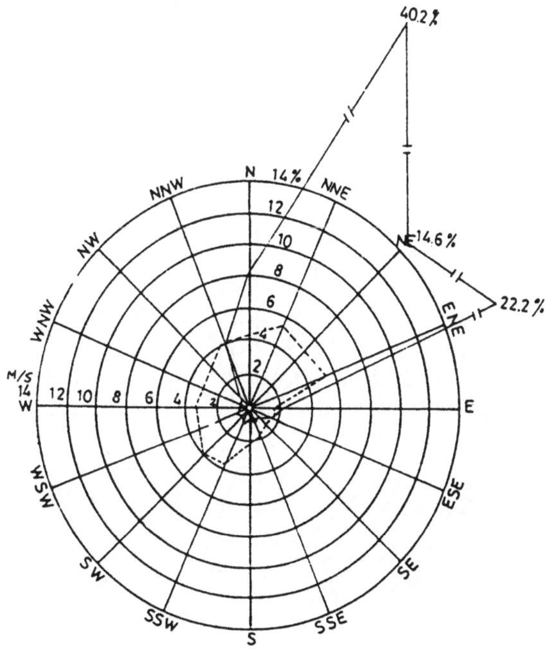

10月～3月 (1963～1965、1968～1972)

Fig. 1.1. Wind Rose of Au-Gu Meteological Station (I)

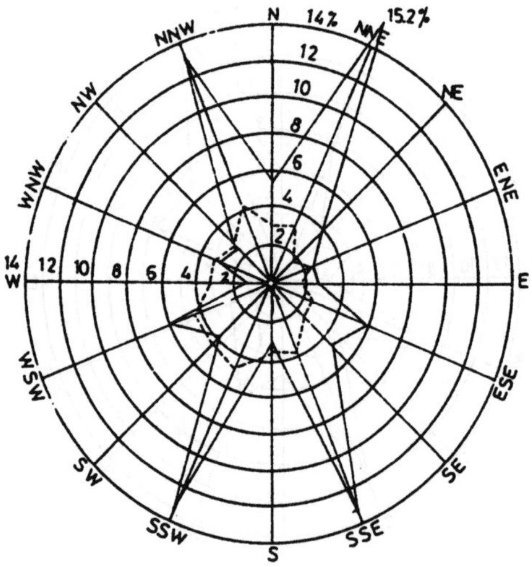

4月~9月　(1973~1974)

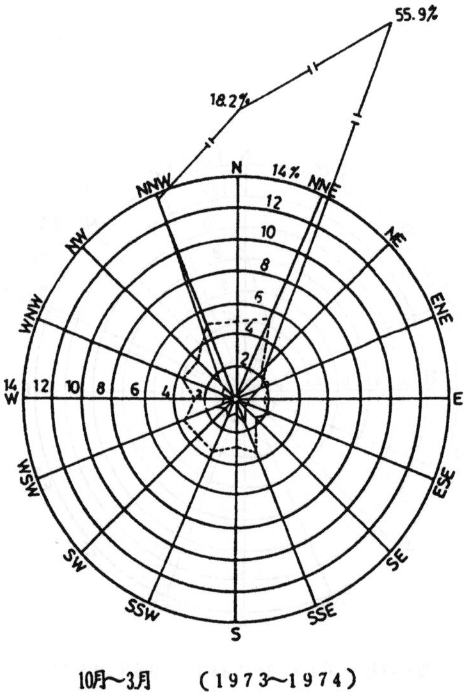

10月~3月　(1973~1974)

Fig. 1.2. Wind Rose of Au-Gu Meteological Station (II)

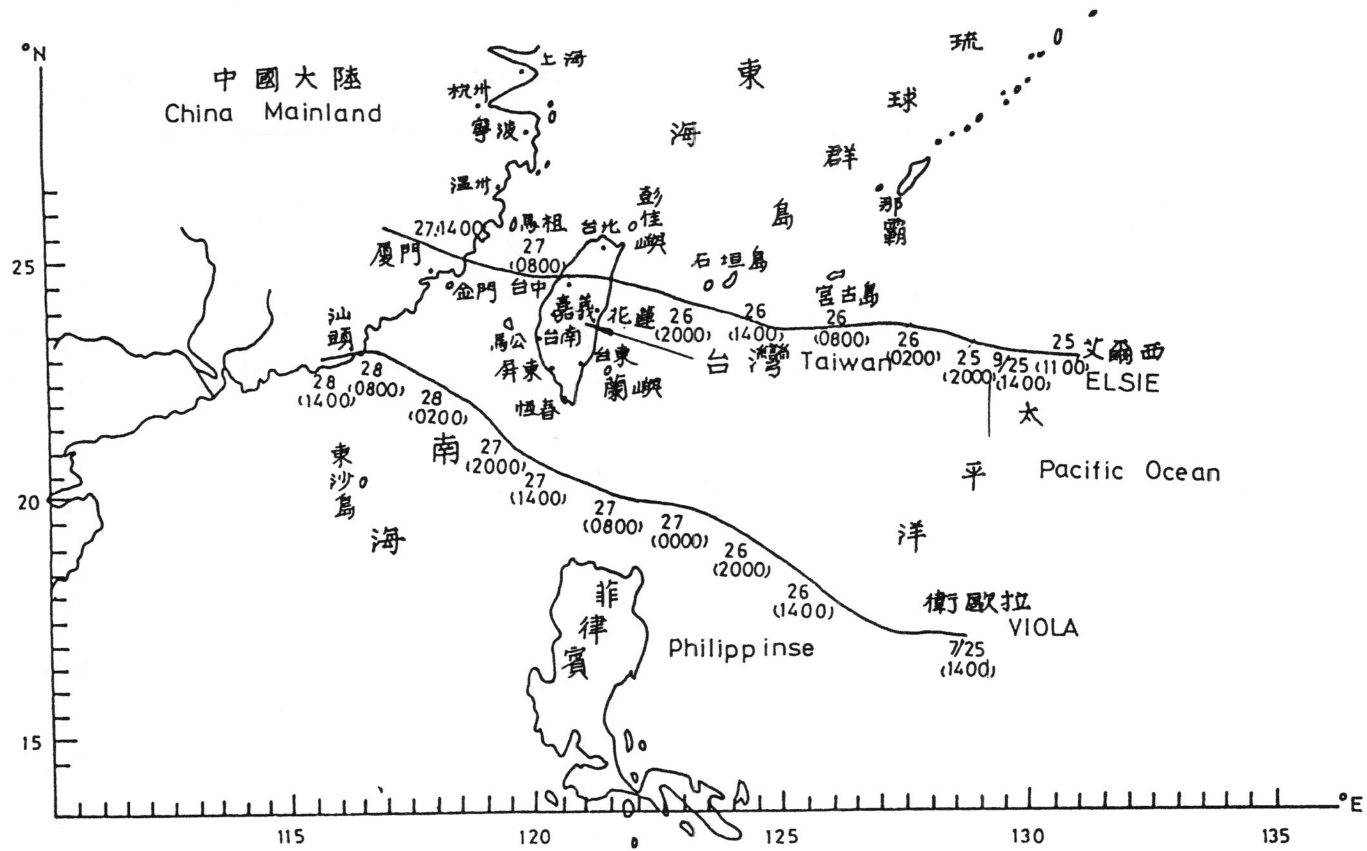

Fig. 2.1. The Passing Tracks of Typhoon Viola and Typhoon Elsie

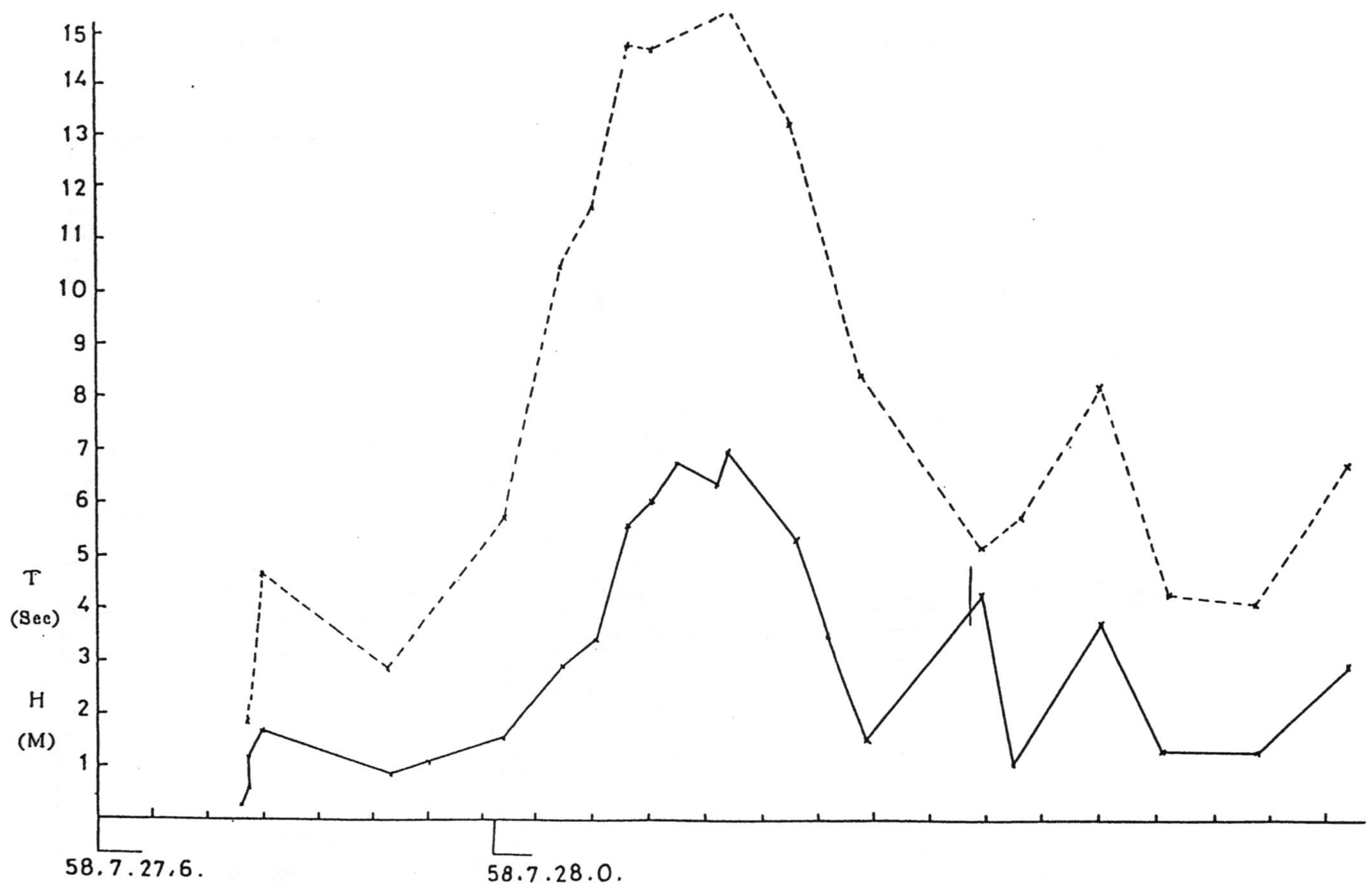

Fig. 2.2. The Hurricane Waves Induced by Typhoon Viola (1969)

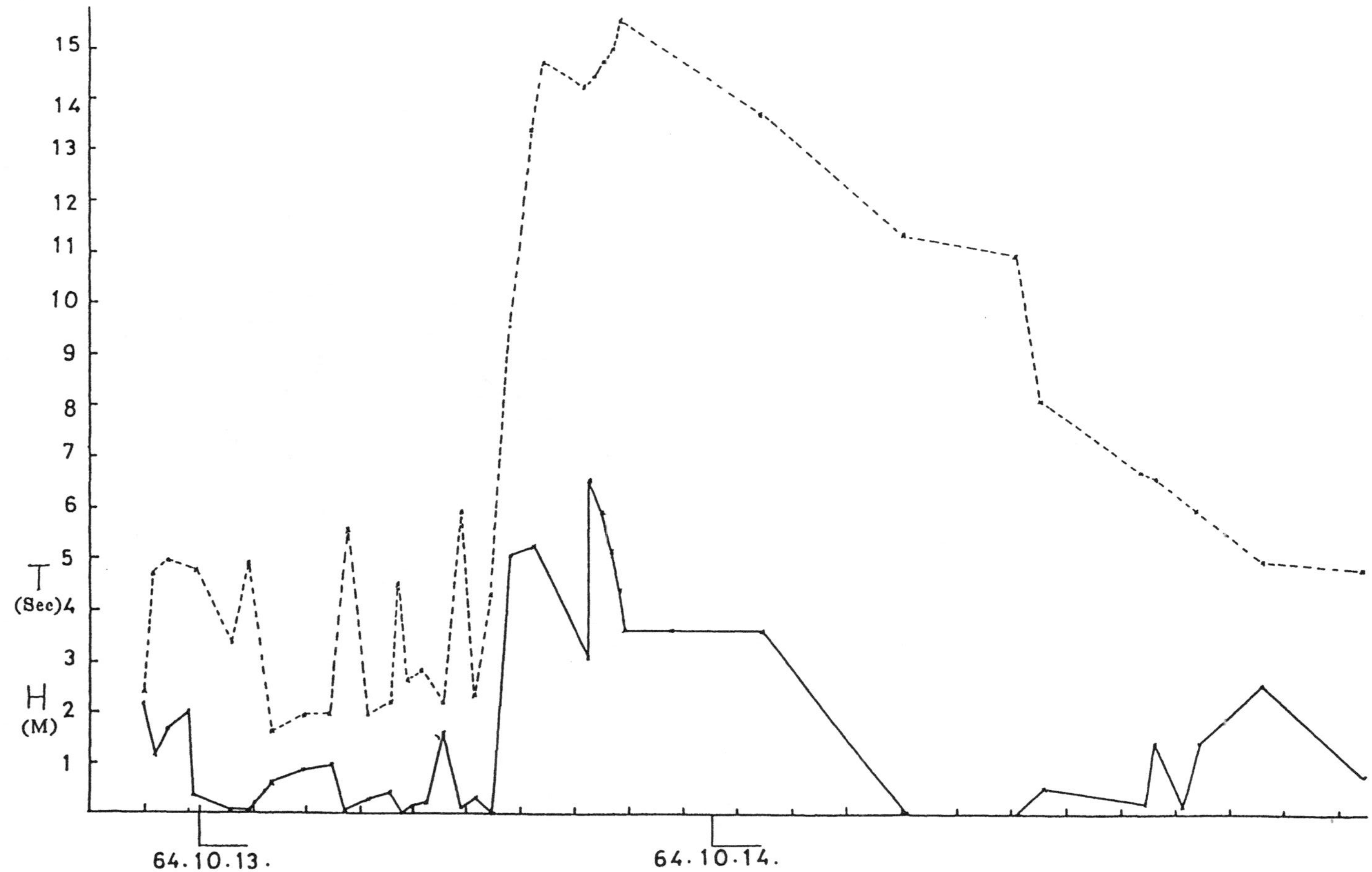

Fig. 2.3. The Hurricane Waves Induced by Typhoon Elsie (1975)

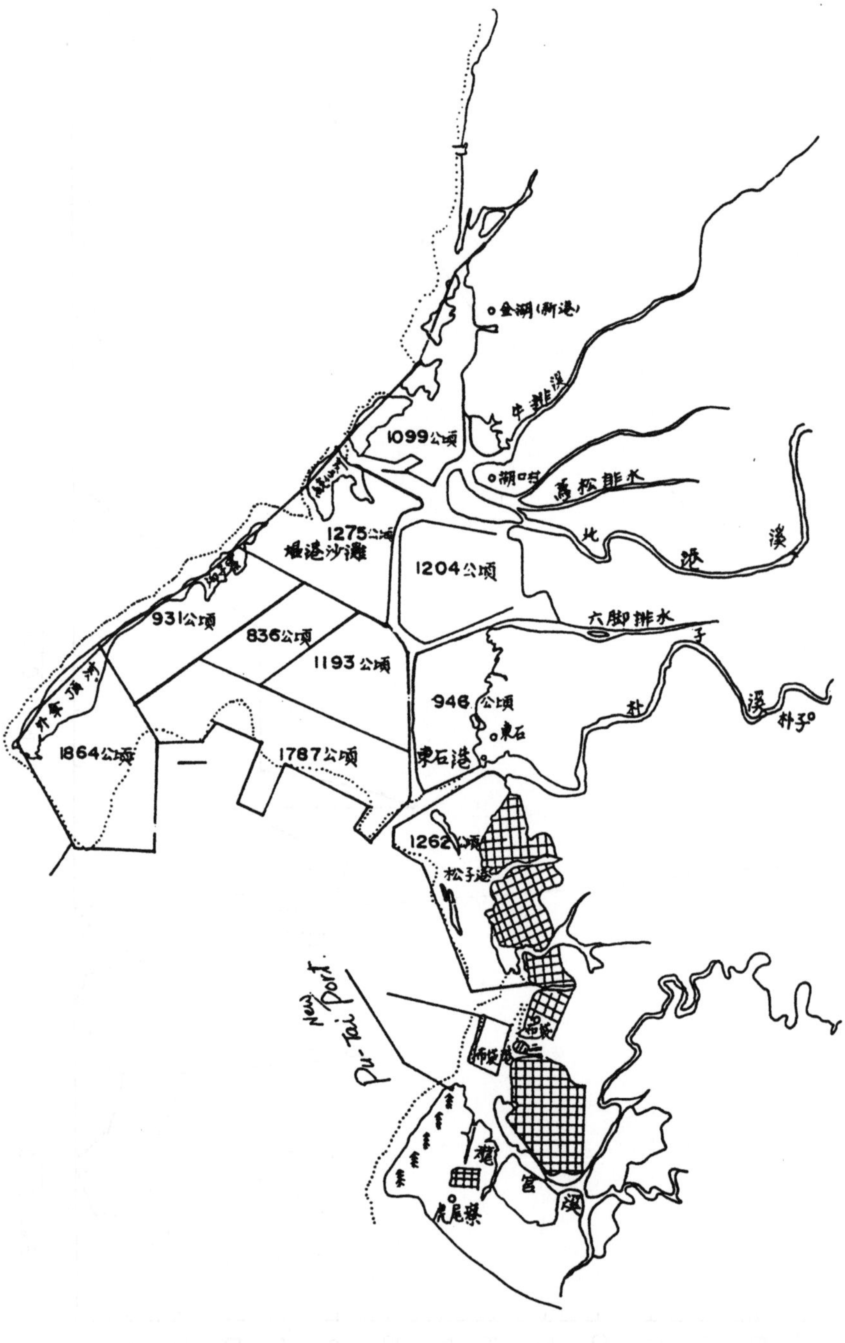

Fig. 3. Layout of Wai-San-Ding Sand Barrier Development

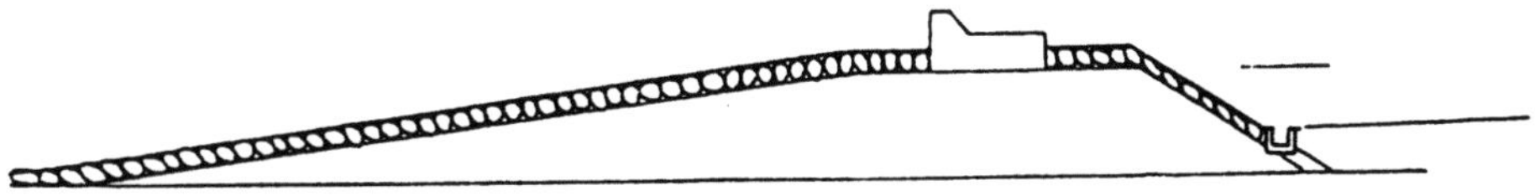

Seawall Section 1

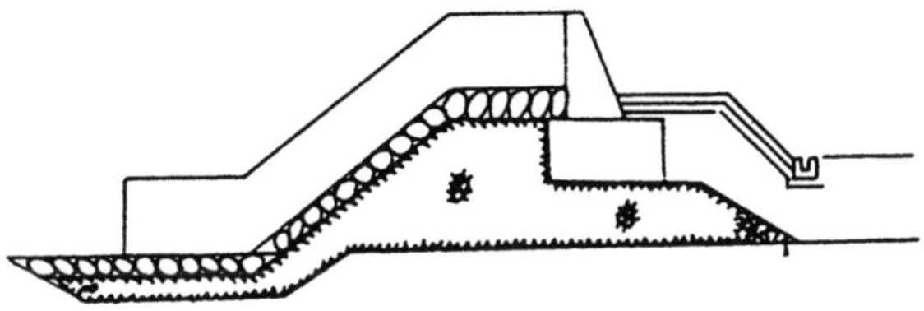

Seawall Section 2

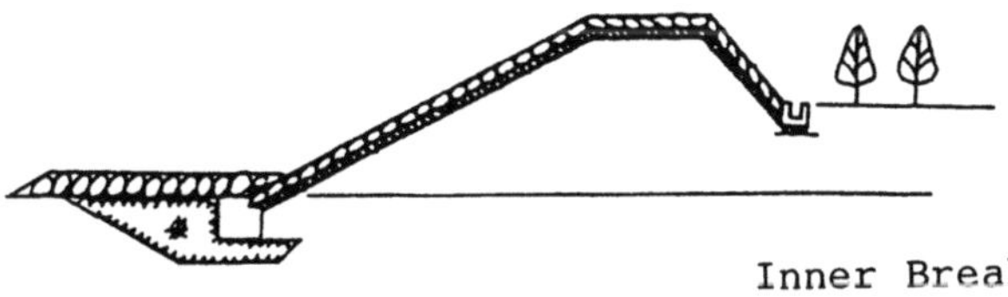

Inner Breakwater

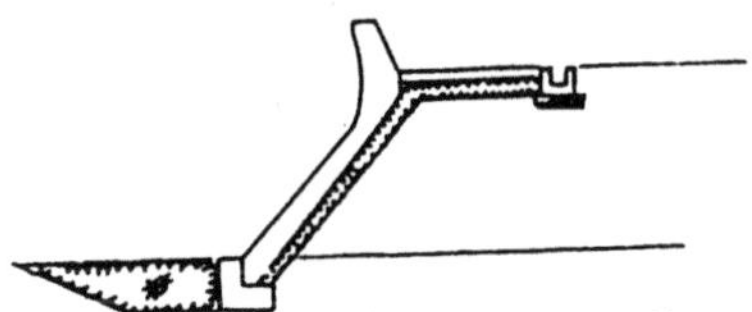

Quay Section 1

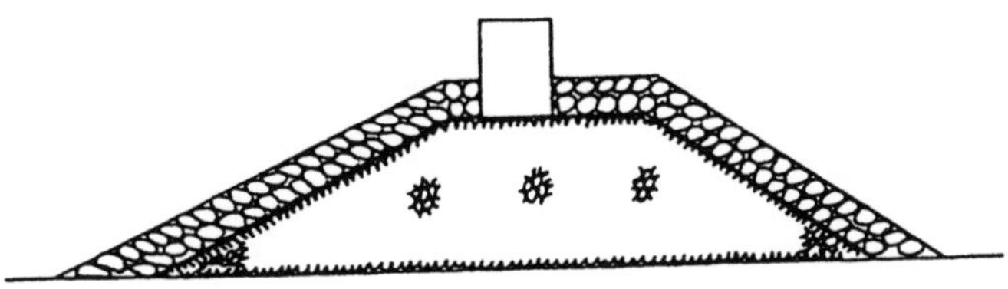

Quay Section 2

Fig. 4. Design Section of the Layout
of Wai-San-Ding Sand Barrier

The Ise Bay Management and Development Plan for the 21st Century

T. Iijima, A. Kaneko and I. Fukuda

The 5th District Port Construction Bereau, Ministry of Transport, Tsukijicho, Minato-ku, Nagoya, 455 Japan

1. Prologue

During the high economic growth period in 1960s, Japan's coastal area especially major bay areas like Ise Bay faced serious problems such as environmental problems and conflict between industry and fishery. As economic and social conditions have changed since the oil crisis, the necessity to propose the new policy to manage Ise Bay and the bay area from the new point of view has been raised in 1970s and 1980s.

Under this circumstance, the association of the port authorities in Ise Bay area was established in 1982. The association published "The Ise Bay Management and Development Plan for the 21st century" which deals with future development, utilization and preservation of the bay through serious discussion among the members of the association.

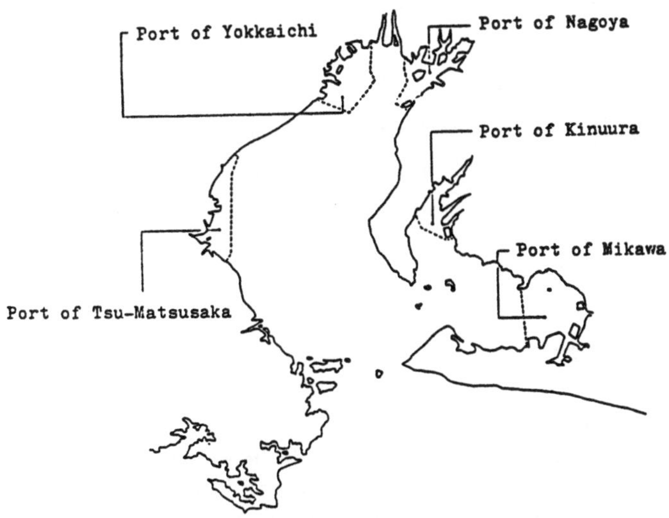

Figure-1 Ise Bay and Ports

This paper presents major future development ideas of Ise Bay from "The Vision".

2. Present State of Ise Bay

Ise Bay located almost in the center of Japan, having an area of 2,100km which is twice wider than Tokyo Bay and one and a half times wider than Osaka Bay. Atsumi peninsula on the east and Shima peninsula on the west jet out into the sea and form the mouth of the bay. Chita peninsula lays in the center of the bay from north to south and divides Mikawa Bay, the sub area of Ise Bay, from the other parts of Ise Bay. This topographic condition makes the bay a typical enclosed water body. It has a natural and geographical character easy for people to use, with relatively shallow depth of an average of -17m and with calm sea condition.

Nagoya, the center of this area, having a population of 200 million at the most inner part of the bay, and surrounding cities such as Yokkaichi, Gifu, Toyoda, Okazaki and Toyohashi, form one of the most important industrial zone in Japan. Table-1 shows the characteristics of the three major bays in Japan with some socio-economic indexes. Ise Bay

Table-1 Comparison of the 3 major bay areas

Term / Name of Bay	Ise Bay	Tokyo Bay	Osaka Bay	Whole Country
Area of Bay				
Length of Coast Line (km)	290	154	160	31,382
Water Area (km²)	2,130	1,160	1,400	—
Volume of Cargo Handled (1981, million ton)	171	502	328	2,909
Number of Vessels Calling the Port (1981, thousand vessels)	319	354	401	9,120
(Land) Total Area (km²)	21,502	13,495	14,833	377,708
Population (March 1981)	9,870,000	28,489,000	15,929,000	117,009,000
Population Density (March 1981, person/km²)	459	2,111	1,074	310
Net Product (1979, billion yen)	16,788	55,994	27,839	189,517
Volume of Industrial Shipment (1979, billion yen)	23,173	46,106	27,923	173,919
Industrial Density (1979, billion yen/km²)	3.4	5.3	5.6	1.4

Note: (1) Land indicates Aichi, Mie, and Gifu Prefecture for Ise Bay, and Tokyo, Kanagawa, Chiba, Saitama Prefecture for Tokyo Bay, and Osaka, Hyogo and Kyoto Prefecture for Osaka Bay.

(2) Industrial Density = Volume of Industrial Shipment / Resident Area

area has much share of industrial shipment compared to that of population and net regional production. It has much land available for residential use, and the bay itself has enough room for freight transportation and navigation.

Ise Bay is also used for fishing industry and coastal and marine recreation. The coastal fishery is very active in Mie prefecture. And it is also noteworthy that Ise Bay is highly valuable for recreation and sightseeing. The Ise-Shima national park in the western part of the bay is the first designated area as a national park, and in the eastern part there is Mikawa Bay quasi-national park. These parks are connected to each other by ferry and sightseeing boat service, and twenty nine big to small marinas are sprinkled along its coast lines.

As for the environmental situation, the bay is farely clean compared to Tokyo Bay and Osaka Bay on COD (Chemical Oxigen Demand) though the water quality of the inner area (between Nagoya and Yokkaichi) and Mikawa Bay are not so clear as the other part of the bay. It should be noticed, however, that the red-tide frequently attacks, and that tremendous improvement of the water quality is not observed despite measures taken in the past such as total pollutant load reduction.

3. Direction of the Development of Ise Bay area

Observing the future course of Japan, it might be possible to imagine the future of Ise Bay and the bay area as follows:

Tokyo metropolitan area, the capital of the country, is strengthening its political and other pivotal functions, and Osaka area, the second largest metropolitan area, is growing as the center of culture, finance, etc.. On the other hand, Ise Bay area with Nagoya has developed as a manufacturing industry and modernized agriculture area, and has been playing an important role as a commodities producing center. The condition of Ise Bay area; accumulation of the industrial technology, sufficient space to utilize, fine natural environment of the bay itself, might be very advantageous for the future progress of industry because the development of industry today is mainly based on positive technological innovation and comfortable living condition.

Taking all these into consideration, by improving the necessary functions, Ise Bay area is expected to be the center of industry not only of Japan but also of the world.

4. Direction of the Utilization of Ise Bay

As pointed out, Ise Bay is not only one of the major industrial zone in Japan, but it has fine natural environment with national and quasi-national parks. Thus it is the precious property not only of Ise Bay area but also of Japan.

The viewpoint for the utilization of Ise Bay with such characteristics, can be imagined as follows.

A. As the nucleus of Ise Bay area, the industrial metropolice of Japan, it is necessary to utilize Ise Bay from nationwide and worldwide viewpoint.

B. Ise Bay area has developed with Ise Bay topographically and historically. Thus it is necessary to take measures as a stage of regional activities.

C. Standing on the position to create and develop the fascinating Ise Bay and to manage it properly for the 21st century, it is necessary to create a fine environment harmonized with a highly technological accumulated society.

In order to make them possible, it is effective to consider comprehensive and organic utilization of Ise Bay noting the environmental preservation, and to promote the specialization and advancement of each region and marine area. The result of our trial discussion is shown in Figure-2.

5. Direction of the Development of the Ports in Ise Bay

For the 21st century, the ports in Ise Bay are required to overcome various kinds of tasks, such as providing transportation system composed with maritime, land and air transportation, taking comprehensive measures as the gateway of internationalization, supplying industrial spaces relevant to the technology leading age, securing space for recreation

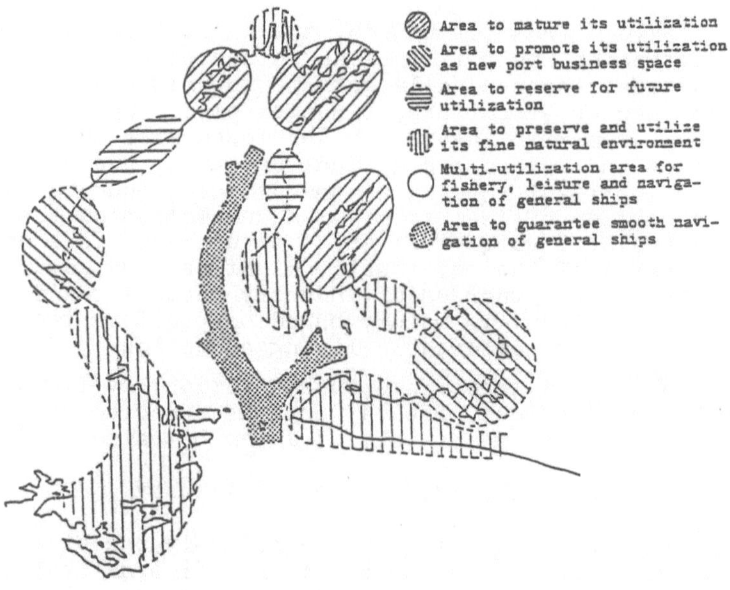

Figure-2 Direction of the utilization of Ise Bay

and culture corresponding to the age of leisure, and preservation and development of marine products.

The following four themes are the generalization of the tasks from the viewpoint of taking into advance the changing functions of ports and harbors.

A. Utilization of coastal area matching with the change of industrial structure

Coastal industries composed mainly with basic material industries are facing stagnation from the viewpoint of both the output and the demand of land. On the contrary, inland industries such as processing and assembling industries are in good condition. Especially, 10 to 20% annual increase of production is estimated in the field of high technology industries such as IC industry and industrial robot industry. Such a change of industrial structure reflects fundamental conditions surrounding us such as limitation of natural resources and energy, increasing severity of the international industrial competition, movement of the internationalization, and it is no doubt that this tendency will continue in the future.

In these circumstances, high technology industries require high speed transportation network system with airline or expressway, well-educated laborer and accumulation of urban function, rather than big space, mass water and a lot of laborer ever shown in basic material industries.

In Ise Bay area, these inland industries such as world famous Toyota Moter Corporation, located in the 30 to 50km zone from Nagoya city. From now on, it is important in the coastal area to cooperate with the inland industries utilizing present accumulation of the industries and to provide good environment to introduce high technology industries.

On the other side, considering the merits of coastal area as industrial zone, and geographical and natural potential of Ise Bay, coastal industries in this area are expected to develop in the long run. Namely, technology leading industries such as space industry, ocean-developing industry and energy industry which are based on highly applied technology, having high probability to be main industry of Ise Bay area. These industries have intention to set up their facilities at coastal area, because they need wide space to work functionally. Thus it is necessary to develop and arrange industrial area steadily and to establish the system for cooperative relations of these area as soon as possible.

B. Promotion of world-wide freight transportation

Structure of freight transportation in Japan is changing due to the advancement of high technology and high added value industrial goods, which lessen the size of cargos. This tendency is also applied to the freight transportation structure in the ports and harbors, the fundamental base of the trade oriented country. For the freight transportation at the port, it is strongly expected to correspond to the less-but-frequent production, just-in-time transportation and shortening of lead-time, reflecting the needs of efficient and rational freight transportation. Containerization is progressing as one of the result of these requirements. It is necessary to further respond to them by improving the function of ports and harbors coordinating with the regional development plan such as "Pilot plan for the Tokai loop-city belt".

On the other hand, commercial ports in Ise Bay, such as Nagoya and Yokkaichi, handles 9 million tons of liner cargo a year. But these ports are still in the developing stage to become commercial ports which supports Ise Bay area, because the volume of cargos feedered to the ports in Tokyo Bay and Osaka Bay are remarkable. This situation is not desirable to keep a good road traffic condition and to preserve natural environment. It has been pointed out that the cause of the problems are weakness of the function of the business house, insufficiency of high speed traffic methods, lack of port sales, etc.. In these days, however, many countermeasures have been gradually excuted.

From these viewpoints, in order to have adequate function of the ports corresponding to the needs of economic activities in this area, it is necessary to take measures from the aspect of both 'hard' and 'soft' and to promote mutual contact and coordination among the ports of Ise Bay.

C. Promotion of international information ports

In these days, the movement of informationalization on transportation is rapidly progressing. It is due to the fact that transportation and information basically have very close relation and this is clear from the examples that online reservation system is employed in early time by the Japanese National Railways and air line companies. Nowadays, the introduction of new media such as VAN(Value-Added-Network) and CATV(Cable TV) are rapidly promoted in the field of information.

At the group of transporters and warehousings in the ports, introduction of efficient network system of freight transportation is progressing because their business is basically a labor-intensed one and they have to follow complicated procedure of foreign trade. On the contrary, the function of international trade information center represented by the international trade fair, and the function of intermodal transportation information center including air cargo and truck cargo have been reinforced in the ports, corresponding to the international trading environment. In the United States and England, as one of these movements, construction of the 'Teleport' has been promoted by the port authorities. By this, the port takes the role as a door of international and domestic information through the artificial satellite.

In these circumstances, Ise Bay should work as a window on international and transportation information for Ise Bay area, and concerning necessary works should be promoted.

D. Creation of the port town

Since the ports and harbors in Ise Bay have grown based on the post-war industrialization; cultural and historical aspects based on trade and exchange of people are rather poor. Therefore, citizens used to have little interest in ports. However, recently, expectations for them are growing from the side of the needs of the region. One of them is an active movement of international association at the ports, and another is a request for sea and coast in their city life. Declaration of conventional city in the port of Kobe and 'Minato-Mirai 21'(MM21; Plan for provision of the port prospecting the 21st century) in Yokohama are the typical examples which link such needs of the region with the cultural aspect of the ports.

Since 1975, the port development works under the slogan of "Creation of Familiar Port" began earnestly in Ise Bay, and development of green zone in the coastal area, construction of the international exhibition hall and the port building (having port observation room, maritime museum and conference halls) are carried out. However, some problems are pointed out, such as lack of harmony with hinterland, international symbolism and connection with foreign countries.

From now on, it is desired to further promote these movements for creation of attractive ports and bring them up to life stage of the industrial capital.

6. Major Projects

Four big and popularly required projects for achieving the tasks of Ise Bay are described in Table-2. They are as follows.

* Plan for provision of international transportation center
* Plan for provision of international high technology industries base
* Plan for formation of cultural port city
* Plan for creative and practical use of Ise Bay area

In this chapter, organization and function of some projects will be summarized based on the tasks mentioned before.

A. Comprehensive freight transportation network plan

* Organization

This network system will be established and managed by port authorities, business houses, etc.. The system is composed with information center for freight transportation, comprehensive freight transportation base and freight distribution center including inland depot, and they are connected by advanced information and communication network, and highways and bypasses.

* Function

This system reduces the cost of cargo transportation and saves time by efficient gathering, packing and distribution of cargos. At the same time, it provides an original

Table-2 Four Basic Plans and Major Projects

Plan for provision of international transportation center
--- Ise Bay connected with the world ---

* International container base
* Comprehensive freight transportation network
* Deep berth center
* New international airport
* Coastal highway

Plan for provision of international high technology base
--- Advanced technology area open to the world ---

* International transportation and processing base
* Technology-leading industry base
* Teleport
* Organization of technology development of coastal industry

Plan for formation of cultural port city
--- Development of towns utilizing sea and coast ---

* Nagoya port town
* Marine-police

Plan for creative and practical use of Ise Bay area
--- Searching for better utilization ---

* Ocean information center
* Establishment of the Ise Bay utilization conference
* Formulation of utilization and management plan

effective system of gathering and distribution in Ise Bay, standarized transportation fee, optimal ship-berth selecting system.

B. Construction of teleport

* Organization

This will be a major international information center utilizing artificial satellites which is connected with pivot facilities, some companies and industries in Ise Bay area by optical fiber. It is composed of communication center, business buildings, etc..

* Function

It will work as a window of the information of world-wide market, banking business, industries and cargo transportation. And information of marine environment, fishing place and ship movement will be exchanged here. Also, it will provide filing, processing and refering service of information.

C. Creation of Nagoya port town

* Organization

The following facilities will be situated continuously from Tsukijiguchi to Kinjo Pier (in the inner part of the

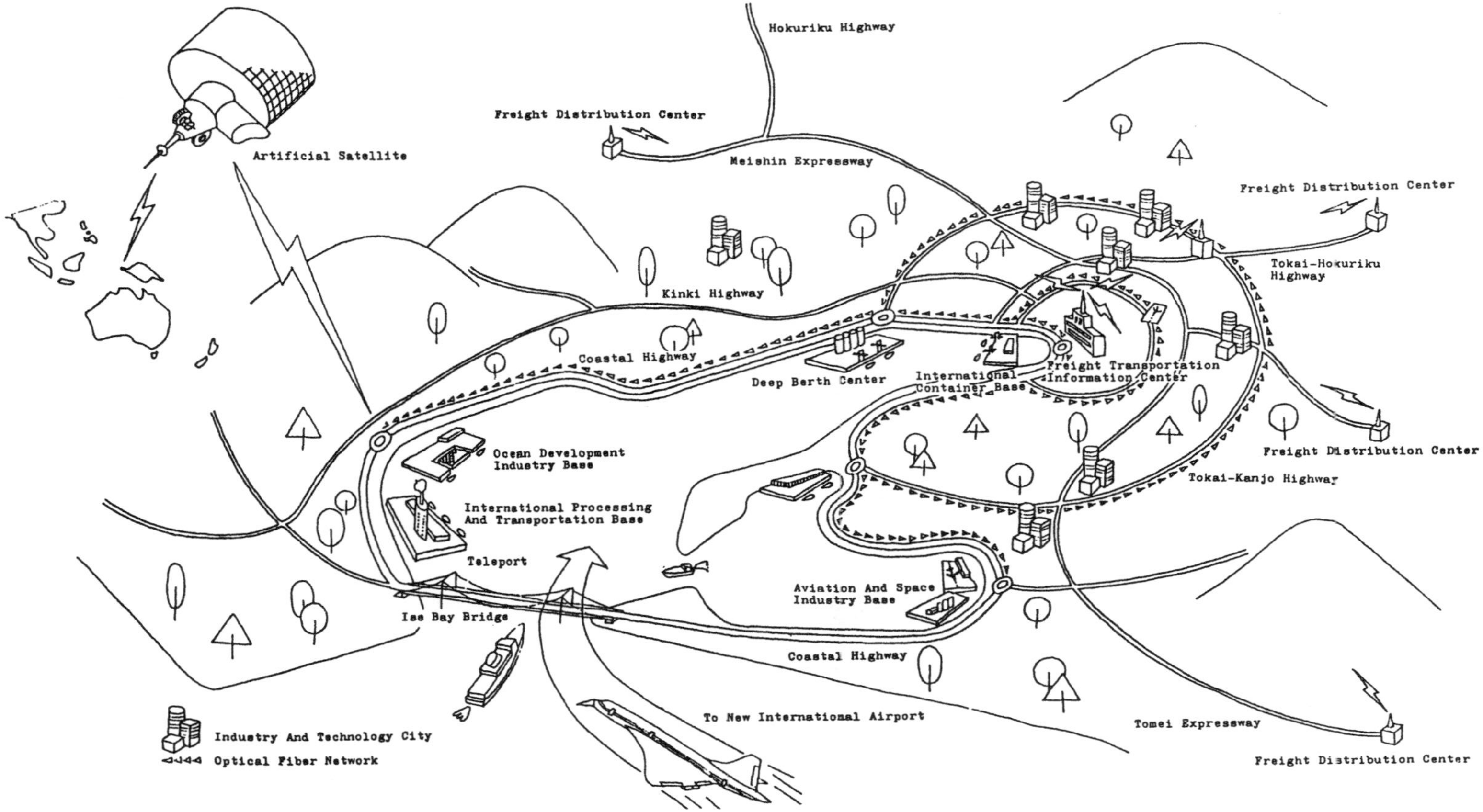

Figure-3 Future of Ise Bay -- Wide Area Complex of Industry and Freight Transportation port of Nagoya).

1. International communication plaza
2. International port-in-service training institute
3. Marine information center)Center for reserch and study of marine and oceans, for collection, stock and exchange of information, and for enlightenment for marine culture)
4. Port viewing promenade with green
5. New traffic system etc.

* Function

The port which has direct connection with abroad, will take a great role as a major international communication base of Ise Bay area. At the same time, the port will be open to city life.

7. Epilogue

This paper is the result of the discussion which consists of the prople concerning the ports, in order to consider deeply about the future of Ise Bay. These results are not based on the economic frame strictly, but to have a vision of Ise Bay for the 21st century, with mutual contact and coordination. There may be some lack of ground or vertifications, but we wish this vision will be the basis for the future development of Ise Bay.

East River Development Project

F.D. Chu[1], T. Omholt[1] and M. O'Keefe[2]

[1] New York Maritime College, State University of New York, Fort Schuyler, Bronx, NY 10465, USA
[2] The Water Club, 500 East 30th Street, New York, NY, USA

ABSTRACT

This paper describes a proposed 30-acre pile-supported development on the East River in New York City. The development is to consist of 32% residential and retail space, 34% park area, and 34% promenade and marina. The base for this development is linked barges, to be built in shipyards and floated into position. Once in position, they are ballasted and come to rest on piles.

The purpose of the paper is to show how in many large and densely populated coastal cities sea expansion can be a viable alternative to land development, and may be the only way to create sound new environments in crowded cities like Manhattan. Waterfront developments have the potential for superior design, as they can provide both intimacy and openness, and a balanced blend of land, water, air, and light.

In the discussion of the East River Development Project, the authors will touch both upon the engineering and design problems of the project, as well as its financial and political aspects.

In the first category, special attention will be given to a unique feature of the project, namely its energy efficiency. The East River, the sun, and seasonal winds will provide the energy for much of the cooling and heating of the buildings, a feature that will not only save money but make the project ecologically attractive.

In the second category, such aspects as cost saving and legal problems will be discussed. Special attention will be given to the advantage of utilizing shipyards to build part of the project, as the depressed shipbuilding industry is presently capable of producing quality at a reasonable price.

INTRODUCTION

A 30-acre waterfront site from 16th to 24th Street, East of the F.D.R. Drive is proposed for development to provide a park of 10.3 acres (34%), and promenade and marina of 10.3 acres (34%). The remainder, 9.62 acres (32%), would be developed to provide 1,300 low-rise large condominium apartments, 175,000 sq. ft. of retail and commercial space, and 250 hotel rooms and parking for 1,600 cars.

In densely populated areas like Manhattan, which is in a booming construction phase, undeveloped land is not only expensive but virtually non existant. A practical alternative to land development here is, therefore, to build on water. Manhattan being an island is surrounded by plenty of that. However, creation of new land by fill is for various economic and political reasons now viable. Thus, the only alternative is to build on floating or pile supported platforms.

Ecological, environmental and aesthetic criteria are central to the design. By incorporating the effects of water and wind patterns, the local microenvironment can be favorably moderated to reduce operating costs. The low rise buildings (none over 12 stories), 120-slip marina, tree-lined sidewalks, and parkland will provide a highly desireable living environment.

The city and its people would be major beneficiaries of the 300 to 400 million dollar project. These would include millions in yearly taxes, hundreds of new jobs, and over 20 acres of beautiful new recreational facilities. It is to generate just this type of general public benefit that the city has for some time being encouraging priviate redevelopment and revitalization of its waterfront areas. Thus, both the economic and political climate seem to favor the proposed development.

ENVIRONMENTAL DESIGN FEATURES

The main project objective is to establish an attractive living and working environment in harmony with nature. To accomplish this, microclimatic control techniques would be used. Several of these are described below.

Buildings are to be arranged around parks and a marina in such a manner as to create a feeling of intimacy and openness. By understanding the spacial relationship between the buildings, and the water, the heat, the light and patterns of the air, it is possible to maintain very favorable environmental conditions at low cost. A general view of the project is shown in Figures 1 and 5.

Vegetation is used to moderate the local climate. Parks to the south will allow south summer breezes coming off the river to cool the area. The deciduous trees to the south will also allow the winter sun to penetrate; but, provide shade in the summer. Rows of evergreens along the northern boundary will block the cold north winter winds. Many pools and recessed terraces will also help to control the climate.

All surfaces on facades are to be staggered at 45° from due south (see Fig. 2). This gives them optimum average exposure to the low winter sun. Also this orientation offers minimum direct exposure to the bothersome low sun which is due west on summer afternoons.

Significant heating and cooling is accomplished by placing the system on water. The East River will act as a giant heat sink or source since its temperature is out of phase with that of the air. This is clear from Fig. 3. which shows the cooling effect of the river during the summer and the warming effect in the winter. Furthermore, because of the proximity of the river, reversible heat pumps can be used for cooling and heating of the buildings.

PLATFORM DESIGN

The base for support will be linked barges as shown in Figure 4. These can also accommodate parking and storage space. The barges would rest on steel or concrete piles.

If the barges are constructed of steel, they could be fabricated in under-utilized shipyards close to New York. There is a potential saving in cost because of this. After construction, the barges are towed into position and ballasted.

Alternatively, the barges could be constructed of concrete. This is the preferred material for heavy loads or long spands. It is more economical than steel. Concrete also has the advantage of being non-corrosive. With concrete, construction could take place on site or at some other location like a shipyard.

The piles would be either steel or concrete. Both types can carry very heavy loads. Steel piles are easier to drive through obstructions and can be easily spliced. They are, however, subject to corrosion and must be protected by special coatings or a cathodic system.

Concrete piles have the advantage of being non-corrosive. There are, however, several disadvantages. They are usually less readily available, splicing should be avoided, and large sizes require heavy driving equipment. If concrete piles are used, the deck structure is usually also concrete.

CONCLUSION

The proposed project would create an attractive new development along the East River in Manhattan. It would strongly aid in the revitalization of the surrounding area. Both the residential, retail and park space created are vitally needed.

Studies have shown the project to be both technically and economically sound. However, because the concept is so new, New York City has yet to support it. Even with city support there are several other serious hurdles.

Since construction would be in navigable waters, a permit from the U.S. Army Corps. of Engineers is needed. Also, because of potential effects on the water quality in the East River both the U.S. Department of Environmental Protection and the State of New York Department of Environmental Conservation would be reviewing the plans. These reviews could take many years; thus, the project's future is very uncertain.

Fig. 1. East River Development Project.

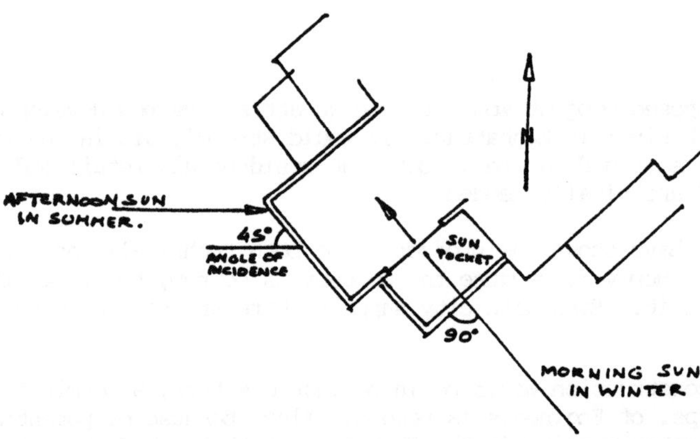

Fig. 2. Orientation of Buildings.

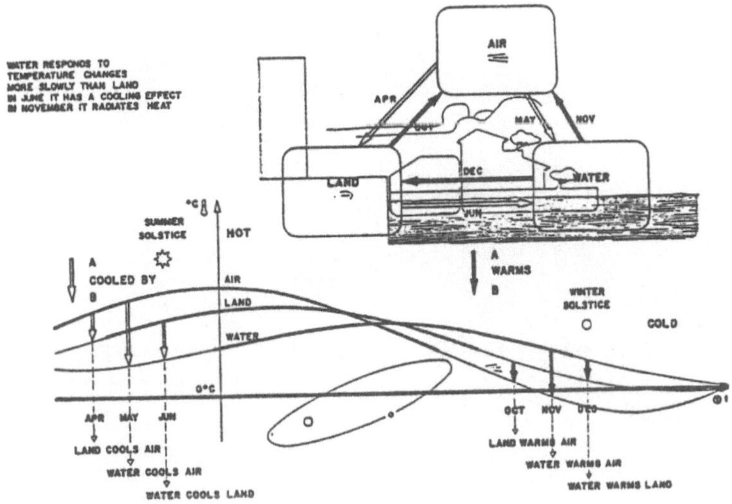

Fig. 3. Seasonal Temperature Differences between water, land and air.

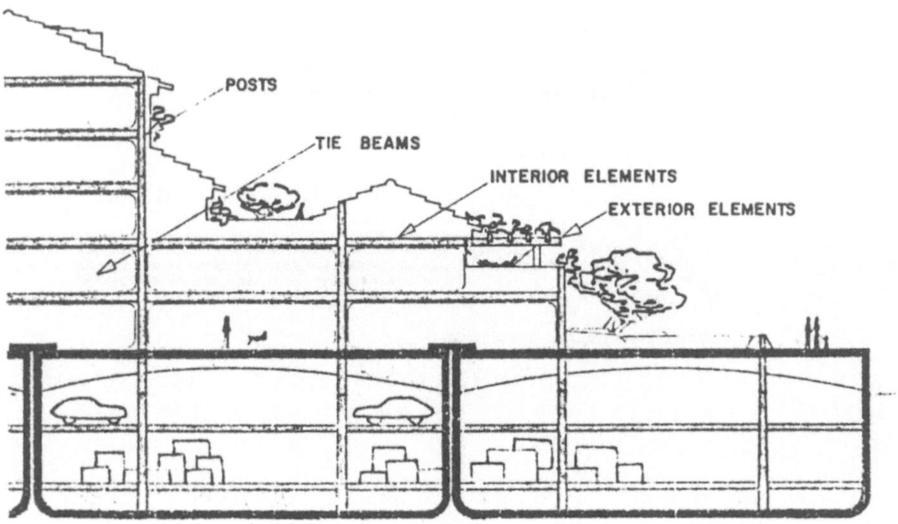

Fig. 4. Linked Barges. Support piles not shown.

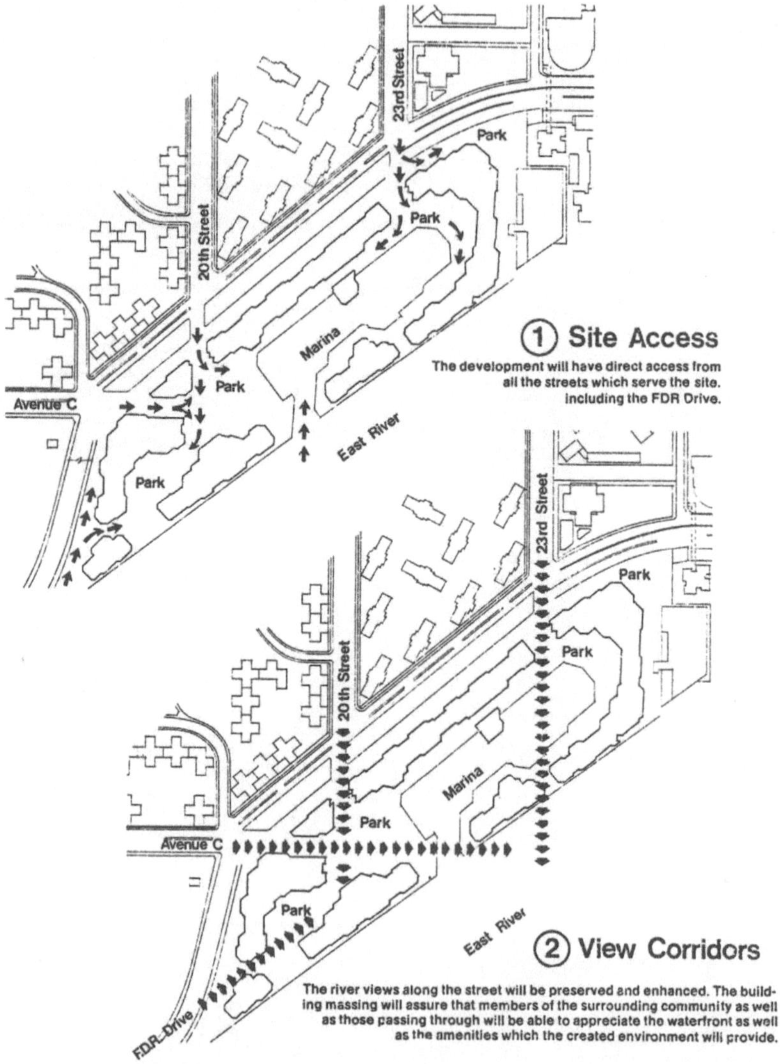

Fig. 5 Site Location

Roles of Fishing Ports and Communities in Use and Control of the Sea

S. Sakai

The Japanese Institute on Fishing Ports and Communities, Akasaka, Minato-ku, Tokyo, 107 Japan

ABSTRACT

Japan is the island country which is made of 3,900 islands of various sizes. Its sea area is far larger than the land area. Along the coast there exist about 4,000 ports and 7,000 fishing communities which have taken part in use and control of the sea through fishery since old times.

These fishing ports and communities have played important roles in spatial use of the sea, in effective use, maintenance and control of living resources and in succession of sea culture.

The aim of this report is to analyze functions of fishing ports and communities in terms of use and control of the sea and thus to clarify that in the future fishing ports and communities will have to play more important roles in a system of use and control of the sea.

1. INTRODUCTION

What is the main reason Japan, which was completely destroyed in W.W. II., has rapidly risen up in the ruins and become as prosperous as it is today? Not a few people suggest, as one reason, industry and good education of Japanese people along with uniformity and equality of the race. But here we should not ignore the fact that Japanese are fish-eating people.

Japan suffered a serious shortage of food after the war as a result of the ruined agriculture — food supply from agriculture is not sufficient even today, though. A lot of people starved to death after the war. But fortunately Japan is surrounded by the sea rich in natural resources, from which people could obtain plenty of food. Japanese people could gather and eat fish and sea weeds, which were the products of the resourceful sea, along with scarce agricultural products. Moreover, marine products contain animal protein of good quality, vitamins and minerals. It was important those marine products didn't need any artificial energy.

Another important thing was that fishing ports and communities were situated all over the country. In result, fishery was possible almost all along the coast in the country and we could take great advantage of marine resources.

Today, Japan holds an important position as a worldwide economic country, and we cannot overlook the role of fishing ports and communities as a basis of Japan's economic growth. These ports and communities are efficiently domestic scattered over the country, supporting food supply as bases of sea food production, making a great contribution to the well-balanced growth of the whole nation.

Japan is expected to keep a stable economic growth in the future. Fishing communities, which face the resourceful sea and hold good human relations, are expected to increase the importance as living spaces as well as production bases.

This, therefore, suggest that in the future fishing ports and communities will play important roles in use and control of the sea.

2. OUTLINE OF FISHING PORTS AND COMMUNITIES

2-1 Outline of fishing ports

A fishing port is a port used as a base of fisheries.

It provides a well-designed gate way for easy entry and exit of fishing vessels and has a multiple function, as the base of fishing activities, such as safe arrival and mooring of fishing vessels, unloading, preserving and marketing handling, processing, of fish catches, maintenance and repair of fishing gears, and offering crew resting places. It also functions as a center of fishing village for their daily life.

Fishing ports in Japan are classified into three categories as follows:

(1) Ports to which the Fishing Port Law is applied — about 2,907 ports
(2) Ports to which the Port and Harbour Law is applied and used by fishing vessels — about 1,000 ports
(3) Anchorages, inlets and seashores which neither law is applied to, but fishing boats actually use — about 100

Therefore, the number of ports used as fishing ports is estimated to be about 4,000.

That is, there exist one fishing port in every eight kilometers of Japan's sea coast on an average.

The state of activity of fishing ports is shown in Table 1. As to the table, look carefully at the figures related to small scale coastal fishery which I discuss later.

2-2 Outline of fishing communities

Though fishing communities can be defined various ways, I here define it rather vaguely as "the communities which are situated along the coast and inhabited by fishermen and their families who exercise or administrate fishery right".

There are 6,865 communities such as defined above (1983 Fishery census). Consequently there are, roughly estimated, 4,000 fishing ports and 7,000 fishing communities.

According to Table 2, the characteristics of present fishing communities can be summarized as follows:

(1) mainly coastal fishery
(2) small scale villages
(3) narrow flatland
(4) faced to open sea; dense villages

2-3 Relations between fishing ports and communities

Fishing communities are, like towns and cities, made up

Table-1 Outline of the Japanese Fisheries (1983)

Items	Statistics	Unit
Fish Catch (Weight)	11,757	1,000 ton
Fish Catch by Coastal Fishing included upper (Weight)	3,179	do
Fish Catch (Value)	27,239	100 M¥
Fish Catch by Coastal Fishing included upper (Value)	12,642	do
Marine Power Vessels (Number)	422,996	vessels
do (Gross Tonnage)	2,855	1,000 G.T.
Marine Power Vessels under 10 G.T included upper (Number)	402,976	vessels
do (Gross Tonnage)	702	1,000 G.T.
Total Fishermen	447	1,000 men
Fishermen engaged in Coastal Fishing included upper	343	do
Population of Fisherman's Family	1,308	1,000 men
Fish Meat Plants	1,218	plants
Employee	95,811	men
Products (Value)	19,448	100 M¥

Table-2 Fishing Communities (1976)

Items		Total Fishing Communities		Major F.C. Fishermans House Ratio 50% and over	
		NO	%	NO	%
Total		5,433	100	1,967	(36)
Communities based on Coastal Fishing		5,130	75	1,873	75
NO of Houses	∿ 50	1,336	25	1,028	52
	50 ∿ 100	1,335	25	537	23
	100 ∿ 200	1,256	23	270	14
	200 ∿ 500	1,003	18	110	6
	500 ∿	503	9	22	1
Geographical Condition	Narrow	3,423	63	1,561	79
	Open sea	3,950	73	1,569	80

on the basis of social overhead capital or public spaces. Among these public spaces the most important one is a fishing ground. A fishing ground is the space of production which supports the fishing community, and it has to be used and managed commonly.

A fishing port is the joint of a fishing ground and a fishing community.

Usually a fishing ground and a fishing community are spatially combined, but in some cases they are situated distantly. Sometimes more than one fishing communities hold a fishing port in common. In these ways the base of fishing activities of a fishing community is a fishing port.

The next important thing in the relation between fishing ports and fishing communities is that a fishing port functions variously as a common space in daily lives of a fishing community. A fishing port is a precious public property of a small fishing community and the area around it usually functions as the administrative, economic and cultural center. Moreover, the open space in a fishing port is used for various purposes such as a festival of the community and recreation of inhabitants.

As related above, a fishing port is a common space for production activity and daily life of a fishing community and functions as the basis of the social structure of a fishing community.

By the way, there are some fishing communities which is scarcely related to fishing ports; those used only in major fishing season or those used only as a refuge for fishing vessels, for example. These are defined as the fishing ports for use and control of the sea.

3. PRESENT STATE OF USE AND CONTROL OF THE COASTAL AREA

3-1 Present state of use of the coastal area

Table 3 shows the present state of use of the coastal water area within 50m depth, in terms of fishery. The figures in Table 3 are related only to the area used only for fishery

Table-3 Fisheries Use Area in the Coastal Water

Items	Depth 0∿20m		20 ∿ 50m		Total	
	Area	%	Area	%	Area	%
TOTAL	29.74	100	50.26	100	80.00	100
Present Use Area	5.41	18	2.22	4	7.63	10
Impossible Use Area	5.62	19	5.14	10	10.76	13
Possible Use Area	18.71	63	42.90	86	61.61	77

Remarks) "Impossible Use Area" is port areas, channels, reclamation areas and so on.

such as
(1) the area whose water plane is used by aqua culture or fixed net fishery
(2) natural or artificial reef ground
(3) the area in which breading and protecting animals and palnts are intended by releasing seeds and other ways

This table doesn't include the areas used as a route of large vessels as well as purse seine or trawler.

So it is remarkable that the percentage of the used water plane amounts to 10. From this we can judge that the water area along the coast in Japan is already used to a large extent.

3-2 Fishery right

One of the characteristics of use of the sea in Japan is the "fishery right" system.

Fishery right is the right entitled to
(1) fishery of space-possession type
(2) fishery of resource control type
on the basis of actual conditions from old times of the use of coastal fishing grounds by the people living in fishing communities. The water area where this fishery right exists covers most of Japan's coastal area with a certain width, and almost all the bay areas are admitted fishery right.

The caracteristics of fishery right are as follows: First, as it is the right entitled to public water plane, Fishery Cooperative Societies, which serves for public interests, are given priority to be endowed the right.

Secondly, it is the right well protected against invasion. The first point functions to reinforce the sense of community in fishing communities. The second point is one of the reasons Japan's fishery has survived against pressure in spite of the scramble between fishery and other use of the sea. This also functions to prevent sea pollution and to preserve marine environment.

There are a variety of fishery in Japan. Other than the one with fishery right, there are "admission fishery" which requires admission by the Minister of Agriculture and Forestry or the Governor, and "free fishery" which is scarcely regulated.

Seeing that the right of these kinds of fishery are also well protected, we can tell that Japan, as a fishing country, has legally supported fishery since old times.

3-3 Control of the sea by fishery

In order to admit fishery right, a plan of fishery ground is necessary. The aim of this is to maintain optimum fishing products from a certain water area by combining several types of fishing. Consequently, what is regulated in this plan is total condition of use and control of the sea in terms of fishery. This regulation includes following articles:
(1) limit of the number and qualification of fishermen
(2) closed fishing ground
(3) closed season
(4) restriction or prohibition of fishing gears and fishing method
(5) restriction on the size and capacity of fishing vessels
(6) restriction on the size and the sex of fish

These articles of restriction or prohibition are discussed in Fishery Cooperative Societies and developed into detailed plans of use and control of fishing grounds.

Coastal fishermen memorize detailed shapes and names of the bottom of the sea in the area they have their own fishery right, and know thoroughly what kinds of living things live in a certain season.

If something unusual happens there, naturally they find it immediately.

Recently they release fish seeds to maintain and increase natural resources. It is no longer a dream that the sea is used and controlled just like a farm land.

Japan has created "the eco-system in which nature and human beings coexist" which means fishermen live a stable life by using and controlling living resources in the sea properly.

4. ROLES OF FISHING PORTS AND COMMUNITIES IN USE AND CONTROL OF THE SEA

4-1 Use of the sea and fishery

Ways of the use of the sea are classified into four categories as follows:
(1) use of marine resources
(2) spatial use
(3) cultural use
(4) space for circulation and purification
As for (4), circulations of air and purification of water have become serious problems recently, but here I explain relations between fishery and items (1) through (3).

Marine resources include minerals, energy and, of course living resources. Living resources are very important for human beings as their food.

As Japanese are fish-eating people, the consumption of marine products in Japan is very large. A Japanese people eat 65 kg of marine animals and 6 kg of sea weeds per year (1982) on an average. Japanese people depend on marine products for about 44% of their animal protein intake.

What produces these marine products is fishery, and fishing ports and communities form bases of fishery. In the near future, developing countries are expected to attach importance to developing plans of fishery. In that case the key to success is whether developing plans of fishery and plans of fishing ports and communities work together.

It should also be noted that marine living resources can be reproduced by the productivity of the sea itself without fertilization or cultivation.

It may be said that fishery is an industry of high productivity which utilizes nature's power.

As for spatial use, we have already seen that most of shallow-sea area has already been used, and the depth of the sea practically used for aqua culture has reached 150m. Shallow-sea area is the most productive area for fishery, and on the other hand it is convenient for spatial use. So it is inevitable that shallow-sea area will be scrambled to be used in various ways.

Cultural use of the sea is based on fishery. In fishing ports and communities, people have inherited from old times their own culture which form foundations of Japan.

To inherit this culture to posterity will lead to the creation of new culture. In Japan's fishing communities, there have been a life cycle, that is, younger fishermen join large fishing vessels and they come back to coastal fishing when they grow older. A fishing community is the place older fishermen keep working and hand down their own culture as long as they live. We must preserve this tradition.

5. ROLES OF FISHING PORTS AND COMMUNITIES

In this chapter I will sum up the roles of fishing ports and communities in use and control of the sea.

Firstly, as a result of the fact that a lot of fishing ports and communities are situated along highly-productive shallow-sea areas, the followings become possible:

(1) Effective use of the productivity of the sea
(2) supply of a great deal of and a variety of marine products
(3) delicate control of resources
(4) observation and preservation of marine environment

In addition, as people depending on fishery live even in the fringe of the country, the followings also become possible:

(1) well-balanced development of the country
(2) sea rescue and prevention of accidents caused by users of recreational facilities
(3) observation of the national boundaries

What stated above hold good as long as the sea and fishery exists. In other words, if fishery becomes impossible, fishing ports and communities will be ruined and the whole country will be exhausted.

Generally, a project for development of the sea tends to eliminate fishery as an obstacle. But such tendency is totally mistaken. It is necessary for a proper project for development of the sea to be able to coexist with fishery. It is important not to abolish the use of the sea which is one of fishery's characteristics, that is, coexisting with nature, and to put it into a system of developing of the sea.

6. CONCLUSION

The sea is no longer the sea if it is polluted or unfit for living things to live in.

The sea must be always clean. Fishery cannot exist without clean water. It is human beings who should keep the sea clean. Fishing ports and communities are the bases where fishermen engage in fishery, maintain and control fishing grounds, and live their lives.

Here I emphasize that to make a proper project for development of the sea, it is imperative that coexisting with fishing ports and communities should be regarded as an important part of the system in a plan of use and control of the sea.

Technical Session:
Materials and Construction

1. Design of Concrete

Flexural Behavior of Ferrocement with Carbon Fiber Reinforced Cement as a Matrix

Y. Ohama and A. Shirai

Department of Architecture, College of Engineering, Nihon University, Koriyama, Fukushima, 963 Japan

ABSTRACT

This paper discusses the flexural behavior of ferrocement using carbon fiber reinforced cement (CFRC) as a matrix, and compares the flexural behavior with that of the ferrocement using ordinary cement mortar (OCM) as a matrix. The ferrocements using CFRC and OCM are prepared, and tested for first cracking load, ultimate load, toughness and cracking resistance. It is concluded from the test results that the first cracking load, ultimate load, toughness and cracking resistance of the ferrocement using CFRC noticeably increase with raising carbon fiber content, and that the occurrence of the cracks and the increase of the crack widths may be considerably restrained by the action of the carbon fibers. Generally, the flexural behavior of the ferrocement can be markedly improved by use of CFRC instead of OCM.

INTRODUCTION

Ferrocement is a thin cement mortar with reinforcement in the form of multiple layers. The ferrocement has been widely used for marine structural applications such as boats, yachts, buoys and pontoons because of its excellent strength, toughness, cracking resistance and impact resistance. However, with the growing interest in the use of the ferrocement in the marine construction industry, higher strength, toughness, cracking resistance and durability than conventional ferrocement have been required. To satisfy these requirements for the ferrocement, the modification of its cement matrix by the addition of carbon fibers to the ferrocement may be considered. The purpose of this study is to investigate the flexural behavior of the ferrocement using carbon fiber reinforced cement (CFRC) as a matrix, and to compare the flexural behavior with that of the ferrocement using ordinary cement mortar (OCM) as a matrix.

In this paper, ferrocements using CFRC and OCM are prepared, and their cracking load, ultimate load, toughness and cracking resistance are tested. Based on the test results, the flexural behavior of the ferrocement using CFRC is discussed in comparison with that of the ferrocement using OCM.

MATERIALS

Materials for Carbon Fiber Reinforced Cement (CFRC)

Ordinary portland cement, silica fume and polyalkyl aryl sulfonate-type water-reducing agent, and low modulus carbon fibers were used in all the mixes of CFRC. The physical properties of the carbon fibers are given in Table 1. The silica fume which was a by-product in the manufacture of ferrosilicon was used as a dispersant for the carbon fibers. The chemical compositions and physical properties of the silica fume are shown in Table 2.

Table 1. Physical Properties of Carbon Fibers

Type of Carbon Fiber	Average Diameter (μm)	Average Length (mm)	Specific Gravity (20°C)	Young's Modulus (kg/mm^2)	Tensile Strength (kg/mm^2)
Low Modulus Carbon Fiber	14.5	3	1.63	3800	80.0

Table 2. Chemical Compositions and Physical Properties of Silica Fume

Chemical Compositions (%)								
ig.loss	SiO_2	Al_2O_3	Fe_2O_3	CaO	MgO	Na_2O	K_2O	Total
3.0	88.9	1.8	2.4	0.5	0.8	0.32	1.61	99.33

Physical Properties		
Specific Gravity (20°C)	Particle Size (μm)	Water Content (%)
2.212	0.1 - 0.3	0.85

Materials for Ordinary Cement Mortar (OCM)

The same cement as stated above and river sand (size, 1.2mm or finer) were used in all the mixes of OCM.

Reinforcement

Reinforcement used was a combination of welded wire fabrics formed from wires of 2.6 mm in diameter and crimped wire cloths formed from wires of 1.5 mm in diameter, specified in JIS A 3551 (Welded Steel Wire Fabric) and JIS A 3553 (Crimped Wire Cloth), respectively. The reinforcement consists of three layers of the welded wire fabrics and two layers of the crimped wire cloths, and the details of the reinforcement are illustrated Fig. 1.

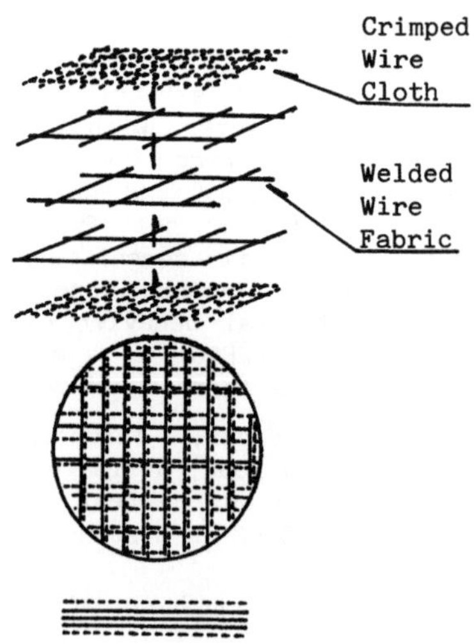

Fig. 1. Details of Reinforcement

PREPARATION OF SPECIMENS

Ferrocement Specimens with CFRC as a Matrix

In conformity with JIS R 5201 (Physical Testing Methods of Cement), CFRC was prepared with the mix proportions given in Table 3. Ferrocement specimens 55 x 25 x 2 cm with CFRC as a matrix were molded, and then precured at 20 °C, 80 % R.H. for 24 hours. Immediately after precuring, the specimens were cured in autoclave where the maximum temperature of 180 °C was maintained under a steam pressure of 10 kg/cm^2 for 3 hours.

Table 3. Mix Proportions of CFRC as a Matrix for Ferrocement

Cement : Silica (by Weight)	Water-Cement Ratio (%)	Carbon Fiber Content (%)	Water-Reducing Agent Content (%)	Flow
1 : 0.4	30	3	5	146
			6	148
		5	5	124
			6	138
	35	3	5	166
			6	176
		5	5	140
			6	150

Ferrocement Specimens with OCM as a Matrix

According to JIS R 5201, OCM was mixed with the following mix proportions ; cement : sand = 1 : 3 (by weight), water-cement ratio of 59.3% and flow of 172. Ferrocement specimens 55 x 25 x 2 cm with OCM as a matrix were molded, and then subjected to a 1-day-20 °C-80 % R.H.-moist, 6-day-20 °C-water and 21-day-20 °C-50 % R.H.-dry cure.

Flexural Tests

The flexural tests were performed on ferrocement specimens with CFRC and OCM as matrices under third-point load (span ; 45 cm). Sensitive LVDT (linear variable differential transformer) was used to measure their central deflection, which was automatically recorded. The widths of the cracks appeared on the surfaces of the specimens were measured by a crack scale.

TEST RESULTS AND DISCUSSION

First Cracking Load

Fig. 2 shows the first cracking load of ferrocements using CFRC and OCM as matrices. The first cracking load of the ferrocement using CFRC generally increases with increasing carbon fiber content because the occurrence of first cracking is restrained by carbon fibers. In particular, this tendency is remarkable at a higher water-cement ratio of 35 %. This is explained to be attributed to an improvement in the workability of CFRC due to an increase in the water-cement ratio. However, the first cracking load of the ferrocement using CFRC is hardly affected by the variation of the water-reducing agent content. The maximum first cracking load of the ferrocement using CFRC attains to about 700 kg, and is approximately 5 times that of the ferrocement using

OCM. Such results suggest that a defect in the flexural behavior of conventional ferrocement, that is, inferior first cracking load can be improved by use of CFRC as a matrix.

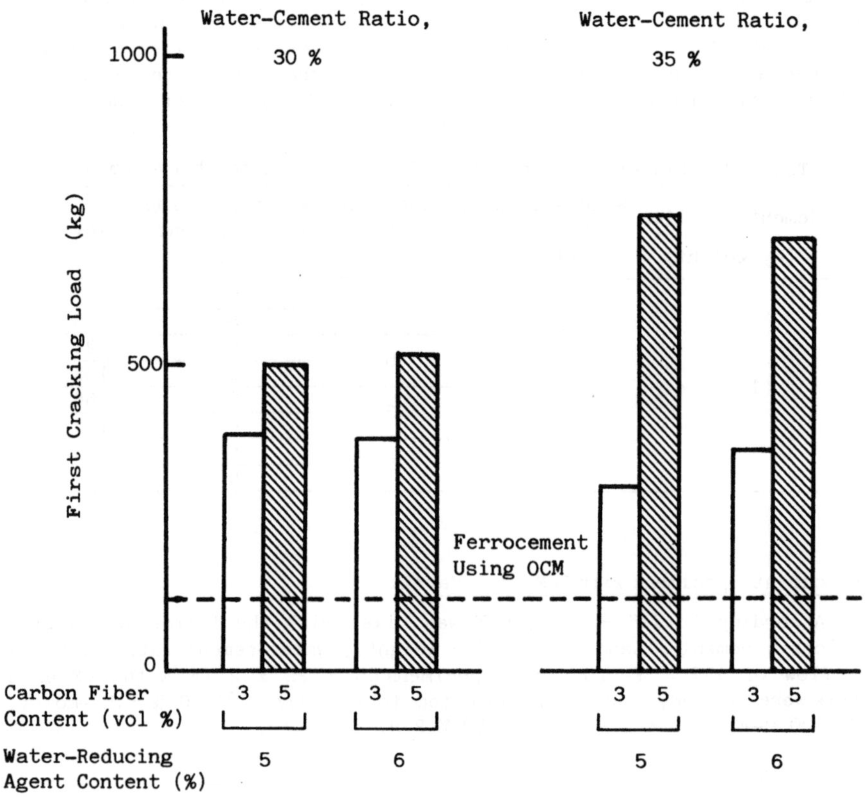

Fig. 2. First Cracking Load of Ferrocements Using CFRC and OCM as Matrices

Ultimate Load

Fig. 3 represents the ultimate load of ferrocements using CFRC and OCM as matrices. Like the first cracking load, the ultimate load of the ferrocement using CFRC tends to enlarge with a rise in the carbon fiber content, and reaches about 1000 kg at a carbon fiber content of 5 vol % regardless of the water-cement ratio and water-reducing agent content. The ultimate load hardly depends on the water-cement ratio and water-reducing agent content. As compared with such a high ultimate load, that of the ferrocement using OCM is only approximately 500 kg.

Toughness

Fig. 4 exhibits the toughness at first cracking of ferrocements using CFRC and OCM as matrices. The toughness at first cracking of the ferrocement using CFRC generally enlarges with increasing carbon fiber content, and this trend is more remarkable at a water-cement ratio of

35% rather than 30%. The toughness is scarcely dependent on the water-reducing agent content. The largest toughness of the ferrocement using CFRC is obtained at a carbon fiber content of 5 vol %, a water-cement ratio of 35 % and a water-reducing agent content of 6 %, and is about 15 times that of the ferrocement using OCM.

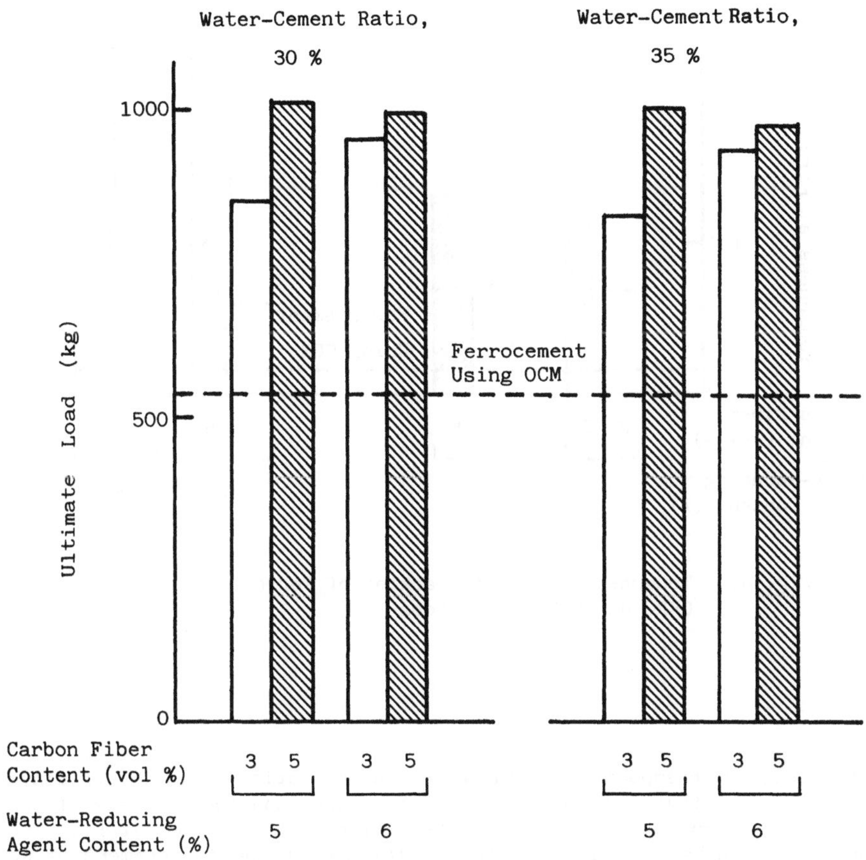

Fig. 3. Ultimate Load of Ferrocements Using CFRC and OCM as Matrices

Cracking Behavior

Fig. 5 illustrates the number of cracks and crack widths with increasing moment of ferrocements using CFRC with a water-reducing agent content of 6 % and OCM as matrices. The number of cracks and crack widths of the ferrocement using CFRC decrease with increasing carbon

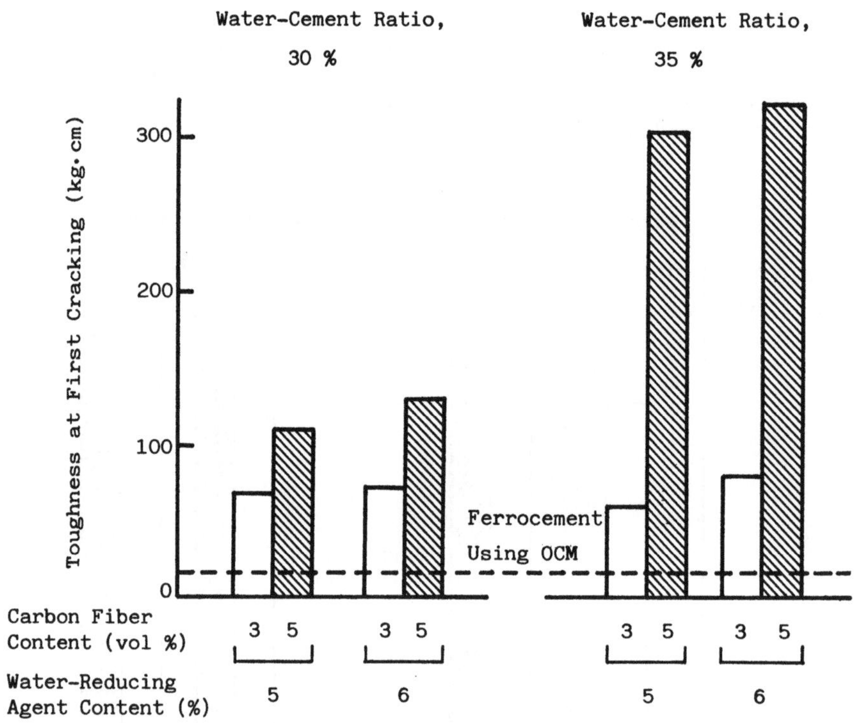

Fig. 4. Toughness at First Cracking of Ferrocements Using CFRC and OCM as Matrices

fiber content irrespective of the water-cement ratio. This is explained by considering that the occurrence of the cracks and the increase of the crack widths may be restrained by the action of the carbon fibers. The number of cracks and crack widths at failure of the ferrocement using CFRC are about 2/3 and 1/2 of those of the ferrocement using OCM, respectively. The above-mentioned tendency of the cracking behavior is similar to that in the ferrocement using CFRC with a water-reducing agent content of 5 %.

Fig. 6 gives the cracking resistance of ferrocements using CFRC and OCM as matrices. In this study, the cracking resistance was estimated as a flexural load at which crack width reaches 0.1 mm. Such crack width limitaion is proposed by ACI Committee 224 [1,2] for water retaining structures. The cracking resistance of the ferrocement using CFRC tends to increase with a raise in the carbon fiber content regardless of the water-cement ratio and water-reducing agent content. The cracking resistance of the ferrocements using CFRC with carbon fiber contents of 3 and 5 vol % provides about 3.4 and 4.3 times that of the ferrocement using OCM, respectively.

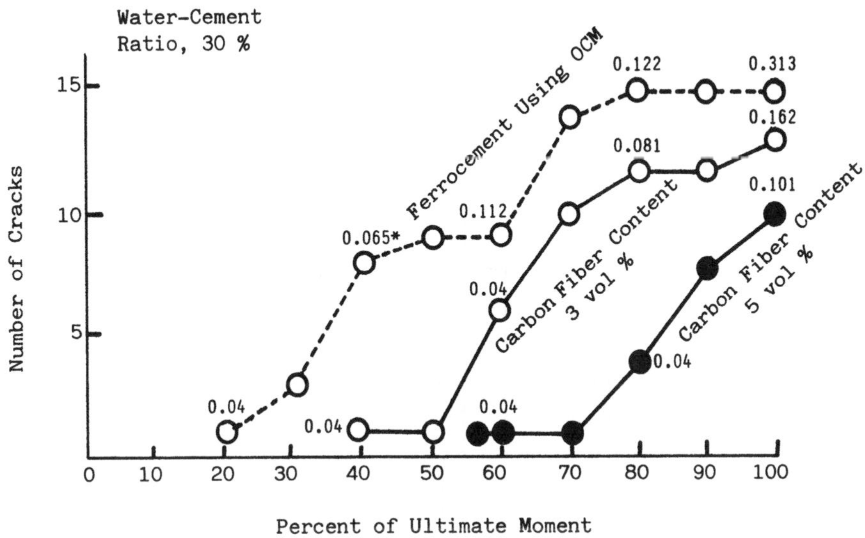

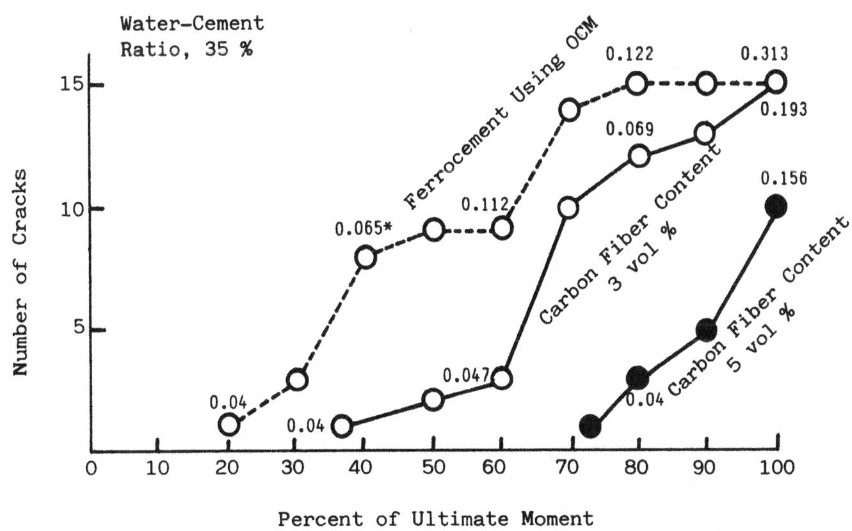

Fig. 5. Numbers of Cracks and Crack Widths with Increasing Moment of Ferrocements Using CFRC with Water-Reducing Agent Content of 6 % and OCM as Matrices

Note ; * Figures indicate average crack widths (unit : mm).

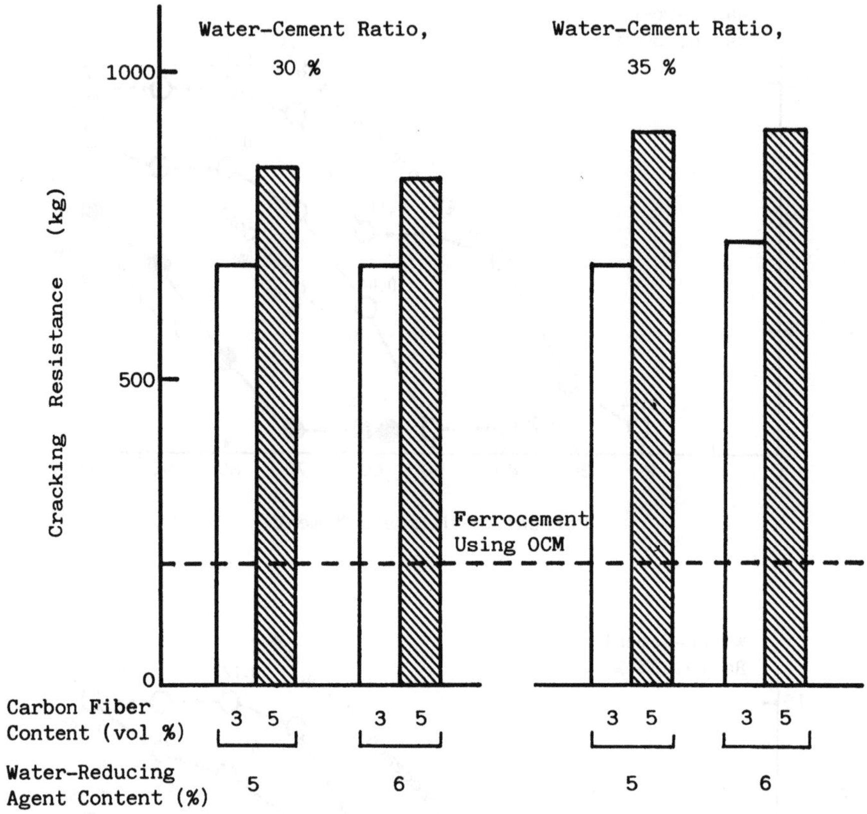

Fig. 6. Cracking Resistance of Ferrocements Using CFRC and OCM as Matrices

CONCLUSIONS

(1) The first cracking load, ultimate load, toughness and cracking resistance of ferrocement using CFRC as a matrix markedly increase with raising carbon fiber content regardless of the water-cement ratio and water-reducing agent content.

(2) The occurrence of cracks and the increase of the crack widths of ferrocement using CFRC as a matrix may be considerably restrained by the action of carbon fibers.

(3) In general, the flexural behavior of ferrocement can be remarkably improved by using CFRC instead of OCM as a matrix.

REFERENCES

1. ACI Committee 224 : Control of cracking in concrete structures. ACI Journal, Proceedings. V.69, No.12 (1972), 717-753.

2. Kobayashi, K. ; Fiber Reinforced Concrete - Properties and Applications (in Japanese). Tokyo : Ohm-sha 1981.

Water Leak Tests of Ferrocement in Tension

Y. Kobayashi

Ship Structure Division, Ship Research Institute, Ministry of Transport, Shinkawa, Mitaka, Tokyo, 181 Japan

ABSTRACT

Watertightness comes up as an important problem for water retaining structures. Watertight crack width is considered to be one of the limit states on design and inspection for use. In this paper, water leak tests in tension were carried out on ferrocement plates. The leak stress, the crack width at each water leaking stage and failure, and the crack spacing were observed experimentally.

INTRODUCTION

Ferrocement formed into thin plates and shells is suitable for boats, pontoons,water tanks and other structures. Crack width and spacings of ferrocement are finer than those of conventional reinforced concrete. Ferrocement made from mortar with a low water to cement ratio(0.35-0.40 by weight) has very low permeability. As low permeable mortar can prevent the corrosion of mesh and delay the growth of deterioration, improved durability can be expected in sea water environment.

Although cracks were found out on surface of ferrocement, water does not always leak immediately. In that case, thin cover becomes the origin of corrosion during a long period. Water leaks only through penetrated cracks. From the above viewpoints, the limit states on design and inspection for use are necessary to be classified by crack width in consideration of water leakage behavior.

Walkus[1] suggested the lower limit crack width of 0.02mm for complete watertightness in ferrocement. Guerra et al.[2] proposed the limiting average crack width of 0.04mm on leak tests of ferrocement cylindrical tanks. Endo et al.[3] defined 150 $\times 10^{-6}$ strain for the watertight limit state where the crack penetrated the plate. Ravindrarajah et al.[4] emphasized that it is necessary for water retaining structures to be designed in the range of the elastic limit. As mentioned above, the crack width for watertightness of ferrocement is narrower than the allowable crack width of 0.1mm in reinforced concrete structures that is reported by ACI committee 214, but the crack width of ferrocement is stable up to the high stress level of reinforcement[2][5].

In this paper, water leakage behaviors, leak stress, crack width of water leaking and spouting, and crack width and crack spacing after failure were revealed and discussed by water leak

Table 1 Mix proportions of mortar

Slumpcone 5x10x15 cm

Exp. Series	F.M.	Flow value (mm)	Slump (cm)	Air (%)	$\frac{W}{C+G}$ (%)	$\frac{C+G}{S}$ (%)	Unit content (kg/m³) Water (W)	Cement (C)	Sand (S)	Expnsive additive (G)	Water reducer (cc/m³)
F	2.84	170	1.1	4.3	40.9	50	261	592	1274	45	3049
G	2.52	182	1.0	4.8							

Table 2 Strength of mortar

(kg/mm²)

Exp. series	Strength Comp. σ_{mc}	Tensile σ_{mt}	Bending σ_{mb}	Elastic modulus (1/3 σ_{mc}) Em
F	6.16	0.49	0.56	2740
G	6.39	0.36	0.46	2860

tests of ferrocement plate in tension.

Table 3 Experimental program

Exp. Series	Specimen Type	Reinforcement	Total No. of Layers	Longi. Volume Fraction	Longi. Specific Surface (mm²/mm³)	Thickness of Plate (mm)
F	A	8x8 Woven	6, 8	.025~.033	.104~.131	19.8~23.8
	A	10x10 Woven	6, 8	.017~.027	.068~.110	20.3~22.5
G	A	8x8 Woven	4,10,12	.030~.041	.120~.165	13.3~29.0
	A	10x10 Woven	4,10,12	.023~.032	.093~.127	13.5~29.7
	B	8x8 Woven,Rod	6, 12	.048	.117, .143	23.5, 35.9
	B	10x10 Woven,Rod	6, 12	.043, .041	.098, .115	23.1, 36.1
	B	12.7x12.7 Weld,Rod	6, 12	.034, .031	.068, .081	23.9, 35.6

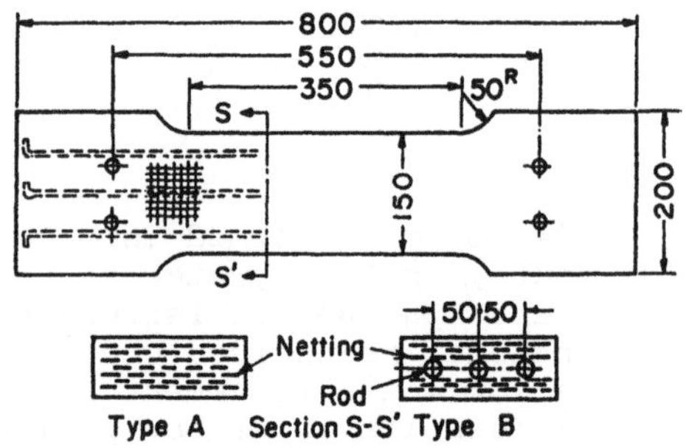

Fig.1 Specimen

EXPERIMENTAL INVESTIGATION

Mix Proportions of Mortar

The mix proportions of the mortar is shown in Table 1. Experiments were classified into two experimental series F and G. Normal Portland cement and river sands with the maximum size of fine aggregate 2.5mm were used. Admixtures such as expansive additive and water reducing agent were added in.

The mortar and specimens were cured by water curing through four weeks after low pressure steam curing. The mortar strength at the age of four weeks is shown in Table 2.

Specimens of Water Leak Test

The size of specimen and two types of reinforcement are shown in Fig. 1. Type A consists of 4 to 12 layers of same type mesh, and Type B consists of one layer of reinforcing rods in the middle of section and 3 or 6 layers of mesh on the both sides. Experimental program is shown in Table 3.

As the reinforcement of ferrocement specimens, Woven Wire Cloth (JIS G 3555), Welded Steel Wire Fabrics (JIS G 3551), and Steel Bars for Concrete Reinforcement (JIS G 3112) were used. Woven meshes were made of galvanized low carbon steel wires after cold drawn in 1mm diameter. The size of square woven mesh openings were 8 and 10mm, respectively. The square welded mesh with 0.9mm diameter and 12.7mm mesh openings was galvanized after welding. Steel rods were hot rolling steel bars; SR-24, 6mm diameter. The ultimate tensile stress and yield point of these meshes were 33-39kg/mm^2, and 29-35kg/mm^2, respectively.

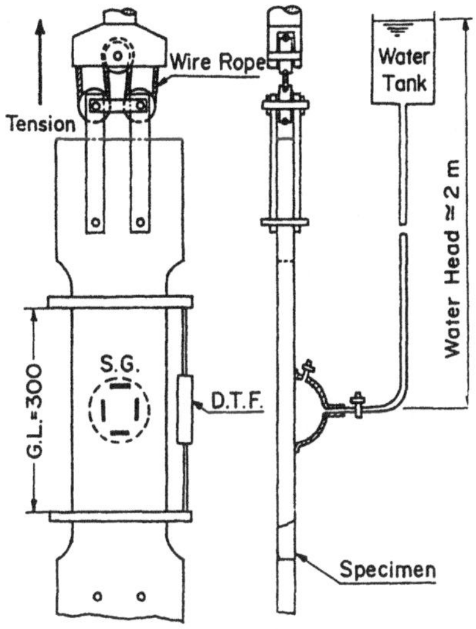

Fig.2 Loading device in tensile and water leak test

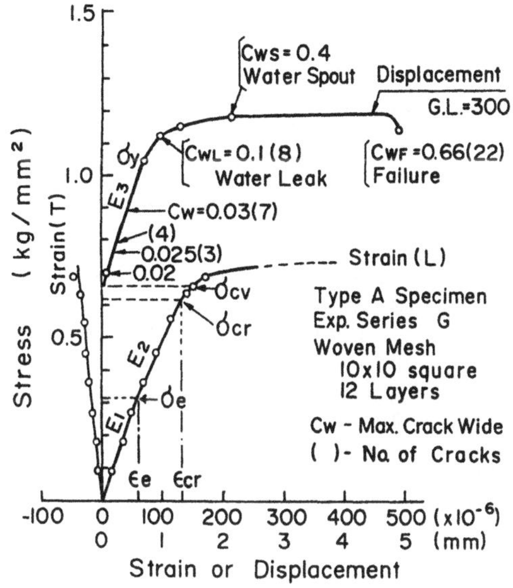

Fig.3 Stress-strain and stress-displacement curves of ferro-cement plate in tension

Testing Procedures

Tensile tests were carried out by a hydraulic testing machine with loading rates 100-130 kg/min. As shown in Fig.2, the both sides of specimen were fixed with wire rope in order to apply uniform tensile load.

Water leak tests in tension were carried out setting up the rubber-made tank on the surface of specimen. The water head of tank was constantly kept at 2m through the test. Water leakage behavior and crack width were observed. Crack width was measured using a 60 times microscope with scale. The gauge length of measurement was 300mm long.

RESULTS AND DISCUSSION

A Few Remarks on the Tensile Tests

Examples of the stress-strain and stress-displacement curves in tension are shown in Fig.3. The tensile test results of water leak tests are as follows. The detailed results and discussion are shown in Ref. 6.

(1) Until first crack was found out, two elastic modulusli were observed in type A specimen, but type B elastic modulus agreed with a law of mixture.

(2) Similar to Ref.7, the stresses of first crack increased proportionally to the specific surface of the reinforcement, irrespective of types of mesh and specimen. But this tendency presented different slopes for different mortar. The first crack did not occur until the surface strain of specimen attained the tensile fracture strain of the mortar in splitting test. First crack stress σ_{cr} can be considered as visible crack stress σ_{cv}.

(3) The elongation at failure increased proportionally to the volume fraction of the reinforcement. The elongation of large mesh openings was larger than that of small mesh openings.

Water Leakage Behavior

The ferrocement did not always begin to leak when cracks were found out on the surface of specimen. For this reason, it is satisfactory to consider that water leaks only through penetrated cracks, and that water spouts through widen cracks.

Therefore, water leakage behavior can be classified into the following three stages in order to characterize. The first stage of leakage or watertight stage is in the state where no leak occurs through cracks on the surface. This state is in danger of leaking, but water is retaining. The second stage or water leaking stage is in the state where water begins to leak and is leaking through penetrated cracks. The third stage or water spouting stage is in the state where leaking water begins to spout.

In the stress- displacement curve of Fig. 3, the first stage changed to the second after the crack width reached 0.1 mm, and the second stage changed to the third after the crack width reached 0.4mm.

Stress Ratio on Water Leakage Behavior

It was shown that leakage behavior presented different stages depending upon applied stress level. The relationship between the ratio of the applied stress σ to the ultimate tensile stress σ_u, σ/σ_u and the volume fraction of the reinforcement Vs is shown in Fig. 4. Where, Vs is calculated by taking into account the reinforcement in the loading direction. $\sigma cv/\sigma u$ is related to Vs by the following empirical equation.

$$(\sigma cv/\sigma u)(Vs+0.006)=0.017 \qquad (1)$$

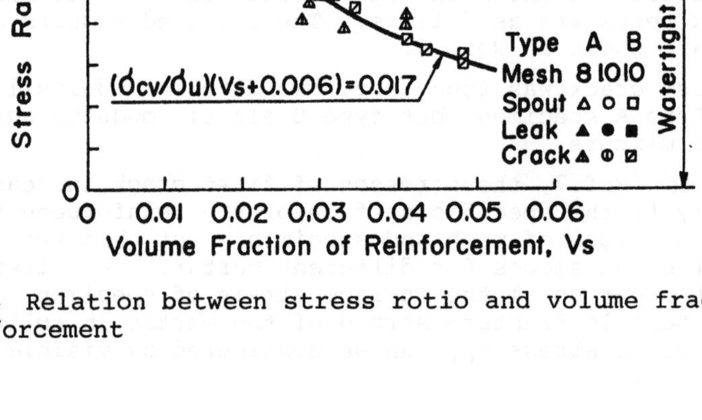

Fig.4 Relation between stress rotio and volume fraction of reinforcement

No water leaks in the range of lower stress ratio than that of eq. (1). This range is completely watertight.

For Vs<0.035, the minimum leak stress ratio was 0.6. For Vs>0.035, leak stress was over yield point; $\sigma y/\sigma u=0.85$. It is considered that cracking, leaking and failure occur at the same time in the range of Vs<0.011. Therefore, the range of the first stage is defined from eq.(1) of Fig.4 until water begins to leak. The third stage was over stress ratio of 0.9, and this stage was seen just before failure in many specimens.

On the other hand, flexural cracking moment was 17 percents of the maximum flexural moment for Vs>0.025[8]. The range of the first stage tends to extent because one side of the plate surface is in compression. In case of the first stage in flexure and tension, it is necessary to consider corrosion of mesh rather than leaking.

Crack Width of Water Leaking and Spouting Stage

The crack width of first crack was almost same as the visible crack width 0.02mm. C_{WL} and C_{WS} were used respectively as the crack width through which water begins to leak and spout. The relationships between these crack width and the specific surface of the reinforcement S_L are shown in Fig.5. Where, S_L was calculated by taking into account the reinforcement in the loading direction.

C_{WL} was inversely proportional to S_L, and the small size of mesh openings showed the large value of C_{WL}. The minimum crack width of C_{WL}=0.025 is indicated in Fig.5. Similar tendency was found out on relationships between C_{WS} and S_L, but C_{WS} of 8mm mesh was much wider than 10mm mesh. And it was seen that the size of mesh openings had effects upon the water leaking stage. The minimum crack width of C_{WS} was 0.05mm.

The correlation of the crack width of yield point C_{WY} and C_{WL} is shown in Fig.6. This correlation shows that C_{WL} is approximately equal to C_{WY} for many specimens: Water leaking is considered to take place at yield point of specimen. However, there were a few specimen in which C_{WL} was narrower than C_{WY}. For this reason, it can be considered that Vs was less than 0.035 as shown in Fig.5 and that position of mesh layers was on one side in the plank section. As shown in Fig.7,

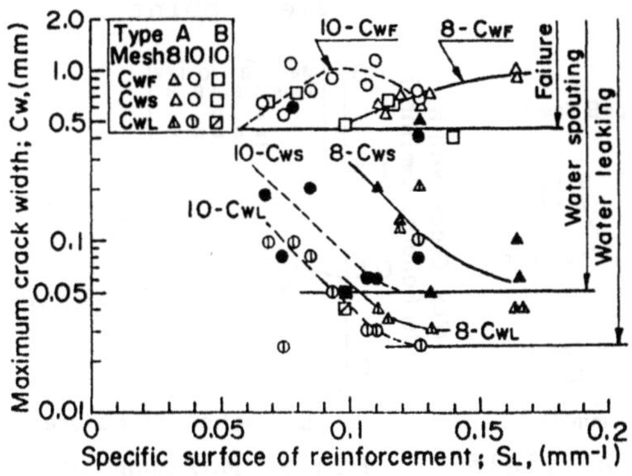

Fig.5 Relation between maximum crack width and specific surface of reinforcement

C_{WS} was wider than C_{WY} in many specimen. It is seen that water leaking state goes on from C_{WY} or less to wider crack width.

C_{WY} of type B with skeletal rods increased with increasing small S_L as in Fig.8. This means that many layers of mesh contribute considerably to crack control.

Maximum Crack Width and Average Crack Spacing after Failure

C_{WF} is the maximum crack width which remains on the surface of specimen after failure. On the process gradually improving to failure, crack width was locally expanded at the part with yielding of mesh because of decreased bond strength. Therefore, as it can be considered that narrow crack has no effect on the elongation at failure, C_{WF} can be assumed to be the crack width at failure.

$$C_{WF} = 0.25\,\delta + 0.45 \qquad (2)$$

As shown in Fig.9 and eq.(2), a linear relationship existed between C_{WF} and elongation δ, irrespective of types of mesh and specimen. C_{WF} increased with increasing δ. For δ=0, C_{WF} is given as 0.45mm. This value was indicated in Fig.5.

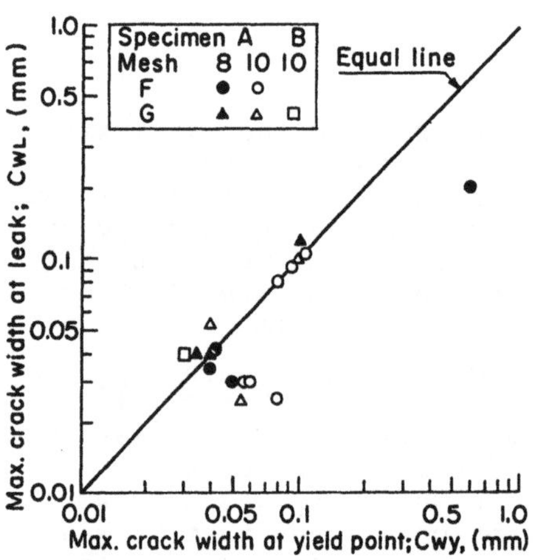

Fig.6 Correlation of maximum crack width at leak and yield point

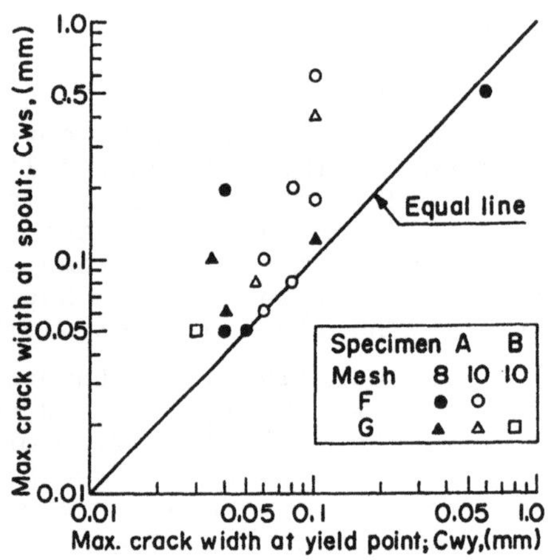

Fig.7 Correlation of maximum crack width at spout and yield point

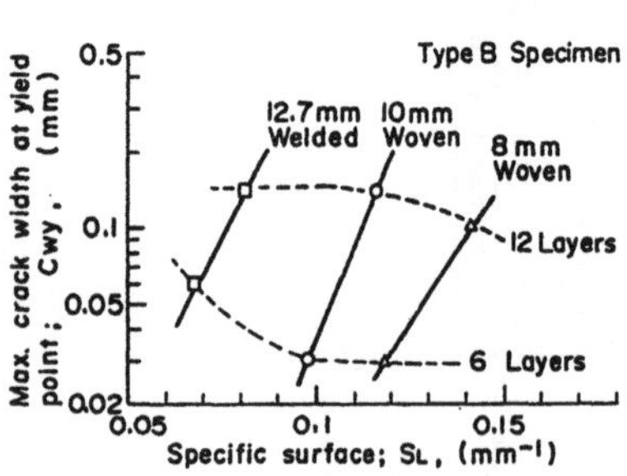

Fig.8 Crack width at yield point vs. specific surface of reinforcement

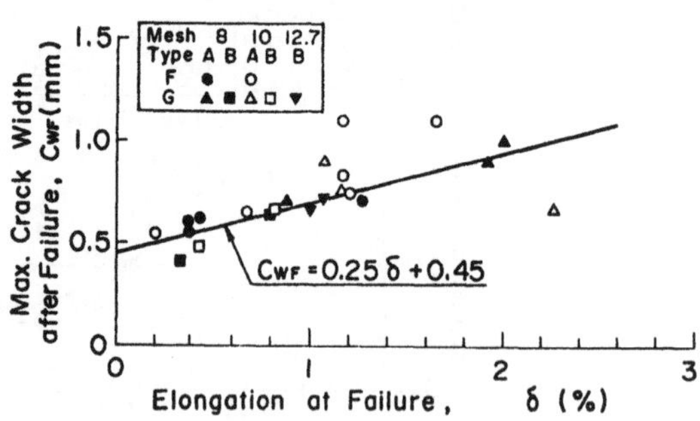

Fig.9 Maximum crack width after failure vs. elongation

Meanwhile, as shown in Fig.10, the free end slip length of 0.5mm was obtained at general yielding in pull-out tests with woven wire embedded in the G series mortar[9]. The value was close to the minimum value of C_{WF}. In Fig.10, the average bond stress τm was almost same for different sizes of mesh openings.

The effective embedded length Le is the length on the border where a wire slips or breaks in pull-out tests. Le can be considered to be the minimum crack spacing where no crack occurs on the surface of ferrocement. Average bond stresses for yield point and failure at Le are shown in Fig.10 as τ_{my} and τ_{mu}. On yield and failure load of specimens, the experimental values agreed well with the calculated values for τ_{my} and τ_{mu}, respectively[9].

The average crack spacing l_{cs} is shown in Fig.11 by the method of Ref.7. lcs was related to S_L by equation (3).

$$lcs \cdot S_L = 1.72 \tag{3}$$

The constant value of the right hand was obtained from τmu and the tensile stress of the mortar τ_{mt}. Experimental data satisfied eq.(3) approximately.

The maximum of average crack spacing [lcs]max is expressed with 1.5 Le, and indicated in Fig. 11. Therefore, the minimum value $[S_L]min$ can be obtained by substituting [lcs]max into eq.(3). This value shows the minimum specific surface required not to fail simultaneously with cracking. Subsequently, on G series experiment, the minimum volume fraction [Vs]min which corresponds to $[S_L]min = 0.04mm^{-1}$ was 0.01.

On the other hand, [Vs]min is defined by the ratio of tessile stress of mortar to the ultimate tensile stress of the reinforcement, and the value of 0.01 had been obtained in Ref.6. The value by two different methods agreed with the above mentioned results. Consequently, the propriety of the effective embedded length in pull-out tests was cofirmed for performance required in ferrocement.

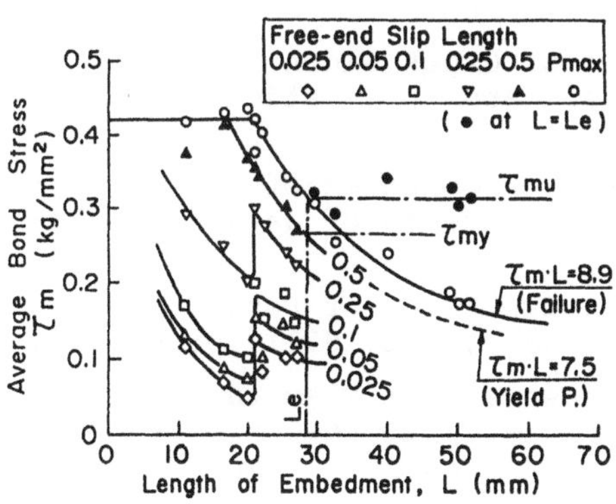

Fig.10 Relation between average bond stress and length of embedment

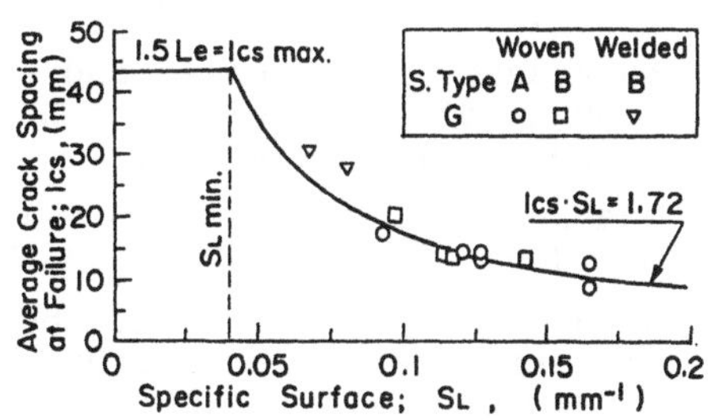

Fig.11 Relation between average crack spacing at failure and specific surface of reinforcement

CONCLUSIONS

The summary of results on water leak tests of ferrocement in tension is indicated below.

1) The water leakage behavior is devided into three stages: watertight, water leaking and water spouting stage. Water does not leak immediately after first crack. Watertight stage is retaining in danger of leaking. Under 3.5 percent of the volume fraction, water does not leak until the applied stress exceeds 60 percents of the ultimate tensile stress. In case of the larger volume fraction than the above value, leak stress is over the yield stress. Water oozes its way through the crack from the state where water begins to leak until the state where water begins to spout.

2) The crack width at which water begins to leak is considered to be close to the crack width of yield point. The minimum crack width for water leaking and spouting were 0.025mm and 0.05mm, respectively. Many layers of mesh contribute considerably to crack control in ferrocement with skeletal rods.

3) The maximum crack width after failure increases proportionally to elongation, irrespective of types of mesh and ferrocement. It is estimated that a load carring ferrocement plate with the crack width of 0.45mm is already getting to failure.
The critical performance required in ferrocement also can be confirmed from the effective embedded length in pull-out tests.

REFERENCES

1. Walkus, B.R.: The behaviour of ferrocement in bending. Journal of Structural Engineering, Vol.3, No.3, (Oct.1975) pp.113-125.
2. Guerra, A.J.; Naaman, A.E. and Shah, S.P.: Ferrocement Cylindrical Tanks; Cracking and Leakage Behavior, ACI Journal, Proceedings, Vol. 75, No.1, (Jan. 1978) pp.22-30.
3. Endo, H.; Matsuoka, K.; Naoi, T. and Nagasawa, H.: Transverse Strength of Ferrocement Boats. Journal of the Society of Naval Architects of Japan, Vol.148, (Dec.1980) pp.303-311.
4. Ravindrarajah, R.S. and Tam, C.T.: Watertightness in Ferrocement. Jounal of Ferrocement, Vol.14, No.1, (Jan.1984) pp.11-20.
5. Naaman, A.E.: Design Predictions of Crack Width in Ferrocement, Ferrocement-Materials and Applications, ACI, SP-61, (1979) PP.25-42.
6. Kobayashi, Y.: Tensile and Leakage Properties of Ferrocement. Journal of the Society of Naval Architects of Japan, Vol.153, (June,1983) pp.386-395.
7. Naaman, A.E. and Shah, S.P.: Tensile Tests of Ferrocement. ACI Journal, Proceedings, Vol.68, No.9, (Sep.1971) pp.693-698.
8. Kobayashi, Y.; Inoue, H. and Nagasawa, H.: Flexural Strength of Ferrocement. Journal of the Society of Naval Architects of Japan, Vol.150, (Dec.1981) pp.371-380.
9. Kobayashi, Y.: On the Bond Strength of Low Carbon Steel Wire and Mortar - Application to Tensile Strength of Ferrocement. Abstract Note of the 41th General Meeting of Ship Research Institute, Ministry of Transport, (May,1983) pp.42-46.

Polymer Concrete Composites for Concrete Ocean Structures

K.G. Babu

Ocean Engineering Centre, Indian Institute of Technology, Madras 600 036, India

ABSTRACT

Recent advancements in concrete materials technology has opened up many avenues for effective utilisation of these materials. Polymer concrete composites being one such has been attracting the attention of many to provide special alternatives for concrete to perform under severe environmental conditions. The present paper gives the methodology for the production of these materials in the laboratory. The characteristics of each of these materials with respect to strength as well as durability have been presented. Lastly, application possibilities with particular reference to ocean structures has been discussed.

INTRODUCTION

In the recent years substantial improvements have taken place in materials technology, particularly both in steel and concrete - two of the most utilized materials for construction. While the advancements in steel has allowed structures to be built in the extremely hostile environments of the oceans to start with, concrete material alternatives are not lagging far behind. In many cases concrete alternatives have shown lower costs of construction and maintenance.

Polymer concrete composites have brought a new dimension in this regard. However, even though the basic technology of the production of these materials has been perfected mostly in the laboratory, its capability is available only to a limited section. Compounded to this, considering the shapes and sizes of the various ocean structural systems, the application possibility of these techniques in field are faced with numerous problems. It is envisaged in the present paper to evaluate the advantages and limitations of these materials along with the application possibilities for ocean structures, in the light of the laboratory and field applications so far achieved and reported along with the author's test results.

POLYMER CONCRETE COMPOSITES

Polymer concrete composites are basically subdivided into three different categories, polymer impregnated concretes - with hardened concrete fully or partially impregnated with polymer, polymer cement concretes - with polymeric admixtures to green concrete in the form of latexes or plasticizers and polymer concretes - with the cement in concrete replaced by a polymeric binder.

Polymer Impregnated Concretes (PIC)

The author has experimented with both full impregnation by means of vacuum-pressure technique as well as partial impregnation by soaking. After 28 days of wet curing the concrete specimens have been dried in a hot air oven at a temperature of 150°C to constant weight. In vacuum-pressure technique the specimens were vacuumed at 30" of mercury for a period of 1 hour in a special chamber fabricated. With the vacuum port closed the monomer is now allowed to flow into the chamber. A pressure of 100 psi is applied now for a period of 6 hours. In soak impregnation the dried specimens are kept immersed in the monomer with at least 1" of the monomer standing over the specimens. All the specimens are cured by thermo-catalytic action in a constant temperature water bath at 55°C. The catalyst used was Benzoyl peroxide. All specimen ranging from 5 cm dia x 10 cm long cylinders to 15 cm x 15 cm x 70 cm beams could be fully impregnated by the vacuum pressure technique adopted.

Impregnations were attempted with polymer and copolymer combinations of Methyl Methacrylate (MMA), Butyl Acrylate (BA), Styrene (Stn) and Acrylo Nitrite (ACN). Table 1 gives some of the typical results obtained from the laboratory experiments. Based on the properties of these materials a theoretical analysis of the behaviour of PIC beams and columns has been presented earlier (1). Results obtained from beam specimens in the laboratory are being analysed and compared with the earlier predictions. An evaluation of the results on polymer impregnated ferrocement specimens has also been reported earlier (2).

Polymer impregnated specimens have shown 3-5 fold increase in compressive strength and corresponding increase in tensile strength. Table 2 gives the results of these materials regarding durability which shows that they can give several fold increase in this direction. Soak impregnated specimens showed increased strength depending upon the depth of impregnation while the other aspects were almost similar to fully impregnated counterparts.

Polymer Cement Concretes (PCC)

Polymer cement concretes are generally prepared by using DOW 460 latex. This has generally resulted in an improvement in the strength of the order of 20-25% apart from good workability. However, trials with laboratory prepared copolymer latexes of MMA/BA and commercially available SBR latex and Natural Rubber Latex have met with little success. Polymeric superplasticizers - Melment from West Germany, Irgament Mighty from Japan and locally available Lomar - have given comparable improvements in strengths and durability (acid attack and sulphate resistance) apart from much better workabilities. A more detailed account of the results of superplasticized concretes has been presented earlier (3,4).

A lot of study has gone into the various aspects of these materials by the author. Particularly the increased workability has been found to be most useful in locations of reinforcement congestion. A part from marginal strength increases these materials have been found to have substantial improvements in durability aspects. Table 2 presents some of the typical results.

Polymer Concrete (PC)

Polymer concretes have been prepared in the laboratory both by using viscous polyester-styrene with MEK peroxide, Dimethyl Aniline and Cobalt Naphthanate as catalysts and promoters as well as normal polymers mentioned earlier. The viscous polymers in general are mixed with dry aggregates and compacted in moulds while the others are allowed to permeate into prepacked columns often aided by vacuum. Later similar methods are utilized to make polymer (polyester-styrene) mortars useful for repair work. These have been used in various problems in field by the author.

These materials have both very good strength and durability. They have also exhibited good bond with normal concrete as well as steel making them very useful for repair work. They can be cured very fast and are ready to function immediately. Table 2 shows that they can be as strong as PICs with even better durability.

Discussion on Polymer Concrete Composites

Comprehensive information on the methodology and production of polymer concrete composites has been presented earlier. Table 2 gives an overview of the properties of these materials as observed in the laboratory. A few of these techniques have already been used for structural applications as well as repairs in severe environments. The observations so far show that they have very good potential where either strength or durability - even under the most aggressive environments - is the primary consideration.

The laboratory tests indicate that very high strengths of the order of 1500 kg/cm^2 can be obtained through impregnation which has been successfully utilized in field (5). Polymeric binder replacing cement can also produce similar strengths. Both these two types have been found to be several fold better than even a high strength concrete(550 kg/cm^2). PCC though with marginal increase in strength has shown better durability against deterioration.

APPLICATIONS IN OCEAN STRUCTURES

During the last few years a lot of interest is evinced in the application polymer concrete composites for structures in severe environmental conditions (5,6). An understanding of the problems faced in concrete marine structures and possible remedial measures can be had from references 7-11. However, in this brief report it may not be possible to even comprehensively discuss all these points.

In a general way it is enough to point out that from laboratory and field studies done so far there is enough knowledge available for a successful utilisation of polymer concrete composites to ocean structures. In particular with minimum modifications field applications of PIC can be done to splash and inter-tidal zone of concrete ocean structures. It may be easier to adopt this at the construction stage rather than in service. Another advantage of PIC - its strength - can be usefully employed in the heavily stressed end blocks of prestressed constructions in ocean structures apart from corrosion protection. As reported in (9) superplasticizers or latexes will be useful in the total construction particularly with congested reinforcements being obvious in large ocean structures. Polymer concretes will be useful materials for repair work and cover coatings of portions exposed to severe corrosion as well as abration.

From the above one can easily say that the alternatives offered by polymer concrete composites being innumerable one has to look only to find a suitable solution for any given problem even in the case of the most severe environments of the oceans. Full impregnation for strength and durability or partial impregnation for corrosion protection alone being possible cost and construction convenience may play an important role on the choice of the best suitable solution.

CONCLUSIONS

1. Polymer concrete composites can give very high strengths of around 1500 kg/cm^2 in PIC and PCs while PCC can give marginal improvements.

2. The durability of all polymer concrete composites is found to show a several fold increase over conventional concrete.

3. Overall cost and production convenience will show that PICs are best suitable for locations requiring both high strength and corrosion resistance (like splash zone of gravity platforms), while PCCs are useful in total structure.

4. PCs will be a great help in fast repairs apart from use in prestressing anchor block etc.

REFERENCES

1. Subrahmanyam, B.V., Ganesh Babu, K. et al., Strength and Plastic Ductility of Polymer Impregnated Concrete Beams and Columns, Second Int. Cong. on Polymers in Concrete, Texas (1978), 229-244.

2. Ganesh Babu, K., Behaviour of Polymer Impregnated Ferrocement, to be presented at Second Int. Sym. on Ferrocement (1985).

3. Ganesh Babu, K., Subrahmanyam, B.V., et al., Strength and Behaviour of Superplasticized Concretes, Indian Concrete Journal,(1982),159-163.

4. Ganesh Babu, K., Subrahmanyam, B.V., Fibre-Reinforced Concretes with Polymeric Superplasticizers, Bridge and Structural Engineer (1981).

5. Mehta, H.C., Manson, J.A., et al. Polymer-Impregnated Concrete: Field Studies, ASCE Transportation Engg. Jl. (1975), 1-27.

6. Dujeiym J.T., Steinberg, M., et al, Concrete-Polymer Materials, Third topical Report (1971).

7. Seki, H., Deterioration of Concrete of Coastal Structures in Japan, Durability of Concrete, ACI Special Publication 47.

8. Browne, R.D.and Geoghegan, M.P., The Corrosion of Concrete Marine Structures, presented at Sym. on Corrosion of Steel Reinforcement in Concrete Construction, London (1978).

9. Anon, Precast Units for North Sea Oil Platform, Precast Concrete Jl. (1977).

10. Gerwick, B., Concrete Sea Structures, Prestressed Concrete Institute (1976).

11. Gerwick, B., Applications of polymers to concrete sea structures, Second Int. Cong. on Polymers in Concrete, Texas (1978), 37-43.

Table 1 : Results of Laboratory Experiments on PIC

Specimen Name	Initial Concrete Strength			Monomer Combination (%)				Impregnated concrete properties		
	Design (Comp)	Actual (Comp)	Tensile (Split)	MMA	BA	Stn.	ACN	Compressive	Split	% loading
MBS3	250	255	35	100	-	-	-	1100	114	7.0
MBS4	500	560	55	100	-	-	-	1600	157	5.5
MBS5	"	"	"	70	30	-	-	1200	152	5.5
MBS6	"	"	"	50	50	-	-	1000	110	6.2
MBS7(F)*	"	"	"	100	-	-	-	1400	180	5.0
SAS8	"	"	"			70	30	1350	-	5.8
SBS9	"	"	"		10	90		1000	96	5.2

*Fibre reinforced concrete (fibre - 1% by volume)

Table 2 : Properties of Polymer Concrete Composites

Property	Conventional concrete	PIC				PCC		PC	
		MMA/BA			Stn-ACN	DOW	Super-plasticizer	Poly-ester Stn.	MMA
		100/0	70/30	50/50	70/30	460			
1. Compressive Strength (kg/cm^2)	550	1600	1200	1000	1300	650	620	1000	700
2. Tensile strength (kg/cm^2)	50	160	150	110	100	70	60	100	-
3. Modulus of Elasticity ($x 10^5$ kg/cm^2)	2.70	4.36	4.34	2.71	-	-	3.46	3.5	-
4. Water Absorption (% wt)	6.00	0.40	0.40	0.50	0.6	4.5	4.80	2.0	2.5
5. Acid Attack* (% wt. loss_	35.0	1.0	1.20	1.30	2.0	7.0	7.50	1.0	1.0
6. " " (% strength loss)	43.0	Nil	Nil	Nil	1.0	5.6	6.80	1.1	1.3
7. Sulphate Attack**(% strength loss)	28.0	Nil	Nil	Nil	1.5	7.2	10.50	Nil	Nil
8. Abrasion Resistance (% improvement)	-	15	15	15	8	4	3	10	8

*Results after 30 days in 15% sulphuric acid
**Results after 30 days of 15% sodium sulphate cyclic exposure test

Research on Resistance of Concrete to Seawater

O. Kato[1] and N. Ohtsuki[2]

[1] R&D Laboratory, The Cement Association of Japan, Toshima, Kita-ku, Tokyo, 114 Japan
[2] Port and Harbour Research Institute, Ministry of Transport, Nagase, Yokosuka, 239 Japan

ABSTRACT

The characteristics of concrete specimens immersed in seawater for 10 years were investigated. In general, concrete immersed in seawater expanded more than in case of standard curing (20°C in fresh water), and it was estimated that expansion would progress even after 10 years. The compressive strength of Portland cement concrete immersed in seawater was lower at 10 years than 3 years, but did not decline very much with blended cement concrete. However, all seawater-immersed specimens were of lower compressive strengths than specimens with standard curing. Surface deterioration of specimens in surface appearance and compressive strength were unrelated. According to the results of chemical analyses of the concretes, the surface portions and the interiors were of different compositions with MgO, SO_3, and Na_2O plentiful at the surface portions with CaO conversely small in quantity. As for Cl, it had penetrated to the centers of the specimens.

1. Introduction

The Cement Association of Japan and the Port and Harbour Research Institute commenced joint tests from October 1969 with the purpose of investigating the resistance to seawater of concretes using Portland and blended cements. With experiments on unreinforced concretes as Part 1 and on reinforced concretes as Part 2, these tests are still going on at present with 20-year age as the target. This report is a compilation of test results up to the age of 10 years for unreinforced concrete specimens immersed in seawater.

2. Materials Used

(1) Cements

The six kinds of cement were used in Table 1.

(2) Aggregates

River gravel and river sand were used in Table 2.

Table 1. Cements

Cement	Blaine's specific surface (cm^2/g)	C_3A (%)	Free CaO (%)
Ordinary Portland cement (OPC)	3,180	9	0.9
High-early strength Portland cement (HPC)	4,290	9	0.9
Moderate heat Portland cement (MPC)	3,260	5	0.5
Portland blast-furnace slag cement, class B (BSC)	3,850	-	0.3
Portland pozzolana cement, class A (PPC)	3,640	-	0.8
Portland fly-ash cement, class B (FAC)	3,740	-	0.6

Table 2. Aggregates

Aggregate	Specific gravity	Absorption (%)	Fineness modulus
Coarse agg.	2.66	1.00	7.38
Fine agg.	2.60	1.00	2.81

Table 3. Mix proportions of concretes

Series	Identification of specimen	Unit weight (kg/m^3) Water	Cement	Fine agg.	Coarst* agg.	AE-agent (ml/m^3)	Air content (%)	Slump (cm)	W/C
A	OPC-A	152	304	715	1,242			7	0.50
	HPC-A	156	312	707	1,228			7.5	
	MPC-A	153	306	713	1,241			8	
	BSC-A	154	308	706	1,230			8.5	
	PPC-A	154	308	707	1,234			7	
	FAC-A	148	296	712	1,245			8.5	
B	OPC-B	152	234	775	1,240			7	0.65
	HPC-B	158	243	764	1,224			8.5	
	MPC-B	154	237	772	1,236			8	
	BSC-B	154	237	768	1,229			8.5	
	PPC-B	154	237	770	1,232			7	
	FAC-B	148	228	774	1,242			8	
C	OPC-C	145	223	736	1,229	60	4.3	7.5	0.65
	HPC-C	147	226	712	1,242	53	4.6	7.5	
	MPC-C	143	220	739	1,233	22	4.2	8	
	BSC-C	141	217	738	1,232	43	4.1	8	
	PPC-C	147	226	731	1,221	70	4.1	8	
	FAC-C	138	212	741	1,237	60	4.3	9	

* Maximum size of coarse aggregate: 40 mm

3. Preparation of Concrete Specimens

The mix proportions of concretes were in three series, A, B, and C, as shown in Table 3. The specimens prepared were ϕ15x30 cm cylinders for compressive strength tests and 10x10x40 cm prisms for length change measurements. The location of seawater immersion was in two levels with specimens set at the upper level submerged in and exposed from seawater twice daily. Specimens at the lower level were submerged in water at all times. The annual temperature of seawater was a maximum of approximately 25°C and a minimum of approximately 8°C.

4. Test Results and Considerations

(1) Appearances of Specimens

From surface appearances, standard-cured specimens showed no abnormalities even after 10 years had elapsed.

With specimens immersed in seawater, after 3 years had elapsed, part of the blended cement concretes had brown surface coloring, while there specimens with textures seen to have become coarse, and especially, the B and C series were conspicuous. From 5 to 7 years, surface coloring and roughening became recognizable for concretes of the A series also. With the B and C series, there was much spalling of cement paste, especially in the cases of BSC and FAC. At 10 years, discoloring and surface roughening was seen with concretes of all of the cements, with surface deterioration particularly advanced for BSC and FAC.[1)]

(2) Chemical Analysis and X-ray Diffraction Tests of Concretes

Fragments were sampled from the surface, center and intermediate portions of ϕ15x30 cm specimens, pulverized, and made into samples of fine powder passing a 105 μm sieve. The results of chemical analysis of these samples are given in Table 4, and the results of X-ray diffraction tests in Table 5. The chemical compositions at the surface and center portions of concrete differed whichever cement was used, and especially, there were differences in the quantities of CaO, MgO, SO_3 and Na_2O. There was less CaO at the surfaces than at the centers, and this trend was especially prominent in case of BSC. According to X-ray diffraction, there were many cases of $Ca(OH)_2$ of the surface portions being small in quantity and calcite being large. MgO, SO_3, and Na_2O were concentrated at the surfaces, and ettringite was detected in X-ray diffraction. In spite of the fact that large quantities of MgO existed, only a slight amount of brucite was recognized. It may have been that large crystals of brucite precipitated at the specimen surfaces were scrubbed off when the specimens were washed. Except for BSC, Cl had penetrated completely into the center of all specimens.[2), 3), 4)]

(3) Compressive Strength

The results of compressive strength tests with ϕ15x30 cm specimens, and the ratios of strengths with immersion in seawater to strength by standard curing are shown in Table 6. In case of immersion in seawater, compressive strengths of Portland cement concretes peaked at 3-year age and were all reduced at 10 years. The compressive strengths of blended cement concretes, in case of the A series with W/C ratio of 0.50, were mostly lowered at 10-year age similarly to Portland cement concretes, but with the B and C series of W/C ratio 0.65, most were not reduced. On looking at the strength ratios of seawater immersion to standard curing, the ratios for Portland cement are small, whereas for blended cement concretes they are not very small. In effect, the reductions in strengths due to seawater immersion for 10 years were larger for Portland cement concretes than for blended cement concretes.

(4) Length Change

The length change rates measured on 10x10x40 cm prism specimens (scale length 250 mm) are shown in Table 7. The length changes of specimens upon performance of standard curing were small, and although all specimens had expanded up to 10-year age they were in a range of 41 to 156×10^{-6}. On the other hand, specimens immersed in seawater had expanded several times more than standard-cured specimens. Those submerged in seawater at all times tended to be slightly more expanded than those in the tidal zone. The difference between W/C ratios of 0.50 and 0.65 was not distinct with Portland cement concretes, but with blended cement concretes expansion was larger for W/C ratio of 0.50 than for 0.65.

5. Conclusions

Tests were carried out on unreinforced concrete specimens immersed in seawater for 10 years and the following conclusions were reached;

(1) The surface portions of specimens had become of chemical compositions differing from center and CaO was of small quantity while MgO, SO_3, and Na_2O were concentrated. Ettringite was detected at surface, but brucite was recognized only in slight amounts.

(2) Compressive strengths of specimens immersed in seawater were high at 3-year age, but at 10 years, they had declined to about the same levels as for 1 year in many cases. On comparison with standard curing, the cases of W/C ratio 0.65 of blended cement concretes showed no difference, but the cases of blended cement concretes at W/C ratio of 0.50 and of Portland cement concretes were all lower for seawater immersion. There was no difference between the cases of seawater immersion at all times and of specimens set at the tidal zone.

Further, a relationship was not seen between deterioration of specimen surface and compressive strength at 10-year age.

(3) Standard-cured specimens did not show large length changes even when 10 years had elapsed, but specimens immersed in seawater showed large expansions both in case of immersion at all times and in case of immersion at the tidal zone, and a trend was seen for expansion to continue even after 10 years.

Acknowledgements

The authors sincerely thank the many members of the Committee on Cement Technology of the Cement Association of Japan who kindly cooperated in carrying out the study.

References

1) Smolczyk, H.G.; Some observations and new aspects concerning seawater action on concrete in the tidal zone, Bul. RILEM No.32 (1966) 299-304.

2) Gjorv, O.E.; Long-time durability of concrete in seawater, J. ACI (Jan. 1971) 60-67.

3) Regourd, M.; Physico-chemical study of cement pastes, mortars, and concretes exposed to seawater, ACI SP-65-4 (1980) 63-82.

4) Gjorv, O.E., Vennesland, ϕ.; Diffusion of chloride ions from seawater into concrete, Cem. Con. Res. Vol. 9 (1979) 229-238.

Table 4. Chemical Analysis of Cements and Concrete Specimens

(%)

Sample	Curing	Sampled position	55°C Loss on igni-tion	Acid soluble component								Cl	CO_2*
				SiO_2	Al_2O_3	Fe_2O_3	CaO	MgO	SO_3	Na_2O	K_2O		
OPC-cement			-	21.9	5.5	3.1	65.2	1.3	2.0	0.30	0.57	-	-
OPC-A	Std. curing	center	32.4	27.5	8.3	4.3	55.5	1.8	1.7	0.3	0.7	0.1	8.2
OPC-A	In seawater	surface	40.7	26.2	6.9	3.7	56.0	3.4	3.2	0.3	0.4	6.0	4.8
		1 cm from surface	31.4	26.2	8.1	5.0	55.4	2.3	1.9	0.3	0.9	5.9	6.2
		3 cm	32.6	26.0	7.3	4.2	58.1	1.7	1.7	0.3	0.7	4.4	5.6
		5 cm	32.6	26.8	8.3	4.8	55.4	2.1	1.7	0.3	0.7	4.2	6.2
		7 cm	30.9	25.2	6.8	4.0	59.6	1.6	1.9	0.4	0.5	4.2	9.8
OPC-B		surface	39.5	27.7	7.5	4.5	51.6	3.9	4.0	0.5	0.3	4.1	17.6
		center	35.3	26.7	8.5	4.3	55.6	1.6	1.4	0.8	1.1	4.5	10.5
OPC-C		surface	38.5	25.5	7.9	4.6	55.2	3.3	2.6	0.5	0.4	7.1	10.0
		center	33.7	26.1	7.5	4.4	57.9	1.8	1.5	0.4	0.4	5.2	9.8
HPC-cement			-	20.9	5.3	2.7	66.4	1.2	2.5	0.38	0.55	-	-
HPC-A	Std. curing	center	31.0	25.3	6.8	3.9	59.6	1.6	2.3	0.3	0.3	0.1	9.2
	In seawater	surface	32.0	24.8	7.2	3.8	58.8	2.0	3.0	0.3	0.3	7.3	7.2
		center	32.7	24.6	6.9	4.0	59.6	1.6	2.5	0.4	0.4	3.5	10.7
HPC-B		surface	44.2	27.8	7.3	4.2	53.9	2.9	2.9	0.5	0.6	6.0	2.9
		center	36.5	27.3	7.5	4.2	56.5	1.6	2.1	0.3	0.5	5.0	6.6
MPC-cement			-	23.8	4.6	3.3	64.1	1.2	1.7	0.28	0.49	-	-
MPC-A	Std. curing	center	36.1	27.3	5.9	4.2	58.2	1.7	2.0	0.5	0.3	0.4	6.3
	In seawater	surface	43.6	27.7	5.7	4.3	55.4	3.2	3.2	0.4	0.2	5.4	3.8
		center	38.5	27.5	5.3	4.0	59.2	1.5	1.9	0.3	0.2	3.3	3.7
MPC-B		surface	26.0	27.2	7.2	5.6	52.0	3.9	2.8	0.6	0.6	5.0	12.6
		center	28.2	28.3	7.2	5.3	55.1	1.6	1.6	0.5	0.5	4.4	8.2
BSC-cement			-	26.2	9.9	2.1	55.1	3.3	2.4	0.57	0.52	-	-
BSC-A	Std. curing	center	41.5	27.9	9.5	2.8	52.6	3.5	2.8	0.5	0.4	0.6	6.0
	In seawater	surface	50.8	25.6	11.2	3.0	35.8	17.8	5.5	1.0	0.3	2.0	7.2
		1 cm from surface	42.8	27.9	9.4	2.7	51.7	4.0	3.6	0.5	0.2	5.8	1.6
		3 cm	42.8	29.2	7.3	2.7	53.5	3.6	3.0	0.4	0.3	4.7	2.2
		5 cm	41.2	28.3	8.4	3.1	52.5	3.8	3.1	0.3	0.3	1.4	6.5
		7 cm	41.2	28.0	9.3	2.7	52.9	3.7	2.9	0.3	0.2	0.5	6.5
BSC-B		surface	46.3	23.8	9.6	2.4	33.6	24.0	5.4	1.1	0.2	2.5	2.5
		center	47.8	29.0	9.8	2.9	50.3	4.3	2.9	0.5	0.3	1.4	2.2
PPC-cement			-	22.1	5.6	2.9	64.7	1.3	2.2	0.57	0.70	-	-
PPC-A	Std. curing	center	30.8	27.2	7.2	3.9	57.4	1.6	1.9	0.5	0.4	0.2	7.6
	In seawater	surface	43.4	28.3	7.5	3.8	53.9	2.4	3.1	0.5	0.4	5.1	8.2
		center	32.3	26.2	7.2	4.0	58.0	1.7	1.9	0.5	0.5	3.4	7.6
PPC-B		surface	31.7	25.5	8.0	5.3	51.2	4.6	3.8	0.9	0.7	4.5	11.2
		center	29.1	28.4	8.4	4.6	53.7	2.1	1.3	0.8	0.8	4.3	6.2
FAC-cement			-	22.0	6.6	3.1	63.2	1.5	2.2	0.77	0.69	-	-
FAC-A	Std. curing	center	32.1	28.3	9.2	4.5	52.8	2.2	1.9	0.4	0.6	1.7	8.2
		surface	33.1	28.5	8.9	4.8	44.4	7.4	4.4	1.0	0.5	2.7	13.4
		center	36.0	28.3	8.9	5.0	52.4	2.0	2.1	0.6	0.8	0.8	11.8
FAC-B		surface	37.0	26.4	10.5	5.2	28.6	24.2	2.9	1.6	0.6	1.7	19.4
		center	30.3	30.0	11.0	5.8	47.3	2.5	2.0	0.5	0.9	2.1	9.5

* CO_2 calculated by following formula

CO_2 (%) = 1,000°C Loss on ignition (%) - [500°C Loss on ignition (%) + Cl (%)]

Table 5. X-ray Diffraction of Concrete Samples

Sample	Curing	Sampled position	Relative intensity of XRD*			
			Ettringite	Friedel's salt	$Ca(OH)_2$	Calcite
OPC-A	Std. curing	center			51	2
	Seawater	surface	7	23		28
		1 cm from surface				6
		3 cm	3	18	49	13
		5 cm		5	15	9
		center	4	25	55	
OPC-B		surface	6			33
		center				
OPC-C		surface	3	38		
		center		56		
HPC-A	Std. curing	center			93	
	Seawater	surface	5	33		
		center	1		37	
HPC-B		surface	8	29		21
		center	7	40	15	15
MPC-A	Std. curing	center			53	
	Seawater	surface	45	111	17	74
		center		62	87	22
MPC-B		surface		15	7	13
		center		18		
BSC-A	Std. curing	center			49	19
	Seawater	surface	34	24	16	145
		1 cm from surface	31	93		24
		3 cm		39	12	17
		5 cm			27	
		center			80	18
BSC-B		surface	36	44		57
		center			19	13
PPC-A	Std. curing	center			73	
	Seawater	surface	6	24		
		center	8			
PPC-B		surface		10		18
		center		16		
FAC-A	Std. curing	center				
	Seawater	surface		8		17
		center			16	
FAC-B		surface	3			
		center				

* α-Quartz used as reference material.

Table 6. Compressive Strengths of Concrete Specimens

Concrete specimen	Compressive strength (kgf/cm^2)									Ratio of strength					
	Standard curing (a)			Immersing in seawater											
				Tidal zone (b)			Immersion zone (c)			(b)/(a)			(c)/(a)		
	1Y*	3Y	10Y	1Y	3Y	10Y	1Y	3Y	10Y	1Y	3Y	10Y	1Y	3Y	10Y
OPC-A	473	501	476	421	512	415	435	432	411	0.89	1.02	0.87	0.96	0.86	0.86
HPC-A	462	483	541	457	490	484	456	482	459	0.99	1.01	0.89	0.99	1.00	0.85
MPC-A	534	491	532	420	462	424	429	456	397	0.79	0.94	0.80	0.80	0.93	0.75
BSC-A	476	470	501	355	432	406	382	431	401	0.74	0.92	0.81	0.80	0.92	0.80
PPC-A	424	474	447	414	440	381	407	433	369	0.98	0.93	0.85	0.96	0.91	0.83
FAC-A	456	499	501	409	475	434	393	459	407	0.89	0.95	0.87	0.86	0.92	0.81
OPC-B	323	306	321	314	314	265	305	303	269	0.97	1.03	0.83	0.94	0.99	0.84
HPC-B	299	327	335	295	305	275	288	290	279	0.99	0.93	0.82	0.96	0.89	0.83
MPC-B	328	333	346	279	301	238	271	271	256	0.85	0.90	0.69	0.83	0.81	0.74
BSC-B	363	374	374	280	342	359	313	363	370	0.77	0.91	0.96	0.86	0.97	0.99
PPC-B	295	334	322	281	309	264	271	277	268	0.95	0.93	0.82	0.92	0.83	0.83
FAC-B	318	326	359	311	337	394	312	322	364	0.98	1.03	1.10	0.98	0.99	1.01
OPC-C	265	275	266	251	248	228	244	241	224	0.95	0.90	0.86	0.92	0.88	0.84
HPC-C	281	315	326	279	280	254	258	250	244	0.99	0.89	0.78	0.92	0.78	0.75
MPC-C	324	332	335	214	282	255	222	277	246	0.66	0.85	0.76	0.69	0.77	0.73
BSC-C	334	324	337	264	309	320	272	312	325	0.79	0.95	0.95	0.81	0.96	0.96
PPC-C	243	270	263	239	245	245	230	217	236	0.98	0.91	0.93	0.95	0.80	0.90
FAC-C	277	318	333	293	314	346	289	293	318	1.06	0.99	1.04	1.04	0.92	0.95

* Y: age (years)

Table 7. Length Changes of Concrete Specimens

($x10^{-6}$)

Concrete specimen	Standard curing (a)							Immersion in seawater													
								Tidal zone (b)							Fully immersed zone (c)						
	1Y*	2Y	3Y	4Y	5Y	7Y	10Y	1Y	2Y	3Y	4Y	5Y	7Y	10Y	1Y	2Y	3Y	4Y	5Y	7Y	10Y
OPC-A	39	48	52	40	44	68	44	65	76	120	120	132	196	311	79	92	120	116	192	180	311
HPC-A	33	65	85	87	91	110	99	83	84	140	152	152	188	269	92	112	152	128	184	144	256
MPC-A	100	84	114	124	74	56	117	55	40	88	116	88	160	225	132	172	192	244	284	348	420
BSC-A	104	88	105	121	92	79	119	51	80	124	140	140	248	432	76	138	196	250	306	416	499
PPC-A	27	40	44	36	39	68	45	81	92	112	104	152	220	339	100	150	180	212	284	376	475
FAC-A	61	115	132	129	135	164	156	87	112	132	136	156	212	264	123	140	186	186	250	304	346
OPC-B	59	76	72	60	69	92	76	103	132	208	204	212	296	397	103	160	232	220	280	296	384
HPC-B	35	83	88	99	94	114	106	104	108	180	196	188	224	311	100	132	240	232	292	284	376
MPC-B	84	72	109	98	60	37	97	84	84	132	161	156	232	284	120	112	152	196	232	256	301
BSC-B	68	64	54	88	57	47	85	35	48	80	116	112	208	293	60	62	120	156	200	280	332
PPC-B	12	28	32	20	21	64	47	32	36	76	72	92	172	243	57	104	152	160	220	264	300
FAC-B	31	73	89	84	83	102	95	71	68	108	76	108	160	188	83	116	140	124	148	188	188
OPC-C	13	40	32	28	24	56	41	28	68	140	144	184	252	340	39	88	136	144	212	196	305
HPC-C	53	77	93	85	93	113	114	85	112	192	208	204	168	324	79	124	176	176	212	204	267
MPC-C	48	28	60	62	28	-11	47	60	76	108	148	156	228	299	75	116	140	180	216	248	347
BSC-C	100	84	62	120	90	68	119	119	132	172	192	184	236	285	103	140	176	192	180	192	219
PPC-C	19	28	28	16	29	52	43	47	80	136	140	184	260	336	60	84	136	128	172	196	228
FAC-C	48	113	121	121	121	148	137	81	88	128	108	120	152	163	92	144	176	128	164	188	189

* Y: age (years)

Specific Behavior of Structural Concrete in the Seas Pore Pressure Effects

A. Ilantzis and L.C. Zaleski-Zamenhof

ARBEM–the French Association for Research on Concrete in the Seas, 75737 Paris Cedex 15, France

ABSTRACT

The authors propose the notion of pore pressure, as generated inside concrete, while immersed and submitted to hydrostatic loads, in a similar (but slightly different) way, to the one applied in soil mechanics. The idea of total and effective stresses is subsequently developed. The analysis reported is based on an extensive research program, run for several years for ARBEM in CEBTP and CNEXO laboratories. Pore pressure effects appear to generate an early swell and a further absence of creep in immersed and water saturated concrete, thus allowing to consider an appropriate increase of E-moduli. An impact of this statement on elastic stability of concrete floaters, often governing the design concept of offshore structures, is then pointed out.

INTRODUCTION

The specific behavior of concrete, as used to build marine and off-shore structures, is due to permanent contact of this material with the environmental seawater and to significant hydrostatic pressure, that the concrete is subsequently submitted to. As concrete is a porous material, albeit of relatively low permeability, water is penetrating inside of pores, thus generating subsequent pore pressures. These pore pressures are assumed to modify somewhat the mechanical behavior of concrete, when saturated. A theory, as commonly applied in soil mechanics, may thus serve, after appropriate adjustments, to evaluate pore pressure effects also in concrete. The adjustments have to account for actual elastic characteristics of the solid skeleton of the investigated material, while soil is a granular medium with no (or low) cohesion between the grains. For quantifying these adjustments, an experimental program on concrete, when subjected to hydrostatic pressures, is carried out for ARBEM in France ; the results of tests, and especially of triaxial creep tests, as have been applied on various concrete specimens, are hereafter presented and discussed.

PORE PRESSURE INVESTIGATIONS IN CONCRETE-STATE OF THE ART

To date, just some limited experiments have been performed to investigate the mechanical behavior of solid porous materials, including concrete, when being saturated. BIOT (1) has developed a general theory dealing with porous media mechanics. The application of this theory appears nevertheless difficult and involves in any case sophisticated experimental verification. BISHOP (2,3) investigated pore pressures inside rocks in undrained conditions. He showed the limitations of effective stress concept application, as used in soil mechanics. MESRI, ADRACH and ULRICH (4) have published some tests on sandstone, marble and limestone with respect to coefficients as defined by BISHOP. CLAYTON (5) has carried out rupture tests on concrete cylinders in saturated conditions. In a publication of a more general nature, MORLAY (6) presents a review of various methods and theories as applied in the field of pore pressure in porous media. A number of tension tests on concrete specimens under various experimental conditions have been performed by BUTLER (7). He concluded, TERZAGHI's expression of effective stress remained still valid, albeit a more precise formulation such as the one given by SKEMPTON (8) could be better applied for concrete. Finally scarce experimental information are only available with respect to hydrostatic pressure effects on mechanical concrete properties. To compensate the subsequent lack of knowledge in this domain, the research programme, as reported and discussed here after, has been undertaken and is still under way.

THE ARBEM RESEARCH PROGRAM

ARBEM is running for several years, on behalf of several French research institutes, oil companies, as well as contractors involved in offshore construction and design, an extensive research program, to investigate concrete behavior, when submitted to significant hydrostatic loads, as pore pressure generating phenomena. These investigations are oriented to :

- an appropriate distinction between total and effective stresses in immersed concrete ;
- hydrostatic pressure effects on concrete compressibility, compressive and tensile strength (in terms of total stresses) ;
- behavior of immersed concrete, when submitted to multiaxial state of stresses ;
- significant parameters affecting pore pressure generation, as concrete permeability, hydrostatic pressure level and duration ;
- methods of evaluating the permeability of concrete, versus its porosity, nature of ingredients, water/cement ratio etc. ;
- evaluation of impregnation level versus concrete permeability as well as versus hydrostatic pressure parameters ;
- evaluation of concrete bulk modulus variation versus impregnation level, versus hydrostatic pressure/compressive strength ratio, multiaxial stress parameters, concrete age and versus hydrostatic pressure duration (creep effects).

The present report is mainly dedicated to this last item, as pore pressure appears to affect creep in an extremely significant way with respect to concrete behavior in offshore structures and, subsequently to the design considerations related to such structures.

Creep tests have been carried out on cylindrical concrete specimens (Ø = 16 cm, H = 32 cm), submitted to a triaxial state of stresses (fig.1). All specimens have been stored in water during six weeks, previously to the application of any load. Thus the test loading has been applied on specimens already impregnated under atmospheric pressure conditions. In parallel with the "free" (able to be impregnated) specimens, some calibration specimens have been tested, after having been "jacketed", to avoid any possible impregnation of concrete. Tests have been carried out under triaxial uniform stresses as well as under axisymmetric ones, the axial stress, generated mechanically, being in that case higher than the lateral stresses, generated by means of hydrostatic pressure applied around the specimen. The applied stress values have been :

σ	= 5 and 10 MPa	for uniform stress tests
σ_l (lateral)	= 5 MPa	for axisymmetric tests
σ_a (axial)	= 10, 15 and 20 MPa	

Three types of concrete have been tested. The respective specimens presented the following properties :

Characteristics		Concrete A	Concrete B	Concrete C
permeability	cm/sec	10^{-14}	10^{-12}	10^{-13}
compression strength	MPa	50,0	36,0	45,0
tension strength	"	4,2	4,8	4,0
modulus of elasticity	"	38000	31000	33000
Poisson's ratio		0,15	0,17	0,15

Fig. 1 : Creep tests run for ARBEM in the CEBTP laboratories at St-Rémy-lès-Chevreuse (France)

The results of tests (9) are summarized on the figures hereafter. On the Fig. 2 the results of creep tests under uniform triaxial state of stress are presented, while the Fig. 3 gives some results due to axisymmetrical stress application.

DISCUSSION OF RESULTS

The curves, as presented on Fig. 2, are illustrating the behavior of "free" (impregnated) specimens, quite different from the conventional behavior of the "jacketed" ones, not submitted to any water penetration. Albeit creep of jacketed specimens under uniform triaxial state of stress is still lower than if submitted to an uniaxial stress, almost no creep appears on the "free" cylinders. Furthermore, for concretes B and C, presenting more usual permeability values, as well as even for concrete A, of exceptionally low permeability, when this latter has been submitted to higher hydrostatic pressure, strain is even decreasing with time (i.e. the volume of the specimens is increasing). Concrete A, under lower hydrostatic load (thus generating lower pore-pressure effects) has shown some creep, but still of an order of 5 to 7 times less than measured on the calibration specimen in absence of pore pressure at all. Results of the triaxial creep tests involving $\sigma_a > \sigma_l$ (fig. 3) show still an attenuation of creep with increase of hydrostatic pressure ratio, and of permeability of concrete. A swell of more permeable specimens, quite spectacular on the Fig. 2, is also still appearing on the Fig. 3, with respect to the concrete B, for σ_l / σ_a ratio higher than 0.5.

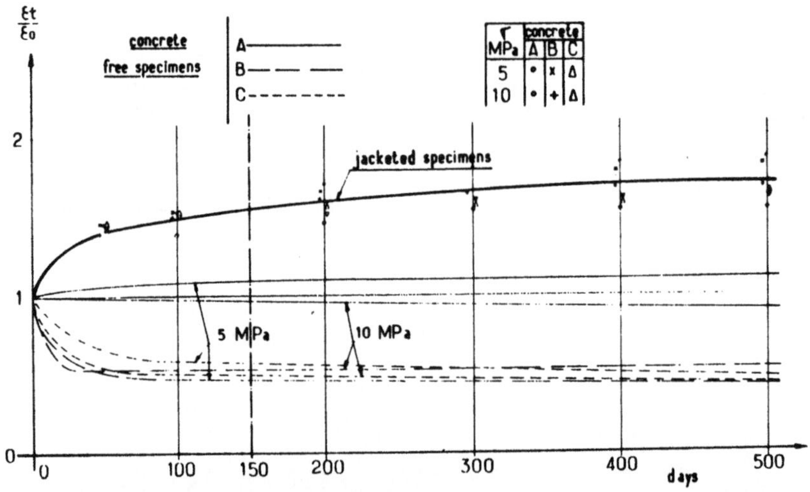

Fig. 2. Strain ratio (creep) tests results under uniform triaxial stress $\sigma = \sigma_1 = \sigma_a$

ε_o - instantaneous concrete strain for the given σ values

ε_t - actual strain at time t

(ε_o and ε_t figures are representing uniform strain, both axial and lateral ones)

Fig. 3.
Strain ratio (creep) tests results at time t = 150 days, under axisymmetric hydrostatic pressure:

σ_1 = 5 MPa

ε_{ao} - instantaneous axial strain for given σ_1/σ_a ratio

ε_{at} - actual axial strain for given σ_1/σ_a

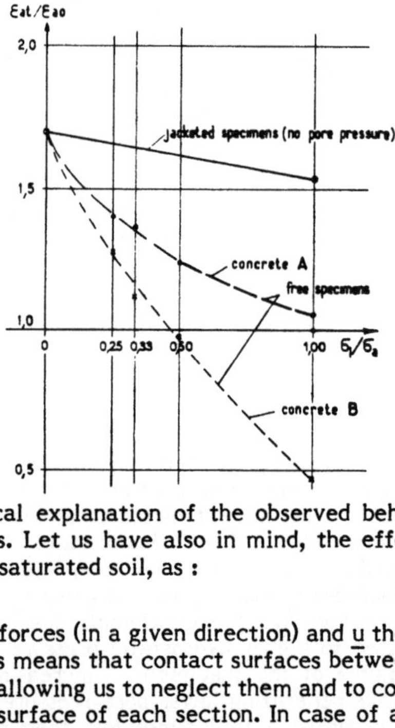

Let us try to present a phenomenological explanation of the observed behavior, involving the notion of effective stresses. Let us have also in mind, the effective stress is defined, in soil mechanics, for a saturated soil, as :

$$\sigma_e = \sigma - u \qquad (1)$$

where σ is the stress due to the applied forces (in a given direction) and u the pore pressure. From physical point of view, this means that contact surfaces between the grains of soil are very small indeed, thus allowing us to neglect them and to consider water pressure, as exerted on the whole surface of each section. In case of a solid porous material of a porosity f, and assuming a pore pressure u, we could a priori expect that the resulting effective stress, in a state of saturation, would be :

$$\sigma_e = \sigma - fu \qquad (2)$$

Various experimental results have actually confirmed such an approach involving, however, an f value higher than the actual porosity of material. This value, having thus to be determined from experimental results, is called "effective porosity". In the formulas hereafter f is consequently representing this effective porosity. When applying such an approach to the case of triaxial uniform state of stresses, the state of equilibrium $\sigma = u$ involves :

$$\sigma_e = (1 - f)\,\sigma \tag{3}$$

the corresponding strain being :

$$\varepsilon = (1 - f)\,\frac{\sigma(1 - 2\nu)}{E} \tag{4}$$

Assuming unchanged E and ν figures, the jacketed specimens ($u = 0$) present a strain :

$$\varepsilon_s = \frac{\sigma(1 - 2\nu)}{E} \tag{5}$$

Subsequently, the strain reduction due to pore pressure effects results in :

$$\frac{\varepsilon}{\varepsilon_s} = 1 - f \tag{6}$$

From creep tests results, such as presented on Fig. 2, ε and ε_s figures may be read, allowing to determine the effective porosity f as well as the effective stress to total stress ratio σ_e/σ. Thus from the Fig. 2, for t = 150 days and σ= 5 MPa, there are, as far as the concretes A and B are concerned :

CONCRETE	$\varepsilon/\varepsilon_o$	$\varepsilon_s/\varepsilon_o$	$\varepsilon/\varepsilon_s=\sigma_e/\sigma$	f
A	1.10	1.55	0.71	0.29
B	0.47	1.55	0.30	0.70

A similar approach, when applied to the cases of non uniform (but axisymmetric) triaxial state of stresses ($\sigma_a \neq \sigma_l$), is resulting in effective stresses formulation :

$$\sigma_{ae} = \sigma_a - f\sigma_l$$
$$\sigma_{le} = (1 - f)\,\sigma_l \tag{7}$$

the axial strain becoming :

$$\varepsilon_a = \frac{1}{E}\,(\sigma_a - (f + 2\nu(1-f))\sigma_l) \tag{8}$$

while for jacketed specimens there is still :

$$\varepsilon_{as} = \frac{1}{E}\,(\sigma_a - 2\nu\sigma_l) \tag{9}$$

Thus :

$$\frac{\varepsilon_a}{\varepsilon_{as}} = \frac{\sigma_a/\sigma_l - (f + 2\nu\,(1 - f))}{\sigma_a/\sigma_l - 2\nu} \tag{10}$$

the corresponding effective stresses ratio being :

$$\frac{\sigma_{ae}}{\sigma_{le}} = \frac{\sigma_a - f\sigma_l}{(1-f)\sigma_l} = \frac{\sigma_a/\sigma_l - f}{1 - f} \tag{11}$$

The strain reduction due to pore pressure effects is hereafter computed from the formula (10) on the example of concretes A and B, at time t = 150 days, for σ_a/σ_l ratio varying from 1 to 4. The effective porosity of these concretes have been assumed unchanged with respect to the values determined here above under uniform triaxial stresses. The computed strain reduction figures are then compared to the ones as read on the experimental curves (see Fig. 3) for assumed σ_{le}/σ_{ae} ratio of effective stresses. The results appear to confirm quite well the suggested analytical methodology (see the table here after) :

Concrete	σ_l/σ_a	σ_a/σ_l	σ_{ae}/σ_a	σ_{le}/σ_l	σ_{le}/σ_{ae}	$\varepsilon_{as}/\varepsilon_{ao}$	$\varepsilon_a/\varepsilon_{ao}$	$\varepsilon_a/\varepsilon_{as}$ measur.	comput.
A	1	1	0.71	0.71	1	1.55	1.10	0.71	0.71
	0.5	2	0.86	0.71	0.41	1.62	1.32	0.81	0.88
f=0.29	0.33	3	0.90	0.71	0.26	1.64	1.43	0.87	0.93
	0.25	4	0.93	0.71	0.19	1.66	1.48	0.89	0.95
B	1	1	0.30	0.30	1	1.55	0.47	0.30	0.30
	0.5	2	0.65	0.30	0.23	1.62	1.28	0.79	0.72
f=0.70	0.33	3	0.77	0.30	0.13	1.64	1.43	0.87	0.83
	0.25	4	0.83	0.30	0.09	1.66	1.50	0.90	0.87

Nevertheless, the slight differences between measured and computed strain ratio make desirable some supplementary research in order to determine more accurately the effective porosity figures and/or its possible variation when non uniform stresses are applied.

ACTUAL APPLICATIONS TO SEA STRUCTURES DESIGN

The characteristic features of concrete gravity off-shore platforms are cylindrical or lobate enclosures, providing buoyancy chambers, as necessary during in-shore construction (see Fig. 4),towing (see Fig. 5) and installation stages, and being available to be used afterwards, if necessary for storage of crude oil extracted.

Fig. 4 :
The NINIAN platform, designed and built by HOWARD DORIS for CHEVRON U.K., during in-shore construction phase in Loch Kishorn (Scotland)

Hydrostatic pressure, as applied during these temporary stages on the concrete walls, are generally governing their structural design and their necessary thickness in particular. As the walls thickness governs their weight, the weight governs the needed volume of buoyancy chamber and the latter governs in turn its weight and, so on, it appears obvious, a most careful dimensioning of marine structures, to fit with hydrostatic pressure applied, is presenting a capital factor of safe and economic design, deciding even in several cases on feasibility of the off-shore structure. Actually the buoyancy tank enclosures have to be designed to ensure (10) an elastic stability involving a safety factor as required with respect to the critical hydrostatic pressure this depending a.o. on actual concrete E-moduli having to be evaluated with respect to :

- concrete properties at its age at time of load application ;

- loading duration ;
- stress/strength ratio involved by load as applied at various time steps (see Fig. 6) ;

and furthermore, in the case of immersed concrete structures :

- the effects of pore pressure as generated by the hydrostatic loads applied (11).

Fig. 5 : The Ninian platform, while towed to the field.

When looking on the example as presented on Fig. 6, there appears the critical pressure of about 1.1 MPa, corresponding to a load duration of some 20 days has been governing the buoyancy tank wall thickness. This wall has been actually designed 120 cm thick, involving, for its curvature radius of 25.4 m, a hoop force of 28 MN and a subsequent critical hoop stress (in terms of total stresses) of 23.3 MPa. Assuming furthermore, the concrete of the structure investigated is comparable to concrete B as described above, there could be read on Fig. 2, for $\underline{t}$ = 20 days :

$$\mathcal{E}/\mathcal{E}_o = 0.64 ; \quad \mathcal{E}_s/\mathcal{E}_o = 1.30 ; \quad \text{therefore } \underline{f} = 0.51.$$

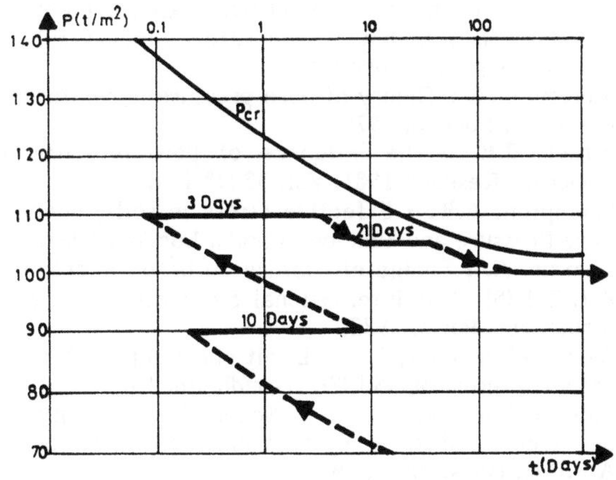

Fig. 6 : Critical load versus loading duration, compared with platform installation scenario

However accounting for a low σ_l/σ_a ratio, and for a subsequent minor E-modulus improvement, this could nevertheless contribute, through some reduction of wall thickness, of curvature radius and of buoyancy tank perimeter, to an eonomy of about 10 % on the total concrete volume of the wall investigated, i.e. 320 x 55 x 1.20 x 0.10 = 2 100 m^3 (!), thus presenting a gain of some two and half million dollars.

An other example (12) : A raft slab of a concrete gravity off-shore platform, of 150 m diameter and 1.50 m thickness had to be prestressed to resist bending moments, especially assumed to occur during installation phase. As this raft had at this stage to be submitted to a radial hydrostatic pressure of 1.40 MPa, due to water depth at this site, the effects of this pressure could be accounted for as a natural (free of charge !) prestressing, involving, with an f figure of 0.5, an effective stress of 0.7 MPa, resulting in an effective compressive force of 1.5 x 0.7 $\cong$ 1.0 MN/m. This could result in an economy of some 160 tonnes of high strength prestressing steel cables to be distributed in both directions over the raft area, i.e. some other half a million dollars more !

CONCLUDING REMARKS

An accounting for pore pressure effects in immersed concrete of marine structures, and especially of the off-shore ones, leads to a better understanding of the actual behavior of such structures and to a significant economy on their design. The beneficial impact of such approach will be evidently still increased when designing structures for deeper and deeper waters ; an accounting for pore pressure effects may even present in such cases a governing factor of feasibility statement. On the other hand, the pore pressure inside immersed concrete should be taken into account on the safe side, having to be subtracted in an appropriate way from the compressive forces generated by hydrostatic pressures, if such forces could contribute to a natural " prestressing" of elements investigated.

References

1. Biot, M.A. ; Mechanics of deformation and acoustic propagation in porous media. Journal of Applied Physics (1962), Vol. 33, N° 4, 1962.
2. Bishop, A.W. ; The influence of system compressibility on the observed pore-pressure response to an undrained change in stress in saturated rock. Geotechnique 1976, Vol. 26, N° 2.
3. Bishop, A.W. ; The influence of an undrained change in stress on the pore pressure in porous media of low compressibility. Geotechnique 1973, Vol. 23 N° 3.
4. Mesri, G. ; Adachi, K. ; Ullrich, C.R. ; Pore-pressure response in orck to undrained change in all-round stress. Geotechnique 1976, Vol. 26, N° 2.
5. Clayton, N. ; Strain of concrete cylinders subjected to fluid pressure. ACI Journal, Sept. October 1980.
6. Morley, C.T. ; Theory of pore pressure in reinforced concrete cylinders. FIP 8th Congress, London, 1978.
7. Butler, J.E. ; The influence of pore pressure upon concrete. Magazine of Concrete Reseach 1981, Vol. 33 N° 114.
8. Skempton, A.W. ; Effective stress in soils, concrete and rocks. Symposium on Pore Pressure and Succion in Soils, London, 1961.
9. Ilantzis, A. ; Fluage et rupture du béton imprégné sous sollicitation triaxiales. ARBEM/ISF 2-nd International Symposium on Behaviour of Offshore Concrete Structures, Paris, 1982.
10. Zaleski-Zamenhof, L.C. ; Limit state approach of buckling capacity of concrete sea structures. BOSS'1976, Trondheim, 1976.
11. Zaleski-Zamenhof, L.C. ; Sécurité des coques en béton précontraint des ouvrages off-shore eu égard à l'instabilité de forme sous compressions élevées. A.F.P.C. Days, Paris, 1978.
12. Zaleski-Zamenhof, L.C. ; Ozanne, P.J. ; Ilantzis, A. ; Problèmes posés par le comportement mécanique du béton saturé dans les structures en mer. Brasil Off-shore'79, Rio de Janeiro, 1979.

Technical Session:
Materials and Construction

2. Durability of Concrete

Durability of Lightweight Concrete for Arctic Concrete Structures

Y. Nojiri, Y. Tazawa and Y. Nobuta

Kajima Institute of Construction Technology, Kajima Corporation, Tobitakyu, Chofu, Tokyo, 182 Japan

ABSTRACT

The experimental results regarding characteristics of lightweight concrete incorporating condensed silica fume are discussed. Many factors such as type and moisture content of lightweight aggregates, dosage of condensed silica fume, air content, water to cement ratio are selected as experimental factors. It was concluded that high strength and durable lightweight concrete satisfying requirements for arctic concrete structures could be obtained by using lightweight aggregate with low moisture content and utilizing condensed silica fume as a mineral admixture.

INTRODUCTION

These days, along with a full scale start-up of the development of marine petroleum, the needs for large-scale concrete structures such as platforms for oil exploration and production and man-made island have been increasing. In case of the anticipated use of those in the arctic region, the concrete to be used needs to maintain enough durability even under the severe environmental conditions of the arctic region. At the same time, since the structures themselves are required to be lightweight from a viewpoint of towing, the development of lightweight concrete having both high strength and superior durability has come to be in demand.

On the other hand, since condensed silica fume, which has been researched mainly in the Northern Europe and the North America, was reported as being substantially effective to both the increase in concrete strength and the improvement of durability of concrete, an application of condensed silica fume as an admixture for marine concrete is being watched with a keen interest.

This paper, based on the above situations, shows the experimental results on characteristics of lightweight concrete incorporating condensed silica fume, and reports the effectiveness of condensed silica fume and strength development while concentrating on the practical investigation in relation to freezing and thawing resistance.

EXPERIMENTAL METHODS

1. Program of Experiments

The experiments consist of the following three series. Factors and levels of experiments are shown in Table 1.

Series I : To investigate basic effects of condensed silica fume on the properties of concrete including freezing and thawing resistance. Lightweight aggregate with high moisture content being widely used in Japan was to be employed.

Series II : To investigate the effectiveness of condensed silica fume in case of using a lightweight aggregate with low moisture content.

Series III : To study freezing and thawing resistance and strength development while setting both the type of lightweight aggregate and its moisture content to be as the main parameters.

2. Materials

Tables 1 and 2 show materials used and the characteristics thereof. Condensed silica fume, a byproduct in the process of manufacturing ferro-silicon and silicon metal, is an ultrafine active silica powder consisting of SiO_2 as its main constituent. In this experiment, employed were three kinds of condensed silica fume with different qualities. Those characteristics are shown in Table 3. As for a lightweight aggregate, there are both non-pelletized and pelletized artificial lightweight aggregates with variable moisture conditions.

As the experimental conditions concerned, being of high cement content and low water to cement ratio, a high range water reducer was used to allow sufficient mixing, and also a superplasticizer from a view point of maintaining workability in actual construction.

3. Test methods

A pan-type mixer was used for mixing. Slump after superplasticizing

Table 1 Program of experiments

Experimental Series			Factors: Type of Condensed Silica Fume	Dosage of Condensed Silica Fume SF/C+SF (%)	Water to (Cement + Condensed Silica Fume) Ratio W/C+SF (%)	Unit Quantity of (Cement +Condensed Silica Fume) C+SF (%)	Dosage of Air-entraining Agent ×(C+SF) (%)	Type of Lightweight Aggregate	Type of Cement	Remarks
I	Levels	1	W	0	30	400	0	AH	High-early Strength Portland	The Orthogonal Array Table L_{27} (3^{13}) was used (27 mixes)
		2	X	5	36	450	0.006			
		3	Y	10	42	500	0.012			
II	Levels	1	W	0	28	417	0.004 ~ 0.007	AL	Normal Portland	The Orthogonal Array Table L_9 (3^4) was used (9 mixes)
		2		5	32	469		BL		
		3		10	36	536		CL		
III			W	10	28	450	0.004 ~ 0.009	AA, AL, BL, CL, AM, BM, CM, AH, BH	Normal Portland	(24mixes)
					30	500				
			X		36	536				

was designed to be 21 ± 2.5 cm. The items and methods of the property tests concerning both fresh and hardened concretes are as follows:

- Air content (fresh concrete) ; JIS A 1128 (Pressure method)
- Air content (hardened concrete) ; ASTM C 457 (Modified point-count method)
- Compressive strength ; JIS A 1108 (specimen ∅10 x h20 cm)
- Freezing and thawing resistance ; ASTM C 666 (Procedure A, rapid freezing and thawing)
- Air void system ; ASTM C 457 (Modified point-count method)

RESULTS OF EXPERIMENTS

1. Compressive strength

(1) Type and dosage of condensed silica fume

Table 2 Materials

Materials		Characteristics	
Cement		High-early Strength Portland Cement ; Specific Gravity=3.14	
		Normal Portland Cement ; Specific Gravity=3.16	
Fine Aggregate		River Sand ; Specific Gravity=2.62, Fineness Modulus=2.8	
Coarse Aggregate (Lightweight)	Moisture Content ; 0%	AA	Pelletized Type ; Sp.Gr. (Oven-dry)=1.30, F.M.=6.51
	Moisture Content ; 4～6%	AL	Pelletized Type ; Sp.Gr. (Oven-dry)=1.29, F.M.=6.39
		BL	Pelletized Type ; Sp.Gr. (Oven-dry)=1.24, F.M.=6.41
		CL	Non-pelletized Type; Sp.Gr. (Oven-dry)=1.25, F.M.=6.51
	Moisture Content ; 10～15%	AM	Pelletized Type ; Sp.Gr. (Oven-dry)=1.28, F.M.=6.58
		BM	Pelletized Type ; Sp.Gr. (Oven-dry)=1.25, F.M.=6.43
		CM	Non-Pelletized Type ; Sp.Gr. (Oven-dry)=1.25, F.M.=6.51
	Moisture Content ; 20～25%	AH	Pelletized Type ; Sp.Gr. (Oven-dry)=1.28, F.M.=6.58
		BH	Pelletized Type ; Sp.Gr. (Oven-dry)=1.25, F.M.=6.43
Condensed Silica Fume		W	Foreign Source ; Spcific Gravity=2.2
		X	Japanese Products ; Specific Gravity=2.2
		Y	Japanese Products ; Specific Gravity=2.2
Admixture	Air-entraining Water Reducer	Hydroxlated Polymer and Calcium Lignosulfonate	
	Air-entraining Agent	Modified Rosin-compound	
	High-range Water Reducer	Polymer of Aromatic Sulfonate	
	Superplasticizer	Polymer of Melamine Sulfonate	

Table 3 Properties of condensed silica fume

Items		W	X	Y
CaO	(%)	0.2	0.4	—
SiO_2	(%)	93.1	87.3	80.3
Al_2O_3	(%)	0.1	0.3	—
MgO	(%)	0.5	0.5	—
Fe_2O_3	(%)	0.1	3.5	—
K_2O	(%)	0.16	0.41	—
Na_2O	(%)	0.07	0.08	—
Total C	(%)	0.21	0.25	—
Loss on ignission	(%)	1.0	2.9	—
Suface (N_2)	(m^2/g)	25.2	24.0	21.8
Mean primary prticle size	(μm)	0.15	—	—

Fig.1 shows the influence of condensed silica fume on compressive strength. Although a slight difference being recognized in relation to some difference in compressive strength due to the difference in the type of condensed silica fume, this was not significant difference.

Excluding the results obtained at the age of 3 days for those in Series II, compressive strength increased with the increase in dosage of the condensed silica fume. In case of Series I, the ratio of compressive strength at the same age, assuming it to be 1.0 for the dosage of 0 %, was given by 1.06 ~ 1.09 for 5 %, and 1.16 ~ 1.20 for 10 %, respectively.

(2) Type of lightweight aggregate

Fig. 2 shows a relationship between the type of aggregates and compressive strength. As for the effect of the shape of the aggregates on compressive strength, there was no clear trend.

As for the influence of moisture content of the aggregate on compressive strength, if the shape thereof being the same, no significant difference was recognized.

(3) Strength development

Fig. 3 shows a development of compressive strength in Series III up to an age of 91 days. According to Fig. 3, high compressive strength could be obtained, for example, about 600 kgf/cm^2 at the age of 28 days, and 650 ~ 700 kgf/cm^2 at the age of 91 days, that means an almost unprecedented high strength lightweight concrete was found to be available with use of condensed silica fume and chemical admixtures. Besides, also in this case almost no influence of difference in moisture content of the aggregate on compressive strength was recognized.

2. Freezing and thawing resistance

(1) Effect of lightweight aggregate with high moisture content

Fig. 4 shows a typical case of the test results concerning freezing and thawing in Series I obtained with the use of the lightweight aggregate with high moisture content, showing that repetition of 100 cycles results

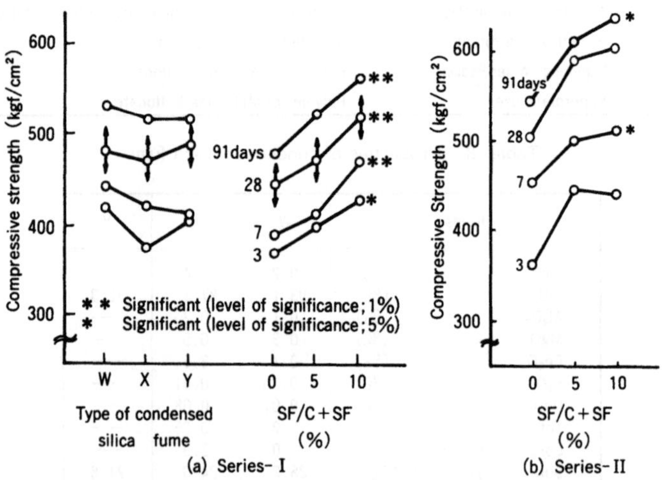

Fig. 1 Relationships between type and dosage of condensed silica fume and compressive strength

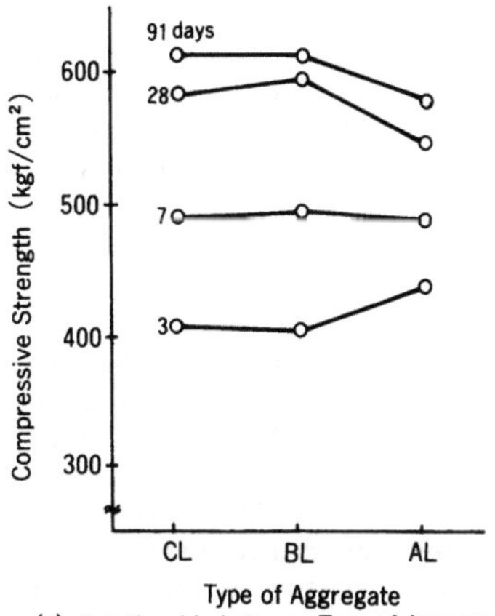

(a) Relationship between Type of Aggregate and Compressive Strength

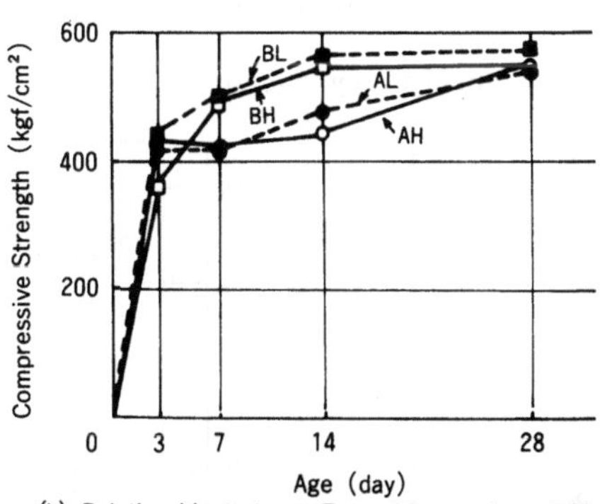

(b) Relationship between Type of Aggregate and Strength Development

Fig. 2 Type of aggregate and compressive strength

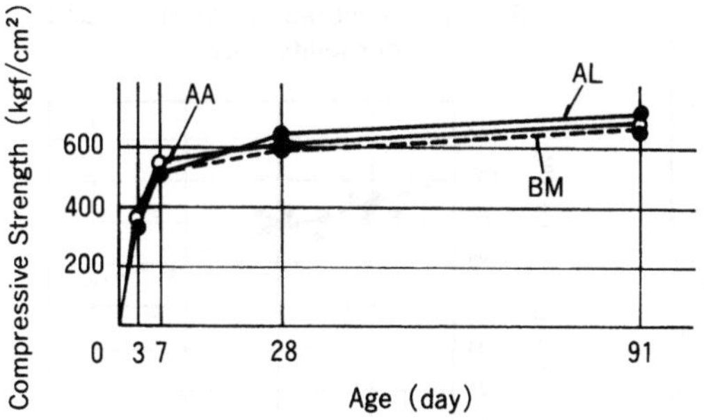

Fig. 3 Strength development

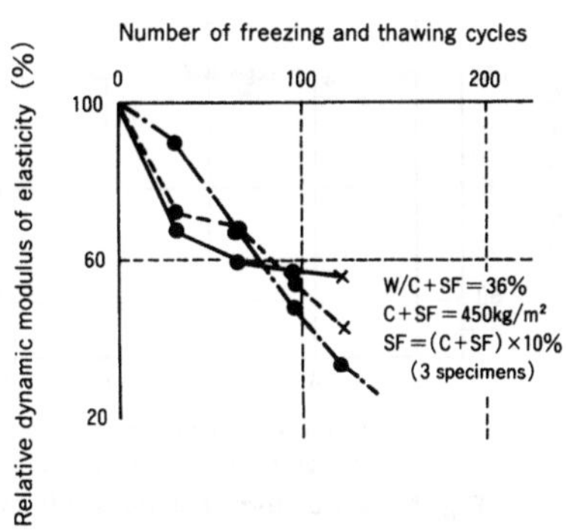

Fig. 4 Results of rapid freezing and thawing tests

in the lowering of relative dynamic modulus of elasticity to the level of less than 60 %. In case of Series I, even under other experimental conditions, similar results could be obtained, thus in spite of the difference in water to cement ratio, the use of condensed silica fume, and so on, it was obvious that the freezing and thawing resistance thereof remarkably fell behind.

(2) Air content

A relationship between the air content and the durability factor is shown in Fig. 5. As shown in Fig. 5 (a), there was a great difference in durability between the results in Series I using the aggregate with high moisture content and those in Series I, II using the aggregate with low moisture content. So far as these results were concerned, no great influence of variance in air content on durability was recognized.

Fig. 5 (b) shows a relationship between the air content of hardened concrete and the durability factor. In case of the aggregate with low moisture content in use, the durability factors were indicated to be more than 90 % when the range of air content was within 2 to 8 %.

(3) Compressive strength and durability factor

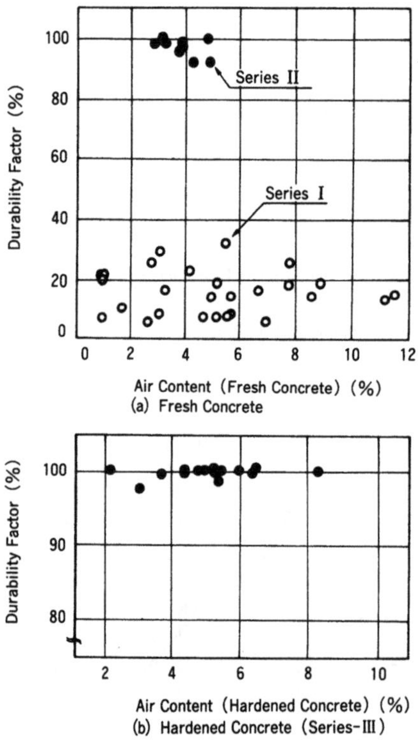

Fig. 5 Air content and durability factor

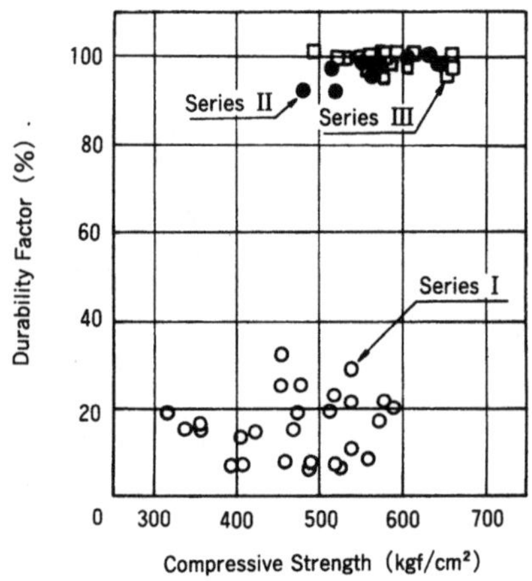

Fig. 6 Compressive strength and durability factor

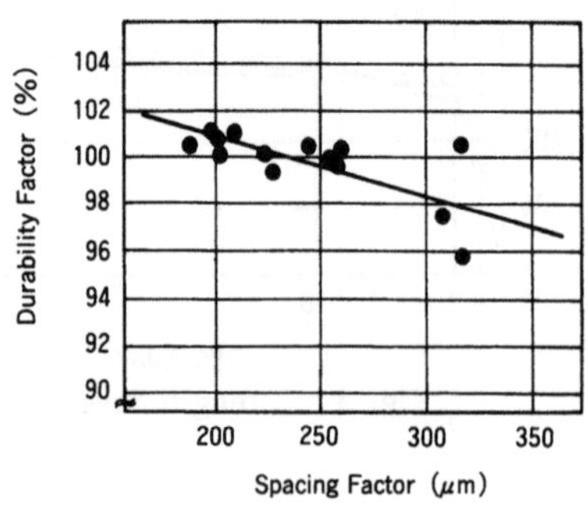

Fig. 7 Spacing factor and durability factor

Fig. 6 shows a relationship between the compressive strength and the durability factor. Similarly as those shown in Fig. 5 (a), a considerable difference between the results in series I and those of Series II, III could be seen. No influence of variance in strength on durability factor was seen under the conditions for high strength exceeding the compressive strength of 400 kg/cm .

(4) Spacing factor and durability factor

Fig. 7 shows a relationship between a spacing factor and a durability factor in Series II and III. There existed a trend that the durability factor be reduced along with an increase of spacing factor. In either case, the durability factor was always greater than 90 % and no remarkable lowering of the durability factor due to fluctuation of the spacing factor was indicated.

(5) Dosage of condensed silica fume and durability factor

Fig. 8 shows an effect of dosage of condensed silica fume on a durability factor. As shown in Fig. 8, it was indicated that the durability factor of concrete incorporating condensed silica fume could be made to be a greater one in value than that of concrete with no dosage. However, no quantitative trend between the dosage and the durability factor was recognized.

(6) Moisture content of lightweight aggregate and durability factor

As mentioned above, the freezing and thawing durability of a lightweight concrete is found to be under the influence of the moisture content of the aggregate. Fig. 9 shows a relationship between the moisture content of aggregate and the durability factor.

According to Fig. 9, if the aggregate with low moisture content of less than about 5 % was used, satisfactory durability having durability factor more than 90 % could be obtained regardless the type of aggregates. When the moisture content went beyond the level of 10 %, the durability varied widely in accordance with the type of aggregates. This is considered to be mainly due to the difference in structural firmness of the aggregates themselves, in diameters and distributions of internal voids. In the case where the moisture content went over 20 %, the freezing and thawing resistance became extremely deteriorated despite the differences of other conditions. Fig. 10 shows the typical results of changes in relative dynamic modulus of elasticity on every feezing and thawing cycle when the low moisture content aggregates were used.

CONCLUSIONS

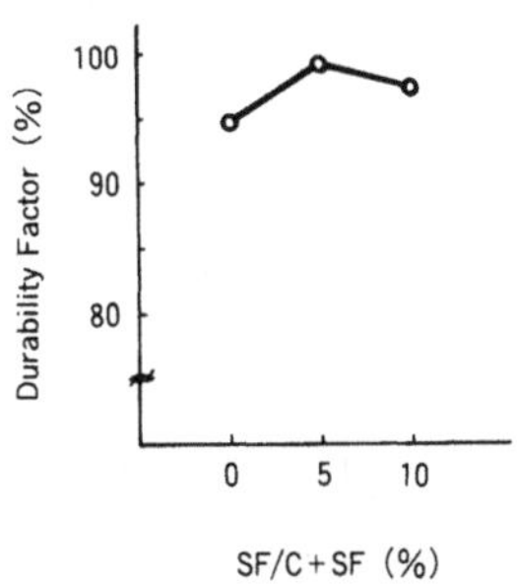

Fig. 8 Silica fume content and durability factor

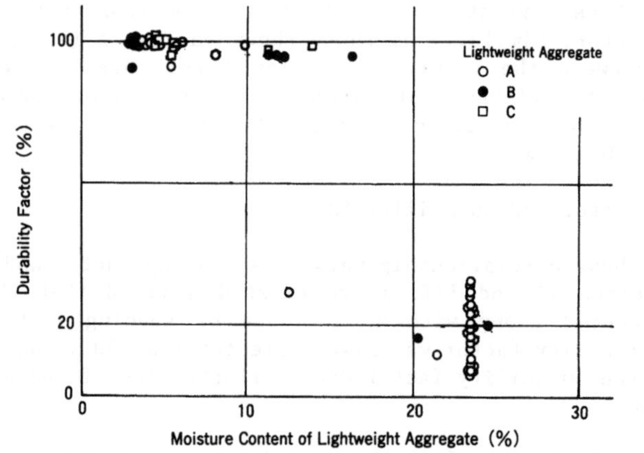

Fig. 9 Moisture content of lightweight aggregate and durability factor

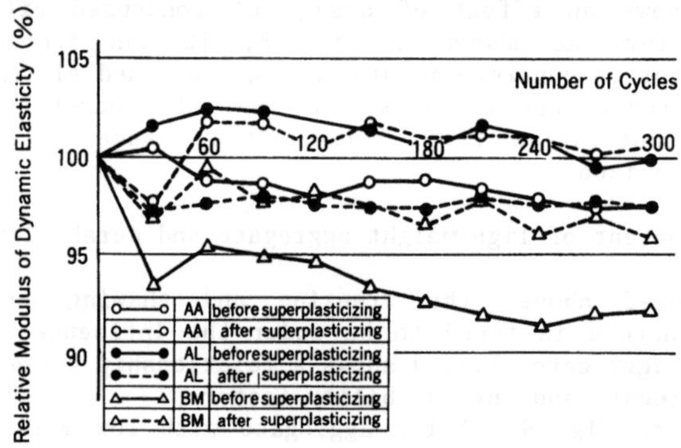

Fig. 10 Results of freezing-and-thawing test

The study on the physical properties of the lightweight concrete incorporating condensed silica fume was carried out for the purpose of utilizing it for the concrete structures in the arctic region that demanded high quality from a view point of strength and durability.

As the result, the following points were indicated:

(a) The use of condensed silica fume is substantially effective to the increase in concrete strength.

(b) The freezing and thawing resistance of lightweight concrete is subject to the moisture content of lightweight aggregate.

(c) The critical moisture content that has an influence on the freezing and thawing resistance is different in accordance with the type of lightweight aggregates.

REFERENCES

1) CANMET/ACI,"Fly Ash,Silica Fume, Slag and Other Mineral By-products in Concrete," Publication SP-79, ACI, JULY 1983

2) Tazawa, Y., Nobuta, Y. and Ishi, A.,"Physical Properties and Durability of High-strength Lightweight Concrete Incorporating Condensed Silica Fume," Proceeding of 6th Annual Conference of Japan Concrete Institute, Vol. 6, pp. 121 - 124, JULY 1984 (in Japanese)

3) Hjorth, L.,"Microsilica in Concrete," Nordic Concrete Research Publication, No. 1, 1982

Static and Cyclic Behavior of Structural Lightweight Concrete at Cryogenic Temperatures

D.E. Berner[1] and B.C. Gerwick, Jr.[2]

[1] Ben C. Gerwick, Inc., San Francisco, CA 94111, USA
[2] Civil Engineering Department, University of California, Berkeley, CA 94720, USA

ABSTRACT

Experimental research conducted at the University of California at Berkeley indicated that plain and prestressed/reinforced lightweight concrete behaved favorably under both static and cyclic loading conditions at cryogenic temperatures. This research indicated: (1) that the mechanical properties of lightweight concrete are significantly affected by moisture content, but not significantly by relatively high-intensity cyclic fatigue, (2) that one-way prestressed concrete specimens subjected to cyclic axial tension and compression as well as cryogenic thermal shock exhibited only minor degradation, and (3) that two-way prestressed concrete specimens subjected to cyclic membrane shear and also to cryogenic thermal shock exhibited only minor surface cracking and reductions in flexural stiffness. These results indicate that prestressed/reinforced high-strength lightweight concrete exhibits excellent cryogenic properties, and may qualify as a material for use as a secondary barrier for offshore liquefied gas containment vessels.

INTRODUCTION

Structural lightweight concrete offers several attractive properties for use in offshore cryogenic containment vessels. Among these properties are lightweight concrete's low thermal conductivity, its low modulus of elasticity, its excellent durability against both freeze-thaw and wetting-drying damage, and its extremely low permeability at low temperatures.

Several offshore concrete cryogenic containment vessels have been proposed however, to date only one offshore concrete cryogenic containment vessel has been fabricated, the ARCO floating liquefied petroleum gas, LPG, production terminal in Indonesia. The ARCO LPG terminal meets existing codes and regulations as it uses separate steel containment tanks for the containment of a relatively warm cryogenic liquid propane, which has a boiling point of -42°C. This vessel does not rely on the cryogenic behavior of the concrete hull of the terminal however, research summarized in this paper indicates that structural lightweight concrete may possess adequate cryogenic properties to act as a secondary barrier for the containment of liquefied natural gas, LNG, (-160°C) even under severe loading conditions of high-intensity cyclic loading and simultaneous cryogenic thermal shock.

TESTING PROGRAM

This paper summarizes the results of an experimental program conducted at the University of California at Berkeley. Details of this study are reported elsewhere (ref. 1). The testing program examined both plain and prestressed/reinforced high-strength lightweight concrete under both static and cyclic conditions. Experimentation for the plain concrete included determination of compressive strength, tensile strength, elastic modulus, and freeze-thaw resistance over a temperature range from 23°C to -196°C. Plain concrete with three different moisture contents, saturated, air-dried, and oven-dried were tested. Experimentation for the prestressed concrete specimens included high-intensity, cyclic axial tension and compression on one-way prestressed slab specimens, and high-intensity, cyclic membrane shear on two-way prestressed slab specimens. Both sets of prestressed/reinforced specimens were tested under conditions of simultaneous cryogenic thermal shock from liquid nitrogen at -196°C as well as the cyclic loading.

A single lightweight concrete mix design containing fly ash was held constant for purposes of comparison throughout the study, for both the plain and the prestressed/reinforced concrete specimens. A superior quality lightweight concrete containing condensed microsilica fume was also included in the final phase of testing on two-way prestressed slab specimens subjected to cyclic membrane shear. Both concrete mixes contained 10-mm expanded shale coarse aggregate, and standard-weight fine aggregate, while the mix containing fly ash had a water/cementitious material, W/C, ratio of 0.35 and a 28-day compressive strength, f_c', of 44.8 MPa, and the mix containing microsilica fume had a W/C ratio of 0.29 and a value of f_c' of 55.2 MPa.

Plain Lightweight Concrete

Results from the studies on plain lightweight concrete with fly ash indicated that moisture content served as a key parameter in determining the cryogenic behavior of the concrete, and that cyclic loading did not significantly degrade the mechanical properties of the concrete. The applied cyclic loading for the plain concrete followed a history of wave-induced membrane stress such as would be encountered by a vessel in a 20-year storm in the North Atlantic (Fig. 1). The histogram indicates a nominal precompression of 37.5% of the 28-day compressive strength, which was introduced by the testing machine in order to approximate the stress conditions within a prestressed concrete hull.

Research conducted by Rostasy, Schneider and Wiedemann (ref. 2) has indicated that 85% relative humidity represents a critical moisture content for the degradation of concrete at cryogenic tmperatures. As the properties of oven-dried concrete change very little at cryogenic temperatures, the work performed at Berkeley focused on moist and air-dried specimens with only a few specimens tested in the oven-dried state. Compressive strength tests were performed at cryogenic temperatures on both cycled and uncycled concrete cylinders. The compressive strength of the uncycled saturated specimens increased approximately 80%, and air-dried concrete increased its compressive strength by over 30% at cryogenic temperatures. Cyclic loading affected the moist concrete slightly more than the air-dried concrete at low temperatures, reducing the strength gain to 58% for moist, and 29% for air-dried concrete. Oven-dried concrete followed a similar trend as the air-dried concrete. The modulus of elasticity for moist concrete increased by approximately 75%, while air-dried concrete increased by about 65% at low temperatures. Cyclic loading had a very minor effect on the elastic modulus at cryogenic temperatures, with a slightly larger influence at room temperature, indicating that the increased strength at low temperatures may help to protect the concrete from cyclic fatigue damage. Splitting-tensile tests were only conducted on uncycled specimens, and the results indicate a sharp increase in tensile strength between -5°C and -90°C, which then gradually decreased with lower temperatures. The tensile strength of moist concrete increased by about 80%, and air-dried concrete increased by about 50% at cryogenic temperatures. Oven-dried concrete showed little or or no change in splitting-tensile strength at low temperatures.

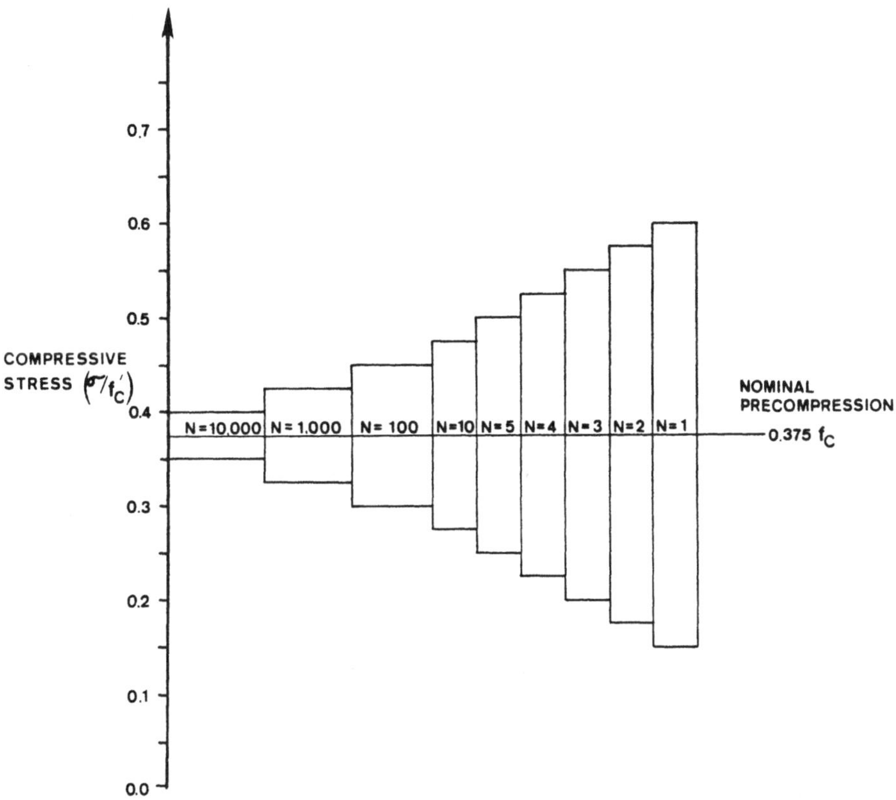

Fig.1 Histogram for cyclic loading of specimens in compression

Air-dried microsilica fume lightweight concrete was also subjected to cryogenic thermal shock freeze-thaw cycles between 21°C and -196°C. Two parameters, air-entrainment, and steel fiber reinforcement were examined. Fig. 2 presents the test results for these freeze-thaw tests, comparing the relative dynamic modulus of elasticity of the concrete to the number of cryogenic thermal shock cycles. The results indicate that for thermal shock, air-entrainment had little influence on either the plain or the fiber-reinforced concrete specimens, while the fiber reinforcement had a very beneficial effect. After as few as five cycles the relative dynamic modulus of the plain microsilica concrete had decrease by approximately 50%, while the fiber-reinforced concrete had decreased by only about 15%. Differential thermal contraction between the steel fibers and the concrete probably enhanced the behavior of the fiber-reinforced concrete by triaxially prestressing the concrete. The air-entrainment probably had little effect on the cryogenic thermal shock resistance of the concrete specimens due to their their relatively low moisture content, approximately 50%, and also due to the difficulty in relieving pore pressure induced by thermal shock.

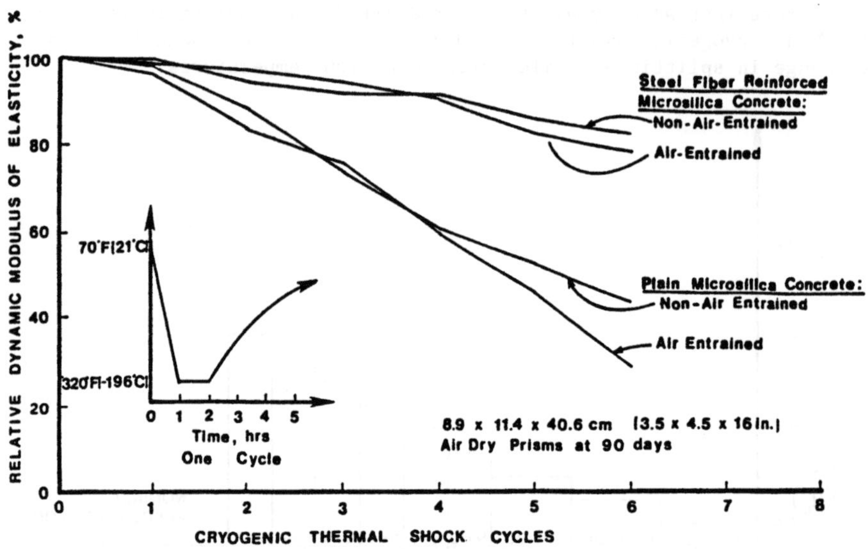

Fig. 2 Cryogenic Freeze-Thaw Resistance of Air-Dried Microsilica Lightweight Concrete

ONE-WAY PRESTRESSED SPECIMENS

Two 2.9 by 0.64 by 0.15-m lightweight one-way prestressed concrete slabs were subjected to a similar cyclic loading history as that shown in Fig. 1, and also cryogenic thermal shock from liquid nitrogen on the top central portion of the slabs. This testing sequence was intended to simulate an accidental release of LNG from the primary steel containment tank on to the concrete hull during storm-wave loading conditions. It is essential that in the event of such an accident that the secondary barrier provided by the concrete hull retain its structural integrity free from cracking, which could release significant portions of volatile gas. The one-way slabs were air-dried, and had 2.25% longitudinal steel, and closely spaced stirrups.

The test results indicated that no through cracking occurred that could release cryogenic liquid. Table 1 summarizes the amount of degradation in flexural stiffness resulting from the various stages of cyclic axial tension and compression loading and cryogenic thermal shock. A third conventionally reinforced slab specimen without prestressing is included in this table for comparison. The results indicate a maximum flexural degradation of the prestressed specimens of 19%, and of 34% for the conventionally reinforced specimen when subjected to less severe conditions than the prestressed specimens. These results demonstrate the superiority prestressed concrete to reinforced concrete for cryogenic containment, and also the ability of prestressed/reinforced concrete to accommodate severe thermal stresses and high-intensity cyclic fatigue loading, and still retain its structural integrity.

Table 1			
Stiffness (EI) from Load-Deflection Tests for One-Way Slabs			
Slab	Condition	Stiffness	
		lb·$in.^2$ x10^9	kg·m^2 x10^5
Prestressed No. 1	Initial	1.63 ±0.1	4.77 ±0.3
	After thermal shock and histogram loading	1.53 ±0.1	4.48 ±0.3
	After 750 additional peak-level cycles (not precracked, no thermal shock)	1.38 ±0.1	4.04 ±0.3
	After 9250 further peak-level cycles (precracked, no thermal shock)	1.33 ±0.1	3.89 ± 0.3
Prestressed	Initial	1.63 ±0.1	4.77 ±0.3
No. 2	After thermal shock and loading histogram	1.43 ±0.1	4.19 ± 0.3
No. 2	After thermal shock and loading histogram (precracked, and thermally shocked)	1.43 ±0.1	4.19 ±0.3
Reinforced No. 3	Initial	0.97 ±0.1	2.84 ± 0.3
	After thermal shock and cyclic loading	0.64 ±0.1	1.87 ± 0.3

TWO-WAY PRESTRESSED SPECIMENS

Four 0.71 by 0.71 by 0.11-m lightweight concrete two-way prestressed slab sections were subjected to cyclic membrane shear and cryogenic thermal shock under conditions more severe than those imposed on the one-way prestressed/reinforced slab specimens. The history of cyclic loading imposed on the two-way prestressed/reinforced specimens was intended to simulate the most severe portion of an 80-year wave-loading history in the North Atlantic. Two of the specimens, 1t and 4t, used fly ash concrete, while specimens Nos. 2t and 4t used microsilica fume concrete. In addition, specimen No. 3t was also subjected to a circulating water bath on the opposite face of the cryogenic thermal shock in order to simulate the ocean water and impose a more severe thermal gradient, also specimen No. 4t was loaded in shear into the post-elastic range thus imposing loading beyond the normal service conditions for such a structure. The two-way slabs contained 2.61% steel along one axis, and 2.25% along the other thus simulating the heavily reinforced concrete sections typically used offshore.

The test results indicated that the specimens performed very well even though they were surrounded by stiff boundary beams in order to simulate the adjoining structural elements in the prototype structure. No through cracks were observed that could release cryogenic liquid, and specimen No. 3t with the circulating water interface showed no surface cracking at all. Table 2 summarizes the amount of flexural degradation that each slab sustained during the test sequence. The results indicate that specimen No. 3t with microsilica concrete and a water interface had the least degradation with a 24% reduction, and that specimen No. 4t with fly ash and also post-elastic loading had the most degradation with about a 30% reduction in flexural stiffness. These results indicate the superiority

TABLE 2			
Deflections Measured 3 inches (7.6 cms) from the Center of the Two-Way Slabs due to a 15-kip (66.7-kNt) Load			
Slab No.	Initial Deflection in.x10^{-3}, (mm)	Deflection after Histogram of Cyclic Loading and Thermal Shock in.x10^{-3}, (mm)	Final Deflection after Additional Cryogenic Thermal Shock in.x10^{-3}, (mm)
1t	5.7 (0.145)	7.4 (0.188)	7.9 (0.201)
2t	5.6 (0.142)	7.3 (0.185)	7.7 (0.196)
3t	5.8 (0.147)	7.2 (0.183)	7.6 (0.193)
4t	5.6 (0.142)	7.5 (0.191)	8.0 (0.203)

of microsilica fume lightweight concrete for cryogenic use, and that biaxial prestressing helps to reduce the effects of cryogenic freeze-thaw damage.

CONCLUSIONS

The findings of this experimental research program indicate that high-strength lightweight concrete may qualify for use as a secondary containment barrier for the offshore storage of LNG. Positive findings from this program include:

1. Prestressed lightweight concrete has the ability to withstand cryogenic thermal shock with only minor reductions in ductility.
2. Concrete containment vessels can be designed to prevent the development of through cracks which can release volatile liquids or gases.
3. Microsilica fume concrete has superior cryogenic behavior to concrete containing fly ash.
4. Biaxial, or triaxial, confinement can reduce the effects of cryogenic thermal shock.
5. Active heating on the opposite face to which a slab section is cryogenically shocked, can help to mitigate the consequences of the shock by limiting the relatively steep thermal gradient to a limited area.
6. The relatively low thermal conductivity of lightweight concrete can limit the structural area affected by the low temperatures from a cryogenic release.

REFERENCES

1. Berner, D.E.; Behavior of Prestressed Concrete Subjected to Low Temperatures and Cyclic Loading. Ph.D. disseration, University of California at Berkeley 1984.

2. Rostasy, F.S.; Schneider, U.; Wiedemann, G.: "Behaviour of Mortar and and Concrete at Extremely Low Temperatures," Cement and Concrete Research. Vol. 9, No. 3 (1979) 365-376.

3. Gerwick, B.C.,Jr.; Mansour, A.; Price, E.; Thayamballi, A.: "Feasibility and Comparative Studies for the Use of Prestressed Concrete in Large Storage/Processing Vessels," Transactions of SNAME. Vol. 86 (1978) 163-196.

4. Mast, R.F.: "The ARCO LPG Terminal Vessel," Proceedings of the Conference on Concrete Ships and Vessels. University of California at Berkeley Sept. 1975.

5. International Maritime Organization (IMO): Code for the Construction and Equipment of Ships Carrying Liquefied Gases in Bulk. Resolution A.328.(IX), IMO, London 1976.

Some Aspects of the Classifications of Pollution by Sea Water Aerosol for Reinforced Concrete Buildings

N. Kashino

Materials Department, Building Research Institute, Ministry of Construction, Tatehara, Tsukuba, Ibaraki, 305 Japan

ABSTRACT

Classification of reinforced concrete buildings standing along seashores was done based on investigations of actual conditions regarding the distributions of sea water aerosol in the atmosphere measured by the distance from the shoreline and the quantity of penetration of the aerosol into concrete. According to the classification, parts of reinforced concrete structures standing up to 200 m from the shoreline directly facing the sea are subjected to salt damage without fail so that it is necessary for some kind of countermeasure to be taken. In case of 1 km from the shoreline, it is necessary for care to be exercised since there is a possibility of salt damage being sustained.

INTRODUCTION

Japan is an island country completely surrounded by the sea. There are cases of reinforced concrete structures standing at the seaboard into the concrete of which sea water aerosol (abbreviated as SWA) carried in from the sea by salt wind penetrates to cause corrosion of reinforcing steel in the interiors. This is generally called salt damage. This paper comprises a consideration regarding the establishment of salt damage classification for reinforced concrete buildings derived from investigations of the quantity of SWA distributed in the atmosphere measured according to distance, and investigations of reinforcing steel corrosion in concrete of reinforced concrete buildings [1].

It should be noted here that a part of this study was reported at an Architectural Institute of Japan convention held in October, 1984 [3].

TREND OF SEA WATER AEROSOL FLIGHT

Sea water aerosol is produced when air bubbles in sea water burst, and borne on updrafts of air above the sea, the aerosol is transported to land by wind. The sizes of droplets differ since they are deformed on the way by evaporation and collapsing. The sizes which had been 3 to 18 μ above the sea become from 0.03 to 0.6 μ by the time they come inland. In effect, there is a trend for sizes of droplets becoming smaller the farther inland they go.

Figure 1 shows the results of measurements made in City A on the Pacific Ocean side of Japan [1]. This is a graph prepared based on maximum value averages of results from monthly measurements by the gauze method at 20 measuring points in City A. Variations in measured values according to season can hardly be recognized. Figure 2 gives the results of measurements along a river at City B on the Sea of Japan side. This also is a graph prepared based on average values of results from measurements, again by the gauze method. The distribution trends of sea water aerosol shown in these figures are similar to those given in a report compiled by ISO [4]. Regarding the vertical distribution of sea water aerosol, it is roughly the same to about 100 m above the ground.

With respect to distribution of sea water aerosol in the atmosphere according to distance from the seashore, the quantity is large up to about 200 m from the shoreline, while it may be seen that the quantity is extremely small more than 1 km inland. However, at a part of the Sea of Japan side the area of aerosol flight is somewhat broad, while in regions where typhoons often hit, at plain areas opened facing the sea, and along rivers, sea water aerosol sometimes flies about 10 km inland [1].

On the other hand, even in the vicinity of the shore, places that are sheltered by enclosures have less sea water aerosol. According to measurements made by the author, in a range up to 100 m from the shore, the quantities of sea water aerosol adhering to buildings directly facing the sea and those sheltered off are 1:0.4 in ratio.

PENETRATION OF SEA WATER AEROSOL INTO CONCRETE AND CORROSION OF REINFORCEMENT AT INTERIOR

Regarding how far inland from the shoreline salt damage can be seen, the results of representative past investigations on quantities of sea water aerosol penetration into concrete according to distance from the shore [1], [2] will be organized and shown. Here, concretes of exterior walls facing the sea of roughly equal water-cement ratios and not previously containing salt are considered.

Figure 3 shows the results of investigations on the sea and at 18 m, 250 m, and 800 m from the shore. All are given in terms of average values. The walls at 18 m and 250 m from the shore were finished with several millimeters of mortar, but the walls at 800 m were of architectural concrete. The ages of the buildings were those shown in parentheses in the figure. Figure 4 shows the results of investigations in 1983. The average values of investigation results at 10, 20, 150, 800, and 900 m from the shoreline are plotted in this figure. The building at 10 m was finished with architectural concrete and the others with thin courses of mortar. The ages of the buildings are again indicated in parentheses in the figure. These data were all obtained by boring several concrete cores each from walls facing the sea and making slices according to distances from the surfaces to analyze individual salt contents.

The penetration amount of sea water aerosol greatly accelerating corrosion of reinforcement at the interior, in ordinary cases, is 0.03 percent converted to NaCl by weight of concrete. According to the example given in Fig. 3, in the building at 250 m from the seashore, the amount of penetration of sea water aerosol is far greater than 0.03 percent even at 3 to 4 cm from the surface of the concrete, and in fact, the reinforcement is extremely corroded. However, at the buildings 800 m or more inland from the shore, the amounts of penetration are small and corrosion cannot be seen on reinforcement.

In effect, within several hundred meters from the seashore, sea water aerosol in sufficient quantity to corrode reinforcement in concrete even at young ages has penetrated, but at 800 m or more inland from the seashore, the amounts of penetration of sea water aerosol into concrete are small even in case of advanced ages and there is hardly any risk of reinforcement corrosion. Similar trends can be seen in investigation data of the Ministry of Construction Building and Repairs Bureau and also in examples of investigations at highway bridges [6]. Further, even at buildings near the sea, in cases where there are sheltering objects the amounts of penetration of sea water aerosol into concrete are generally less than 0.03 percent [1].

ESTABLISHMENT OF SALT DAMAGE CLASSIFICATIONS FOR REINFORCED CONCRETE BUILDINGS

Upon organization of data obtained so far, the relations between distributions of sea water aerosol in the atmosphere from the shoreline toward inland areas and amounts of penetration into concrete are as shown in Table 1. And as shown in Table 2, the salt damage classifications with distances from the seashore to reinforced concrete buildings as indices can be considered.

As indicated here, the amounts of sea water aerosol transported more than 1 km inland from the seashore are small in Japan and it can be said there would be no concern about salt damage, but care is needed since there have been regions where salt damage was seen in the past due to strong seasonal winds and typhoons.

CONCLUSION

Salt damage classifications for reinforced concrete buildings were established as a result of detailed investigations of sea water aerosol distributions in the atmosphere and of many investigations on amounts of penetration of aerosol into concrete. The heavy salt damage area designated in the classifications is an area requiring measures against salt damage to be taken at walls facing the sea without fail, the salt damage area an area where measures should be taken as necessary, and a semi-salt damage area an area where measures are necessary in cases.

REFERENCES

1. Ministry of Construction Comprehensive Technology Development Project: Development of technologies for improving durability of building structures, 1983 Building Research Institute Direct Research Project.

2. Kawakami, H.: Summaries of Technical Papers of the Architectural Institute of Japan, Materials Session, Sept. 1983.

3. Kashino, N.: Summaries of Technical Papers of the Architectural Institute of Japan, Materials Session, Oct. 1984.

4. ISO/TC 156: Classification of corrosivity of atmospheres, 1983.

5. Ministry of Construction Building Research Institute: Building Research Institute Research Report No. 90, Mar. 1980.

6. Oshiro, T.; et al.: Roads and Concrete, No. 60, June 1983.

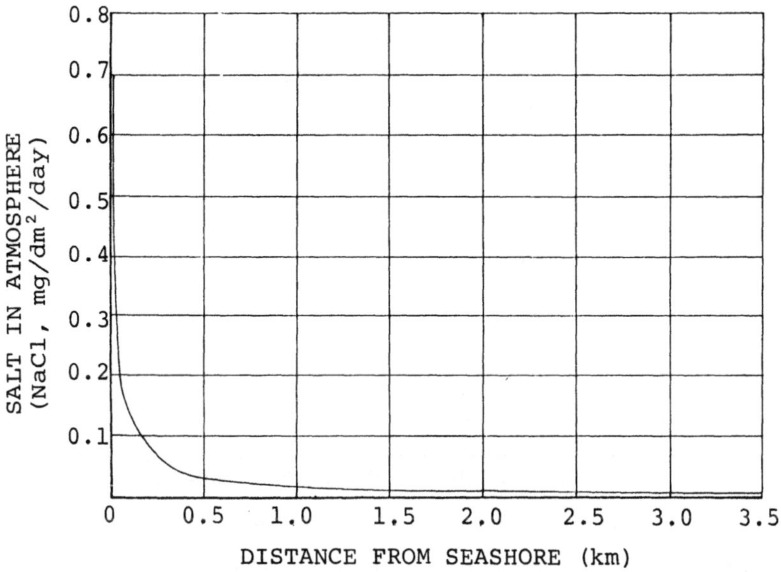

Fig.1. Measurements at City A. (Remarks: 1. Measurements by gauze method at site opened out toward sea (maximum value averages). 2. Measuring points City A at 20 locations at different distances from seashore.)

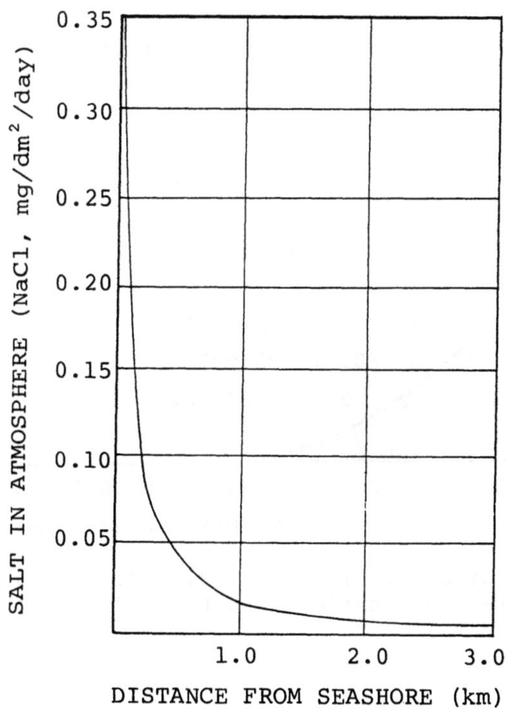

Fig.2. Measurements at City B. (Remark: Measurements by gauze method along river at City B on Sea of Japan side (averages of all data).)

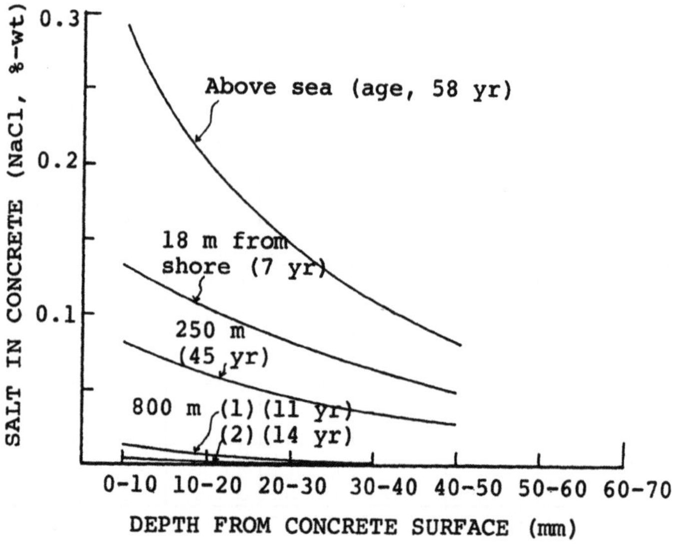

Fig.3. Quantity of salt penetrated--investigations of influences according to different distances from seashore [1],[2]. (Compared with concretes of water-cement ratios 0.60 to 0.70 and not previously containing salt.)

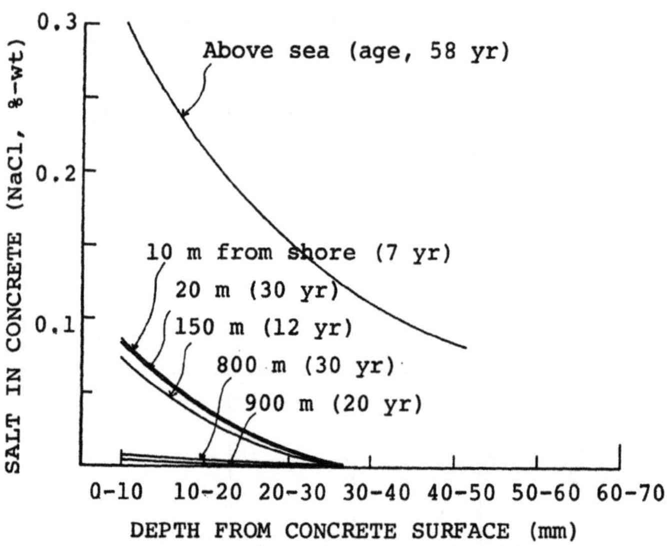

Fig.4. Investigations in 1983 [1],[2]. (Compared with concretes of water-cement ratios 0.60 to 0.70 and not previously containing salt.)

Table 1 Salt Contents in Atmosphere vs, Amount of Salt Penetration into Concrete

Distance from shore	Up to about 200 m	Up to 1 km	Up to 10 km	More than 10 km inland
Salt content in atmosphere (Nacl)	0.3- 0.4 mg/dm^2/day (sea water droplets adhere directly in cases.)	0,2 mg/dm^2/day	Salt exists to an extent under special conditions such as during typhoons.	Almost no salt exists in atmosphere.
Salt penetrated in at surface part of concrete	Far exceeds 0.03%*	Exceeds 0,03% at 500 m from shore. Does not exceed 0.03% at 800 m.	-	-

Note: *Salt content to cause corrosion of reinforcing steel (NaCl %-concrete weight)

Table 2 Salt Damage Classifications of Reinforced Concrete Buildings with Distances from Seashore as Indices

Salt damage classification	Heavy salt damage area	Salt damage area	Semi-salt damage area	General area
Distance from shore	0- 200 m	Up to 1 km	Up to 10 km	Other than the left
Condition	Subject at times to direct spraying of sea water under normal conditions. (Includes atmosphere at sea--landing pier, etc.)	Area adjacent to heavy salt damage area. Area liable to suffer salt damage from seasonal winds and minor typhoons.	Area along river or plain open toward sea, where salt damage suffered in past due to strong seasonal wind, typhoon, etc. (Inland parts of Okinawa, etc.)	Area thought to be practically free of salt damage.
	(Excludes areas surrounded by structures and high land with no fear of salt damage.)			
Necessity for countermeasure	Effective finishing materials and corrosion-resistant reinforcement needed to be used without fail.	Finishing materials to be selected as necessary.	Suitable finishing materials to be used in cases.	No countermeasure necessary.

Durability of Reinforced Concrete Structures in Marine Environment

T. Sakugawa[1], M. Sakuta[1] and K. Kishitani[2]

[1] Department of Oceanic Architecture and Engineering, College of Science and Technology, Nihon University, Narashinodai, Funabashi, Chiba, 274 Japan
[2] Department of Architecture, Faculty of Engineering, The University of Tokyo, Hongo, Bunkyo-ku, Tokyo, 113 Japan

ABSTRACT

This report presents the result of the investigation of corrosion of reinforced steel and concrete deterioration.

The corrosion of reinforced steel is the most critical problem to affect the durability of reinforced concrete in marine environment. When reinforced steel rust, the corroded products generally occur considerably more volume than that of the original steel. This expansion depend on corrosion of reinforced steel causes cracking and spalling of the cover concrete which accelerates the rate of corrosion.

On the result of this investigation, the reinforced concrete structures in marine environment, the corrosion of reinforced steels are affected by the contented salt (use of beach sand and the penetration of chloride) in concrete. In particular, the structures built in the coastal zone suffered heavy damage due to the penetration of sea-salt particles.

The relation of the sea-salt particles and the penetration of chloride into concrete is, when the sea-salt particles volume is greater than 0.12mg/dm²/day, the penetration of chloride volume tends to increase suddenly.

INTRODUCTION

Reinforced concrete structures in marine environment is different from those on land and used under the severe corrosive conditions of sea where the durability of structure due to the deterioration of materials is a problem. In particular, on the reinforced concrete structure in marine environment, corrosion of steel bar are affected by the permeation of salt from outside (Cl^- in particular) and the behavior salt in concrete.

Chloride ions may be present in concrete structures either as constituents of the original mix or by diffusing into hardened concrete from the environment. In the latter case, it is recognized that the chloride level is dependent on factors difficult to difine in practical situations (permeability of the concrete, effect of cracks, proportion of chloride ions trapped by the concrete), and high levels of chloride can penetrate to the steel reinforced even in high quality concrete, in relatively short periods of time.

This report presents the result of the investigation of corrosion of reinforcing steel and salt attack on reinforced concrete in marine environment.

INVESTIGATED OBJECTIVES

Investigated objectives are 176 buildings of 76 schools in Okinawa Prefecture, the primary and middle schools. Investigations were carried out for 19 buildings of seven schools in 1980 and 26 buildings of 12 schools in 1981 and 112 buildings of 47 schools in 1982, and 30 buildings of 12 schools in 1983 respectively. Ages of objectives are 8 to 29 years when investigated and which are all reinforced concrete.

There are various locations in coastal zone 0.6 km and inland zone 3.0 km from beach. They are, however, not always affected by sea salt particles. Further, there is less influence of polluted atmosphere containing sulphur dioxide. The locations of objectives are shown in Fig. 1.

WEATHER OF OKINAWA PREFECTURE

Okinawa Prefecture is located in the southernmost place of Japan, consisting of many islands. The weather is featured subtropic, being warm during the year at an average temperature of 22.2°C. Days below 10°C are counted less even in wintertime. Humidity is high, recording 77% as the yearly average. Rainfall is more than the average amount of Japan but not particularly much. Compared with that of Tokyo the amount is about 1.6 times more. Large amount of rainfall at a time is a feature according to the squall and typhoon which are particular in the subtropic zone.

INVESTIGATION METHOD

Investigation is made mainly to observe concrete of main structural members of each building and deterioration of internal reinforcement. For places of investigation main structural members were selected enabling the grasp of deterioration of entire building. On the members other than main structural members, the places on finished mortar and wing walls where the deterioration was to have thought clearly progressed were recorded.

Condition of Concrete

1) Carbonation Depth of Concrete

By spraying alcoholic solution containing phenolphtalein of 1%, the depth from concrete surface up to colored part is measured and recorded. Average carbonation depth on each investigation place is recorded as a carbonation depth of the place.

2) Estimation of Compression Strength of Concrete

On the place same to that of 1), the test tc estimate compression strength is carried out by means of a Schmidt concrete test hammer. The hammer is used for 20 times at an interval of 3cm, on upper, lower, right and left places, after making the concrete surface smooth. The compression strength was estimated according to the following equation.

Compression strength : $F(kg\ f/cm^2) = -110 + 10.0\ Ro$

Ro : standard hardness $Ro = R + \Delta R$
R : average value of measured hardness
ΔR : correction value of impact direction
(0 for the present investigation provided
ΔR of slab = - 6)

3) Execution Degree of Concrete

In execution of concrete on the investigation place, pit corrosions and cavities are recorded, if any.

Corrosion on Reinforcement

Concrete on each investigation place is scraped off until the reinforcement is completely exposed and the corrosion on exposed reinforcement is checked according to the corrosive forms. In this case, recording is to be made by classifying roughly the corrosion into type of slow and even corrosion and type of quick and deep pit corrosion progressing in a part. Both types are indicated by the rate of corrosive area. Indication is made by the scale mentioned in Table 1 for convenience.

Degree of corrosion is indicated as A: no corrosion at all, B: even corrosion slightly and C: pit corrosion slightly.

Investigation on Building Interior and Exterior by Visual Inspection

On the interior and exterior of each building, the phenomena proving the deterioration such as lifting, exfoliation and crack are observed with the naked eye and the positions and sizes of remarkable corrosions are recorded by means of a camera.

Analysis of Salt Content in Concrete

Salt content in concrete is measured with a piece scraped out as in 1) according to the Volhard method.

Sampling of Sea Salt Particles

Sampling of sea salt particles were carried out by use of gauge exposure method and chemical analysis of sea salt particles is measured with Volhard method.

INVESTIGATION RESULTS

Internal Reinforcement

1) On a place where the internal reinforcement is corroded, crack appeared in concrete according to the corrosive expansion of internal reinforcement, and in the remarkable case many places without concrete, being separated and exfoliated, were observed.

2) On poor concrete execution place with pit corrosions and cavities, the internal reinforcement offers particularly the obvious corrosive forms like layer or exfoliation.

3) As concrete contains much salt, corrosion on the internal reinforcement is formed evenly pit corrosion. On the remarkable place, a partly corrosive form is regarded being considered the complex of nearby pittings.

4) On the internal reinforcement slab, concrete is already exfoliated and the internal reinforcement is being weatherbeaten. There are many places where the reduction of cross section of reinforcement is clearly observed.

5) In the relation of carbonation of concrete and rate of corrosive area, the rate of corrosive area of reinforcement becomes larger according to the increase of the carbonation.

6) As to the covering depth of concrete and the rate of corrosive area of reinforcement, there is a tendency that the rate of corrosive area of internal reinforcement becomes smaller accordingly as the covering depth becomes is thicker.

Compression Strength of Concrete

1) For aggregate of concrete, beach sand is used as fine aggregate and crushed calcareous stone as coarse aggregate where concrete itself is brittle like plaster wall.

2) Compression strength of concrete is measured with a Schmidt concrete test hammer. Letting the design strength when constructed be 180 kg/cm² as standard, it is 80 150 kg/cm² on poor execution place and around 200 kg/cm² on ordinary execution place. Particularly, measurement is not pos-

sible on the place of insufficient filling of mortar and considerable number of pit corrosions and cavities.

Salt Content in Concrete

1) Salt content in concrete largely exceeds the Institute Standard, being in the range of about 0.09 1.1% by weight of sand.

2) Salt content tended to decrease according as the objective building leaves beach towards inland. (See Fig. 2.)

3) On the objective buildings of 0.5 and 1.0 km from beach, Fig. 3. gives relation between the elasped year of objective building and salt content. Salt content in concrete tends to increase with age and decrease according to distance from beach.

4) The data in Fig. 4. shows the relation of sea salt particles volume and the penetration of chloride ion volume. The sea salt particles volume is greater than 0.12 $mg/dm^2/day$. The penetration of chloride volume tends to increase suddenly.

Carbonation of Concrete

1) Carbonation of concrete tended to increase with age. Concrete is carbonated in about 25 mm after 20 years of age.

2) Existence of finishing agent largely affects the carbonation of concrete. Types of finishing agent are mortar painting and painting as well as concrete work without finish. Progress of the carbonation of concrete increase in the order of mortar painting finish, painting finish and concrete work without finish.

CONCLUSION

On the reinforced concrete structure in marine environment, concrete and internal reinforcement are affected by the permeation of salts from outside (Cl^- in particular) and the behavior of salts contained in concrete. Moreover, the effects from marine environment to reinforced concrete construction and metals appeared as deteriorations such as carbonation of concrete, corrosion on internal reinforcement, crack in concrete and corrosion on metal. These phenomena cause the important problems of deciding materials, execution management and durability performance of the construction as well as lifetime.

As above, the reinforced concrete structure in marine environment requires sufficient investigation, and the anti-corrosive method and repairing method are necessary to be studied. In addition, when the reinforced concrete structure are to be built in future, the local climate must fully be considered as one of the external environmental factors.

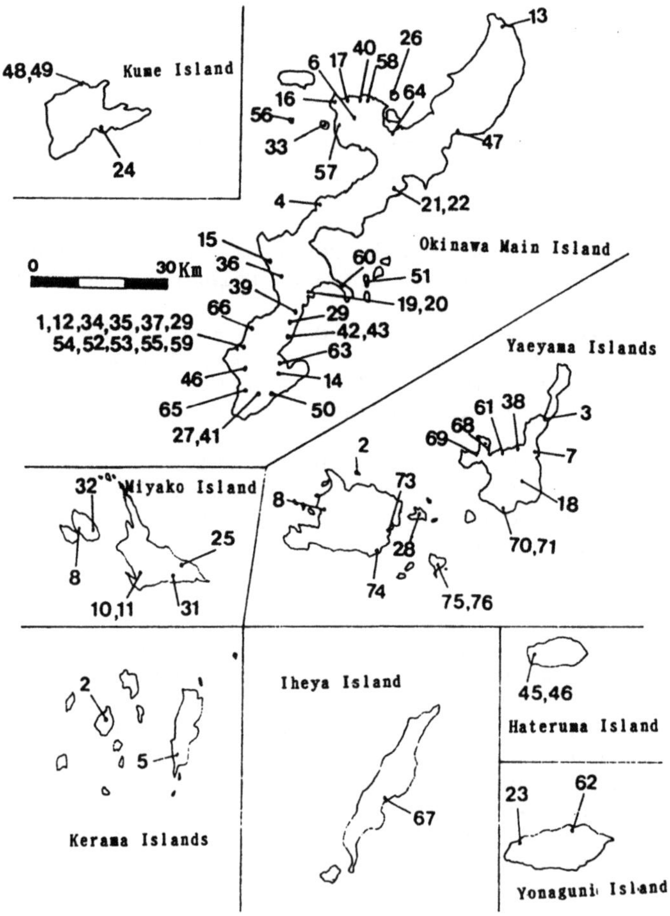

Fig. 1 The location of objectives

Table 1 Corrosion Scale of Reinforcement

Rate of corrosive area, Scale (%)	Condition of corrosion
0	No corrosion at all
5	Infintesimal corrosion. Pit corrosion
15	Corrosion estimated between 5% and 25%
25	Corrosion estimated about 1/4 of surface area (of an objective)
50	Corrosion estimated about 1/2 of surface area
75	Corrosion estimated 50% and 100%
100	Corrosion extended entirely

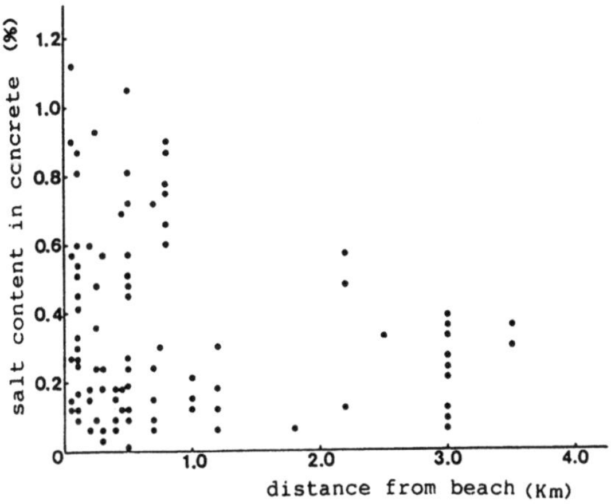

Fig. 2 The relation between distance from beach and salt content in concrete

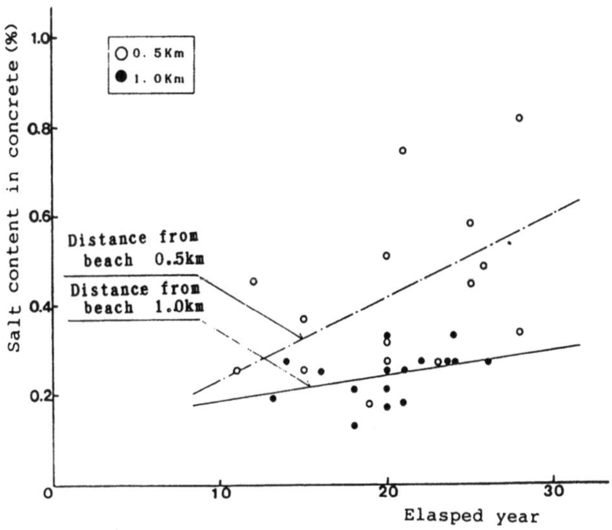

Fig. 3 The relation between elasped year and salt content in concrete

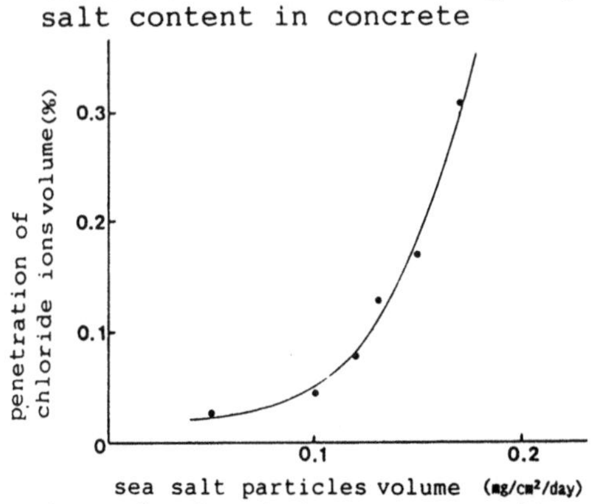

Fig. 4 The relation between sea salt particles volume and penetration of chloride ions volume

Life Time Test of Prestressed Concrete

Various Measures of Protection against Corrosion in Splash Zones

M. Ota[1*], A. Numata [2*] and S. Kondo[3*]

[1] Technical Division, P.S. Concrete Co., Ltd., Tokyo Branch, Kudan-Kita, Chiyoda-ku, Tokyo, 102 Japan

[2] Technical Division, P.S. Concrete Co., Ltd., Marunouchi, Chiyoda-ku, Tokyo, 100 Japan

[3] Technical Division, Oriental Concrete Co., Ltd., Gobancho, Chiyoda-ku, Tokyo, 102 Japan

ABSTRACT

Japan is a long and narrow seagirt country with an oceanic climate, stretching from subarctic to the subtropical zones.

Since the country is apt to be struck by typhoons in the seasons from summer to early autumn and the districts facing the Japan Sea in particular are affected by seasonal winds, splashes of sea salt particles fall on the areas about 10 km from the coastal lines.

Concrete structures have been regarded as semipermanent structures having high durability, but it has become known that concrete also deteriorates the same as other materials, with the result that its durability is also posing a problem from the social point of view.

Paying attention to the salt damages which are recently much talked about, regarding the deterioration of concrete structures, Japan Prestressed Concrete Contractors Association (JPCCA) have decided to carry out a developmental study of highly advanced corrosion preventive technology intended for oceanic structures, especially the concrete structures constructed in splash zones, as well as long-term durability evaluation.

The present study will be carried out as a joint work by Public Works Research Institute, Ministry of Construction and JPCCA with Research And Development Laboratory The Cement Association of Japan, it has been decided to make an overall study by sharing the roles making effective use of the respective strong points. As is evident from Table 1 showing concrete study items and contents, the study mainly consists of experiments.

BASIS OF CORROSION PREVENTIVE TECHNOLOGY FOR CONCRETE

Cover Depth and Mix Proportion

The corrosion preventive effect on concrete with cover depth will be investigated classified by cement type and water-cement ratio. This is the standard of the corrosion preventive technology for concrete, we named it Standard System.

Three types of cements (ordinary portland cement, high-early-strength portland cement and blast furnace cement Class

---

* Members of Technical Committee of Japan Prestressed Concrete Contractors Association

Table 1. Study Content

Major Category	Middle Category	Minor Category	Study content
Corrosion preventive technology of concrete structures in splash zones	(1) Basis of corrosion preventive technology for concrete	1 Cover depth and mix proportion	Identification of the effects of corrosion protection varying with the differences in cement type, cover depth and water-cement ratio.
		2 Identification of internal structure and corrosion preventive performance of concrete	Study of corrosion preventive effects by porosity and curing.
		3 Development of corrosion checking and repairing technology	Research on time dependence of progressive corrosion and repairing materials.
	(2) Corrosion preventive technology for steel in concrete	4 Practical application of epoxy coated steel bars	Performance on corrosion protect of epoxy coating.
		5 Development of corrosion preventive materials for PC tendon	Development of polyethylene sheath and corrosion preventive performance.
	(3) Development of coating materials for coastal structures	6 Practical application of water proof membranes.	Facilitation of development of paint or water proof membranes.
		7 Development of impregnating agents.	Facilitation of development of impregnating agents.
		8 Development of salt-free concrete.	Development of internally sealing mixture.

B) were selected, and water-cement ratios of 40% and 50% were used.

For the protective covering, 4 different thicknesses (2.5 cm, 3.5 cm, 5.0 cm and 7.0 cm) were adopted (See Table 2).

Table 2. Test Beams of Standard System

Cement / Water-cement ratio / Test years / Cover depth		Ordinary cement						High-early strength cement						Blast furnace cement Class B		
		40%			50%			40%			50%			50%		
		2	5	10	2	5	10	2	5	10	2	5	10	2	5	10
Small size test beams	2.5cm	A001	A002	A003	A004	A005	A006	A007	A008	A009	A010	A011	A012	A013	A014	A015
	3.5cm		A016	A017		A018	A019		A020	A021		A022	A023		A024	A025
	5.0cm		A026	A027		A028	A029		A030	A031		A032	A033		A034	A035
Medium size test beams	7.0cm		A036	A037		A038	A039		A040	A041		A042	A043		A044	A045
Types		PC			RC			PC			RC			RC		

Identification of Internal Structure and Corrosion Preventive Performance of Concrete

This test is intended to examine the corrosion preventive effects of concrete resulting from porosity and curing in relation to the standard system. Three types of cement are used. They are ordinary portland cement, high-early-strength portland cement and blast furnace cement Class B in accordance with the standard system, where water cement ratios of 40%, 50% and 60% are used.

Four types of curing methods are employed, i.e. no wet curing, 7 days, 14 days and 28 days wet curing and then stored in air.

Investigation of Corrosion and Development of Repairing Methods

To improve the nondestructive inspection and repairing methods, test beams have been prepared to meet the technological improvement expected in the future. Consequently the test pieces have been classified into 3 groups.

Group 1 consists of the test beams in which steel stresses are 800 kgf/cm^2, 1000 kgf/cm^2, 1200 kgf/cm^2 and 1400 kgf/cm^2. This is intended to test the differences in the damage of corrosion due to stress level. Furthermore this group will be used for accelerated corrosion tests due to cracks in the test beams.

Group 2 consists of the test beams of full-sized bridge girder (girder length : 5.00 m). This gruop will be mainly used to examine the degree of corrosion for various cross section of bridge girder and furthermore to improve the repairing methods.

Group 3 consists of the test beams for investigation of corrosion by comparison with the test beams which are given corrosion preventive measures.

CORROSION PREVENTIVE TECHNOLOGY FOR STEEL IN CONCRETE

Practical Application of Epoxy-Coated Steel Bars

This test is intended to test the corrosion preventing performance of epoxy-coated steel bars and how to apply practically.

Epoxy-coated steel bars can be expected to prevent corrosion because they themselves have an electrically insulating property even if surrounded by the chlorid ions.

Galvanized steel bars will also be tested for comparison.

Development of Corrosion Preventive Materials for PC Tendon

This test is intended to develop protecting sheaths to replace the conventional steel sheaths having no corrosion resistance. Paying attention to the quality of polyethylene, attempts will be made to test the corrosion preventing effect and applicability of polyethylene sheaths. In the present test, polyethylene sheaths having a nominal diameter (inside diameter) of 30 mm, outside diameter of 42 mm, thickness in the range from 0.8 mm to 1.2 mm and spiral pitch of 9.5 mm are used as the post-tension members.
For the erection bars, epoxy-coated steel bars are used.

DEVELOPMENT OF COATING MATERIALS FOR COASTAL STRUCTURES

Practical Application of Water Proof Membranes

To prevent or inhibit the permeation of chlorid ions into concrete by applying a coating on the concrete surface, the test is carried out to research the effect of inhibiting permeation of chloride ions by the coating membrane with respect to various types of coating materials.

The coating membranes are classified into three groups as described below.

Group 1 is for prestressed concrete members, and the coating membrane is not required, so high crack followability (1% or more).

Group 2 is for reinforced concrete members, and the membranes required crack followability (4% or more).

Group 3 is for long term corrosion preventive coatings intended for application on the concrete structures in which particularly severe corrosion is anticipated.

For these three groups, it has been decided to carry out the test with respect to 15 types of coating.

Development of Impregnating Agents

The purpose of application of the impregnating agents is same as water proof membranes described in Practical application of water proof membranes.

Because of their characteristics, the impregnating agents intended for application on the prestressed concrete members of which crack followability is not required, and it has been decided to carry out the test on 7 types of impregnation.

Development of Salt Free Concrete

To prevent or inhibit penetration of chloride ions into concrete members by making the concrete structures fine by adding an internally sealing mixture.

In the present test, silica fume has been adopted as the mixture.

CONCLUSION

A total of 154 test beams were prepared in April 1984. In addition, a total of 264 cylinders (ø150 x 300) also were prepared to provide an auxiliary role.

These both types of test pieces were set up in the Ministry of Construction's Ohigawa Research Station installed off River Ohi in Suruga Bay, and exposure tests have already started.

This exposure test has just started and the period of study is scheduled for ten years at present.

The social needs call for urgent estabrishment of protecting measure from corrosion. However, since the state of development varies with study item, much depends on the future progress of technology and development of new materials.

Under these circumstances, we plan to modify the test items and test methods conforming to the technological progress.

We should feel greatly rewarded for our effort, if the present study should prove helpful to the improvement of durability of concrete structure.

Developments of Protection Technologies for Concrete Structures in Marine Environment

K. Katawaki and M. Makita

Public Works Research Institute, Ministry of Construction, Toyosato, Tsukuba, Ibaraki, 305 Japan

INTRODUCTION

Concrete structures have been wide-spreadly constructed since concrete structures are recognized to be reliable and minimum maintenance structures. Onshore concrete structures, a part of marine coast highways, bay bridges and artificial islands are placed in offshore areas around the Japanese coast.

Meanwhile it was found out that some concrete road structures on sea coast were deteriorated by the corrosion of reinforcemnnt in concrete. The improving durability becomes the center of interest. It is considered that the corrosion of reinforcement is caused by the chloride which are brought from salt paricles in oceanic wind.

It can be declared that the permeation of salt particle into concrete is the main reason of the corrosion of reinforcement, because it was made sure that the surface layer of the deteriorated concrete structures had remarkably high salt contents.

The corrosion of reinforcing steel in concrete structures built in an environment subjected to the influence of sea water as in or on the sea, induces cracks and delamination of the concrete. However, the corrosion behavior of reinforcing steel in concrete as related to the cement paste quality, coverage over steel, and presence of cracks have been not clarified. To prevent such deterioration, the reinforcement should be covered with good quality and thick cement paste, and the generation of crack must be controlled perfectly, or directive protective methods. The protection technology for concrete structure is required.

This paper summarizes mainly previous PWRI works related to concrete deterioration and describes the brief current status of the corrosion and protection technology for improving durability of concrete structures.

CORROSION OF REINFORCING STEEL

Steel in passive state is well effectively protected from corrosion, since the alkalinity around steel is 12.5 caused by the saturated hydroxide calcium in pore water in good quality concrete and a protective layer can be formed on the steel surface.

The corrosion of steel in concrete is an electrochemical process and the reaction mechanism can be described by electrochemical diagram.

Electrochemical technique is applied for measuring corrosion behavior. Potentiodynamic polarization technique can identify the type of corrosion; pitting corrosion, general corrosion or passivated state. Its technique can allow characterization of the active or passive region of steel surface. This technique involves measurement of log current as the potential of the steel specimen is scanned slowly from active to passive potential regions.

Corrosion tests were performed on the solutions in which chloride content, oxygen concentration and alkalinity were simulated to those in concrete. Based on simulated corrosion tests, the conclusion can be made that the reinforcing steel even corroded in high alkaline when the chloride content was above threshold values which were between 0.7 and 2.0 kg/m^3. Corrosion current increased as concrete got more chloride.

Macro cell corrosion is dictinctive in actual marine concrete structures as shown in Fig.1. The part above the tidal zone have noticeable corrosion produced, and this part corresponded to valley (anode) in the distribution of potential in the Figure. Irregularity in the potential distribution was measured corresponding to corrosion of reinforcing steel. For the cause of such irregularity, it is considered that the initial un-uniformity of concrete or crack of concrete caused a difference in the permeation of chloride and oxygen from the outside and the irreguarality of the electric resistance. With an irregularity produced in the distribution of the potential, a macro cell current flows into that part which is of lower potential (valley part) to localized corrosion. Current from passive area to the another area produces localized corrosion on steel surface. Usually a flow of current in the steel from an anodic area to cathodic area is observed. Reinforcing steel is apt to corrode with a current flow. As the part above the tidal zone have noticeable corrosion produced, the reinforcing steel having a constant current is between below and above sea water.

PENETRATION OF CHLORIDE

A result of typical chloride analysis of concrete specimen is shown in Fig.2.

The chloride concentration decreased sharply from the concerete surface to the inner of the concrete in marine environment. This phenomenon shows thrt the chloride in concrete is mainly penetrated and diffused from the surface. The rate of penetration of the chloride into the inner concrete varies with the concrete quality, especially the water-cement ratio, being greater in the case of the water-cement ratio of 0.70 than in the case of 0.40. Penetrated chloride to the steel deteriorates the passivity film locally and the corrosion cell is produced on the surface even among high alkalinity.

The chloride content was in the surface layer at the splash zone rather than the atmospheric or submerged zone. It is considered that the chloride was concentrated by wet and dry cycle at concrete surface.

DETERIORATION MECHANISM

Deterioration process involves many phenomena: chloride penetration, chloride accumulation, start of steel corrosion, crack initiation and crack propagation. The deterioration process is proposed in Fig.3. A chain of deterioration process can be explained chronologically. Although crack propagation seems to be the most influential to the structure failure, the chloride penetration is the first stage of the tragic chain reaction.

COVER DEPTH AND CRACK WIDTH VERSUS STEEL CORROSION

Results of the visual examination of the reinforcing steel taken out of the concrete specimens which were exposed in Tokyo Bay for 1000 days, are shown in Fig. 4. With the cover depth of 2 cm, both corrosion length ratio and corrosion degree of the reinforcing steel were large in splash zone, and without exception the specimens had a crack of maximum 0.3 mm due to corrosion of the steel, with particularly severe corrosion of reinforcing steel noted at the place of crack. With the cover depth of 7 cm, the specimen with the water-cement ratio of 0.55 had no corrosion of reinforcing steel noted whether they were mixed with service water or sea water. However when the water-cement ratio was 0.70, a degree of corrosion of about several percent in the corrosion length ratio was noted in the specimen. With the cover depth of 12 cm, no corrosion of the reinforcing steel was observed.

Results of the visual examination of the reinforcing steels taken out from the concrete specimens, exposed for 1000 days with the crack width kept constantly, is shown in Fig. 6. When mixed with service water, greater crack width tends to give greater degree of corrosion. To prevent the generation of noticeable corrosion in the reinforcing steel under such exposure condition the crack width (width of a crack located on the concrete surface) must not be greater than 0.1mm. When mixed with sea water, the specimens with a crack width of about 0.3mm had no corrosion, while those with a crack width of about 0.05 mm had great corrosion produced. Thus no relationship is observed between the crack width and the corrosion of reinforcing steel. Mixing with sea water involves a possible risk since such a small crack is apt to produce appreciable corrosion.

It is seen that both the cover thickness and water-cement ratio have a great influence on the corrosion of reinforcing steel, and also the crack width of concrete have much influence under thick coverness.

PROTECTION OF REINFORCING STEEL

Concrete structure built in an environment subject to the influence of sea water or sea wind induce cracks and deterioration of concrete. To prevent such corrosion, it is necessary that care is taken the design and the mixing and placing of the fresh concrete and to take crack generation precaution. At some cases, corrosion of the reinforcement may be avoidable without additional direct protective measures.

Many kinds of protection methods have been developed recently and they show specific features and application area. They are classified according to the protective function.

1) coated reinforcing steel
    - epoxy coated reinforcing steel
    - galvanized steel bar
2) concrete surface sealing
    - polymer impregnation
    - concrete surface sealer
3) concrete surface coating
4) cathodic protection

The evaluation of these protective methods and material improvement tests have been performed in PWRI. For instance in Tokyo Bay field exposure tests on the corrosion preventive effects of epoxy coating bar, galvanized bar, concrete surface coating material and cathodic, protection have been experimented.

The results of experiments of the concrete specimen exposed shows as belows;

The epoxy coated steel bars showed good corrosion resistant performance. The best in protective performance among the epoxy coating is the powder type epoxy. For protective performance, a coating thickness of minimum 150 m, but for good bonding to concrete the thickness is less than 150 m. The coating thickness of 150 m is considered to be optimum. Galvanized steel bar gave protective performance but was not always satisfactory in the concrete exposed to very corrosive marine atmosphere. The galvanized steel bar had not only white rust of zinc but also spot rust of steel in those portions of specimen since galvanized treatment had little cathodic protection effects in splash zone. Urethane coating had no deterioration. The result of cathodic protection or reinforced concrete showed good protective effects in the tidal and in sea water. The optimum potential to be applied is -1.0 to -1.2V. When it is more noble than -0.8 V, the protection effect is not satisfactory, and when lower than -1.5 V, over protection may result. From the measurement of the current density of reinforcing steel having cover of 2 cm and 7 cm thickness, were obtained values of current density required for cathodic protection under submerged conditions and near the water line; 5 mA/m for steel with a cover depth of 7 cm; 30 mA/m for steel with a cover depth of 2 cm; 30 mA/m for steel with a cover depth of 7 cm in cracked concrete.

A GUIDELINE TO PREVENT CORROSION DAMAGE

In Feb. 1984, "Tentative guideline of countermeasure of road bridges in sea side area to prevent corrosion" was published from Japan Road Association. The brief contents of the guideline are shown as follows;

Japan is devided to three areas according to the severity of the environment and classification of countermeasure is ruled.

Three methods are adopted in the guideline, which are the increasing cover depth on reinforcement, the use of epoxy coated reinforcement, and the coating on concrete.

a) Cover thickness

The method, which will be considered at first, is the increase of the cover thickness on reinforcement. The minimum cover thickness are as shown table, according to the countermeasure classification.

minimum cover depth (cm)

Class	Superstructure			Substructure	
	deck railing cap	beam		beam	column
		pretention precast PC	other		
I	5.0	5.0	7.0	7.0	7.0
II	4.0	3.5	5.0	5.0	5.0
III	3.0	2.5	3.5	3.5	4.0

b) Epoxy coated reinforcement or coating on concrete

When the minimum cover depth is not adopted, the use of exposy coated reinforcement or the coating on concrete must be chosen. When epoxy coated reinforcement is used, allowable bond stress for design is 80% of normal one, and it is required to have enough workability and durability in situation, enough bond strength to concrete and alkali-resistnace. When the coating on concrete is used, coating material must be chosen among the following three kinds;

A Epoxy and polyurethane
B Soft type epoxy or soft type polyurethane
C High build type epoxy or glassflask polyester

A is applied to the prestressed concrete structure, B is applied to reinforced concrete and C is coated for important concrete structure or the one in which repairment will be difficult.

FURTHER RESEARCH ON GOINT ON

The new exposure tests in the following research on protection methods of marine concrete structure have been started from May 1984 at the Suruga marine exposure test station, which was constructed by the national research fund for promoted by The Science and Technology Agency.

1) Protective performance and durability of paints on concrete
2) Protective performance and durability of concrete using with epoxy coated reinforcement
3) Materials development and durability of concrete sealer and polymer impregnation
4) Durability of chloride resistant concrete using with silica-fume

REFERENCES

(1) Tentative guideline of countermeasure of road bridges in sea side area to prevent corrosion, The Japan Road Association (Feb. 1984)
(2) Makita, Mori, Katawaki, Marine corrosion behavior of reinforced concrete exposed at Tokyo Bay. Performance of concrete in marine environment (1980)
(3) Makita, Mori, Katawaki, Performance of typical protection methods for reinforced concrete in marine environment, Performance of concrete in marine environment (1980)

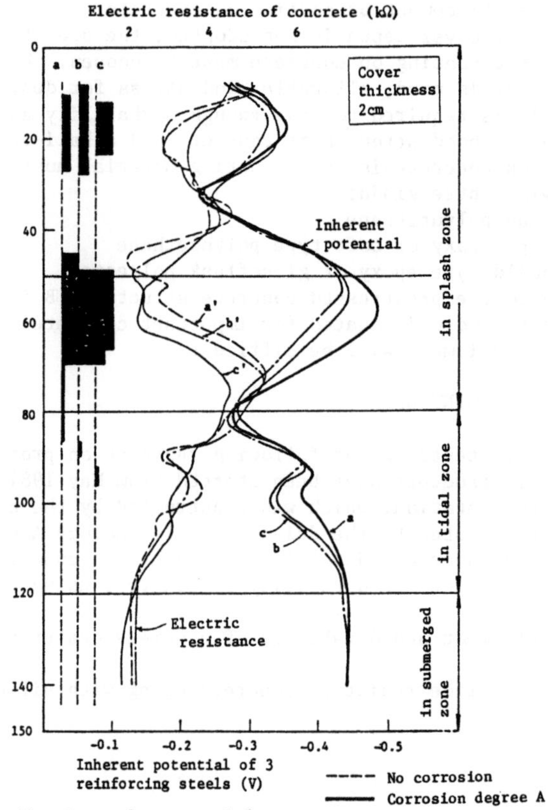

Fig. 1 Distribution of potentials of concrete specimen exposed on/in sea water

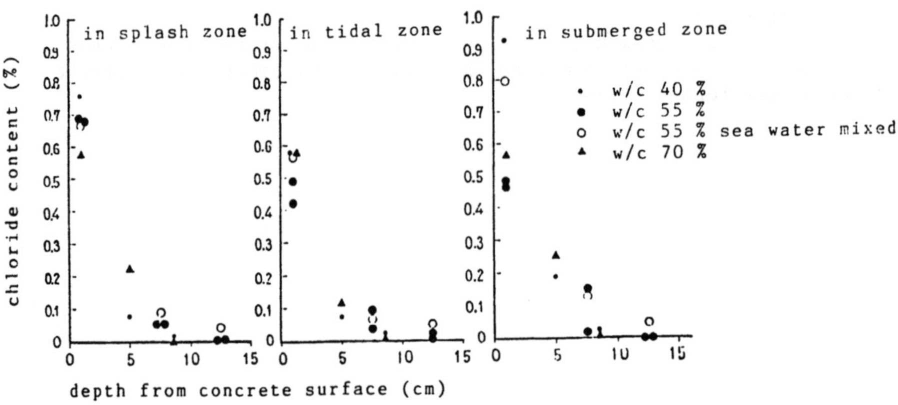

Fig. 2 Chloride analysis of concrete specimen exposed on/in sea water

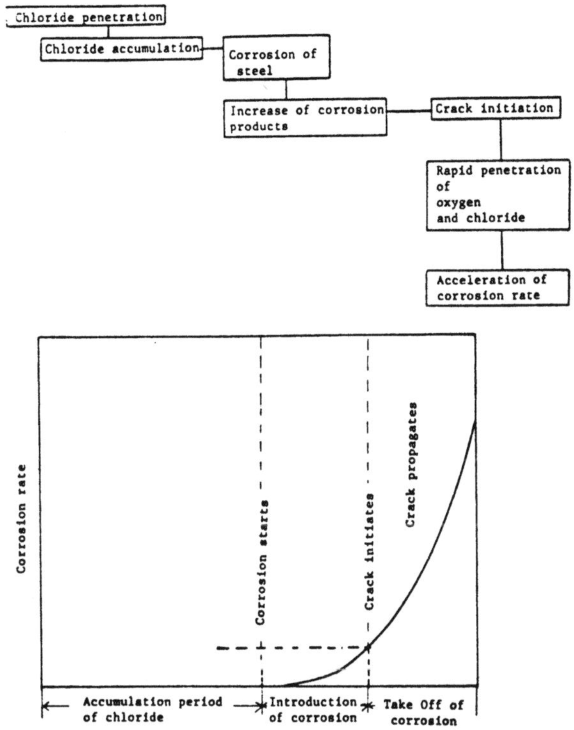

Fig. 3 Deterioration mechanism of reinforced concrete in chloride atmosphere

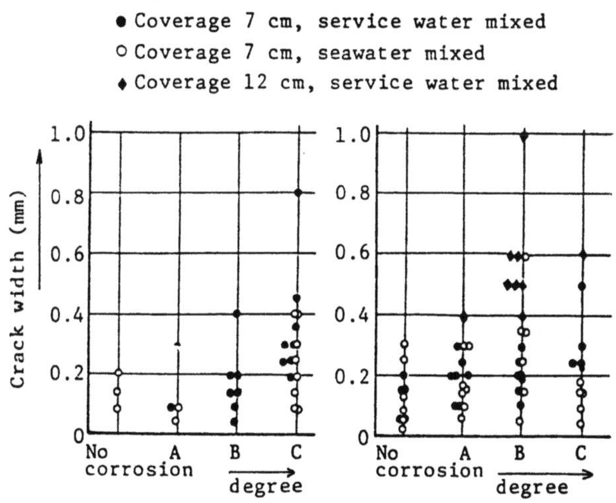

(a) Specimens exposed in splash zone

(b) Specimens exposed in tidal zone

Fig. 4 The relation between crack width and corrosion degree of reinforcing steel

Technical Session:
Materials and Construction

3. Construction

The Strength of the Drill Pipe for the Offshore Drilling

N. Tsukada

Department of Mechanical Engineering, College of Industrial Technology, Nihon University, Izumicho, Narashino, Chiba, 275 Japan

ABSTRACT

The study aims to clarify the influence of deformation of drill pipe inside the raiser pipe due to axial load and torque for the off-shore boring technique.[1] To estimate the deformation of practical pipe of alternately stepped cross sections jointed with screwed thread's fittings is so trouble-some that the ideal equivalent pipe with uniform cross section is introduced. The deformation of this ideal equivalent pipe is discussed. The most important points of this problem are made a proposal to predict the ideal equivalent pipe with sufficient effects of the drill pipe with alternately stepped cross sections in practical use.

INTRODUCTION

The drill pipe for the off-shore boring consists of a long jointed pipe with alternately stepped cross sections. There happens to occur the trouble of deformation of the drill pipe inside the raiser pipe due to axial load that is bit load for high efficient boring and torque. Accordingly,it would be pointed out that the longer pipe has the more serious problems on the boring. In the off-shore boring,there are too many uncertain factors to the trouble of deformation of drill pipe. This paper discusses on the deformation of the drill pipe inside the raiser pipe due to simultaneous action of axial load and torque on boring. The drill pipe is treated with a pipe with uniform section.[1,2] In this paper,the validity of the introduction of ideal equivalent pipe with uniform cross section has been critically discussed instead of a pipe with alternately stepped cross sectional area. The ideal equivalent pipe has the characteristics of the two equivalent pipes with uniform sectional area to axial load and torque.
The first one is provided with bending rigidity of Euler's buckling in equivalence of a pipe with alternately stepped cross sectional area at the same length and axial load. And the second is provided with torsional rigidity on torsinal free vibration in equivalence of alternately stepped cross sectional pipe at the same length and frequency.
The relations of the ideal pipe are given with the expressions in making use of formulae. With this ideal equivalent pipe, the influence on the drill pipe due to axial load and torque is studied.

CONCEPT OF ASSUMPTION AND METHOD FOR ESTIMATION OF DRILL PIPE

ASSUMPTION OF DRILL PIPE

1. Rotational rate of drill pipe is low enough to be negligible.
2. The top and bottom of drill pipe are fixed.
3. Except axial load and torque to drill pipe,another external force is not acted on it,and the raiser pipe stands still.[2] And the circulating water has not.

METHOD FOR ESTIMATION

IDEAL EQUIVALENT PIPE AND EQUIVALENT PIPE

To estimate the strength of drill pipe judging from deformation of the pipe due to axial load and torque, ideal equivalent pipe is introduced. Ideal equivalent pipe consists of one equivalent pipe characteristic of bending rigidity EIm of Euler's buckling and another pipe characteristic of torsional rigidity GJm with uniform cross section. The one equivalent pipe with uniform cross section is provided with bending rigidity EIm of Euler's buckling in equivalent of a pipe with alternately stepped cross sections at the same length L and axial load P. The other is provided with torsional rigidity GJm of torsional free vibration in equivalent of an alternately stepped cross sectional pipe at same length L and frequency fn. With such ideal equivalent pipe,the influence of deformation on drill pipe due to axial load P and torque T is studied.

PROCEDURE OF CONCEPT ON METHOD FOR ESTIMATION

The procedure of estimation for deformation of drill pipe due to axial load and torque is summarized with the flow chart on the above concept in Table-1.

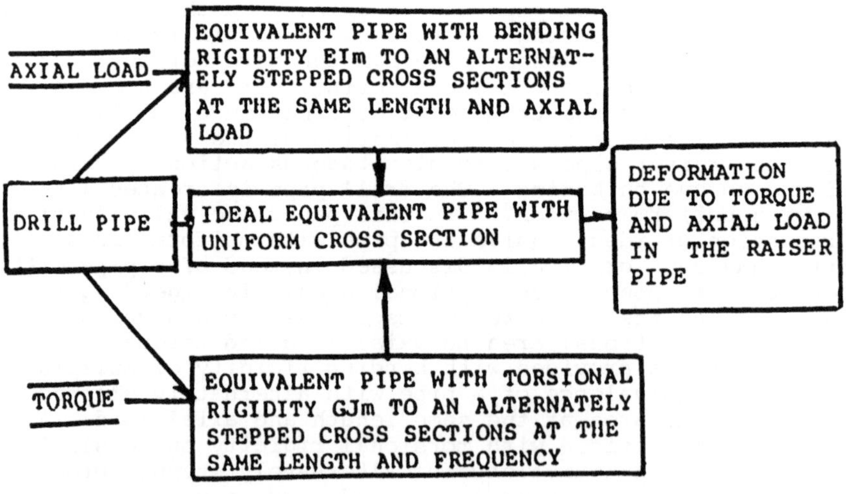

Table-1 METHOD FOR ESTIMATION OF DRILL PIPE

OUTLINE OF DRILL PIPE AND BIT LOAD

CONFIGURATION OF DRILL PIPE/IDEAL EQUIVALENT PIPE

The drill pipe stands inside the raiser pipe, and it is jointed alternately with screwed thread's fittings for connections of its pipe what is called tool joint shoulder. The configuration of drill pipe for the off-shore boring is shown in Fig 1. The relations between equivalent and ideal pipe with uniform cross section to drill pipe are shown in Fig-1, and Fig-2.

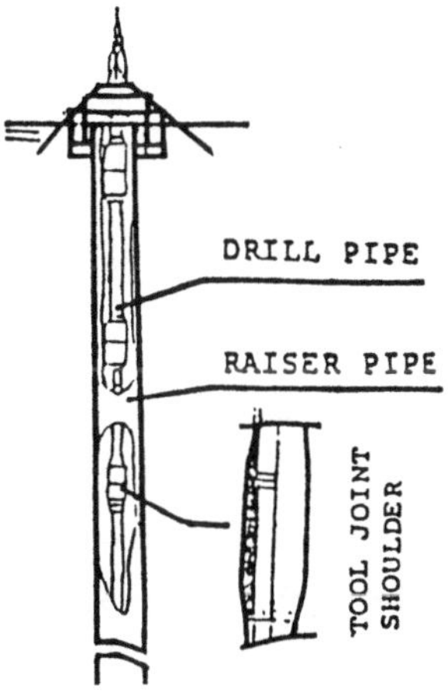

Fig-1 OUTLINE OF DRILL PIPE AND RAISER PIPE

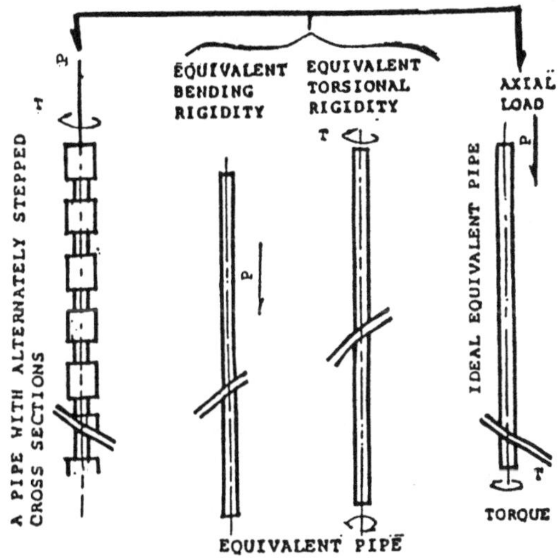

Fig-2 EQUIVALENT PIPE AND IDEAL EQUIVALENT PIPE

OUTLINE OF BIT LOAD/AXIAL LOAD OF DRILL PIPE

The drill pipe would be loaded "compression" and "tension " to the boring in order to rid the dead weights of itself, and keep on constant bit load because of prevention against buckling due to axial load. (refer Fig 3)

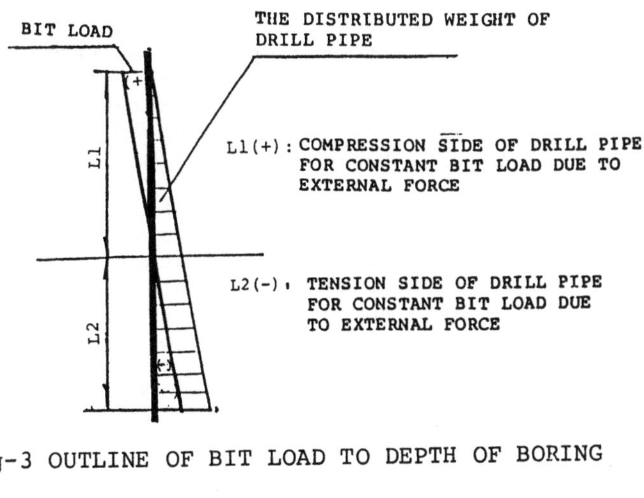

Fig-3 OUTLINE OF BIT LOAD TO DEPTH OF BORING

SIMULATIONS FOR THE ESTIMATION OF DRILL PIPE

EQUIVALENT BENDING RIGIDITY

It is given with the following expressions[3)4)] to relation that equivalent pipe with uniform section is provided with bending rigidity EIm of Euler's buckling in equivalence of a pipe with alternately stepped cross sections at the same length L and axial load P. The characteristics of Euler's buckling load due to alternately stepped cross sections' increment give asymptotical buckling load Pcr. With Pcr,equivalent rigidity EIm in equivalence of a pipe with alternately stepped cross sections can be obtained.

In case of odd number of total small pipe n

$\eta_1 = D$

$\eta_2 = (AD+B)/D$

$\eta_3 = [(A^2+B)D+AB]/(AD+B)$

$\vdots$

$$\eta n = \frac{[I]nD + [II]n}{[III]n + [IV]n}$$

In case of even number of total small pipe n

$\eta_1 = F$

$\eta_2 = (AF+B)/F$

$\eta_3 = [(A^2+B)F+AB]/(AF+B)$

$\vdots$

$$\eta n = \frac{[I]nF + [II]n}{[III]n + [IV]n}$$

(General terms)

$$\eta n = T+C \quad \cdots\cdots 1)$$

Where the relation of the above expressions is given with $A=-(U+V)/(h+k)-(h+k)$, $B=-UV/(h+k)^2$, $C=-U/(h+k)-h$, $D=V/(Q-k)-k+C$, and $F=S+C$. And where $\alpha=\sqrt{P/(EI_1)}$, $\beta=\sqrt{P/(EI_2)}$ $k=1/(\beta\tan\beta a)$, $h=1/(\alpha\tan\alpha s)$ are given. $S=-1/(\beta\tan\beta a/2)$, $T=-\tan(\alpha s/2)/\alpha$, $Q=-1/(\alpha\tan\alpha s/2)$, $U=-1/(\alpha^2\sin^2\alpha s)$, $V=-1/(\beta^2\sin^2\beta .a)$ are given. And where [I],[II],[III] and [IV] are described on the reference 4. From 1) T,C and ηn are function of P. Accordingly,Pcr buckling load is gotten. As result, equivalent bending rigidity of buckling load of a pipe can be obtained by relation $Pcr = \pi^2 EIm/L^2$. That is $EIm = Pcr.L^2/\pi^2$.

EQUIVALENT TORSIONAL RIGIDITY

It may be expressed in the following relation[5)] that equivalent pipe with uniform section is provided with torsional rigidity GJm at torsional free vibration in equivalence of a stepped cross sections at the same length L and frequency fn. The characteristics of the first mode of torsional free vibration due to alternately stepped cross sections'increment give asymptotical frequency fn. With fn,equivalent torsional rigidity is introduced. In relative expressions above described buckling,rearranging $\alpha=(GJp1/s)\sqrt{Ip1/K1}\cdot\omega$, $\beta=(GJp2/a)\cdot\sqrt{IP2/K2}\cdot\omega$, $h=1/[(\alpha/GJp1)\tan\Gamma s]$, $k=1/[(\beta/GJp2)\tan\Theta a]$, $S=-1/[(\beta/GJp2)\tan\Theta a/2]$, $T=-(GJp2)\cdot\tan(\Gamma s/2)/(\alpha/GJp2)$, $Q=-1/[(\alpha/GJp1)\tan\Theta s/2]$, $U=-1/[(\alpha/GJp1)^2\sin^2\Gamma s]$, and $V=-1/[(\beta/GJp2)^2\cdot\sin^2\Theta a]$, A,B,C,and F above described expression would be function of torsional natural angular frequency ω. Accordingly, from equation 1) ω can be obtained along axial length of alternated stepped cross sections of a pipe,L. The modulus of elasticity in shear G of a pipe with uniform section is obtained due to the obtained ω. Where $\Gamma=\sqrt{Ip1/K1}(\omega/s)$, $K1=GJp1/s$, and $\Theta=\sqrt{Ip2/K2}(\omega/a)$, $K2=GJp2/a$ are given. The torsional relation is $fn=[\lambda/(2\pi L)]\sqrt{G\cdot g/\gamma}$ on equiv.pipe. Accordingly,equiv.torsional rigidity GJm is able to obtain. Where $\lambda=\pi/2$ is given and g is acceleration of g and γ is specif.weight.

APPLICATION TO ESTIMATION FOR EQUIVALENT RIGIDITY OF IDEAL PIPE

EXAMPLE OF DRILL PIPE

Example of drill pipe is shown in Table-2. The estimations above the method of description are shown in Fig-5 on the moment of inertia of the cross-sectional area Im for equivalent bending rigidity and in Fig-6 on angular frequency ω of the first mode for torsional free vibration with Jm, the polar moment of inertia of the cross sectional area for equivalent torsional rigidity. Judging from asymptotical buckling load Pcr with alternately stepped cross sections in Fig-4, the Im due to changes of stepped cross sectios to buckling load Pcr is shown in Fig-5. The curve of Im to length is reducing gradually with aysmptote. Accordingly, Im for equivalent bending rigidity and the length of equivalent pipe are set on the stable range of asymptote in Fig-5.

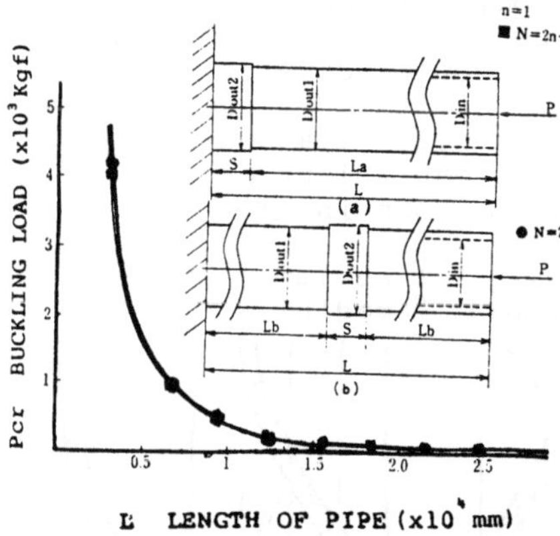

Fig-4 BUCKLING LOAD AND LENGTH OF ALTERNATELY STEPPED CROSS SECTIONS

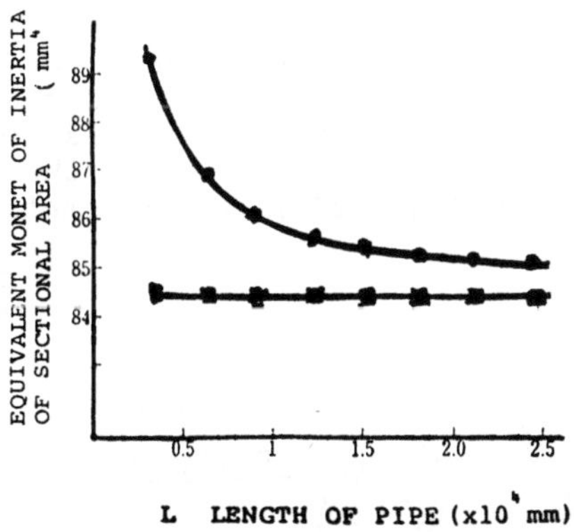

Fig-5 EQUIVALENT MOMENT INERTIA AND LENGTH OF ALTERNATE STEPPED SECTIONS

Similarly, the inclination of ω of the first mode for torsional free vibration with Jm to equivalent length is asymptotical from Fig-6. Accordingly, ω angular frequency and the length are set on the stable range of asymptote.

(mm)

		Length	Din	Dout	I (mm⁴)	
Small pipe	La	3000-24000	65.1	77.0	I_1	84.39
	Lb	1500-12000				
Large pipe	S	300	65.1	88.9	I_2	218.44
L Total		3300-24300				
E		21000 (Kgf/mm²)				

E: Young's Modulus
Small pipe Dout→Dout 1
Large pipe Dout→Dout 2

Table-2 ALTERNATELY STEPPED CROSS SECTIONS

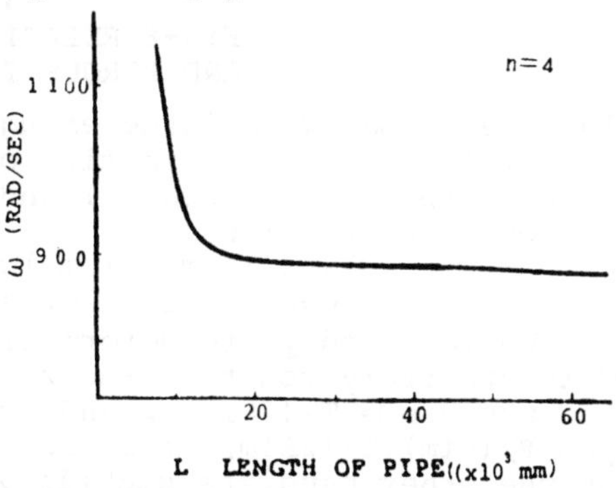

Fig-6 TORSIONAL FREQUENCY AND LENGTH OF ALTERNATELY STEPPED CROSS SECTIONS

BUCKLING OF IDEAL EQUIVALENT PIPE

Drill pipe inside the raiser pipe associates complicated deformation[6] with axial load and torque. Ideal equivalent pipe with the properties of bending rigidity EIm and torsional rigidity GJm of equivalent pipe is most easily expressed as model of drill pipe. The axial load P and torque T would deform its pipe to helical tortuosity without sufficient equivalent rigidity, EIm and GJm. Consequently ideal equivalent pipe would touch the inner wall of raiser pipe. This relation is represented by the following formulae.
The requisition of touching slightly on its inner wall with helical tortuosity of ideal equivalent pipe due to torque T without axial load P (refer Fig-7 and Fig-8) is that if the relations $T > C \cdot \sin\beta / r$ and $T = b \cdot \cos\beta / r$ are satisfied ideal equivalent pipe will be formed helix due totorque T and touch its inner wall slightly. Where r is deviation due to tortuosity of ideal equivalent pipe, O-O′ inside raiser pipe, that is pitch radius, and β is pitch angle. If T is great enough and β is very small, the pipe would scrub its inner wall. Where b=EIm and C=Gjm are rigidities of ideal equivalent pipe. And where E is Young′s modulus and G is equivalent modulus in shear, of the ideal equivalent pipe.
The configuration of deformation is shown in Fig-8

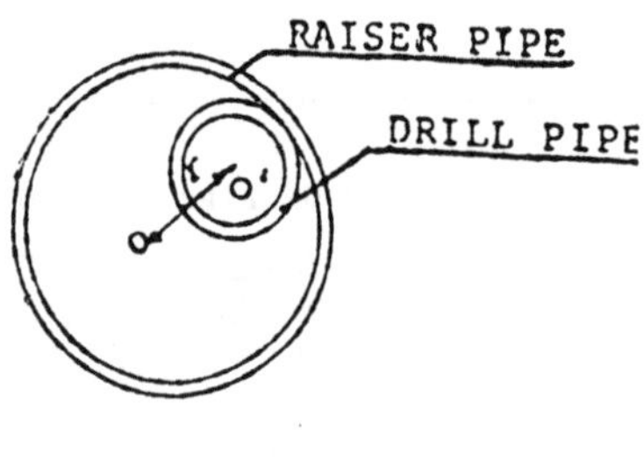

Fig-7 TOUCH OF IDEAL PIPE THE INNER WALL OF RAISER PIPE

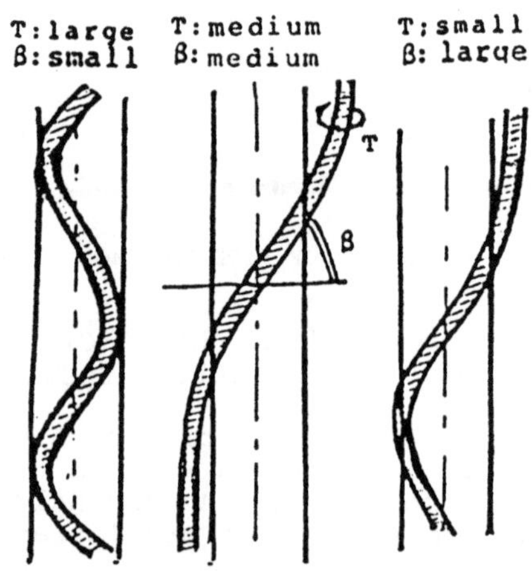

Fig-8 RELATION TO CHANGES OF PITCH ANGLE β AND TORQUE T

In case of GJm with large enough equivalent torsional rigidity, the relations of deformation of the pipe in space inside raiser pipe due to torque T and axial load P are given in following expressions.[7]

$$\left.\begin{aligned} x &= C\cdot(1/m_1 + 1/m_2)\cos\mu z \\ y &= C\cdot(1/m_1 - 1/m_2)\sin\mu z \end{aligned}\right] \quad \ldots\ldots 1)$$

Where x and y are deformation of direction of radius, and C is arbitrary constant, and z is distance of axial direction. And where $m_1 = T/(2EIm)+\mu$, and $m_2 = T/(2EIm)-\mu$ are given and $\mu^2 = P/(EIm)+T/(2EIm)$ is given.
On the other hand, the stability of the pipe under twisting couple is given with the following relation.[8]

$$\xi=C[(\lambda/2)(1-\cos\lambda\xi)-(\lambda\xi-\sin\lambda\xi)] \quad \ldots\ldots\ldots \; 2)$$
$$\eta=C[(1-\cos\lambda\xi)-(\lambda/2)(\sin\lambda\xi+\lambda\xi^2-\lambda\xi)]$$

Where relation $\lambda=\pm 2*4.934$ is given and $\zeta=z/L$ is given. And where $\xi=X/L$ and $\eta=Y/L$ are expressed. By using of the requisition of tortuosity,and the equations of 1) and 2),the deformation is studied.

EXAMPLE OF DEFORMATION OF IDEAL EQUIVALENT PIPE

From the obtained equivalent rigidities,the deformation of ideal equivalent pipe is shown with making use of formulae above description in Fig-9. And the obtained equivalent rigidities are shown in Table-3.

	IDEAL PIPE
LENGTH	25000 (mm)
INNER DIA	65.1 (mm)
OUTER DIA	77.0 (mm)
E	21000 (kgf/mm^2)
EQUIV. BENDING RIGIDITY EIm	0.01785 (kgfmm^2)
EQUIV. TORSIONAL RIGIDITY GJm	17722.5 (kgfmm^2)

Table-3 IDEAL EQUIVALENT PIPE AND EQUIVALENT RIGIDITIES

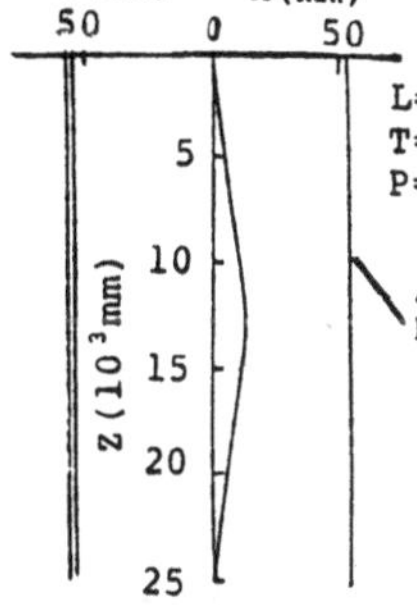

Fig-9 DEFORMATION OF IDEAL EQUIVALENT PIPE DUE TO RELATION 1)

TEST ON BORING OF PRACTICAL DRILL PIPE AND IDEAL EQUIVALENT PIPE

The photograph in Fig-10 shows texting on boring of practical drill pipe. And test on boring of ideal equivalent pipe is shown in Fig-11.

Fig-10 ON TESTING OF BORING WITH PRACTICAL DRILL PIPE

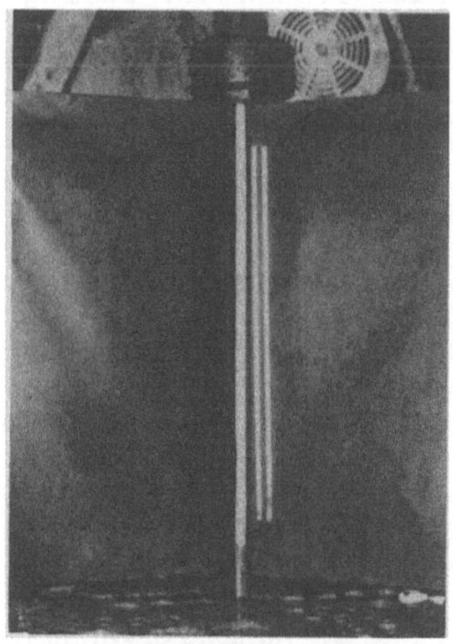

Fig-11 ON TESTING OF IDEAL EQUIVALENT PIPE

In Fig·12, the white stripe is set on ideal pipe. And the inclination of deformation of it is studied with making use of stroboscope. The parallel white stripe on black face is standard line for test of deformation. The inclination of the test of ideal equivalent pipe is almost identifiable with the relations above description. In Fig·12, the result of equation, 2) is shown.

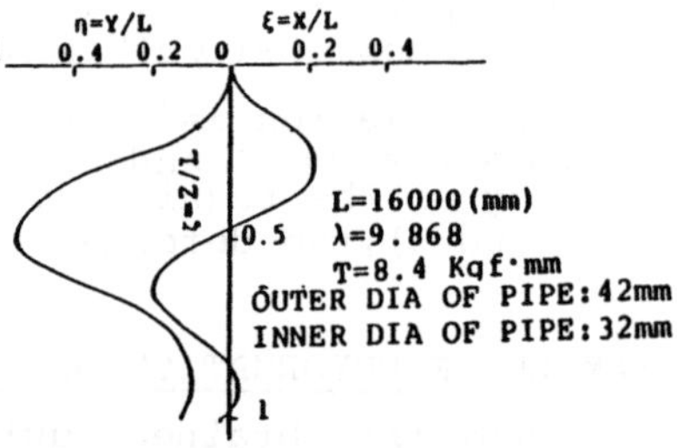

Fig-12 DEFORMATION DUE TO RELATION 2)

CONCLUSION

Practical drill pipe is able to be in equivalence of a long pipe with alternately stepped cross sections. Such a long pipe of equivalence can be in equivalence of pipe with uniform section provided with two kinds of characteristics of equivalent rigidities with bending due to buckling load and futher more with torsion. In this paper, a pipe with uniform cross section having above mentioned two characteristics is defined as the ideal equivalent pipe. The behavior and action of statics and dynamics on drill pipe are possible to clarify with making use of analysis with the ideal equivalent pipe. By using the ideal equivalent pipe, to get the fruitful result of calculation for off-shore boring, further study is needed.

AKNOWLEDGEMENTS:

The author wishes to express their sincere thanks to Dr. Masato, and Yoshino of Tone boring CO.Ltd, and to Dr.Owada, Mr. M.Yoshizawa, and T.Horie.

References:

1. Kazuo Yuwamatsu: "Boring Handbook", Morikita press, pp3-38, 1976.
2. Kazuo Aso, and Katsushige Kan : "On vibration of rod for deep ocean boring", Trans. of the Japan Society of Mech. Engg, vol 46 No 408, pp 894-903, 1980. Pt.C
3. Walter D.Pilkeys $ Pin Yu Chang: "Mordern Formulas for statics and Dynamics", McGraw-Hill, pp245-281, 1978.
4. N.Tsukada: "The Critical Load of a long shaft for drilling with Periodically Stepped Cross-Sections", J. of The College of Industrial Technolgy, Nihon University, vol 17, No 1 pp.11-20 1984.
5. N.Tsukada: "The Characteristics of Torsional Vibration of a Shaft with Alternately Stepped Cross Sections", Lecture Meeting Note of the Japan Society of Precsion Engg. at Mie Prefecture, No.823-5, pp24-26, 1982.
6. A.E.H.Love: "A treatise on the Mathematical Theory of Elasticity"'Dover Press, pp414-419.
7. A.G.Green Hill: "On the strength Of shafting When exposed both to torsion and to end thrust", Inst.Mech, Proc, pp.182-209. April 1883.
8. Atsubumi Okumura: "Supplemental Note on the Stability of Shafts under Twisting Couples", Trans. of the Japan Society of Mech.Engg, vol-18, No 65 pp.30-36 1952.

Falling Behavior of a Pipe in the Sea

T. Shimogo and T. Kawashima

Department of Mechanical Engineering, Faculty of Science and Technology, Keio University, Hiyoshi, Kohoku-ku, Yokohama, 223 Japan

ABSTRACT

To set up a cold water pipe at the OTEC plant, a long pipe floating initially on the sea surface is sent to the sea bottom or vertically suspended in the sea. In this paper, dynamical behavior of a falling pipe in the sea were studied for two methods of setting up. At the first method, a pipe is initially suspended by many floats distributed along the pipe and pulled by a wire rope. One end of the rope is fastened at the shore and the other end is carried by a tugboat. By cutting off the floats in turn from the shore-side or the offshore-side, the pipe is sent to the sea bottom. The maximum bending moment occured at the pipe, a collision velocity of the pipe with the sea bottom, and a required time to laydown the pipe were examined for various cut off speeds of the floats and various tensions of the rope by a theoretical analysis and a numerical simulation. At the second method, a pipe, which is initially filled with air, is sunk by pouring water into the pipe through an orifice provided at the end of the offshore-side and by discharging air out of the pipe through an orifice provided at the end of the shore-side. In this case, the end of the shore-side is supported by a pinjoint. An inclination of the pipe, a water level in the pipe, a reaction at the supporting end, and a bending moment occured at the pipe were investigated with the lapse of time by a theory and an experiment. The effects of a pipe length and an orifice area on the falling behavior of the pipe were clarified.

INTRODUCTION

In order to transport many kinds of resources to the shore or cold water to a thermal energy conversion plant on the shore, it is necessary to install a very long pipeline on the sea floor. In this study, a falling behavior of this sort of pipeline during the installation is analysed.

In this problem, there are a few studies of the stress which occured in the pipe during its installation by a laybarge for the marine pipeline on the sea floor [(1)~(3)], or by a subsurface tow [(4)]. However, these studies dealt with the static problems.

In the installation of a very long pipeline on the sea floor, a long pipe, which is assembled at the shore, is pulled out from the shore to the offshore under the sea surface at first, and is sunk on the sea floor by loosing its buoyancy.

One of the idea to pull out the pipeline to the installation location is to tow a pipe with many buoys by a barge, and it is named the subsurface tow method. Although this method is an economical convenient method without a large crane barge and a lay barge, it has been used in few cases for the sinking behavior of the pipeline is not clear. In this study, the sinkin behavior of the pipeline with a horizontal tension from the subsurface of the sea to the sea floor by removing the buoys is analysed, and the maximum bending moment produced in the pipe and a required time to laydown the pipe are examined for various cut off speeds of the buoys and various tensions of the rope.

The other idea is to fill the pipe with air initially and to pour water into the pipe gradually to loose buoyancy. It is called the water inflowing method. Although this method is very simple and economical, a large bending moment may be produced at the pipe due to the buoyancy of air discharged through an exhaust port at the shore-side and due to the gravity of water poured through an intake port at the off-shore-side. However, if the pipe is relatively short or very flexible, this idea may be available in practice. In this study, on the assumption of a rigid pipe, an inclination of the pipe, a water level in the pipe, a reaction at the supporting end of the pipe, and a bending moment occured at the pipe are investigated with the lapse of time for various pipe lengths and various port areas.

THE SUBSURFACE TOW METHOD

Analysis

To simplify an analytical model of the subsurface tow method, we put the following assumptions:

(1) As shown in Fig.1, the pipe has a uniform cross section, and its motion is two-dimensional in a vertical plane.

(2) The pipe has a small inclination, and is the Euler-Bernoulli's beam. And a horizontal displacement of the pipe-line is neglected.

(3) The buoys are distributed uniformly and continuously along the pipe.

(4) The constant horizontal tension is acted on the pipeline by a barge.

(5) The ropes affect the pipe ends as shearing forces.

(6) The buoy of the shore-side rope is removed at the same time as the removal of the buoys of the pipe begins. The buoys of the pipe are removed continuously with a constant rate. The buoy of the offshore-side rope is removed at the same time as the removal of the buoys of the pipe finishes.

(7) Since the pipe, the ropes and the buoys have been under the sea surface at first, there is no change of buoyancy during the sinking of the pipeline.

(8) The pipeline is initially straight and horizontal, and has no initial sinking velocity.

(9) Effects of the sea wave and the ocean current are neglected.

From these assumptions, the equation of motion of the pipe is given as follows:

$$\rho_p \partial^2 z(x,t)/\partial t^2 + (C_D \rho_w D_p/2)\,|\partial z(x,t)/\partial t|\,\partial z(x,t)/\partial t$$
$$+ EI\partial^4 z(x,t)/\partial x^4 - T\partial^2 z(x,t)/\partial x^2$$
$$- w_p + bH(x - vt) - V_1\delta(x) - V_2\delta(x - L) = 0 \qquad (1)$$

where z = vertical displacement, x = horizontal coordinate, t = time, ρ_p = mass per unit length of the pipe including added mass of water, C_D = fluid drag force coefficient, ρ_w = density of the sea water, D_p = outer diameter of the pipe, EI = bending stiffness of the pipe, T = horizontal tension, w_p = gravitational force per unit length of the pipe including buoyancy, b = buoyancy per unit length due to the buoys (b = w_p), v = removing rate of the buoys, L = length of the pipe, $H(\ )$ = Heaviside's step function, $\delta(\ )$ = Dirac's delta function, V_1, V_2 = shearing forces at the pipe ends given by the shore-side rope and the offshore-side rope, respectively.

$$V_1 = (1/2)w_r L_1 H(vt) - T\{z(0,t)/L_1 - \partial z(x,t)/\partial x|_{x=0}\}$$
$$- (1/8)C_D\rho_w D_r L_1\,|\partial z(0,t)/\partial t|\,\partial z(0,t)/\partial t$$
$$- (1/3)\rho_r L_1 \partial^2 z(0,t)/\partial t^2 \qquad (2)$$
$$V_2 = (1/2)w_r L_2 H(vt - L) - T\{z(L,t)/L_2 + \partial z(x,t)/\partial x|_{x=L}\}$$
$$- (1/8)C_D\rho_w D_r L_2\,|\partial z(L,t)/\partial t|\,\partial z(L,t)/\partial t$$
$$- (1/3)\rho_r L_2 \partial^2 z(L,t)/\partial t^2 \qquad (3)$$

where w_r = gravitational force per unit length of the rope including buoyancy, L_1 = length of the shore-side rope, D_r = outer diameter of the rope, ρ_r = mass per unit length of the rope including added mass of water.

The initial conditions and the boundary conditions of the pipe are represented as follows:

$$z(x,0) = 0, \qquad \partial z(x,t)/\partial t|_{t=0} = 0 \qquad (4)$$
$$\partial^2 z(x,t)/\partial x^2|_{x=0,L} = 0, \quad \partial^3 z(x,t)/\partial x^3|_{x=0,L} = 0 \qquad (5)$$

The nonlinear drag force terms of Eq.(1),(2) and (3) are linearized by using the equivalent linearization technique. To minimize the mean squared error between the nonlinear and the linearized drag forces, the equivalent linearized coefficient is represented as a functional of the solution, which is determined by solving the linearized equation. To simplify the computation, the linearized coefficient is assumed to be constant by approximating the solution to a simple time function. The linearized equation can be solved by using the modal analysis. In this problem, two rigid modes and three elastic modes were taken into consideration.

As a numerical example, the following data of the Tokunoshima OTEC test plant are used: L = 1500 m, EI = 1.40×10^6 Nm^2, T = 1.96×10^5 N, v = 5.0 m/s, L_1 = 150 m, L_2 = 800 m, D_p = 0.55 m, w_p = 147 N/m, ρ_p = 503 kg/m, D_r = 0.03 m, w_r = 39.2 N/m, ρ_r = 5.45 kg/m, ρ_w = 1030 kg/m^3, C_D = 1.1.

The buoys are generally removed from the shallow shore-side of the pipeline to make sure of the laydown position of the pipe. In this case the results of numerical example are shown in Fig.2 (broken lines) . If the buoys are removed from the offshore-side, the results are shown in Fig.3 (broken lines). In these figures an ordinate is enlarged by 100/15.

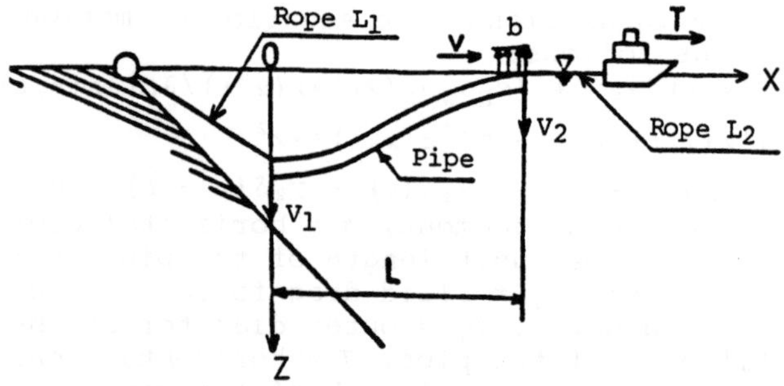

Fig.1. Illustration of model for the subsurface tow method

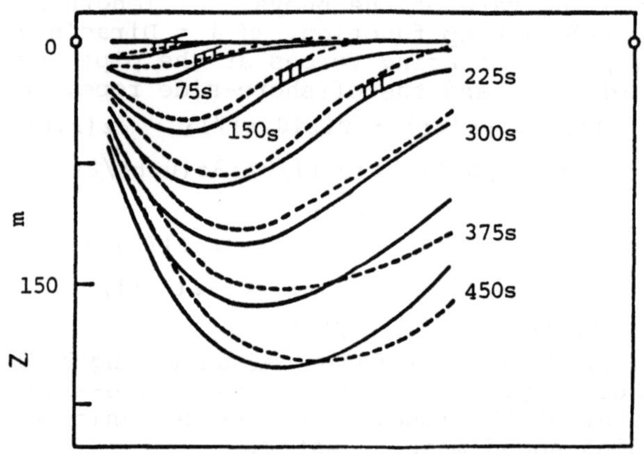

Fig.2. Sinking behavior of the pipe
(The buoys are removed from the shore-side)
--- analytical solution, —— finite difference solution

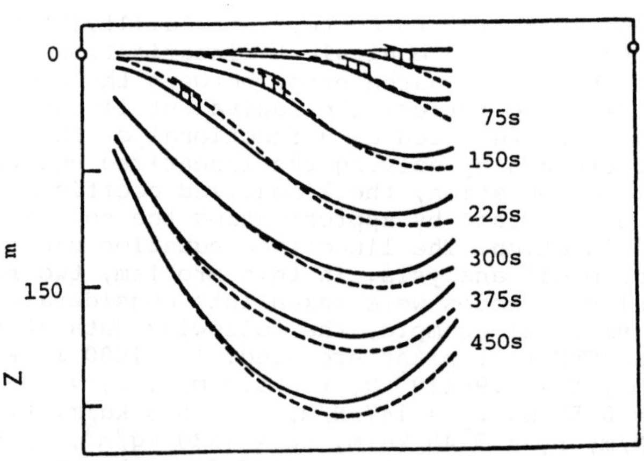

Fig.3. Sinking behavior of the pipe
(The buoys are removed from the offshore-side)
--- analytical solution, —— finite difference solution

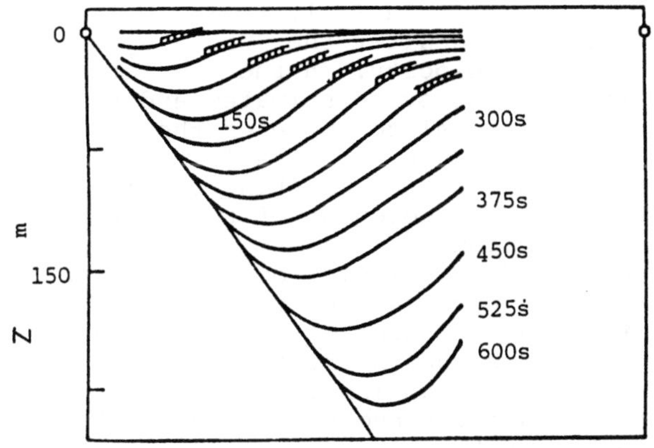

Fig.4. Sinking behavior of the pipe including the landing
(finite difference solution)
(The buoys are removed from the shore-side)

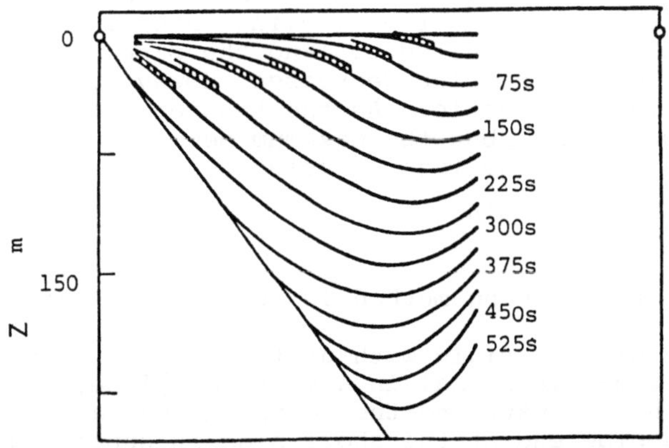

Fig.5. Sinking behavior of the pipe including the landing
(finite difference solution)
(The buoys are removed from the offshore-side)

Numerical Simulation

To confirm the analytical results, a numerical simulation is carried out by applying the finite difference method to the nonlinear equations (1),(2) and (3), in which the elastic terms are omitted because they are relatively much smaller than the other terms in our numerical example. The results are shown in Fig.2 and 3 (solid lines).

The landing of the pipe on the sea floor is calculated on the assumption that the pipe does not rebound and does not slide horizontally at the landing. The results are shown in Fig.4 and 5, assuming the inclination of the sea floor is 1/5.

From Fig.2 and 3, it is seen the analytical solution is roughly in good agreement with the finite difference solution. In the analysis, the equivalently linearized drag force is over-estimated just after the beginning of sinking because the sinking velocity is relatively small, while the drag force is under-estimated after the long elapsed time because of the large sinking velocity. In the simulation, the curvature of the pipe is over-estimated and the inclinations of the pipe ends are equal to those of the ropes after removing all buoys, because the elasticity of the pipe was neglected. In the case of the constant tension, the pipeline sinks till the vertical component of the gravity is balanced with that of the tension.

The maximum landing velocity occures at the first landing point of the pipe, which is usually the shore-side end. The maximum bending moment occures near the pipe end just after the buoys of the rope is removed or just after the landing begins. The maximum landing velocity and the maximum bending moment in case the buoys are removed from the off-shore-side are larger than those in case the buoys are removed from the shore-side. In practice, the offshore-side rope is slowly pulled out from a barge to complete the landing, and the landing velocity and the bending moment can be controlled by pulling out the rope. In case of Fig.4, the maximum bending moment is 1940 Nm, and the maximum landing velocity is 0.147 m/s. As the removing rate of the buoys increases, the time required to land the pipe decreases and the deformation of the pipe decreases. However, the removing rate is limited in practice. As the tension decreases, the landing becomes fast, but the maximum bending moment increases for the large landing velocity.

THE WATER INFLOWING METHOD

Analysis

The water inflowing method is analyzed on the following assumptions:

(1) The pipe is rigid with uniform cross section, and its motion is a two-dimensional in a vertical plane, further the effects of the sea wave and the ocean current are neglected.

(2) The pipe end of the shore-side is supported by a pin joint as shown in Fig.6, and the other end has a lumped mass such as a water intake portion.

(3) The air-filled pipe is initially at an equilibrium state in water, and the lower edge of the pipe at the supporting end coincides with the water surface at first. In case the pipe cross section is partly above the water surface, a fluid drag force and an added mass are reduced by a ratio of the cross section under the water surface to the whole cross section.

(4) Pressure drops through the air exhaust port at the shore-side and the water intake port at the offshore-side are given on the assumptions of compressible and incompressible flows, respectively.

(5) A state of the fluid in the pipe is quasi-steady, and the air in the pipe changes isothermally, and the water keeps a horizontal surface.

(6) Effects of Coriolis force and centrifugal force are neglected.

From these assumptions, the equation of motion of the pipe is given as follows:

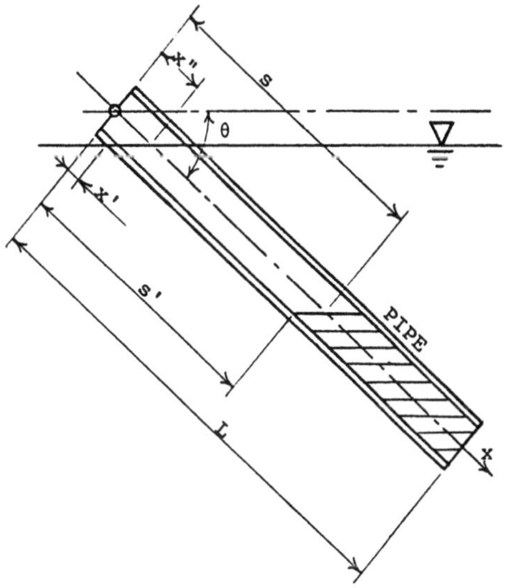

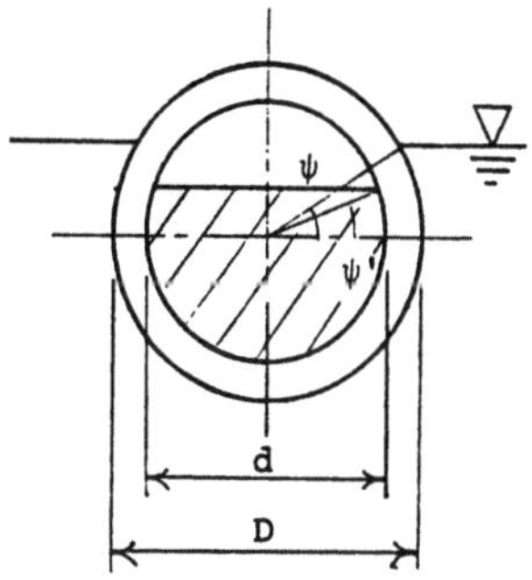

Fig.6. Analytical model for the water inflowing method

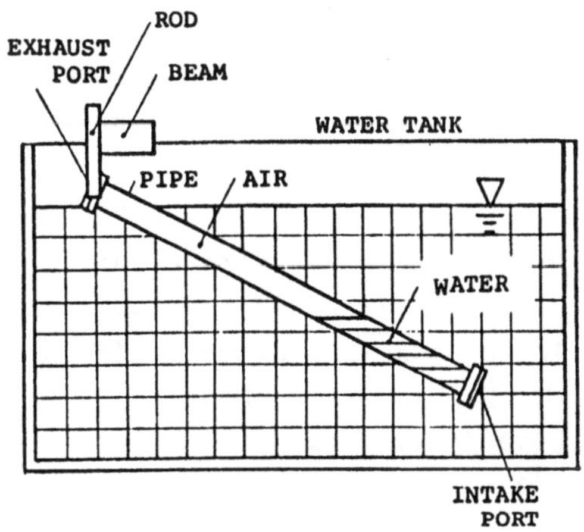

Fig.7. Experiment model for the water inflowing method

$$\begin{aligned}
&\{(1/3)\rho_p L^3 + m_c L^2 + \textstyle\int_{s'}^{L} \rho_w(\pi d^2/4)k_2(x)x^2 dx \\
&+ \textstyle\int_{x'}^{L} \rho_w(\pi D^2/4)k_1(x)x^2 dx\} d^2\theta/dt^2 \\
&+ \textstyle\int_{x'}^{L} (1/2)C_D\rho_w D k_1(x)x^3 dx\,|d\theta/dt|\,d\theta/dt \\
&- \{\rho_p g L^2/2 + m_c' g L + \textstyle\int_{s'}^{L} \rho_w(\pi d^2/4)g k_2(x)x dx \\
&- \textstyle\int_{x'}^{L} \rho_w(\pi D^2/4)g k_1(x)x dx\}\cos\theta = 0 \qquad (6)
\end{aligned}$$

where θ = angular displacement of the pipe from the water surface, t = time, x = distance along the pipe axis from the supporting end, x',s' = distances between the supporting end and the outer water surface or the inner water surface, respectively (see Fig.6), L = length of the pipe, d = inner diameter of the pipe, D = outer diameter of the pipe, ρ_p = mass per unit length of the pipe, m_c = lumped mass at the tip of the pipe, $m_c'g$ = weight of the tip in water, ρ_w= density of the sea water, C_D = fluid drag force coefficient, g =

acceleration of gravity, $k_1(x)$ = ratio of the outer cross section immersed in water to the whole outer cross section of the pipe, $k_2(x)$ = ratio of the inner cross section filled by water to the whole inner cross section of the pipe.

The equations of water and air in the pipe are given as follows:

$$ds/dt = -(\alpha_w A_w/A_p)\sqrt{2}\{(s'\sin\theta + (d/2)\cos\theta - D/2)g + (P_o - P_a)/\rho_w\}^{1/2} \tag{7}$$

$$dP_a/dt = -(P_a/s)(ds/dt + (\alpha_a A_a/A_p)(P_o/P_a)^{1/\kappa} \cdot\{2\kappa/(\kappa-1)\cdot(P_o/\rho_{ao})((P_a/P_o)^{(\kappa-1)/\kappa} - 1)\}^{1/2}) \tag{8}$$

where s = distance between the supporting end and the inner water surface (see Fig.6), A_p = inner cross section of the pipe, A_w, A_a = cross sections of the water intake port and the air exhaust port, respectively, α_w, α_w = discharge coefficients of the water intake port and the air exhaust port, respectively, P_o = atmospheric air pressure, P_a = air pressure in the pipe, ρ_{ao}= atmospheric air density, ρ_a = air density in the pipe, κ = specific heat ratio (=1.4).

The initial equilibrium position of the pipe is determined by the following expression:

$$\rho_p g L^2/2 + m_c' g L = \int_{x'}^{L} \rho_w(\pi D^2/4) g k_1(x) x dx \tag{9}$$

$\theta(t)$, $s(t)$ and $P_a(t)$ are obtained from equations (6), (7) and (8), which are solved by the Runge-Kutta method.

The vertical and horizontal reactions occured at the supporting end of the pipe are given by the expressions

$$\begin{aligned} R_V = &-\{\rho_p L^2/2 + \textstyle\int_{s'}^{L} \rho_w(\pi d^2/4) k_2(x) x dx \\ &+ \textstyle\int_{x'}^{L} \rho_w(\pi D^2/4) k_1(x) x dx\} d^2\theta/dt^2 \cos\theta \\ &- \textstyle\int_{x'}^{L} (1/2) C_D \rho_w D k_1(x) x^2 dx\, |d\theta/dt|(d\theta/dt)\cos\theta \\ &+ \{\rho_p L + \textstyle\int_{s'}^{L} \rho_w(\pi d^2/4) k_2(x) dx \\ &- \textstyle\int_{x'}^{L} \rho_w(\pi D^2/4) k_1(x) dx\} g \end{aligned} \tag{10}$$

$$\begin{aligned} R_H = &\{\rho_p L^2/2 + \textstyle\int_{s'}^{L} \rho_w(\pi d^2/4) k_2(x) x dx \\ &+ \textstyle\int_{x'}^{L} \rho_w(\pi D^2/4) k_1(x) x dx\} d^2\theta/dt^2 \sin\theta \\ &+ \textstyle\int_{x'}^{L} (1/2) C_D \rho_w D k_1(x) x^2 dx\, |d\theta/dt|(d\theta/dt)\sin\theta \end{aligned} \tag{11}$$

where the lumped mass at the tip of the pipe was neglected. The maximum bending moment is approximated as follows:

$$\begin{aligned} M = &(f - \rho_p g)\xi^2/2 - f\xi^3/6L + \rho_w(\pi D^2/4) g (L - x'') \\ &\cdot(\xi - (L - x'')/2) - \rho_w(\pi d^2/4) g (L - s)(\xi - (L - s)/2) \\ &\text{for } 0 \le \xi \le L - s \\ M = &\{f - \rho_p g + \rho_w(\pi D^2/4) g\}\xi^2/2 - f\xi^3/6L \\ &- \rho_w(\pi d^2/4) g (L - s)(\xi - (L - s)/2) \\ &\text{for } L - s < \xi \le L - x'' \\ M = &\{f - \rho_p g + \rho_w(\pi D^2/4) g - \rho_w(\pi d^2/4) g\}\xi^2/2 - f\xi^3/6L \\ &\text{for } L - x'' < \xi \le L \end{aligned} \tag{12}$$

where

$$\begin{aligned} f = &(3/2)(\rho_p + \rho_w(\pi d^2/4)\{1 - (s/L)^2\} \\ &- \rho_w(\pi D^2/4)\{1 - (x''/L)^2\}) g \end{aligned} \tag{13}$$

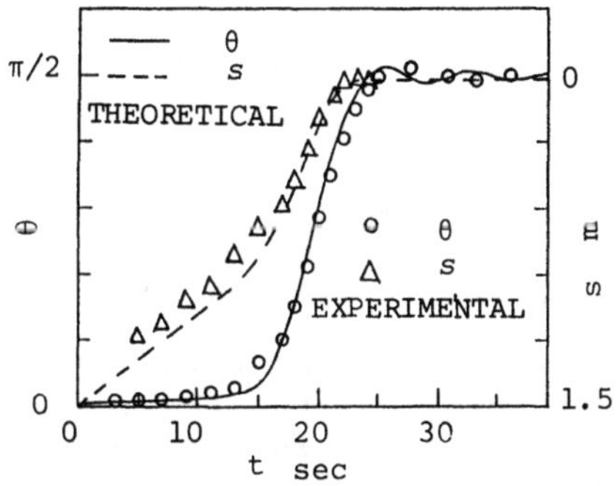

(a) pipe length = 1.5 m, outer diameter = 34 mm, inner diameter = 28.2mm, intake port diameter = 8.80 mm, exhaust port diameter = 3.10 mm

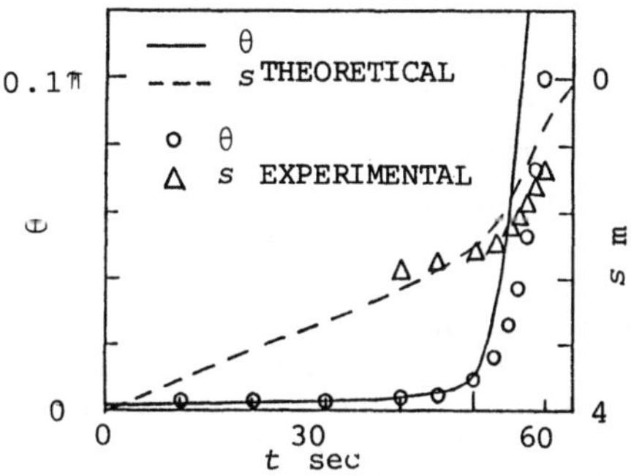

(b) pipe length = 4.0 m, outer diameter = 34 mm, inner diameter = 28.2 mm, intake port diameter = 8.80 mm, exhaust port diameter = 3.10 mm

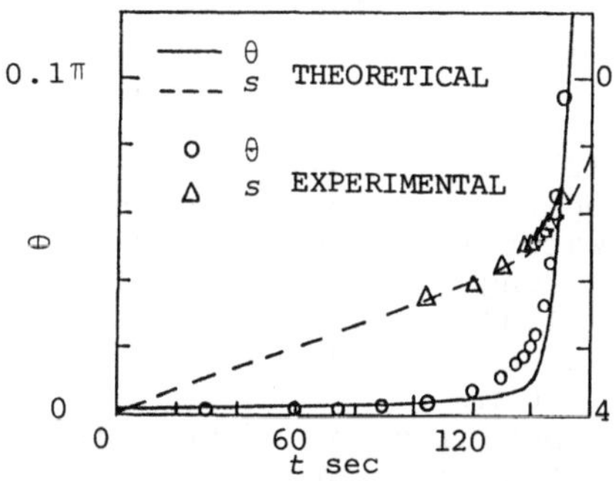

(c) pipe length = 4.0 m, outer diameter = 34 mm, inner diameter = 28.2 mm, intake port diameter = 4.95 mm, exhaust port diameter = 3.10 mm

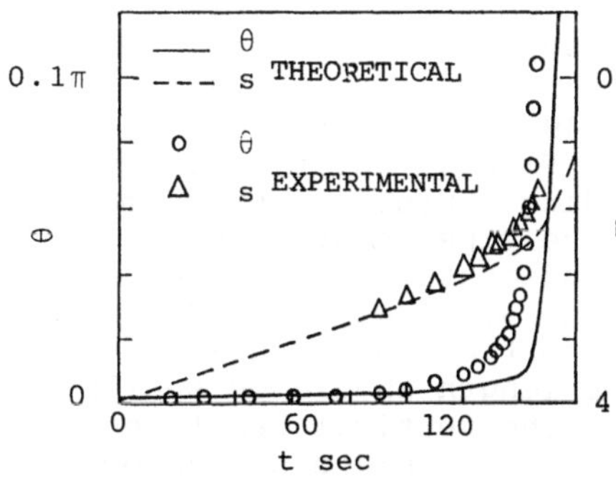

(d) pipe length = 4.0 m, outer diameter = 34 mm, inner diameter = 28.2 mm, intake port diameter = 8.80 mm, exhaust port diameter = 3.10 mm

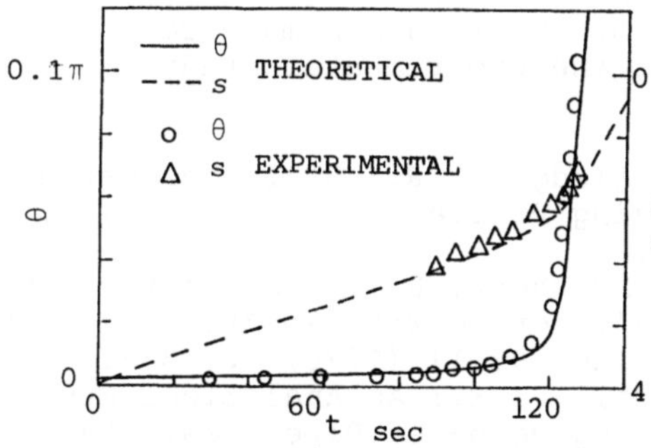

(e) pipe length = 4.0 mm, outer diameter = 48 mm, inner diameter = 40.8 mm, intake port diameter = 8.80 mm, exhaust port diameter = 3.10 mm

Fig.8. Sinking behavior of the pipe at the water inflowing method where x'' = distance between the supporting end and the outer water surface (see Fig.6), and the inner water surface was assumed to be vertical to the pipe axis.

Experiment

A scheme of an experimental model of the water inflowing method is indicated in Fig.7. In this model, the dimension of a water tank is 5x3x1.5m and a side plate is transparent for observation.

The experimental results are shown in Fig.8(a),(b),(c),(d) and (e), and the corresponding theoretical results are plotted in this figure. From these results, it is seen the above mentioned analysis is sufficiently effective to estimate the behavior of the pipe at the water inflowing method.

The angular displacement of the pipe suddenly increases after a certain critical time elapsed. Therefore, the water inflowing method is not adequate to laydown the pipe on the sea floor owing to the large landing velocity. However, this method is available to set up a vertically suspended pipe such as a riser tube by reason of its simplicity. The critical time elapsed increases as the water intake port or the air exhaust port is small, especially for the long and big pipes, as shown in Fig.8. The total time is roughly proportional to the length.

From calculation of equation(12), it is seen the maximum bending moment is approximately proportional to squared value of the pipe length. Therefore the water inflowing method is not adequate to set up a relatively long pipe, for example, if the steel pipe has dimensions D= 1.845 m, d = 1.8 m and the port diameters are same as the inner pipe diameter, the pipe length should be smaller than 200 m.

The vertical reaction at the supporting end of the pipe may occure upward at the beginning of sinking process owing to the bending moment. From calculation of equation(10), it is seen the maximum upward reaction is roughly proportional to the pipe length. It is important to check the amount of this reaction to prevent an overturn of a barge.

ACKNOWLEDGMENT

The authers are grateful to Tokyo Kyuei, Inc., for the offer of the data of the Tokunoshima OTEC test plant and for the cooperation of the experimental study.

REFERENCES

1. Dixen, D.A.; Rutledge, D.R.: Stiffened Catenary Calculations in Pipeline Laying Problem. Trans. ASME., J. Eng. Ind. 90-2 (1968) 153-160.
2. Dareing, D.W.; Neathery, R.F.: Marine Pipeline Analysis Based on Newton's Method with an Arctic Application. Trans. ASME., J. Eng. Ind. 92-11 (1970) 827-833.
3. Healey, A.J.; Hall, J.E.: An Analytical Study of Controlled Buoyancy in Laying Deep-Sea Pipelines. Trans. ASME., J. Eng. Ind. 97-2 (1975) 266-273.
4. Kan, W.C.; Healey, A.J.: Finite Element Analysis with the State Variable Transfer Matrix and Geometric Nonlinearity for Marine Pipelines in Subsurface Tow. Trans. ASME., J. Energy Resour. Technol. 103-3 (1981) 26-31.

Free Falling Installation Test of Artificial Steel Reef

N. Sakaki and T. Okamoto

Civil Engineering Laboratory, Structure Research Department, Technical Research Center, Nippon Kokan K.K., Minami-Watarida, Kawasaki, 210 Japan

ABSTRACT

In this paper, the free falling installation of the large-size Artificial Steel Reef (ASR) will be discussed in order. Specifically, a prediction of its landing point on the sea floor, the force of the impact to evaluate the stress of the members of the ASR upon landing and an actual impact acceleration for evaluating impact force will be explored.

The accuracy of the method for predicting the landing point is studied through experimental results of a laboratory model-test and a full-scale test in the field. The measured stress of the members of the ASR in the drop test on to the ground, is compared with one which was obtained by the Finite Element Method (FEM) which applies the impact force, evaluated by measured accelerations. The impact accelerations, measured upon the landing on the sea floor, is compared with that recommended by the design guide.

INTRODUCTION

The 200-nautical mile exclusionary economic zone in foreign waters has stimulated the demand for large-size artificial reefs which enhance the productivity of the fisheries in home waters. Until now, concrete reefs have been used mainly for artificial reefs. Recently, however, steel has been used for artificial reefs due to its superior flexibility, lightness, ease of fabrication, etc.

The free falling installation of ASR is economical, but the following problems still need to be solved.

(a) Prediction of the landing point of the ASR on the sea floor
(b) Safety precautions against the force of impact
(c) Control of the posture of the ASR when it lands on the sea floor

The first two problems will be discussed in the following sections by the results of the laboratory test, the full-scale test in field and the analytical approach.

PREDICTION OF LANDING POINT

Laboratory Model Test

The horizontal motion of an object in a uniform flow is expressed by Eq. (1).

$$(\sigma V + C_M \rho V)\frac{\partial u}{\partial t} = C_D A_H \frac{\rho (U-u)^2}{2} \qquad (1)$$

where σ= the density of the object, ρ= the density of the fluid, V= the volume of the object, A_H= the projected area of the object in the horizontal direction, U= the velocity of the flow in the horizontal direction, u= the velocity of the object in the horizontal direction, C_D= the drag coefficient, C_M= the inertia coefficient and t= time. The left-hand tefm of the Eq.(1) represents the inertial force and the right-hand term represents the drag force. Solving Eq.(1) based on the initial conditions that u=0 and X=0 at t=0, the travel length X in the horizontal direction is found as

$$X = Ut - \frac{a}{c} \ln(at+b) + \frac{c}{a} \ln (b) \tag{2}$$

where $a=C_D A_H \rho U$, $b=2V(\sigma+C_M\rho)$, $c=2UV(\sigma+C_M\rho)$.

A laboratory test was performed in order to verify the accuracy of the theoretical travel length (Eq.(2)). In this test, the 1/30 scale model (see Fig.1) was dropped into the water. Its depth is 180cm. The velocity of the flow is shown in Fig.2. The mean velocity $\bar{U}$ of the flow was 21.5cm/s. The locus of the model in the water was recorded by a photograph.

The measured elapsed time (T), from the dropping until the landing, the locus of the model in the water and the travel length (L_H), from the dropping point to the horizontal landing point are also shown in Fig.2. The elapsed time (T), calculated by the equation for free falling motion in a fluid, is given in Ref.1. The calculated elapsed time was 3.7s. The measured elapsed time was 3.8s. The locus of the model was calculated by Eq.(2) where $U=\bar{U}$ the vertical travel length is given in Ref.1. Fig. 2 shows that the measured locus compatible with the calculated one. A_V shown in Fig.2 represents the projected area of the model in the vertical direction. The calculated travel length (L_H) agrees within 3% error with the results of twelve types of experiments.

Installation Test of the Full-Scale Model

A full-scale, free fall installation test of the ASR (see Fig.3) was also conducted in order to test the theoretical treatment in the field of the elapsed time (T), and travel length (L_H) offshore at Nakiri in Mie prefecture.

The velocity of the current measured immediately before the dropping of the ASR at the installation site is shown in Fig.4. The measured travel length (L_H), the travel time (T), the calculated (T) and (L_H) are also shown in Fig. 4. The theoretical value of travel length (L_H) was calculated by the numerical integration of Eq. (1), varying the current velocity at each water depth. Only the calculated travel length is shown in the circle (see Fig.4) because, unfortunately, the current direction was not measured. As shown in Fig. 4, similar values were obtained from the experiment and the calculations. The calculated elapsed time (T = 22.4s.) coincides well with the experimental result (T = 21.0s.) obtained through acceleration records when landing on the sea floor. Therefore, from the laboratory and the field test, it was found that using the present theory, we can accurately predict the landing point of the ASR.

SAFETY AGAINST AN IMPACT FORCE

Impact Force

We performed a drop test of the real ASR (see Fig.3) on the ground in order to compare the measured stresses of ASR at landing with the calculated ones. The locations of accelerometers and strain gauges are shown in Figs. 5 and 6.

In order to calculate the stresses of the ASR, statically, we assumed that the impact force (F_i), which acts on the ASR can be expressed as (Ref.2)

$$F_i = m\alpha \qquad (3)$$

where m=the mass of the members of the ASR and α= the measured maximum acceleration. Applying body force obtained from Eq. (3), the impact stress of all the members of the ASR upon landing was calculated by the structural analysis program "NASTRAN" (Ref.3). The three-dimensional, finite element model is shown in Fig.5. The statically calculated stresses per unit of gravity acceleration (1G) and the measured ones are shown in Fig.7. Good correlation can be seen between the calculated and experimental values. Therefore, it has been confirmed that we can accurately estimate the stress of the ASR upon landing through static analysis. When we applied the impact force for design to the ASR, both the measured stresses and the calculated ones of all the members of the ASR were less than the allowable stress.

Impact Acceleration

We measured the impact accelerations of the ASR (see Fig.3) in the field installation test. The locations of the accelerometers were the same as shown in Fig.5. The total weight of the ASR, including the concrete anchor weight, was 8 tons and the projected area in the vertical direction is $18.2m^2$. The depth of the water at the test site is 52m. The sea floor of the test site consists of coarse sand. Its uniformity coefficient was 1.9. The angle of internal friction was 40 degrees and the coefficient of viscosity was 0. the impact accelerations were measured by strain gauge type accelerometers (AS-20B) at four points A,B,C and D (see Fig.5). The values of acceleration ranged from 1.7G to 2.2G in which the gravity acceleration (1G), was included (see Fig.8). Therefore, the values of 4G-5G recommended by the Design Guide (Ref.2) as the impact acceleration for sand were found to be conservative values for the design of the ASR.

CONCLUSIONS

(1) The present theoretical method can predict the landing point of the ASR with good accuracy.
(2) The stresses of the members of the ASR upon landing on sea floor can be calculated by statically applying the inertial force to the ASR.

(3) When we applied the impact force for design to the ASR, both the measured stresses and the calculated ones of all the members of the ASR were less than the allowable stress.
(4) The measured impact accelerations of the large-size ASR upon landing on the sea floor was one half of that recommended by the Design Guide.

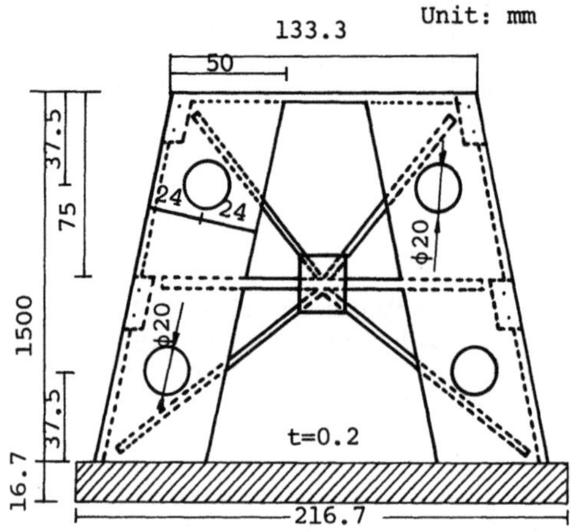

Fig. 1 A side view of 1/30 scale model of ASR

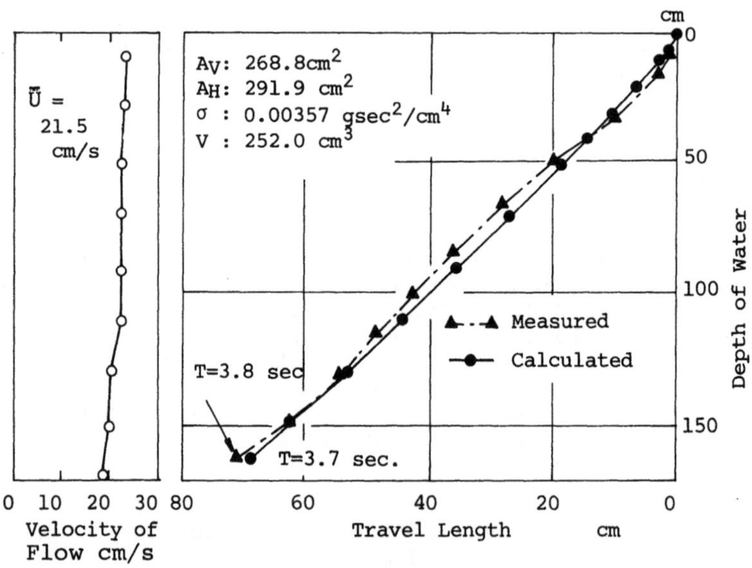

Fig. 2 Travel length, travel time T and locus of 1/30 scale model of ASR in laboratory test

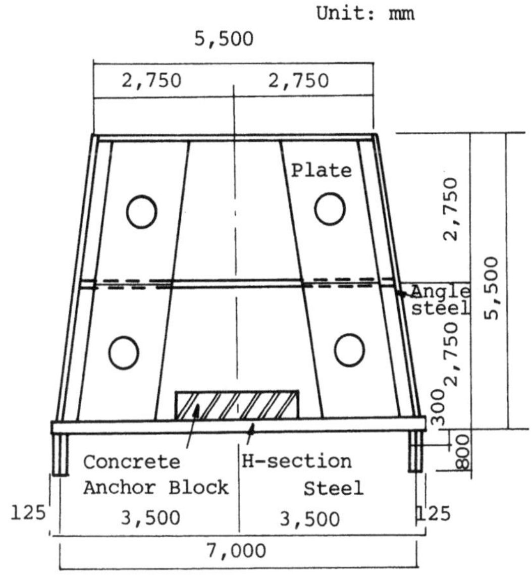

Fig. 3 A side view of large-size ASR

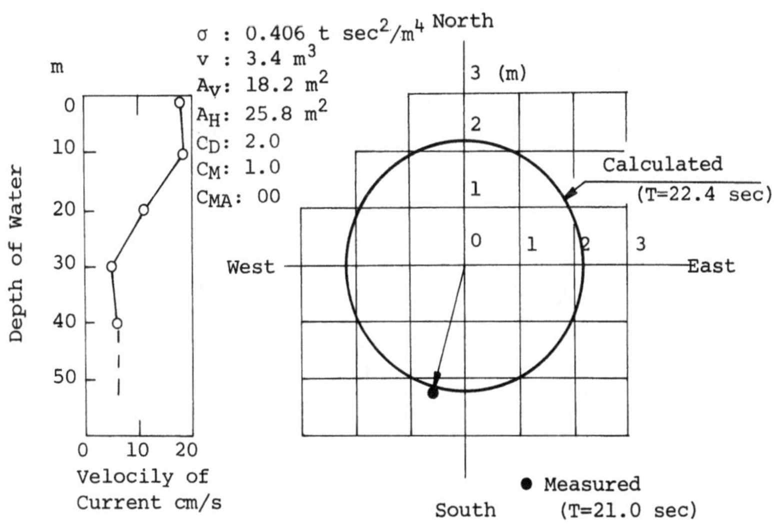

Fig. 4 Travel length L_H of the real ASR in the full scale free falling installation test in field

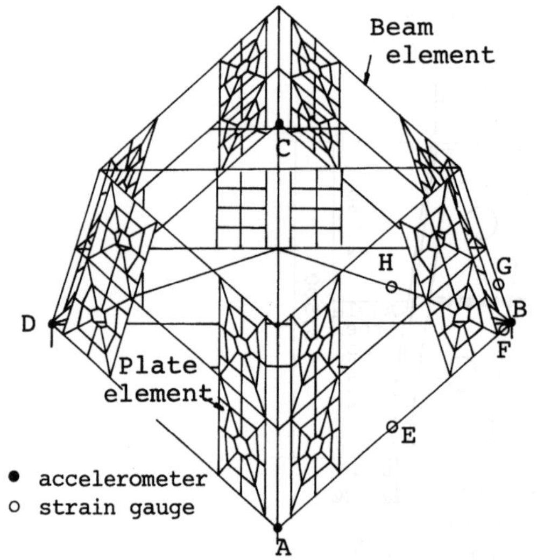

Fig. 5 Finite element model that consist of beam and plate elements, and the location of accelerometers and strain gauges

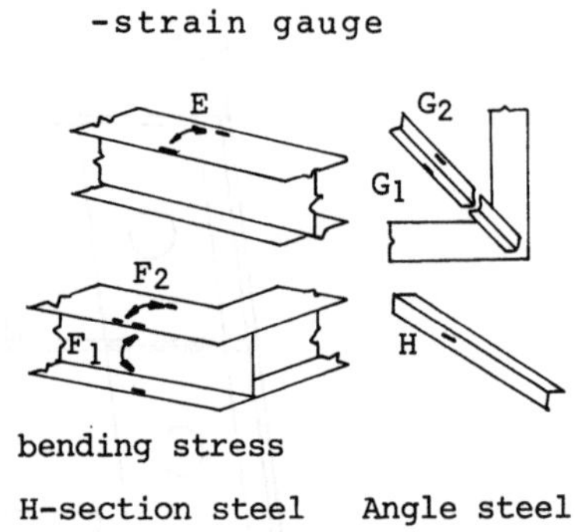

Fig. 6 Measured points of stresses

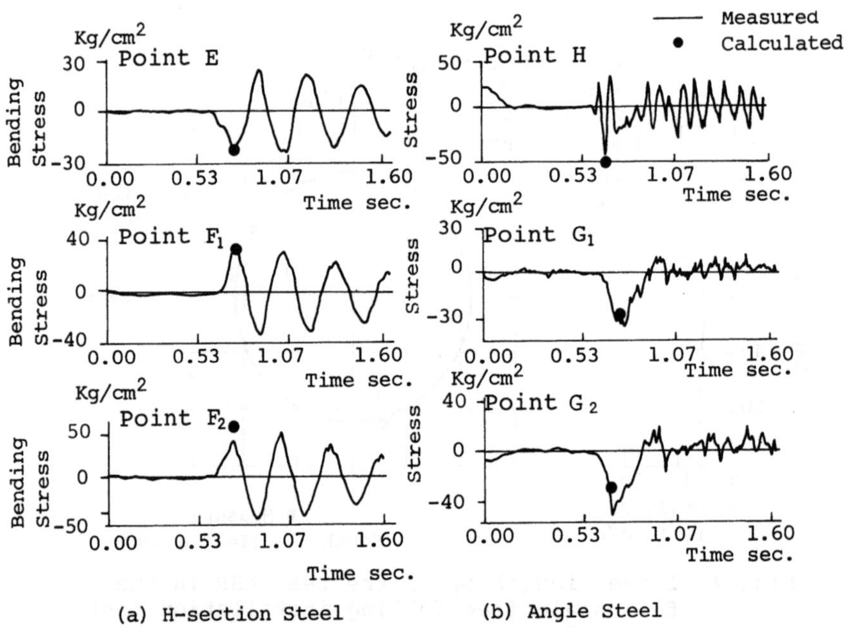

Fig. 7 Measured and calculated stresses of members of ASR upon landing

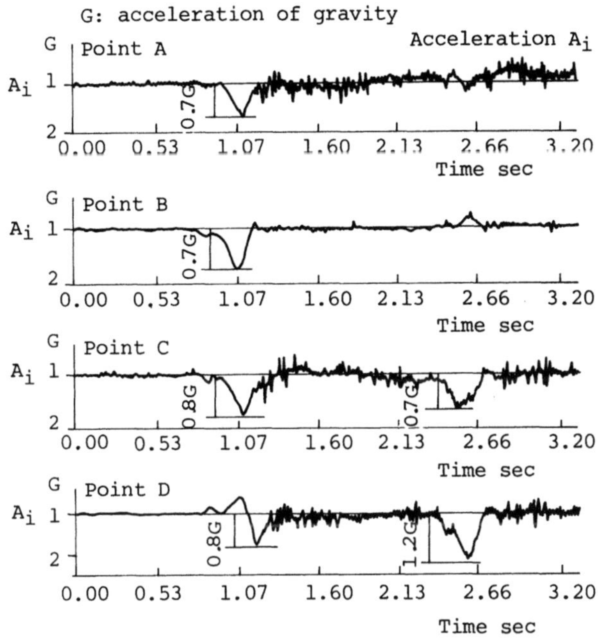

Fig. 8 Measured accelerations of ASR upon landing on the sea floor

References

1. Makoto Nakamura.: "SUISAN DOBOKU GAKU", 1979, pp 401-456.
2. An imcorporated association Japan Fisheries Resources Preservation Association: The Structural Design Guide for Coastal Fisheries Consolidation and Development Enterprise, 1978, pp 7-18.
3. Richard H. MacNeal, Editor.: MSC/NASTRAN Handbook For Linear Static Analysis, MSC/NASTRAN Version 60, The MacNeal-Schwendler Corporation, 7442 North Figueroa Street, Los Angles, California 90041, 1980.

Technical Session:
Materials and Construction

4. Corrosion and Protection

Corrosion Surveys of Steel Structures under Marine Environment

H. Tsuchida, T. Yokoi and M. Abe

Materials Laboratory, Structures Division, Port and Harbour Research Institute, Ministry of Transport, Nagase, Yokosuka, 239 Japan

ABSTRACT

For studying the corrosion of steel structures under marine environment and for examining the effectiveness of corrosion protection, 229 port facilities with steel members were surveyed. In some of them, concentrated corrosions were found just below the mean low water level and the rate of corrosion where the severest concentrated corrosion had taken place was estimated to be larger than 0.93 mm/yr. Except such places, the corrosion rates of steel in sea water and bottom soil were about 0.1 mm/yr. In the structures to which the corrosion protection with the electrode of alluminum alloy was applied, the protection potentials were maintained below the corrosion potential and no concentrated corrosion was found.

INTRODUCTION

For effective utilization of ocean space in the future, the steel is being considered as one of the major structural materials. The reason is that the steel has many advantages as a construction material, such as relatively high strength per unit cross section, uniform quality, wide varieties in shape and size, and easiness in cutting and jointing. On the contrary, the steel has a disadvantage to be corroded rather rapidly under certain marine environments. For instance, some of the open-type wharves with steel pipe piles constructed in 1960's in Japan had to be reconstructed, because of the serious corrosion of the piles. The facilities lasted for about 25 years. Therefore, to secure serviceability of the steel structures for the ocean space utilization and to save the steel and other related resources, knowledge and technology on the corrosion and the protection of the steel under marine environments are very important.

The Port and Harbour Research Institute (PHRI) has been conducting extensive researches in this field. Subjects of the researches at PHRI are as follows:

1) Development of thickness meters for steel materials in sea water,
2) Corrosion surveys of existing port facilities,
3) Performance surveys of cathodic protection applied to existing port facilities,
4) A field corrosion test of proto-type steel piles,
5) Study of painting techniques to be used as corrosion protection and a performance survey of existing facilities where piles were protected by painting,
6) Laboratory study of the corrosion mechanism,
7) Tidal pool experiment to study the corrosion mechanism and the effectiveness of corrosion protection against the concentrated corrosion.

This paper presents the summary of the corrosion surveys and the performance surveys.

The corrosion survey conducted by PHRI can be divided into two phases. The first phase began in 1959 and 470 sheetpiles selected from the sheetpile bulkheads in 11 ports were surveyed. The second phase began in 1967, and 205 facilities (162 sheetpile bulkheads, 30 pile structures, and 13 structures of other types) in 71 ports were surveyed by 1983. For the performance survey, structures which had been under the protection more than 10 years were selected. As of March 1984, 24 facilities were surveyed. The details of the surveys have been published as the technical notes of PHRI.

CORROSION SURVEY

Survey Procedure

Two types of survey procedure were organized, namely, the detailed survey and the simplfied survey.

(1) Detailed Survey

The detailed survey can be divided into two sub-surveys and they consist of the following investigation items:

Structural Survey

1) Collection of information on structural conditions,
2) Measurement of corrosion (= Measurement of remaining thickness by the thickness meter in situ and/or by the micrometer in laboratory on specimens cut out from structure).

Environment Survey

1) Measurement of seawater quality,
2) Measurement of bottom mud quality,
3) Observation or information collection on climatic condition.

A standard survey procedure was established and all the detailed surveys were carried out based on the standard survey procedure.1)

(2) Simplified Survey

The simplified survey consists of the following investigation items:

1) Collection of information on structural conditions,
2) Observation by eye during low tide on sheetpiles or piles at mean low water level.

Example of Corrosion Survey on Sheetpile Bulkhead

As an example of the detailed corrosion survey on the sheetpile bulkhead, results from the survey on a sheetpile bulkhead with water depth -5.5 meter in Toyama Port will be explained.2) The bulkhead was constructed in two year time, 1965 and 1966. The cross section of the bulkhead is shown in Fig. 1. No corrosion protection was executed. The survey was made in March 1981. At the time of survey, it had been about 16 years since the construction. Four sheetpiles were chosen and the thickness were measured at 13 levels. At each level, measurements were made at 6 point at each level, as shown in Fig. 2. The results are shown in Fig. 1. By the observation by eye, pits through entire thickness of sheetpile were found at corners of 13 sheetpiles. Photo 1 is showing the pits through which the water was flowing out.

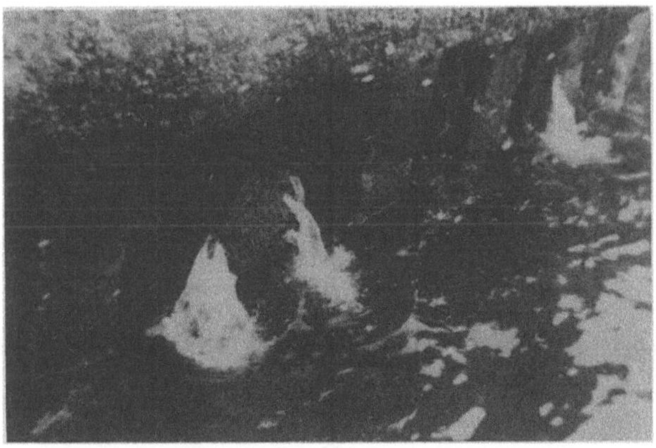

Photo 1 Sheetpile bulkhead and pits through which water is flowing out

The corrosion rate at the places of pits was estimated as 0.58 mm/yr. The water qualities were measured on pH, DO, BOD, and SS at 4 sites near the bulkhead.

Corrosion Tendency

Based on the surveys which had been performed on the sheetpile bulkheads, it was found that there were 5 different cor-

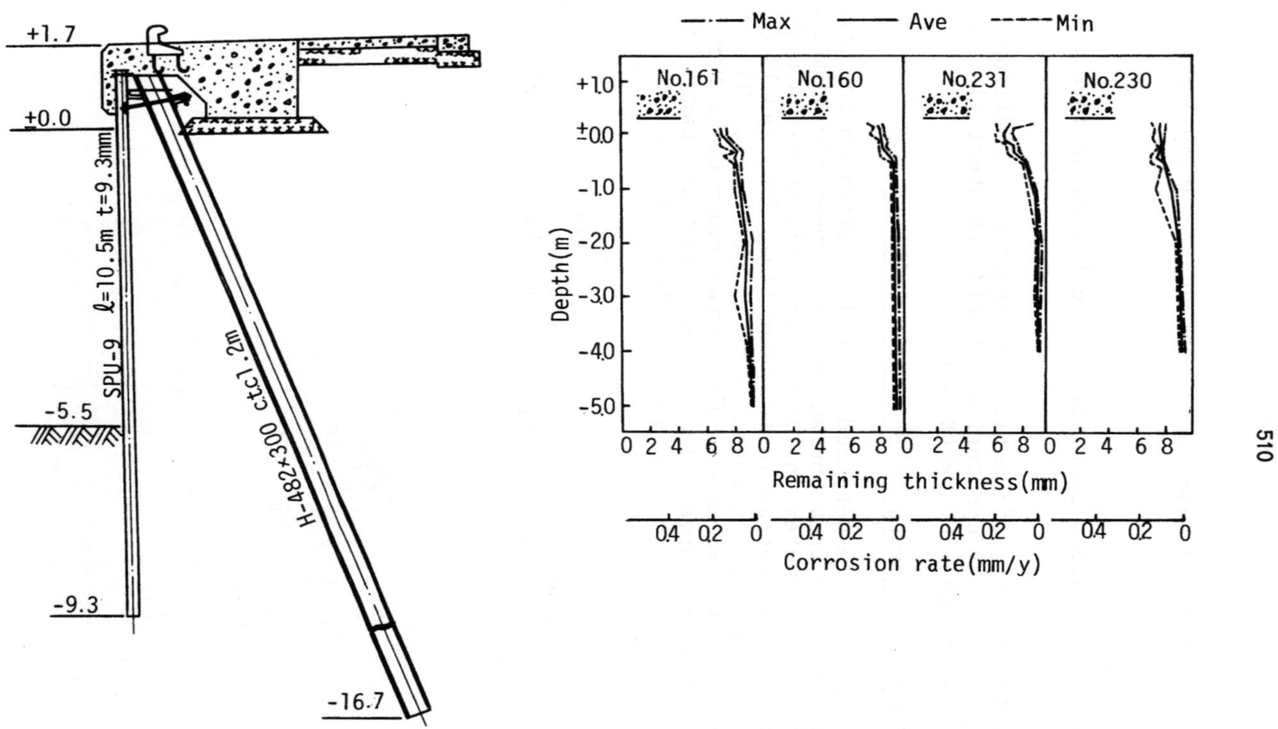

Fig. 1 Sheetpile bulkhead in Toyama Port

rosion tendencies, as shown in Fig. 3. The rate of corrosion of the structural steel under low water level was about 0.1 mm/yr, except the cases in which the concentrated corrosion had developed.

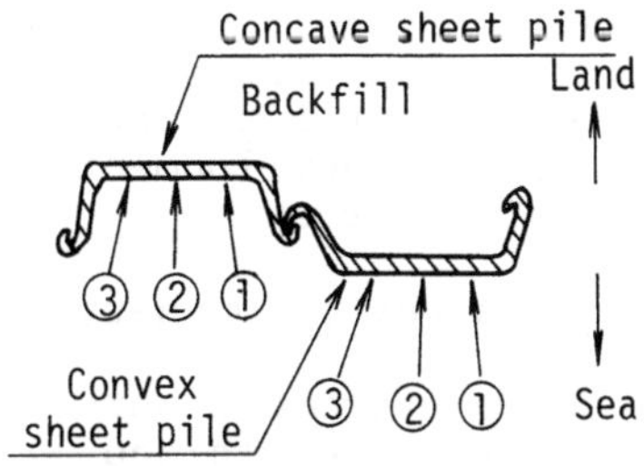

Fig. 2 Points of thickness measurements at each level

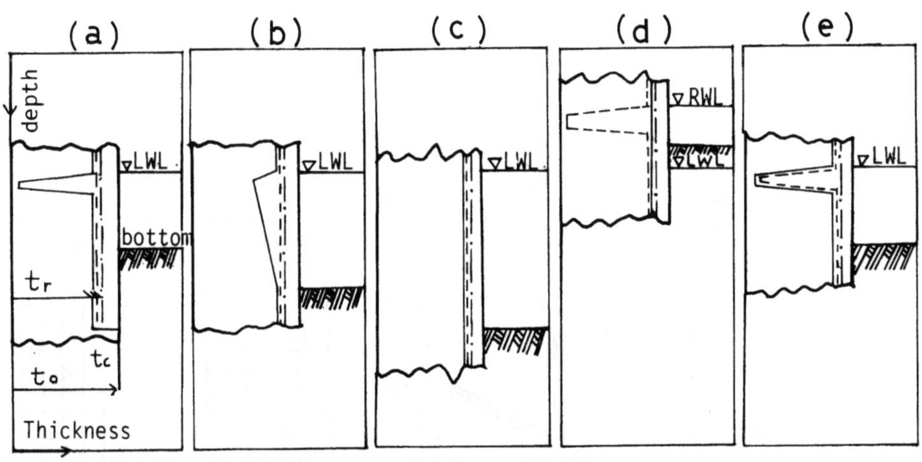

t_o: Total thickness of sheetpile
t_r: Residual thickness of sheetpile
t_c: Corroded thickness of sheetpile

Fig. 3 Corrosion tendencies of steel sheetpile

In one of five corrosion tendencies, the tendency (a), the corrosion of high speed develops under the mean low water level. This is the concentrated corrosion and the largest corrosion rate estimated from the corrosion surveys carried out during 1978 through 1981 was "larger than 0.93 mm/yr."

The corrosion tendency of steel piles are not much different from that of the sheetpile bulkhead.

PERFORMANCE SURVEY OF CATHODIC PROTECTION

Survey Procedure

The performance survey can be divided into three sub-surveys, namely, the electrode survey, the structureal survey, and the environmental survey. The electrode survey consists of the following investigation items:

Electrode Survey
1) Measurement of corrosion protective current,
2) Measurement of electrode potential,
3) Measurement of remaining electrode weight.

The structural survey in the performance survey includes all the items of that in the corrosion survey. In addition, it includes measurement of electric potential (protective potential) of the structure. The envitonment survey is identical with that of the corrosion survey.

Example of Performance Survey on Open-type Wharf

As an example, the performance survey on the open-type wharf in Maizuru Port will be explained.3) Cross section of the wharf is shown in Fig. 4. The structure was constructed 11 year before the survey, and it has been under the cathodic protection since the completion.

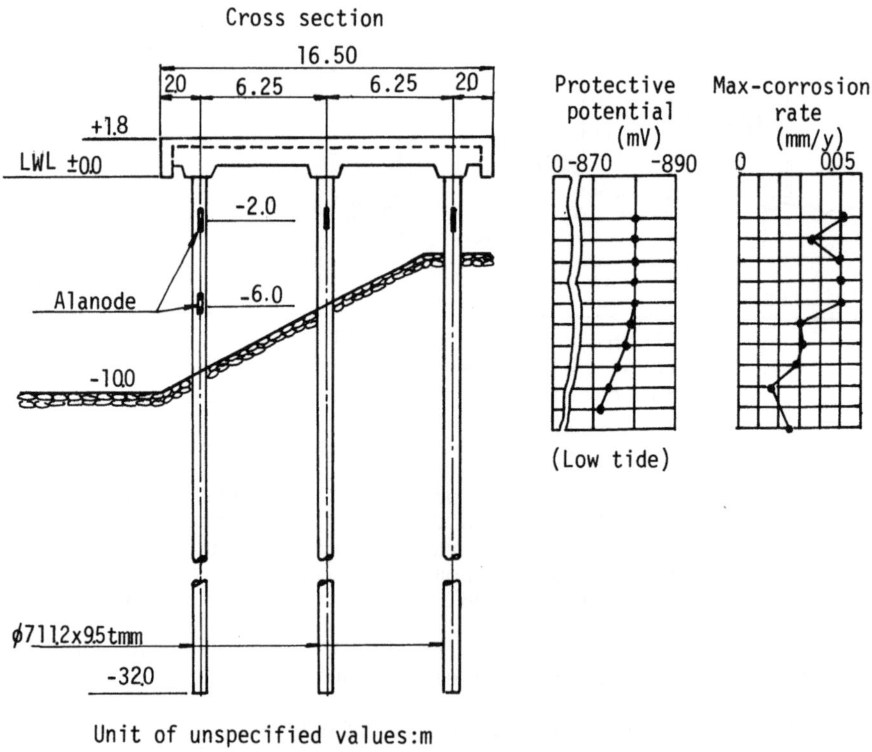

Fig. 4 Open-type wharf in Maizuru Port

The protective potential of the structure was measured at 11 levels, when the tide was low and high. The results for low tide are shown in Fig. 4. From the remaining weight of the electrode and the protective current measured, the total life of the electrode was estimated as 26 years. The total life considered in the design was 20 years. The thickness of pile was measured at 10 levels. The maximum corrosion rate at each level are shown in Fig. 4.

Major Results

The protective potentials of all the facilities surveyed are shown in Fig. 5. As it is seen in the figure, the structures, except two, are kept below the design protective potential (-780 mV with reference to silver-silver chloride-seawater electrode) which is indicated by the broken line in the figure.

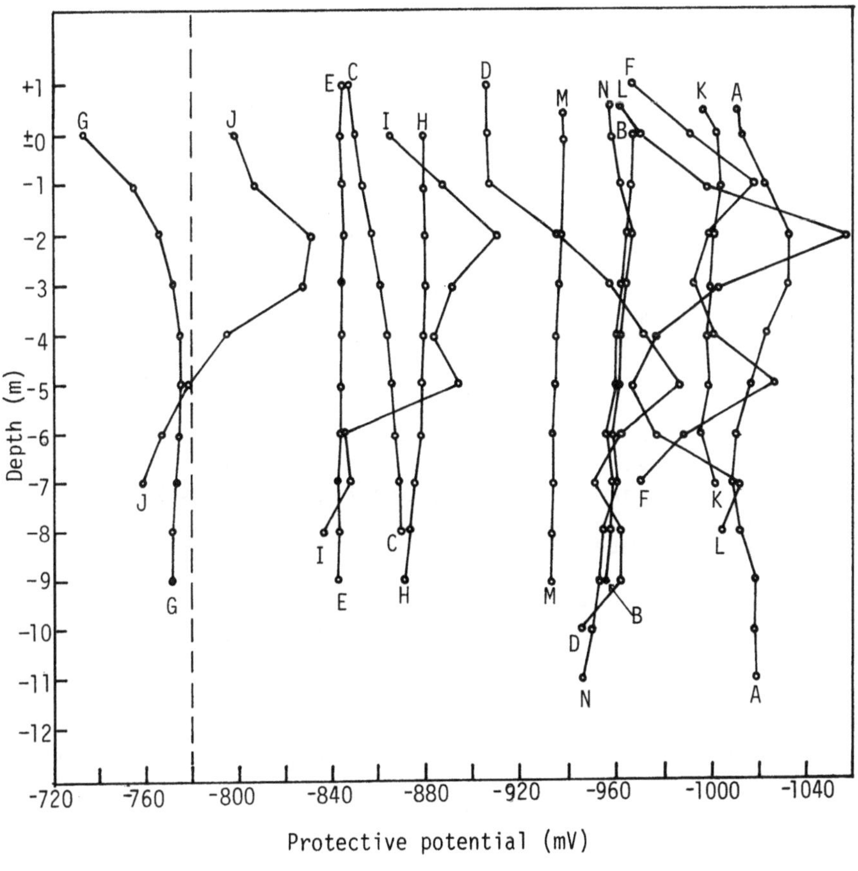

Fig. 5 Protective potentials of 16 facilities

The cathodic protections for the 16 facilities in Fig. 5 were designed with the same design procedure. However, the protective potentials of structures varied widely. At present, it is not possible to explain the variations in relation to the environmental conditions. At least, it may be said that the current design procedure of the cathodic protection in Japan seems adequate or conservative in most of the cases.

CONCLUSIONS

From the data obtained by the corrosion surveys on 205 facilities and the performance surveys on 24 facilities, the followings are concluded:

1) In some facilities, the concentrated corrosions occured at mean low water level and its rate of corrosion was 0.93 mm/yr, in a case.

2) Without the concentrated corrosion, the rate of corrosion of steel in seawater and bottom soil was about 0.1 mm/yr.

3) The cathodic protections surveyed were functioning properly in most of the cases. The current design procedure of the cathodic protection seems to be adequate or conservative.

ACKNOWLEDGEMENT

Dr. Kazuaki Zen, formaly Senior Research Engineer of Structures Division, PHRI, presently Director of Development Division, Nakagawa Corrosion Protecting Co. Ltd., established the procedures of the corrosion survey and the performance survey and carried out most of the corrosion surveys refered in this paper, while he was working at PHRI, together with one of the authors of this paper, Abe. Dr. Zen's contribution is highly acknowledged.

REFERENCES

1. Zen, K.; Abe, M.: Corrosion Survey of Steel Structures at Harbours (1st Report), Corrosion Surveys of Steel Structures at Hirohata, Osaka, Shibushi, Sasebo, Shimonoseki, and Toyama Harbors, Technical Note of the Port and Harbour Research Institute, No. 49, June 1968.

2. Zen, K.; Abe, M.: Corrosion Surveys of Steel Structures in Marine Environments (1978-1981 Surveys), Technical Note of the Port and Harbour Research Institute, No. 413, March 1982.

3. Zen, K.; Yokoi, T.; Abe, M.: Performances of Steel Structures under Marine Environment, being Applied Cathodic Protection (Part 1), Technical Note of the Port and Harbour Research Institute, No. 475, March 1984.

Electrochemical Measurements of Corrosion and Biofouling Films on Simulatcd OTEC Hcat Exchangers

P.K. Sullivan[1] and B.E. Liebert[2]

Departments of Ocean Engineering[1] and Mechanical Engineering[2], University of Hawaii, Honolulu, HI 96822, USA

ABSTRACT

Electrochemical measurements of corrosion and biofouling films have been performed using a recently developed technique called Electrode Impedance Spectroscopy (EIS). EIS is based on the hypothesis that corrosion and biofouling films can be characterized by observing the steady-state response to a small amplitude sinusoidal perturbing potential imposed on the heat exchanger material.

Copper-nickel, aluminum, and titanium alloys were evaluated with specially designed EIS corrosion cells placed in flowing seawater. Results were compared with the weight loss of coupons in line with the corrosion cells.

The results from these tests indicated that instantaneous corrosion rates can be determined at low frequencies, changes resulting from biofouling occur at high frequencies, the occurance of localized corrosion and multiple fouling films can be identified, and the effect of cleaning the surface of the material can be readily determined.

INTRODUCTION

Continued interest in Ocean Thermal Energy Conversion (OTEC) has identified a number of engineering problems in need of attention. Since the cost of heat exchangers for closed-cycle OTEC has been estimated to be of the order of 50 percent of the total system cost, the performance of heat exchangers and the selection of heat exchanger materials is of considerable importance to the designer.

A number of techniques have been used to evaluate materials for both corrosion and biofouling resistance, such as weight-loss and heat-transfer measurements [1-3]. Although heat-transfer measurements can be obtained in real time, and indirectly the rate of biofouling, the determination of corrosion rates and mechanisms has required the removal of specimens and subsequent analysis. Clearly, a technique that would allow the evaluation of both corrosion and biofouling to be made in situ would be desirable.

A technique that offers this capability is Electrode Impedance Spectroscopy (EIS, or ac impedance) [4]. Electrodes made of the material under test are placed in seawater flowing at 2 m/s, typical of OTEC conditions. A 10 mVpp sinusoidal voltage at a fixed frequency is applied to the electrodes and the current response is measured. From this information, the electrical impedance of the metal/fouling film/seawater system can be determined. If the measurement is made over a wide frequency range, typically 20 μHz to 120 kHz, the complete frequency response, in the form of a Bode plot (log impedance and phase vs log frequency), can generally be obtained [5].

From these results, an electrical equivalent circuit consisting of resistors and capacitors can be used to model the system. The values of the individual components can be determined from graphical and/or nonlinear regression techniques. From a knowledge of the expected fouling behavior of the metal, specific properties of the system, such as the thickness, dielectric constant, and resistivity of biological and corrosion films, can be determined.

For example, a metal completely inert to the formation of biofouling and corrosion films can be modelled as a single resistance in series (Rs) with a parallel resistance-capacitance (Rp-Cp); where Rs is the resistance of the electrolyte and leads, Rp is the charge-transfer resistance, and Cp is the double-layer capacitance. In this case, Rp is inversely proportional to the corrosion rate [6].

For each type of film that develops on the metal's surface, an additional parallel RC circuit, in series with the above model, is present and is characteristic of the film's thickness and the resistance of the metal to corrosion. Fig. 1 shows the equivalent circuit and Bode plot for such a system. The high frequency limit of the impedance is Rs, the low frequency limit is Rs+Rf+Rct, and at intermediate frequencies where the impedance is constant, the impedance is Rs+Rf. The capacitances can also be determined at the intersection or extrapolation of the impedance with log f = -log (2π).

If the corrosion rate is controlled by the rate of charge transfer at the metal/seawater interface, the corrosion rate can be calculated as shown in Eqn. 1. If the impedance of the film is not negligible, both Rct and Rf must be taken into account. In general, the loss in thickness due to corrosion is given by [7]:

$$\text{loss (um/yr)} = \frac{Ba*Bc*M*K}{2.303*Rct*(Ba+Bc)*p*n*F} \qquad (1)$$

where Ba and Bc are the Tafel constants obtained from a separate experiment or the literature, M is the average atomic mass, K is a conversion factor, p is the average metal density, n is the average number of electrons transferred, and F is Faraday's constant. The film thickness can be determined from the measured capacitance if the dielectric constant of the film is obtained from a separate experiment or the literature.

$$\text{thickness} = \varepsilon*\varepsilon_o \frac{A}{C} \qquad (2)$$

where ε is the dielectric constant of the film, ε_o is the permittivity of free space, A is the area of the electrode, and C is the measured capacitance.

EXPERIMENTAL METHODS

Facilities at the Natural Energy Laboratory of Hawaii, Keahole Point, Hawaii were used as a continuous source of warm seawater for nearly one year. During this time, specially-designed corrosion cells and an automated data acquisition system recorded EIS spectra for copper-nickel (CA706), aluminum (alclad 3003), and titanium (commercially-pure). The computer-controlled system multiplexed among the three materials, taking a complete EIS measurement for each material once every three days. Instrumentation was controlled with an HP 85 micro-computer and data were processed with nonlinear least squares regression using Marquardt-Levenburg optimization of the magnitude of the impedance.

RESULTS AND DISCUSSION

Example results are provided to illustrate the utility of EIS to monitor various material characteristics of interest. All materials were found to form biofouling films that result in an increase in the heat-transfer resistance.

Titanium

Since titanium in known to be chemically inert under these conditions [1] due to a protective oxide film, changes in the impedance spectrum are due solely to the presence of a biofilm. Fig. 2 shows the EIS spectrum in the presence of a two-layer biofilm, as discussed above.

Fig. 3 is the EIS spectrum of titanium after brush cleaning. The EIS spectrum suggests that brushing removes a loosely-adhering biofilm; however, a tightly-adhering film of negligible heat-transfer resistance remains after brushing.

Aluminum

Aluminum has a tendency to corrode by localized attack (pitting) in seawater [8], although recent results indicate that pitting is unusual under OTEC conditions [3]. It has been found [9-10] that the magnitude of the impedance at low frequencies decreases with decreasing frequency if localized attack is present. As seen in Fig. 4, this behavior is evident below 0.1 Hz. However, after several brush cleanings the areas of localized attact repassivate and uniform corrosion is found, as shown in Fig. 5. Notice also the low-frequency impedance increased two orders of magnitude after repassivation, indicating a substantially lower corrosion rate.

Copper-Nickel

Although copper-nickel is incompatible with ammonia, the most likely candidate for a closed-cycle OTEC working fluid, copper-nickel is of interest because of its frequent use in other marine applications [11]. Corrosion rates predicted from the weight loss of sample coupons are compared to rates calculated from EIS spectra and shown in Fig. 6. The points with error bars were determined from the weight loss of individual sample coupons while the other points were obtained from EIS data. The agreement between the two techniques is very good.

SUMMARY

The EIS method has been shown to be a powerful technique for the in situ monitoring of the condition of heat exchangers. It is certainly not limited to OTEC applications; indeed, any electrochemical or biological process can be monitored with this technique. It is useful for testing materials for design purposes, such as the effect of brush cleaning versus chlorination and the influence of water quality on biofouling and corrosion rates.

The support of The Seagrant Program and the State of Hawaii is gratefully acknowledged. The authors would like to thank Prof. L.R. Berger, Dept. of Microbiology at the University of Hawaii, for many fruitful discussions. This work was done for partial fulfillment of the Ph.D. degree in Ocean Engineering for one of the authors (PKS).

REFERENCES

1. Liebert, B.E.; Berger, L.R.; White, H.J.; Moore, J.; McCoy, Wm.; Berger, J.A.; and Larsen-Basse, J.: The Effects of Biofouling and Corrosion on Heat Transfer Measurements. Proc. Sixth OTEC Conf., Washington, D.C. (1979) 12.7-12.16.
2. Liebert, B.E.; Larsen-Basse, J.; Berger, J.; and Berger, L.R.: Biofouling and Corrosion Studies at Keahole Point, Hawaii (1978-1980). Proc. Eighth OTEC Conf., Washington, D.C. (1981) 421-430.
3. Munchmeyer, FC.; Berger, L.R.; and Liebert, B.E.: Five OTEC Biofouling and Corrosion Experiments at Keahole Point, 1976-1980. Ocean Sci. Engr. 7 (1982) 1-20.
4. Macdonald, D.D.: Transient Techniques in Electrochemistry. Plenum Press, NY (1977).
5. Sullivan, P.K.: Electrochemical Measurements of Corrosion and Biofouling Films on Simulated OTEC Heat Exchangers. Ph.D. Dissertation, Dept. Ocean Engineering, University of Hawaii (1985).
6. Mansfeld, F.: Recording and Analysis of AC Impedance Data for Corrosion Studies. I. Background and Methods of Analysis. Corrosion 36 (1981) 301.
7. Stern, M. and Geary, A.: A Method for Determining Corrosion Rates from Linear Polarization Data, Corrosion, 14 (1958) 440.
8. Godard, H.P.: An Insite Into the Corrosion Behavior of Aluminum, Materials Performance 20 (1981) 9.
9. Bertocci, U.: Applications of a Low Noise Potentiostat in Electrochemical Measurements. J. Electrochem. Soc. 127 (1980) 1931.
10. Freeman, R.S.: Weight Loss, Potentiostatic Polarization, and Impedance Measurements of Three Aluminum Alloys. M.S. Thesis, Dept. Mechanical Engineering, University of Hawaii (1984).
11. Lewis, R.O. and F.L. LaQue: Evaluation of Copper-Nickel Alloys for OTEC Applications, Report by Argonne National Laboratory, Argonne, Illinois, ANL/OTEC-BCM-019 (1981).

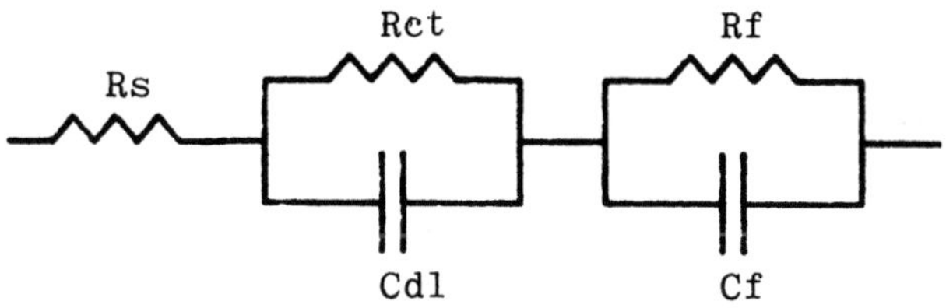

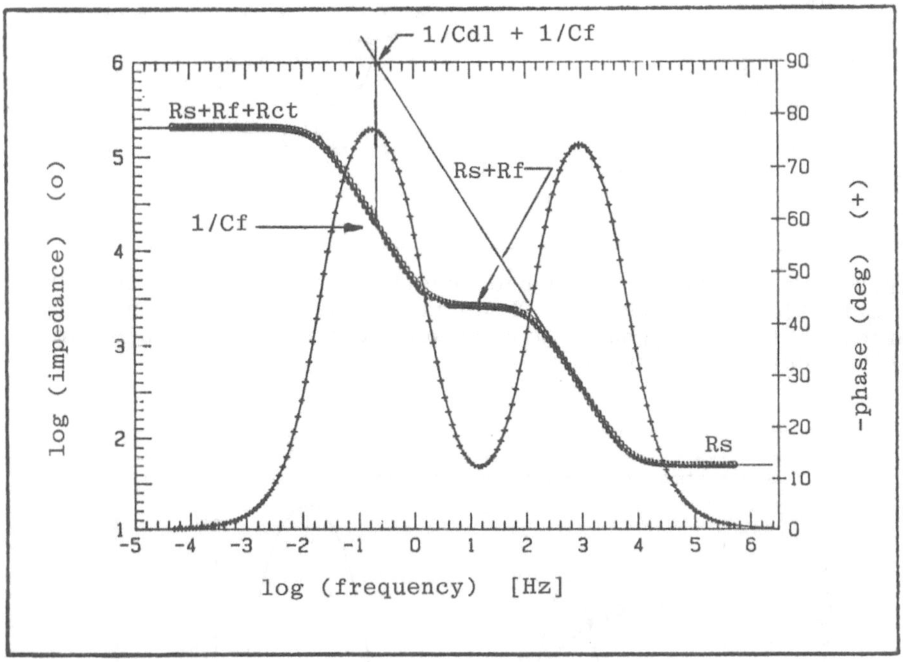

Fig.1. The equivalent circuit and corresponding Bode plot for a metal with a single fouling film. Rs, Rct, and Rf are the series, charge-transfer, and film resistances. Cdl and Cf are the double-layer and film capacitances. The high, intermediate, and low frequency impedances are Rs, Rs+Rf, and Rs+Rf+Rct, respectively. The capacitances can be obtained at the intersection or extrapolation of the impedance with log f = $-\log(2\pi)$, as shown, or by nonlinear regression.

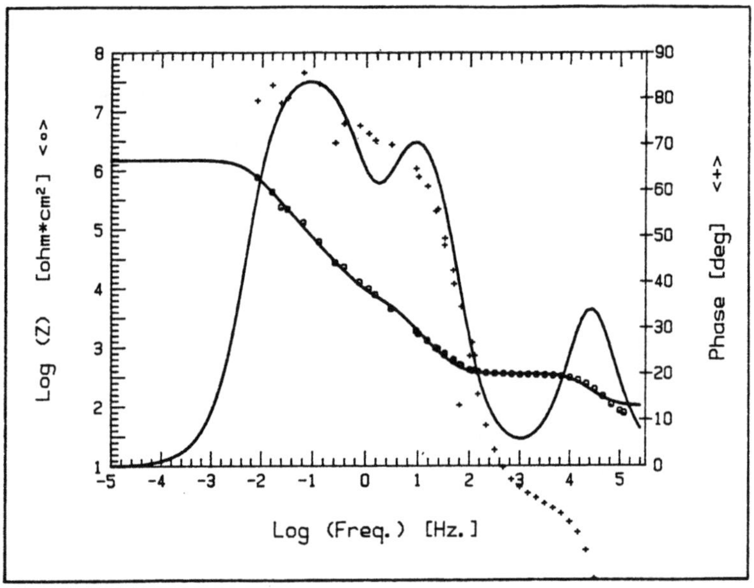

Fig.2. The electrode impedance spectrum of titanium after 206 days in flowing seawater. Three time constants, evidenced by three peaks in the phase, are present indicating two biofilms are present on the surface.

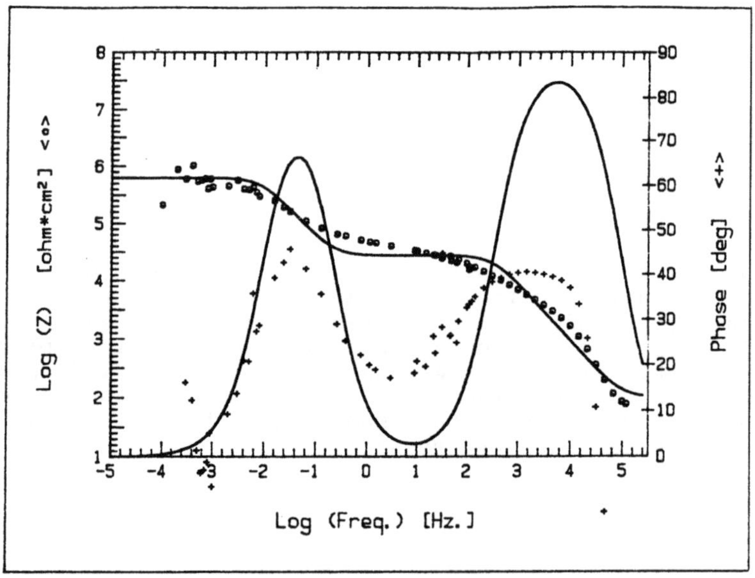

Fig.3. The electrode impedance spectrum of brush-cleaned titanium after 274 days in flowing seawater. Two time constants are present suggesting that brush cleaning removed only one of the biofilms.

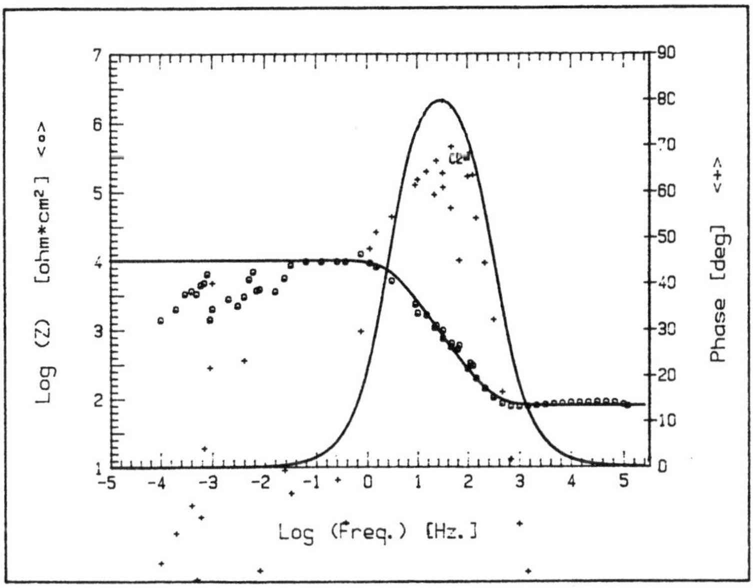

Fig.4. The electrode impedance spectrum of aluminum after 102 days in flowing seawater. Pitting is evident from the decrease in the impedance with decreasing frequency. Pitting was visually confirmed upon removal from the system.

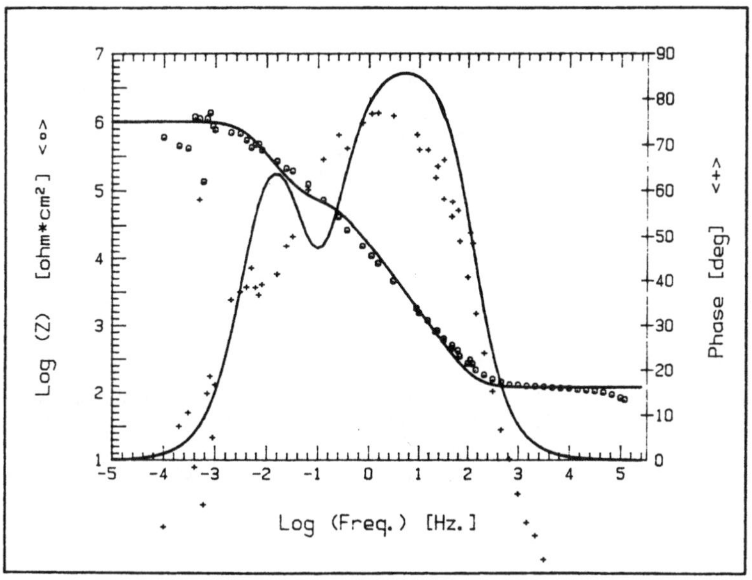

Fig.5. The electrode impedance spectrum of aluminum after 310 days in flowing seawater. Pitting attack has ceased and the low-frequency impedance is two orders of magnitude higher than the same sample at an earlier time shown in Fig. 4 indicating a substantially lower corrosion rate due to passivation.

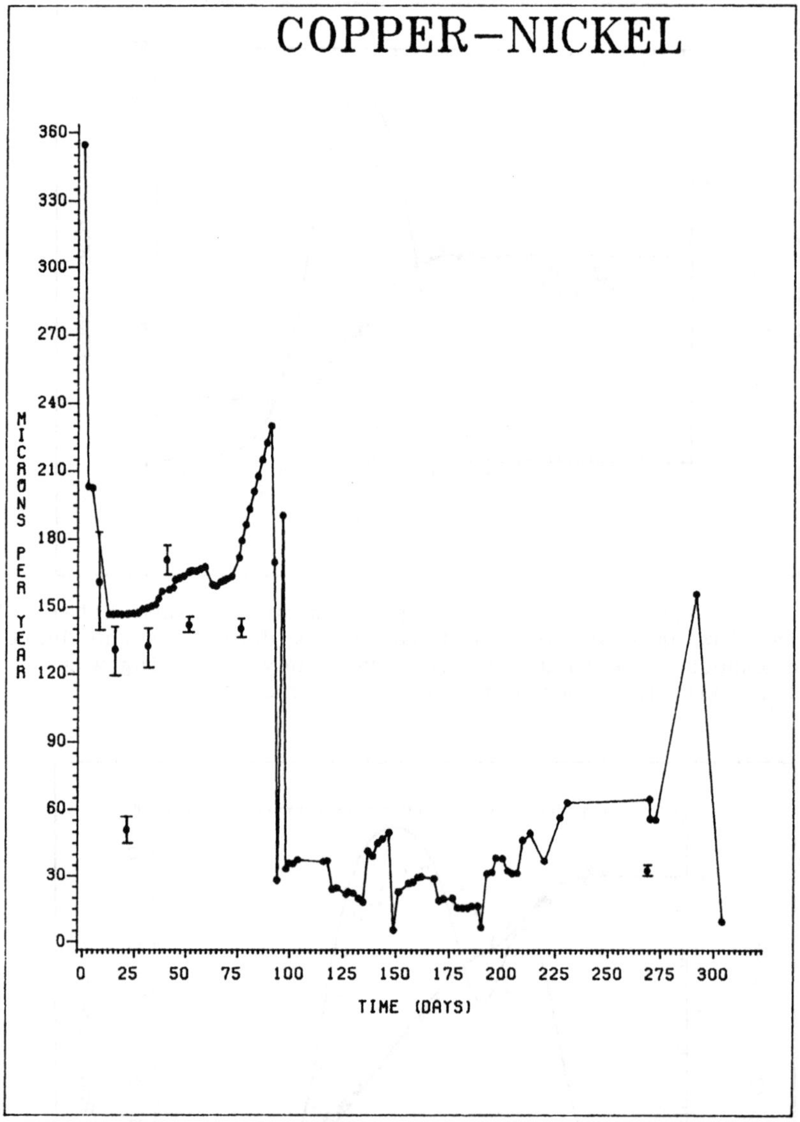

Fig.6. The corrosion rate of copper-nickel in flowing seawater over a 310 day period. The eight points with error bars were obtained from weight-loss measurements. The other points connected by a straight line were obtained from the EIS data. There is a general tendency of a decreasing corrosion rate with time, typical of film-forming metals.

Atmospheric Exposed Test vs. Simulated Acceleration Test on Painted Test Piecc

T. Nishijima[1] and R. Tanaka[2]

[1] Public Works Research Institute, Ministry of Construction, Toyosato, Tsukuba, Ibaraki, 305 Japan
[2] Public Works Research Center, Taito, Taito-ku, Tokyo, 110 Japan

1. Foreword

In cooperation with test equipment and paint manufacturers, as well as other companies and associations involved in the handling of anti-corrosive agents, we at the Public Works Research Institute, Ministry of Construction, have developed and put into trial use an apparatus which simulates environmental conditions for the purpose of testing at an accelerated rate the durability of anti-corrosive products used on steel structures in coastal areas. The apparatus is so built that environmental conditions such as light, salinity, the presence of acidic gas, humidity and temperature can be controlled at will, the test pieces being subjected to these simulated conditions in cycles which can be repeated at pre-determined intervals. Since 1979, we have been conducting simulated accele ration tests to investigate the durability of corrosion protective paints. In this report, we would like to discuss some of the findings which these tests have yielded. This research was conducted jointly by the Public Works Research Institute, Ministry of Construction, the Public Works Research Center, and the Japan Steel Bridge Paint Specialists Association. Research planning and the compilation of the research results were executed by the Committee for Research on Simulated Acceleration Testing of Corrosion Protective Materials, set up internally in the center and chaired by Dr. Harumitsu Tamano.

2. Outline of Research

2.1 Research Objectives

It is well known that paint coverings on steel structures in coastal areas tend to deteriorate relatively faster, their deterioration being promoted by the severe influence of salt particles in the environment to which they are subjected to. The deterioration tendencies of the paints also differ greatly with the position on the structure of the areas to which they are applied; areas subjected to dampness or located where there is much salt accumulation, such as the undersides of flanges and such, tend to deteriorate more readily than others.

Therefore, the first objective of our research was to find a way to re-create the environmental conditions which promote the deterioration of paints applied to the undersides of steel bridges, whereby we could accelerate the speed of paint deterioration. Experiments were also undertaken with the purpose of discovering test conditions corresponding to other environmental conditions for testing and evaluating the durability of newly developed paints.

2.2 Research Methods

In developing tests calling for the acceleration of the deterioration process in a simulated environment, it is essential to devise a method which yields test results correlating closely to the deterioration of paints in an actual environment. It was therefore necessary to first collect data on the deterioration process in a natural environment of paints applied to test pieces of the same nature as those to be used in the simulated acceleration tests. Since the results of simulated acceleration tests conducted in accordance with JIS standards would be necessary for comparison, salt water immersion and saline mist tests were executed parallel to the exposure and the simulated environment tests. The methods and the specifications of the painted test pieces used in these tests are listed in the following test summaries.

(1) Exposure Tests
1) Okinawa Region
<1> Underside of the Ufuganiku Bridge, North Okinawa prefecture
<2> Okinawa City, Mid Okinawa Prefecture
2) Kanto Region
<1> Ichikawa City, Chiba Prefecture
<2> Ajigaura, Nakaminato City, Ibaraki Prefecture
(2) JIS Standard Tests
1) Salt Water Immersion Tests
2) Saline Mist Tests
(3) Simulated Environment Tests
1) Outline of Test Equipment

The structure of the test apparatus is shown in Figure 1.

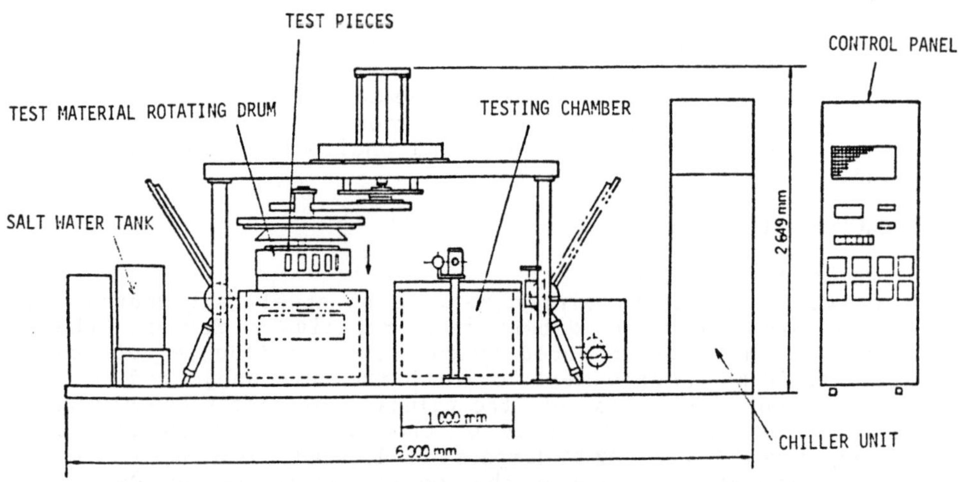

Figure 1. Simulated Environment Test Apparatus

2) Test Conditions

The effects of salt and repeated inundation and drying are

thought to be the strongest factors promoting the deterioration of paint coverings on the undersides of bridges in coastal regions. Subsequently, the test conditions listed below were set up:

A: One cycle = Saline mist, 2 hours (@ 35°C)→ Drying, 1 hour→ 90% RH @50°, 20 hours.
B: One cycle = Saline mist, 2 hours (@ 35°C)→ Drying, 1 hour→ 80% RH @ 50°C, 15 hours→ 50% RH @50°C, 5 hours.

<4> Preparation of the Test Pieces
Finding test conditions which would yield results correlating closely to the results of the exposure tests being our purpose, and wishing to obtain test results in a relatively short time, thin-film coats of long-term exposure etching primer, epoxy zinc primer, inorganic zinc-rich primer and zinc anti-corrosive paint were applied in one application to the test pieces according to the specifications shown in Table 1.

Table 1. Specifications of the Test Pieces

Film Thickness (μm) / Paint	Target thickness	Actual thickness upon measuring	
		Smallest	Greatest
Long-term Exposure Etching Primer	15 ~ 20	16	22
Epoxy Zinc-rich Primer	15 ~ 20	14	24
Inorganic Zinc-rich Primer	15 ~ 20	17	20
Cyanamide Zinc Anti-corrosive Paint	30 ~ 35	26	33

3.2mm thick SS41 blast steel sheets measuring 150 X 70mm were used as test pieces.

<5> Test Piece Deterioration Evaluation
The progress of test piece deterioration was assessed by visual observation. If no abnormalities could be seen, the evaluation "{circle}" was given; with the appearance of spot rust or other light signs of defects in the paint film, the evaluation mark "{triangle}" was given; when signs of clear defects could be observed, an "X" was given. White rust was not considered an object of evaluation, therefore the extent of its appearance was merely recorded with the comments, "little," "medium," or "much."

3. Research Results and Observations

3.1 Exposure Test Results
Comparing the severity of the environmental conditions at each of the exposure test locations based on the deterioration tendencies of the test pieces, the following order can be ascertained: Okinawa Ufuganiku Bridge underside > Ajigaura, Ibaraki Pref.≠ Okinawa City > Ichikawa City, Chiba Pref., in descending order of severity.

Comparing the different paints, one finds the long-term exposure type epoxy primer displaying the most defects, followed by the single-coat cyanamide zinc anti- corrosive paint. Next to these two paints, the zinc-rich primers are more durable, however the inorganic type seems to be slightly more durable than the epoxy type, these results more or less verifying common knowledge. The degree of white rust formation {in the zinc-rich types} was greater on the underside of Ufuganiku Bridge and in Okinawa city than in, not only Ichikawa City, but even Ajigaura, where the tests were conducted over sea water. This fact demonstrates that, more than the effects of salt particles, the effects of humidity and temperature play an important role in promoting paint deterioration in Okinawa.

3.2 Results of the JIS Tests

Results of salt water immersion tests differed from those of the exposure tests concerning the ranking of the different paints according to deterioration progression, and there were even greater differences between the two testing methods concerning conditions surrounding rust formation. In contrast, the results of the saline mist tests were the same as those concerning the ranking of the paints according to deterioration progression in the expo sure tests, lending credence to the suitability of this test for comparing corrosive tendencies in coastal areas. There were, however, no similarities in the rust formation conditions.

3.3 Results of the Simulated Environment Tests

The results of the simulated environment tests are shown in Table 2. In comparing the results of the tests conducted under the two different test conditions, A and B, one discovers that the formation of white rust is extensive in zinc-rich paints subjected to long term, high temperature exposure under test condition group A. However, the ranking of paint type according to deterioration progression and the deterioration tendencies remain the same {under the different test conditions}, i.e. long-term etching primer> cyanamide zinc anti-corrosive paint>> epoxy zinc-rich primer = inorganic zinc-rich primer, in descending order of deterioration progression.

Comparing these test results with those of the exposure tests, we find that the degree of accelerated deterioration was the highest with etching primer and cyanamide zinc anti-corrosive paint. Compared to deterioration underneath the Ufuganiku Bridge in Okinawa, where the most severe natural conditions prevail, the test pieces in the acceleration tests using these two paints showed the equivalent of three to four months deterioration after six cycles (six days). After the same period, they displayed deterioration equivalent to four to six months exposure at Okinawa City and Ajigaura, and to one year's exposure in Ichikawa City, where the environmental conditions were the least severe. However, in separate tests spanning 130 cycles (more than four months) conducted on metallic paints containing a high density of metallic zinc, like the zinc-rich paints, there was no occurrence of defects, with the exception of white rust, in the paint films. We think this may be due to the formation of an oxidized zinc film on the surface of zinc-rich paints which has a protecting effect.

The ranking of the different paints according to deterioration progression is: long-term etching primer> cyanamide zinc anti-corrosive paint>> epoxy zinc-rich primer = inorganic zinc-rich primer. With the exception of the zinc-rich types, the ranking of the paints according to deterioration progression, the progression of paint deterioration and the conditions of the rust formations in the simulation tests resembled very closely those in the exposure tests.

On comparing the simulated acceleration test results with those of the JIS test results, the degree of deterioration acceleration was clearly greater in the simulation tests than in the salt water immersion tests, and slightly greater than in the saline mist tests, on organic coverings such as the etching primer. However, due to an insufficiency of data, no such conclusions can be drawn concerning the zinc-rich primers.

4. Conclusion

Summarizing these test results, the following statements can be made:

<1> In testing the durability of organic paints with the simulated environment test, the progression of paint deterioration, the ranking of the different paints according to deterioration progression, and the conditions surrounding the formation of rust are very similar to those in an actual environment, thus demonstrating that this method of testing is superior to contemporary methods for re-creating the conditions which promote paint deterioration.

<2> The rate of paint deterioration was greatly accelerated in the simulation tests in comparison to its rate under conditions of natural exposure. The same holds true when comparing the results of the simulation tests with those of the salt water immersion and the saline mist tests; it seems to us that, by devising additional combinations of factors promoting deterioration and exposure time distribution, these simulation tests could be executed even more rapidly.

Judging from the above-mention facts, we conclude that the simulated environment test is an effective method for accelerating the deterioration of paint films. Based on these test results, it is expected that the same kind of good results can be obtained by shortening the length of time the test pieces are exposed to high temperatures, thus shortening by that much the time required for a single cycle. In the future, we plan to conduct tests under these revised test conditions as well as combining conditions of drying by light waves into the test scheme. We hope to use methods of testing devised in this manner for investigating and evaluating the durability of newly developed anti-corrosive materials and paints.

In conclusion, we wish to express our sincere gratitude to the members of the research committee for their guidance, as well as to those persons who worked so hard in executing the tests and compiling the results thereof.

Table 2 Simulated Environment Test Results

Test conditions	Paint \ Number of cycles	4	5	6	11	19	25
A One cycle=Saline mist, 2 hours (@35°C)→Drying, 1 hour→90% RH @50°, 20 hours.	Long-term Exposure Etching Primer	△	△	×	×	×	×
		△	△	×	×	×	×
	Epoxy Zinc-rich Primer	○ white rust: little	○ white rust: little	○ white rust: medium	○ white rust: much	○ white rust: much	○ white rust: much
		○ white rust: little	○ white rust: little	○ white rust: little	○ white rust: much	○ white rust: much	○ white rust: much
	Inorganic Zinc-rich Primer	○ white rust: little	○ white rust: medium	○ white rust: medium	○ white rust: much	○ white rust: much	○ white rust: much
		○ white rust: little	○ white rust: little	○ white rust: medium	○ white rust: much	○ white rust: much	○ white rust: much
	Cyanamide Zinc Anti-corrosive Paint	○	○	△	△	×	×
		○	○	△	△	×	×
B One cycle=Saline mist, 2 hours (@35°C)→Drying, 1 hour→80% RH @50°C, 15 hours 50% RH @50°C, 5 hours	Long-term Exposure Etching Primer	△	△	×	×	×	×
		△	△	×	×	×	×
	Epoxy Zinc-rich Primer	○ white rust: little	○ white rust: little	○ white rust: little	○ white rust: medium	○ white rust: much	○ white rust: much
		○ white rust: little	○ white rust: little	○ white rust: medium	○ white rust: much	○ white rust: much	○ white rust: much
	Inorganic Zinc-rich Primer	○ white rust: little	○ white rust: little	○ white rust: medium	○ white rust: much	○ white rust: much	○ white rust: much
		○ white rust: little	○ white rust: little	○ white rust: little	○ white rust: much	○ white rust: much	○ white rust: much
	Cyanamide Zinc Anti-corrosive Paint	○	○	△	△	×	×
		○	○	△	△	×	×

(NOTE) Observations made continuously on two test pieces per type of paint (observations limited to front-side)

Table 3. Results of Exposure Tests

Location of Exposure	Term of exposure / Type of paint \ Observed side	2 Months		3 Months		4 Months		6 Months		12 Months	
		F	B	F	B	F	B	F	B	F	B
Ichikawa City (Chiba Prefecture)	Long-term exp. etching pr.	○	○	–	–	–	–	○	△	×	×
	Epoxy zinc-rich primer	○	○	–	–	–	–	○L	○L	○L	△
	Inorganic zinc-rich primer	○	○	–	–	–	–	○L	○L	○L	○L
	Cyanamide zinc anti-cor.pt	○	○	–	–	–	–	○	○	×	○
Ajigaura (Ibaraki Prefecture)	Long-term exp. etching pr.	–	–	–	–	×	×	×	△	×	×
	Epoxy zinc-rich primer	–	–	–	–	○	○	○L	○L	×	△
	Inorganic zinc-rich primer	–	–	–	–	○	○	○L	○L	○M	○M
	Cyanamide zinc anti-cor.pt	–	–	–	–	○	○	×	×	△	△
Okinawa City (Mid-Okinawa Prefecture)	Long-term exp. etching pr.	–	–	–	–	△	×	△	△	×	×
	Epoxy zinc-rich primer	–	–	–	–	○	○M	○M	○M	○M	○M
	Inorganic zinc-rich primer	–	–	–	–	○	○M	○M	○M	○M	○M
	Cyanamide zinc anti-cor.pt	–	–	–	–	○	△	△	△	×	△
Ufuganiku Bridge, underside, (North Okinawa Prefecture)	Long-term exp. etching pr.	–	–	△	×	–	–	×	×	×	×
	Epoxy zinc-rich primer	–	–	○	○	–	–	○M	○M	△	△
	Inorganic zinc-rich primer	–	–	○	○	–	–	○M	○M	○M	○M
	Cyanamide zinc anti-cor.pt	–	–	○	○	–	–	×	×	×	×

Table symbol meanings: ○ No abnormalities observable ○L: Little white rust F: Front side
○M: Much white rust B: Reverse side
△ Small amounts of spot rust, etc., minor deterioration
× Rust, etc., advanced deterioration

Note: Test pieces showing the most deterioration were removed at the time of obsersvation after each term of exposure

Full names of corrosion protection coatings: Long-term exposure etching primer
Epoxy zinc-rich primer
Inorganic zinc-rich primer
Cyanamide zinc anti-corrosive paint

Development and Progress of Anti-Corrosive Coating in the Marine Atmosphere

M. Fukushima*, S. Hashimoto** and T. Tsuji***

Technical Committee, Public Works Research Center, Taito, Taito-ku, Tokyo, 110 Japan

ABSTRACT

The transition of rust preventive coating in the Japanese marine atmosphere is presented focusing on coating system, coating technology and evaluation technique. The paper reports the research projects and the data measured in the first year regarding the research of durable rust-preventive coating technology in the marine atmosphere carried our in Research Facilities of Marine Test installed in Suruga Bay.

INTRODUCTION

It has been conceived that coating is the most effective measure for rust prevention in the marine atmosphere. The coating system suited for the marine environment has been developed as the most important subject in Japan. The development of the coating system has been supported by the advanced coating technologies and evaluature techniques. The coating systems in the Japanese marine atmosphere have significantly changed in the past two decades. Coating Standards for Steel Bridges and the annexed manual for the Main Island-Shikoku Island Connecting Bridges revised in 1980 have been standardized.[1] In order to utilize the ocean more effectively in the future, however, it is required that the technology of higher-performance, more reliable and durable rust-preventive coating will be developed.

CORROSIVE ENVIRONMENT

Since Japan is an island country where the zones of life are massed at the coastal areas due to mountainous regions in the centre, most steel structures are constructed in the coastal areas. Severe corrosion of structural steel is caused by sea salt particles adhering to it. The nearer the structure is located to the coast, the more the salt adheres to the structure which is very much corroded. Even coated steel is much more deteriorated in the marine atmosphere than in the others. According to the investigation made in 1982, after the elasped times in which 25% of the bridges investigated have been so much deteriorated has to be repaired are halve in the marine environment compared with those in the mountainous environment.[2]

---

* Development Group Manager, Nippon Paint Co., Ltd.
** Senior Engineer, Kansai Paint Co., Ltd.
*** Senior Engineer, Dainippon Paint Co., Ltd.

TRANSITION OF ANTI-CORROSIVE COATING

Changing of coating system

Rust preventive coating is applied to structures such as a cross-channel bridge, oil-drilling equipment, a sea berth and an observation tower in the marine atmosphere. Since the past records of the cross-channel bridges have been well preserved, a tracking survey can be easily run on them. The changes of the coating system in the marine atmosphere are stated based on the past records of the cross-channel bridges hereafter (See Table 1). Coating had been sometimes made on mill scale (the larger bridge of Choshi) until 1960 and since then, steel surface has been coated after blasting. Alkyd resin coating was applied on anticorrosive oil paint early in 1960.

Table 1. The past records of the cross-channel bridges

Name of bridge	Date of construction	Painting systems	No.of coats
The Choshi bridge	1960	Oil base primer/Alkyd paint (Over mill scale)	4
The Wakato bridge	1962	Oil base primer/Alkyd paint	5
The Tenmon bridge	1966	Oil base primer/Alkyd paint/ Aluminum paint	5
The Onomichi bridge	1968	Oil base primer/Alkyd paint	
The Keihin bridge	1971	Inorganic zinc rich/Epoxy paint/ Urethane paint	5
The Jonan bridge	1971	Inorganic zinc rich/Epoxy paint/ Urethane paint	5
The Hiroshima bridge	1973	Inorganic zinc rich/ Chlorinated rubber paint	5
The Kanmon bridge	1973	Tower:Oil base primer/Alkyd paint :Zinc spray/Wash primer/ Zinc chromate primer/ Phenolic micaceous iron ore/ Chlorinated rubber paint	5
The Arakawa-wangan bridge	1974	Epoxy zinc rich/ Chlorinated rubber paint	5
The Minato bridge	1974	Oil base primer/ Phenolic micaceous iron ore/ Chlorinated rubber paint	5
The Omishima bridge	1978	Inorganic zinc rich/wash primer/ Zinc chromate primer/ Phenolic micaceous iron ore/ Chlorinated rubber paint	6
The Into bridge	1982	Inorganic zinc rich/Epoxy paint/ Urethane paint	5

Even cross-channel bridges were coated by the same coating system as that, or only by repeating the coating (the larger bridge of Wakato) or with aluminium paint as a top-coat at the most (Temmon Bridge). Zinc-rich paint, chlorinated rubber coating, epoxy resin coating and polyurethane coating had been applied on steel structures in around 1965. Since these coatings were approved to be well rust-preventive, they had been applied on cross-channel bridges in around 1970 (the larger bridge of Keihin, the larger bridge of Jonan, the larger bridge of Hiroshima and the coastal bridge of Arakawa). In 1970's, cross-channel bridges had been increasingly constructed and zinc metal spraying and micaceous iron oxide paint had been applied on them (Kanmon Bridge and the larger bridge of Minato). Since then, the coating system had been studied for application to the Main Island-Shikoku Island Connecting Bridges which would connect both Islands approximately 38 kilometers apart from each other across channels from island to island. The coating system was investigated during a committee established by Japan Civil Engineering Institute as requested by the Main Island-Shikoku Island Connecting Bridges Construction Corporation. The goal of the rust preventive coating was to lengthen its life on the sea up to a decade or longer. The fundamental principles of the rust preventive coating submitted from the committee in 1974 are as follows:

1) Zinc layer approximately 75μm thick shall be applied on the steel surface by zinc metal spraying or by coating with zinc-rich paint.
2) Another thick film coating approximately 200μm thick shall be applied to the zinc layer (See Table 2).

Coating Standards for Steel Bridges and the annexed manual for coating of the Main Island-Shikoku Island Connecting Bridges established in 1977 and revised in 1980 have been standardized into the current bases of coating in Japanese marine atmosphere.

Table 2. Painting systems for Main Island-Shikoku Island Connecting Bridges

Surface preparation	Blasting	
	High build inorganic zinc rich, 75μm or Zinc spraying, 75μm	
paint designation, film thickness	Wash primer	Mist coat
	Phenolic zinc-chromate primer, 30μm	Epoxy anti-corrosive 60μm
	Phenolic micaceous iron oxide paint, 60μm	Epoxy anti-corrosive 60μm
	Phenolic micaceous iron oxide paint, 60μm	Polyurethane paint 30μm
	Chlorinated rubber paint, 35μm	Polyurethane paint 30μm
	Chlorinated rubber paint, 25μm	
Total film thickness	285μm	255μm

Changes of coating technology

The technology of coating in the marine atmosphere has been studied in research installations off the coast of Iwaya in Awaji Island, on the coast of Tarumi, Hyogo Prefecture and on the pier of Ajigaura, Ibaragi Prefecture. As a result, it was found that there were the following technical problems in the field of durable rust preventive coating in the marine atmosphere:

(1) to develop a water-resistant coating with high-barrier properties,
(2) to establish the control criterion for pretreatment of coating,
(3) to strengthen the rust-preventive weak spots at the edge and joint of structural steel
(4) requirement of keeping up appearance of film long also accompanied by improving the durability of rust preventive coating,
(5) to solve the problem of sea salt particles adhering to the substrate in painting operation in the maine atmosphere and
(6) to develop a reliable repainting system.

It was confirmed that thick-film coating with water-resistant epoxy resin coating possessing high-barrier properties is effective for coating in the marine atmosphere and MIO and glass flakes are effective fillers to improve the barrier properties.

Although SSPC and SIS have been utilized as the standards for pretreatment of coating so far, new standards have been established assuming the possible practical conditions.

Splice plates sprayed with zinc and aluminium and high-strength bolts plated with cadmium were used at the joints of Kammon Bridge completed in 1973. Later on, abrasion-resistant inorganic zinc-rich paint has been applied on the splice plates and newly developed special lubricating primer has been applied on the high-strength bolts of cross-channel bridges.[3] The treatment and the coating at the edge of structural steel were studied. Long durable colour and gloss of film were investigated to follow the durability of rust preventive coating. Binder, pigment and colour with low susceptibility of discoloration were selected by extended exposure test and accelerated test.[4] Adhesion of sea salt particles to the substrate is unavoidable in painting operation in the marine atmosphere. It was confirmed that coating can be applied without hindrance on the substrate to which sea salt particles have adhered at a density of 100mg/m^2 or less.[5] Although the analytical manual for adhering salt is available, the determination may be difficult depending on the kinds of film.[6]

It is hard to recover the same degree of rust preventing properties as the new coating by repainting on the sea. Any of the several methods of repainting in the marine atmosphere proposed so far is not available. It is a problem of awaiting solution to develop a reliable repainting system.

Changes of evaluation technique

Visual inspection of the film coated on steel structures has been made for judging the degree of deterioration for more than two decades by Japanese National Railways. This is the criterion for judging according to the total points of the five items including peeling, blistering, cracking, rusting and chalking. Various inspection methods have been investigated so far. The current method applied to inspection of steel structures was published as the manual of JSSC in 1982. Research is being carried out at present for estimating the deterioration and clarifying the deterioration mechanism of steel structures.[7] The method of evaluating the rust preventing properties of film by alterating impedance developed in around 1960 is widely used on-site.

Some reports on the method evaluating the rust preventing properties of film by polarization characteristics of steel surface below the film have been published since 1975. The methods called corrosion-rate meter method and current interruptor method have been developed for measuring polarization characteristics.[8,9] An accelerated test to simulate the complicated natural corrosive environment is being studied. JSSC reported the results of test made by combined test apparatuses, while Building and Public Works Research Institute is making research by a prototype of composite environmental test apparatus.[10] New test apparatuses have been announced by another test machine manufacturer.

RESEARCH ON THE TECHNOLOGY OF DURABLE RUST PREVENTIVE COATING IN THE MARINE ATMOSPHERE

The coating system maintaining the rust preventing properties for a decade or longer in the marine atmosphere has been developed. More durable coating systems are required to be developed to effectively utilize the ocean in future. Start has been made with the research in Research Facilities of Marine Test installed in Suruga Bay in 1984 to develop coating systems maintaining the durability for two decades or longer in the marine atmosphere and to study the evaluation method and the repainting technology for improving the reliability and practicality of coating system.

Research Facilities

- Installation spot: 250 meters out in Suruga Bay off Shizuoka Prefecture
- Facilities: the structure consists of the main body over the level two meters above the surface of the sea and the exposure rack installed on the deck about fourteen meters above the sea level.

Subjects of study

Measurement of ambient corrosive environment

The corrosive environmental conditions in the facilities such as surface temperature, surface wetting time, amount of airborne salt, atmospheric temperature, humidity, velocity and direction of wind, amount of precipitation and amount of corroded structural steel were measured to understand the marine environment.

Coating on main body of facilities

Twelve kinds of coatings were applied on the main body of the facilities. The specification of the coating system and the initial measurements of the film were listed in the table (See Table 3). These coating systems are those that have been already applied and will be applied in the near future in the marine atmosphere. Although the facilities were taken in tow on the sea as far as one thousand and several hundreds kilometers to the site from the manufactory where the coating had been completely finished, any noticeable scratches were not observed on the film.

Table 3. Coatings on the facilities and the initial measurements

Surface preparation	Paint designation, No. of, film coat thickness	Initial measurements Film thickness	Gloss
Blasting	High build epoxy zinc rich, 1 coat, 75μm Glass flake polyester, 2 coats, 700μm Fluoric resin paint, 2 coats, 40μm	946	40
	Polyester primer, 1 coat, 40μm Glass flake polyester, 2 coats, 700μm Polyurethane paint, 2 coats, 60μm	1022	22
	Glass flake epoxy, 2 coats, 1500μm Polyurethane paint, 2 coats, 60μm	1756	11
	Glass flake epoxy, 2 coats, 600μm Polyurethane paint, 2 coats, 60μm	901	18
	Epoxy zinc rich, 1 coat, 25μm Modified polyester primer, 1 coat, 100μm Modified polyester paint, 2 coats, 100μm	414	61
	High build inorganic zinc rich, 1 coat, 75μm Epoxy anti-corrosive, 2 coats, 120μm Polyurethane paint, 2 coats, 60μm	311	41
	High build inorganic zinc rich, 1 coat, 75μm Epoxy anti-corrosive, 1 coat, 150μm Polyurethane paint, 2 coats, 60μm	308	57
	Epoxy zinc rich, 1 coat, 15μm High build epoxy paint,1 coat, 2000μm Polyurethane, 1 coat, 30μm	2513	2
	High build inorganic zinc rich, 1 coat, 75μm High build vinyl paint, 1 coat, 150μm Vinyl paint, 1 coat, 30μm	263	36
	High build inorganic zinc rich, 1 coat, 75μm Fluoric resin primer, 1 coat, 100μm Fluoric resin paint, 3 coats, 90μm	416	70
	High build inorganic zinc rich, 1 coat, 75μm Tar epoxy paint, 2 coats, 400μm	890	7
Power tool cleaning	Modified epoxy paint, 2 coats, 150μm Polyurethane paint, 1 coat, 30μm	184	25

Development of new coating system

Approximately 350 kinds of test pieces were placed on the exposure rack installed on the first deck to develop coating systems standing for two decades or longer on the sea.

Study on repainting system

It is necessary to establish a repainting system for using marine structures semipermanently. These coatings are requited to possess the following performance:

1) Applicability on rusty surface
2) Quick-drying properties
3) Ability to finish coating by a few times of application based on sufficiently thick film provided by one-application.

Exposure test of repainting test pieces was made in the current year.

Research on explication of deterioration mechanism of coated structural steel and technology estimating deterioration

A technology estimating the duration of life is required to effectively develop a durable rust-preventive coating system. Although some reports on explication of deterioration of structural steel have been published, any effective technology concerning thick-film coating for durable rust-prevention has not been reported yet. The purpose of the research is to clarify the deterioration mechanism of thick-film coated steel plate in the marine atmosphere and to successively obtain the technique estimating deterioration by tacking its process. The measurement items are as follows:

1) Permeabilities and their changes of water, oxygen and chlorine ion for examining the performance of film related to the rust-preventing properties
2) Changes of binder and bond of binder with filler for examining qualitative changes of film
3) Changes of adhesive strength and steel surface below the film for examining the changes between the film and the structural steel

REFERENCES

1. Coating standards for steel bridges and the annexed manual for the main Island-Shikoku Island Connecting Bridges. The Honshu-Shikoku Bridge Authorities. 1980.
2. R. Tanaka, Investigation of steel bridges, Steel Bridge Painting, Vol.9, No.4 (1981)
3. M. Fukushima, Coating materials up-to-date (1), Steel Bridge Painting, Vol.11, No.3 (1983)
4. Y. Mori, Durability of film appearance, Steel Bridge Painting, Vol.5, No.3 (1977)
5. T. Takagaki, K. Takatsu, K. Hasegawa, Sea saer particles under film and adhesion, Rust Prevention & Control, Vol.28, No.9 (1984)
6. S. Moriya, Study on sea saer particles density, Steel Bridge Painting, Vol.11, No.4 (1983)

7. Manual for the investigation of paint film on steel bridges, JSS IV 03-1982
8. T. Yamamoto, H. Amako, Y. Oyabu, Studies on corrosion Phenomena of steel plates coated by paint films with very high electric resistance (I) ∿ (III), J. of the Japan Society of Colour Material, 48. 49 (1976), 50 (1977)
9. Y. Sato, Electro-chemical measurment on under-film corrosion, corrosion Engineering, Vol.32, No.3 (1983)
10. Y. Sakae, O. Kawada, Final report of the Accelerated Weathering test of sub-committee, JSSC, Vol. 10, No.107 (1974)

Exposure Tests of Corrosion-Protected Steel Pipe Piles in Marine Environment

M. Makita[1] and R. Tanaka[2]

[1] Public Works Research Institute, Ministry of Construction, Toyosato, Tsukuba, Ibaraki, 305 Japan
[2] Japanese Association for Steel Pipe Piles (JASPP), Nihonbashi Kayabacho, Chuo-ku, Tokyo, 103 Japan

1. INTRODUCTION

The degree of corrosion of offshore structures affected by the marine environment is far greater than that of onshore structures in a land environment. It is indispensable therefore that appropriate corrosion-protection measures be taken when building offshore structures to correspond with their designed service life. In recent times in particular, there is a strong demand for the development of corrosion protection technology which can secure long-term durability of offshore structures from fifty to a hundred years.

2. SORTING THE PROBLEMS CONNECTED WITH EXPOSURE TESTS

2.1 CLASSIFICATION OF MARINE ENVIRONMENTS

The conditions of corrosion of steel structures depend largely on the exposure positions of these structures in relation to their environments. We therefore classified the marine environmental effects by dividing the offshore structures' zones as follows: the sea air, splash, tidal, seawater and bottom soil zones.

2.2 CHARACTERISTICS OF THE SPLASH AND TIDAL ZONES

Fig 1 shows that the difference in the test methods will result in the difference of the tendencies of corrosion.
Tests using test pieces (micro tests) revealed that a considerable degree of corrosion appeared in the tidal zone, while tests using continuous test piles (macro tests) showed that large amounts of corrosion took place in the splash and seawater zones.

2.3 EXPOSURE TEST USING STEEL PIPE PILES

We conducted exposure tests using actual steel pipe piles (ø: 406 mm, t: 9 mm, l: 54 mm) at Off-Kawasaki in Tokyo Bay between 1974 and 1976.
Fig. 2 shows the conditions of corrosion that occurred on the unprotected steel piles. On the basis of the test results, we planned another series of corrosion protection tests with steel pipe piles for the splash and tidal zones.

3. EXPOSURE TEST OF CORROSION-PROTECTED STEEL PIPE PILES

3.1 TEST PROCEDURE

In order to secure long-term durability of offshore steel structures, the establishment of corrosion protection technology is indispensable. Particularly, the corrosion conditions are severest at the splash and tidal zones, where maintenance and repairs are difficult, and it is absolutely necessary to design a corrosion protection technology which can be applied in this environment.
Therefore, as part of the Japanese Government Ministry of Constructions' project for "Development of Construction Technology for Offshore Structures," exposure tests of corrosion-protected steel pipe piles with the objective of developing a new, long-lasting and maintenance-free corrosion protection technology were started at two locations on the Pacific Ocean coast of Japan with a 15-year schedule from 1973 to 1987 in the form of joint research of the Ministry of Construction Public Works Research Institute and the Japanese Association for Steel Pipe Piles.
This paper is a summarization of the test results for the 10-year period from 1973 through 1983.

3.2 TEST MATERIALS

Two types of test materials were used. One was steel pipe piles of 508 mmø diameter. Various types of corrosion protection were executed for a length of 6 m including the splash and tidal zones on each pile. (The piles were actually driven by pile-drivers and then their heads were isolated and connected to serve as exposure test materials.)

The other type used was test pieces in a shape of a pipe pile with a size of 100 mø x 2000 mm. After having been corrosion protected, they were attached to the splash/tidal zones of the exposed steel pipe piles mentioned above. These pieces were inflicted in advance with 3 types of artificial defects before being attached to the piles for investigation of expanding rate of deterioration from the damaged parts. The types of defects inflicted prior to exposure were abrasion, scratch and impact. Types of corrosion protection executed on the test pieces are shown on Table 1.
The exposure test sites selected were Off-Chiba in Tokyo Bay and Off-Ajigaura facing the Pacific Ocean. The former was selected as a standard marine corrosion environment of Japan and the latter as a site with rough waves.

3.3 TEST ITEM

a. Corrosion-protected steel pipe pile

Protected steel pipe piles were visually inspected every year, and corrosion maps of test pipe piles were formed on them. In the 4th year since the exposure started, partial exfoliation of the protection was performed so that the adhesive strength of coatings and residual amount of salt on the surface of the base material could be measured. After the measurements, the exfoliated portions were repaired, and observation of the repaired portions was continued to check for deterioration caused by aging.

b. Test piece

Every year, the test pieces were observed. Also, they were retrieved at intervals of a certain period of time for detailed inspections of corrosion, including destructive tests. Chief items of investigation when they were retrieved included impact resistance by drop hammer tests, adhesive strength of the coating, electrical resistance of coating, number of pinholes, and analyzing of salt infiltration using EPMA and corrosion of steel around the given defects.

3.4 TEST RESULTS

(1) Visual inspection

Table 2 shows the results of visual inspection of external views of the corrosion-protected steel pipe piles that have been exposed for 8 years. Marks A - E in the table indicate relative evaluations of the exterior visual conditions of the corrosion protection, and have a slightly different meaning from the durable life of corrosion protection.

(2) Results of partial exfoliation test

In the 4th year of exposure, a partial exfoliation test of the corrosion-protected steel pipe piles was performed. Table 3 shows the results of investigations. The items of investigation included the corrosion of the base steel material, adhesive strength of coating, amount of residual salt on the steel surface, number of pinholes, etc.
The table shows a large residual amount of salt on the steel surfaces. There is, however, a possibility that this measurement is not reflecting the conditions under the coatings in the strict sense. So, simply the results are given here.

(3) Secular changes of electrical resistance

Fig.3 shows examples of the secular changes of electrical resistance measured on the coating. The oblique lines in the charts show the external insertion performed to estimate the values of electrical resistance at the 10th year of exposure. For effective protection of steel materials by painting, it is understood that an electrical resistance level of $10^6 \Omega\ cm^2$ or higher is required. By the charts, it is clearly seen that the resistance values of the protection systems which manifested outstanding signs of deterioration recognizable by sight fell below this critical value quite early after the exposure had begun. The secular changes of electrical resistance and the performance of corrosion-protection of the painting method correspond with each others comparatively well.

(4) Distribution profile of salt with EPMA

Fig.4 shows the results of investigations performed on the distribution profile of salt with EPMA concerning the protection systems whose electrical resistance is relatively low and amount of the residual salt is high. It can clearly be inferred from the chart that the movement of the corrosive factors through film is extremely brisk in the systems where the deterioration of protection is great.

(5) Natural potential

Fig. 5 shows the natural potential of resin mortar (epoxy). Cases where the potentials are nearly the same as those of unprotected steel pipes indicate that the sea water has penetrated through the resin mortar lining, causing a decrease in the resistance of the coating.

(6) Attachment of biological matter

Table 4 shows the percentage of the areas where biological matter (especially shells) have attached to the test pieces that have been applied with contamination-protection paintings.

Bibliography:

1) Development of corrosion protected steel pipe piles and the results of their exposure tests in marine corrosion environments (Part 10), Compiled by JASPP (1984)

2) Technical Memorandum of Public Works Research Institute No. 1245 (1977)

3) Report of the Port & Harbor Research Institute Vol. 22, No. 2, Dr.K.Zen

---

Figures and Tables

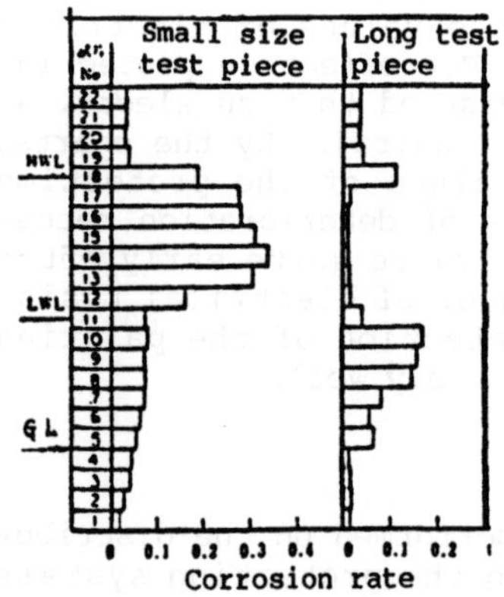

Fig. 1 Differences in corrosion under experimental conditions

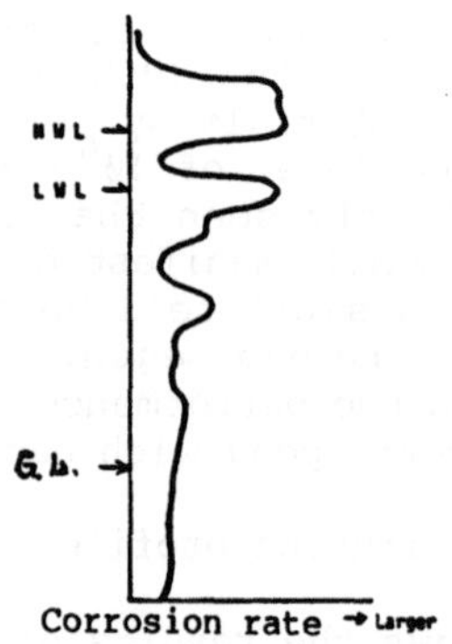

Fig. 2 Condition of corrosion of steel pipe piles

TABLE 1: TYPES OF COATING FOR CORROSION-PROTECTED TEST PIECES

No. of Test piece	Types of coating
1	Stainless steel (25Cr - 13Ni), sheet wrapping (3 mm)
2	Low-alloy corrosion-resistant steel (Cu-Cr-Al), sheet wrapping (6 mm)
3	Inorganic zinc-rich paint (75 μm) + epoxy with glass cloth (300 μm x 3)
4	Flame-sprayed aluminum (100 μm) + coal-tar epoxy with vinylon cloth (300 μm x 2)
5	Monel sheet wrapping (2 mm)
6	Flame-sprayed aluminum (100 μm) + coal-tar epoxy (300 μm x 2)
7	polyethylene (cohesion type, extrusion process 4 mm)
8	Inorganic zinc-rich paint (75 μm) + coal-tar epoxy (300 μm x 3)
9	Flame-sprayed aluminum (200 μm)
10	Flame-sprayed aluminum (100 μm) + coal-tar epoxy (300 μm x 2)
11	Butyl rubber (4 mm)
12	Epoxy resin mortar (4 mm)
13	Glass-flake filled polyester (1 mm)
14	No treatment
15	Titanium (cladding, 0.5 mm)
16	Underwater-curing epoxy (5 mm)
17	Glass-flake filled epoxy (2 mm)
18	polyethylene (adhesive type, extrusion process, 4 mm)
19	Fusion bonded epoxy (0.3 mm) + polyethylene (extrusion process, 4 mm)
20	Glass fiber reinforced cement mortar (20 mm)
21	Epoxy-saturated nonwoven fabric tape (3 mm)
22	Glass-flake filled polyester (2 mm)
23	Epoxy resin mortar (5 mm)
24	Polyethylene (fusion bonded process 3 mm)
25	Shrinkable polyethylene tube (3 mm)
26	Coal-tar epoxy mastic (14 mm)
27	Chloroprene sheet (5 mm)
28	Petrolatum tape (2mm) + FRP cover (3 mm)
29	Foamed polyurethane (10 mm)
30	Asphalt mastic (10 mm)
31	Inorganic zinc-rich paint (75 μm) + coat-tar epoxy with glass cloth (1000 μm) + ORG-ANOMETALLIC COPOLYMER TYPE antifouling paint (75 μm x 2, applied to lower portion of steel pipe, 2500 mm) (Test piece length: 4 m)
32	Inorganic zinc-rich paint (75 μm) + epoxy (500 μm x 2) + (Copper oxide + organic tin compounds) antifouling paint (50 μm x 2, applied to lower portion of steel pipe; 2500 mm) (Test piece length: 4 m)
33	Glass flake-filled vinylester (100 μm)
34	Inorganic zinc-rich primer (20 μm) + Glass flake-filled isophthalic acid polyester (1400 μm)
35	Glass flake-filled visphenolic polyester (1400μm)
36	Underwater curing epoxy for repairing (5 mm, Non-surface finishing, works executed in the atmosphere)
37	Underwater curing epoxy for repairing (5 mm, Non-surface finishing, works executed in the seawater)
38	Organic zinc-rich primer (20μm) + epoxy resin mortar lining (5 mm)
39	Aluminum (No backing-up steel material, 4 mm) (4m length)
40	Stainless steel (SUS 316L cladding, 2 mm) (4m length)
41	90Cu-10Ni (cladding, 3 mm) (4m length)
42	Inorganic zinc-rich paint (75μm) + epoxy (100μm x 3) + epoxy (35μm x2, applied to upper portion of steel pipe: 1500 mm) + (cuprous oxide + organic tin compounds) anti-fouling paint (80 μm x 4, applied to lower portion of steel pipe: 2500 mm)
43	Inorganic zinc-rich paint (75 μm) + epoxy (300 μm) + (cuprous oxide + organo-metallic copolymer) anti-fouling paint (100 μm x 2, applied to lower portion of steel pipe: 2425 mm)
44	Inorganic zinc-rich paint (75 μm) + epoxy (300 μm) + (cuprous oxide + triphenyltin: high content type) anti-fouling paint (60 μm x 3, applied to lower portion of steel pipe: 2425 mm)
45	Inorganic zinc-rich paint (75 μm) + epoxy (300 μm) + (cuprous oxide + organo-metallic copolymer) anti-fouling paint (20μm x 6, applied to lower portion of steel pipe: 2425 mm)
46	Inorganic zinc-rich paint (75 μm) + epoxy (300 μm) + (Organo-metallic copolymer) anti-fouling paint (1 mm, applied to lower portion of steel pipe: 2425mm)

TABLE 2: RESULT OF EVALUATION OF CORROSION-PROTECTED STEEL PIPE PILES

Types of coating	Marine air splashzone						Tidal Zone						
	3.5 yrs	4.5 yrs	5.5 yrs	6.5 yrs	7.5 yrs	8.5 yrs	3.5 yrs	4.5 yrs	5.5 yrs	6.5 yrs	7.5 yrs	8.5 yrs	
1. Stainless steel (25Cr-13Ni sheet, 3mm)	A	A	A	A	A	A	A	A	A	A	A	A	
2. Flame sprayed aluminum (200µm)	B	B	B	B	C	C	D	D	D	D	E	E	
3. Flame sprayed aluminum (100µm) + coal-tar epoxy (300µm x 2)	C	D	D	D	D	D	D	D	D	E	E	E	
4. Flame-sprayed aluminum (100µm)+ coal-tar epoxy with vinylon cloth (300µm x 2)	B	C	C	C	C	C	C	C	C	C	C	C	
5. Flame-sprayed aluminum (100µm)+ coal-tar epoxy (300 mµx 2)	AB	B	B	B	B	B	C	D	D	D	E	E	
6. Inorganic Zinc-rich paint (75µm)+ coal-tar epoxy (300µm x 3)	AB	B	B	B	B	B	B	C	C	C	C	C	
	AB	AB	B	B	B	-	AB	AB	B	B	B	-	*
7. Inorganic Zinc-rich paint (75µm)+ epoxy with glass cloth (300µmx 3)	B	B	B	B	B	B	A	A	A	A	B	B	
	AB	AB	B	B	B	-	A	A	A	A	AB	-	*
8. Polyethelene (cohesion-type, extrusion process, 4mm)	A	A	A	A	A	A	A	A	A	A	A	A	
	A	A	A	A	A	-	A	A	A	A	A	-	*
9. Epoxy resin mortar (4mm)	BC	C	C	C	CD	CD	BC	BC	BC	C	C	C	
	AB	AB	B	BC	BC	-	A	A	A	B	B	-	*
10. Steel fiber reinforced concrete (100mm by casing, 50 mm by spraying)	B	B	B	B	B	B	B	B	B	B	B	B	
11. Polyurethane Mastic (3mm)	C	C	C	C	C	-	C	C	C	CD	D	-	*
12. Cement Mortar (47 mm)+ Polyester FRP cover (3mm)	A	A	A	A	A	-	A	A	A	A	A	-	*
13. Polyurethane rubber (4mm)	A	A	A	A	A	-	A	A	A	A	A	-	*

Results marked * are the Ajigaura site evaluation

Table notation: A, B, C, D, and E: Evaluation by visual inspection

TABLE 3: RESULTS OF PARTIAL REMOVAL OF CORROSION PROTECTED STEEL PIPE PILES (Off-Chiba; 4th year of exposure)

Type of coating	Condition of removed and base material Position of removed portion: Splash Zone	Adhesion strength (kg/cm2)	Salt Amount on covering /base steel surface (mg/m2)	Pinholes	Repair method
Inorganic zinc-rich paint (75μm) + coal-tar epoxy (300μm x 3)	Adhesion of film is good No rust has occurred on base material. (Defect portion: Rust occurred on base material)	28	1776/ 69	None	Coal tar epoxy
Inorganic zinc-rich paint (75μm) + epoxy with glass cloth (300μm x 2)	Adhesion strength between epoxy and cloth is weak. Occurrence of rust on the base materal	13	2168/ 27	None	see note 1.
Flame-sprayed aluminum (100μm) + coal-tar epoxy (300μm x 2)	Adhesion of coats is good. No occurrence of rust is observed	32	1044/ 42	1 pt. dia. 2 mm	Coal tar epxy
Flame-sprayed aluminum (100μm) + coal-tar epoxy with vinylon cloth (300μm x 3)	Adhesion str. of cloth to paint is weak. No abnormality in aluminum. Defect ptn: Rust on base mater'l	28	1355/ 54	None	see note 2.
Flame-sprayed zinc (100μm) + coal-tar epoxy (300μm x 2)	Adhesion of film is good No occurrence of rust is observed	28	1219/ 42	None	Coal tar epxy
Epoxy resin mortar (Epoxy 20wt%, 4mm)	Almost no bonding strength observed. Black rust covering whole surface of base material	-	2249/ 204	Countless	Resin see nt.3
Polyethelene (cohesion-type, extrusion process, 4mm)	Weak bond strength between base material and lining. Rust observed along the spiral bead	-	542/ 81	None	see note 4.
Steel fiber reinforced concrete (50 mm by spraying)	Fibers near the surface have rusted. But 10 mm from surface, fibers exhibit metallic gloss. No abnormality in the base material observed	-	1192/ 56 (Intermediate 667)	-	Cement mortar

Note 1: Underwater curing epoxy + thick membrane-type epoxy (4.7 mm)
Note 2: Underwater curing epoxy + coal-tar epoxy (5.7 mm)
Note 3: Underwater curing resin mortar
Note 4: Petrolatum tape (2 mm) + polythelene sleeve wrap

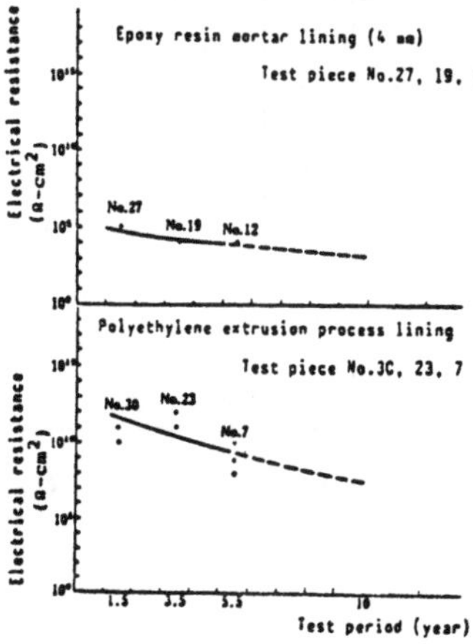

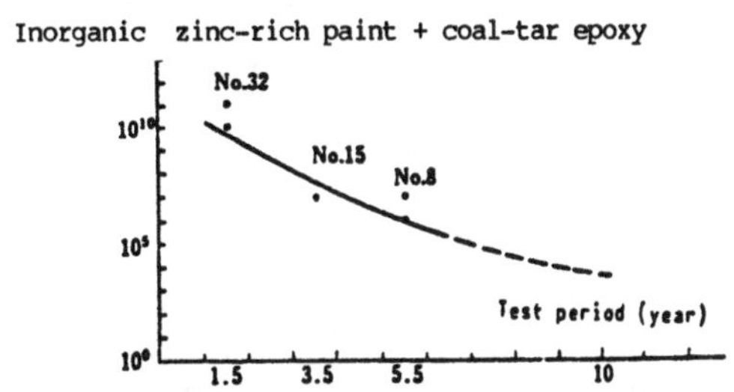

Fig. 3 Secular changes of electrical resistance (Off Chiba)
Dotted line is a 12-year forecast

Glass flake filled polyester film

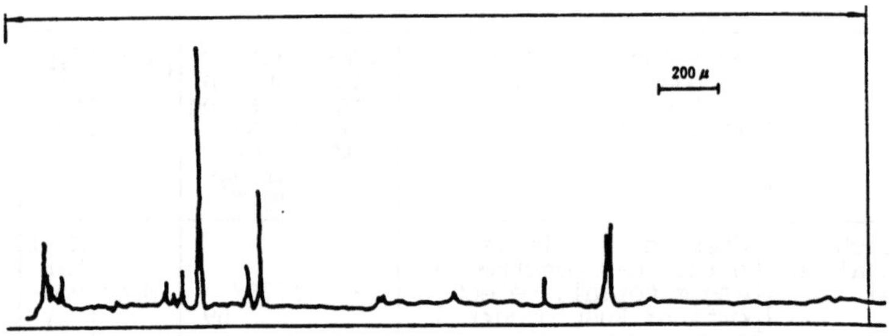

Glass flake filled polyester (2mm)
Epoxy resin mortar film

Base steel

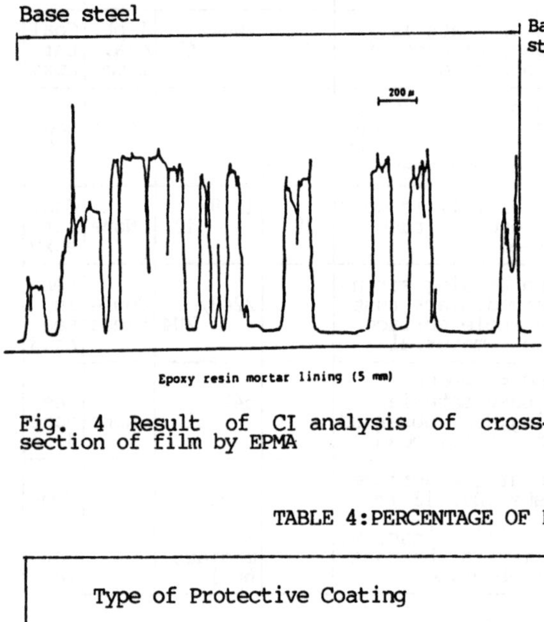

Fig. 4 Result of CI analysis of cross-section of film by EPMA

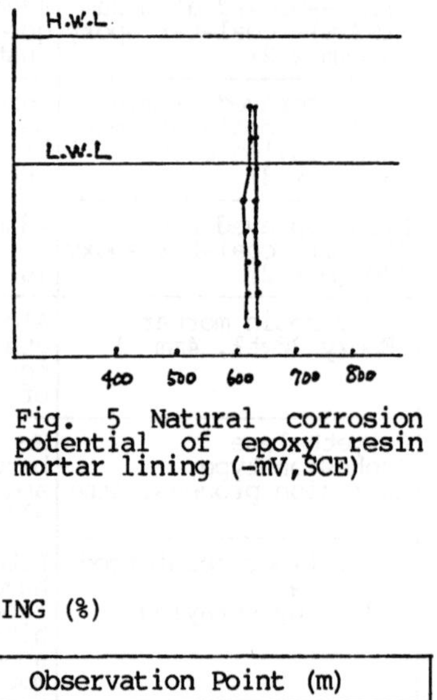

Fig. 5 Natural corrosion potential of epoxy resin mortar lining (-mV,SCE)

TABLE 4:PERCENTAGE OF FOULING (%)

Type of Protective Coating	Observation Point (m)					
	+1.5	+0.9	+0.5	LWL +0.5	-0.5	-0.9
Organo-Metallic Copolymer type, anti-fouling paint (75 m x 2)	70	20	30	30	30	80
Cuprous oxide + organic tin type anti-fouling paint (50 m x 2)	20	100	40	40	80	80
Cuprous oxide + organic tin type anti-fouling paint (80 m x 4)	6	6	0	0	0	0
Cuprous oxide + triphenyl tin type paint (high content, 60 m x 3)	90	30	2	0	0	0
Cuprous oxide + organic-metallic co-polymer anti-fouling paint (80 m x 6)	100	100	80	30	0	0
Organo-metallic copolymer type anti-fouling paint (1 mm x 1)	100	70	70	0	0	0

Note: Observations of half the circumference, a 30 cm width band on the ocean side

New Corrosion Control System for Steel Pipe Piles of the Oversea Bridge Footings

T. Kakehi, Y. Imakita and Y. Hoshino

Nakagawa Corrosion Protecting Co., Ltd., Kajicho, Chiyoda-ku, Tokyo, 101 Japan

ABSTRACT

The new corrosion control system named PTC system was applied to the tidal and splash zone of steel pipe piles of the Ube-Bridge in Yamaguchi Pref. The footings of the bridge consist of 84 steel pipe piles with diameter 2.5m and length 40m. The PTC system is made up of petrolatum tape for corrosion prevention and plastic cover for protecting the tape from mechanical damage and for long life service. Construction method and characteristics of the PTC system will be presented using the example of the Ube-Bridge.

Fig. 1 Whole view of the Ube-Bridge

INTRODUCTION

In 1981, the Ube-Bridge was constructed by Ube Industries Ltd. as a road bridge for transportation of industrial materials. Just after the construction, the cathodic protection system with galvanic anodes was applied to the immersed and mud zone of the bridge and they were protected enough up to the mean sea level (MSL). On the other hand, the part above mean sea level was left as it was and exposed to corrosion damage. But after two years of the construction, the protective covering system was applied to the tidal and splash zone.

CORROSION OF STEEL PILE

The tidal and splash zone of marine structures are always pounded by waves and splashed with sea water. Those areas are also subjected to marine growth, weathering action, tidal action, driftage, and alternate wetting and drying out.

Therefore, those areas exhibit severe corrosion. According to our marine test result, the splash zone of steel pile of the Ube-Bridge was also severely corroded. The maximum corrosion rate of the test pile was 1mm/y.

SELECTION OF PROTECTION SYSTEM

The Ube-Bridge is bearing an important role for transportation of industrial materials, and is required to withstand long use. Consequently, it is a necessary condition that the steel pile of the footings are protectively lasting from corrosion. For that reason, the excellent performance of corrosion protection and long term durability are rquired from the protection system. Furthermore, being capable of installing under bad site condition and without water pollution is required. Regarding the above mentioned, various protective covering sytems are examinied. Thereafter the PTC system was selected.

SUMMARY OF PTC SYSTEM

The PTC system is an abbreviation of Petrolatum Tape and Covering System which is a combination of petrolatum corrosion protecting tape and FRPP (glass fiber reinforced polypropylene) cover. Composition of coating layer of the PTC system consisted of a paste layer, petrolatum impregnated tape and protective cover (FRPP cover) backed with a foamed resin sheet. The section of the coating layer is shown in Fig. 2.

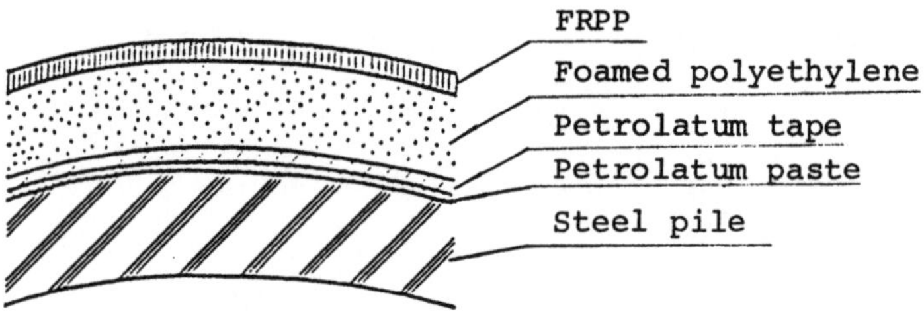

Fig. 2 Section of coating layer of PTC system

PTC system was completed by solving many difficult problems utilizing our long year's research and ample experiences, and it has various characteristics as follows.

(1) It can be installed by simple preparation of steel surface of an existing steel structure.
(2) It can be installed in sea water and on splash zone.
(3) It exhibits an excellent corrosion protection effect.
(4) It shows superiority in durability and can be expected of a long life.
(5) It has an excellent strength, flexibility, and shock durability.
(6) It can be installed for various forms of structure.
(7) It can be installed without fire.
(8) It is lightweight, and does not make a load to a structure.
(9) It is not poisonous and causes no pollution.

DESIGN OF THE PTC SYSTEM FOR THE UBE-BRIDGE

In previous investigation, it was proved out that the effect of cathodic protection was effective enough up to mean sea level (CDL + 0.98m). So, the protection area was settled from mean sea level to lower side of concrete footing (CDL + 3m)

(1) Objects to be protected
Area of the objects to be protected is shown in table 1.

Table 1

pier	diameter (m)	quantity (pieces)	area (m^2)
P2	2.5	12	188.4
P3	2.5	16	251.2
P4	2.5	20	314.0
P5	2.5	20	314.0
P6	2.5	16	251.2
total		84	1318.8

(2) Specification
The specification of the PTC system is shown in table 2.

Table 2

Process	Articles	Standard	Specification
Cleaning	Hand tool	-	Remove all loose material
Priming	Nakabo peto paste	WA	Gloved hand coating
Tape wrapping	Nakabo peto tape	100 HT	55% over wrapping
Protective covering	Nakabo multi cover (FRPP cover)	PAW	FRPP 2.7mmt, flange type foamed PE 10 mmt
	FRPP band	PAW	FRPP 2.7 mmt 100 mmW
Filling up	Polyurethane resin	NS12	30mm wide filling up
Fixing	(Upper part) FRPP band	PAW	FRPP 2.7mmt 100mmW
	(Lower part) Steel band	LS	30mm x 8 part welding

(3) Physical properties of main materials
Physical properties of the main materials of PTC system is shown in table 3 and table 4.

Table 3 Physical properties of petrolatum tape*

Item	Unit	Value
Thickness	mm	1.1
Weight	kg/m^2	1.4
Elongation	%	15.0
Water absorption ratio	%	0.1
Tensile strength	kg/25 mmW	8.0
Resistivity	Ω·cm	10^{14}

* Nakabo peto tape

Table 4 Physical properties of FRPP

Item	Unit	Value
Tensile strength	kg/mm^2	7.5 - 9.0
Bending strength	kg/mm^2	13 - 16
Elastic modulus for tension	kg/mm^2	450 - 490
Elastic modulus for bending	kg/mm^2	500 - 560
Anti-shock energy*	kg·cm	5325

* Anti-shock ability of FRPP cover tested by dropping 21.3 kg steel weight.

INSTALLATION

The installation of the protective coating by PTC system was carried out in the following order.

(1) Cleaning of steel surface
Shells and seaweeds, loose or rising rusts and paint films on the steel surface of protective range were removed fully.

(2) Priming
Petrolatum paste was spread evenly on the whole surface of protective range of steel pile by groved hand or trowel.

(3) Wrapping of petrolatum tape
Petrolatum tape was spirally wrapped over the fully cleaned and primed surface commencing at the lowest point and allowing a 55% overlap of the tape. Sufficient tension is used to ensure a good bond.

(4) Fixing of FRPP cover
FRPP covers were covered on the surface of petrolatum tape wrapping and fixed tightning fully.

(5) Upper part fixing
The upper part of FRPP cover was fixed by FRPP band, and filled up with polyurethane resin to the opening space of concrete and FRPP cover.

(6) Lower part fixing
The lower part of FRPP cover was fixed by steel band welded to steel pile.

Fig. 3 Petrolatum tape wrapping

Fig. 4 Steel pile protected by PTC system

CONFIRMATION OF PROTECTIVE EFFECT

For the confirmation of protective effect of the PTC system, a lid was made on a FRPP cover per every pier. The pile surface can be watched by opening the lid.

EPILOGUE

The PTC system work for this bridge was completed in September 1983. Since the completion, respective pile sections have been satisfactorily protected by the PTC system in tidal and splash zones, and also by the cathodic protection system in sea water and sea mud. Thus the Ube-Bridge will be securely and indefinitely protected from corrosion. The author wishes to express his appreciation to Ube Industries Ltd for permission of publication of this article.

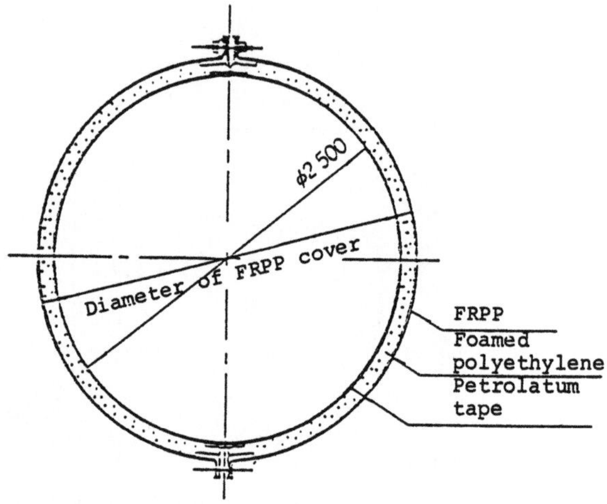

Fig. 5 Section of PTC system for steel pile

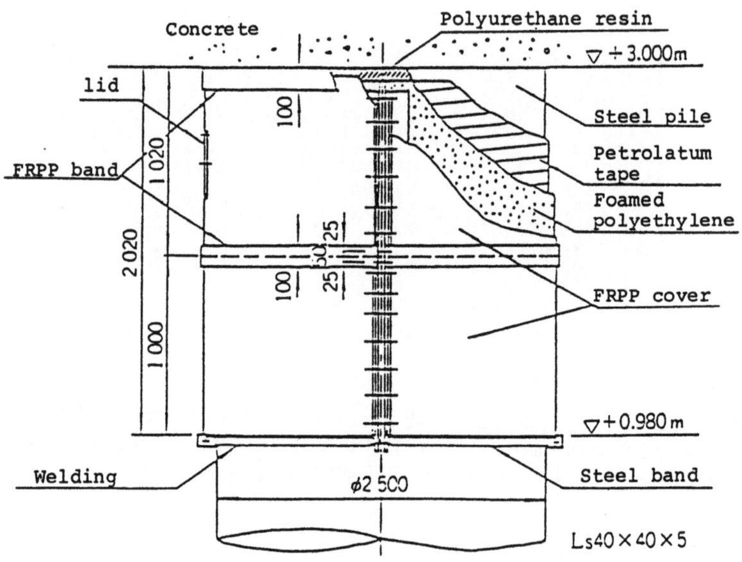

Fig. 6 Side view of steel pile protected by PTC system

A New Evaluation Method of Anti-Corrosive Properties of Heavy Duty Coating Systems for Offshore Structure Using Thermal Shock Technique

M. Arita, K. Matsuoka, H. Inoue, S. Naito, T. Shibata and K. Ohnaga
Ship Research Institute, Ministry of Transport, Shinkawa, Mitaka, Tokyo, 181 Japan

ABSTRACT

Two important factors in the flow, determining necessary conditions of heavy duty anti-corrosive coating systems for offshore structure use, are discussed. One is a distribution of coating thickness, which is closely related to defects in coating system. A relation between a characteristic of a coating thickness distribution and a shape of objects, on which the coating system is applied, is discussed. The other factor discussed is an accelerated deterioration method of coating systems which is called the thermal shock test using two water bathes. It is shown that the test has a possibility to evaluate anti-corrosive properties of coating system within a relatively short period of time.

INTRODUCTION

Ocean is a corrosively agressive environment for offshore steel structures. And these structures have no possibility of dry docking as carried out in shipping. Therefore the necessity of the development of heavy duty coating systems for them is very high.

Fig.1 shows the flow chart, which determines necessary conditions of heavy duty anti-corrosive coating systems for offshore structure use. In the flow, steps of B, Ib, IIb and IId are characteristics of these coating systems, and are different from usual coating. In the step B, pinhole, porosity, crevice, rumple and also ununiformity of coating thickness are considered to be defects. Because shortage of coating thickness may increase risk of easy access of the outside environment to the metal surface, and an excess of coating thickness may increase separations within coating.

On the other hand, an accelerated deterioration method of coating system is an indispensable tool to be developed to estimate the deterioration rate or corrosion rate in the marine environment stipulated in steps IIb and IId in the flow.

The knowledge of the characteristics of coating thickness distribution and the development of accelerated deterioration method of coating system may lead to a better estimation of anti-corrosive properties of coating systems developed for offshore structure use.

COATING THICKNESS DISTRIBUTION

Anti-corrosive properties of coating systems deeply depend on those of used paints and combination of them, and also on conditions of spraying. Improper selections of items mentioned above lead to several kinds of defects.

Recently the wide variation of coating thickness itself is considered to be a kind of defects[1,2]. But there are few quantitative informations about it. Therefore influences of the shapes of painted objects on the distribution of coating thickness were experimentally examined.

Experiment

Five coating systems were used. They are shown in Table 1. They were applied to flat plate specimens(specimen A in Fig.2) and the Spec. #5 was applied to hemisphere(specimen B).

Total coating thickness was measured by using the electromagnetic thickness meter named MINITEST FD-digital made by Electro-Physik Co. Ltd..

For flat specimens, total coating thickness was measured at three points on both top and back sides of every test specimen. Total number of thickness data were 1188 (198 specimens multiplied by 6 points) for Spec. #1 and 300 (50 specimens multiplied by 6 points) for Spec. #2 through #5.

For hemisphere specimens, total coating thickness was measured at cross points of every 10 degree of longitudinal and latitudinal line. Therefore total number of thickness data for one specimen was 325 (36x9 + pole) .

Results and Discussion

(1) Coating Thickness on Flat Surface

Summary of experimental data of coating thickness on flat plate specimens are shown in Table 2. An example of histogram and normal distribution fitting is shown in Fig.3. In spite of very careful coating by a specialist of painting, the coating thickness varied very widely. Not a little numbers of thickness data were thinner or thicker than the target thickness t*.

As shown in Table 2 and Fig.3, normal distribution fittings were considered to be practically appropriate for coating thickness distribution on flat plates as follows;

$$f(t) = \mathrm{Exp}\{-(t-t_m)^2/2\sigma^2\} \,/\, \sigma\sqrt{(2\pi)} \qquad (1)$$

where, f(t)=distribution function, t=coating thickness, t_m= mean coating thickness, σ=standard deviation. Values of coefficients of variation σ/t_m were in the range of 8.18 to 13.2%. The values were fairly large compared with those measured in case of tensile strength of mild steel [3] which was usually used material for offshore structures.

(2) Coating Thickness on Curved Surface

In order to investigate the relation between radii of curvature of objects painted and distributions of coating thickness on them, seven hemispherical models were used.

Fig.4 shows examples of rough sketches of coating thickness for hemisphere. Fig.5 shows histograms of coating thickness along an equal latitude girth and that of total data for D=355mm. Fig.6 shows relations between diameter of hemisphere and coefficient of variation.

As shown in Fig.4 and Fig.5, there were such general tendency that coating thickness was relatively large at latitude θ_ℓ=30° to 60° compared to that at θ_ℓ=0° to 20° or 70° to 80°. As for tendency of distribution (see Fig.5), range of thickness distribution was wide at low θ_ℓ and that became narrower and narrower with increase of θ_ℓ.

Let's assume that the width of the thickness distribution can be expressed roughly by the coefficient of variation σ/t_m. And σ/t_m is considered to be a function of a radius of curvature D/2, an absolute angle of inclination to holizontal θ , an area per one datum a_d , number of data N and other factors X as a kind of coating system and so on. Then σ/t_m can be expressed as follows;

$$\sigma/t_m = f(D/2, \theta, a_d, N, X) . \qquad (2)$$

Here in this study N=const. and X=const., and there exists following relation.

$$a_d \propto (D/2)^2$$

Then Eq. (2) becomes,

$$\sigma/t_m = f(D/2, (D/2)^2, \theta) . \qquad (2')$$

The contribution of D/2 to f in Eq.(2') is considered to be as a decreasing function of D and $(D/2)^2$ is considered to be increasing function of D. Then Eq.(2') becomes conceptual as shown in Fig.7. By this conception, relations between σ/t_m and D can be well understood in Fig.6.

Level of σ/t_m values for θ_ℓ =const. were rather low as compared to those for flat plates (see Table 2). But there were differences in sample numbers and an area per one datum. Levels of σ/t_m for all values of θ_ℓ were equal or somewhat higher than those for flat plates. Sample numbers of hemisphere (N=325) were nearly equal to those for Spec.#2 through #5 in Table 2 (N=300). But the area per one datum for any case of hemisphere (a_d=0.23 to 12 cm^2) was quite smaller than that of flat plates (A=100 cm^2). It is possible to say from these results that the range of coating thickness distribution on objects with curvature becomes considerably wider than that on flat plate and that the danger of defects due to excessively thin or thick coating thickness becomes higher.

ACCELERATED DETERIORATION METHOD OF COATING SYSTEM

Many accelerated deterioration method have been developed [4,5] in order to test anti-corrosive properties of coating systems. But almost all are used to test generations of spots of rust or other defects in coating system after certain period of exposure in some accelerating atmosphere. Weak points of these are little informations obtained about process of deterioration. In order to get multiple informations about deterioration of coating systems within relatively short period of time, we have developed a thermal shock test of coating system using two water bathes.

Experimental Method

Test specimens are shown in Fig.2(A). They are same ones

used to measure coating thickness distributions. Test specimens were immersed to a hot bath for a certain period of time and then they were pulled up to be immersed to cold bath. This hot-cold cycle was repeated untill the numbers of cycles settled were reached.

Changes of mechanical and electric properties due to the thermal cycles were examined by three methods. The first was a mesurement of an adhesion force by using Model 106 Adhesion Tester made by Elcometer Instrument Ltd. England. The second was a three point bending test in order to know the relation between an amount of deflection and an initiation of pinholes. Each span between supports was 120 mm. Pinholes were detected by the pinhole detector TRE-10A made by Sanko Electronic Co. Ltd. Japan. Voltage of the detector was settled to be 5 kV. The third was a measurement of an impedance by Dielectric Loss Measuring Set TR1-C made by Ando Electronic Co. Ltd. Japan, assuming that a coating system was a circuit consisted of a resistance and a condenser connected parallel with each other.

Experimental conditions of two siries of thermal shock tests were as following;

	ΔT (°C)	N	t_{in} (sec)
Exp. Ser.#1	81 to 95	10 and 20	20 (hot), 30 (cold)
Exp. Ser.#2	55 to 63	10, 20, 30 40 and 50	10 (both bathes)

where, ΔT=the difference of temperature between the hot bath and the cold bath, N=the numbers of thermal cycles, and t_{in} =the immersion time in bathes.

Results and Discussion

Initial values of virgin test specimens were measured by the adhesion test and three point bending test. Impedance was measured for every specimen before thermal shock test.

Experimental results are summarised in Table 3 (Exp. Ser. #1) and Table 4 (Exp. Ser.#2). Values in the tables show deviations from initial ones. Measurements of resistance and capacitance were carried out for a frequency range 0.1 to 10 kHz, but values only for 1.0 kHz are shown in tables.

Numbers 0, I, II in Table 3 mean types of behaviors of coating system after initiation of the first detected pinhole at three point bending test as follows;

- 0 - no pinholes were detected all through the test
- I - no rapid increase of pinholes detected after the initial detection of them
- II - rapid increase of pinhole and successive peeling of coating after the initial detection of pinhole.

There are at least two stages of deterioration for coating systems. The one is the initial strengthening stage (stage 1) and the other is the weakening stage (stage 2). These stages were observed for test pieces exposed to marine environment and also for those exposed to an activated environment [4]. At stage 1, both adhesion force and resistance have tendencies to increase and the capacitance to decrease with the lapse of time. These tendencies were realized by Spec.#3 of the experimental series #1 shown in Table 3.

The types of pinhole behavior at three point bending test for Spec.#3 were also well corresponding with deterioration stages. In the cases of Spec.#1, 2, 4 and 5 of the experimental series #1, tendencies mentioned above were also observed, though they were not clear as compared with those of Spec.#3.

In the experimental series #2 (Table 4) , the behavior of Spec.#2 showed a typical process of deterioration. For others, the tendencies were same as in the series #1, but they are somewhat obscure.

CONCLUSIONS

(1) The distributions of coating thickness on flat surface roughly obeyed to the normal distribution. The deviation range of coefficients of variation was 8 to 13%, which was fairly wide compared with that of the tensile strength of mild steel.

(2) The influence of radius of curvature of painted objects on the distribution of coating thickness was examined. Data obtained well coincided with the analysis.

(3) The coefficients of variations of coating systems on curved surfaces were fairly larger than those on flat surfaces. It may increase the generations of defects in coating systems.

(4) Thermal shock test of coating systems using water medium was proposed. Deterioration stages were well observed by the method, and it may have a possibility to be a rapidly accelerated deterioration test of heavy duty coating systems.

ACKNOWLEDGMENT

This study formed a part of the research program named the Utilization of Marine Space by Coastal and Offshore Structures, which was performed through Special Coodination Funds for Promoting S&T of the Science and Technology Agency of the Japanese Government.

REFERENCES

1. Bailey,G. and Deacon,D.H.: Results of Offshore Tests of Selected Coating Systems. Trans. I. E.(C), Vol.91 (1979) 43-50.

2. Kumanotani,J.: Degradability of Organic Coating Films - Chemical Reaction in Degradation. Jr. of Metal Finishing Soc. of Japan, Vol.32, No.11 (1981) 35-42.

3. Yokobori,T.: Strength, Fracture and Fatigue of Materials. Gihodo (1955) 4 .

4. Shipbuilding Res. Assoc. of Japan : Res. Rep. of SR182, No.341, 347, 357 and 369 (1981 to 1984) .

5. Sato,Y.: Anti-Corrosive Properties Testing of Paints. Jr. of Metal Finishing Soc. of Japan, Vol.16, No.5 (1965) 29-34.

6. Amako,H.: Anti-corrosive Evaluation of Coated Steel by Electrochemical Method. Jr. of Metal Finishing Soc. of Japan, Vol.32, No.9 (1981) 38-48.

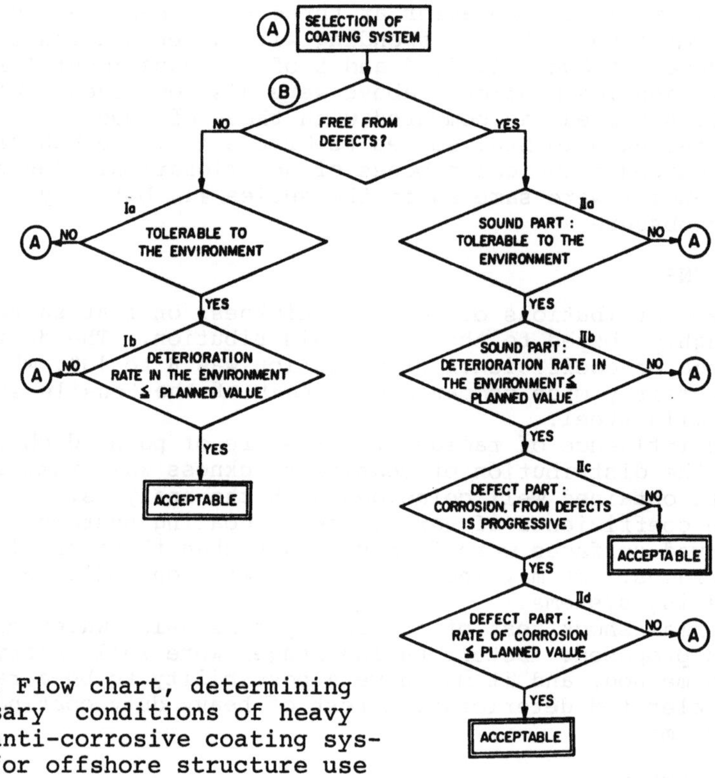

Fig.1. Flow chart, determining necessary conditions of heavy duty anti-corrosive coating systems for offshore structure use

Table 1. Description of coating systems

SPEC NO.	DESCRIPTION OF COATING (THICKNESS IN μm) 1ST COAT	2ND COAT	3RD COAT	4TH COAT	5TH COAT	6TH COAT	TOTAL THICKNESS (μm)
1	IZ (75)	PE·A/C MIST COAT	PE·A/C (100)	PE·A/C (100)	PE·IM (40)	PU·T/S (35)	350
2	IZ (75)	TE MIST COAT	TE (250)	–	–	–	325
3	PE·A/C (100)	PE·A/C (100)	PE·A/C (100)	PE·T/S (40)	PE·T/S (40)	–	380
4	TE (200)	TE (200)	–	–	–	–	400
5	IZ (75)	PE·A/C MIST COAT	PE·A/C (100)	V·IM (60)	V·T/S (40)	–	275

SURFACE PREPARATION = SHOT BLAST + IZP

IZP = INORGANIC ZINC RICH PRIMER
IZ = HIGH-BUILT INORGAMC ZINC RICH PAINT
PE·A/C = PURE EPOXY ANTI-CORROSIVE PRIMER
PE·IM = PURE EPOXY INTERMEDIATE PAINT
PU·T/S = POLYURETHANE TOPSIDE PAINT
TE = TAR EPOXY PAINT
V·IM = VINYL INTERMEDIATE PAINT
V·T/S = VINYL TOPSIDE PAINT

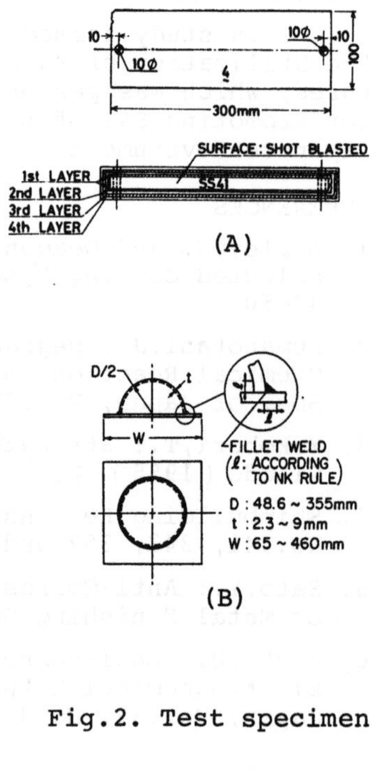

Fig.2. Test specimen

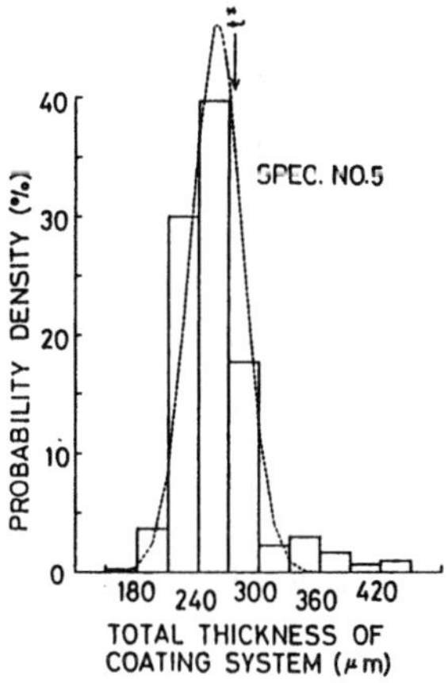

Fig.3.Coating thickness distribution of flat plate and normal distribution fitting

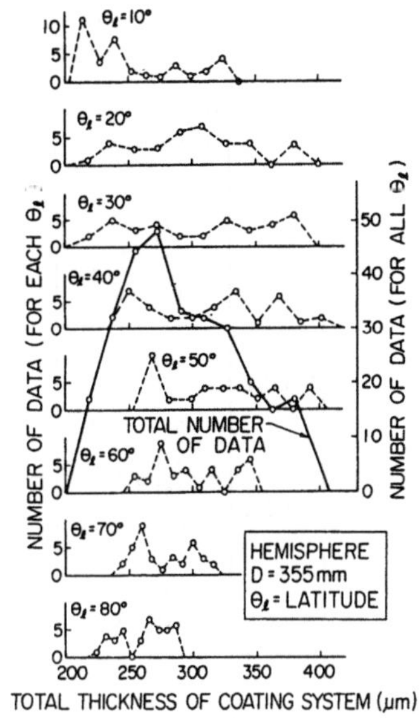

Fig.5. Coating thickness distribution for hemisphere D = 355 mm

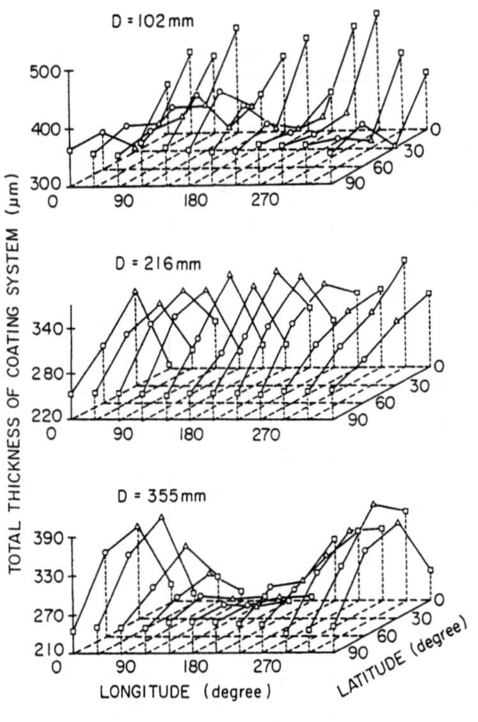

Fig.4. Rough sketches of coating thickness on hemispheres

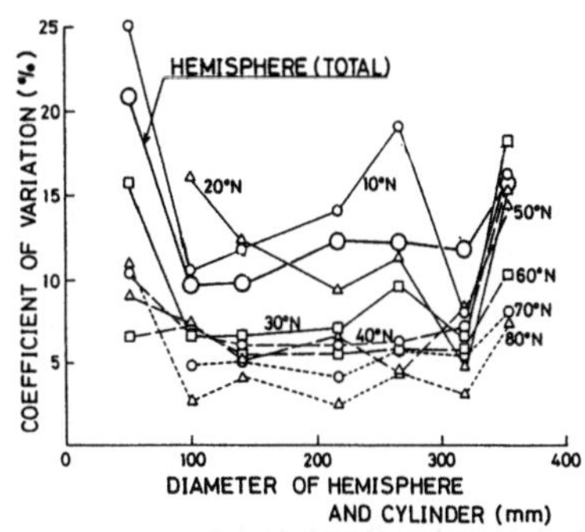

Fig.6. Relation between diameter D and coefficient of variation σ/t_m for hemisphere

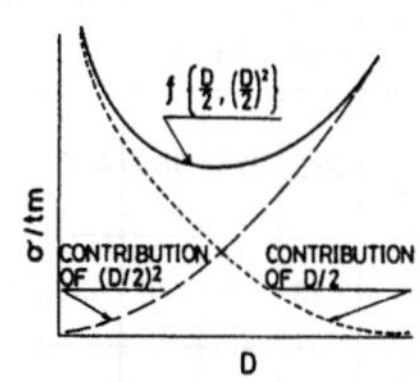

Fig.7. Conceptional view of relation between D and σ/t_m

Table 3. Summary of thermal shock test (Exp. Series #1)

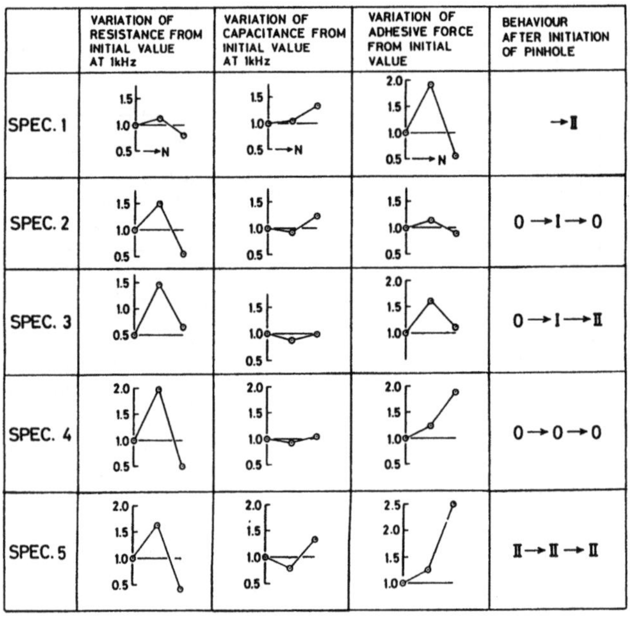

Table 4. Summary of thermal shock test (Exp. Series #2)

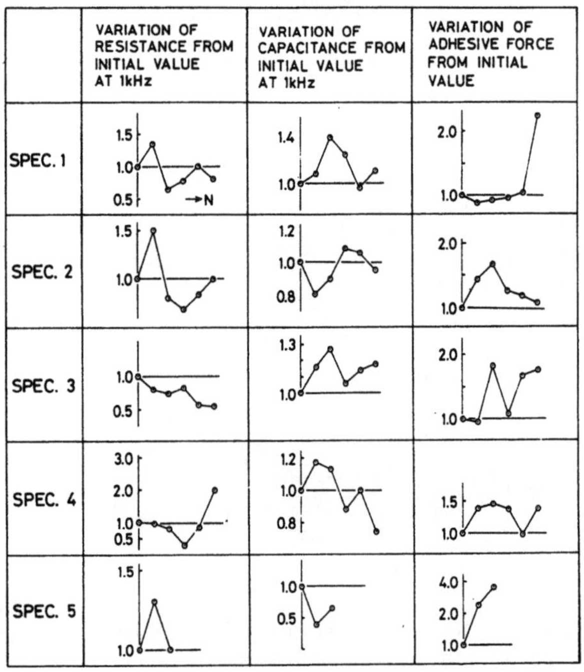

Table 2. Parameters of normal distribution fitting of data for flat plate

SPEC. NO.	1	2	3	4	5
SAMPLE NUMBER	1188	300	300	300	300
TARGET OF TOTAL THICKNESS t^* (μm)	350	325	380	400	275
MEAN THICKNESS t_m (μm)	428	376	410	436	258
STANDARD DEVIATION σ (μm)	41.5	44.5	33.5	57.5	26.0
t_m/t^*	1.22	1.16	1.08	1.09	.937
σ/t_m	97.1×10^{-3}	118×10^{-3}	81.8×10^{-3}	132×10^{-3}	101×10^{-3}

Study of Corrosion-Protection Technologies in Splash and Tidal Zones

Establishment of Corrosion-Protection Technologies to Provide Long-Term Durability at Low Cost for Offshore Steel Structures by Use of Methods for Wrapping of Highly Corrosion-Resistant Metallic Materials

Y. Suzuki*, K. Doi*, T. Kyuno** and Y. Shimizu***

The Kozai Club, Nihonbashi Kayabacho, Chuo-ku, Tokyo, 103 Japan

ABSTRACT

To meet the requirement for a long service life of offshore structures, experiments and research on new corrosion-protective wrappings of highly corrosion-resistant metallic materials (austenitic stainless steel, dual-phase stainless steel, cupronickel, monel metal, titanium, etc.) were started in April 1984, using a part of general research installation for ocean technology (off Ohigawa-cho in Shizuoka Prefecture, Japan).

INTRODUCTION

Offshore steel structures sometimes suffer unexpected corrosion damages. Conventionally, various kinds of organic coatings have been applied to protect these structures in the splash and tidal zones. However, organic coatings are usually damaged sooner or later. In addition, on-site repairing of a damaged area and maintenance of protection measures are very difficult. More than a decade has passed since the outset of the construction and installation of full-scale offshore structures in Japan. Recently, as the problems associated with corrosion of these structures came to the fore, the significance of corrosion-protection technologies has steadily been considered a priority by the engineers concerned.

So far, no specific measures of corrosion protection of steel structures in the splash zone, regarded as the most critical service environment for steel structures, have been developed. Presently, various kinds of organic coatings are widely used, the reliability of which deteriorates with time. Taking into consideration the permeability of oxygen and water, deterioration due to ultraviolet rays and other factors, the employment of this method, which is good only in

---

* Nippon Steel Corporation
** Kawasaki Steel Corporation
*** Nippon Kokan K.K.

the early stages of construction, does not suffice to achieve an expected service life of several decades for offshore structures. Accordingly, additional works will be involved in maintenance as partial repairs or reconstruction of the damaged areas. When such works become inevitable, the expenses incurred will be quite costly due to the reason that such works will be carried out in marine environments. Hence, the total cost for corrosion protection during the period of expected service life will add up to an appalling amount.

Therefore, the argument is that high-performance corrosion protection approaches that dispense with maintenance works are more viable, regardless of the relatively high initial cost. Based on this concept, the wrapping of a costly but extremely corrosion-resistant metal in marine environments is highly conceivable. This approach, however, has hardly been attempted yet, as described in the report "Current Study of Corrosion Protection Technologies for Offshore Structures in Japan"[1], prepared by the Kozai Club in 1975. In some countries abroad, there have been many cases where monel metal was used as a protective covering[2], but the adoption of other metals, including stainless steel, has rarely been reported.

Indeed, the high initial cost factor and the absence of very effective lining technologies, together with the lack of proper data concerning corrosion-resistant properties of these metals, are three reasons largely responsible for the slow progress of the above method. Although some data on individual metallic materials are available from various tests made, including an exposure test off the coast of Chiba Prefecture conducted by the Japanese Association for Steel Pipe Piles[3] and another exposure test made by INCO of the U.S.A.[4], comprehensive and systematic tests covering chemical composition of wrapping metals and wrapping methods have not been fully attempted.

Under such circumstances, the Kozai Club has decided to carry out a study on the subject entitled "Establishment of Corrosion-protection Technologies to Provide Long-term Durability at Low Cost for Offshore Steel Structures by Use of Methods for Wrapping of Highly Corrosion-resistant Metallic Materials", as the main theme of the development task assigned to the Club as a joint study with the Public Works Research Institute of the Construction Ministry.

OUTLINE OF THE EXPERIMENTAL PLAN

This study constitutes a part of the "Study of Effective Utilization of Offshore Space by Building of Offshore Structures", and is subsidized by the Scientific Technology Promotion Coordination Expenditures of the Japanese Government for completion in five years from 1982. The study has been conducted jointly by the Kozai Club and the Public Works Research Institute of the Construction Ministry, using a general research installation for ocean technology, off Ohigawa-cho in Shizuoka Prefecture.

Although the entire plan is scheduled for completion by 1987, the study will be continued in the event that a positive result could not be achieved within the allocated period. (Total time required is estimated at twenty years.)

DETAILS OF THE STUDY PLAN

Basic Concepts for the Study

1) As for the corrosion-protective materials under study, emphasis is being placed on stainless steel which not only excels in corrosion resistance, but also has recently been making rapid progress in the reduction of manufacturing cost. Also under study, partly for comparison purpose, are monel metal and various other non-ferrous metals, which have been considered to have potential for cost reduction and improved performance in the future.

2) The study of the methods of corrosion protection focuses on wrapping ordinary steel, which is considered the most practical and feasible material. Also used in the study are clad material and dual-phase stainless steel pipe. Hot-dip galvanized and flame-sprayed zinc and aluminum alloys are also employed in combination with heavy-duty coatings.

3) All test pieces are in the shape of pipes, the corrosion-resistant properties of which are assessed and studied in the plane, at weld joints and at parts having an artificial crevice. These crevices are formed using a polyethylene shrinkable tube covering for lined materials, and by welding a flat sheet to the outer surface of pipe for titanium test pieces.

Test Pieces and Test Schedules

Table 1 shows the corrosion-protection systems for the individual test pieces and projected years for dismantling test pieces for destructive tests. An appearance observation is carried out every year.

Exposure Test Methods

1) Artificial crevice forming

For stainless steel test pieces, artificial crevices are formed by heating and shrinking a polyethylene shrinkable tube with an adhesive (150A, thickness 3 mm, length 200 mm) at estimated positions as shown in Fig. 1. In the case of a titanium test piece, the artificial crevices are formed by welding with a titanium alloy sheet.

2) Test-piece settlement positions

Half of the part of the research installation on the south, in the splash and tidal zones, was selected as a site for the exposure stand to which test pieces were settled. The stand was secured inside the cross beams of the structure.

Table 1. Specification of test pipes

No.	Corrosion protection system	Specification	Dimension (mm)	Time for destructive test (years) 5	10	15	20
A-1	Metallic wrapping	SUS 316L(3mmt)	171.2ø x 3500 x 14t	◎	◎*	◎	◎
A-2	Metallic wrapping	22Cr-26Ni-5Mo-Ti(3mmt)	171.2ø x 3500 x 14t		◎*		◎
A-3	Metallic wrapping	25Cr-50Ni-6Mo-Cu-Ti(3mmt)	171.2ø x 3500 x 14t		◎*		◎
A-4	Metallic wrapping	26Cr-4Mo-Ni(3mmt)	171.2ø x 3500 x 14t		◎*		◎
A-5	Cast cladding	SUS 316L(3mmt)	171.2ø x 3500 x 14t		◎		
A-6	Flame spraying + Coating	Flame sprayed Zn + 13Al + Urethane (2500 μm)	171.2ø x 3500 x 14t		◎		
A-7	Dual-phase stainless steel pipe	25Cr-7Ni-3Mo-Cu-W-N	114.3ø x 3500 x 10t		◎		◎
A-8	Dual-phase stainless steel pipe	25Cr-6Ni-3Mo	89.1ø x 3500 x 9.6t		◎		◎
A-9	Dual-phase stainless steel pipe	22Cr-6Ni-3Mo	139.8ø x 3500 x 12.7t		◎		◎
A-10	Nonferrous metal wrapping	9Cu-1Ni (2mmt) (cupronickel)	171.2ø x 3500 x 13t		◎		◎
A-11	Nonferrous metal cladding	31Cu-65Ni(monel)(3mmt)	171.2ø x 3500 x 14t		◎		◎
A-12	Galvanizing + Coating	Hot-dip galvanizing + Glass fiber filled vinyl ester (1500 μm)	171.2ø x 3500 x 14t		◎		
A-13	Nonferrous metal wrapping	Titanium(3mmt)	171.2ø x 3500 x 14t		◎*		◎

*: Test pipe equipped with artificial crevice.

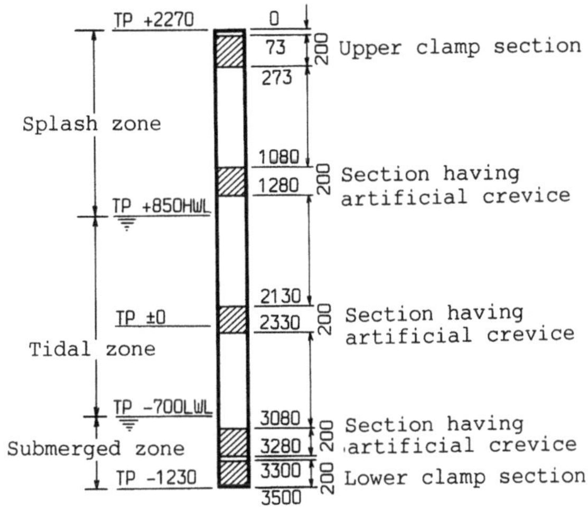

Fig. 1. Dimensions of test pipes and positions for artificial crevices

3) Test-piece settlement method

Test pipes are settled into specially-arranged fixed beds and clamps. Neoprene rubber is stuck on the inside of each clamp and each fixed bed to isolate the test pipes electrically from the pipe stand. Good insulating properties are proved by the determination of electrical-insulation resistance conducted after the test-piece settlement.

Test Procedure

1) Test items to be measured prior to exposure

a) Material inspection ... Mill sheet checking
b) Color inspection ... Checking for cracks in weld joints in accordance with JIS Z2324
c) X-ray inspection ... Making certain that weld joints are of class 2 or higher, in accordance with JIS Z3104
d) Dimensional inspection ... Checking of outside diameter tolerances (+4 mm, -0 mm) and straightness to ensure that pipe end lines run parallel to each other
e) Thickness measurement ... Measurement of thickness of corrosion-protection layer for lined materials, and plate thickness before pipe-making for clad materials, respectively, using a micrometer
f) Surface roughness measurement ... With a surface roughness meter
g) Film thickness measurement ... Measurement of thickness of paint film, galvanized layer, etc., using an electromagnetic micrometer

h) Appearance observation ... A close-up photographing of overall view, welds, singular parts and crevices, etc.
i) Others ... Adhesive force of coated film and AC impedance, etc.

2) Evaluation of corrosion resistance during exposure
   a) Appearance observation ... The same procedure as that for the advanced measurement
   b) Localized corrosion rate ... Only when localized corrosion occurs
   c) Measurement of immersion potential

3) Destructive test
   a) Determination of an average corrosion loss and a pitting depth for individual parts, by means of a depth gauge, section microscopic examination, surface roughness measurement and the like (determination of the corrosion loss of base materials, according to circumstances)
   b) Analysis of rust composition
   c) Other items peculiar to each type of test piece

SUMMARY

To meet the requirements for a long service life of offshore structures, experiments and research on new corrosion-protective wrapping of highly corrosion-resistant metals were started, with the ultimate aim of developing and establishing a new corrosion-protection system that can dispense with any maintenance works. The exposure test using a part of the general research installation for ocean technology was started in April 1984 and will be continued until the year 2004.

References

1. Current study of corrosion protection technologies for offshore structures in Japan: The Kozai Club 1975

2. NACE-Standard RP-01-76: Recommended practice: Control of corrosion on steel, fixed offshore platforms associated with petroleum production

3. Development of corrosion-resistant steel pipe piles and record of exposure tests (part 10): Japanese Association of Steel Pipe Piles 1984

4. E.R. Greamer (International Nickel Co.): OTC paper 1274, 1970

5. Report on maintenance technologies for offshore structures: Public Works Research Institute, Construction Ministry 1982

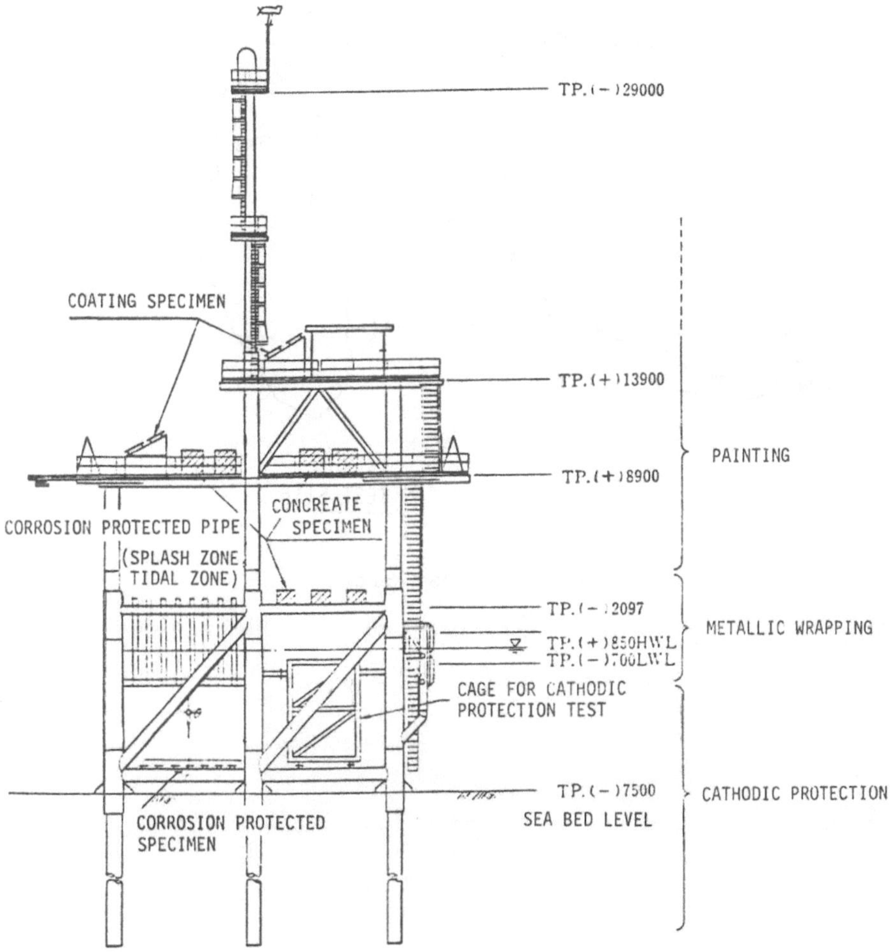

Fig. 2. General research installation for ocean technology

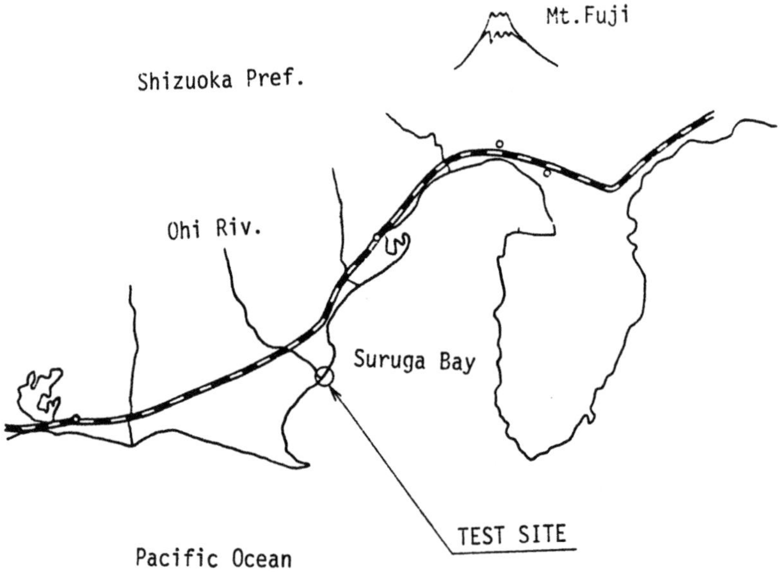

Fig. 3. Test Site

Photo 1. Out view of installation

Experiments on Repair of Corrosion Protective Coverings Applied to the Splash and Tidal Zones of Steel Pipe Piles

Y. Mori[1] and T. Tomura[2]

[1] Chemical Division, Public Works Research Institute, Ministry of Construction, Toyosato, Tsukuba, Ibaraki, 305 Japan
[2] Technological Section, Nakagawa Corrosion Engineerings Co., Ltd., Kanda Kajicho, Chiyoda-ku, Tokyo, 101 Japan

Foreword

The splash and tidal zones of steel structures constructed in coastal areas are protected against corrosion through the application of thick film coverings and/or linings. However, the deterioration of organic covering materials is severe in splash zones due to the effects of the sun rays and the repeated wetting and drying to which they are exposed, and the protective coverings on the tidal zones are very easily damaged by littoral drift, the effects of waves, and organic organisms which attach themselves to the piles. Despite their relative susceptibility to damage, the repair of these coverings still often remain a difficult task.

Therefore, the Public Works Research Institute, Ministry of Construction, in cooperation with private firms involved in corrosion protection manufacturing, has been conducting research and on-site experiments on corrosion protective covering repair technology.

In this report, we will discuss two such repair case studies.

The Experiments

1. Experiments on Corrosion Protective Coverings Applied to the Supportive Steel Pipe Piles of the Drift Sand Observation Pier

1.1 General Condition of the Steel Pipe Piles

(1) Location

The piles in the experiments are supportive steel pipe piles on a pier used for the observation of drift sand and waves by the Coastal Research Division, Public Works Research Institute, Ministry of Construction, . The pier is located on the shore of Ajigaura in Nakaminato, Ibaraki Prefecture. Experiments were conducted in the splash and tidal zones on 16 of the support piles in a 97 m long section constructed in 1972; the total length of the pier is 200 m.

(2) Corrosion and Placement of the Piles

The piles are placed as shown in Figure 1. In 1976, four years after their construction, the piles were inspected. Results showed that there was an average of 1 mm reduction in thickness, with corroded holes of up to 4.8 mm in depth.

1.2. Types of Corrosion Protective Coverings and Areas of Application

Corrosion protective coverings were applied to the areas of the piles shown in Figure 2. The area between TP+2.0m and the top of the piles is the splash zone, while the area between TP+1.0m and TP+2.0m is considered the tidal zone. Corrosion protective coverings appropriate to the conditions of each area were selected and applied, as shown in Table 1 and 2.

1.3 Survey of Protective Covering Durability

(1) Survey Method

Surveys to check the durability of the coverings are generally carried out by visual observation. The coverings are checked for the occurrence of peeling, rust, cracks, bubbles and so forth, in accordance with the thickness and properties of each type of covering.

(2) Survey Results

Table 5 shows the results of the surveys conducted over the first four years after construction. The results of the survey conducted eight years after construction is as follows:

1) Vinyl Coverings

Separation of the film layers begin rather early, leading us to conclude that the durability of vinyl type coverings is not very good.

2) Epoxy Coverings

Despite the appearance of rust spots in some of the corroded holes of the steel surfaces, there seem to be few changes in the insolvent epoxy coverings, thus their durability has been judged as being excellent. However, their adhesion to the polyurethane surface layer is poor.

With the denatured epoxy and (other) epoxy coverings, the advance of rust spot over the years is negligible, and their adhesive qualities are also excellent. However, there is a lot of discoloration in the surface coat of the denatured type due to chalking.

Thick-film epoxy types have both excellent corrosion protective and adhesive qualities.

The corrosion protective qualities of organic zinc + thick-film epoxy coverings are just as good as those of the thick-film epoxies, but their adhesive qualities are somewhat inferior.

The corrosion protective qualities of denatured epoxy coverings are also on par with the other epoxy types, however problems remain with chalking and their adhesive characteristics.

3) Polyetser Coverings (Containing Glass Flakes)

Of the two piles to which polyester coverings were applied, some spot rust appeared on one of them, while the other shows absolutely no signs of rust and is still in excellent condition today, eight years after initial application.

4) Thick-film Tar Epoxy Coverings

The piles to which this type of covering was applied are located where the waves break, and work was also carried out at a time when the waves were high, thus the desired film coating couldn't be obtained. The results, being entirely contrary to expectations, were also unsatisfactory.

5) Water Curable Epoxy Coverings

Water curable epoxy A coverings were applied to four of the piles with spatulas, but shortly after application, they began to show signs of peeling, cracking and bubbles, phenomenon which tended to increase with the passage of time.

Water curable epoxy B coverings were hand-painted onto the tidal zones of three of the piles. About two years after application, cracks, peeling and bubbles occurred, showing considerable early deterioration of the coverings applied to these three offshore piles.

A water curable epoxy C covering was applied to pile number 15 of the tidal zone using the earlier described method of pouring the epoxy into forms, however, after the forms were damaged four years later, deterioration advanced quickly and the forms and protective coatings had all exfoliated eight years later.

Water curable epoxy was applied to pile number 16 of the splash zone with a trowel and, perhaps because the sea was calm at the time, it is still intact and in excellent condition eight years later.

6) Petrolatum Anti-Corrosive Tape + Protective Cover

We covered the petrolatum tape with a protective cover of EPT (Ethylene Propylene Tar-polymer) rubber sheeting, however this gradually lost its ability to retain its proper shape and became more and more loose, also rolling up from the bottom due to the effects of waves. However, when the sheet and the tape were removed, there was a thin, black film on the surface of the steel pile, and no corrosion could be discerned. The EPT sheet was replaced with an FRP protective cover after four years of service due to its ever worsening rolling. With the application of this new sheet, we used a method employing the parallel implementation of a closed cell foam protective tape to the pile. Today, four years after the replacement of the old sheet, the new covering is still in excellent condition.

7) Cement Mortar + FRP Cover

Four years after its application, small cracks which are thought to be caused by steel pipe pile corrosion, appeared in the top section of the mortar covering, and the formation of (a) gap(s) between the mortar and the FRP cover became obvious. After seven years, the FRP cover was severely damaged by breakers and it fell off the pile.

8) Chloroprene Rubber Sheet Coverings

The adhesive power of the chloroprene rubber sheet to the surface of the steel pipe was 2.5 - 4.5 kg/cm after one year, 2.0 - 3.5 kg/cm after two years, and 1.0 - 3.0 kg/cm after three years; thus it can be seen that its adhesive strength tends to decline with the passage of time. However, as no corrosion could be seen in places where the cover had separated from the pile, its corrosion protective qualities have been judged as excellent. After passage of about six years, it began to peel from the pile beginning at the bottom, and after eight years most of the cover had peeled away.

2. Experiments on Repairing Corrosion Protective Coverings of (Corrosion Protective) Steel Pipe Piles

For the purpose of determining the durability of Corrosion Protective Coverings applied to the steel structures of the splash and tidal zones which stand in sea water, exposure experiments have been conducted since 1974 on 24 corrosion protective steel pipe piles, each treated with a different protective coating, erected in in Tokyo Bay offshore from Chiba City. In 1979, after 4.5 years of exposure, tests on separation were conducted on nine piles (types of protective covering), together with tests where sections of the piles were cut away to enable inspection of the protective qualities of the coverings, as well as to monitor the extent of any corrosion on the insides of the steel pipe piles. At the same time, the tested spots were used for experiments on the repair of such protective coverings.

2.1 Lineup of Piles and Types of Coverings Used in Experiments

The lineup of the piles used in the experiments on the repair of protective coverings is shown in Figure 4. The types of protective coverings subjected to the mentioned experiments are listed in table 3.

2.2 Repairs

Repairs on the protective coverings were carried out in accordance with the different types of coverings, as described in Table 3. For the repair work, a day was selected on which the sea surface was calm.

2.3 Durability Surveys on Repaired Sections

(1) Method Employed in Surveys

Surveys to check the durability of coverings applied to repaired sections were carried out three years after the repairs were conducted, both by visual means, i.e. examining the exteriors of the coverings for rust, bulging, cracks, peeling, etc., and by subjecting bubbles in the coverings to a tapping test.

(2) Survey Results

The results of these surveys are shown in Table 3. As can be ascertained from the table, defects can only be established in certain repaired areas; most of them continue to be in good condition.

The areas which had begun to show defects were the ones which were repaired with tar epoxy on an aluminum flame coating anti-corrosive base. Areas where repairs were conducted with zinc flame coating or thick-film inorganic zinc-rich paint bases are still all intact. Therefore, it must be concluded that aluminum flame coating is unsuitable as a corrosion protective base.

Figures and Tables

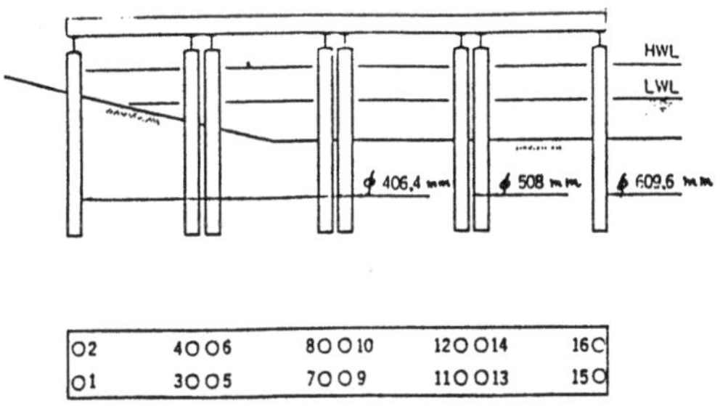

Figure 1. Distribution of Steel Pipe Piles

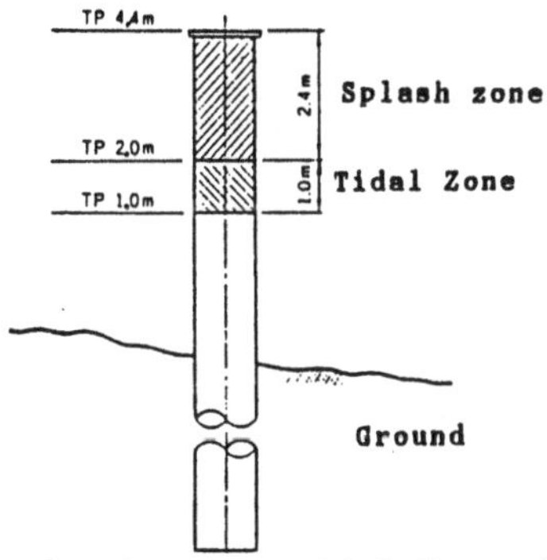

Figure 2. Areas to which Corrosion Protective Coverings were Applied

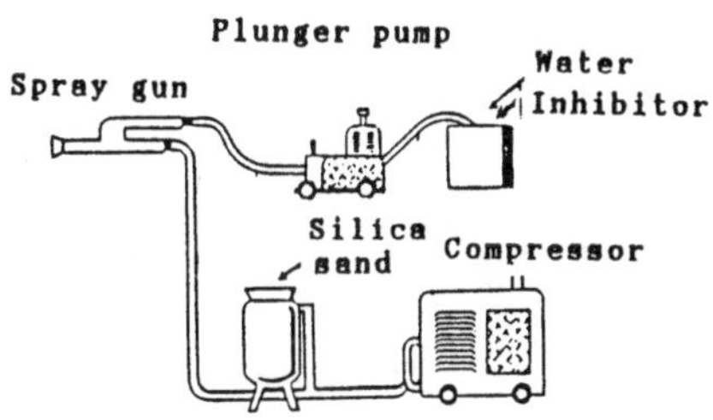

Figure 3. Water Jet Blast Setup

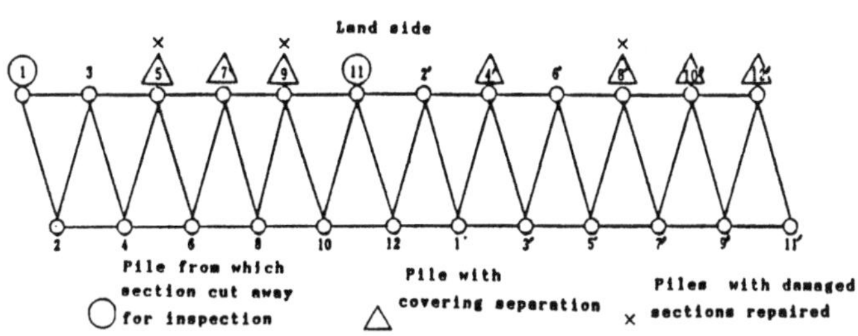

Figure 4. Lineup of Piles Used in Experiments

Table – 1 Protective Coverings Applied to Splash Zones and their Method of Application

Pile Number	Primary Layer	Secondary Layer	Tertiary Layer	Total Film Thickness		Application Method	Environmental Conditions at Time of Application
				Target	Actual		
1	Organic Zone	Thick-film type Vinyl		275 μm	283 μm	Airless Spray	Piles mostly out of water;little wave influence
2	Organic Zinc	same as above		275 μm		same as above	same as above
3	Insolvent Spray	Urethane		330 μm	344 μm	2-head airless gun/brush (2nd lyr)	Due to strong wind, changed to brush for part of the job
4	Denatured Epoxy	Epoxy		300 μm	333 μm	Brush	same as above
5	Epoxy Primer	Thick-film type epoxy	Thick-film type epoxy	440 μm	373 μm	Airless spray/brush (1st layer)	Great deal of wave spray at time of application
6	Organic Zinc	Thick-film type epoxy	Thick-film type epoxy	415 μm	383 μm	Airless spray	Great deal of wave spray at time of application
7	Denatured Epoxy	Denatured Epoxy	Denatured Epoxy	390 μm	380 μm	Airless spray	Great deal of wave spray at time of application
8	Petrolatum paste	Petrolatum paste	EPT sheet* (2 mm)	4 mm	-	Manually wound	Great deal of wave spray time of application
9	Denatured Epoxy	Denatured Epoxy	Denatured Epoxy	390 μm	380 μm	Airless spray	Great deal of wave spray at time of application
10	Cement mortar	FRP cover ** (3 mm)		103 μm	-	Forms attached, mortar poured in	Great deal of wave spray at time of application
11	Polyester w/glass flakes			1,000 μm	1,100 μm	Pressure pump airless spray	Stands where waves break so subject to effects of wave spray
12	Insoluble thick-film epoxy			500 μm	1,180 μm	Airless spray	Stands where waves break so effects of spray severe
13	Polyester w/glass flakes			1,000 μm	1,100 μm	Pressure pump airless spray	Stands where waves break so subject to effects of wave spray
14	Insoluble thick-film epoxy			300 μm	330 μm	Airless spray	Stands where waves break so effects of spray severe
15	Primer (2 liquids)	Adhesive	Cold vulcanized rubber sheet (3 mm)	3 mm	-	Stretch application	Sea surface calm
16	Water curable epoxy D w/glass flakes			4 mm	4.3 mm	Trowel	Sea surface calm

* EPT: Ethylene Ppropylene Tar Polymer

** Fiber Reinforced Plastic Cover. Foundations prepared by removing rust to Sa 2 by means of water jet blast.

Table 2. Protection Coverings Applied to Tidal Zones and their Means of Application

Pile Number	Primary Layer	Secondary Layer	Tertiary Layer	Total Film Thickness		Application Method	Environmental Conditions at time of Application
				Target	Actual		
1 − 4	Water curable epoxy A			5 mm	2.5 mm	Spatula	Effects os spray minimal
5 − 7	Water curable epoxy B			5 mm	4.7 mm	Hand painted	Lower portion subject to influence of seawater
8 − 9	Petrolatum paste	Petrolatum tape	EPT sheet (2 mm)	4 mm	-	wound manually	Subject to waves and wave spray
10	Cement mortar(100mm	FRP cover (3 mm)		103 mm	-	Forms attached mortar poured	Subject to waves and wave spray
11-14	Petrolatum paste	Petrolatum tape	EPT sheet (2 mm)	4 mm	-	wound manually	Subject to waves and wave spray
15	Water curable epoxy C			5 mm	-	Poured into form	Sea surface calm
16	Water curable epoxy D w/glass flakes			4 mm	4.3 mm	Trowel	Sea surface calm

Foundations prepared by removing rust to Sa 2 by means of water jet blast.

Table 3. Survey of Removal and Repair Work   Date of Removal and Repair: 1979 (4.5 yrs. after construction)
Date of Survey: 1982 (3 yrs. after completion of repairs)

Pile Number	Type of Protective Covering	Method of Removal	Method of Repair	Surface condition before beginning Repairs	Condition of Repaired Surfaces
4	Aluminum flame spray (100 μm) + tar epoxy coat (300 μm)	Electric removal machine	Tar epoxy coating (1500 μm)	No appearance of rust on foundations	Corresponding with spreading rust from surrounding areas, about 1/3 of the covering separated from the pile. Red rust and red rust bubbles, swellings and bulges are recognizable
5	Aluminum flame spray (100 μm) + Vinylon cloth reinforced tar epoxy coating (300 μm x 2)	same as above	Water curable epoxy + tar epoxy (5.7mm)	Rust on foundation of damaged areas; removed with disc sander	Brown discoloration due to rust water on surface layer, 5mmφ red rust spot found, otherwise no abnormalities
7	Zinc flame spray (100 μm) + tar epoxy coating (300 μm x 2)	same as above	Tar epoxy coating (1750 μm)	No appearance of rust on foundations	Light discoloration in the form of spots on the surface layer. No abnormalities in the covering film.
8'	Inorganic zinc-rich paint (75 μm) + tar epoxy coating (300 μm x 3)	same as above	Tar epoxy coating (1750 μm)	Rust formation on foundation of damaged areas; removed	In excellent condition; no abnormalities.
9	Inorganic zinc-rich paint (75 μm) + fiberglass epoxy (300 μm x 3)	same as above	Water curable epoxy lining + thick film epoxy coating(4.7mm)	Rust formation on foundation of damaged areas; removed	Staining of surface layer by rust water seen on upper section.Light chalking seen in places,but condition of film excellent.
10'	Extruded polyethylene lining (4 mm)	Cutter, chisel and hammer	Petrolatum tape + polyethylene sleeve	Rust formation in spiral bead section; removed	Portions of exterior of shrink tube fastener have slipped.Bulges can be seen in one section, but condition of cover is normal.
11	Epoxy resin mortar lining (4 mm)	Cold chisel and hammer	Water curable resin mortar	Light rust formation on all foundation; removed	Formation of red rust noticeable from exterior wrap section; otherwise normal.
12'	Fiber concrete lining (sprayed on: 50 mm)	Cold chisel and hammer	Concrete mortar lining	No appearance of rust on foundation	No abnormalities

Studies on the Paint System Applied to the Marine Steel Structure Simultaneously with Cathodic Protection

T. Tsuji*

R&D Center, Dai Nippon Toryo Co., Ltd., Daitocho, Tsurumi-ku, Yokohama, 230 Japan

1. PREFACE

The paint system which is applied to the submerged surfaces of a marine steel structure is generally a combination of tar epoxy coating, epoxy coating, and glass flake polyester coating and organic or inorganic zinc rich primer.

To investigate the effect of cathodic protection, such coated steel panels were tested. Various paint systems were thus checked for film performance, current density, and protection potential.

2. EXPERIMENT

A panel of JIS G 3101 SS41 sandblasted sheet steel measuring 100 x 100 x 3.2 mm and having a surface roughness of 30S was used in the experiment. The sample was prepared by providing a steel panel with the paint system which was a combination of organic or inorganic zinc rich primer and any of the coatings referred to below.

Samples with tar epoxy coating were the greatest in number because this coating is in a relatively wide use. Other samples with glass flake polyester, tar urethane, epoxy, and vinyl coatings were also tested.

The film performance of the tar epoxy resin is considered to largely depend on the kind of hardener used. Consideration was therefore given to the kind of hardener only when the JIS K 5664-1 Type 1 (epoxy resin content, 30% or more) was tested.

One coated side of some samples was given a cut which was 5 cm long at its center and deep enough to reach the base metal.

A potentiostat was used to keep the protection potential at -900 mV and -1,500 mV. The current which changed with the lapse of time while flowing through the sample was investigated. At the same time, the paint film was checked for appearance.

The amount of flowing current was calculated from the measurements of reduction in voltage of shunt.

3. EXPERIMENTAL RESULT AND DISCUSSION

The current density necessary for keeping the protection potential of coated steel panels at -900 mV and -1,500 mV (vs. SCE) is under the influence of various factors. The study

---

* member of Technical Committee of Public Works Research Center

on those factors was conducted and the results obtained are shown below.

Fig. 1 shows the time-lapse change in current density in the sample with tar epoxy coating (JIS Class 1, polyamide) when a protection potential of -900 mV was to be maintained. The data on a protection potential of -1,500 mV is indicated in Fig. 2.

At -900 mV, an increase in current density was noted only in scratched samples which had films 100 μm and 200 μm thick and were free of zinc rich primer. The same phenomenon occurred only to unscratched samples which had a 100 μm-thick film without zinc rich primer. The paint films on those samples were observed to develop blistering. The current density and conditions of film, however, were found to remain unchanged when samples with other paint systems were tested.

While a protection potential of -1,500 mV was being maintained, the current density greatly increased immediately after the onset of the test regardless of the film thickness and the presence or absence of zinc rich primer. The current density was about 100 mA/m^2 and its time-lapse variations were negligible.

Generally speaking, the current density increased more markedly in unscratched samples at -1,500 mV than at -900 mV. A current density of 0.1 to 1 mA/m^2 was measured in the samples with zinc rich primer. This result indicates beneficial effect of zinc rich primer.

In view of the fact that the steady current density in noncoated steel panels is usually about 100 mA/m^2 at -900 mV, it has become evident that this density decreases to between 1/100 and 1/1,000 of normal value at -900 mV after paint application. On the other hand, it has turned out that the current density drastically decreases in unscratched panels with zinc rich primer.

To elucidate the optimum protection potential of zinc rich primer, the samples coated with inorganic or organic zinc rich primer were separately tested.

As is clear from Fig. 3 which indicates the result of the above test, the optimum potential of zinc rich primer is in the range more noble than -1,200 mV and the durability of zinc rich primer sharply decreases as the potential drops from this level.

Evidence has been thus established that the film durability decreases in the paint system with cathodic protection when the protection potential is less noble than -1,200 mV.

The desirable effect obtained from zinc rich primer is indicated in Fig. 1. In the absence of this primer, the current density gradually increased regardless of whether the samples were scratched or not when the film thickness was 100 μm. When the scratched sample with 200 μm-thick film was tested, the current density increased to the detriment of film. In the case of a film thickness of 400 μm, little or no increase in current density was observed.

As in contrast with the instances mentioned above, the current density did not increase even in the samples with a 100-μm thick film and the paint film was protected against deterioration regardless of the presence or absence of scratch when organic or inorganic zinc rich primer was present.

Zinc rich primer itself proved to produce desirable effect. Difference in effect between organic and inorganic zinc rich primers, however, was not clear.

As is seen from Figs. 1 and 2, the larger the film thickness, the smaller the current density. It follows that the effect of the film thickness on the current density is great.

Fig. 1 indicates that the current density increases in the scratched samples with 200-μm thick film in the absence of zinc rich primer and that an increase in current density was hardly noticeable in the samples with 400-μm thick film regardless of the presence or absence of zinc rich primer.

The test results with various coatings are shown in Figs. 4 and 5. As is clear from Fig. 4, tar epoxy coating permits a slight increase in current density regardless of whether or not zinc rich primer is present. When tar epoxy coating was used, the current density never increased.

No increase in current density was noted in the application of either glass flake polyester coating or vinyl coating. In this case, the film did not deteriorate.

The acceleration test was made at a protection potential of -2,000 mV to determine the minimum possible film thickness of glass flake polyester coating. As a result, this paint was found sufficiently durable when it had a film more than 600 μm thick.

As far as tar epoxy coating is concerned, the relationship between the epoxy content and the kind of hardeners on one hand and the current density on the other was investigated. No increase in current density was observed at -900 mV in any kind of tar epoxy coating regardless of film thickness.

At -1,500 mV, the current density was found to decrease with increasing epoxy content of tar epoxy coating.

The current density was greater when polyamide was used as a hardener than with amine adduct.

4. CONCLUSION

The conclusion to be drawn from these results is as follows.

(1) The current density in coated steel panels with undamaged paint film is below 0.5 mA/m^2 which is 1/100 to 1/1,000 of that of uncoated panels.

(2) The paint film on steel panels with zinc rich primer and film 200 μm thick offers excellent durability at -900 mV irrespective of the presence or absence of a scratch.

(3) At a protection potential of -1,500 mV, the paint system applied to unscratched steel panels having zinc rich primer and paint film 200 μm thick offers excellent durability. In the presence of a scratch, however, any paint system quickly deteriorates.

(4) The paint systems which are greatly durable when used with cathodic protections are listed below.

Zinc rich primer	Type of coating
High build type organic zinc rich or high build type inorganic zinc rich (film thickness, 75 µm)	Tar epoxy (JIS K 5664-1/2/3), epoxy, tar urethane, vinyl (film thickness, 200 µm)
	Glass flake polyester (film thickness, 600 µm)
Not provided	Tar epoxy (JIS K 5664-1/2/3), epoxy, tar urethane (film thickness, 400 µm)
	Glass flake polyester (film thickness, 600 µm)

The protection potential must be more noble than -1,100 mV (vs. SCE).

NOTE
This study was carried out as a joint project of the Public Works Research Institute, Ministry of Construction, and Piblic Works Research Center.

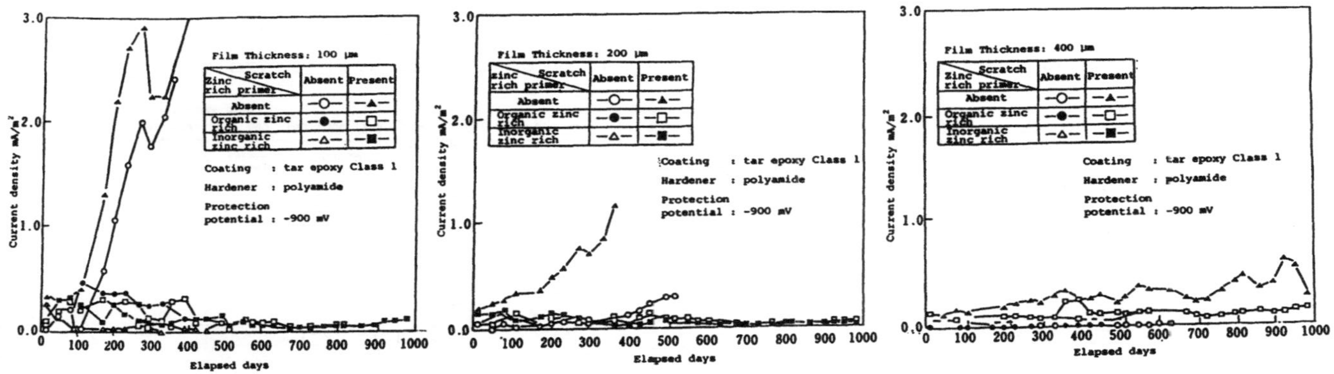

Fig. 1 Effect of various factors on current density (protection potential: -900 mV)
Coating: tar epoxy JIS K 5664-1; Hardener: polyamide

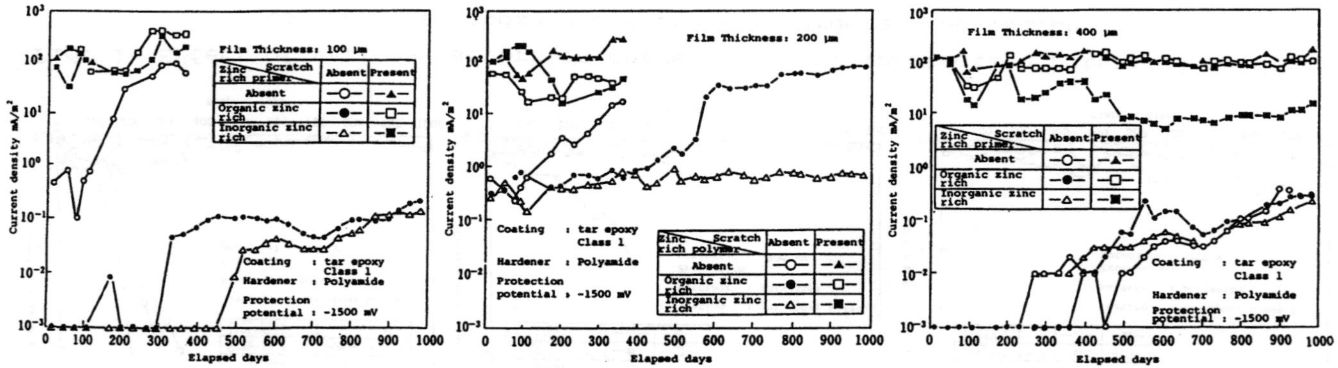

Fig. 2 Effect of various factors on current density (protection potential: -1500 mV)
Coating: tar epoxy JIS K 5664-1; Hardener: polyamide

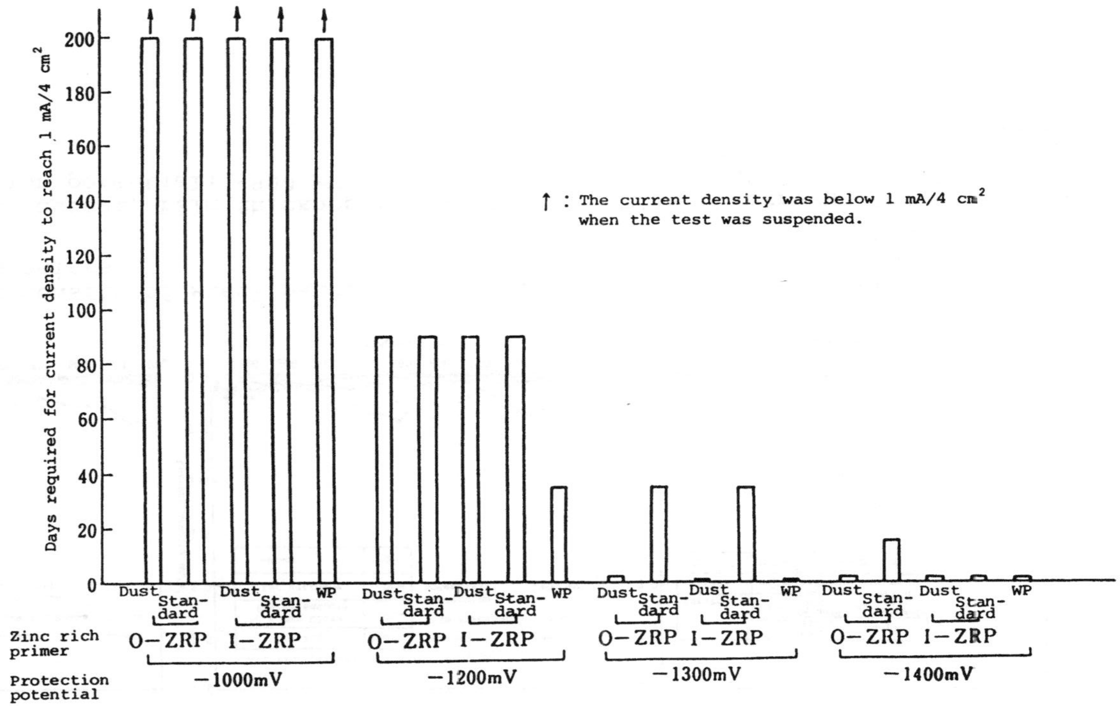

Fig. 3 Time-lapse changes in current density of steel panels coated with zinc rich primer

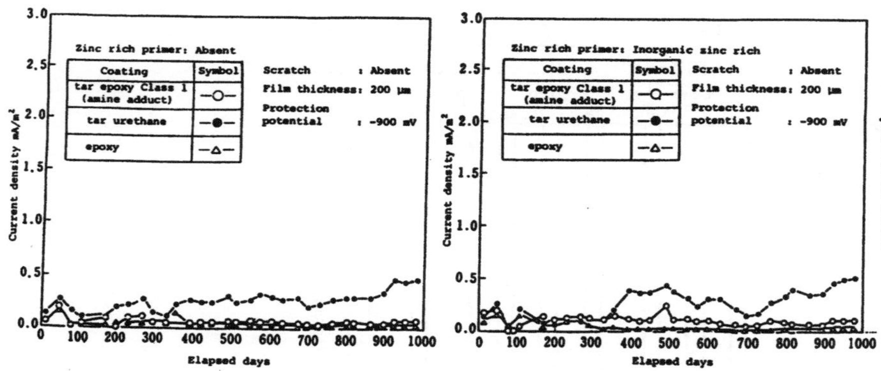

Fig. 4 Current density of steel panels coated with various paints

Scratch: absent; Film thickness: 200 µm; Protection potential: -900 mV

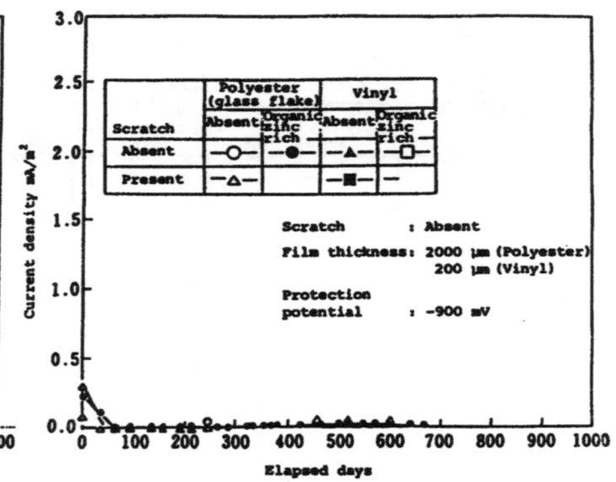

Fig. 5 Current density of steel panels coated with various paints

Film thickness: Polyester (glass flake) 2000 µm Vinyl 200 µm; Protection potential: -900 mV; Scratch: absent; Film thickness: 200 µm

Study on Cathodic Protection in Various Seawater Environments

J. Tomura* and K. Shimizu**

Public Works Research Center, Taito, Taito-ku, Tokyo, 110 Japan

1. ABSTRACT

Cathodic protection is quite effective for corrosion prevention of marine steel structures in seawater and on the sea bottom. Design standards for the application of cathodic protection are still inadequate in Japan. Therefore, the authors carried out experiments on cathodic protection to obtain data in various envinronments such as normal seawater, low temperature seawater, low salinity seawater, polluted seawater, and flowing seawater.

As a result, basic data for cathodic protection design, including the relationship between cathodic potential and corrosion prevention rates for steel materials in various seawater environments, appropriate cathodic protection current density, etc., were obtained. The results are reported as follows.

This study was carried out with the cooperation of the Public Works Research Institue of the Ministry of Construction and the Juridical Foundation of the Public Works Research Center.

2. EXPERIMENTAL METHOD

2.1. Experimental Environment

All experiments were carried out in a laboratory. The experimental environments were as follows.

Normal environment:	Water temperature 25°C, Natural seawater (oxygen saturated)
Low temperature environment:	Water temperature 5°C, Natural seawater (oxygen saturated)
Low salinity environment:	Water temperature 25°C, Seawater diluted so as to have a resistivity of 200 Ω.cm (oxygen saturated)
Polluted environment:	Water temperature 25°C, Tokyo Bay bottom soil, pH 7.1, oxidationreduction potential +68 mV, resistivity 63.2 Ω.cm, sulfate reducing bacteria $1.1x10^4$ MPN/ml (nitrogen supplied)
Flowing water condition test:	Water temperature 25°C, Natural seawater, velocity 0.5-8 m/sec.

* Nakagawa Corrosion Protecting Co., Ltd.

** The Nippon Corrosion Engineering Co., Ltd.

2.2 Test Samples

For test samples other than those used for the flowing water condition test, mild steel pieces of 100x100x3.2 t mm were used. Two types of surface finishing were used for the test pieces. One surface finish was achieved by sand-blasting, and the other by sand-blasting, exposure to the atmosphere for 30 days, and immersion in seawater twice a week during the test period in order to rust the test samples.

For the flowing water condition test, sand-blasted mild steel test pieces of 50x60x5t mm were used.

2.3 Potentiostatic Test

This test was carried out using water tanks in which natural seawater was circulated very slowly and by keeping the test samples at fixed potentials of -850, -900 and -1000 mV (S.C.E.) for 2 to 7 months. By measuring the potential and current during the test period, and after the completion of the test, corrosion rates and corrosion prevention rates were determined from the weight reduction of the test samples.

In the case of the flowing water test, potentials of -800 mV and -900 mV were used and at specific flow velocities, potentials higher than the noble potentials of -770mV were also used.

2.4 Galvanostatic Test

This test was carried out at a current density range of 20-100 mA/m2 intervals using the same tanks in which the potentiostatic test was conducted.

In a polluted environment, 200 and 400 mA/m2 were used and tests at various flow velocities were carried out at 100, 200, 300 and 400 mA/m2.

2.5 Test with an Anode

By connecting a resistor between an aluminum anode of 20øx48 mml and a test piece, currents and potentials were measured. Resistors of 0.03-1.0 kΩ were used.

3. EXPERIMENT RESULTS

3.1 Corrosion Rate

Corrosion rates were 0.027 mm/yr in the normal environment, 0.0456 mm/yr in the low temperature environment and 0.0566 mm/yr in the low salinity environment. The corrosion rate in airsaturated still seawater was estimated to be 0.05-0.06 mm/yr.

In the polluted environment, the corrosion rate showed a high value of 0.0831 mm/yr. In polluted environments, the corrosion rate varies, depending on the degree of pollution.

Test results under various flow velocity conditions are shown in Fig. 1. As velocity increases, the diffusion rate of oxygen increases and the corrosion rate of steel increases rapidly. At flow velocity of a m/sec. the corrosion rate showed a high value of 0.5 mm/yr and increased further as the velocity was increased, but the rate of increase decreased gradually.

3.2 Potential and Corrosion Prevention Effect

In all of the normal, low temperature and low salimity environments at noble potentials lower than -800 mV, the corrosion rate displayed lower values of 1/20 that of the natural corrosion rate. In the normal environment test, -760 mV and -770 mV were supplied and, as shown in Fig. 2, a corrosion prevention rate over 905 was obtained at -770 mV, showing that the same result was obtained at -770 mV, the generally recommended corrosion prevention potential in seawater.

In the polluted environment, the corrosion rate was 0.0136 mm/yr and the corrosion prevention rate was 84% at -800 mV. Compared with normal environment, the corrosion rate was higher. At -900 mV, however, very little corrosion was noticed. Therefore, in an environment where sulfate reducing bacteria exist, the corrosion prevention potential must be about -900 mV.

In the flowing water condition tests, the corrosion rate showed values on the order of 0.003 mm/yr at potentials less noble than -800 mV, regardless of flow velocities, showing that the corrosion prevention effect was very good. High corrosion prevention rates of 98% at 1 m/sec and -720 mV, and 82.3% at 4 m/sec even at -620 mV were obtained. In flowing water condition tests, corrosion prevention seems to be possible at potentials higher than the normal protection potentials. The relationship between set potential and corrosion prevention rate is shown in Fig. 3.

3.3 Current Density

3.3.1 Normal, low temperature and low salinity Environments

For both sand-blasted steel and rusted steel, when maintained at constant potentials, the protective current densities were rapidly reduced respectively, and they reached steady states in 30-100 days. The lower the set potential, the shorter the period required to reach a steady state. The time required was shorter for sand-blasted steel than for rusted steel.

Although the period and the current density required to reach -800 mV by flowing a current at a constant current density varied slightly depending on the type of environment, the period required to reach -800 mV was less than one month when the current density was 80 mA/m2 larger. Data for the low salinity environment are shown in Fig. 5.

A current density above 80 mA/m2 was the initial protection design current density required to reach -800 mV, but for commonly used cathodic protection using an aluminum anode, the protection current is reduced over time and reaches a steady state. As a result of experiments for determining this protection current density in the steady state using an aluminum anode, it was found that the current density reached a steady state at about 40 mA/m2, half the initial value in each environment. Data for the low salinity environment are shown in Fig. 6.

3.3.2 Polluted Environment

In the polluted environment, as shown in Fig. 7, high polarization was observed in the early stages of the test, but after that, the potentials approached the noble potentials and then stabilized. Few differences were found between the sand-blasted steel and rusted steel. To obtain a potential less noble than -900 mV, a current density of 100-200 mA/m2 is required, and when the potential fluctuates, a current density of almost 100 mA/m2 is required to maintain current density.

3.3.3 Flowing Water Condition

At the beginning of water flow, the current density showed a very high value, and the higher the flow velocity the higher the current density. The relationship between flow velocity and current density after 15 days is shown in Fig. 8. In the low flow velocity range of 0.5-2 m/sec, the current density increased rapidly with the increase of flow velocity, but the rate of increase decreased in the high flow velocity range above 2 m/sec. At -800 mV especially, this tendency was very pronounced.

The results of flowing current through painted steel showed a protective current density of 1-2 mA/m2 after 60 days. At high flow velocity, current density can be reduced greatly by painting the steel.

4. CONCLUSION

(1) Basic data for protection potentials and protection current densities used in designing cathodic protection in normal, low temperature, low salimity and polluted environments and in flowing water conditions were obtained.

(2) Although the normal, low temperture and low salimity environments show similar tendencies, in the polluted environment and under the flowing water conditions, the corrosion rate is high and current density varies with the conditions. Therefore, attention must be paid to these various conditions.

(3) As this report reflects experimental results achieved in a laboratory, it will be necessary to carry out experiments in actual sea environments in the future to obtain more practical data.

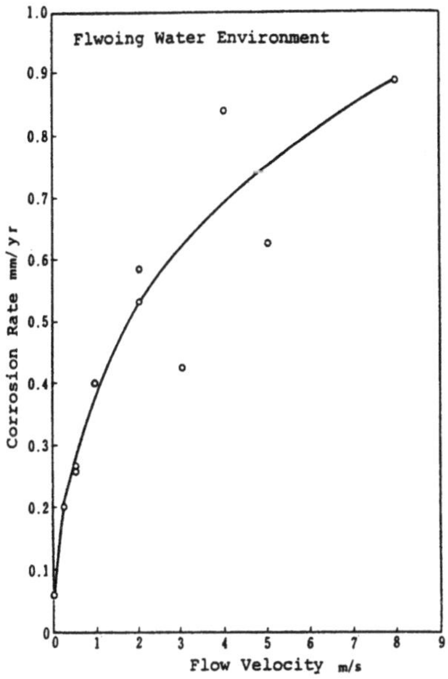

Fig.1. Corrosion rate at each flow velocity

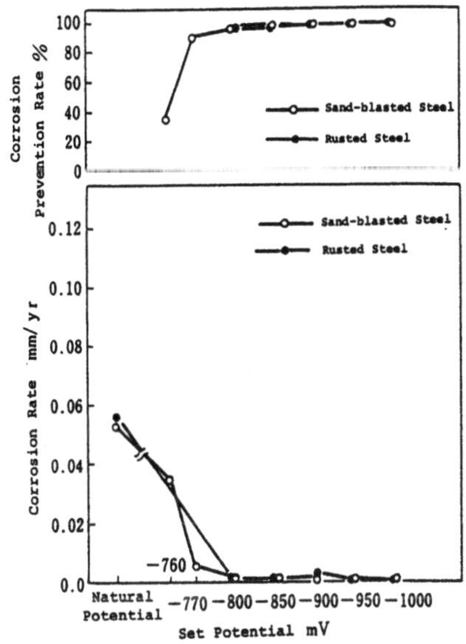

Fig.2. Corrosion rate and corrosion prevention rate

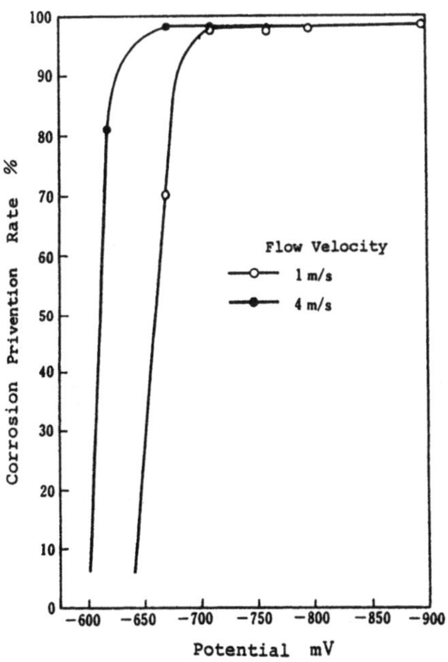

Fig.3. Set potential vs corrosion prevention rate

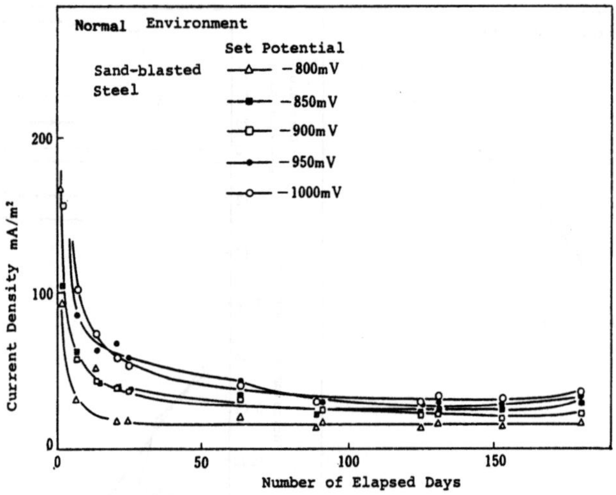

Fig.4. Relationship between set potential and current density

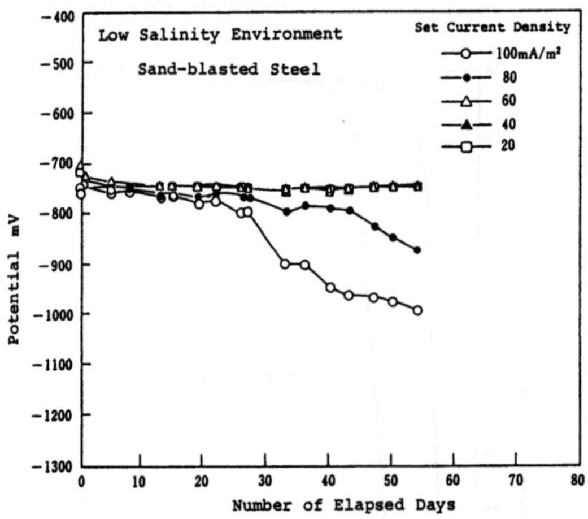

Fig.5. Relationship between current density and potential

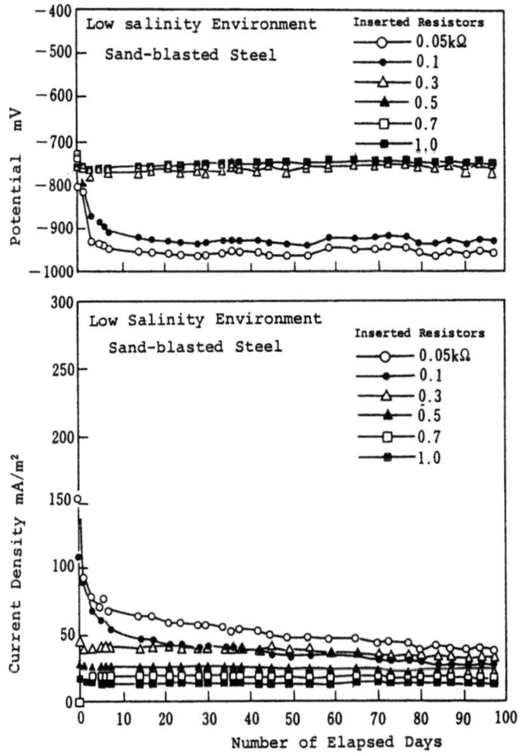

Fig.6. Potential and current density changes over time

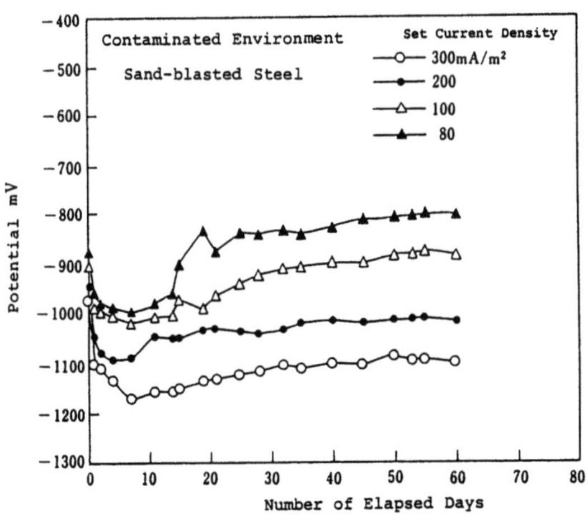

Fig.7. Potential and current density changes over time

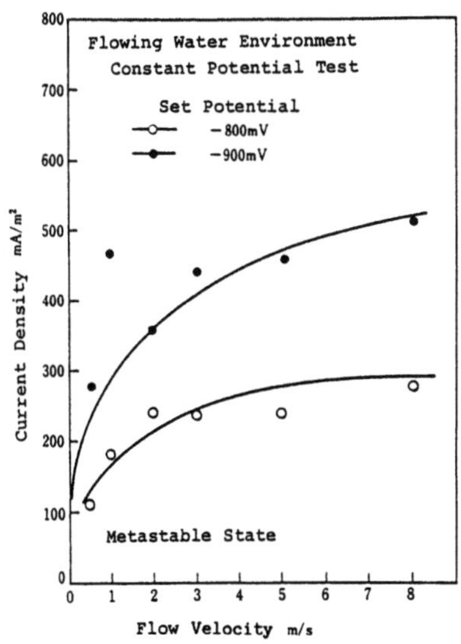

Fig.8. Relationship between flow velocity and current density

New Design Code for Cathodic Protection System of Maritime Steel Structures

K. Katawaki[1] and H. Tamano[2]

[1] Public Works Research Institute, Ministry of Construction, Toyosato, Tsukuba, Ibaraki, 305 Japan
[2] Metropolitan Expressway Public Corporation, Kasumigaseki, Chiyoda-ku, Tokyo, 100 Japan

ABSTRACT

Maritime structures which will be exposed in severe corrosive condition, the corrosion protection is essential problem to be solved urgently. Cathodic protection is most expected to protect submerged area of steel structures.

Public Works Research Institute and Public Works Research Center jointly have developed new design system of cathodic protection through labolatories experiments and in actual fields in cooperation with private companies. Finally we proposed the new design code which contains design rules and criterias for cathodic protection system of maritime steel structures.

INTRODUCTION

Many kinds of maritime structures, channel over bridges, artificial island, fixed offshore platforms are being placed in offshore areas around Japanese coast. These structures are being in deep waters, and become larger, more complex, in severe corrosive condition, most of them is expected over 100 years life.

Corrosion of maritime structures can be devided into three major areas; the submerged area, the splashed area, and the atmospheric area. The submerged area includes that portion below waterline and below the mudline.

Cathodic protection will be expected to be most reliable and rational protection method, so, it is worth proposing a new code on cathodic protection related to the maritime structures. The protection design for these long life structures will be improved and design policy to be well established. With new cathodic protection code, the expected life, maintenance, economics will be strictly testified and approved.

This new design code provides guidelines of corrosion on steel surface for corrosion control of maritime steel structures with cathodic protection.

TASK COMMITTEE

Public Works Research Institute and Public Works Research Center have jointly studied "the new design for cathoidic protection in maritime structure" which had been executed with the active committee of cathodic protection design chaired by Dr. Harumitu Tamano, and have got much knowledge of the design items; suitable potential, protective current which varys with marine condition. Basis on this study, this code was summarized the essential design items when the structural designer draws protection design of maritime structure without any difficulty.

This code was prepared by the task committie which consists of the members listed in Table-1, and is issued by PWRI and PWRC.

Table-1 Task committee on the new design code of cathodic protection

Kiyoshi Katawaki	Public Works Research Institute
Yoshinori Mori	Public Works Research Institute
Kazuo Shimizu	The Nippon Corrosion Engineering Co.,Ltd.
Toshio Shinohara	Dai Nippon Toryo Co., Ltd.
Ryunosuke Tanaka	Japanese Association for Steel Pipe Piles
Juichi Tomura	Nakagawa Corrosion Protecting Co., Ltd.

FEATURES OF THE NEW CODE

This code has some distinctive features as follows,

1) Criteria in this code have been developed through laboratory experiment for four years. This experiment composes many different phases related cathodic protection. Each criteria have been testified in simulated test condition and accelerated condition.

2) This code classifies the maritime condition into three zones; normal sea water zone, polluted sea water zone and high velosity sea water zone. Low temperature sea water zone is included in normal sea water zone. These zones are classified with locality on the basis of water quality and velosity.

3) The fundamental of cathodic protection is to control minimumly corrosion rate, therefore, the essential point at design and maintenance is to control potential. The potential at which corrosion is controlled is a function of the oxygen content and environmental conditions. Since the protection design corresponds to corrosiveness of sea water, appropriate potential must be selected and suitable protective current to maintain the potential is varied.

4) When marine condition is must severe and more protective current is needed, it may be profitable to choose the combination with cathodic protection and coating system. The combination system is situated one of cathodic protection. Also the effectivenes of this combination system is described and suitable coating system is suggested.

5) This code gives presentation of the protection procedure, so the designer and maintenance people can easily acceses the standard format of cathodic protection design.

OUTLINE OF DESIGN CODE FOR CATHODIC PROTECTION

This code recommends procedures for designing cathodic protection systems that will provide effective corrosion control of maritime structures. The design code book is 47 pages long, the content is briefed in this paper.

Section 1 General

1.1 Application area

This code can be applied for cathodic protection design of steel structure submerged area including below mudline.

1.2 Definition

cathodic protection: a electrochemical technique to reduce corrosion rate of structure surface by electrochemical polarization.

combinated cathodic protection: a variation of cathodic protection in conjunction with steel surface coating for the purpose to reduce protective current.

current density: the current flowing to or from a unit area of an steel surface to maintain protective potential. protective potential: potential to reduce corrosion rate negligibly of a metal surface by inducing flow of current. Essential item for protection design and maintenance of structure.

zone: classified area by the environmental condition.
Corrosion of structure in submerged area is influenced by oxygen supply and water quality, therefore, zone is devided into three zones, normal sea water zone which contains no polluted substance nor high water velosity, polluted sea water zone which sea water and sea mud are polluted with accumulated chemical substance, for example, inlet bay accumulated with polluted sludges of industrial waste. High velosity sea water zone means that water velosity exceds 0.5m per second.

Section 2 Design of cathodic protection

2.1 General

The objective of using cathodic protection is to reduce corrosion rate of steel surface effectively, so design code supply systematic criteria including preliminary survey, decision making of potential and current density according to marine condition, quantity and specification of anodes and selection of cathodic protection system. Standard procedure of cathodic protection design is described in Figure 1.

Fig. 1 Standard procedure of cathodic protection design

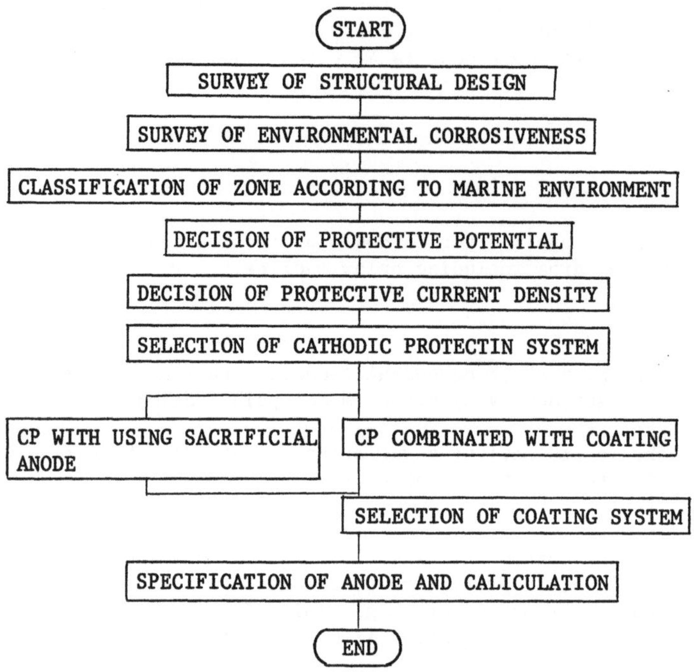

2.1.1 Selection of cathodic protection system

Cathodic protection usually means galvanic anode system. In normal sea water less than the specific resistivity 30 ohm-cm, galvanic anode system is efficient and economical, so this system is used widely because a galvanic anode system requires substantial initial capital investment but little maintenance cost. It is worth to compare feasibility of cathodic protection in conjunction with coatings in special conditions, submerged structure which is placed for example, in high velosity sea water condition, otherwise complex structure which the current distribution is not un-uniformly. Cathodic protection in conjunction with coating can reduce current density and less the galvanic anode is required at the same condition, which results in easy operation.

2.1.2 Calculation of anode number

Galvanic anode system should be designed to provide sufficient current flow thoughout steel surface to produce potential for a life of the anode system. The anode system must also provide sufficient size and weight of anode material to supply current for the desired number of years of protection. The number of the galvanic anode to be required is calculated in equation(1).

$N = \bar{I}/Ig$ ----- (1)

N : number of galvanic anode
I : required current
Ig: initial output capability of anode

2.1.3 Specification of anode

Galvanic anode is aliminum alloys. Aluminum alloy can be obtained in a variety of shaped and weight to deliver protective current with optimum current distribution. The typical sizes to be specified in this code are listed in Appendix 1.

2.2 Field survey

Field survey work prior to designing cathodic protection is required.

2.2.1 Construction specification

Information necessary for designing related to construction specification and practice must be surveyed, construction drawing, size and locations of structure, anode installation accessibility and availability of monitoring equipment. Interference from neighborhood structure is to be monitored.

2.2.2 Site condition

Information on corrosiveness of offshore site is collected if necessary. Oxygen content, velocity temperature, water resistivity and turbulence give useful information, sometimes corrosion test and protection test data on site is recommended. These test data are available for estimating current requirement.

2.3 Potential

Standard potential to maintain good corrosion control is described in Table 2.

Table 2 Protective potential

Marine condition	Potential V vs S.C.E	
	Without coating	With coating
Normal sea water zone	-0.77	-0.77 < < -1.10
Polluted sea water zone	-0.90	
Night velosity sea water zone	-0.99	

2.4 Protective current density

Design current density at initial stage varies according to corrosiveness of sea water and subsoil contacted with steel surface. Typical design current density under each zone is based on Table 3.

Table 3 Typical current density

Marine condition		Protective current density (mA/m2)	
		Initial	Constant
Normal sea water zone	in sea water	100	50
	in subsoil	20	10
Polluted sea water	in sea water	100K	50K
	in sludge	150	100
	in subsoil	20K	10K
High velosity sea water	2m/s	230	115
	5m/s	270	135

K varys 1.2 to 1.5 depends on the degree of polution

When cathodic protection in conjunction with coating are applied, protective current is multiplied protective current of bare steel by damaged paint surface area ratio.

This numerical values are concluded from laboratory tests performed during four years. Constant current density decrease after polarization of steel surface. This decrease tendency is affected from potential and availability of calcium solved in sea water. Initial current density is ruled mainly with corrosive agents, principally oxygen and pollutants. Protective current density is affected by fluctuation of environmental condition, for example, in polluted condition, current density needs more in summer because of higher water temperature. Therefore it is hopeful to decide initial current density and constant current density in good condition via corrosion test over one year. When primary test is difficult to be performed, typical current density in Table 3 can be applied through the classification of zone after field survey. Fig. 2 shows the relationship between potential and current density. This figure indicates that environmental influence on current density but the protective current density in normal sea water zone. Normal sea water zone can include the low temperature sea water and diluted sea water from the experiments because the required current density of these conditions is similar to normal sea water.

2.5 Coatings

Coatings which is combinated with cathodic protection is especialy durable under cathodic protection condition. Conventional paint should not be used for this purpose since high alkalinity caused polarization on steel surface under cathodic protection.

Many coatings are available but only a few are suitable. Long life porotective and water resistent paint is specified in this code.

Coating system in use is as below.
Inorganic zinc rich primer + mist coat + tar epoxy top coat (2 coats)

Inorganic zinc rich primer + mist coat + high build epoxy top coat (2 coats)

Organic zinc rich primer + glassflake polyester top coat (2 coats)

These coating system have been developed and shows excellent performance through laboratory experiments. From the laboratory tests, the paint film on steel with zinc rich primer and paint film offers durability at -0.9V irrespective of the presence or absence of a scratch. At a potential of -1.5V, scratched paint steel deteriorates, it is concluded that the protection potential must be more noble than -1.1V to protect coating system.

Table 4 Coating system

Zinc rich primer	Type of coating
High build type organic zinc rich or high build type inorganic zinc rich	Tar epoxy (JIS K 5664-1/2/3) epoxy, tar urethane, vinyl (film thickness 200 μm) Glass flake polyester (film thickness 600 μm)
Not provided	Tar epoxy (JIS K 5664-1/2/3) epoxy, tar urethane, vinyl (film thickness 200 μm) Glass flake polyester (film thickness 600 μm)

FURTHER RESEARCH

Based on this cathodic protection design code, we applied the cathodic protection to the Suruga marine test station constructed newly in May 1984, and continue to testify the effectiveness of this protection system.

ACKNOWLEDGEMENT

We wish to express our sincere gratitude to the members of the committee for their cooperation, as well as to those persons who worked so hard in executing the experiments and compiling the test results.

REFERENCES

1) The design code for cathodic protection system of marine steel structures, Public Works Research Institute/Public Works Research Center (1984)

2) The study reports for cathodic protection design of maritime structure (1980, 1982 and 1983) Public Works Research Institute/Public Works Research Center.

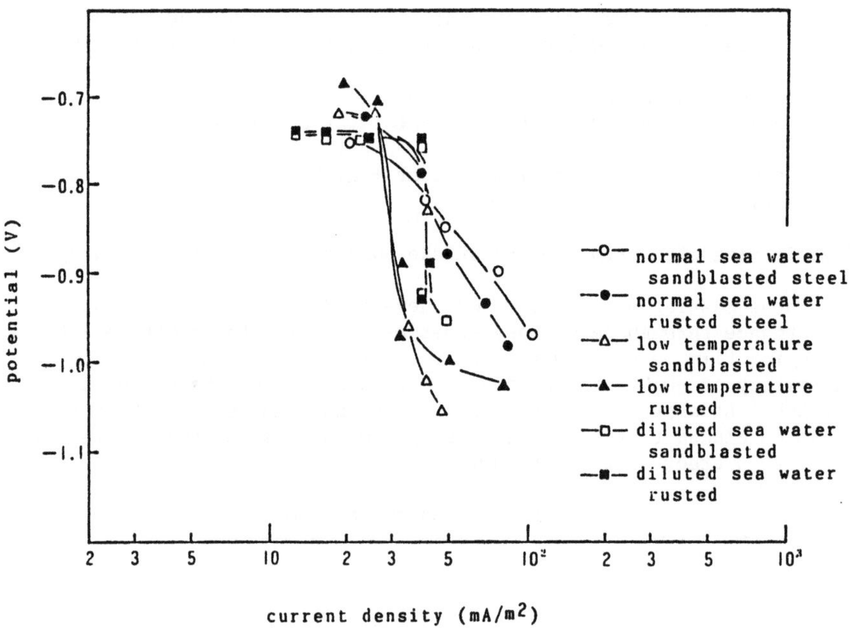

Fig.2 The relationship between potential and current density

Technical Session:
Materials and Construction

5. New Concept

Tremie Process Non-Contacting with Water

G. Shimizu

Department of Oceanic Architecture and Engineering, College of Science and Technology, Nihon University, Narashinodai, Funabashi, Chiba, 274 Japan

ABSTRACT

The present investigation suggests feasibility of new tremie process of non-contacting with water for high qualities of underwater concrete. Experimental works in the laboratory based on the new concept were performed and defined successfully from the test results of lucid cylindrical vessels of 63ϕ×300cm. At the same time, it was found that the possibilities of vibrating compaction and no water pollution were realized in the new process.

INTRODUCTION

It is evident that the qualities of underwater concrete greatly depend on not only the design of concrete and high technique based on long time experiences in the field, but also the construction methods themselves.
As for the process of concrete placement, many researches and various improvements have been introduced to obtain reliable qualities of concrete. In any cases, however, they are limited to minimize the contact with fresh concrete and water. Therefore, whichever all kinds of past methods may be used, such problems are still remained as the washing of fresh concrete brings damage to the qualities as well as pollution to the aqueous environment in early stage of concrete placement. Farthermore, vibrating compaction and moving the mouth of tremie pipe under construction are largely limited from the points of view as mentioned above.
Recently, the author has a new concept to solve positively those problems by embeding a boundary layer at the face between fresh concrete and water in tremie process. This paper presents the outline of new tremie concrete placement non-contacting with water, and some related experiments.

PRINCIPLE OF PROCESS

Fig.1 (a)-(c) shows the new process of tremie concrete placement. What the new process differs from the ordinary one is to embed the boundary liquid layer at the face between concrete and water in tremie process. First of all, it is necessary that the liquid which has larger specific gravity value than that of water is placed on the bottom of caisson before the placement of concrete, and then the concrete

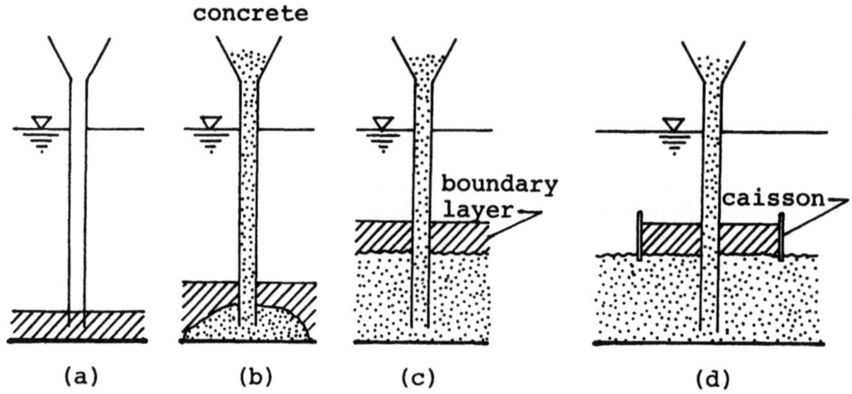

Fig.1. Definition sketches for tremie process non-contacting with water, (a),(b);early stage, (c);standard state and (d);partial use

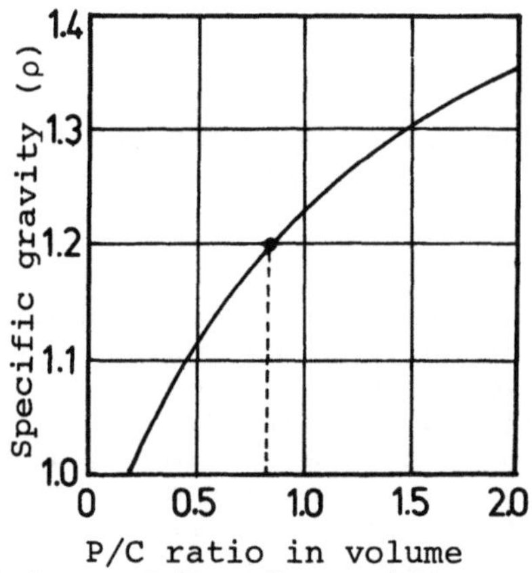

Fig.2. Relation between P/C and specific gravity of solution P; Liquid Paraffin (ρ=0.89),C; Carbon Tetrachloride (ρ=1.59)

placement due to general method may be performed. Fresh concrete from the mouth of pipe,which has the largest value of specific gravity, dives by itself under the liquid layer and becomes standard state as shown in Fig.1.(c) when the bottom is filled with concrete. The boundary liquid layer is required for satisfying the principles as follows;

(1) The liquid must be nonhydraphilic and middle value of specific gravity between concrete and water.
(2) The liquid has chemical and physical stabilities when processing.
(3) Less influence to the qualities
(4) Less sea-water pollution of environment
(5) Less harm to works and easy to treat

Various experimental works by adopting the solution of liquid paraffin and carbon tetrachloride were performed in this investigation,finally almost satisfying results are obtained.

Fig.2. shows P/C-ρ curve calculated from each value of specific gravity. General value of ρ may say to be 1.1-1.2 for practical use. In here, the boundary liquid layer used before is possible to be re-used by withdrawl after placement. As for the applications to underwater concrete, the partial use for large cross-section as shown in Fig.1 (d), the maintenance as well as the perfect compaction with vibrator at the joint of precast concrete members may be adopted.

CONSTRUCTIBILITY OF NEW PROCESS NON-CONTACTING WITH WATER

Discussing the realization and efficiency of the present process based on the new concept,the experiment of the following laboratory model was carried out. Two lucid cylindrical caissons,which were 63φ×300cm each,consisted of three sheets of vinyl chloride resin (200×100×0.3cm) and compacted with concrete at each bottom,were stalled up.
Each of the caissons was reinforced by thick steel bands at intervals of 50cm. Moreover,each of them was fixed with arranged eight steel pipes to keep them in security.
As shown in Table 1.,two kinds of ordinary ready-mixed concrete, proporting strength 210kg/cm^2,slump 21cm (210-21) and 180kg/cm^2, 21cm (180-21) were used. First of all,solution (ρ=1.2) of boundary layer was poured with the depth of about 20cm from the bottom of caisson filled with water. After that,the placement was begun with the plastic pipe of which the diameter of 12.5cm. A view of experimental works is shown in Fig.3.
On the beginning of placement,silvery concrete out of the mouth in clear solution flowed horizontally without any washing as would be expected. It was observed that there were no problems at all against any removal of the mouth of tremie pipe if within the head of solution,and also concrete in place could never sack solution because of its nonhydrophric.
Therefore,actual facts were allowed subsequent concrete to place on the concrete embedded. The head of concrete perfectly separated from water kept on arising as the quantity of fresh concrete. Clear solution of boundary layer is found to keep fully in a stable state against lowering during placement. Vibrator was applicable to less flowable concrete such the poor mixture as (180-21). In such a case,entrapped air bubbled at the face between concrete and boundary layer and popped out themselves when the buoyancy became stronger. Segregation,washing or water pollution could not be seen in any case except bubbling. So,this process based on a new concept was recognized to be technically feasible and also effective to resolve various problems.

Table 1. Mix proportions of tremie concrete

Mix.no	Maximum size of G	W/C	Slump	S/A	Air	Unit weight (kg/m^3)			
(F_C-slump)	(mm)	(%)	(cm)	(%)	(%)	W	C	S	G
210-21	25	59	21	49.6	4	179	303	884	910
180-21	25	64	21	50.2	4	179	280	905	910

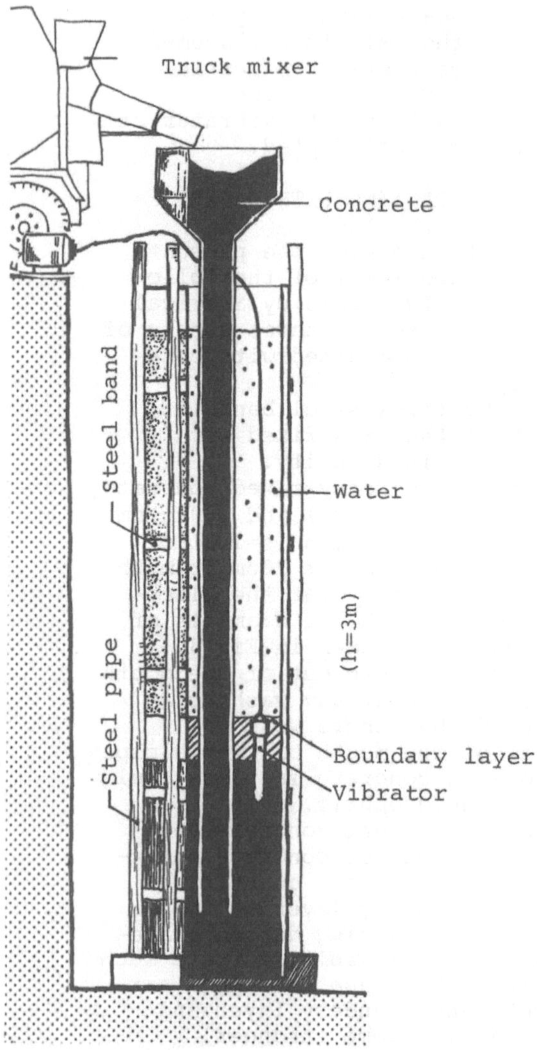

Fig.3.Illustrationofcross-sectional view and experimental picture

QUALITY OF TREMIE CONCRETE

Concrete specimens for compressive strength were chosen as the three kinds of subject,the one is the test cylinder(15ϕ× 30 cm) for comparison obtained from the ready-mixed concrete,the other is cored elements(15ϕ× 30cm) and the another is the large size cylinders of which the length 126cm cut the upper and lower sections of hardened tremie concrete(63ϕ× 126).
The test of the large size cylinders was performed by the universal type testing machine of capacity 3000ton of Nihon University. A summary of the test results is given in Table 2. Comparing the compressive strength obtained by the present

works with those of the standard specimens,it could be known that there is no any great difference between them. Compactibility which has no fault is,here by,certificated from the observed results of cross section,cored elements and even if the bottoms of concrete.
Some experiment was performed to discuss the maximum content and influence to the quality when concrete and boundary liquid were mixed each other by the pan type mixer in the laboratory. It is resulted that the solution content was,at most,less than 2 percent of volume for each batch of concrete slump 21cm. Compressive strength and some related values are summarized in Table 3 . The reduction of compressive strength compared with plain concrete is within 1-19 percent. Those values have tendency of the richer the mixture,the larger the difference of the compressive strength. That is the reduction of compressive strength is almost equivalent to the ratio brought about by incresed air in AE-concrete. But, in actual process,the possibility contained the solution in the static state is so few that the influence to the quality of concrete may be disregarded.

Table 2. Compressive strengths of tremie concrete

Mix.no (F_C-slump)	Size of specimen (cm)	Curing	Age (days)	Compressive strength (kg/cm^2)	Modulus of elasticity (10^5kg/cm^2)	Note
210-21	15ϕ× 30	20℃Water	268	310	3.01	Ready-mixed
	15ϕ× 30	20℃Air	268	-	-	
	15ϕ× 30	Cored	268	265	3.07	Tremie
	63ϕ×126	Outdoor	268	285	3.03	
180-21	15ϕ× 30	20℃Water	409	333	3.03	Ready-mixed
	15ϕ× 30	20℃Air	409	253	2.43	
	15ϕ× 30	Cored	409	318	2.49	Tremie
	63ϕ×126	Outdoor	409	305	2.62	

Table 3. Characteristics of concrete added boundary liquid

W/C (%)	Liquid	Compressive strength (kg/cm^2)		Modulus of elasticity (10^5kg/cm^2)		Tensile strength (kg/cm^2)	
	Age	4	13	4	13	4	13weeks
45	non	337	405	3.08	2.40	36.8	30.6
	added	274	338	3.04	2.34	-	30.2
55	non	303	278	2.63	2.50	32.2	23.2
	added	266	261	2.48	2.07	-	23.4
65	non	215	267	2.81	2.74	30.1	21.7
	added	212	251	2.92	2.62	-	23.6

(max.containable volume,2%)

Fig.4. Cross section and cores of tremie concrete specimen

Fig.5. Large size test specimen at failure (63ϕ×126cm)

REINFORCED CONCRETE MEMBER

When the newly developed tremie process is applied to the reinforced concrete member(structure),the discussion of the performance of the member is required because,under construction,reinforcements are forced to be coated by the mineral oil as the boundary liquid. Here,the beam specimens,in which reinforcement-bars were precoated sufficiently with mineral oil by brush before placement of concrete,were prepared and tested as skew-symmetric loading type as shown in Fig.6. The details of bar arrangements and test results are summarized in Table 4. Fig.7 shows load-deflection curves which were measured at middle of test specimen. In the results of the experiments,the great differences with respect to ultimate load, failure mode and load-deflection curve cannot be certified as compared precoated specimens with normal ones.

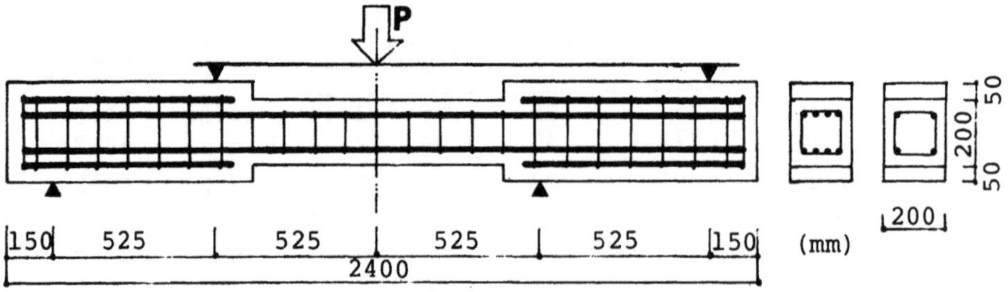

Fig.6. Skew-symmetric loading type

Table 4. Bar arrangement in beam specimen and observed results

Top and bottom reinforcement			Stirrup			Liquid	Ultimate load (ton)				Failure mode
Dia. (mm)	No.	p_t (%)	Dia. (mm)	Spa. (mm)	p_t (%)		P_{bc}	P_{bsc}	P_y	P_{max}	
D13	2	0.64	4φ	125	0.1	non	4	11	11.8	13.6	shear
						coated	4	8	12.0	13.2	
D13	4	1.27	6φ	50	0.6	non	5	12	22.0	26.6	flexure
						coated	5	15	23.0	26.0	
D13	2	0.64	6φ	50	0.6	non	4	13	11.2	14.8	flexure
						coated	4	12	11.4	14.8	
13φ	4	1.33	6φ	50	0.6	non	4	16	18.1	19.4	flexure
						coated	6	18	17.8	19.4	

P_{bc} :Load at flexured cracking
P_{bsc}:Load at flexured and sheared cracking
P_y :Load at yielding
P_{max}:Maximum load

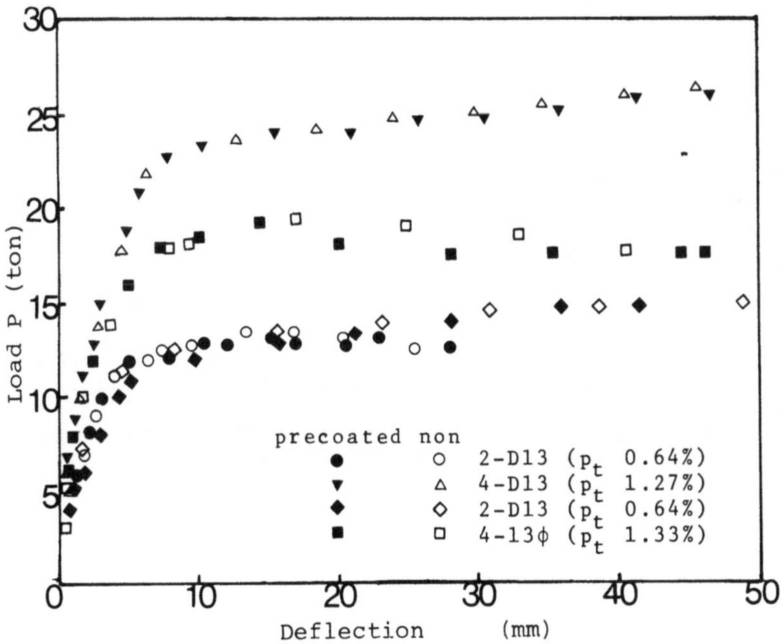

Fig.7. Load-deflection curves at the middle of test specimens

CONCLUSION

A newly developed tremie process was described and recognized to be technically feasible through this research. The characteristics of the results obtained from the new process are that tremie concrete is never washed by the water even if under vibrating,and not contaminated in the aqueous environment because the boundary liquid layer separates perfectly the water and the fresh concrete. The application of tremie process non-contacting with water to the field may be possible through the present experimental investigation. As for the history about the present time,it is still not long and there are few data to discuss about the corrosion of reinforcements in the concrete placed by the new process. However,from one year's test results,no corrosion is found on the reinforcements precoated by the boundary liquid in concrete and also steel bars embedded in the solution as boundary liquid. In the near future,it will be required to discuss about the durability of the interaction between reinforcements and harden concrete.

ACKNOWLEDGEMENT

The author would like to express the thanks to the members of concrete structure research laboratories of Dept. of Oceanic Architecture and Engineering for structural experiments,and also the staffs of the Institute of Science and Technology in Nihon University for performing the large size test.

References

1. Gerwick,B.C.Jr;Placement of tremie concrete.ACI Special publications"Concrete construction in aqueous environments" paper NP.2.1966.pp.9-20
2. ACI Committee 304:Proposed revision of ACI 614-59;Recommended practice for measuring.mixing.transporting and placing concrete.ACI Journal.Jul.1972
3. Schoewert,L.C;Hillen,H.F.;Underwater transporting of concrete with the hydro-value.ACI Journal.Sep.1972
4. Holland,T.C.;Gerwick,B.C.Jr;Turner,J.R.;Tremie concrete placement."Wolf creek dam cutoff wall"ACI Journal.Oct.1982. pp.26-34
5. Hiroshi,S.;Tadasuke,O.;Recent development of underwater concreting.JCI Journal.Apr.1983.pp.4-13

A Concept of Large-Scale Artificial Ridges Using a New Hardened Product Made from Coal Ash

T. Suzuki

Engineering Division, Hazama-Gumi, Ltd., Kita-Aoyama, Minato-ku, Tokyo, 107 Japan

ABSTRACT

This report describes one concept for developing a new type of fishing ground. The idea is to form a large-scale artificial ridge on the bottom of the sea, using coal ash which is produced in large quantities by coal-fired thermal power plants. The artificial ridges are expected to help promote coal-fired thermal power generation and improve fishing resources.

INTRODUCTION

Japan is facing a problem of decreasing fish resources due to restrictions on fishing within 200 mile-zones in many countries and smaller catches in coastal fishing grounds due to water pollution from coastal reclamation and industrial effluent and due to overfishing. The situation calls for the development of new coastal fishing grounds.

At the other end, as the second highest energy consuming country next to the USA, Japan is rapidly converting from oil to coal power generation to avoid too much dependence on a single energy source and thus ensure a stable energy supply. As a result, the electric power output of coal-fired thermal power plants has already reached 6.6 GW, and that level is expected to triple within 10 years.

At the same time, coal-fired thermal power plants are producing coal ash at an annual rate of approximately 360,000 tons per 1 GW of power, which must be disposed of in some manner. At present, most coal ash is used as a raw material for reclamation of coastal areas.

Although more effective uses for coal ash have been developed, there is no market for it in such large quantities.

The concept of artificial ridges is, therefore, designed to solve both of these problems. These ridges feature a newly developed hardened product made from coal ash. They are constructed on the sea bottom in areas with poor fishing resources, and thereby improve these fishing grounds.

UPWELLING FISHING GROUNDS

Coastal areas offshore from California, Peru and Algeria are well known as good upwelling fishing grounds. The upwelling is, in many cases, generated by offshore winds. As shown in Fig. 1, the offshore wind generates a surface layer current in an offshore direction, and this causes sea water at the bottom to move up toward the surface following the topography. Through this process, the upwelling carries nutrients up towards the surface. Seawater containing rich nutrients produces plantplankton when exposed to sunlight. A large amount of plantplankton so generated becomes the foundation of a food chain in the sea, which increases fish resources.

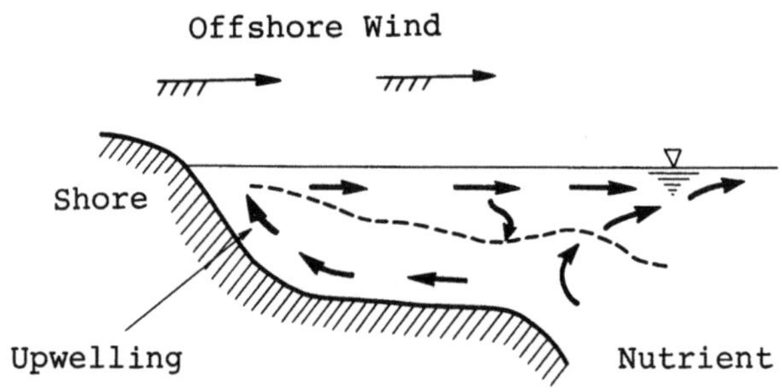

Fig. 1. The model of upwelling generated by the offshore wind

In addition to wind-generated upwelling fishing grounds as above mentioned, there are many smaller fishing grounds generated by current in local areas.
In Japan, current-generated upwellings are found around islands, including Hachijojima, and in channels at Naruto and other locations.

A detailed study was conducted on Shiomaki Sho(reef), Yamaguchi Prefecture by Uekita, Nakamura and Hideshima et al 1), and it reports various interesting findings on current-generated upwellings. The study indicates that this type of upwelling is generated by differences in pressure between a slow water flow in the bottom layer and a fast water flow in the surface layer when the current flows over the reef.

ARTIFICIAL UPWELLING FISHING GROUNDS

The study on Shiomaki Sho implies that an upwelling can be generated by an artificial structure located on the sea bottom, which has a shape similar to the two humps of a camel.

Naturally, the exact shape and size of the structure have to be designed in consideration of the speed and direction of currents and other local conditions in a particular sea area.

Another example of an artificial structure is an upwelling generator currently being developed by Matsumoto and Kanari, et al 2). It has a structure as shown in Fig. 2, and is designed to bring bottom water up to the surface layer using differences between static and dynamic pressures in the sea-water due to the presence of the structure.

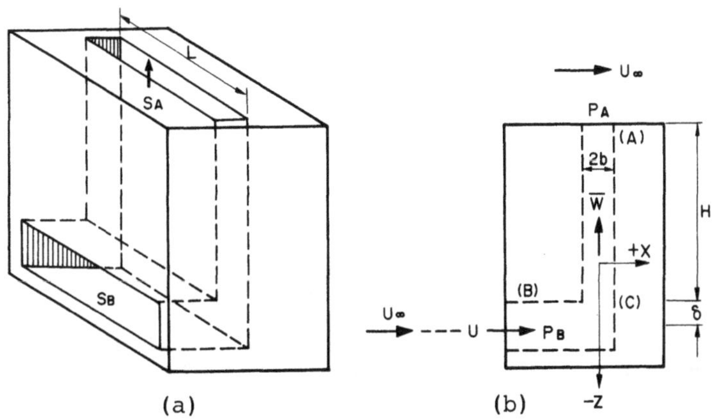

Fig. 2. Schematic view of an underwater structure with L-shaped flow-path upwelling generator (a), and definition sketch for theoretical consideration (b).

The velocity of upwelling $\overline{W}$ can be expressed as

$$\overline{W} = \frac{U_\infty}{\sqrt{\left(\frac{S_B}{S_A}\right)^2 + 2\left(\frac{H}{b}\right)\gamma^2}} \qquad (1)$$

where U_∞ is the velocity of flow, S_A, S_B are the areas of outlet and inlet, b is the width of flow-path, γ is a constant.

A structure made on this principle has been installed off the coast of Ooigawa, Shizuoka Prefecture, for experimental purposes.

NEW HARDENED PRODUCT

We have successfully developed a new hardened product using coal ash, characterized by low cost and high strength. The new hardened product is made of coal ash, cement, water and a newly developed admixture. As shown in Fig. 3, the new hardened product has a completely different composition than concrete, and has a lower specific gravity of 1.7.

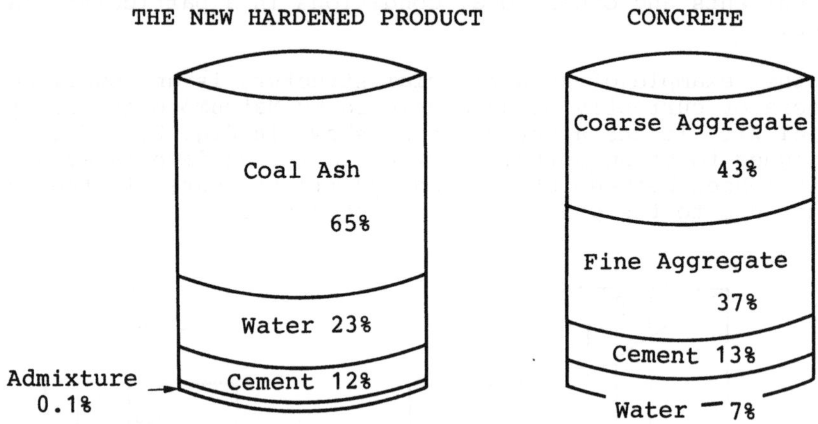

Fig. 3. The comparison of the new hardened product and concrete

The new hardened product has high compressive strength which is due to the new admixture as shown in Fig. 4.

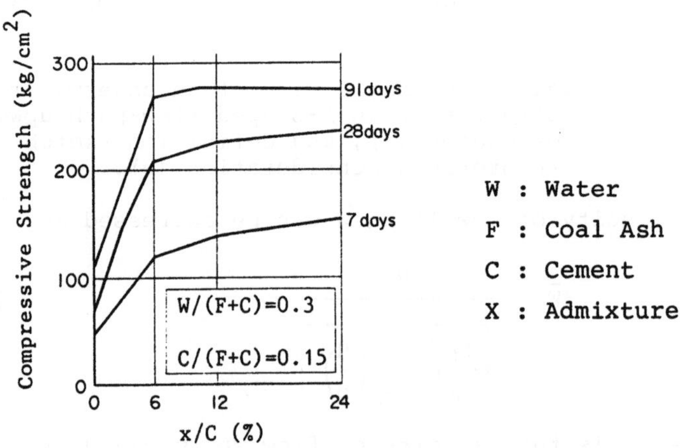

Fig. 4. Relation between the percentage of admixture and the compressive strength

The relation between the compressive strength of the hardened product and the percentage of cement is shown in Fig. 5. This figure shows that any strength can be produced by controling the percentage of cement.

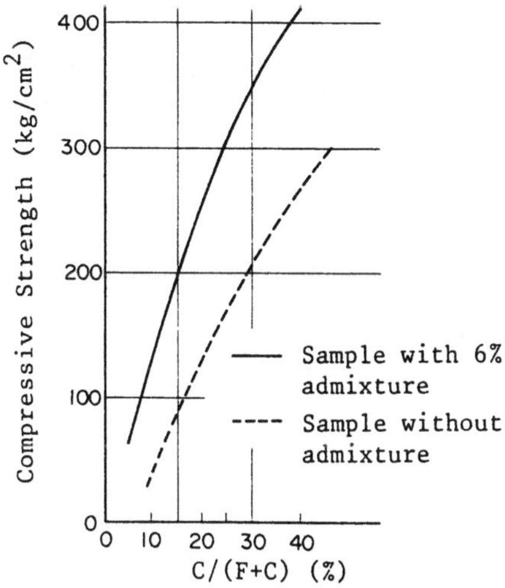

Fig. 5. Relation between the percentage of cement and the compressive strength

Also, the new hardened product can be produced much cheaper than concrete, since its raw material cost is about one third the cost of concrete with 200 kg/cm^2 strength as shown in Fig. 6.

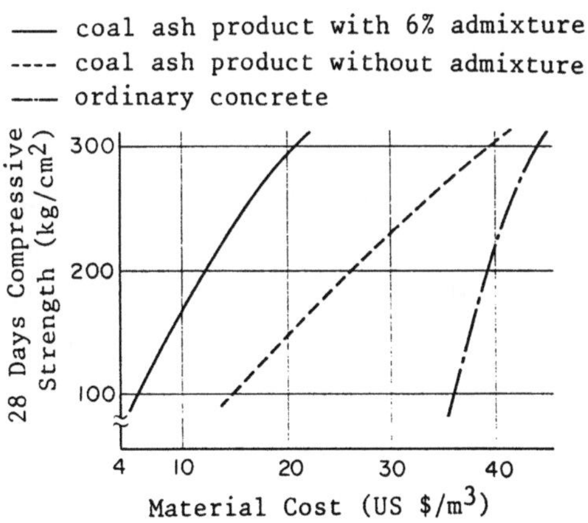

Fig. 6. The relation between the compressive strength of the product and the cost of raw materials per cubic meter

Using the new hardened product, we have developed a large artificial fishing reef, 52 tons in weight, 2.5 m in height and 7.5 m in diameter. The fishing reef has the shape shown in Fig. 7, and was proven to be highly effective in attracting many kinds of fishes. In particular squid and octopus lay eggs in the reef.

This clearly demonstrate that the new hardened product effectively creates excellent fishing grounds.

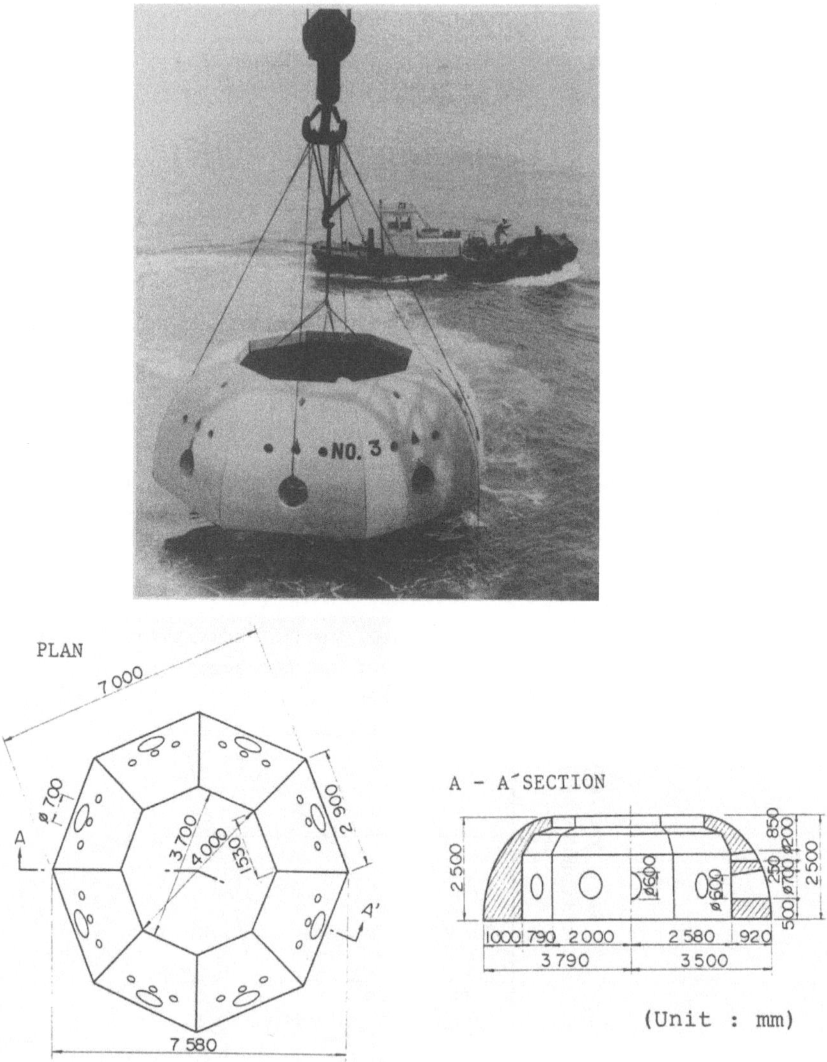

Fig. 7. The photograph shows the installation of the fishing reef, and the sketch shows the dimensions of the reef

LARGE-SCALE ARTIFICIAL RIDGES EMPLOYING THE NEW HARDENED PRODUCT

To illustrate the process of constructing artificial ridges using coal ash, let us assume that a 1 GW coal-fired thermal power plant is located on the coast, and that the area has a sand/mud bottom with limited fishing resources. Further assuming that all coal ash produced by the plant is used as the raw material for a structure shaped like the two humps of a camel as shown in Fig. 8.

Fig. 8. The image of artificial upwelling ridges

In this case, several construction methods are possible, but let us assume that blocks, 100 tons each, are continuously manufactured and stacked close by. The power plant produces 1,000 tons of coal ash per day, which is mixed with cement, water and the admixture to produce 17 blocks per day. Each block has a volume of 120 m^3 if its percentage of void is assumed to be approximately 50 %. Manufactured blocks are temporarily stacked, and later sunk and piled up on the bottom using special barges to form two pyramid shaped structures, width of 100 m, length of 200 m and height of 35 m, after a period of one year. The estimated cost of construction is around 2 billion yen.

The artificial ridges are considered to have the following advantages;

(1) A flat sand/mud area with poor fishing resources can be converted into an excellent fishing ground relatively cheaply.

(2) A shallow water area, being a very good place for marine life reproduction, can be preserved by reclamation using coal ash such areas would otherwise be destroyed.

(3) Coal ash, an industrial waste product, can be effectively used to increase food production.

To actualize the concept, we would like to propose that a feasibility study be carried out with the participation of the government, universities and industrial circles.

References

1. Uekita; Nakamura; Hideshima: Observation of the upwelling occurring behind the large reef in case of Shiomaki. Bulletin of national institute of fisheries engineering. NO.5.1984

2. Matsumoto; Kanari; Myose: Hydraulic model experiments on underwater enforced-flow structures for applications to artificial fish shelters and fish farms. Extrait de la mer (Bulletin de la Société franco-japonaise d'océanographie) Tom 19, N°2 Mai 1981

Effective Uses of Pressed, Packed and Sealed Waste Blocks for Rubbles of Man-Made Fishing Strands in Shallow Seas

Y. Nagasaka[1], S. Hori[1], H. Fujita[1] and S. Sakai[2]

[1] Kisokougaku Consultants Co., Ltd., Kanda Tacho, Chiyoda-ku, Tokyo, 101 Japan
[2] Department of Civil Engineering, College of Science and Technology, Nihon University, Narashinodai, Funabashi, Chiba, 274 Japan

SUMMARY

A great amount of solid wastes are issued in urban area. At the dumping site, they generate pollutant gas, offensive odor and filthy water. Adding these pollutions, there takes place long-lasting settlement of dumping site.

Because of this, it becomes very difficult to find new place for dumping in urban area recently, and we cannot use the site effectively for long period after dumping.

By pressing, packing with steel-mesh and sealing with concrete, it can decrease the volume of wastes and prolonge the decomposing process. Also the generation of pollutant gas, ill odor and filthy water from wastes are prevented, and the settlement of dump site do not appear.

It can prolonge the life-span of dumping site by decreasing the volume of wastes. For the strength of concrete seal, the dumped ground has very high quality, for example high bearing capacity, then we can start to use the dumped grounds for other purposes, such as residential quarters immediately after finish of dumping.

The concrete sealed blocks can be used as materials on construction works such as retaining walls, bubble mound, filling of caisson, groundsill of water way and rubbles for man-made fishing strand in shallow sea or lake.

1. INTRODUCTION

The treatment and disposal methods of solid wastes and garbage of urban area in Japan have progressed from raw dumping to incineration, crushing, segregation of recyclable materials, pressing, packing and sealing disposal.

Incineration residues of garbages and solid wastes are from 10 to 20 percents of original weight and volume.

By compressing the domestic waste with high pressure (100 - 150 kg/cm^2) the volume can be reduced from 30 to 80 percents of original one and in case of the solid waste, from 40 to 90 percents.

Treatment of urban waste in Japan is legally ruled by "the Act on the Waste Disposal and Public Cleansing" enforced in 1970. In this Act, wastes are classified into two categories; domestic waste and industrial waste.

Domestic waste is treated by municipalities and industrial wastes are treated by the waste generating industrial establishments themselves.

Several other laws were establishd for environmental protection.

The total amounts of urban domestic wastes is about 100,000 tons/day in Japan. Out of this 70 percent are incinerated, 25 percent are dumped (raw dumping) and 5 percent are recycled. The amounts of industrial waste is about 700,000 tons/day.

Final disposals of raw domestic waste (garbage and solid waste) and incineration ash are usually dumped at land-fills. Most of dumping site are located at inland valleys. For environmental protection, the method of dumping fills are standerdized in the following aspects;

a) Application of impermeable base layers to protect the ground water and public surface water from pollutant leachates through the wastes.
b) Reduction of volume of the waste to prolong the life span of dump site.
c) Stabilization of biochemical and mechanical properties.
d) Sanitary considerations of surroundings.

Table 1.1 Laws for Environmental Protection in Japan

Public Water Contamination Control Law	1958
Industrial Drainage Control Low	1958
Basic Law for Environmental Pollution Control	1967
Noise Prevention Law	1968
Act of Anti-air Pollution	1968
Act of the Waste Disposal and Public Cleansing	1970
Law of Prevention of Soil Pollution in agricultural Land	1970
Law of Prevention of Water Contamination	1970
Law of Self Burden Cost for Prevention Works of Pollution	1970
Law of Preservation of Natural	1972

Many regional regulations were enforced in 1970-1975, based on these Laws and Act and varied according to the regional requirements.

Some standardized techniques are shown for these purposes by the Ministry of Health and Welfare as follows,

1) Sealing of the surface of bottom layers.
2) Solidification of pollutant materials with cement.
3) Pretreatment by crushing, dehydration, incineration or pressing, according to the characters of wastes.
4) Covering with earth layer; 50 cm in thickness for each 3 meters of waste fills.

Even if treated with these techniques mentioned above, some problems remains in the subsequent use of the site, such as the existence of offensive odour, inflamable gas, long-lasting settlement, etc.

We try to introduce a new method for treatment of domestic wastes, especially of solid waste for this ten years. The method is composed of three steps of treatment, pressing, packing and sealing.

Decomposition of waste is very much prolonged by this method. To be sealed the pressed and packed block of waste by concrete, it has very high strength.

Then the blocks can be used for some construction materials. We show some examples of effective uses of the blocks, especially the use for rubbles of man-made fishery ground in shallow sea. This usage is more suitable for environmental protection against sea muddiness than the natural rubble method.

We try this at Uchiura bay for five years successfully.

2. PHYSICAL PROPERTIES OF DOMESTIC WASTES

Physical properties of domestic wastes vary widely by time, season and region of investigation, and other factors.

Moisture content in weight distributes from 40 to 80 percents; higher in summer (50 to 80%) and lower in winter (40 to 50%) (Table 2.1).

Unit weight also widely varies from 0.15 to 0.4 g/cm^3 (Table 2.2).

It can be said that the incineration method, which is presently the most commonly practiced in Japan, is not always suitable for high content of incombustible solid and plastics unfit for incineration furnace.

Table 2.1 Moisture content of domestic waste by weight.

No.	Max.	Min.	Mean
1	69.7	32.1	56.0
2	6.7	30.5	52.0
3	70.0	23.4	55.0
4	61.6	39.9	53.1
5	59.8	43.5	51.4

Source: Tokyo Metropolitan Bureau of Cleansing

Table 2.2 Weight per cubic meter and breakdown proportion of components of 4 samples of domestic wastes.

Classification	1	2	3	4	Average	Volume percent
Wooden furniture	0.185	0.150	0.157	0.144	0.159	7.6
Fibers	0.101	0.093	0.114	0.083	0.098	7.4
Papers	0.048	0.070	0.071	0.044	0.050	14.3
Wood fragments	0.095	0.115	0.082	0.2000	0.123	19.9
Scum of Sundries	0.019	0.094	0.043	0.229	0.121	2.9
Garbage	0.269			0.091	0.180	1.2
Total					0.101	53.0
Electric goods	0.185	0.133	0.176	0.196	0.173	2.2
Cans	0.091	0.169	0.105	0.090	0.114	15.9
Glass	0.241	0.469	0.349	0.355	0.354	5.2
Fragments of china	0.424	0.684	0.425	0.232	0.441	3.4
Total					0.202	26.7
Plastics	0.043	0.041	0.055	0.046	0.046	15.6
Others (soil, etc.)	0.427	0.202	0.393	0.548	0.393	4.8
Total					0.128	100.0

Source: Cleansing Bureau of Nara City. Unit, kg.

3. TREATMENT OF DOMESTIC WASTE BY PRESSING, PACKING AND SEALING

3.1 Problems of Domestic Waste Treatment by Crushing and Incineration

Treatment of waste is operated with the intention of reduction of its volume, preventing pollutions and improvement of chemical and physical stability.

Treatment by crushing improves reduction of volume, increase of incinerating efficiency and makes it easier to segregate useful materials. Treatment by incineration aims to prevent the pollutions, to improve the chemical, biochemical and physical stability and to reduce its volume.

But it is difficult to complete all these requirements only by these treatments.

Furthermore, there are other problems around the incineration plant, for examples, noise, vibration, exhaust gas, ill odour, traffic accident on the roadside, etc.

Reduction ratio by incineration in volume is about 1/10 ∿ 1/20 in volume, 1/5 ∿ 1/30 in weight.

3.2 Problems of Land Fill Site by Raw or Crushed Waste and Incineration Ash.

Raw or treated waste is finally brought to disposal in land fill site. It is disposed with earth sandwich method, called sanitary land fill in Japan, where the waste is decomposed under aerobic or anerobic condition and it takes very long time.

According to the decomposition it takes place settlement of the site and exhaust of guas, ill-odour and exudation of filthy water.

This prevents the effective uses of the site for a long time after finish of disposal.

3.3 Pressing, Packing and Sealing Treatment of Domestic Waste

For the purpose of prevention of some problems mentioned above, we try to introduce a new method of treatment for this ten years.

This treatment method is composed of three processes. At first, waste is compressed mechanically with high pressure (100 ∿ 200 kg/cm^2). In this process, the volume of waste is reduced.

Reduction of volume is shown in Table 3.1.

Table 3.1 The reduction of the volume with 100kgf/cm^2 in compression pressure

Type of waste	Before Compression weight kg	Before Compression volume m3	Before Compression unit weight t/m3	After compression weight kgf	After compression volume m3	After compression unit weight t/m3	Reduction ratio in volume
Refregerator	1000	10.3	0.1	1000	0.55	1.82	1/18.2
Electric washing machine	980	11.0	0.09	920	0.54	1.72	1/18.9
Metalic can	2810	38.7	0.073	2780	1.44	1.93	1/31
Television	750	6.8	0.11	730	0.68	1.07	1/9.6
Wooden furniture	1000	11.2	0.09	965	1.86	0.52	1/5.8
Mixed Materials	940	7.0	0.134	910	0.71	1.28	1/9.6
Plastics	2050	8.2	0.25	2.022	0.63	3.20	1/12.8
Garbage	10000	24.3	0.25	9800	9.70	1.02	1/4.0

Secondly the pressed waste is packed by wire net. Then it does not expand. The weight of wire net is about 3 kg for one cubic meter of pressed waste volume. These two processes are usually practiced several stations in a town. Packed waste is easy to transport.

Finally the packed block is sealed with asphalt or concrete.

Photo. 3.1 Concrete-Sealed Blocks

Asphalt seal is deteriorated in a few years after sealing, while concrete seal does not, unless subjected to mechanical destruction. Decomposition of waste proceed very slowly in the sealed block, then it keeps its form for over 10 years. The waste in the sealed block will most-likely undergo the carbonation process.

Table 3.2 Size of block and the requirement of sealng materials

Size of block mm	Asphalt weight kg/per blocks	Concrete Thickness mm	Concrete Volume m3/per block
500x500	4 - 5	75	0.15
600x600	6 - 7	75	0.21
650x650	7 - 8	75	0.24
700x700	8.5 - 9.5	100	0.39
800x800	11 - 12	100	0.49
900x900	14 - 15	100	0.60

Concrete sealed block has high strength, then we use this block for some construction materials and use the disposed site with such blocks immediately after the completion of filling.

In this treatment, generation of secondary pollutions such as vibration, noise, ill-odour, exhaust gas and exudation of filthy water, are decreased. Many kinds of refuses can be treated by this method. An amount of 20 to 100 tons of waste can be treated in 8 hours by this method now.

3.4 Evaluation of Pressing, Packing and Sealing Treatment

According to a survey conducted right after landfill was completed with asphalt sealed blocks of urban wastes, the coefficient of volume compressiblity (mv) was from 0.027 to 0.029 cm^3/kg which is about half as much as usual peat or clay and the modulus of deformation is 35 to 37 kg/cm^3.

The N-value by standard penetration test is 2 to 18. In the case of concrete sealed block, there is no problem in strength and large durability of concrete itself.

The merits of this method are summarized as follows:

1) The method is sanitary because it prevents malodor and harmful insects by dehydration and pressing.
2) There is no fear of secondary pollution, because the wastes are sealed by concrete.
3) Life span and capacity of dump site can be improved by reduction of the volume of waste, especially of solid wastes.
4) The block is non-pollutant mass and easy to transport.
5) The block can be used for some construction materials.

Furthermore, we can build this plant on a limited site and decrease the generation of secondary pollutions around the plant compared with the one of incineration plant.

4. EFFECTIVE USES OF CONCRETE SEALED BLOCKS

The concrete sealed block is durable and strong. Therefore it can be used as some construction materials as shown in Fig. 4.1, Photo 4.1 ∿ 4.4.

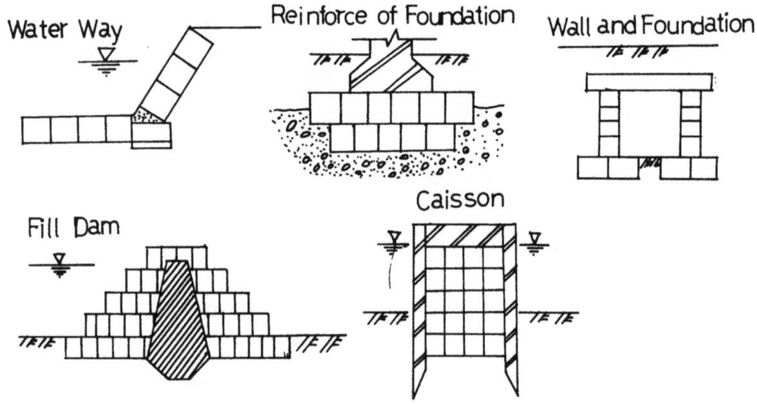

Fig. 4.1 Examples of effective uses of concrete sealed blocks

Photo. 4.1 Retaining Wall (1)

Photo. 4.2 Retaining Wall (2)

Photo. 4.3 Architectural Piller (Under construction)

Photo. 4.4 Architectural Pillar (Accomplished)

It becomes necessary to get an enormous amount of material for creation of fish-gathering place in Japan. The use of natural rocks is not economical and tends to pollute the sea water by fine material adhered to the rock when sunk into the sea.

Concrete sealed blocks are used for this purpose as shown in Fig. 4.2.

They are free from adhered soil and other pollutant materials, and will not pollute the sea water.

Fig. 4.2 Use for man-made fishery ground rubbles

To confirm the certainty of non-pollution of the sea-water, the leaching test from concrete-sealed block was performed in a water tank filled with marine water for a year. A little materials were leached, these are solved only from the concrete surface.

Full size experiment was performed in Uchiwa bay, central part of Japan. Test point is about 200 m off the coast, and it is 30 m of depth with fine sand deposit. Twenty-five blocks were sunk on Dec. 1978.

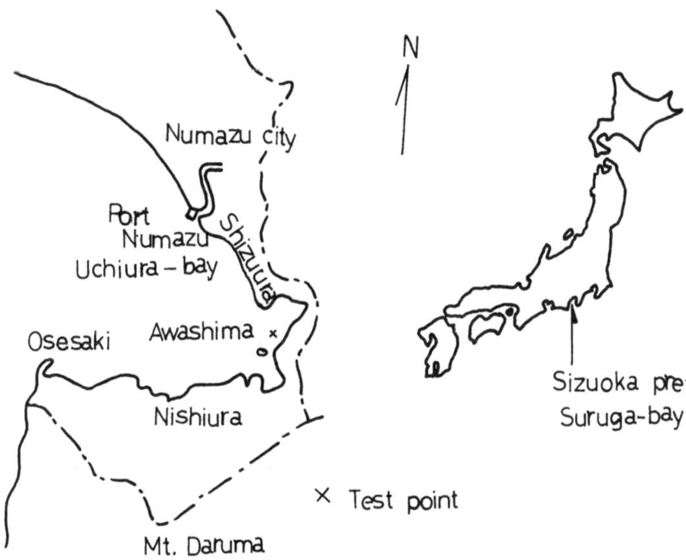

Fig. 4.3 Location of experiment

Each five blocks were connected by a chain (Photo. 4.5). We explored the states of affairs of blocks in the sea every year. After one year, some kinds of marine plants and benthoses were grown on the surface of the blocks and many kinds of fishes were found in and around the mound of the blocks.

Photo. 4.5 Sunken Concrete Blocks

On Dec. 1982, five blocks were pulled up and broken. Contented wastes were seen almost same as fresh one. Strenth of concrete did not go down. Weight of block increased about 100 kg per one block.

This increae of weight corresponds to sea-water weight permeated into the blocks proportionally to the water pressure.

Table 4.1 Classification of Marine Plants and Benthoses on the Surface of Block

Chlorephycas	Ulva pertusa
Rhodophycophyceae	Gelidium amonsii, Grateloupia filicina, Ceramium Rondi
Annelida	Hydroides norvegica
Sipwnculoidea	Phascolosoma scalops
Arthoropoda	Balanus trigonus
Ma acostraca	Caprolla acutifrons, Allorchestes plumicornis
Decapoda	Acanthophrys longispinus, Etisus laevimanus
Echinodermata	Hemicentrolus pulcherrimus
Mollusca	Modiolus flavidus
Gastropoda	Callopomella excellenus, Tristichotrochus haliarchus, Morulina musiva, Ergalatax constrictus, Bedevina birileffi, Euplica scripta, Siphonalia fusoides,

We considered that the decomposition of sealed waste was prolonged under the saline water and the water pressure. On this matter we must borrow much knowledge from chemistry and bio-chemistry and must check up on the truth of the matter.

We will pull up the sunken blocks in 1985 and 1988.

ACKNOWLEDGEMENTS

The authors wish to express their appreciation to Mr. Suzuki S., Chief Engineer of Tezuka Kosan Co., Ltd. and members of the Environmental Engineering Branch for their kind advices and help on this study, also to Mr. Miyagishima T. on the field experiment.

REFERENCES

Sower, G.F. (1977) Disposal and Utilization of Solid. Waste in the U.S. Panel discussion of 9th lSMFE.

Maeda, K. et al. (1970) Charactors of Waste Fill and Effective Uses of Fill Site (in Japanese).

Ueshita, K. et al. (1981) Study on Environmental Geotechnique (in Japanese).

Technical Session:
Materials and Construction

6. Ocean Structures

Large Column-Supported Floating Platforms

R.J. Seymour and F.N. Spiess

Institute of Marine Resources at Scripps Institution of Oceanography, La Jolla, CA 92093, USA

ABSTRACT

Deep draft spar buoys provide the maximum decoupling from surface waves of any floating structures. The research vessel FLIP, developed at the Scripps Institution of Oceanography (SIO), demonstrated how the problems of transporting these very long buoys could be solved by towing them in a horizontal attitude and then ballasting them to the vertical position at the operational site.

As slightly larger stable platforms were required, particularly for offshore petroleum exploration, the semi-submersible concept was developed. By placing the major axis of the flotation volume in the horizontal, rather than the vertical direction, the need for rotation was eliminated at the cost of greater response to waves and higher structural loads.

As the size of the floating platform and its anticipated time on station are increased, and the need for mobility decreased, cost and reliability considerations suggest that the vertical column supported floating platform may again become the design of choice. The elements of such a large structure are amenable to serial production using highly automated shipyard techniques. They are then towed, or alternately barge transported, to the construction site and erected. The remarkable stability of the column structures facilitates joining them at sea into very large structures.

The development of this concept over a twenty year period at SIO is described. Following the success of FLIP, a multiple column unit was designed. Numerical and wave tank modelling preceded a 10 m draft (1/8 scale) system test at sea in which multiple elements were transported and joined to form a stable platform. Various applications are envisioned including airfields or sites for at-sea processing of sea floor minerals.

BACKGROUND

The cylindrical spar or pole buoy has been employed at sea long enough that its origins are unknown. It is characterized by a small waterplane diameter compared to its length. Therefore, a change in the waterline position produces only a minor change to the buoyant force. If the inertia of the buoy is large enough, its response will be small to this forcing. In addition to the volume related force associated with the mean hydrostatic pressure, the passing waves impose an oscillatory hydrostatic component, which decays exponentially with increasing depth, on all submerged surfaces. The forces resulting from both of these pressure fields are cancelled in the horizontal, but not necessarily in the vertical direction. The integrated pressure on the bottom of the spar produces a vertical force which is in phase with the water elevation. However, if the buoy has a deep enough draft, the wave-caused variation in base fluid pressure will be small. The vertical stability of this configuration has made it useful for visual estimations of wave height.

It is possible to modify the design of a simple cylindrical spar to futher increase its stability to heave forces. By increasing the diameter of the spar well below the waterline, some counterbalancing of wave forcing can be achieved. The force from integrating the oscillatory part of the wave-induced pressure over the base can be more or less balanced by a higher pressure acting downward over the smaller annular area where the diameter increase is effected. It should be noted that exact balance occurs at one wave period and one draft, or loading condition for the spar. However, the response at drafts or periods other than this null point is substantially decreased from that of a straight cylinder of the same displacement. [For a more detailed treatment see Spiess (1968).]

DEVELOPMENT OF A SINGLE COLUMN PLATFORM

In 1960, one of us (F.N.S.), in conjunction with F.H. Fisher and P. Rudnick, initiated a study of a very stable platform for making acoustical measurements in the ocean at a depth of about 100 m. [see Fisher and Spiess (1963)] The attempt was to measure fine-scale fluctuations in acoustical propagation. A single spar with a transition to a larger diameter at the base was chosen as the basic configuration. The draft was selected at 90 m, largely on the basis of acoustical measurement needs.

Because the primary purpose for the platform was acoustical research, no machinery was to operate during experiments. This required the platform either to drift or to be moored rather than to use propulsion for station keeping. To facilitate access to ports and increase the towing speed while moving to the experiment site, the platform was arranged so that it would float with its long axis horizontal. At the experiment site, it could then be moved into the vertical, stable position and back into the horizontal mode again by adjusting seawater ballast.

The final configuration chosen is shown in Figure 1. The overall length is about 108 m. A reduction in cross-sectional area of 60% was achieved above a depth of 45 m by means of a 27 m long conical transition section. Towing characteristics and heave and pitch response were physically modelled at scales of 1/100, 1/25, and 1/10. The heave response curve is shown in Figure 2. The null point, at which heave response is nil, is clearly shown at a period of 21.5 seconds. [see Rudnick (1967)]

The acronym FLIP (for FLoating Instrument Platform) was selected. FLIP was launched in 1962 and put through a series of sea trials including a number of flipping operations. Subsequently, it was determined that the platform could be towed and flipped in seas of 2.5 m without problems. In the more than two decades that followed, FLIP has continued to perform as a unique instrument platform in a variety of physical oceanographic experiments. [e.g., see Regier and Davis (1977)]

The stability of FLIP was severely tested by waves of 24 m maximum height, with peak energies between 16 and 20 seconds, on 1 December, 1969 about 500 km north of the Hawaiian Islands [see Rudnick and Hasse (1971).] At these periods, the response of the platform was still less than 10% of the incident wave height. Therefore, the platform acted as a wave measuring pole. FLIP had been designed with a first-deck clearance of slightly more than 6 m above the mean waterline. As could have been anticipated, 12 m wave crests above mean sea level resulted in flooding through an open hatch 12 m above the still water line.

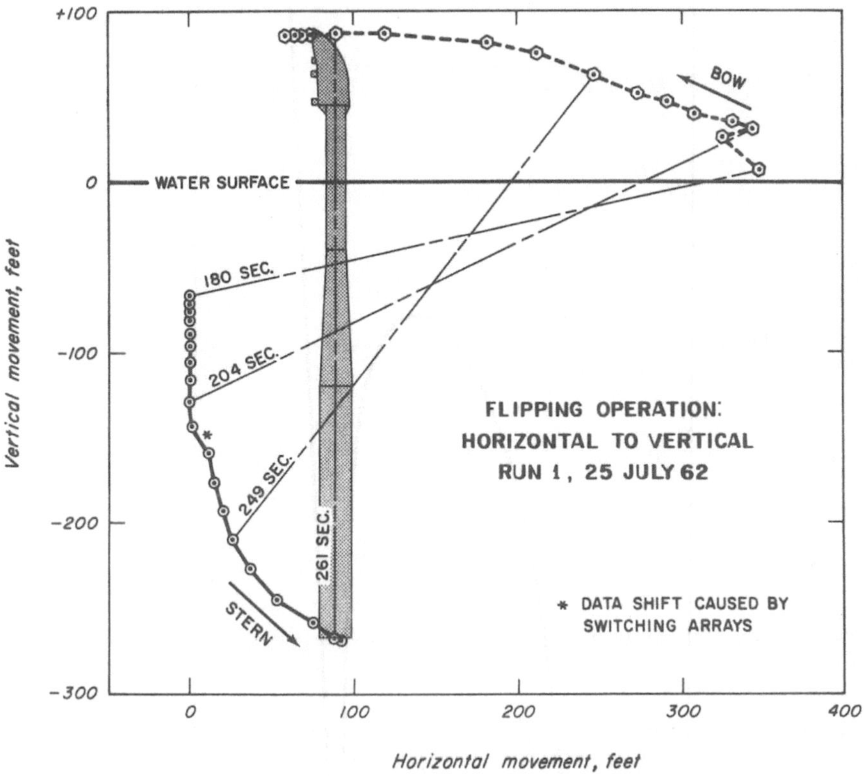

Figure 1. Erecting Maneuver of FLIP

It is interesting to note that Fisher and Spiess (1963) reported that the resonant period of 25 seconds was well outside of the then widely-accepted 18 second maximum for peak wave energy. Earle et al. (1984) describes a very energetic storm in February, 1983 that passed north of Hawaii close to the area where FLIP was operated in the storm of 1969. The period of peak energy in this storm was 25 seconds, perhaps the longest on record in the northern hemisphere. Wave heights are predicted to have exceeded 24 m. With wave forcing at close to the resonant period, vertical response would be in quadrature with the surface elevation, resulting in wave impact elevations even higher on the superstructure of the platform than had been observed with the same wave heights at lower periods. During the same storm, a pipe laying barge secured in a nine-point moor designed for 18 second maximum wave periods, broke all of its moors and went aground at San Francisco with damage exceeding $10 million. Seymour et al. (1984) shows that the global climate anomalies associated with the El Nino and Southern Oscillation phenomena can be expected to produce periods longer than 22 seconds many times each century. Therefore, the resonant period of 25 seconds which appeared safe in the 1960's might be replaced today by a much larger value for operations in the North Pacific.

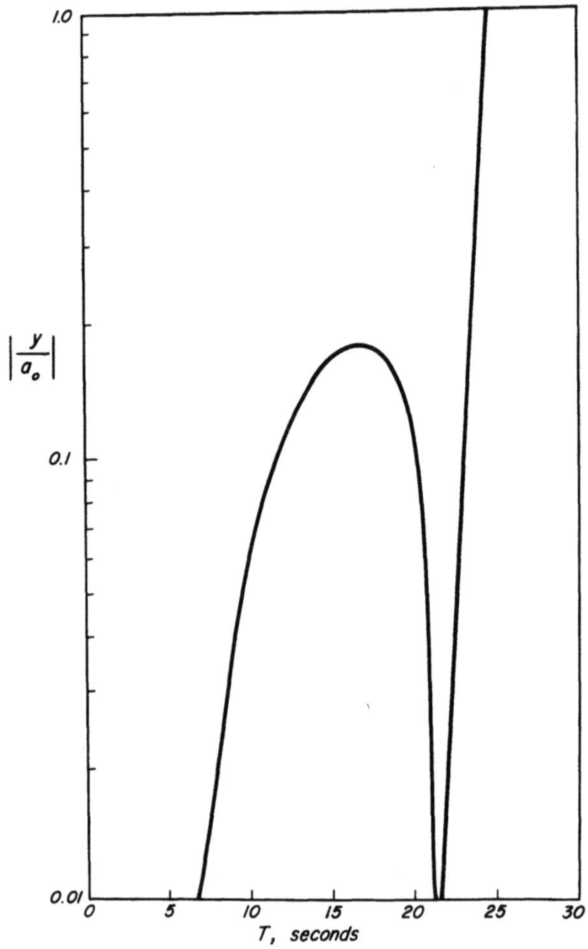

Figure 2. Response Characteristics of FLIP

DEVELOPMENT OF STABLE DRILLING PLATFORMS

During the last decade or so, exploratory drilling for petroleum has moved out of protected coastal waters into very deep water. Particularly in the North Sea operations, this has resulted in the necessity for operations under wave conditions that would have been unthinkable only a few years earlier. Exploration implies relatively short intervals on station and economic pressure to reduce the time required to move to the next site. Multiple-column stabilized structures could have met the stability requirements for drilling in large waves without question. However, in the vertical position their draft would have precluded them from almost all harbors and their towing resistance would have resulted in slow and expensive movement between jobs. Flipping such a multiple-legged structure into a towable configuration is certainly not straightforward. Therefore, development followed a different course.

The universal result was the well known semisubmersible platform. In this configuration, the long axes of the major displacement elements are horizontal, rather than vertical. When towed, the unit is deballasted so that these large horizontal, cylindrical hulls are at the surface, much like FLIP in its towed attitude. For stability, the hulls are ballasted down and the waterplane is broken by a series of vertical struts which offer only small response to the buoyant forces of passing waves. However, the horizontal area acted upon by the pressure field remains very much greater than would be experienced with the axes of the displacement cylinders vertical. In short, the stability - particularly to heave - was sacrificed to some extent in order to achieve mobility.

MULTI-LEGGED COLUMN-STABILIZED PLATFORMS

In 1970, just ten years after the initiation of the FLIP development, a second major program was initiated at SIO by one of us (F.N.S.) The goal of the Stable Floating Platform (SFP) project was the development of the technology required for large platforms capable of carrying substantial payloads or spanning large areas. The first approach investigated anticipated building catamaran type modules made of spar buoy pairs which could be towed to the station in the horizontal mode, flipped to the vertical, and coupled together. As a specific example, a detailed design was produced for a four-legged research platform made from two modules, forming an approximate square about 30 m on a side. The draft was to be 85 m and the payload about 500 tons. There was to be a 12 m clearance above the waterline to allow the free passage of 24 m waves. The design was to employ post-tensioned concrete to the maximum possible extent.

There were substantial differences from FLIP. First, the use of concrete would double the primary structural weight, compared to steel. Second, the payload and the height of the center of gravity were proportionately much greater. The configuration chosen initially was a catamaran barge, shown in Figure 3. Two of these modules would be towed to the site and flipped to the vertical. They would then be joined at their upper ends to form the four-legged SFP. This concept was tested at 1/100 scale in a wave tank. A 1/8 scale model was successfully flipped 26 times in San Diego Bay.

Following this phase, attention was directed toward large area platforms capable, for example, of performing as airfields. It was noted that a barge, then in service carrying pipe for the Alaska pipeline, was capable of carrying 12 columns the size of FLIP (900 tons, 6 m diameter) at 9 knots with a 3400 HP seagoing tug. With this in mind, the concept was developed of carrying a number of supporting columns to station in a group and assemblying them as a large stable frame on which deck sections could subsequently be placed (Figure 4.)

In this concept, the barge used for transportation would be "flippable" with its payload aboard. The barge itself would be incorporated into the overall structure and would contain the necessary machinery and logistical support functions for both initial assembly and subsequent utilization. Because the heave response of both the barge and the columns (after flipping) would be both small and about the same, the problems of assemblying the total structure in a seaway would be minimal.

The columns were to be joined with rigid members - one above the water and one near the base - for each connection. Diagonal guys were to provide desired rigidity. The rigid members would be attached to the spars as part of the barge deck load.

In order to investigate the operational problems, a 1/8 scale model barge and two columns with their associated struts were built and operated at sea off San Diego, CA. The structure was successfully

Figure 3. Catamaran Barge During Flipping Test.

assembled in the scale equivalent of 5 m waves, using pre-rigged lines, and with small craft performing the tug functions. Figure 4 shows the structure fully assembled at sea.

As part of the SFP project, design studies were performed on a 150 m by 1.8 km floating platform which could be used to construct an airport. Two such platforms could be arranged in a cruciform shape to allow for random orientation of the platform relative to the wind. The leg spacing was to be 60 m in the transverse and 60 or 100 m in the longitudinal direction. The other design variables considered included both reinforced concrete and steel columns, simple cylindrical columns and columns with stepped diameters, and drafts of 45, 60 and 80 m.

The deck clearance above the waterline was increased to 15 m. The center strip of the surface was designed for the landing loads imposed by a C-130 cargo aircraft. Other areas were designed to standard barge and drill platform requirements. Moderately high strength steel was proposed for the deck structure to reduce weight.

The total platform cost was found to be less with steel columns than with concrete, at any draft. Columns without stepped diameters were found to range from no difference to 20% less than those with steps, depending upon draft and column spacing.

The lowest cost system, in 1973 dollars, was estimated to be about $500 per sq.m of deck area. This is the constructed cost, only, and contains no allowance for towing or assembly costs.

Figure 4. Ocean Test of 1/8 Scale Three Leg Platform

The Stable Floating Platform program demonstrated several important concepts.

1. Standard seagoing cargo barge technology can be readily adapted to carrying multiple column elements rapidly to the assembly site and to erecting them into the vertical mode.
2. Multiple-legged column-stabilized platforms provide exception stability to heave forcing.
3. Platform elements can be reliably joined at sea under realistic service conditions of wind, waves and currents to form structures of unlimited size.

PROSPECTS FOR THE FUTURE

The single column stable platform, as epitomized in FLIP, provides superb stability and good mobility, but a severely restricted deck area. The semisubmersible platform meets needs for moderate deck areas, for short periods of time on station, and is capable of being moved to any location on the globe. It sacrifices stability to some extent to achieve this mobility. It seems reasonable to reconsider the simplicity of the column stabilized mode for very large deck areas in applications where extreme mobility is not a requirement.

The original impetus for the SFP project included aircraft servicing facilities in the deep ocean, far from land. There have been many studies in Japan and in the United States concerning the advisability of constructing major airport facilities offshore to minimize noise, reduce airspace conjestion, improve safety, and release valuable coastal land employed for buffer zones to more intensive uses. In many locations, such as the Pacific coast of North America, the narrow shelves result in deep water relatively close to shore. Floating structures may well prove to be very economical compared to bottom resting systems in these applications.

Recent exploration has disclosed the existence of potentially very valuable mineral resources in deep water. As opposed to the manganese nodule discoveries of 25 years before, some of these findings are within the Exclusive Economic Zones of Pacific nations. With assurance of national sovereignty, the potential of these resources is being evaluated agressively. If they are determined to be of mineable quantity, their exploitation will be given serious consideration. One restriction that might exist is the onshore environmental impacts of massive movements of ore and tailings to and from island locations that depend upon natural beauty to attract tourism. A solution is processing of the ores at sea, with direct return of tailings. Economical processes require very large areas. Many employ gravity separation and cannot tolerate the vertical accelerations of conventional ship forms. A very large stable floating platform, assembled in the operating area, would serve as a central processing plant converting ores to high value concentrates for direct shipment to markets. Such a facility could be moved occasionally, if the need arose. It might be possible to position it relative to known resources such that it could be utilized in a single location throughout its useful lifespan. In many locations where mineral crust deposits have been located, the thermal resources for ocean thermal energy conversion (OTEC) exist. A recent study [see Marine Board (1982)] found that all of the technology was at hand for a floating OTEC plant except the power cable connection to shore. Using the power on board a floating processing plant was recognized to be presently technically feasible, if it could be economically justified.

REFERENCES

Earle, M.D., K.A. Brush, and G.D. Hamilton: 1984. "High-Height Long-Period Ocean Waves Generated by a Severe Storm in the Northeast Pacific Ocean during February 1983," Journal of Physical Oceanography, Vol. 14, No. 8, pp. 1286-1299.

Fisher, F.H. and F.N. Spiess: 1963. "Flip - Floating Instrument Platform," The Journal of the Acoustical Society of America, Vol. 35, No. 10, pp. 1633-1644.

Marine Board: 1982. "Ocean Engineering for Ocean Thermal Energy Conversion," National Academy Press, Washington, D.C. 69 p.

Regier, L.A., and R.E. Davis: 1977. "Observations of the Power and Directional Spectrum of Surface Waves," Journal of Marine Research, Vol. 35, pp.433-451.

Rudnick, P.: 1967. "Motion of a Large Spar Buoy in Sea Waves." Journal of Ship Research, Vol 11, No. 4, pp. 257-267.

Rudnick, P. and R.W. Hasse: 1971. "Extreme Pacific Waves, December, 1969," Journal of Geophysical Research, Vol. 76, No. 3, pp. 742-744.

Seymour, R. J., R. R. Strange III, D.R. Cayan, and R. A. Nathan: 1984. "Influence of El Ninos on California's Wave Climate," Proceedings, 19th International Conference on Coastal Engineering, Houston, Texas, September 3-7, 1984. 1984.

Spiess, F.N.: 1968. "Oceanographic and Experimental Platforms" in Ocean Engineering, J. Brahtz, Ed., John F. Wiley, New York, pp. 553-587.

Design and Experiments of a Concrete Barge for LNG Storage

K. Noguchi[1] and R. Kobayashi[2]

Concrete Barge Section[1] and Offshore Structures Design Team[2], Technical Development Department, Taisei Corporation, Nishi-Shinjuku, Shinjuku-ku, Tokyo, 160-91 Japan

INTRODUCTION

Natural gas is a pure energy that is stored in the earth, its deposits are almost comparable to those of petroleum, and it is a valuable resource for mankind. As is the case with petroleum, offshore recovery of natural gas is flourishing markedly these days. For a small gas field with less deposits, the liquefaction and storage of recovered gas either on the sea or beach near to that gas field is possible: the famous example of this is the LPG storage barge at thc Arjuna oil field off Java. When storing LNG, the storage facility demands a high standard of safety with respect to the prevention of pollution and disaster. This paper describes the design of offshore storage of LNG by using concrete barge and investigates its safety experimentally by using members and a model barge. As a result of these experiments, this basic design concept has been officially approved by NIPPON KAIJI KYOKAI, the Japan shipping register association.

1. A SUMMARY OF THE PLAN

The present plan is for building domestically an LNG storage barge, towing it up to the destination where it will be moored or grounded and secured backfilled on the seabed. In other words, it functions as a barge while afloat and during towage and, therefore, as a floating tank or semi-underground tank.

The LNG storage facility is shown in Fig. 1 and 2. It mainly consists of a prestressed concrete barge, 6 tanks and insulation. The principal dimensions of the barge are 175m length, 52m breadth and 22.5m depth with double hull construction. The tanks are semi-membrane type and made of stainless steel. Its total storage capacity is 110,000 Kl. The insulation made of polyurethane foam is installed with single-layer joist.

Design consideration involves three stages: building, towage and operation after being settled. In operation, cargo loads are supported by the concrete barge structure through the insulation. The spaces between inner and outer hulls are heated not to be cooled below a temperature 0°C. The inner hulls have the structural strength which will act as secondary barriers in case of the failure of tanks.

Fig.1 Bird's-eye view of LNG storage barge

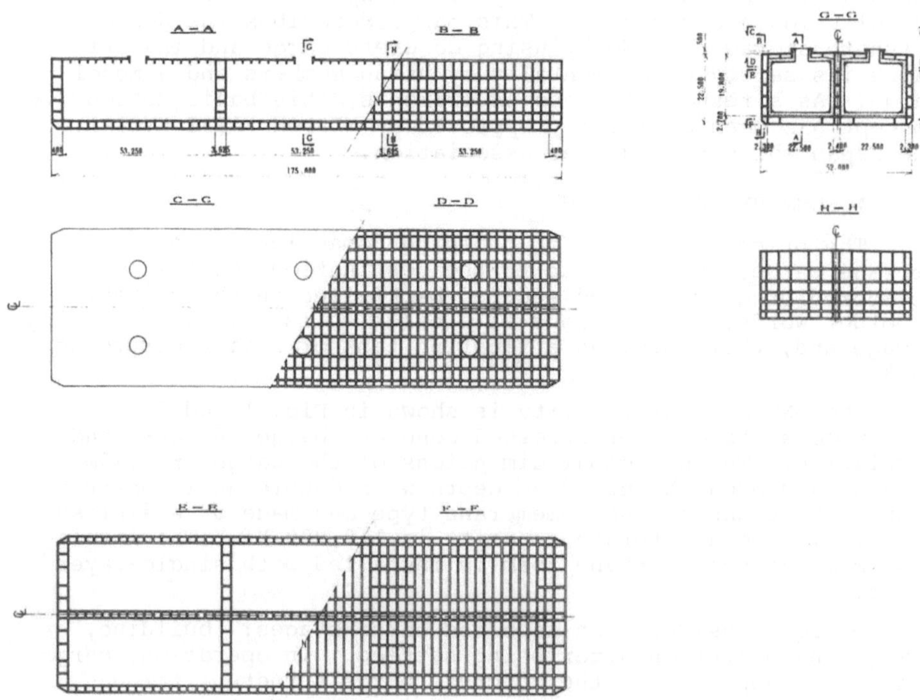

Fig.2 Outline of concrete barge

2. TECHNICAL PROBLEMS

Mechanical characteristics of concrete are excellent on cryogenic condition. "Provisional Rules for Prestressed concrete barges" established by NIPPON KAIJI KYOKAI sets forth that inner hulls of a concrete barge may be used as secondary barriers for LNG tank as long as having liquid-tightness and permitting no leakage of the harmful quantity of the gases.

The water-tightness and liquid-tightness of concrete differ largely in respect to the concrete material and cracking. Material-wise, water penetration can be prevented by densely compacting concrete of small water-cement ratio.

In order for the inner hull to function as a secondary barrier, it is necessary to assure the strength of the structural members and to retain the liquid in the secondary barrier at the time of liquid leakage through the tank.

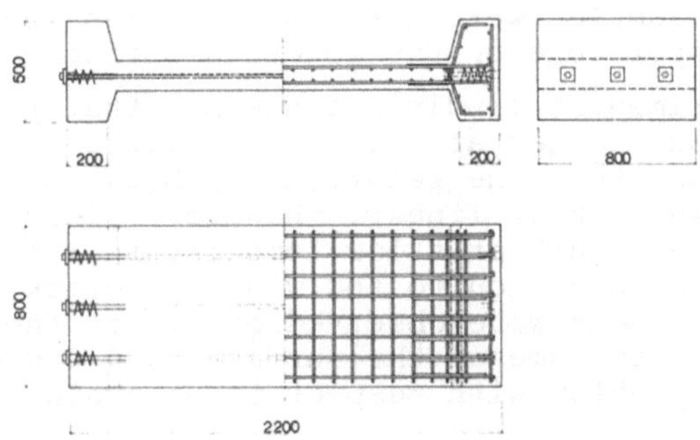

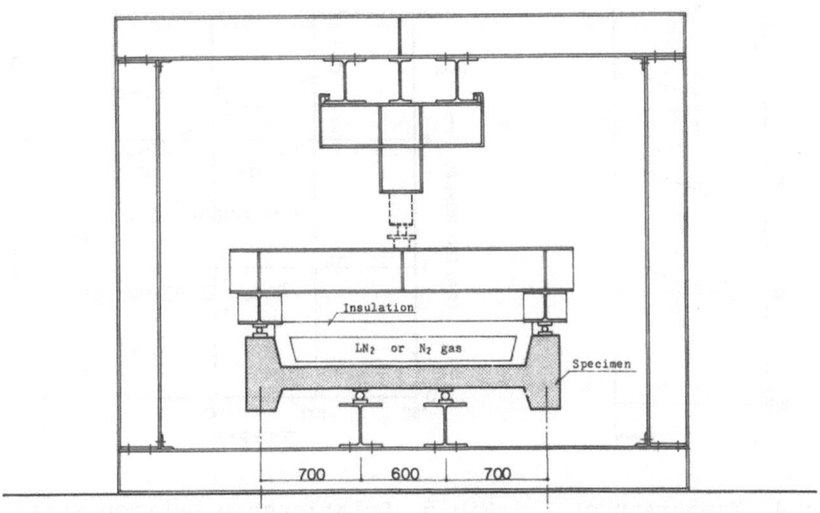

Fig.3 Cooling and loading apparatus for PC panel

In this connection, it was necessary to confirm the bending strength characteristics of prestressed concrete members at the time of one side being cooled, and also to consider the probability that damage to concrete members caused by a collision from outside may affect the inside. For the purpose of clearing up these technical problems, some experiments using prestressed concrete members and a model barge were carried out.

3. EXPERIMENTS ON BENDING STRENGTH OF PRESTRESSED CONCRETE PANELS AT LOW TEMPERATURE

In this experiment, in order to check changes in the strength characteristics of the panel which forms the double hull, when cooled due to liquid leakage through the tank or when thermal insulation material was damaged, one side of the prestressed concrete panel was cooled, and loaded after the panel temperature came to a constant level, to obtain the relationship between the temperature and bending strength characteristics of the panel.

Four samples were used and each was loaded with a different temperature on the concrete surface.

Experimental findings confirmed that the panel's bending strength was larger at a low temperature than at normal temperature and that the relationship between temperature and strength was almost linear. However, since cracking tended to be of larger width at a low temperature, it would be necessary to give consideration to the width of cracking. Also, fine mesh-like cracks were observed partly at the time of cooling, but limited to some of the surface, and would not constitute a serious problem with respect to strength.

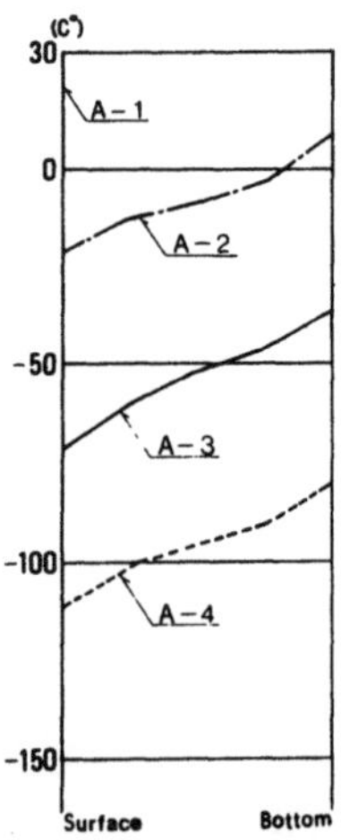

Fig.4 Temperature distribution at center cross section

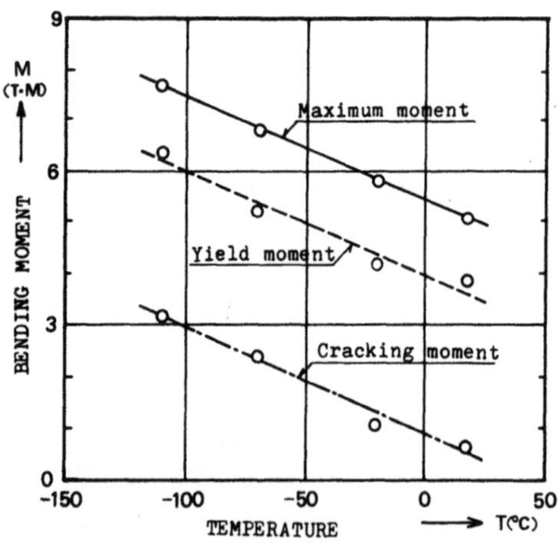

Fig.5 Relationship between strengths and surface temperatures

4. EXPERIMENT ON REINFORCED CONCRETE PANEL'S IMPACT RESISTANCE

This experiment was intended to clarify how the reinforced concrete panels supported their periphery would behave when some rigid object collided against the panels, and to obtain the limit of impact resistance.

Specimens were square, reinforced concrete panels, each measuring 125cm on each side and 10cm and 15cm in panel thickness, with reinforcing steel bars arranged in double layers and equidistant longitudinally and transversely. Their number totaled 17 units.

As the extent of damage was checked mainly by visual observation and it is impossible to be explicit as to the extent, the levels of damage are here defined as follows:

Slight damage ..... bottom surface suffers cracks, but no separation is observed.

Half-damage .... A considerable extent of bottom surface separation is observed, but almost no water leak.

Serious damage .... Bottom surface separation is great with water leak.

Fig.7 shows the center cross sections of panels having

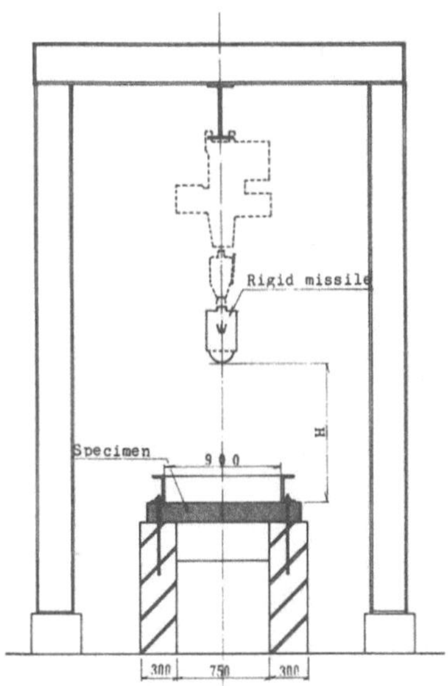

Fig.6 Impact experimenting apparatus

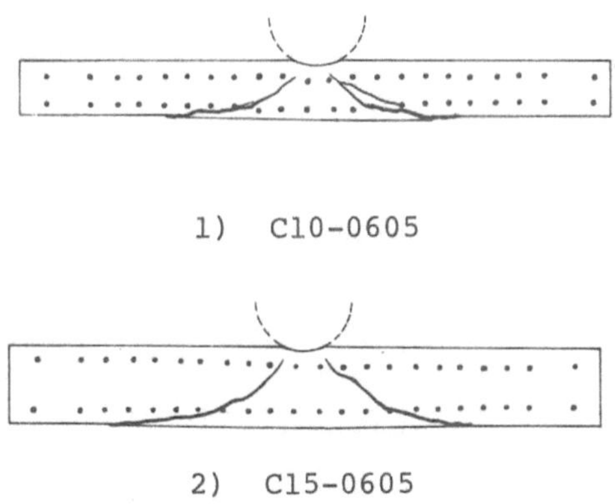

Fig.7 Cross sections of after impact experiment

the different panel thickness but arranged with reinforcing steel bars in same patterns. Both panels suffer ring-like cracks on the bottom surface, and one having a 10cm panel thickness suffers a bottom surface separation which is about to take place, but the cracks did not develop to the surface where collision impact occurred.

Fig.8 shows the relationship between the weight of the impact object, impact speed and damage. Seeking the border between slight damage and half-damage from the graph, a curve will be represented by $WV^{2 \cdot 3}=2.2 \times 9.8$.

5. EXPERIMENT FOR SAFETY CONFIRMATION OF RESTRESSED CONCRETE LNG STORAGE BARGE

The object of this experiment was to clarify through an experiment the problems remaining unsolved from the viewpoints of safety and disaster control concerning the concrete LNG storage barge, such as cracks in the concrete barge and the effect of concrete permeability as affected by the cracks on the tank and secondary barrier, and to prove the safety of the concrete LNG storage barge.

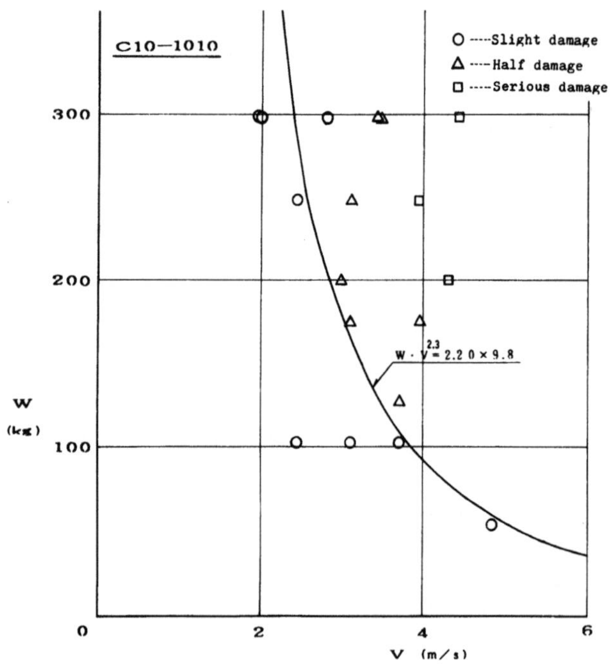

Fig.8 Impact damage of reinforced concrete panels

As shown in the figure, a model barge used for experiment was constructed of prestressed concrete with outside dimensions of 5m x 1.8m x 1.5m and with prestress of 25 kg/cm^2 applied to only the longitudinal direction. This model was set in an experimental basin simulated a marine environment.

The experiments were carried out on the following conditions:

1) The primary tank fails and the liquified gas leaks to the secondary barrier.
2) Thermal insulation material is partially damaged.
3) A floating object collides with the barge.

In these experiments, liquid nitrogen was used as a heat coolant. The concrete from which the thermal insulation material was partially removed was cooled down to -98.5°C at the center of the cross section, but no anomaly was observed except fine cracks on the part of inside surface of the concrete which came into direct contact with the liquid nitrogen.

Even with the occurrence of shearing cracks, no rapid drop in the strength was observed. A drop in airtightness was caused but to an extent that the drop did not affect the liquid tightness at all. When the external forces were removed, almost all cracks closed themselves, and the safety of barrier was confirmed.

Cracks due to impact occurred in the intermediate floor plates connecting the outer and inner hull, but the surface of the inner hull showed no cracks even though 340 μ of dynamic strain was measured. The safety of the double hull was thus proved.

Findings from the airtightness tests and the measurements of the oxygen density in the voids indicated a drop in airtightness of the barrier, but no problem at all with liquidtightness. Referring to the NK's rules, "(1) The boundaries of the tank holds are not to be ruptured at the low temperature in case of the failure of tanks assumed, and are to be as

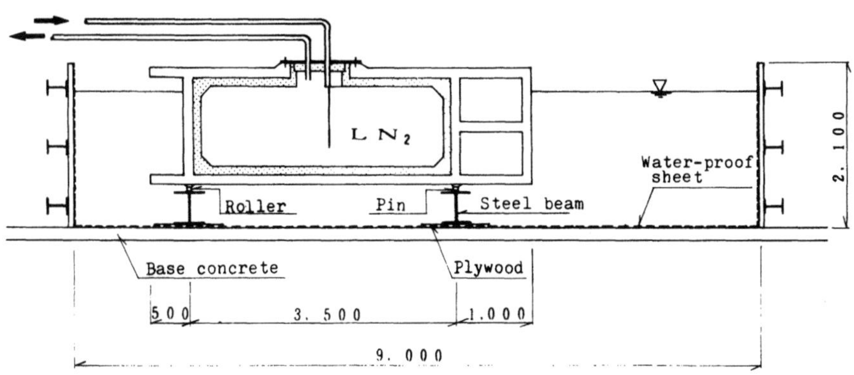

Fig.9 Model barge and experimenting water basin

the structure having liquid tightness and permitting no leakage of the harmful quantity of the gases. (2) It is to be confirmed by the model test for the boundaries of the tank holds to satisfy the preceeding (1)".

In the light of the above provision, the amount of leak was small with this experiment, and the provisions could be met.

6. CONCLUSION

Findings from the present experiments showed a drop in the concrete secondary barrier's airtightness and the necessity to quantitatively elucidate the extents of airtightness according to various cases. Also, a small amount of gas leak could be dealt with by a sufficient monitor and ventilation system. But in order to improve safety, it is effective if the inner hull wall is lined with a very tough material. It is presumed that fluctuation of the thermal stress at the time of the thermal insulation material being notched off partially covers a large area, and it would be necessary to do, in the future, a detailed thermal stress analysis with additional consideration given to concrete cracks. Our design concept was approved by NIPPON KAIJI KYOKAI on the precondition that the data from the experiments will be submitted and a detailed analysis will be made.

ACKNOWLEDGEMENT

We extend our sincere appreciation to the members of the Hull Department, NIPPON KAIJI KYOKAI who gave us valuable advice and to the Japan Ocean Industries Association which gave us a precious opportunity to conduct experiments by means of a model barge.

REFERENCE

NIPPON KAIJI KYOKAI; Provisional rules for prestressed concrete barges. Tokyo, 1975

Flexibility of TSG Platform to Project Requirements

E. Gnone

Marine Structures Department, Tecnomare S.p.A., Venezia 2091-30124, Italy

ABSTRACT

A study has been conducted to evaluate the applicability of the Tecnomare Steel Gravity (TSG) platform concept for developing offshore oil fields. This evaluation includes both the technical feasibility and economic attractiveness of the concept.

To aid in evaluating the question of economic attractiveness, a computer program was written to perform preliminary sizing of a TSG platform and to prepare an estimate of weight and towing requirements. Using this program, a parameter study was performed and a set of weight estimating curves was developed. Using these estimating curves in combination with up-to-date material and construction costs, a designer or operator can quickly estimate the cost of a TSG platform for a range of environmental conditions and operating requirements.

INTRODUCTION

The steel gravity platform (TSG) concept developed by Tecnomare has been the object of increased industry interest during the past few years. The concept has economic advantages over steel piled platforms for a number of applications due to its minimal offshore installation and hook-up time, integral oil storage capability, and ease of removal. In addition, it can have advantages over concrete gravity platforms in that a sheltered, deepwater fabrication site is not required and may be easily installed over a predrilled wells template.

To date, four light TSG platforms have been installed in the Loango Field off the Congo and one larger TSG in the Maureen Field in the North Sea.

THE TSG CONCEPT

Platform Description

The TSG is shown in Fig. 1. The main structural components of the TSG structure are three steel buoyancy cylinders connected by a triangular spaceframe, an hexagonal framed tower that is supported by the triangular spaceframe, and three foundation bases (1).

The cylinders provide towing buoyancy, naval stability, and oil storage capacity. Each cylinder has two or more hemispherical domes, to provide compartmentation for the towing and stability requirements. The cylinders are ring-stiffened shell structures with the rings located on the interior or exterior of the cylinders.

The triangular spaceframe is a tubular, highly redundant three-dimensional structure connecting the tower to the cylinders and to the foundation bases. The main structural joints are reinforced by increased chord and stub thickness.

The framed tower is a tubular three-dimensional structure, composed of vertical or slanted braced plane faces, internally stiffened by horizontal bracing. The exagonal shape can be modified into a rectangular or square shape at the upper level to accommodate a rectangular or square deck arrangement.

The foundation of the TSG structure consists of three circular steel bases, one beneath each cylinder. Stiffened skirts extend from the bottoms of the bases concentrically around the perimeter. Concrete or heavy aggregate is added to the interior of the bases to provide ballast weight for towing in addition to on-bottom stability.

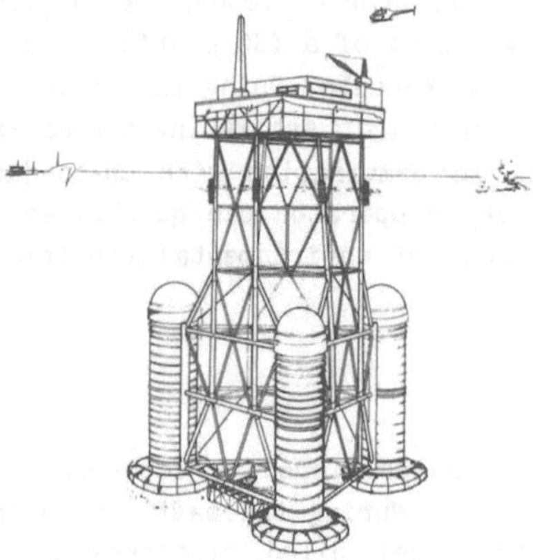

Fig. 1. TSG configuration

Fig. 2. Maureen platform

The TSG structure can be adapted to support a wide variety of deck configurations. These include the conventional modular deck used primarily on steel jacket platforms and the integrated deck configurations that have seen extensive use with concrete gravity structures in the North Sea.

Platform Constructions

The TSG platform has a configuration that will permit prefabrication in large structural blocks that can be assembled using several procedures.

Among these are:

. Construction and assembly in dry dock.

. Construction on land and assembly in wet dock.

These two procedures can be used together to better fit the available yard and harbor facilities and to reduce platform completion time (2).
Using dry dock fabrication, normal construction methods for tall structures can be used. After the structure is assembled and equipped, the dry dock is flooded and the structure is towed to the mating or installation site. This technique has been used for the Maureen platform (see Fig. 2).
The wet dock fabrication procedure allows for the floating assembly of the main structural blocks. The structural components are towed to the wet dock and assembled by a heavy lift barge and/or mating barge. This procedure reduces the heights of the lifts and the draft requirements near the dock. The wet dock procedure was used for the construction of the Loango platforms (Fig. 3).

Transportation

When platform fabrication has been completed with the appropriate equipment and the deck support frame and/or deck has been installed, the TSG platform is ready for transportation to the installation site (Fig. 4)
The towing fleet is composed of a proper number of tugs with adequate power for the selected towing speed. Normal towing speeds range from 1.5 to 3 kt. The required power for the tow is generally achieved with the minimum number of tugs in the available tug classes ranging from 5000 to 25,000 hp (1). The structure is designed to withstand damage in the upper part of each cylinder in the water plane zone. For long distance transportation, special compartmentation is implemented to guarantee stability in the event of cylinder damage.

Installation

One of the key features of the TSG platform is its ability to be installed without preparation of the seabed. The platform can be positioned with the assistance of the towing spread. In most environments, special mooring systems are not required (1). The platform installation is performed by means of a water ballasting system located on the platform.
The control of the immersion operations is automatically performed by the water distributor which is a mechanical self-stabilizing device. This system guarantees the vertical position of the platform during immersion.
Possible differential penetration of the foundation skirts may be controlled by properly water ballasting the cylinders. The effect of soil uneveness can be minimized by underbase grouting, performed soon after platform setting. The complete platform installation can be performed in a very short period of time. For each of the four Loango platforms, 23 to 34 hr. were required from platform positioning to final setting.
The TSG can be installed over a seabed template accommodating predrilled wells. This has been the case of the Maureen platform where the final installation accuravy has been 5 cm (3) and the duration was 48 hr.

Fig. 3. Loango platforms

Fig. 4. Platform ready for tow

THE PARAMETRIC STUDY

General

The primary purpose of the parametric study is to determine the effect of design criteria and environmental data on the preliminary design characteristics of the TSG platform. This would lead to the development of estimating curves that could be used to predict the structural weight, towing requirements, ballast weight and ancillary equipment requirements of the platform.

A specially developed computer program allows the preliminary evaluation of those characteristics (geometrical, structural, and operational) that have the most effect on the sizing of a TSG concept (4), (5) by means of simplified calculations of naval stability, environmental forces, and foundation design.

Input data

One of the most important features of the program is its input flexibility. The program has the ability to operate with limited data, when incomplete information is available, and when there are specific project requirements.

Program Description

After reading the input data, the program calculates a minimum platform width (center-to-center spacing of the bases) based on the water depth. Beginning with the minimum platform width, the environmental forces under storm conditions are calculated. A naval stability and foundation

check are made for the storm conditions. Ballast requirements are computed. When all of these requirements are satisfied, the foundation check is repeated under the operating conditions in order to verify the overturning stability just after installation. If any modifications are made during these checks, the procedure starts again at the beginning of the detailed analysis. After the last check is made, the solution is stored for the final total weight comparison. The routine increases the platform width and the procedure continues.
When this procedure has been completed for all values of the platform width, the minimum total weight (or cost) solution is determined and another, more accurate weight estimate calculation is repeated for the minimum weight (or cost) solution (1). Additional calculations pertaining to the cathodic protection system, installation forces, towing requirements, hydrostatic properties, static stability, and foundation characteristics are performed for the minimum weight (or cost) solution.

Output Results

In the complete form, the output consists of weight and cost breakdowns by component, component dimensions, and ancillary equipment and ballast requirements for each solution step (base spacing). Finally, the complete characeristics for the minimmm-weight solution are presented.

Procedure for Parameter Study

The purpose of the parameter study is to determine the effects of design criteria and environmental conditions on TSG platform characteristics (total structural weight, ballast requirements, base spacing). To simplify the presentation and subsequent use of the results, the environmental conditions are divided into three categories. These three categories represent areas of mild, moderate, and severe environmental conditions. Geographical areas associated with these environmental conditions are given in the following table.

Environment	Geographical Area
Severe	North Sea
Moderate	Gulf of Mexico, Argentina, Bass Straits of Australia
Mild	West Africa, Indonesia

The key parameters affecting platform characteristics include water depth, towout load, soil conditions, maximum operating payload, environmental conditions (design wave height), and storage requirements. To analyze the effect of one variable (e.g. water depth) on the platform characteristics in each base case, the single variable is changed while the rest are held constant. For example, to analyze the effect of water depth on total structural weight in the mild environment, the program was run four times once each for water depths of 70 m, 114 m, 160 m, and 180 m.

THE RESULTS

One of the most significant platform characteristics is the structural weight which is responsible for a sizable portion of the total platform fabrication cost. Much of the following discussion is directed to the influence of input parameters on platform structural weight.

Environmental Conditions

Environmental conditions greatly affect TSG platform characteristics, particularly structural weight. Figure 5 presents the base cases plotted with respect to water depth and structural platform weight.
The effect of environmental conditions on structural weight, as shown in fig. 5, is due primarily to design wave height. For the base cases, the design current velocity is less than 1.5 m/s. This current has little effect on the platform structural weight.

Water Depth

As expected, water depth is an important factor influencing TSG structural weight. Figure 5 shows the effect of water depth on structural weight for the base cases. As the height of the structure increases, the structural weight increases. In the mild environment, the effect is nearly linear as water depth decreases from 70 to 180 m. However, in the more severe environments, structural weight increases nonlinearly with water depth.

Towout Load/Storage

Cylinder dimensions are dependent on towout load (buoyancy requirements) and storage capacity. Figure 6 presents the towout load/storage relationship at different water depths for severe environments. The curves show structural weight and storage capacity as a function of towout load.
The following example illustrates the use of the curves from Fig. 6.
In a severe environment at a water depth of 90 m, a towout load of 13,300 tonnes would require a structural weight of 20,000 tonnes. The inherent storage capacity of the cylinders would be 64,000 m^3 (402,000 bbl).

Maximum Operating Load

The maximum operating load includes deck, facilities, expandable materials (drill pipe, cement, drilling mud, potable water and drill water). Figure 7 demonstrates the effect of maximum operating load on structural weight for three environmental conditions. As shown, the effect is larger for milder environments. For the severe environment, maximum operating load has little effect on structural weight due to the design storm conditions controlling the platform design. In general, changes in maximum operating load have smaller effect on structural weight than towout load.

Ancillary Equipment

The ancillary equipment for the TSG structure consists of the floatation system, conductors, risers, piping (including ballast and oil storage piping), boat landings, miscellaneous equipment (stairs, ladders), ballast piping, and oil storage piping (if internal storage is required).
Ancillary equipment requirements are mainly dependent on platform height, storage requirements, and subsea template requirements (1). Figure 8 shows the ancillary equipment requirements for the three environments as function of structural weight.
In general, ancillary equipment requirements increase as platform structural weight and environment severity increase.

Cathodic Protection

The computer programs calculates the platform wetted surfaces on the basis of historical design data for the framed structure, cylinders and bases of TSG platforms. Internal surfaces of the cylinders are calculated only in the case of oil storage requirements. A rough calculation of cathodic protection is performed for naval stability checks. Figure 9 shows cathodic protection requirements for three environmental conditions based on structural weight. As structural weight increases, the weight of the cathodic protection system also increases.

Towing Requirements

The horsepower requirements for the tow of the completed structure depend on the structure size, towing speed, wind and current. The program calculates optimal towing draft and power requirements for four different towing speeds and three towing drafts. Figure 10 shows the horsepower requirements for the severe environment at water depths of 114 m for different cases of deck load and wind speed.Horwepower requirements for the platform increase significantly as towout loads (and structural weights) increase. Towing speed (current can be included in towing speed), has a large effect on the power requirements, especially at towing speeds greater than 1.5 m/s.

Soil Conditions

Soil conditions have a major impact on base size and skirt lengths. As the unconfined compressive strength q_u of the soil decreases, the structural weight increases only slightly until q_u falls below 10 tonnes/ m^2. Below q_u = 10 tonnes, structural weight increases rapidly as foundation components become larger. Figure 11 shows the structural weight requirements for the three environmental cases as a function of unconfined compressive soil strength.

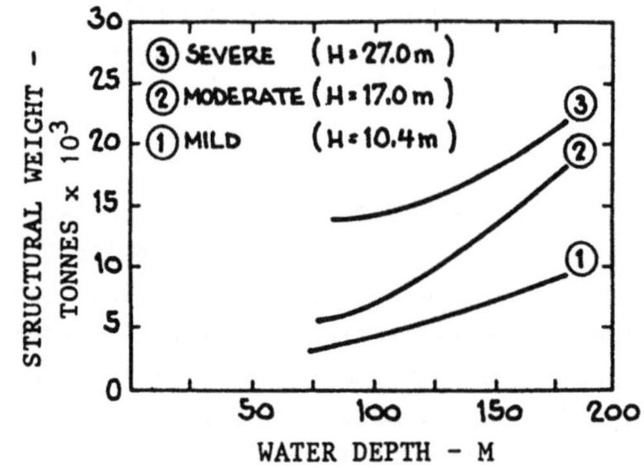

FIG. 5

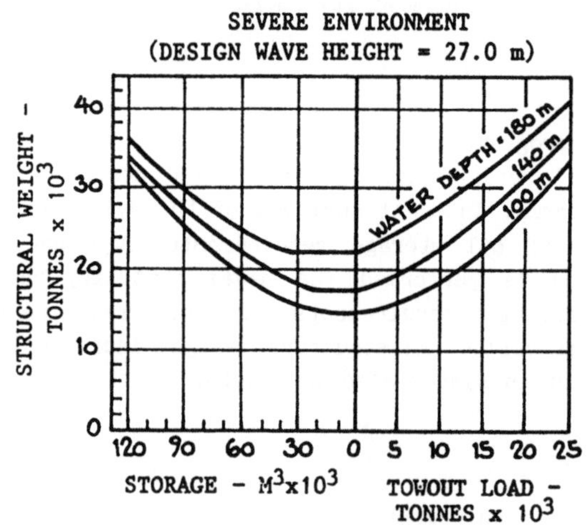

FIG. 6

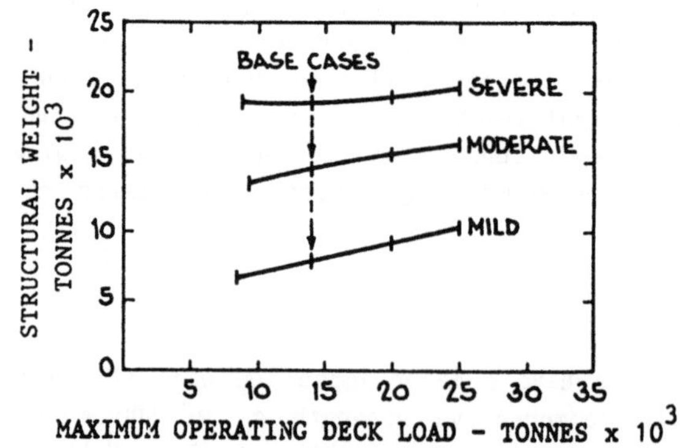

FIG. 7

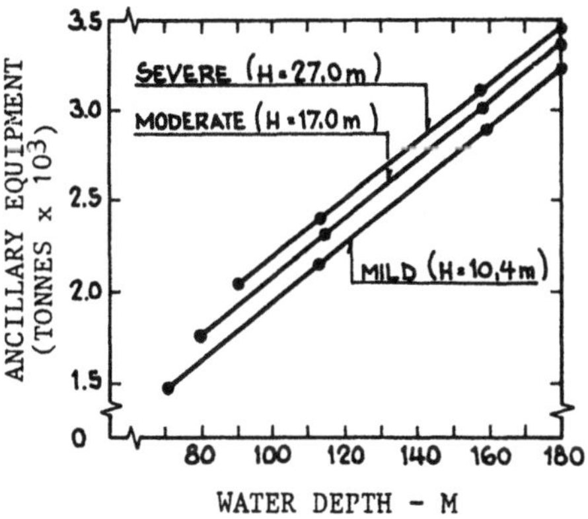

FIG. 8

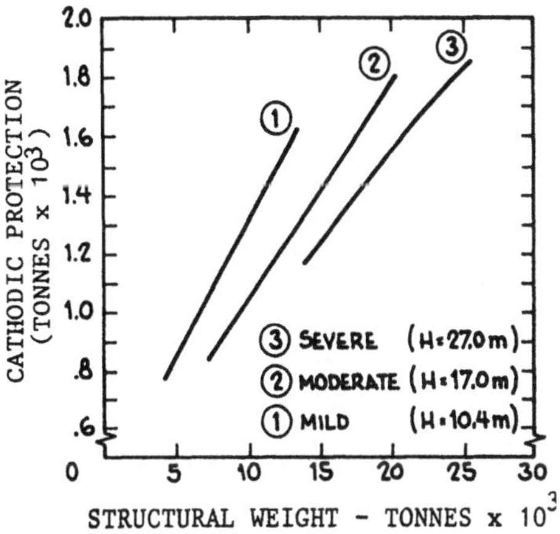

FIG. 9

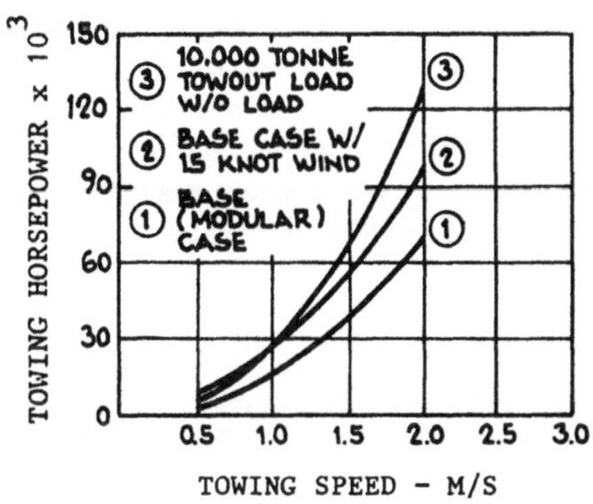

FIG. 10

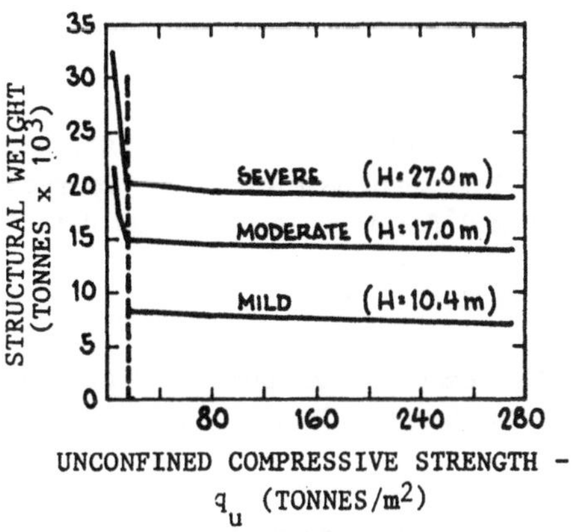

FIG. 11

CONCLUSIONS

Within the scope of this study the design storm wave height is the most critical environmental parameter affecting the TSG platform characteristics.

Large towout loads and storage requirements can significantly affect platform structural weight due to increased cylinder size and structural framing.

Water depth is an important factor in determining platform weight, especially in more severe environments.

Soil strength is not a significant factor in determining platform weight until the unconfined compressive strength of the soil falls below 10 tonnes/m^2.

Maximum operating load does not have a strong influence on platform weight, particularly for severe environments.

In general, more total ballast is required as water depth, towout load, and design wave height increase.

Towing speeed and towout load produce the largest increases in power requirements.

These parameter study curves should be adequate for project screening and development cost estimates where the information is limited to general location, field type, and unconfirmed estimates of field size. When more site-specific information is available (such as environmental surveys, reconnaissance-type soil borings, shallow seismic surveys) and a design basis has been developed, it is recommended that a computer analysis be performed for more accurate and detailed weight and cost estimates.

REFERENCES

1. "Tecnomare Steel Gravity Platform - The preliminary estimate of platform characteristics". Tecnomare S.p.A., Venezia - Italy, Sept. 1980

2. Lalli, D.: New concept of pile-less fixed platform for deep water drilling and production operations. Fifth Annual Offshore Technology Conference, Paper OTC 1916, Houston, 1973.

3. Gnone, E.: Maureen, the first steel gravity platform in the North Sea. C.I.S.M., Udine - Italy, Sept. 1983.

4. Alloni, C.; D'Agostino, A.; Priarone, R.: Design analysis and construction of the Loango steel gravity platforms. Eighth Annual Offshore Technology Conference, Paper OTC 2437, Houston, 1976.

5. D'Agostino, A.; Di Tella, V.; Gnone, E.; Sebastiani, G.: TSG - Integrated storage platform for early production in the North Sea. Twelfth Annual Offshore Technology Conference, Paper OTC 3880, Houston, 1980.

Study on the Fatigue Strength of Pressure Vessel Subjected to Internal Fluctuating Pressure

In the Case of 3% Saltwater as Pressure Medium

M. Kanno, M. Nagao and Y. Sato
College of Engineering, Nihon University, Koriyama, Fukushima, 963 Japan

ABSTRACT

The cylindrical pressure vessel of metallic materials was subjected to internal fluctuating pressure in saltwater, and fatigue strength of the materials was determined experimentally as the effect of saltwater. The cylinders used in this experiment are mild steel (S45C), stainless steel (SUS310S), aluminium (A5056), copper (C1100) and brass (C3604) metal. The cylinder was filled with 3% saltwater similar to seawater. Saltwater is used as a medium in addition to water, turbine oil and castor oil. Fatigue behavior of the cylinder in the saltwater is obtained and depends on the materials. Fatigue strength of copper and brass metal is the highest of all materials, but aluminium metal is the lowest of all. We explain the effect of saltwater on the fatigue strength of all materials. We relate corrosion behavior of metals to the free energy change (ΔG) by the Gibbs. Corrosion and free energy change of copper and brass metal are less than other materials.

INTRODUCTION

As the development of sea space machinery and mechanical elements will be used in seawater more than before, it is of great importance to investigate mechanical properties of materials of pressure vessels under saltwater [1].

The cylinder has rupture possibilities due to the effect of various environments [2,3]. The rupture strength of the cylinder under the influence of the environment such as saltwater, is particularly important under the fluctuating pressure rather than static pressure [4]. In this experiment, the hollow cylinder was subjected to internal fluctuating pressure, and the internal pressure medium was 3% saltwater in addition to water, turbine oil and castor oil. The cylindrical specimens are mild steel, stainless steel, aluminium alloy, copper and brass metal. Fatigue characteristics of the materials were determined experimentally for the effect of saltwater and so on. It seems that these data provide useful clues for the designing of machinery used in sea spaces.

EXPERIMENTAL PROCEDURES

Material and Specimen

Materials of hollow cylinders are ferritic-pearlitic structure steel (0.44%C), austenitic stainless steel (24.76%Cr-19.49%Ni), Al-Mg structure aluminium alloy (4.60%Mg), copper (>99.90%Cu) and brass (59.4%Cu-Zn). Mechanical properties of

Table 1. Mechanical properties of virgin materials at room temperature.

Material	0.2% proof stress $\sigma_{0.2}$ (MPa)	Tensile strength σ_B (MPa)	Elongation δ (%)	Reduction of area ϕ (%)	Vickers hardness HV
Mild steel (0.44%C)	σ_y=392	726	13	47	208
Stainless steel (24.76%Cr-19.49%Ni)	207	521	67	75	169
Aluminium alloy (4.60%Mg)	145	366	18	44	102
Copper (>99.90%Cu)	164	261	31	74	94.2
Brass (59.4%Cu-Zn)	184	438	29	43	132

σ_y : Yield stress

these materials are given in Table 1. Configurations and dimensions of the test specimen are shown in Fig.1. The specimen is a relatively thin cylinder of inner diameter d_1=10mm to outer diameter d_0=12mm.

The hoop stress σ_h acting on the cylinder subjected to internal pressure may be given for a thin cylinder,

$$\sigma_h = \frac{1}{\kappa_0 - 1} P$$

Where κ_0 is the ratio κ_0 of outer diameter d_0 to inner diameter d_1, $\kappa_0 = d_0/d_1$. The value of pressure (P) is maximum internal fluctuating pressure and the range of test pressure was from 20MPa to 90MPa.

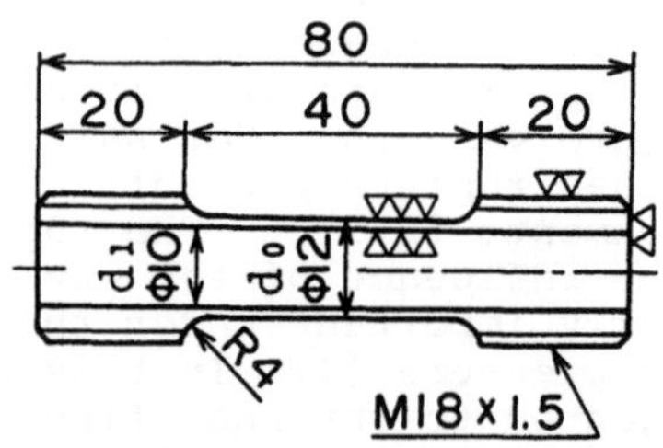

Fig.1. Configurations and dimensions of the test hollow cylinder (mm).

Pressure Medium or Test Environment

Various pressure media including saltwater and their kinematic viscosity are shown in Table 2. The viscosities of saltwater and water were less than other media, and the viscosity of castor oil was the largest of all. Viscosity ratio ν/ν_0 was also shown as the ratio of viscosity ν of a medium to that of water ν_0. Saltwater is 3 percent NaCl solution obtained by dissolving NaCl in water, turbine oil is industrial liquid lubricants of ISO VG 32 and castor oil is

Table 2. Test pressure media and its viscosity (40°C).

Medium	Kinematic viscosity (mm^2/s)	Viscosity ratio ν/ν_0
Water	0.7	1
3% saltwater		
Turbine oil	32	46
Castor oil	244	349

the reagent oil. The temperature of testing media ranged from room temperature to 40°C.

Fatigue Testing Apparatus

Experimental apparatus as shown in Fig.2 was used. In this figure, the specimen (a) was a hollow cylinder as shown in Fig.1, and from the piping (f) of the test apparatus the internal fluctuating pressure was given to the specimen. The pressure is so high that the cylinder presses the sealed packing (d) by the two disk (b) and (c). Thus it prevents leaking of the pressure medium. Air loophole (e) of left end side is provided so as to let air out at the exchange of a media, and during the test, the air loophole was sealed by the plug (g). Internal pressure measurement in the specimen was made using a strain gauge. Cycle of the pressure amplitude was constant f=250 cpm. We continued the test until $N_f=1X10^6$ cycles. The fatigue test was carried out in the stress range above fatigue endurance limit. Fatigue strength of cylinder in saltwater is evaluated as the number of failed cycles.

RESULTS OF EFFECT OF PRESSURE MEDIUM (S-N CURVE)

Mild Steel (S45C)

S-N curves of mild steel for different pressure media as

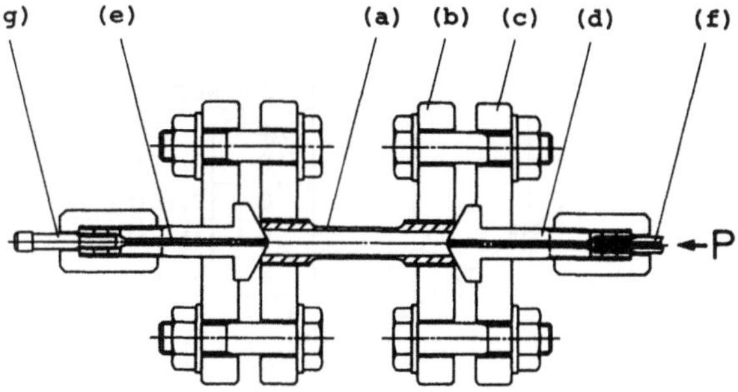

(a) Specimen
(b) Disk
(c) Disk
(d) Sealed packing
(e) Air loophole
(f) Pressure piping
(g) Plug

Fig.2. Fatigue testing apparatus of internal fluctuating pressure.

well as saltwater are shown in Fig.3. As is evident from Fig.3, fatigue strength of the mild steel for saltwater and water was less than other media. It has been known that fatigue strength in saltwater is inferior to that in water [5], but in this experiment no apparent difference between the two media appeared. The stress difference($\Delta\bar{\sigma}_h$) between the fatigue strength for saltwater and castor oil as a given number of cycle is about $\Delta\bar{\sigma}_h$=50MPa. Turbine oil and castor oil were shifted respectively, and as given in Table 2, fatigue strength increases with increase of viscosity.

Stainless Steel (SUS310S)

S-N curves for stainless steel are shown in Fig.4. Fatigue strength of this cylinder is given by two S-N lines, one is for saltwater and water, the other for turbine oil and castor oil. The fatigue strength for fats and oils is larger than the fatigue strength in saltwater and water, but there was no apparent effect of viscosity grade of the turbine oil and castor oil. The fatigue strength under the pressure medium of saltwater has a tendency to be similar to that of S45C.

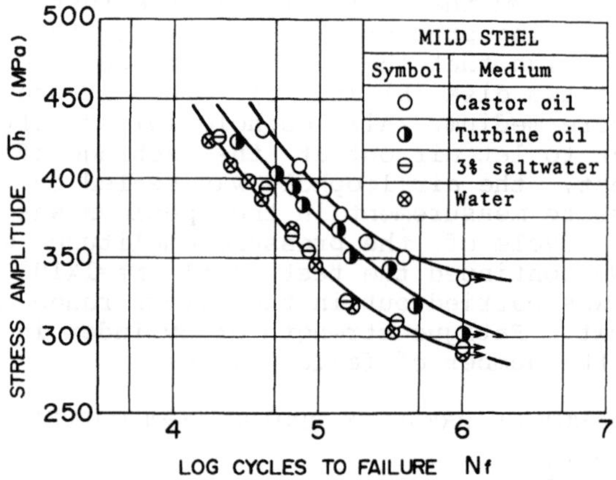

Fig.3. Relation between σ_h and N_f of mild steel in various pressure media.

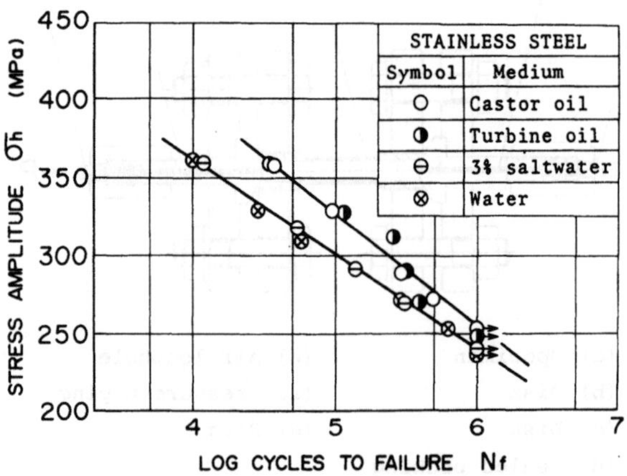

Fig.4. Relation between σ_h and N_f of stainless steel in various pressure media.

In stainless steel, the stress difference was about half that of mild steel, $\Delta\bar{\sigma}_h$=24MPa.

Aluminium Alloy (Al-Mg Structure)

The S-N curves of aluminium cylinder under the pressure medium are illustrated by the three lines in Fig.5. The fatigue strength for turbine oil and castor oil is qualitatively the same as that for SUS310S. However, the fatigue strength for saltwater is less than for water. The stress difference between the oils and saltwater is about $\Delta\bar{\sigma}_h$=63MPa. Aluminium metal has most unfavorable properties as a material subjected to stress amplitude in seawater. For the materials of aluminium and stainless steel,fatigue strengths of cylinder did not always depend on the viscosity of media. Therefore, it seems that the effects of pressure media are surface phenomena for their respective metallic property.

Copper Metal

Figure 6 shows the relation between σ_h and N_f for various test media. The S-N curves for various media are almost on same line. In particular, fatigue strength for saltwater and water is the same as fats and oils. The effect of saltwater on the fatigue strength of copper metal was extremely

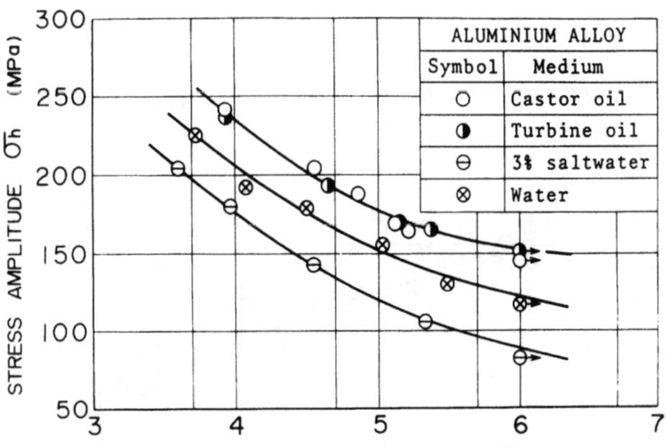

Fig.5. Relation between σ_h and N_f of aluminium alloy in various pressure media.

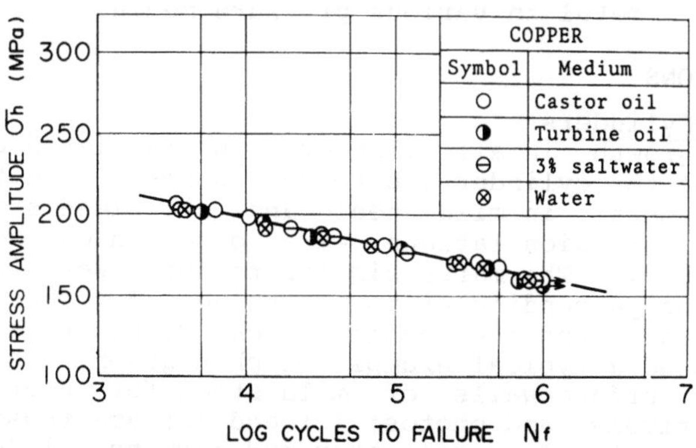

Fig.6. Relation between σ_h and N_f of copper metal in various pressure media.

different from that of cylinders of S45C, SUS310S and Al-Mg alloy. It has been well known that the fatigue strength of copper for rotary bending and pure bending of plates has little difference in saltwater, water or air [5]. Copper metal is found to be insensitive in comparison with other materials for various conditions of pressure medium or environment. This material has excellent saltwater resistance, and is useful for pressure vessels and machinery subjected to internal fluctuating pressure under saltwater or seawater.

Brass Metal (Muntz Metal)

In the case of brass metal S-N curves are shown in Fig.7. The curves of fatigue strength under the various media are obtained at random. For water the relation between σ_h and N_f is a straight line. The S-N curves for saltwater and oil media are approximately parallel to the axis of abscissa for higher stress, $N_f \leqq 2X10^4$ cycles. Brass metal has characteristics fractured for the specialic environments. These are stress corrosion cracking and dezincing [5]. The fatigue strength for saltwater is the greatest of all media. In the case of copper metal the fatigue strength is affected little by a medium, but this material is not applied to structure components. Brass metal is a useful material for pressure vessels in seawater.

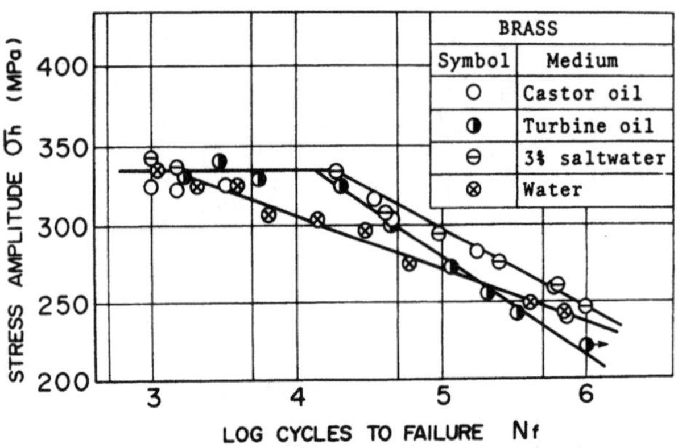

Fig.7. Relation between σ_h and N_f of brass metal in various pressure media.

CONSIDERATIONS

Surface of Cylinders

Crack growth in all cylinders is initiated parallel to the axial of the cylinder due to hoop stress. On the surface of ferrous metal in mild steel and stainless steel, marks peculiar to corrosion fatigue were observed on the inside of cylinders [6,7]. The marks similar to notch were initiated in the direction perpendicular to the principal stress.Many marks were observed around the opening crack. Figure 8 (a), (b), (c) and (d) showed a typical example of photograph. Photo.(a) and (b) are the surface walls of mild steel for water and castor oil, respectively, and photos; (c) and (d) are those of stainless steel fractured in saltwater and turbine oil.Judging from surface condition,the component of the pressure medium changed

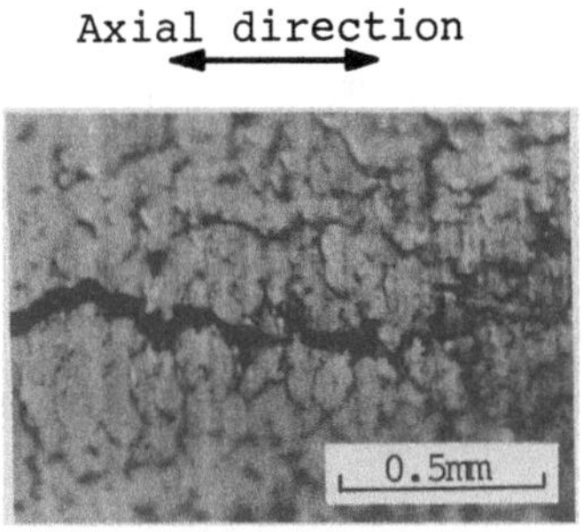

(a) S45C : Water
σ_h=390MPa
N_f=3.80X10^4cycles

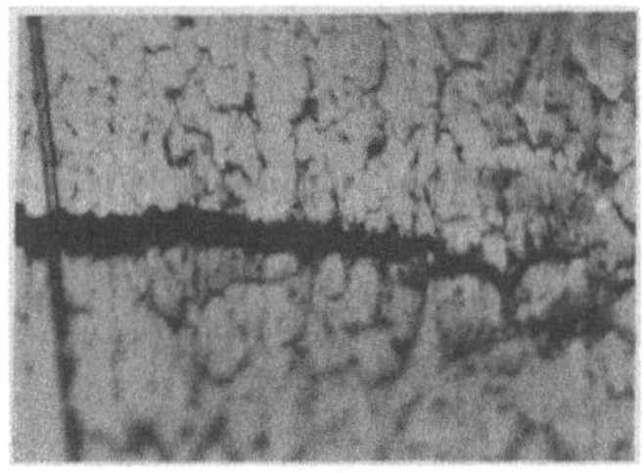

(b) S45C : Castor oil
σ_h=392MPa
N_f=10.47X10^4cycles

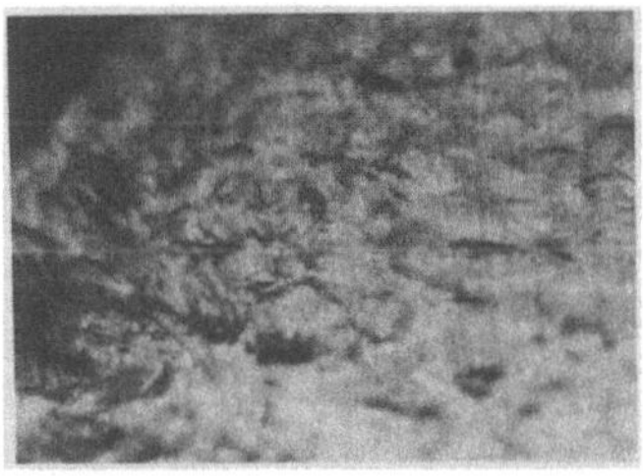

(c) SUS310S : Saltwater
σ_h=317MPa
N_f=5.37X10^4cycles

(d) SUS310S : Turbine oil
σ_h=328MPa
N_f=11.33X10^4cycles

Fig.8. Optical micrograph examples around the opening crack in the inner surface wall of the test cylinder.

the mechanical properties of metals surface. The phenomena of fatigue of materials under various media is considered to be due to the effect of corrosion and fluctuating stress.

Stress Difference due to Metal and Medium

For an explanation of the effect of fatigue strength in saltwater, we considered corrosion behavior of metals for saltwater. Therefore, free energy change given ΔG by the Gibbs are adopted as a parameter of reactivity in the relation of metal and environments. ΔG value can be obtained in corrosion tests of hydrogen generation (pH=0) with reference to measurement value ΔG by the Tomashov [8]. The free energy change ΔG is shown as a function of the stress ratio $\Delta\bar{\sigma}_h/\sigma_y$ in Fig.9. $\Delta\bar{\sigma}_h$ is determined as an average of stress difference $\Delta\sigma_h$. σ_y is the yield stress of a material of 0.2% proof stress. It is clear that corrosion of aluminium metal is the largest of the test materials. That is, the ΔG value of copper and brass metal is smaller than the absolute value ΔG for others. The fatigue strength under the repeated internal pressure does not decrease for saltwater. The stress differences of these metal are substantially $\Delta\bar{\sigma}_h$=0 MPa. The fatigue strength of cylinders for saltwater has a correlation to the free energy change ΔG of metal.

CONCLUSIONS

The cylindrical pressure vessel of mild steel, stainless steel, aluminium, copper and brass metal were subjected to in-

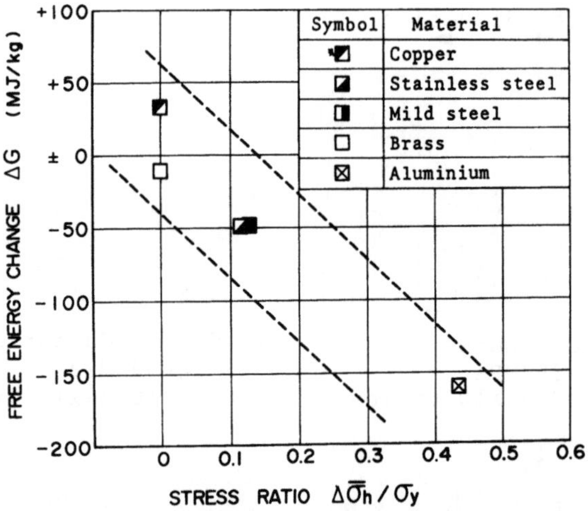

Fig.9. Experimental relation between free energy change of test metal, ΔG, and stress ratio, $\Delta\bar{\sigma}_h/\sigma_y$.

ternal fluctuating pressure by using the pressure medium of saltwater. The fatigue behavior of these materials for saltwater experimentally was clarified. It was found that the phenomena of fatigue was different due to the properties of materials.

Copper and brass metal are useful for the pressure vessel in saltwater, and aluminium metal is most useless material as pressure vessel. Copper metal has strength for water as well as saltwater.

The stress ratio $\Delta\bar{\sigma}_h/\sigma_y$ for the pressure medium of saltwater indicates the effect of saltwater on the fatigue strength of metals. The stress ratio of materials is respectively linear as a function of the free energy change ΔG relating to corrosion of metals. As free energy change is large in the positive region and small in the negative region, metal behavior is insensitive to saltwater. As a result,copper or brass metal is useful for machinery used in the seawater.

REFERENCES

1. Horn, R.M., Kass, J.N., and Ranganath, S., Trans.ASME,J. Press.Vessel Technol., Vol.106, 1984, p.201.
2. Nagai, F., Matsukura, T., and Sato, Y., Trans. Jpn.Soc. Mech.Eng., Ser.A, Vol.48, No.433, 1982, p.1217.
3. Nagai, F., Nagasawa, A., and Sato, Y., Trans.Jpn. Soc. Mech.Eng., Ser.A, Vol.50, No.450, 1984, p.174.
4. Kanno, M., Nagao, M., Prepr.28th.Jpn.Con.Mater.Res., 1984, p.1.
5. JSME Data Book : Fatigue of Metals.
6. Endo, K., Miyao, Y., Trans.Soc.Mech.Eng., Vol.24, No.139, 1958, p.167.
7. Likhtman, V.I., Rebinder, P.A., and Karpenko, G.V., "Effect of a Surface-Active Medium on the Deformation of Metals," Dep.Sci.Ind.Res., London, 1958.
8. Itoh, G., "Corrosion Science and Engineering," Corona Publishing Co., Tokyo Jpn., 1979.

Fatigue Strength of Tubular Joints

K.G. Babu and M. Nasser
Ocean Engineering Centre, Indian Institute of Technology, Madras 600 036, India

ABSTRACT

Parametric equations developed earlier for stress concentration factors (SCF) and strain concentration factors (SNCF) for tubular joints are mainly dependent upon the increase in stress and strain due to geometric, notch and weld effects. Fatigue failure criteria were also formulated from the above. However, previous loading history to levels beyond working loads due to abnormal environmental forces will result in permanent stresses which will have a large bearing on the design of the structure both for static and fatigue loadings. The paper presents some relevant information from the authors' tests to bring out the importance of permanent stresses and their relation to the fatigue strength of tubular joints. The influence of a defect (preformed) on the performance of a tubular joint has also been discussed.

INTRODUCTION

Fixed offshore steel platforms are fabricated entirely from tubular sections because of the many advantages offered. However, lack of realistic data on such complex structures led to the extrapolation of data available from land based structures leading to highly conservative designs. Particularly,having to adopt unprecedented dimensions for the various members, a fundamental and detailed understanding of the behaviour of the tubular joints has become essential.

In such structures, even under static loading, stress concentrations occur due to the following reasons. The 'nominal stresses' arising due to basic structural response need to maintain compatibility between tubes at the joint leading to 'geometric stress concentration'. The highly localised deformations on the part of the tube walls lead to 'notch stresses'. All these are invariably superimposed upon the 'welding stresses'. So much so, the local stresses at the joint and hence the'stress concentration factors' have been hitherto analysed as the combined effect of all these causes both under static and fatigue loading.

Apart from these, as suggested by some of the recent investigators, the previous loading history to levels beyond normal working loads due to unprecedented or abnormal environmental forces will result in local weld toe stresses going well beyond the design levels. Even after the removal of this load, 'permanent stresses/strains' are retained by the weld toe in the form of locked up stresses.

However, the environmental loads are basically of a dynamic nature in ocean structures. After the loss of 'SEA GEM' in 1965, the discovery of cracks on other mobiles and fatigue cracking on southern North Sea gas platforms in the early 70's, fatigue has become a major consideration in the design of offshore platform.

So a better understanding of the above aspects and formulation of more reliable criteria for the fatigue analysis of welded tubular joints has become essential. The present study envisages to evaluate the effect of these parameters, particularly with reference to 'permanent stresses/strains' induced due to previous loading history.

Experiments on the influence of a 'preformed defect' to compressive loading on tubular 'T' joints has also been presented. These defects are assumed to be resulting from minor weld defects which have not been detected or from the omnipresent weld cracks which are deep enough to effect the behaviour of the joint

REVIEW OF LITERATURE

Prior to 1970, fixed offshore steel platforms were essentially designed for member strength only. Later investigators have tried to predict the complex stress distribution that exists at the weld toe around the intersections between tubular members, basically under static loading, in terms of parametric equations. These were based on thin shell finite element analysis and acrylic model data. The validity and limitations of these equations were checked by some of the later experimental results. However, there were significant variations between the predictions of these equations with joint size, attributed generally to weld geometry.

The presentation of fatigue data has been traditionally in the form of S-N curves. The first S-N curves used for offshore designs were developed from onshore welded fabrication design curves by Marshall, justified by results of model tests of tubular joints and pressure vessels. The use of these for onshore structures was simple, that it took account of both weld stress concentration and weld imperfections using member nominal stresses. But in the case of tubular connections of large diameters it is complex due to the presence of high local stresses produced by local wall bending of the chord and brace intersections. Marshall proposed 'T' and 'K' curves to take this into account, based on the concept of punching shear; 'T' for single brace joints and 'K' for simple K-type joints. Both curves had a stress concentration factor built in, but the brace nominal stresses had to be modified to obtain an equivalent acting stress range (1). These emperical punching shear curves are still current with AWS and DOE. But API (1) uses the punching shear concept and gives a single S-N curve 'KK' that overcomes the problem of joint selection faced by the earlier methods.

EXPERIMENTAL INVESTIGATION

A comprehensive series of experimental investigations are in progress presently. Though it may not be possible to give the complete scope of this, relevant information for a better understanding of the present paper is reported hereunder.

Tests were basically designed to study the fatigue strength of tubular 'T' joints, particularly the effect of residual stresses and their possible effect on fatigue, both under low frequency (0.25 Hz) and high frequency

(15-17 Hz). Some of the tests were performed on joints with preformed defect on one side of the saddle of a 'T' joint. The 'T' joints had a chord of 120 mm diameter and 4.8 mm thickness with a length of 500 mm. The brace dimensions were 89 mm Ø x 3.9 mm thick with 250 mm length in all specimens other than ST2(D) and ST3 which had 51 mm Ø x 2.2 mm thick tube as a brace. All specimens with a preformed defect are denoted with a suffix (D) to the specimen name.

As per present practice strain gauges were pasted at a distance from the weld not influenced by the notch stress and geometric stresses are obtained by linear extrapolation (2). ECSC and UKOSRP practice has been adopted in fixing the strain gauges so as to extrapolate the measured strain values to weld toe excluding notch effect. The various stages of fatigue failure process have been according to Radenkovic (3)/ But the test specimens having been loaded in compression all these different stages were not clearly observed. As the change in strain measurements continues very small cracks can be detected (4). Application of soap solution to weld areas helps to detect crack initiation, that even the smallest cracks will show pumping action as minute bubbles on the metal surface when air is pumped (5). Even otherwise a 15% change in strain output can be assumed to this point, which was followed in the present experimentation. It has also been observed earlier (6) that increasing stress ratio in fatigue tests on small tubular joints reduced the life to failure.

DISCUSSION

A set of typical observations for a test specimen (FT8) have been presented in Table 1. A comprehensive picture of the behaviour of all the reported specimens has been presented in Table 2.

In all the tests of the authors' large residual strains were noticed. These were high for large braced specimens (ST1, ST2, FT5 and FT8) compared to small braced specimens (ST3 and ST4), obviously due to the fact that the earlier were subjected to much higher total loads. This resulted in very high SNCF values at low loads after the specimen has been subjected to higher loads earlier.

ST1 recorded very large SNCF values (ranging from 53 at low loads to 16 at higher loads) with an average of 31 towards the end of the test (Stage III). However, when this specimen was tested again after 24 hours, SNCF values have gone to a more consistent average value of 6.1 from the previous 5.5 (Stage IV). It is to be noted that values of 53 and 16 for SNCF are noted taking into consideration the effect of residual strains at lower loads at this stage. In the test that was repeated after 24 hours this effect of residual strains, which is still present, was neglected and hence the lower values. Similar results were obtained from ST3, though to a lower extent because of the smaller loads involved. However, ST1 was subjected to loading a number of times prior to actual test while the strains were monitored from start. This might have led to very large residual strains resulting in large SNCF values.

In high frequency mode, specimens were tested upto 2×10^6 cycles measuring and computing SNCF's and residual strains at specified intervals. Even here residual strains have been found to be substantially increasing with number of cycles. It can be seen that residual strains increased from 110×10^{-6} to 875×10^{-6} for FT5 and from 524×10^{-6} to 1125×10^{-6}

for FT8 at chord saddle, which was the critical location for these specimens. Also the SNCF values have increased from 4.3 to 6.4 for FT5 and 5.5 to 8.7 for FT8 at the same location. This reflects the importance of the effect of residual strains.

Also the specimens with theformed defect, ST2(D) and ST4(D) have shown marked reductions in both SNCF values as well as residual strains (as much as half) compared to their counterparts ST1 and ST3. This reduction has been observed to be the same at both the saddles, with and without defect.

To date, the general fatigue analysis takes into consideration only the nominal, geometric and notch stresses. Residual stresses can be built up in the structure during rolling and welding, during fabrication or while launching and upending in erection. Many have studied about welding stresses in particular and remedial measures like heat treatment were suggested. However, substantial residual stresses may result from environmental forces the structure has been subjected to after installation. Since the residual stresses could be built up during the life time of the structure, its fatigue life would be in most cases be different from that predicted from the present S-N curves as they do not onsider this effect. Fatigue analysis experiments todate usually neglect the effect of residual stresses. Some reports pointed out that this may be the reason for the failure of many of the North Sea structures.

CONCLUSIONS

1. Load cycling, both at high frequency as well as low frequency, has resulted in very large residual (permanent) strains depending upon the level at which the cycling is done.

2. These permanent strains will have a large influence on the SNCF values, which may be the reason for the large variations reported even in experimental results.

3. Once the effect of the earlier load history in the form of permanent strains is neglected the specimen appeared to behave normally under static load. But the fatigue strength will be substantially effected.

4. A preformed defect simulating a through crack in the 'T' joint saddle had the effect of reducing residual strains as well as SNCFs by approximately half.

REFERENCES

1. American Petroleum Institute, 'Planning, Designing and Constructing Fixed Offshore Platforms', API RP-2A, Ed.11, January 1980.

2. De Back J., 'Testing of tubular joints' Repporterteur's notes, European Offshore Steels Research Seminar, Cambridge, November, 1978.

3. Dragos Radenkovic, 'Stress analysis in tubular joints', Steel in marine structures (Int. Conference), Paris, Oct. 1981.

4. Gibstein, M.B., 'Fatigue failure of T-joints: The Det Norske Veritas fatigue testing programme', European Offshore Steel Research Seminar, Cambridge, Nov. 1978.

5. Mc Donald, A., Thomson, J.F., 'The Fatigue strength of large scale welded T-joints', European Offshore Steels Research Seminar, Cambridge, Nov. 1978.

6. Martin, T., Mc Gregor, J., 'An integration into the stress distribution and fatigue strength of welded tubular joints, Ocean Technology Conference, OTC 2856 Houston, Texas, 1977.

TABLE 1 : TYPICAL RESULTS OF A TEST SPECIMEN (FT8)

Load (t)	No. of cycles	Chord				Brace			
		Saddle		Crown		Saddle		Crown	
		SNCF	Res. Stn	SNCF	Res. Stn.	SNCF	Res. Stn.	SNCF	Res. Stn.
STATIC TEST PART:									
0.5	0.0	4.0		3.4		3.8		2.0	
1.0	"	5.0		4.2		5.0		1.2	
1.5	"	5.2		4.2		5.2		1.1	
2.0	"	5.6		4.2		5.5		1.0	
2.2	"	5.4		4.4		5.3		1.0	
2.4	"	5.4		4.4		5.3		1.0	
2.6	"	5.5		4.2		5.2		0.9	
2.8	"	5.4		4.3		5.2		0.8	
3.0	"	5.6		4.5		5.1		0.9	
3.4	"	5.3		4.4		5.4		0.8	
3.6	"	5.0		4.5		5.2		0.9	
3.6	10	5.5		4.5		5.2		0.8	
3.0	"	6.0		5.0		5.1		0.9	
2.0	"	5.6		4.6		3.7		0.8	
1.0	"	6.0		4.6		3.9		0.9	
0.0	"	-	500	-	39	-	314	-	-78
2.0	"	5.9		4.5		4.0		0.8	
3.6	"	6.2		4.2		5.1		0.7	
Max. SNCF		6.2		4.6		5.2		0.9	
Ave. SNCF		5.5		4.4		5.0		0.9	
FATIGUE TEST PART									
3.6	5,000	6.2	515	4.5	43	5.3	301	0.8	-91
"	10,000	6.5	640	4.4	57	5.8	343	0.5	-114
"	50,000	7.0	621	4.6	61	6.0	415	0.3	-107
"	0.1×10^6	7.1	650	4.5	74	6.7	621	0.3	-115
"	0.5×10^6	7.5	715	4.4	63	6.1	584	0.4	-231
"	1.0×10^6	8.5	914	4.6	89	5.9	571	0.5	-272
"	1.5×10^6	7.8	950	4.5	114	6.1	615	0.1	-240
"	2.0×10^6	8.7	1125	4.5	102	6.1	641	0.0	-316
Max. SNCF		8.7		4.6		6.7		0.8	
Ave. SNCF		7.4		4.5		6.1		0.4	

TABLE 2 : SALIENT RESULTS OF AUTHOR'S EXPERIMENTAL INVESTIGATIONS

Spec. Name	Load(t)		SNCF at*		Cycles		Residual strains after cycling (x 10^{-6})					
							Chord			Brace		
	Max.	Min.	Max.	Ave.	No.	Freq. (Hz)	Saddle (defect) (1)	Saddle (non-defect) (2)	Crown (3)	Saddle (defect) (4)	Saddle (non-defect) (5)	Crown (6)
ST1	2.2	0.0	5.7(2)	5.5(2)	10	0.25	-	-	-	-	-	-
	2.8	0.0	5.9(2)	5.4(2)	30	"	-	503	74	-	115	-14
	3.6	0.0	28.1(2)	31.0(2)	10	"	-	1010	240	-	400	-18
	3.6	0.0	6.2(2)	6.1(2)	-	"	-	22	10	-	15	-14
ST2(D)	2.2	0.0	1.8(1)	2.0(1)	10	"	35	-	11	12	10	-
	2.8	0.0	1.8(1)	2.1(1)	30	"	130	-15	89	-	-10	43
	3.6	0.0	2.4(1)	3.2(1)	10	"	404	-25	11	-85	-38	34
ST3	0.7	0.0	5.9(5)	6.5(5)	10	"	-	29	15	-	114	11
	0.9	0.0	6.9(5)	6.1(5)	30	"		9	9	-	510	8
	1.1	0.0	5.8(5)	10.0(5)	10	"	-	-	-	-	569	-
ST4(D)	0.7	0.0	3.9(1)	3.5(1)	10	"	-	-	-	8	8	-
	0.9	0.0	4.3(1)	4.1(1)	30	"	8	16	-	-	-	-51
	1.1	0.0	3.9(2)	3.6(2)	10	"	88	11	40	-85	24	-22
FT5	2.2	0.0	4.8(2)	4.3(2)	30	"	-	110	-21	-	87	-35
	2.2	1.1	6.4(2)	5.5(2)	2x10^6	15.2	-	875	-42	-	211	-93
FT8	3.6	0.0	6.2(2)	5.5(2)	10	0.25	-	524	44	-	302	-83
	3.6	1.8	8.7(2)	7.4(2)	2x10^6	15.2	-	1125	102	-	641	-316

*Maximum SNCF at the particular load and average SNCF at all load stages in the cycle observed at locations(1)-(6)

Constructibility Engineering for Large Offshore Structures

B.C. Gerwick, Jr.

Civil and Ocean Engineering Department, University of California, Berkeley, CA 94720, USA

ABSTRACT

The very large and complex structures now being considered for deployment in deep water and in adverse environments such as the Arctic are increasingly demanding fuller consideration of the entire construction process, in order to reduce the time and cost of construction, to ensure optimum quality and long term behavior, and to integrate more effectively with the other systems involved in the total field development.

These considerations have led to the concept of "Constructibility" as an essential aspect of the entire planning, engineering, and construction process. This concept involves such considerations as to where and how to build, the procedures to be followed, the equipment and facilities required, material procurement and handling, quality control, personnel training and supervision, safety, scheduling, and cost control. While each of these elements has been individually considered in the past, Constructibility wraps these up into a coherent system.

The process starts during the initial feasibility stage where it may affect the decision as to the type of structure, the basic materials to be used, and the way in which the structure will be integrated into the overall field development. Consideration must be given to the fabrication site, the availability of land, and protected inshore harbor areas, the available water depths for exit to the sea, and the environmental, political and social conditions, both those affecting construction and those affected by the construction.

During construction, these huge structures pass through many stages, during which full consideration must be given to draft, freeboard, stability, damage control, structural loads and stresses, deformations and deflections, dynamic response, and access for further construction activity.

Various techniques of construction need to be evaluated. Typical schemes that have application in offshore structures include prefabrication, modular construction methods, steel-concrete construction in its several combinations, and temporary ballasting and grounding, etc. Special construction equipment such as floating gantries may be considered.

Because of the rigors to which the structure will be subjected in service, with consideration of fatigue, impact, low temperature, and corrosion, and recognizing the limitations and cost of in-service inspection and repair, the ability to control quality during construction assumes a major role.

Scheduling also becomes of more than normal importance because of the financial costs attached to any delay and the need to integrate so many complex and sophisticated systems.

Finally, Constructibility requires an evaluation of risks and the establishment of contingency plans to deal with such events as storms during installation, changed soil conditions, accidental flooding, strikes and other delays.

This paper will be illustrated with examples of offshore projects from the North Sea and Arctic. It will propose an initial methodology suitable for Constructibility Planning in Offshore Structures.

INTRODUCTION

The construction of offshore platforms has been a heavily cyclic industry, responding almost frantically to the discovery of new oil provinces such as the North Sea, or significant changes in price level, such as that which followed the OPEC oil embargo of 1973-74. During the intervening periods, as the market stabilized, the industry matured and became more orderly construction-wise and cost-wise. By the time the next boom arrived, technology had also changed, requiring a learning curve before the industry as a whole was geared to the new demands.

The result has been very significant cost over-runs and schedule delays in the early periods of the new development, followed by a gradual steadying as estimates rose to meet actual data and competitive forces brought costs down closer to their targets.

This has most recently been shown in the North Sea. In the late 1970's cost over-runs of almost 100 percent occurred. Today, a stable industry is turning out both steel and concrete platforms within the estimated costs and time schedules.

The cost of structures has a significant influence on the viability of the offshore development, largely because it is an early capital expenditure. Similarly, the time lag between structure expenditure and oil income has tended to increase rather dramatically as the projects have moved to environmentally more demanding areas: deeper water, more remote locations, and greater ecological, social, and political constraints. This is especially true in the Arctic where the cost of exploratory drilling platforms is very high, and regulations plus winter ice limit the number of holes that can be drilled each year.

Engineers and constructors involved in the design and construction of structures can influence the cost of their portion of the development by sound design, construction planning, and construction management. These latter two aspects are germane to the subject matter of this paper and can be encompassed within the term, "Constructibility Engineering."

The offshore structures per se are not usually the major cost of field development. However, they are a significant part of that total cost and they are usually on or close to the critical path in schedule. They are one area where sound innovative design, competent construction planning, and competent construction management can achieve meaningful savings.

In the following sections, specific aspects of constructibility will be addressed.

CONSTRUCTION STAGES

An offshore structure goes through a series of very distinct stages as it moves from fabrication to off-loading (or float-out), to completion afloat, to transport, to installation, and to module erection and hook-up. The stages pertinent to each of the various types of structures have been described in some detail in the previous chapters.

In constructibility planning, it is essential to formally set these stages forth, by title, description, and schematic drawing.

The stages should be further portrayed by a series of appropriate drawings or sketches; isometric drawings have been found extremely useful. The drawings should be essentially outline in character, with key items pertinent to that stage shown in heavy lines. The purpose is to eliminate aspects not essential to that stage, so that the key elements can be clearly recognized. Thus, while they are based on engineering design drawings, they differ from them in emphasis, clarity, and use. Computer-aided-design (CAD) is especially effective in enabling three-dimensional portrayal of the successive stages.

Experience in the preparation of such descriptions and drawings has shown that serious errors have occurred due to jumping past intermediate stages, which have been incorrectly assumed to be unimportant or self-evident. The entire purpose of constructibility planning is negated when this happens, because it is just these skipped stages that turn out so often to be critical.

Once the constructor is satisfied that all the stages have been set forth, then engineering evaluations can be made of each such stage to ensure proper structural, geotechnical, mechanical and hydrodynamic performance. Many elements are subjected to higher forces and stresses during these constrution stages than under the design environmental loads.

For many of the stages, the key issue will involve the interaction of two or more disciplines. For example, ballasting by means of mechanical systems is intimately related to the structural capacity under differential heads, the stability performance afloat, and the instrumentation with its real-time read-out.

Key considerations which have been inadequately addressed in the planning of previous structures include:

1. Draft, with relation to available water depth.

2. Stability during all stages of installation.

3. Hydrodynamic response of structure during tow, especially acceleration forces acting on mechanical installations.

4. Effect of pressure and temperature changes on function of instrumentation, valves, and minicomputers.

5. Initial contact with seafloor and the interactive effects of trapped water trying to escape.

6. Break-out forces from the seafloor, and resultant draft and stability when relocating or removing the structure.

7. Snap loads of mooring lines due to stored energy from long-period excursions.

8. Effect of shallow water and minimal underkeel clearance on wave characteristics, structure, or vessel response, squat, yaw, and seafloor scour.

9. Control of draft and stability in event of ruptured ballast line, jamming of valve, or carrying away of bulkhead, allowing internal flooding.

10. Human error in ballasting control: adoption of controls, training, and system isolation as needed to prevent catastrophe.

11. Arrangement of lines and umbilical control cables to prevent fouling during critical operations.

12. Inadequate weight and tolerance control during fabrication.

13. Inadequate consideration of tolerances in differential heads of ballast water in compartments.

PRINCIPLES OF CONSTRUCTION

Some of the principles which can be beneficially applied to reduce the time and cost of construction are:

1. Subdivision into as large components and modules as it is possible to fabricate and assemble.

2. Concurrent fabrication of major components, in the most favorable location and under the most favorable conditions applicable to each component.

3. Planning the flow of components to their assembly site.

4. Providing adequate facilities and equipment for for assembly. This may include such items as synchrolifts, heavy-lift cranes, both land-based and barge-mounted, drydocks, basins, etc.

5. Simplification of configurations.

6. Standardization of details, grades, and sizes insofar as practicable.

7. Avoid excessively tight tolerances. Provide for flexibility and adjustment in connections, especially in mechanical system piping.

8. Select structural systems that will utilize skills and trades on a relatively continuous and uniform basis.

9. Avoid intermittent peaks in the demand for the labor force. Select construction methods that involve relatively uniform demand.

10. Avoid procedures that are overly sensitive to weather conditions. Arrange for shop prefabrication of elements which are very sensitive.

11. Mechanical systems to be incorporated in or on the structure should be modularized into the largest possible components, even if this requires additional structural support or interruption of the construction of the structure proper.

12. Construction methods should be selected which are appropriate to the specific structure. Avoid fixation on only one method, e.g., concrete pumping, slipforming, welding, barge launching. Be versatile in choice of methods.

FACILITIES

In almost all cases, the early stages of construction are carried out at a shore base. This base may be purpose-built for this one project or may be a relatively permanent facility.

The area for such a facility must be adequate to accommodate not only the structure and/or components themselves but also storage of materials, access roads, support buildings and infrastructure facilities.

MATERIAL SELECTION

The design will, of course, have determined the specifications for the materials, based on their performance in service. Constructibility considerations will now go further, as the constructor addresses the practicability of building the structure to meet the specifications.

With steels, for example, welding procedures and materials are intimately related to the ambient temperature at which the work will be carried out. The constructor has an opportunity to optimize these.

a. He may elect to carry out the majority of the welding within protected, heated, and dry enclosures.

b. He may elect to use pre-heat and/or post-weld treatment to attain the required results.

c. He may elect to purchase specially processed steels which are less sensitive to the conditions.

With concrete structures, the Constructor has even more alternatives available. These include the concrete mix, selection of aggregates, cement, and admixtures, cooling of the mix, placement procedures, and curing methods.

CONSTRUCTION PROCEDURES

Within the context of each construction stage, suitable procedures have to be developed so as to meet the following criteria:

1. Compliance with the specifications and drawings.

2. Assurance of meeting quality requirements.

3. Adaptability to equipment, facilities and skills available.

4. Ability to meet schedule requirements.

5. Economy in overall performance: lowest possible costs consistent with items 1, 2, and 4.

6. Minimum risk of accident or delay.

Since the two largest expenditures within the control of the contractor are the fabrication of the structure and the hook-up of mechanical facilities, the principal attention insofar as efficiency is concerned is directed to these two phases.

However, the phases involving heavy lifting, load out, launching, delivery and site installation, while not heavily labor intensive, are controlling from a technical viewpoint, so attention must be directed to them also to ensure technical performance and safety.

Thus, the focus in the procedures for the different stages of the work differs, in one case being directed to efficiency; in the other, to equipment selection and technical performance.

Because of the immense amount of work in fabrication, whether steel or concrete, the approach should follow the same logic and patterns as those used in the Japanese shipbuilding industry. There the work is broken down into as many sub-units as practicable. Each is then fabricated in the most favorable attitude (often upside down), and under the most favorable conditions.

With regard to the fabrication of steel tubular components, the decision must be made as to where to place the junctures, whether at nodes or in mid length. Since the nodes are three dimensional, fit up is usually much more difficult there. If the juncture is in the mid length of a tubular, the nodes can be first erected to their correct position, the tubular "pup" cut to exact length as measured in the field, and the girth weld readily made.

With regard to a jacket or large module frame, how should subassemblies be selected? Should the jacket be split into its several faces for component fabrication, as is extensively practiced at the McDermott and Brown and Root yards on the Gulf of Mexico, or should it be split into three-dimensional space frames, as used by NKK for the North Rankin platform?

With concrete structures, several decisions have to be made.

Will all elements be cast in place, or will some or any be precast? For the Ninian Central Platform, several hundred concrete shell units were precast in southern England, then transported to Northwest Scotland and erected by sheer legs crane barge. Precast cellular internals were combined with cast-in-place external walls in the Super CIDS platform.

While precasting offers opportunities for dispersion and suboptimization, it then requires consideration of lifting capabilities and joint details.

The next decision is in regard to cast-in-place concrete. Should slip forming be used, or panel forms? Slip forms have been very successfully used on the Condeeps, for example, but require a large surge in manpower during short periods of time. Will the climbing rods and yokes interfere excessively with embedment and reinforcement placement? Panel forms enable the reinforcing installation to be carried out at different location from the concrete placement, facilitating disperson and equalization of manpower requirements, but requiring more construction joints.

A third major decision is the method to be adopted for concrete delivery. Will it be by pumping, as used on the Seatank and DORIS structures, or by bucket, as used on the Condeeps?

Most of the manpower requirements on a concrete offshore structure relates to the reinforcing installation. Should the bars be handled individually, as is practiced on slipforming operations; or should preassembled cages be used? How should splices be made: by lap, weld, or mechanical connectors? Will color coding of reinforcing and pre-packaging of the reinforcing bars required in individual zones save time for the placing crew?

In particular, should the stirrups, of which so many are required in the typical offshore structure, be bent bars, closed pre-welded loops, or mechanically-headed bars?

A very important role can be placed in decision making for critical fabrication operations by use of mock-ups. Full-size sections of the structure, whether intersecting tubular nodes or the juncture of concrete shells, are selected. They are then fully fabricated, with all inserts, post-tensioning ducts, reinforcing steel, stiffener plates, etc. This mock-up enables the visualization of the interaction of the many details, and the practicability of welding, steel placement, and concreting.

The mock-ups also serve a value purpose in training the workmen, especially if they are shown the results of their work. For example, if placing concrete among congested reinforcement results in honeycomb and rock pockets, they will visually see the need for vibration. If welding of studs with excessive heat results in plate warping, they will understand the reason for the apparently excessive shifting of location and imposed time lags.

One lesson from Japanese shipbuilding practice that has been demonstrated repeatedly in the U. S. nuclear power plant construction is that teams of workers, comprised of several trades, and assigned to a specific task group with the objective of completing all work within a specific zone, are much more productive than a highly centralized organization by trades, with each trade then responsible for all work within its classification throughout the structure.

The task group teams will, of course, be reassigned and reconstituted as the needs demand. Even where specialist subcontractors are involved, e.g., post-tensioning, the task group organization appears to give greater overall efficiency and reliability.

ACCESS

A much overlooked aspect of constructibility is that of providing access for men and equipment to the areas where they must work. Men need safe and convenient access. Studies have shown that up to 50 percent of a man's workday is associated with moving. It is inefficient and expensive to have men climb ladders, thread or force their way through congested reinforcing steel, climb across scaffolds and walk planks. Proper and safe access needs to be engineered.

Construction procedures offshore are planned with primary consideration being given to the sea states and weather conditions under which they will be performed. They must be organized around the available equipment. Logistics and support facilities must be considered. Availability of skilled personnel and supervision must be taken into account.

Chen and Rawstron in their paper, "Systems Approach to Offshore Construction Project Planning and Scheduling," Marine Technology Journal, October 1983, make use of advanced simulation techniques in the planning of offshore construction operations. Limiting sea states are determined for various operations such as module lifting, piledriving, pipelaying, and saturation diving, and the effects of vessel motions on operations are evaluated.

TOLERANCES

Offshore structures are not only among the largest structures built by man but they are structures which must be moved, floated, and rotated. They are subject to a wide variety of external loads during construction. Tolerances, therefore, become of far more than normal importance.

Weight control is critical. Steel jackets typically are launched and are then supposed to dynamically return to float on an even trim with only one meter or so of freeboard.

Concrete structures, when ballasted down to receive the deck, have typically only one-half of one percent or less of reserve displacement.

Weight control procedures are therefore instituted and an organization set up to control the weights during the entire construction period.

Geometry control is important to ensure against buckling under external hydrostatic head, and for subsequent fitting of other components, e.g., embedments, piping, precast units, etc.

SURVEY CONTROL

Survey control is, of course, intimately associated with geometry control but extends beyond it to guide the fabrication and erection process before and during construction.

Where components must fit to others, it is often difficult to establish the proper reference line. One must be selected, e.g., a line connecting the center of two best fit circles. The other points must be properly related, in all three planes. Templates will often be found to be the best method of transferring complex interactive dimensions.

Match-casting of precast concrete members has been highly successful in assuring later fit. It was utilized effectively for the breakwater segments of the Ninian Central Platform. Care has to be taken in such match casting to avoid distortions due to thermal effects, e.g., warping during steam curing.

Similar match fitting and templating can be used with steel fabrications, again recognizing potential distortions due to welding.

Proper survey control procedures must also be set up for erection of space frames, of which the steel jacket is the most common example. Distances are large, 50 meters or so, points are high in the air and, hence, of limited accessibility. The sun's heat may cause significant elongation of upper members during the afternoon, while the lower members on the ground are partially restrained by friction, as well as perhaps seeing lesser temperature rise due to shade. Dead weight deflections may account for even greater distortions, since the jacket is usually fabricated in a different attitude from that in which it is installed. Diagonal measurements often provides the best check.

When structures are afloat, there is always difficulty in establishing reference lines, especially the vertical line. Lasers can be rigidly mounted at the base, accurately set normal to the base. They can then project a normal line, called "vertical" even if the structure is slightly listed due to ballasting or dead weight.

Even a relatively rigid structure such as a concrete gravity-base structure undergoes significant deflections during construction due to dead weight and ballasting eccentricities. Thus, the shafts may deflect outward during ballasting down to receive the deck, the ballasting typically producing a hogging moment and deflection in the structure as a whole.

QUALITY CONTROL AND ASSURANCE (QA/QC)

The establishment of a Quality Control Manual and a quality Assurance Program is an essential aspect of constructibility.

The first task is to set up what the requirements are. They, of course, include those specified by the designer. If the designer has used only general requirements, such as "compliance with ASTM Specification --," it is necessary to determine which elements of that specification are important, and further, which will be determined or measured in the construction process.

To these, the constructor must add those requirements necessary to enable him to carry out his work in accordance with the materials selected and procedures adopted. Examples are pre-heat temperature, humidity control for painting, and early strength gain for concrete.

In establishing these lists, every effort should be made to reduce the number to the bare essential minimum. The Nuclear Reactor syndrome that "everything that can be measured must be measured" must be avoided. Paperwork must not become more important than the structure.

Where defects can be immediately corrected, they should be.

Agreement should be reached before construction starts as to what records are to be recorded and which data, e.g., radiographs, are to be kept. Those which are so kept must be properly identified and stored.

Only that should be tested and inspected which is essential for the proper performance of the structure. Only that number of tests should be made which is necessary to ensure maintenance of quality on a statistically-defensible basis.

SAFETY

The engineering of a safety plan for the large offshore project requires careful job-specific study by the construction and engineering personnel responsible for executing the project. They should develop a manual to apply to their project, in which various safety risks are identified and then appropriate preventive or mitigating measures adopted.

Of the many safety precautions and equipment required by various regulatory agencies, which are important to implement on this job? Which are irrelevant or nonapplicable? Which may be detrimental and, hence, require a special exemption?

Is additional scaffolding required? Will life lines and snap-on belts be of value on some high work? Should safety nets be provided; and if so, is their purpose to save a man who falls or the protection of men working underneath, or both? The design of the net and its supports should then be appropriate to the purpose. What provision is made for men to stand aside while loads of steel or concrete are being lowered? Will walkways or recesses be usable at such times? Red plastic caps on projecting reinforcing bars will prevent deep scratches, protect eyes, and prevent puncture wounds. They should be attached before the bars are delivered.

What about man overboard? Arctic waters are at minus $2^{o}C$: a man can live only a few minutes in such water. At low temperatures, boats will not always start instantly. Should engines be left running and the boat be manned at all times?

A continuously manned and operated life boat is now required around the platforms being constructed afloat in Norwegian waters.

Even if a man can swim; and even if he has a life jacket on, to what does he grab hold in a choppy sea and strong surface current? Fiber lines, floating in the water, can be trailed out from the structure. They should be well marked by buoys, to prevent fouling boats' propellers. Are there searchlights that can be used to illuminate the man in the water?

Fire is the scourge of the sea and especially so in the Arctic and sub-Arctic when piping and valves become frozen, and intakes clogged with frazil ice. What secondary means are available for fighting fire?

If a man is injured, what means are available for evacuating him from a congested location inside the structure to a shore-side hospital? As the structure nears completion, there will probably be excellent facilities; but in the early stages, temporary means must be planned.

Finally, training. Major emergencies such as fire, collision, explosion, or imminent overturning require the coordinated action of several hundred men, many of whom are not offshore oriented and trained. Evacuation may need to be carried out under conditions of darkness, wet decks, loss of power, high winds, and a stormy sea. The diverse groups of men abroad need to be organized into crews, and the crews need instruction and rehearsal. This matter is especially difficult where numerous specialist subcontractor personnel, e.g., x-ray technicians, are on board on a temporary basis and, hence, unfamiliar with the organization and the vessel or platform.

CONTROL OF CONSTRUCTION

An offshore structure is a major undertaking on two fronts: first, the effect of the sheer size, complexity, and interdisciplinary aspects; the second, the dynamic movement, transport, and locations, etc., carried out on a grand scale, sometimes involving over half a million tons, and a structure the size of our largest high-rise buildings.

Construction management will have carefully planned each operation. Now, as the work goes on, how is the success or lack of it monitored? What warning signals will be sent , and how will they be recognized in time for corrective action?

It is not enough to try to control by flagging exceptions; the 10 percent overrun or underrun may apply to an insignificant item or one which will soon be completed and, hence, beyond correction. Rather, the major components of the work need to be identified; schedules and budgets assigned, with consideration of the learning curve, the special conditions, etc. The key items are then closely monitored, usually on a crew-day basis.

Constructibility planning must, of course, include and interface with the Critical Path Scheduling. CPM is a valuable technique to evaluate and control the various operations. Critical path schedules are, of course, constantly updated.

While most attention goes quite naturally to the items that lag, consideration must also be given to the opportunities that present themselves when work goes faster than scheduled. The Statfjord C platform was ahead of schedule on several early items; others were then accelerated to enable the completed structure to be placed several months ahead of target.

CONTINGENCY PLANNING

Murphy's Law postulates that "What can go wrong, will go wrong, and at the worst possible time."

For each detailed planning phase, a list of credible potential accidents and errors needs to be listed, including especially those due to human error. Each of these is then examined in detail. What can be done to prevent them? The preventive step may be physical (structural or mechanical), or it may be the assignment of a specially trained man.

Examples are manifold. To offset a stripped valve stem or jammed gate, valve position indicators may be installed, with a remote readout at the control station, so as to verify that the valve really is open, or closed. Valves may be arranged in series, with a space between, so as to provide a back-up in case some foreign object gets in. External screens may be provided over intakes. To prevent snagging and ripping off from a boat line, guards may be installed over the screens.

The above series of steps are now standard in the Norwegian North Sea, ever since a wire line got sucked into the Frigg ballasting line and kept two valves from closing. Fortunately, this occurred near the end of the installation and did not result in serious damage to the structure, but it could have been catastrophic.

We learn primarily from past mistakes, so the advice of experienced personnel is invaluable in preparing and reviewing contingency lists. However, the initial listing can be quite fully prepared even by a less experienced engineer by extrapolation from more conventional problems on land and in surface vessels, and by addressing his imagination to the situation.

However, some of the contingent accidents cannot readily be prevented. For these, back-up equipment should be provided. Especially, care must be taken to prevent progresssive collapse, e.g., the crane boom collapses, falls across the falsework, which allows structural elements to drop, holing the bottom. Some contingent events will be judged so serious as to indicate a major change in construction procedure, even at an increase in cost or time.

MANUALS

From the previous sections of this chapter , manuals are now prepared, covering each major stage and each important or critical component of the construction process.

The preparation of each of these manuals requires the participation of all involved parties, contractors, and subcontractors, and all disciplines. Thus, it turns out to be an effective means of communication, and of making each group aware of the others' needs and concerns at that stage.

A draft of the manual is then circulated for review to management, design engineering, field construction supervisors, consultants, key subcontractors, and insurance surveyors. They are asked to review in detail and comment. Not only do constructive suggestions for improvement arise, but this review makes each party even more fully aware of the operation and enables each to focus on critical aspects.

ON-SITE INSTRUCTION DRAWINGS

While during construction, applicable drawings from the design and those showing temporary construction will be in the construction site office, as will the manuals, these are hardly suitable for use out on the platform structure itself.

Complementary sets of construction drawings are therefore prepared, using the above documents as a source. One set is prepared for each sub-stage or major operation. Unlike design drawings, these drawings show only those elements which are essential for the construction phase. Isometric drawings may be used for certain steps.

These drawings are then issued on waterproof paper, for actual use in the field by the construction personnel. These drawings are especially useful in the rigging, lifting, launching, etc., operations. For these, successive drawings can show the different positions of the load, the booms, and the lines as the load or structure is transferred from one location to the next.

RISK AND RELIABILITY EVALUATION

Risks associated with the various construction stages and procedures can be identified and a qualitative evaluation, at least, made of their reliability and safety involved. The word "qualitative" seems appropriate, even where some effort is made at quantification, because each operation has many unique aspects and because the data base is generally inadequate.

Risk and reliability evaluation is obviously closely-related to contingency planning. The latter, however, is intended to establish specific procedures to prevent or mitigate risks after the overall plan has been established. Conversely, risk and reliability evaluation is intended to serve as a broad guide, and overview, to ensure that suboptimization techniques have not led to adoption of excessively risky procedures, and that areas of high risk will be reinvestigated so as to reduce their probability and mitigate the consequences.

CONCLUSION

Constructibility Engineering offers a significant opportunity to improve safety, reliability, schedule, and costs. While the individual principles expounded have long been employed in piecemeal fashion, often informally, the complexity and importance of large offshore structures justifies a more rigorous and thorough approach.

Concrete Platforms in the North Sea and Some Related Research on Fatigue

R.J. Lenschow

Division of Concrete Structures, Norwegian Institute of Technology, 7034 Trondheim-NTH, Norway

ABSTRACT

The number of concrete platforms on the continental shelf in the North Sea is increasing steadily and has reached 15 today. Most of them are produced in Norway where the conditions are particularly advantageous for such production starting in dry dock and continuing in deep fjord. The concrete mix has gone through more faces, and strength has been increased each time. The experience, now 15 years, with concrete structures in the North Sea has been excellent.

The longtime stochastic loading makes it necessary to consider the fatigue behavior of the concrete platforms. The dynamic strain of concrete may be greater than the timedependent creep. Recent research results are applied in simplified rules for fatigue life of reinforced concrete structures. These rules are discussed.

INTRODUCTION

In the last 10-15 years, the discoveries of oil on the continental shelves offshore the continents have created a new way for oil production and many small and great innovations have been introduced, not least in structures, both steel structures and concrete structures. The rough, hostile sea has been the challenge, and new development has been necessary to master the requirements.

With structures exposed to waveloads year after year, the question of fatigue life is immediately brought up, and research on fatigue both for steel and concrete has been intensified in several countries to meet the challenge from the ocean.

The results of this new development are in the next round used on land, in bridge building, tall buildings, television towers exposed to wind, etc.

This paper will focus on marine concrete structures and also some findings of general interest for reinforced concrete structures.

CONCRETE STRUCTURES IN THE OCEAN

Most concrete platform structures are of the Condeep type, fig. 1. The deepest water for a concrete platform, as yet, is approximately 150 m, which means the entire structure is

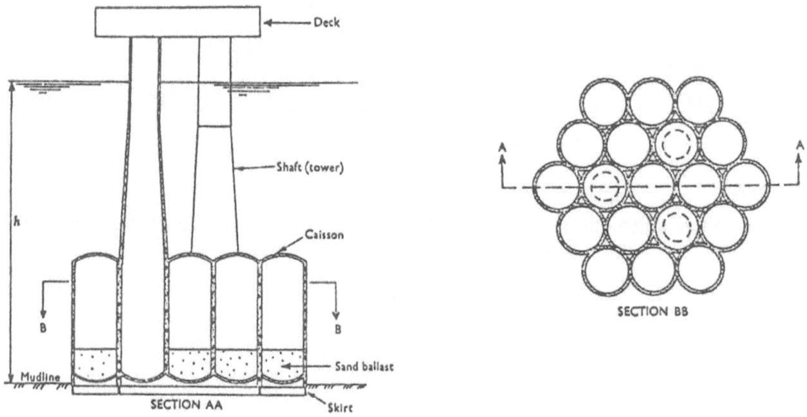

Fig. 1. Concrete Platform. The Condeep type.

approximately 200 m tall. The shallowest sea is about 70 m.

The lower part of a Condeep platform consists of a number of cells. They may be used as storage for oil, for example when deliveries are uneven. Three or four of the cells continue up through the surface and carry the deck. The deck with equipment on top of it may weigh as much as 60 000 tons.

Fig. 2 is a list of concrete platforms in the North Sea. They are placed on the British and Norwegian sectors. Most of the platforms are produced in Norway where deep fjords and high quality sand deposits at the coast are great advantages.

PLATFORM	DESIGNER/ CONTRACTOR	CONSTRUCTION SITE	DELIVERED	CONCRETE VOL. m^3
EKOFISK I	DORIS/SELMER-HØYERE	STAVANGER NORWAY	1973	80 000
BERYL A	NORWEGIAN CONTRACTORS	STAVANGER NORWAY	1975	52 000
BRENT B	NORWEGIAN CONTRACTORS	STAVANGER NORWAY	1975	64 000
FRIGG CDP 1	HOWARD DORIS/ NORW. CONTRACTORS	ANDALSNES NORWAY	1975	60 000
BRENT D	NORWEGIAN CONTRACTORS	STAVANGER NORWAY	1976	68 000
FRIGG MP 2	HOWARD DORIS/ SKÅNSKA DORIS	STRØMSTAD SWEDEN	1976	60 000
FRIGG TP 1	SEA TANK/ MC ALPINE	ARDYNE POINT U.K.	1976	49 000
STATFJORD A	NORWEGIAN CONTRACTORS	STAVANGER NORWAY	1977	87 000
FRIGG TCP 2	NORWEGIAN CONTRACTORS	ANDALSNES NORWAY	1977	50 000
DUNLIN A	ANDOC	ROTTERDAM NETHERLANDS	1977	90 000
BRENT C	SEA TANK/ MC ALPINE	ARDYNE POINT U.K.	1978	105 000
CORMORANT A	SEA TANK/ MC ALPINE	ARDYNE POINT U.K.	1978	120 000
NINIAN CENTRAL	HOWARD DORIS	LOCH KISHORN U.K.	1978	140 000
STATFJORD B	NORWEGIAN CONTRACTORS	STAVANGER NORWAY	1981	140 000
STATFJORD C	NORWEGIAN CONTRACTORS	STAVANGER NORWAY	1984	130 000
			SUM	1 295 000

Fig. 2. Concrete platforms in the North Sea. (1)

The production of a platform starts in a dry dock like an artificial pond, on land near the sea, and emptied for water. When the structure has been cast 6-7 m tall, the wall between the sea and the dry dock is removed, the water floats in and lifts the structure off the ground. Then the floating structure is towed to a fjord nearby for further construction. The lower part of the structure, the storage cells, is slipformed in such a way that the slipform has constant elevation above sea level while the bottom of the structure is lowered deeper into the sea.

At the same time as the platform is built, the deck structure with equipment is made ready. While the price of the concrete structure is about 150 million dollars, the complete deck costs about one billion dollars. The deck is constructed across two ships. At completion, the ships sail on top of the lowered concrete structure and the deck is fixed to the concrete structure.

MATERIALS

The quality requirements for concrete in North Sea platforms are given in regulations and codes and are essentially (1):

- Sound constituent materials
- High compression strength, minimum 40 MPa
- Maximum w/c ratio of 0.45 - or 0.40 in the splash zone

Typical concrete mix proportions and test results from production at Norwegian Contractor's yard in Stavanger are given below:

Condeep Beryl A (1973-75)		Condeep Statfjord C (1981)	
Cement (SP30)	430 kg	Cement (SP30-4A)	380 kg
Sand, Årdal 0-10 mm	900 kg	Sand,Tytlandsvik 0-5 mm	1030 kg
Coarse aggregate, Årdal, 10-32 mm	900 kg	Coarse aggregate, Tytlandsvik (5-20 mm)	845 kg
Water	175 l	Water	160 l
Betokem LP	4 l	Betokem PA(B)	4 l
Slump	120 mm	Slump	220 mm
w/c ratio	0,41	w/c ratio	0,42
Obtained mean 28 day compressive strength	55 N/mm^2	Obtained mean 28 day compressive strength	62 N/mm^2

It can be seen from the above data that an increased strength (by some 12%) is achieved at a reduced cement content (by some 12%). At the same time a more workable mix is obtained and the setting time of the concrete is less dependent on the dosage of admixture. The improved performance is due to development of the main constituents of the concrete, cement, aggregates and admixture. The use and effect of admixture during the last decade at Norwegian Contractor's yard can be considered in three stages: no admixture - lignosuphonate LP - superplast PA(B), 100% compressive strength - 110%-120%, 100% water requirement - 94%-88%, respectively.

The next stage will include even higher specified concrete strength. Other large marine concrete structures have already been constructed with more than 70 MPa characteristic 28 day cube strength.

A major part of the reinforcement has been deformed bars, minimum yield strength 50 MPa and 32 mm bar diameter, 3 bars in one bundle is common practice. The shafts have been prestressed vertically. The cells are not prestressed, except some hoop prestressing at the intersections between the wall and dome. Various prestressing systems have been used with cables with capacity around 4000 kN, and yield stress 1600-1700 MPa. All prestressing ducts have been injected with high quality cement mortar.

Adequate cover to the reinforcement is set to minimum 50 mm for ordinary reinforcement, and twice as much for prestressing steel.

The experience, now up to 12 years with concrete platforms has been excellent. Durability and suitability of concrete for this purpose have been unanimously confirmed so far. Findings from numerous inspections have been negative, and consequently it seems justified to reduce the frequency of the regular inspections of today. Impacts from vessel collisions have occurred several times. The magnitude of the impacts is difficult to quantify. However, people have fallen off their chairs and large indentations on the vessel have been observed, while no damage has been found on the concrete (2).

FATIGUE

Much of the experimental research on reinforced concrete structures which is performed today in various laboratories around the world is connected to marine structures and is often sponsored by the oil industry.

The waveloads are almost at all times acting on a concrete platform at sea. Therefore, the dynamic response - both deformations and strength - may be critical for these structures.

If we let a dynamic load act on concrete, the material will change, most in the beginning of the loading period and if the loads are so heavy that the specimen finally fails, the change of the material are also dramatic before failure. The change can be measured by imposing a pulse, by observing the acoustic emission and by observing the strain-development.

The development of strain has important practical consequences. If we have a beam subjected to static load, the load-deformation diagram is the well known. If we stop the

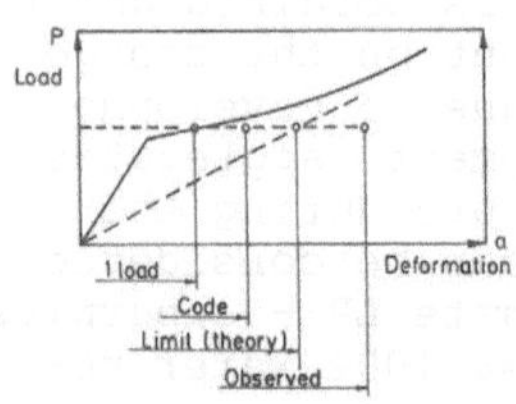

Fig.3. Beam under repeated loading

Fig.4. Crackdevelopment under 1, 10^3, 10^4, 10^5 loadings

static load and start pulsing the load, the deformation will increase further, and it may not stop at the so-called stadium II-line, which is considered to represent the maximum deflection in most textbooks, fig. 3. The cracking zone increases, and more remarkable are the cracks at the top of the compression zone, fig. 4. The behavior of the beam can be explained by considering stress and strain at the top beam, and by taking into account the deformation of the material, the so-called dynamic creep. The compressive stresses decrease by more loads under maximum loads and when unloading the top fibres will come in tension and finally crack, fig. 5. Because of the dynamic creep the stress distribution of a cross-section will change as the dynamic loading continues. The diagram in fig. 6 indicates how dramatically the concrete changes under high dynamic loading. In this case the plastic, permanent deformations are close to 2% and the modulus of elasticity is significantly reduced.

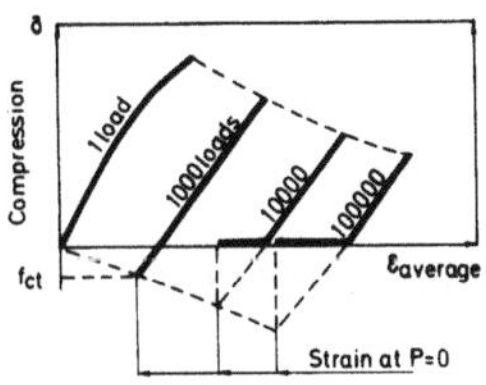

Fig.5. Stress and strain at the top of the compression zone

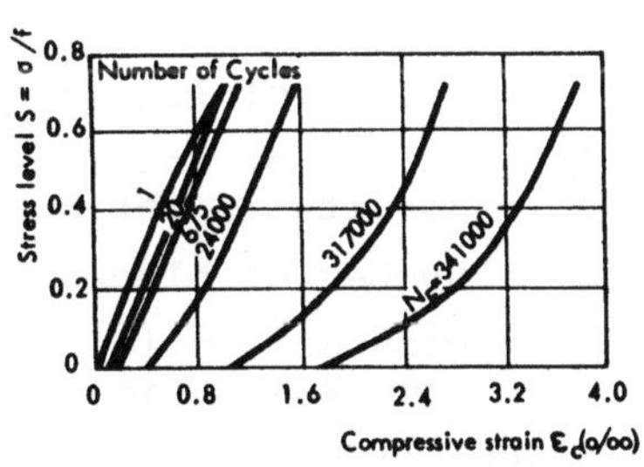

Fig.6. Effect of repeated load in concrete strain

The modulus of elasticity changes under the dynamic loads as indicated in fig. 7. During the lifetime of a structure the dynamic loading causes both dynamic creep and time-dependent creep.

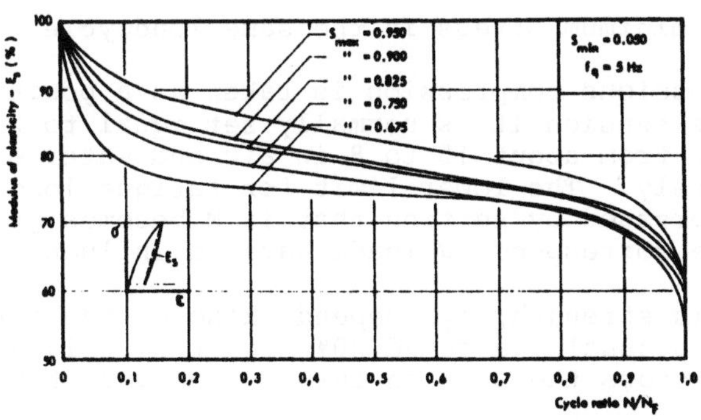

Fig. 7. Percentage reduction of the secant modulus with the cycle ratio. Mean curves for different stress levels

The ordinary time-dependent creep is added to the dynamic creep. The ordinary time-dependent creep is based on (a+RMS) where a = average stress,

$$RMS = \sqrt{\frac{1}{T}\int_{o}^{T} (f_x - a)^2\, dt}, \quad f_x = \text{stress}, \quad T = \text{total time.}$$

The fatigue strength or the lifetime of a structure or part of a structure is usually based on a so-called S-N-diagram or a Wöhler diagram.

To deal with fatigue and loads of various magnitudes Palmgren-Miner's law is applied with some adjustments to determine the fatigue life. The loads are presented in several blocks with constant load amplitude in each block.

$$\sum_{i=1}^{k} \frac{n_i}{N_i} < \eta$$

where

n_i = number of loadings in block i

N_i = number of loadings with the same amplitude which results in fatigue failure

k = number of blocks

η = constant, original 1,0, but later modified in some regulations

The fatigue life of concrete is set:

$$\log_{10} N = C \frac{1 - \frac{\sigma_{max}}{F_n}}{1 - \frac{\sigma_{min}}{\sigma_{max}}}$$

f_r = static strength of concrete in the structure divided by the partial material factor

σ_{max} = maximum stress

σ_{min} = minimum stress in the same loadcycle

In these expressions compression is taken as a positive value, and if σ_{min} is tension it is normally set equal to zero. The value C varies from about 15 to 8 in air and water environment respectively. The lower limit for fatigue loads may be indirectly expressed by setting that if N becomes greater than about 10^7, the corresponding loads have no influence on the fatigue load.

The static strength, f_r, depends also on the stress gradient and is usually assumed 20% higher at triangular stress distribution across the section than a constant stress distribution.

The compression of concrete due to shear is controlled in the same way as shown above by setting

$$\frac{\sigma_{max}}{f_r} = \frac{V_{\gamma max}}{V_{dc}}$$

$$\frac{\sigma_{min}}{\sigma_{max}} = \frac{V_{\gamma min}}{V_{\gamma max}}$$

where V_{dc} = static shear capacity based on concrete compression capacity in the shear zones

$V_{\gamma max}$ = maximum shear load

$V_{\gamma min}$ = minimum shear load

The fatigue life of reinforcement is set (4):

$$\log N = C - a \log \; - b\sigma_{min}$$

where C = 6.5 for straight bars
c = 5.5 bend bars
a = 2.65
b = 0.002

After having calculated the N-values and the stress variations in a period are known, the Miners sum can be made up for the period. If the Miners sum is η, this period is equal to the fatigue life of the structure. The Miners sum is very sensitive to the S-N-curve. For example, a reduction of S-values by 10% may reduce the η-value by 80-90%. In regulations the η-value varies from 0.1 to 1.0.

Much of the discussion about η-values may stem from the fact that Miner's law does not model the behaviour of the material particularly well:

Miner's law says it does not matter whether a high load precedes a low load or the other way around. However, if a low load is applied first, the Miner's law applies well which is indicated by Miner's sum equals 1.0. If we apply the greatest loads first and then the smaller loads, we may obtain a Miner's sum which is only approximately 1/2, while we should, according to the hypothesis by Miner, have had 1.0. This is one of several examples that Miner's law is not representative for the behavior of a material or a structure.

Load effects from waves are naturally stochastic, and only test with stochastic loading on the respective material can give reliable information on the fatigue life. The influence of the environment may be significant. For example, the fatigue life of normal concrete is reduced by moving it from air into water.

At the Norwegian Institute of Technology, Division of Concrete Structures and FCB, experimental fatigue studies on concrete has been carried out the last 15 years. The Miner sum under simulated North Sea loadings has been studied. Usually the Miner's sum range is 0.60-0.85 dependent on the loading response histogram.

ACTIVITY IN THE NEAR FUTURE

The next great step in the North Sea is the construction of the production platform on 350 m depth. New concepts have been presented and designed. For a concept, called "Tripod", with three legs from the bottom joining at mid-height, one inclined hollow leg has partly been test-cast on land with variable diameter and wallthickness, see fig. 8.

Fig. 8

In principle the researchers try to be in front to be able to provide the designers with needed information. Particularly research on durability in a wide sense, comprising corrosion and fatigue, takes time. However, in practice, concepts, materials and construction methods are developed further to be utilized so soon that the researcher and designer work so to say hand in hand. Therefore research, currently and in the near future, is connected to concrete platform on 350 m depth and improvements of today's renowned type.

REFERENCES

1. Moxnes, Jan: "Offshore concrete - recent developments in concrete mix design". Nordisk Betong 2-4, 1982.

2. "Experience from design, fabrication and installation of concrete platforms in the North Sea". Norwegian Contractor paper 1984.

3. Holmen, J.O.: "Fatigue of concrete by constant and variable amplitude loading". Bulletin 79-1, Division of Concrete Structures, NTH-Trondheim, Norway, 1979, 218 p.

4. Det Norske Veritas: "Rules for the Design, Construction and inspection of offshore structure", 1977.

Author Index

MIX
Papier aus verantwortungsvollen Quellen
Paper from responsible sources
FSC® C105338

If you have any concerns about our products,
you can contact us on
ProductSafety@springernature.com

In case Publisher is established outside the EU,
the EU authorized representative is:
Springer Nature Customer Service Center GmbH
Europaplatz 3, 69115 Heidelberg, Germany

Printed by Libri Plureos GmbH
in Hamburg, Germany